OPTICS OF NANOSTRUCTURED MATERIALS

Wiley Series in Lasers and Applications

D. R. Vij, Editor

Optics of Nanostructured Materials
Vadim A. Markel and Thomas F. George (Editors)

OPTICS OF NANOSTRUCTURED MATERIALS

Edited by

Vadim A. Markel
Washington University, St. Louis

Thomas F. George
University of Wisconsin-Stevens Point

A WILEY-INTERSCIENCE PUBLICATION

JOHN WILEY & SONS, INC.

New York • Chichester • Weinheim • Brisbane • Singapore • Toronto

This book is printed on acid-free paper. ♾

For ordering and customer service, call 1-800-CALL-WILEY.

Library of Congress Cataloging-in-Publication Data:

Markel', V. A. (Vadim Arkad'evich)
Optics of nanostructured materials / Vadim Markel, Thomas George,
p. cm.—(Wiley series in lasers and applications)
ISBN 0-471-34968-2 (cloth)
1. Nanostructure mateirals—Optical properties. 2. Optoelectronics—Materials.
3. Nanowires. I. George, Thomas F., 1947- II. Title. III. Series.

TA418.9.N35 M517 2000
621.36–dc21 00-025715

Printed in the United States of America.

10 9 8 7 6 5 4 3 2 1

CONTRIBUTORS

Alexander A. Balandin, Device Research Laboratory, Department of Electrical Engineering, University of California—Los Angeles, Los Angeles, CA 90095

Supriyo Bandyopadhyay, Department of Electrical Engineering, University of Nebraska—Lincoln, Lincoln, NE 68588

Timothy A. Birks, Optoelectronics Group, Department of Physics, University of Bath, Claverton Down, Bath BA2 7AY, UK

Rana Biswas, Ames Laboratory, U.S. Department of Energy, Departments of Physics, and Astronomy and Microelectronics Research Center, Iowa State University, Ames, IA 50011

Sergey I. Bozhevolnyi, Institute of Physics, Aalborg University, DK-9220 Aalborg Øst, Denmark

Yulia E. Danilova, Institute of Automation and Electrometry, Siberian Branch of the Russian Academy of Sciences, 630090 Novosibirsk, Russia

V. P. Drachev, Insititute of Semiconductor Physics, Siberian Branch of the Russian Academy of Sciences, 630090 Novosibirsk, Russia

Thomas F. George, Office of the Chancellor, Departments of Chemistry, and Physics and Astronomy, University of Wisconsin-Stevens Point, Stevens Point, WI 54481

Bernd Hanewinkel, Department of Physics and Material Sciences Center, Philipps-Universität, Renthoff 5, D-35032 Marburg, Germany

Kai-Ming Ho, Ames Laboratory, U.S. Department of Energy, Departments of Physics, and Astronomy and Microelectronics Research Center, Iowa State University, Ames, IA 50011

Alexei A. Kiselev, Physics Research Center, Department of Atmospheric Physics, St. Petersburg State University, Petrodvoretz Ulyanovskaya 1, 198904 St. Petersburg, Russia

Jonathan C. Knight, Optoelectronics Group, Department of Physics, University of Bath, Claverton Down, Bath BA2 7AY, UK

Andreas Knorr, Department of Physics and Material Sciences Center, Philipps-Universität, Renthoff 5, D-35032 Marburg, Germany

Stephen W. Koch, Department of Physics and Material Sciences Center, Philipps-Universität, Renthoff 5, D-35032 Marburg, Germany

Frank L. Madarasz, Center for Applied Optics, University of Alabama in Huntsville, Huntsville, AL 35899

Vadim A. Markel, Department of Electrical Engineering, School of Engineering and Applied Science, Campus Box 1127, Washington University, One Brookings Drive, St. Louis, MO 63130

Eugene F. Mikhailov, Physics Research Center, Department of Atmospheric Physics, St. Petersburg State University, Petrodvoretz Ulyanovskaya 1, 198904 St. Petersburg, Russia

Arkadiusz Orłowski, Instytut Fizyki, Polska Akademia Nauk, Aleja Lotników 32/46, 02-668 Warszawa, Poland

S. V. Perminov, Institute of Semiconductor Physics, Siberian Branch of the Russian Academy of Sciences, 630090 Novosibirsk, Russia

Marian Rusek, Commissariat à l'Energie Atomique, DSM/DRECAM/SPAM, Centre d'Etudes de Saclay, 91191 Gif-sur-Yvette, France

Philip St. J. Russell, Optoelectronics Group, Department of Physics, University of Bath, Claverton Down, Bath BA2 7AY, UK

Vladimir P. Safonov, Institute of Automation and Electrometry, Siberian Branch of the Russian Academy of Sciences, 630090 Novosibirsk, Russia

Andrey K. Sarychev, Center for Applied Problems of Electrodynamics, 127412 Moscow, Russia

Vladimir M. Shalaev, Department of Physics, New Mexico State University, Las Cruces, NM 88003

Mihail M. Sigalas, Ames Laboratory, U.S. Department of Energy, Departments of Physics, and Astronomy and Microelectronics Research Center, Iowa State University, Ames, IA 50011

Costas M. Soukoulis, Ames Laboratory, U.S. Department of Energy, Departments of Physics, and Astronomy and Microelectronics Research Center, Iowa State University, Ames, IA 50011

Mark I. Stockman, Department of Physics and Astronomy, Georgia State University, Atlanta, GA 30303

Frank Szmulowicz, Air Force Research Laboratory, Materials and Manufacturing Directorate (AFRL/MLPO), Wright Patterson AFB, OH 45433

Peter Thomas, Department of Physics and Material Sciences Center, Philipps-Universität, Renthoff 5, D-35032 Marburg, Germany

Segey S. Vlasenko, Physics Research Center, Department of Atmospheric Physics, St. Petersburg State University, Petrodvoretz Ulyanovskaya 1, 198904 St. Petersburg, Russia

Kang L. Wang, Device Research Laboratory, Department of Electrical Engineering, University of California—Los Angeles, Los Angeles, CA 90095

CONTENTS

PREFACE

One of the most important features of nanostructures, from the point of view of an optical scientist, is that they have characteristic length scales much larger than atomic sizes but small compared to the wavelength of light. In a sense, nanomaterials can be thought of as an intermediate phase between bulk samples, which have characteristic sizes much larger than the wavelength in question, and individual atoms and molecules. The vast majority of optical phenomena involving macroscopic samples can be understood from the purely classical description based on Maxwell's equations. Of course, this statement is not entirely true, since the optical constants of matter are treated as phenomenological while, in fact, they must be calculated microscopically; nevertheless, it is possible to understand the physical optics of macroscopic continuous media from a purely classical point of view. By contrast, the interaction of light with atoms and molecules is generally described only by quantum mechanics. In the case of nanostructures, both approaches are used, which is illustrated by the selection of chapters in this volume. As a rough rule, the quantum description is more appropriate when the optically active electrons (or holes) are free but confined in nanostructured potentials, as in the case of quantum dots and wires (Chapters 11 and 12).

Although there exists a vast literature devoted to the fabrication of nanostructured materials and their physical and chemical properties and applications, very few books, apart from the literature on photonic band crystals [1–4], focus specifically on optical properties [5–10]. Even less attention is paid to the *classical* description of the optical properties of nanostructures. This volume is an attempt to fill in these blanks. Although far from being exhaustive, which is, perhaps, impossible in such a rapidly developing and vast field as physics, it collects under one cover review chapters written by physicists who actively work in the field of optical properties of nanostructured objects, including materials that are not designed or fabricated on purpose (such as carbonaceous soot).

Due to volume limitations, we have had to restrict the scope of this book, and, as a result, it tends to naturally tilt toward the scientific interests of the editors. Therefore, a relatively large number of chapters is devoted to the classical description of *disordered* nanostructured systems, including fractals, nanocomposites, and random metal–dielectric fields. However, several chapters deal with {ordered} systems. The book opens with two chapters on photonic band crystals and photonic crystal fibers. Chapters 3 and 4 are devoted to near-field optical studies of nanostructures objects. Localization of light in disordered dielectrics and metal–dielectric films is discussed in Chapters 5 and 6. The next two chapters discuss in detail nonlinear optical properties of fractal clusters built from nanometer-sized particles from the experimental (Chapter 7) and theoretical (Chapter 8) points of view. Chaos of dipolar eigenvalues and nonlinear susceptibilities in nanocomposites are also considered in Chapter 8. Chapters 9 and 10, devoted to the optics of fractal carbonaceous soot, are closely related to the previous chapters (Chapters 5–8) in the way the electromagnetic interaction is described. However, they discuss distinctly different objects, in which this interaction is off-resonant, in contrast to resonant interactions described in Chapters 6–8. Finally, the two closing chapters (Chapters 11 and 12) review the physics and applications of quantum wires and quantum dots.

We would like to express our deep appreciation of the time and effort that the authors spent in writing their excellent chapters. We are very impressed by the authors' genuine interest in this publication and by their promptness and co-operation. We are also grateful to Maggie Kuhl at the University of Wisconsin—Stevens Point who helped with the editorial work.

VADIM MARKEL

Washington University in St. Louis, St. Louis, MO 63130

THOMAS GEORGE

University of Wisconsin—Stevens Point, Stevens Point, WI 54481

REFERENCES

1. C. M. Soukoulis, Ed., *Photonic band gaps and localization*, Plenum, New York, 1993.
2. C. M. Soukoulis, Ed., *Photonic band gap materials*, Kluwer Academic Publishers, Dordrecht, The Netherlands, 1996.
3. J. D. Joannopoulos, R. D. Meade, and J. N. Winn, *Photonic crystals: molding the flow of light*, Princeton University Press, Princeton, NJ, 1995.
4. T. F. Krauss, *Photonic crystals in optical regime: Past, present and future*, Elsevier Science, New York, 1999.
5. F. Henneberger, S. Schmitt-Rink, and E. O. Gobel, Eds., *Optics of semiconductor nanostructures*, Akademie Verlag, Berlin, 1993.
6. J.-P. Leburton and C. S. Torres, Eds., *Photons in semiconductor nanostructures*, Kluwer Academic, Boston, 1993.
7. T. Goto and Y. Segawa, Eds., *Proceedings of the International Conference on Optical Properties of Nanostructures*, Tokyo, *Jpn. J. Appl. Phys.* (1995).
8. S. V. Gaponenko, *Optical properties of semiconductor nanocrystals*, Cambridge, New York, 1998.
9. M. Ohtsu and H. Hori, *Near-field nano-optics: From basic principles of nano-fabrication and nano-photonics*, Kluwer/Plenum, New York, 1999.
10. V. M. Shalaev, *Nonlinear optics of random media: Fractal composites and metal dielectric films*, Springer-Verlag, Berlin, 2000.

OPTICS OF NANOSTRUCTURED MATERIALS

CHAPTER ONE

Photonic Crystals

MIHAIL M. SIGALAS, KAI-MING HO, RANA BISWAS, and COSTAS M. SOUKOULIS

Ames Laboratory-US DOE, Departments of Physics and Astronomy, and Microelectronics Research Center, Iowa State University, Ames IA 50011

1.1. INTRODUCTION

Photonic crystals are a novel class of artificially fabricated structures that have the ability to control and manipulate the propagation of electromagnetic (EM) waves. Properly designed photonic crystals can either prohibit the propagation of light, allow it only in certain frequency regions, or localize light in specified areas. They can be constructed in one, two, and three dimensions (1D, 2D, and 3D) with either dielectric and/or metallic materials.

The concept of photonic band structure arises in analogy to the concept of electronic band structure. Just as electron waves, traveling in the periodic potential of a crystal, are arranged into energy bands separated by band gaps, we expect the analogous phenomenon to occur when EM waves propagate in a medium where the dielectric constant varies periodically in space. There is particular interest in structures that can produce a forbidden frequency gap in which all propagating states are prohibited: Such materials are called photonic band-gap materials and is the topic of intensive studies by many groups theoretically and experimentally [1–4].

Photonic band gaps can have a profound impact on many areas in pure and applied physics. Due to the absence of optical modes in the gap, spontaneous emission is suppressed for photons with frequencies in the forbidden region. It has been suggested that, by tuning the photonic band gap to overlap with the electronic band edge, electron–hole recombination can be controlled in a photonic band-gap material, leading to enhanced efficiency and reduced noise in the operation of various optoelectronic devices [5]. Likewise, the suppression of spontaneous emission can be used to prolong the lifetimes of selected chemical species in catalytic processes [6–8]. Photonic band-gap materials can also find applications in frequency-selective mirrors, band-pass filters, and resonators. Besides technical applications in various areas, scientists are interested in the possibility of observing

Optics of Nanostructured Materials, Edited by Vadim A. Markel and Thomas F. George
ISBN 0-471-34968-2

the localization of EM waves by the introduction of defects and disorder in a photonic band-gap material [9–12]. This will be an ideal realization of the phenomenon of localization uncomplicated by many-body effects present in the case of electron localization. Another interesting effect is that, zero-point fluctuations, which are present even in vacuum, are absent for frequencies inside a photonic gap.

There has been a rapid development over the past several years in the fabrication of photonic band-gap materials. Unlike the case of electron waves, which usually have wavelengths on the atomic scale, the wavelengths of EM of interest are several orders of magnitudes larger, varying between hundreds of nanometers for visible light to meters and centimeters for radio- and microwaves. Thus, while for electron waves the periodic lattice is constrained by the crystal structure, the periodic dielectric structures for photonic band-gap materials are artificial structures that can be designed and fabricated to provide a desired electromagnetic response. Therefore, there is a lot of interest in theoretical calculations for these systems and, over the past few years, advances in the field have been characterized by a close collaboration between theorists and experimentalists.

Photonic crystals in 1D have been well known for more than 50 years and are the basis of many devices, such as dielectric mirrors, Fabry–Perot filters, and distributed feedback lasers [13–15]. A 1D photonic crystal is shown in Fig. 1.1(a). It consists of a superlattice of alumina (dielectric constant, $\varepsilon = 9.61$) layers with thickness of 0.4375 mm and air (dielectric constant of 1) layers with thickness of 1.3125 mm. The transmission of EM waves incident on the structure at three different incident angles (0°, 30°, and 60°) is shown in Fig. 1.1. For normal incidence (solid lines in Fig. 1.1), there is a drop of the transmission from 36 up to 77 GHz. The transmission at the center of the gap is almost five orders (almost −50 dB) of magnitude less than the incident wave. The gap is created by the interference of waves due to the periodicity

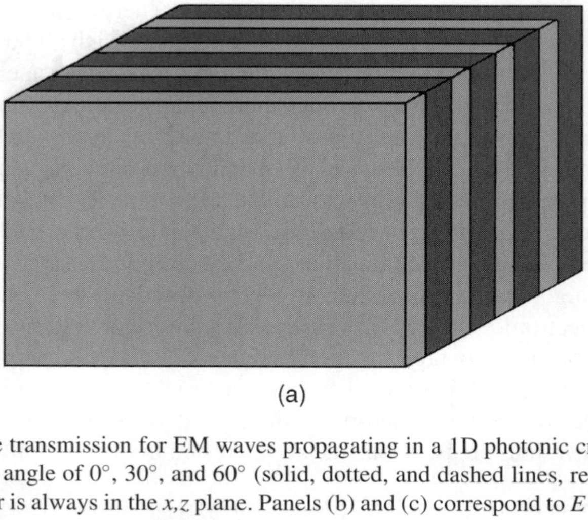

(a)

Figure 1.1. The transmission for EM waves propagating in a 1D photonic crystal [see panel (a)] for incident angle of 0°, 30°, and 60° (solid, dotted, and dashed lines, respectively). The incident **k** vector is always in the x,z plane. Panels (b) and (c) correspond to E fields parallel to the y axis and in the x,z plane, respectively.

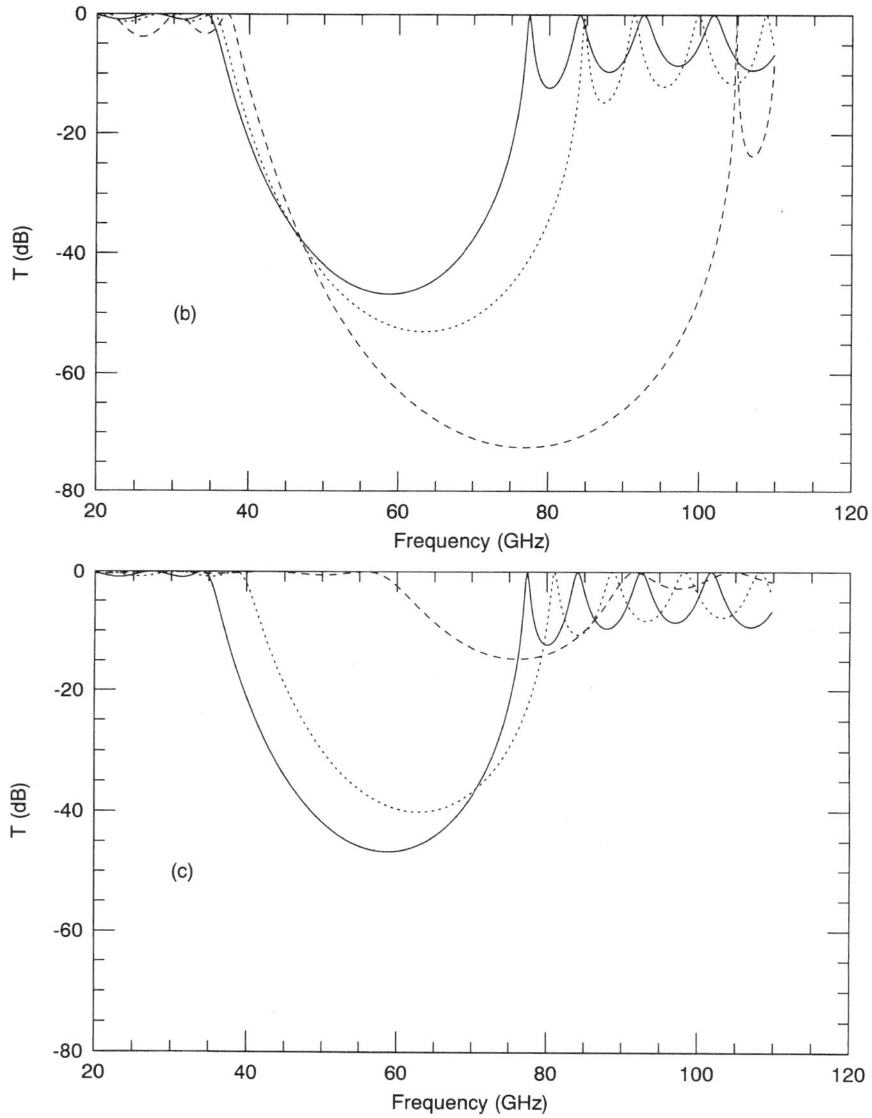

Figure 1.1. (*Continued*)

of the structure along the z axis. One expects that the gap will disappear for other incident angles. Indeed, by increasing the incident angle, the gap increases for the polarization with the electric field out of the plane of incidence *but* the gap tends to disappear for the wave polarized in the plane of incidence.

A realization of a 2D photonic crystal is constructed using infinitely long dielectric cylinders arranged in the so-called 2D triangular lattice [1–4]. The cross-section of this structure is shown in Fig. 1.2(a). The transmission for EM waves with incident **k** vector in the *x,z* plane is shown in Fig. 1.2. We use air cylinders with radius 0.805 mm surrounded by a dielectric with $\varepsilon = 12.25$ (approximately the dielectric constant of GaAs); the distance between the center of the cylinders is 1.75 mm and the total thickness of the system along the *z* direction is 9.1 mm. For waves with the *E* field parallel to the cylinders [Fig. 1.2(b)], there is a small gap at ~48 GHz and a much wider gap at ~80 GHz. As the incident angle increases and the **k** vector is perpendicular to the axis of the cylinders, the second gap moves to smaller frequencies. For the polarization with the *E* field in the *x,z* plane [Fig. 1.2(c)], there is a gap at ~70 GHz for all the angles and for **k** vectors perpendicular to the axis of the cylinders. So, for both polarizations and **k** vectors in the *x,z* plane, there is a gap from 70 to 80 GHz. As in the 1D case, we expect that the gap is going to disappear as the **k** vector moves out of the *x,z* plane since the system is homogeneous along the *y* axis.

It is clear from the previous discussion that we need a 3D structure with periodicity along three directions in order to have a *complete photonic band gap*, that is, a photonic band gap for all the polarizations and all the incident directions. Intense research in the beginning of the 1990s showed that there are some specific structures that poses a *complete photonic band gap* (PBG) [1–4,16,17]. In fact, the first 3D photonic crystal build by Yablonovitch and Gmitter [18] did not have a *complete PBG*. This structure consisted of air spheres embedded in an Al_2O_3

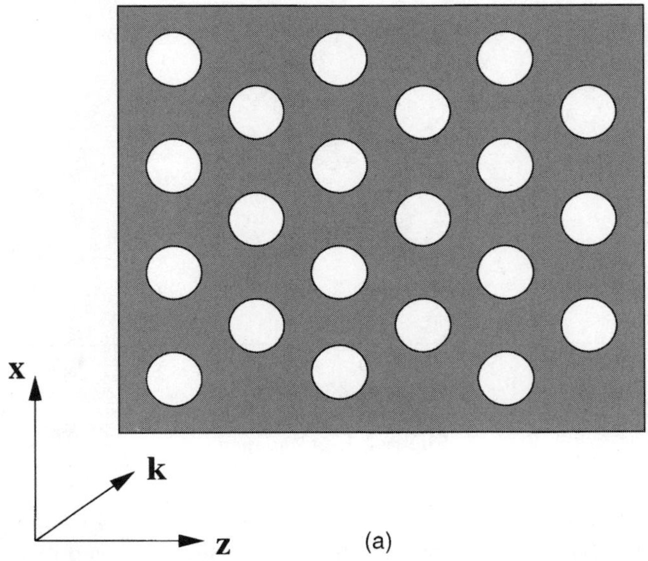

Figure 1.2. The transmission for EM waves propagating in a 2D photonic crystal. The cross-section of the structure is shown in panel (a). The incident angle is 0°, 30°, and 60° (solid, dotted, and dashed lines, respectively). The incident **k** vector is always in the *x,z* plane. Panels (b) and (c) correspond to *E* fields parallel to the *y* axis and in the *x,z* plane, respectively.

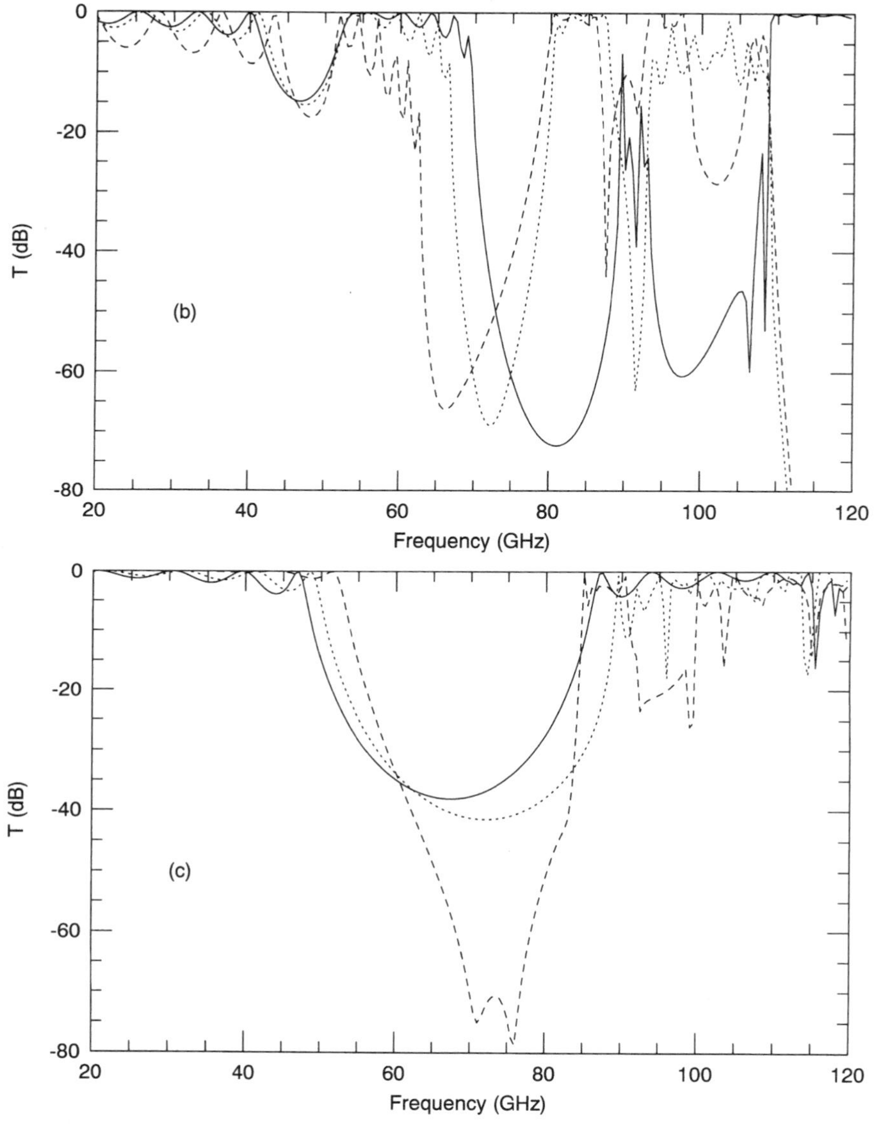

Figure 1.2. (*Continued*)

material forming a face-centered cubic (fcc) lattice. This crystal can be visualized by placing the spheres at the edges and at the center of the faces of a cube. It was constructed by drilling hemispherical cavities on dielectric plates that were stacked together. The whole structure can be constructed by periodically displacing the cube into the space. In contrast to the transmission measurements that showed a *complete PBG* for this structure, subsequent theoretical studies showed that there is not a *complete photonic band gap* for this structure due to a degeneracy of modes at a particular direction [19–21].

More interestingly, Ho et al. [21,22] theoretically proved that the diamond structure consisting of air or dielectric spheres poses a *complete PBG*. A diamond structure is similar to the fcc structure but instead of placing one sphere in each fcc lattice point, we place one more sphere in each lattice point displaced parallel to the body diagonal of the cube by one-quarter of the length of the diagonal. The first photonic crystal with a *complete PBG* was build by Yablonovitch et al. [23]. It has an fcc lattice, but instead of spheres there are cylindrical voids in each lattice point. The so-called "three cylinder structure", can be constructed by drilling holes on the surface of a materials' slab, which are penetrating throughout the whole slab. The holes are forming a triangular array. Three drilling operations are conducted through each hole, 35.26° off normal incidence and spread out 120° on the azimuth. The structure had a *complete PBG* centered at 14 GHz and the forbidden gap width was 19% of its center frequency.

More recently, our group designed and fabricated the layer-by-layer structure shown in Fig. 1.3 [24]. The structure is assembled by stacking layers consisting of parallel rods with a center-to-center separation of a. The rods are rotated by 90° in each successive layer. By starting at any reference layer, the rods of every second neighboring layer are parallel to the reference layer, but shifted by a distance $0.5a$ perpendicular to the rod axes. This results in a stacking sequence that repeats every four layers. This lattice has face-centered tetragonal (fct) lattice symmetry with a

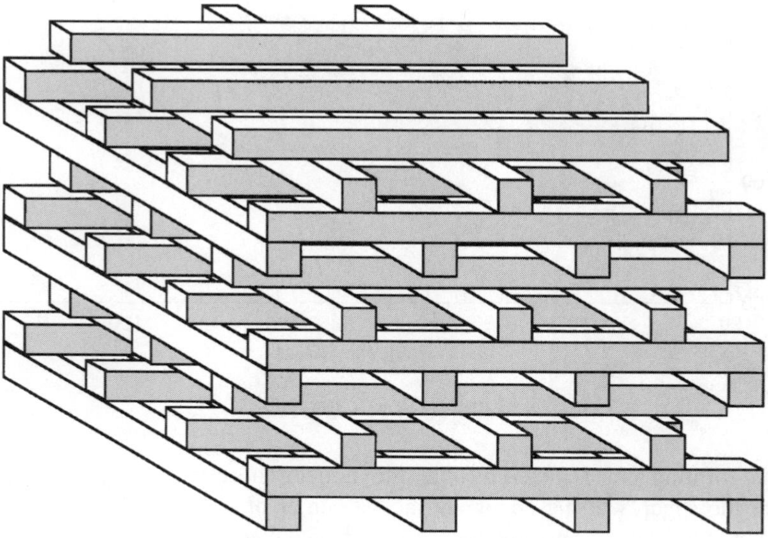

Figure 1.3. Layer-by-layer structure constructed by orderly stacking of dielectric rods. The periodicity is four layers in the stacking direction. Layers in the second neighbor layer are shifted by $a/2$ in the plane.

basis of two rods. This structure has a robust photonic band gap when both the filling ratio and the dielectric contrast meet certain requirements. The photonic band gap is not sensitive to the cross-sectional shape of the rods. Several different structures have been constructed with midgap frequencies at 13, 100, and 450 GHz using etching techniques and Al_2O_3 or Si as materials [25–30]. Recently, the same structures have been fabricated with a measured PBG at the far- and near-infrared (IR) wavelengths using laser-induced direct-write deposition from the gas phase and advanced silicon-processing techniques [31].

Figures 1.4 and 1.5 show the transmission for a layer-by-layer structure with rods having a circular cross-section: The radius of each rod is 20 μm, the in-plane separation of the rods is 160 μm, and the dielectric constant of the rods is 9.61. The crystal contains 12 layers of rods (3 unit cells). For propagation along the stacking direction (**k** parallel to the *z* axis), there is a gap between 0.9 and 1.25 THz and the transmitted intensity at the center of the gap is more than six orders of magnitude smaller than the incident intensity (−60 dB). By increasing the incident angle, the gap becomes smaller and the transmission at the center of the gap increases, but there is a complete PBG for all the angles and polarizations between 0.9 and 1.05 THz.

By creating small distortions (defects) in the photonic crystals discussed in the previous paragraphs, we can also create defect states inside the PBG that give rise to sharp peaks of the transmission inside the PBG. This is analogous to donors or acceptors in semiconductors that introduce impurity states in the electronic energy gap. Such systems can be used as filters. Let us first study a 1D photonic crystal with a defect (Fig. 1.6). The structure is the same as the one studied in Fig. 1.1, with one defect. In the third unit cell, the air slab has a thickness 0.8 times its original thickness, and the dielectric slab has a thickness 1.2 times its original thickness. As a result of that difference, the transmission for normal incidence has a peak inside the gap at 71 GHz (cf. the solid lines in Figs. 1.1 and 1.6). However, when the incident angle increases, the transmission peak moves to higher frequencies and it appears at different frequencies for each polarization (cf. the different lines in Fig. 1.6). Even for the 30° angle, the transmission peak appears at 76 GHz (for *E* field perpendicular to the plane of incidence), 5 GHz higher than its value at normal incidence.

The strong dependence of the defect mode on the incident angle is again related to the fact that the 1D photonic crystal is actually homogeneous along the *x* and *y* directions. It has been shown theoretically and experimentally that 3D photonic crystals can solve this problem [29,30,32–34]. For simplicity, we will illustrate this with a 2D photonic crystal similar to the one in Fig. 1.2. One should keep in mind though that 2D photonic crystals suffer from the same disadvantage as the 1D photonic crystals discussed in the previous paragraph. The reason for this problem is the homogeneity of the structure along the *y* axis [see Fig. 1.2(a)]. For this reason, we will show results only for **k** vectors in the *x*,*z* plane [see Fig. 1.2(a)] and for *E* field parallel to the axis of the cylinders (*y* axis). The defect is introduced by decreasing the radius of one cylinder in the center of the structure. The distorted radius is 0.7 times its original value. For normal incidence (solid line in Fig. 1.7),

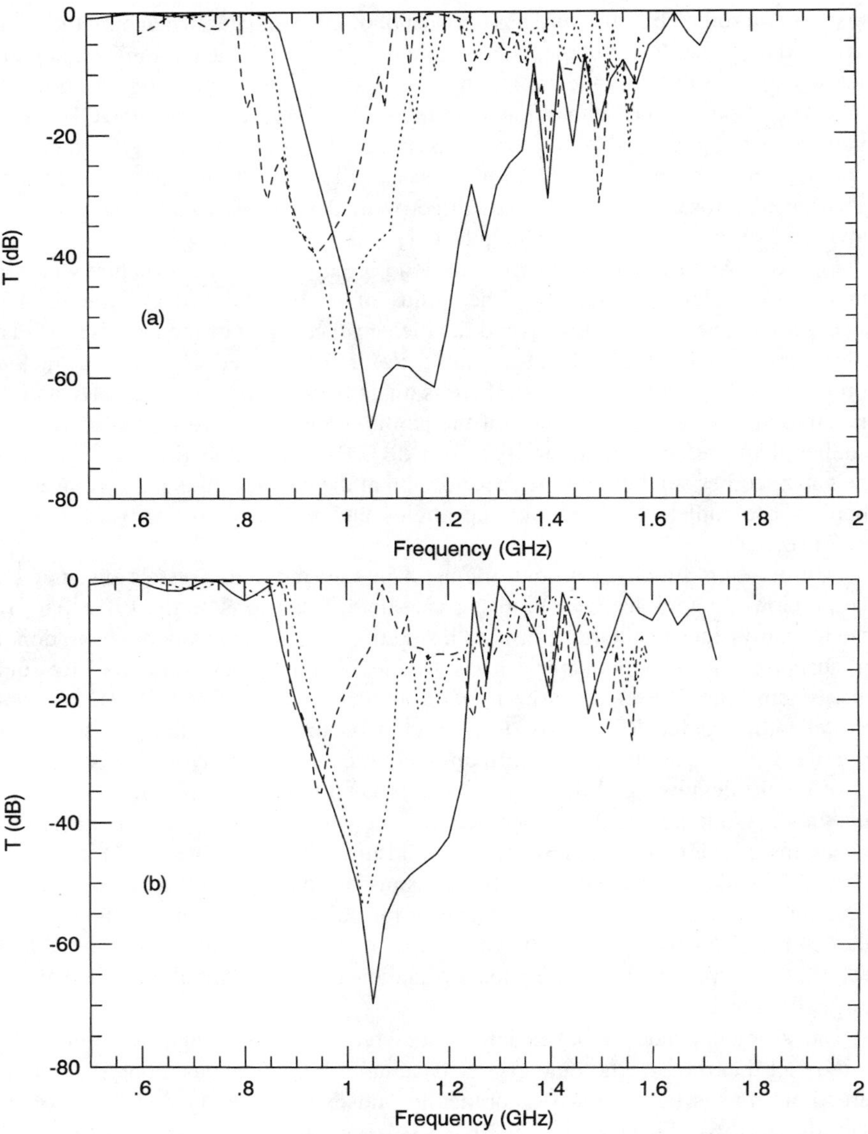

Figure 1.4. The transmission for EM waves propagating in a 3D photonic crystal similar to the one shown in Fig. 1.3. The incident angle is 0°, 30°, and 60° (solid, dotted, and dashed lines, respectively). The incident **k** vector is always perpendicular to the *y* axis. Panels (a) and (b) correspond to *E* fields parallel to the *y* axis and in the *x*,*z* plane, respectively. The rods of the first layer are parallel to the *y* axis.

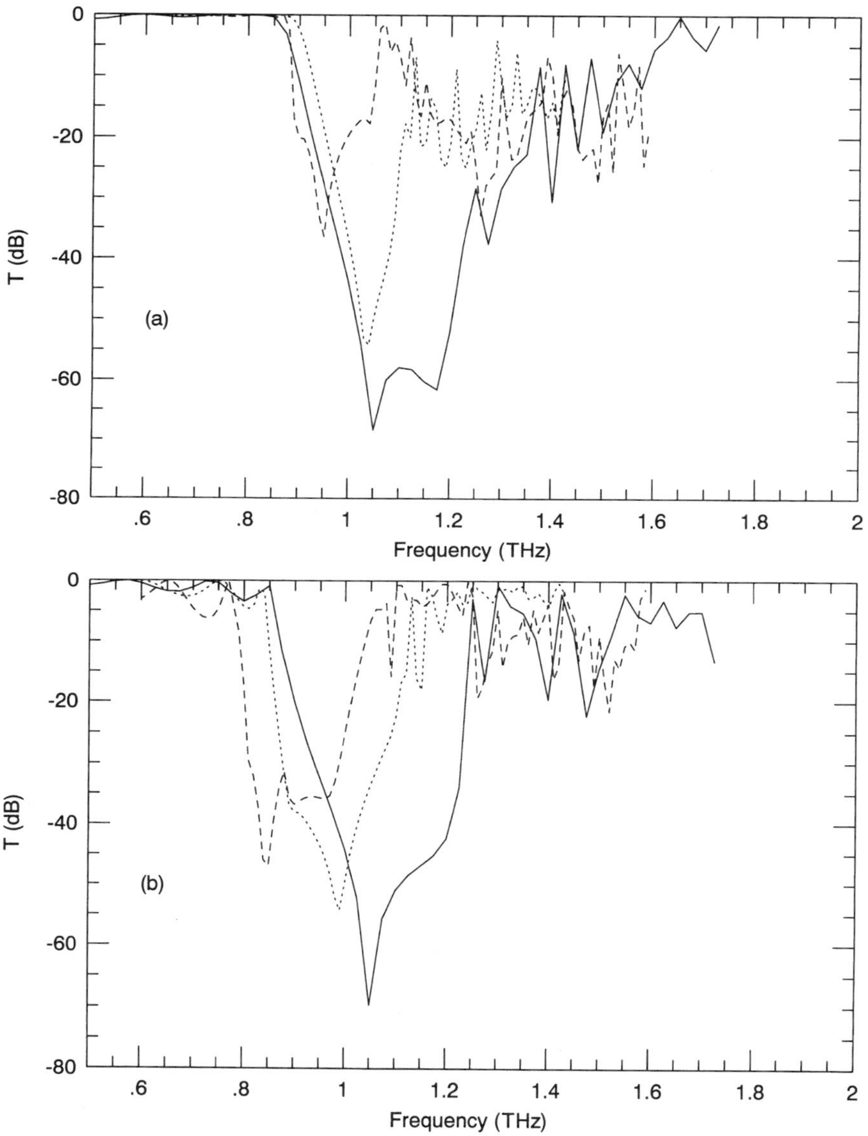

Figure 1.5. The same as in Fig. 1.4 except, the **k** vector is always perpendicular to the x axis.

there are three transmission peaks inside the gap, at 71, 76, and 78 GHz. For an incident angle of 30°, the peaks remain at almost the same frequencies. The small changes in the frequency are actually an artifact of the calculations. We will return to this point when we discuss the computational method (TMM, Section 1.2.2). One can tune the position of the transmission peak inside the gap by changing the radius

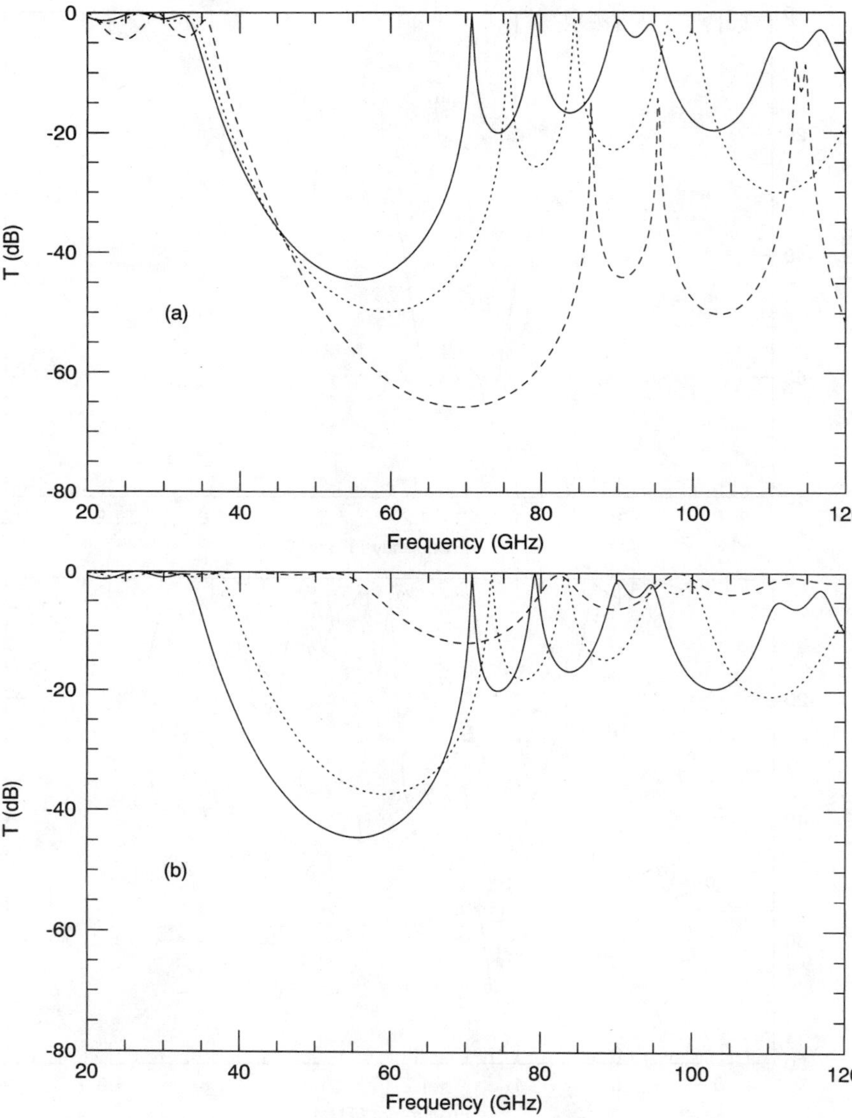

Figure 1.6. The transmission for EM waves propagating in a 1D photonic crystal (similar to the one in Fig. 1.1) with a defect for incident angle of 0°, 30°, and 60° (solid, dotted, and dashed lines, respectively). The incident **k** vector is always in the *x,z* plane. Panel (a) and (b) correspond to *E* fields parallel to the *y* axis and in the *x,z* plane, respectively.

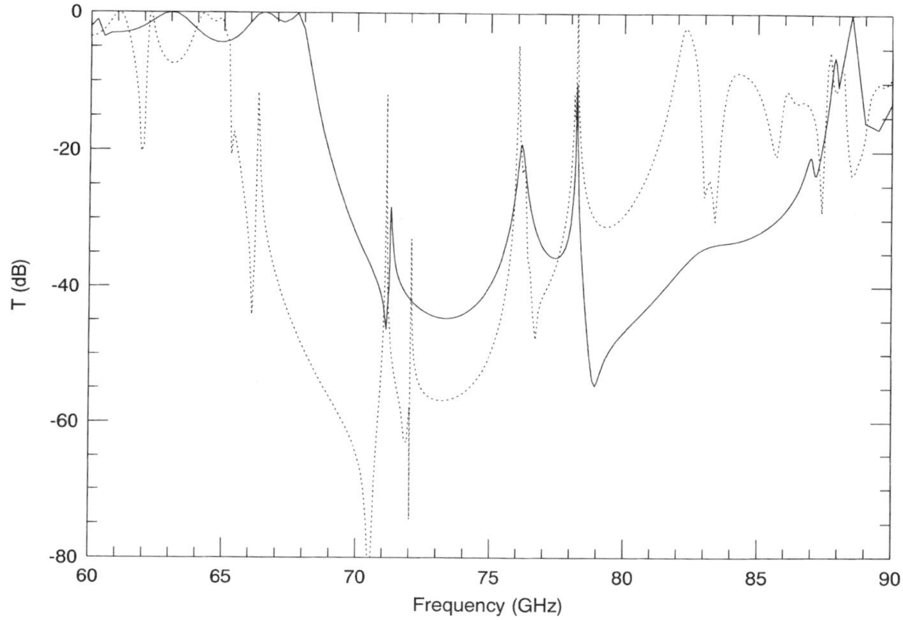

Figure 1.7. The transmission for EM waves propagating in a 2D photonic crystal (similar to the one in Fig. 1.2) with a defect. The incident angle is 0° and 30° (solid and dotted, respectively). The incident **k** vector is always in the *x,z* plane. The *E* field is parallel to the *y* axis.

of the distorted cylinder. The flexibility in tuning defect modes makes photonic crystals a very attractive medium for the design of novel types of filters, couplers, laser microcavities, and so on [1–4].

1.2. THEORETICAL METHODS

1.2.1. Plane Wave Method

To study the behavior of EM waves in photonic band-gap materials, we have to solve Maxwell's equations for a media characterized by a spatially varying dielectric function $\varepsilon(\mathbf{r})$:

$$\nabla \cdot \mathbf{D} = 0 \tag{1.1}$$

$$\nabla \cdot \mathbf{H} = 0 \tag{1.2}$$

$$\nabla \times \mathbf{E} = i(\omega/c)\mathbf{H} \tag{1.3}$$

$$\nabla \times \mathbf{H} = -i(\omega/c)\mathbf{D} \tag{1.4}$$

$$\mathbf{D}(\mathbf{r}) = \varepsilon(\mathbf{r})\mathbf{E}(\mathbf{r}) \tag{1.5}$$

These may be decoupled to generate an equation only in the magnetic field,

$$\nabla \times (\varepsilon^{-1}(\mathbf{r})\nabla \times \mathbf{H}) = (\omega/c)^2\mathbf{H} \tag{1.6}$$

and

$$\nabla \times (\nabla \times \mathbf{E}) = (\omega/c)^2\varepsilon(\mathbf{r})\mathbf{E} \tag{1.7}$$

At this point, we note that the vector nature of the wave equation is of crucial importance. Early attempts [35] adopting the scalar wave approximation led to qualitatively wrong results. The simplest case happens when $\varepsilon(\mathbf{r})$ is a real and periodic function of $\mathbf{r}$ and we assume that it is frequency independent in the range of interest. We also assume the magnetic permeability μ is 1. In this case, the solution of the problem scales with the period of $\varepsilon(\mathbf{r})$: For example, reducing the size of the structure by a factor of 2 will not change the spectrum of EM modes other than scaling all frequencies up by a factor of 2.

Because of the periodicity of the problem, we can make use of Bloch's theorem to expand the electric and magnetic fields in terms of Bloch waves:

$$\mathbf{H}(\mathbf{r}) = \sum_{\mathbf{K}} \mathbf{H}_{\mathbf{k}} \exp(i\mathbf{K}\cdot\mathbf{r}) \tag{1.8}$$

where $\mathbf{K} = \mathbf{k} + \mathbf{G}\cdot\mathbf{k}$ is a vector in the Brillouin zone and $\mathbf{G}$ is a reciprocal lattice vector.

The solution for the magnetic field has the form of an eigenvalue problem

$$\sum_{\mathbf{K}} \mathbf{K} \times \varepsilon^{-1}_{\mathbf{K},\mathbf{K}'}(\mathbf{K} \times \mathbf{H}_{\mathbf{K}'}) = -(\omega/c)^2\mathbf{H}_{\mathbf{K}} \tag{1.9}$$

The corresponding equation for the E field does not have the form of a simple eigenvalue problem since the dielectric function enters into the frequency dependent right-hand side

$$\mathbf{K} \times \mathbf{K} \times \mathbf{E}_{\mathbf{K}} = -(\omega/c)^2 \sum_{\mathbf{K}'} \varepsilon_{\mathbf{K},\mathbf{K}'}\mathbf{E}_{\mathbf{K}'} \tag{1.10}$$

Hence, we obtain photonic band structure by solving Eq. (1.9) for the magnetic fields.

Here $\varepsilon_{\mathbf{K},\mathbf{K}'} = \varepsilon(\mathbf{G} - \mathbf{G}')$ is the Fourier transform (FT) of the dielectric function. Dielectric functions with sharp spatial discontinuities require an infinite number of plane waves in the Fourier expansion. To avoid this problem, we smear out the interfaces of the dielectric objects in the unit cell. For example, for modeling a cylinder of radius a and dielectric ε, we employ the smeared dielectric function

$$\varepsilon(\mathbf{r}) = 1 + (\varepsilon - 1)/(1 + \exp((r - a)/w) \tag{1.11}$$

where the width w of the interface is chosen as a small fraction of the radius a ($\approx 0.01 - 0.05a$). In practice, we incorporate the smearing and define the dielectric function $\varepsilon(\mathbf{r})$ over a grid in real-space. The FT of the dielectric function in our finite plane wave basis set is computed to obtain $\varepsilon(\mathbf{G}-\mathbf{G}')$. The dielectric matrix in Fourier space is then inverted to obtain $\varepsilon^{-1}(\mathbf{G}-\mathbf{G}')$. This procedure yields much better convergence than the alternative method of determining $\varepsilon^{-1}(\mathbf{r})$ in real-space and then performing a FT to obtain $\varepsilon^{-1}(\mathbf{G}-\mathbf{G}')$.

The transverse components of the magnetic field are $h_{\mathbf{K},\lambda}$, that is,

$$\mathbf{H}_{\mathbf{K}} = h_{\mathbf{K},1}\mathbf{e}_1 + h_{\mathbf{K},2}\mathbf{e}_2 \tag{1.12}$$

where the unit vectors $\mathbf{e}_1$ and $\mathbf{e}_2$ form an orthogonal triad $(\mathbf{e}_1, \mathbf{e}_2, \mathbf{K})$.

The solution (1.9) for the magnetic field reduces to the eigenvalue problem

$$\sum_{\mathbf{K}',\lambda'} M(\mathbf{K},\lambda;\mathbf{K}',\lambda')h_{\mathbf{K}',\lambda'} = (\omega/c)^2 h_{\mathbf{K},\lambda} \tag{1.13}$$

The matrix M is defined by

$$M(\mathbf{K},\lambda;\mathbf{K}',\lambda') = |\mathbf{K}||\mathbf{K}'|\begin{pmatrix} \mathbf{e}_2\cdot\mathbf{e}'_2 & -\mathbf{e}_2\cdot\mathbf{e}'_1 \\ -\mathbf{e}_1\cdot\mathbf{e}'_2 & \mathbf{e}_1\cdot\mathbf{e}'_1 \end{pmatrix}. \tag{1.14}$$

In practice, the photonic band structure given by the frequencies $\omega(\mathbf{K},\lambda)$ are computed over several high symmetry points in the Brillouin zone or on a grid in the Brillouin zone if the density of states is needed. Plane wave convergence is closely checked.

The first structure [19–21] considered by researchers was the fcc structure composed of low dielectric spheres in a high dielectric (ε) background. This simple structure with close-packed spheres has the band structure shown in Fig. 1.8. There is no fundamental gap between the second and third bands—the bands are degenerate at the W point of the zone. There is a region of low densities of states between bands 2 and 3—the pseudogap, which may have interesting consequences. Another very interesting feature is a sizable complete gap between the 8 and 9 bands (8–9 gap), which exists over the entire zone, that is, for all directions of propagation of the EM wave. The size of this gap is $\sim$8% for a refractive index contrast of 3.1. The direct fcc structure (high dielectric spheres in a low contrast background), however, does not possess the dip in the photonic DOS.

The diamond structure has been the subject of much investigation [1–4,21,22], since it has a full 3D photonic band gap between the fundamental bands (2–3 gap between the second and third bands). This gap exists for (1) high dielectric spheres on the sites of the diamond lattice (Fig. 1.9), (2) the diamond structure with low dielectric spheres on the diamond sites [Fig. 1.9; conjugate of (1)], and (3) the diamond structure connected by dielectric rods (Fig. 1.10).

The best performing gap (29%) is reached for the diamond structure with 89% air spheres, that is, a multiply connected sparse structure. A similar large gap (30%) is also found for the diamond structure connected with dielectric rods with $\sim$30% dielectric filling fraction. These gap magnitudes are for a refractive index contrast of 3.6, appropriate for GaAs.

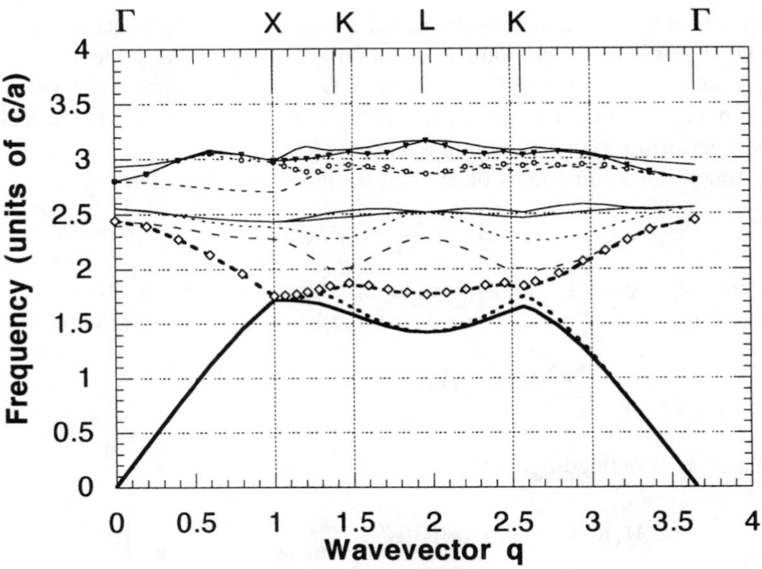

Figure 1.8. Photonic band structure for the fcc structure composed of air spheres in a high dielectric background (dielectric constant $e = 9.61$). The geometry is for close-packed spheres, that is, 74% filling ratio. The bands are shown along the 110 axis of the Brillouin zone.

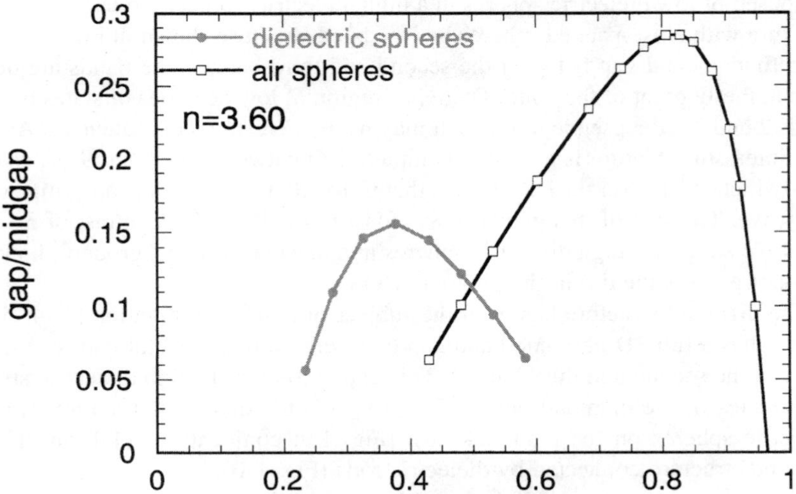

Figure 1.9. Size of the 3D photonic band gap measured by the gap/midgap ratio for a diamond structure with spheres on the diamond sites and its conjugate structure. Band gaps are plotted as a function of the filling ration. The dielectric contrast of 12.96 is used.

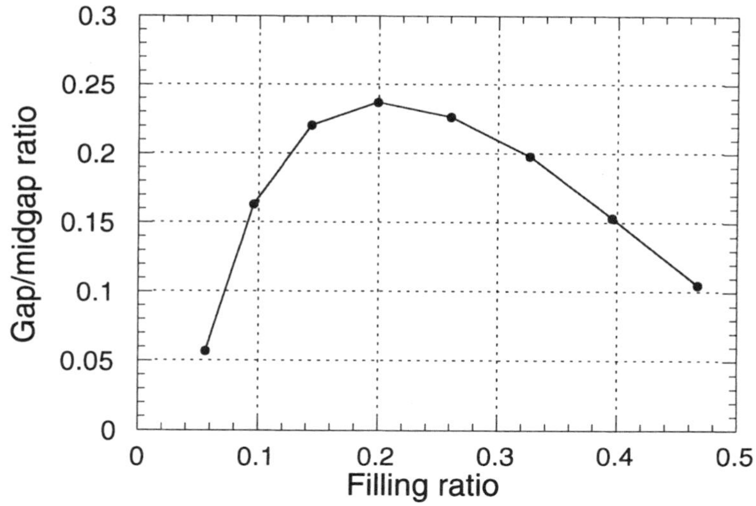

Figure 1.10. Size of the 3D photonic band gap when the diamond structure is connected by dielectric rods ($e = 12.96$). Rectangular rods connect the sites in 111 planes whereas cylindrical rods connect the rods along the 111 axis.

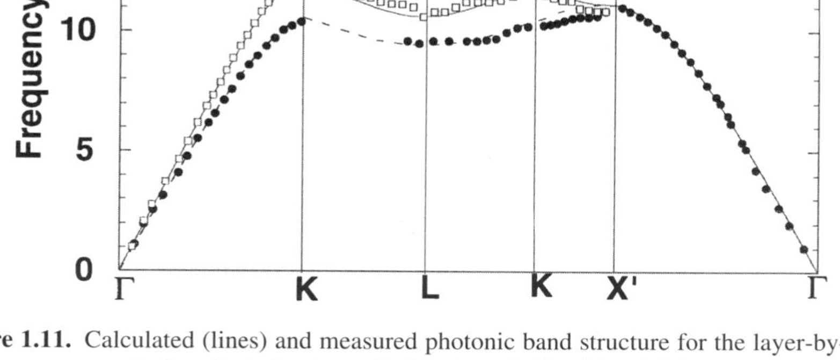

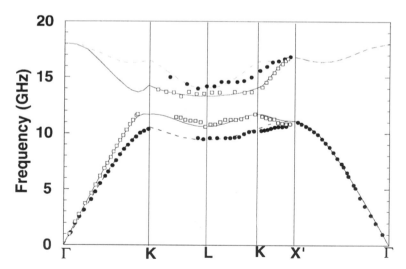

Figure 1.11. Calculated (lines) and measured photonic band structure for the layer-by-layer structure composed of stacked alumina cylinders ($e = 9.61$) with a full band gap between 12 and 14 GHz. The circles represent measurements for the E-field polarization parallel to the rod axes while squares represent measurements performed with an E-field perpendicular to the rod axes.

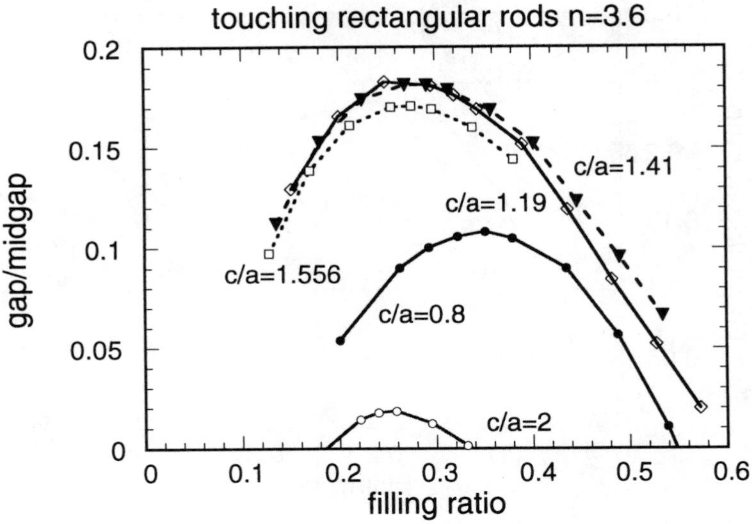

Figure 1.12. Size of the gap in the layer-by-layer structure as a function of filling ratio for different *c/a* ratios. The parameter *c* is the repeat distance and *a* is the rod separation. A dielectric contrast of 12.96 has been used to facilitate comparison.

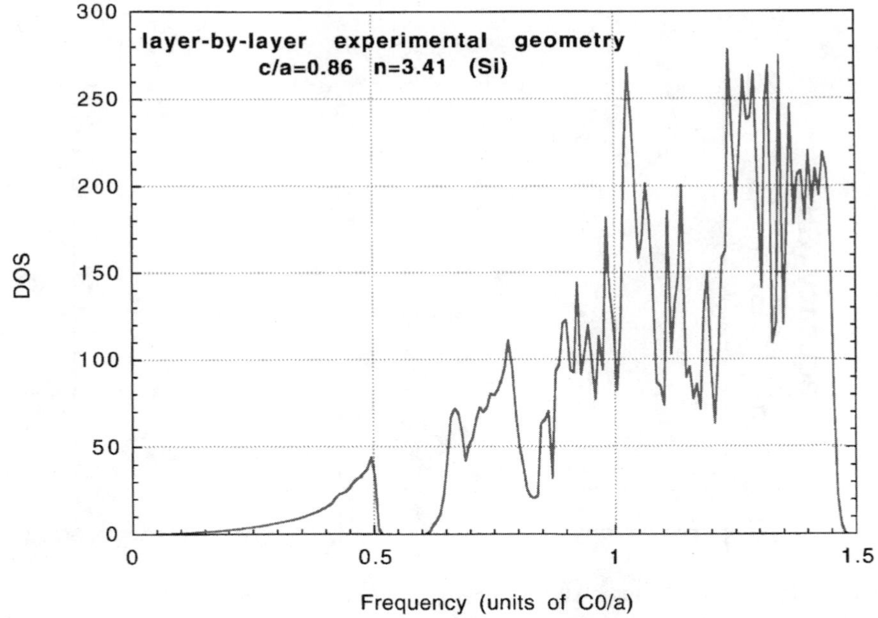

Figure 1.13. Photonic densities of states for the layer-by-layer structure using the experimentally fabricated geometry with a filling ratio of 0.26 and silicon ($e = 11.67$) as the dielectric material.

A novel layer-by-layer structure was designed and fabricated at Iowa State University [24–30] (Fig. 1.3) that has a full 3D fundamental PBG (Fig. 1.11). The agreement between the calculated bands and the experimental measurements are excellent (Fig. 1.11) for both EM wave polarizations. This particularly robust structure has been fabricated at length scales providing gaps ranging from 13 to 500 GHz [24–30]. As is typical for the diamond structure, the magnitude of the gap is maximized at $\sim$25% (Fig. 1.12) for the contrast of $n = 3.6$. The densities of photon states for the experimentally fabricated structure with silicon micromachining (Fig. 1.13) provides a picture of both the fundamental gap and other frequency regions that display depleted or enhanced DOS.

The photonic band gap depends on (1) the local connectivity of the dielectric structure, (2) the contrast between the two media, and (3) the filling ratio. A minimum dielectric constant ($\varepsilon > 4$) is usually needed to observe the band gaps. The photonic band structure method is a systematic way to search for the existence of band gaps in dielectric structures [3,36].

1.2.2. Transfer Matrix Method

While the method described in Section 1.2.1 focuses on a particular wavevector, there are complementary methods that focus on a single frequency. In the transfer matrix method (TMM), first introduced by Pendry and MacKinnon [37], Eqs. (1.3 and 1.4) are discretised and the z components of the fields can be eliminated, so, we derive the following equations:

$$
\begin{aligned}
E_x(i,j,k+1) = {} & E_x(i,j,k) + ic\omega\mu_0\mu(i,j,k)H_y(i,j,k) \\
& + \frac{ic}{a\omega\varepsilon_0\varepsilon(i,j,k)} \\
& \times [a^{-1}\{H_y(i-1,j,k) \\
& - H_y(i,j,k)\} - b^{-1}\{H_x(i,j-1,k) - H_x(i,j,k)\}] \\
& - \frac{ic}{a\omega\varepsilon_0\varepsilon(i+1,j,k)} \\
& \times [a^{-1}\{H_y(i,j,k) - H_y(i+1,j,k)\} - b^{-1}\{H_x(i+1,j-1,k) \\
& - H_x(i+1,j,k)\}]
\end{aligned} \tag{1.15}
$$

$$
\begin{aligned}
E_y(i,j,k+1) = {} & E_y(i,j,k) - ic\omega\mu_0\mu(i,j,k)H_x(i,j,k) + \frac{ic}{b\omega\varepsilon_0\varepsilon(i,j,k)} \\
& \times [a^{-1}\{H_y(i-1,j,k) \\
& - H_y(i,j,k) - b^{-1}\{H_x(i,j-1,k) - H_x(i,j,k)\}] \\
& - \frac{ic}{b\omega\varepsilon_0\varepsilon(i,j+1,k)} \\
& \times [a^{-1}\{H_y(i-1,j+1,k) - H_y(i,j+1,k)\} \\
& - b^{-1}\{H_x(i,j,k) - H_x(i,j+1,k)\}]
\end{aligned} \tag{1.16}
$$

$$\begin{aligned} H_x(i,j,k+1) = {} & H_x(i,j,k) - ic\omega\varepsilon_0\varepsilon(i,j,k+1)E_y(i,j,k+1) \\ & + \frac{ic}{a\omega\mu_0\mu(i-1,j,k+1)}[a^{-1}\{E_y(i,j,k+1) - E_y(i-1,j,k+1)\} \\ & - b^{-1}\{E_x(i-1,j+1,k+1) - E_x(i-1,j,k+1)\}] \\ & - \frac{ic}{a\omega\mu_0\mu(i,j,k+1)}[a^{-1}\{E_y(i+1,j,k+1) - E_y(i,j,k+1)\} \\ & - b^{-1}\{E_x(i,j+1,k+1) - E_x(i,j,k+1)\}] \end{aligned} \tag{1.17}$$

$$\begin{aligned} H_y(i,j,k+1) = {} & H_y(i,j,k) + ic\omega\varepsilon_0\varepsilon(i,j,k+1)E_x(i,j,k+1) \\ & + \frac{ic}{a\omega\mu_0\mu(i,j-1,k+1)} \\ & \times [a^{-1}\{E_y(i+1,j-1,k+1) - E_y(i,j-1,k+1)\} \\ & - b^{-1}\{E_x(i,j,k+1) - E_x(i,j-1,k+1)\}] \\ & - \frac{ic}{a\omega\mu_0\mu(i,j,k+1)}[a^{-1}\{E_y(i+1,j,k+1) - E_y(i,j,k+1)\} \\ & - b^{-1}\{E_x(i,j+1,k+1) - E_x(i,j,k+1)\}] \end{aligned} \tag{1.18}$$

The parameters $\varepsilon(i,j,k)$ and $\mu(i,j,k)$ are the dielectric constant and the magnetic permeability at the subcell (i,j,k). The dimensions of each subcell along the *x,y,z* directions are *a,b,c*. Equations (1.15)–(1.18) are connecting the fields at the $k+1$ plane with the fields at the k plane. By using TMM, the band structure of an infinite periodic system can be calculated, but the main advantage of this method is for the calculation of transmission and reflection properties of EM waves of various frequencies incident on a finite thickness slab of PBG material.

Such calculations are extremely useful in the interpretation of experimental measurements of transmission and reflection data. The TMM method can also be applied to calculate PBG structures containing absorptive and metallic materials. The TMM has previously been applied to defects in 2D PBG structures [38], photonic crystals with complex and frequency dependent dielectric functions [39], metallic PBG materials [40,41], and angular filters [42]. In all these examples, the agreement between theoretical calculations and experimental measurements was very good.

At this point, we would like to return to the discussion of Fig. 1.7. In that case, we used a rectangular conventional unit cell consisting of 15–26 subcells. In order to create the hexagonal lattice, we put cylinders at the corners and at the center of the rectangular unit cell. The system is finite along the *z direction* [see Fig. 1.2(a)] having 3 unit cells thickness. Along the *x* direction, we use a supercell consisting of 3 conventional unit cells and we assume periodic boundary conditions at the edges of the supercell. So, there are infinitely many defects along the *x* direction with separation of $3a$. This is the reason for the small change in the frequency as we change the angle. Calculations with a bigger supercell (consisting of 5 conventional unit cells) show negligible angular dependence.

1.2.3. Finite Difference Time Domain Method

While the preceding transfer matrix method is employed for steady-state solutions, the finite difference time domain (FDTD) method is used for general time-dependent solutions including transient behavior. In this method, the Maxwell curl equations are numerically solved

$$\nabla \times \mathbf{E} = -(1/c)\frac{\partial \mathbf{H}}{\partial t} \tag{1.19}$$

$$\nabla \times \mathbf{H} = (1/c)\varepsilon(\mathbf{r})\frac{\partial \mathbf{E}}{\partial t} \tag{1.20}$$

The derivatives in the Maxwell equations are approximated with finite differences and the electromagnetic field components are located on a Yee cell [43]. In the Yee cell, the E-field components at time $n\Delta t$ are located on the sides of a cube. The magnetic field, H, components at times $(n + 1/2)\Delta t$ are located at the face-centered points of the Yee cell. This results in both spatial and temporal offsets of the two fields when the Maxwell curl equations are solved on each face of the cube. The system is described by a spatial grid. The time step is chosen such that an EM wave will propagate less than a grid spacing during the time step.

$$E_x^{n+1}(i,j,k) = E_x^n(i,j,k) + \frac{\Delta t}{\varepsilon(i,j,k)}\left[\frac{H_z^{n+1/2}(i,j+1/2,k) - H_z^{n+1/2}(i,j-1/2,k)}{\Delta y} - \frac{H_y^{n+1/2}(i,j,k+1/2) - H_y^{n+1/2}(i,j,k-1/2)}{\Delta z}\right] \tag{1.21}$$

$$E_y^{n+1}(i,j,k) = E_y^n(i,j,k) + \frac{\Delta t}{\varepsilon(i,j,k)}\left[\frac{H_x^{n+1/2}(i,j,k+1/2) - H_x^{n+1/2}(i,j,k-1/2)}{\Delta z} - \frac{H_z^{n+1/2}(i+1/2,j,k) - H_z^{n+1/2}(i-1/2,j,k)}{\Delta x}\right] \tag{1.22}$$

$$E_z^{n+1}(i,j,k) = E_z^n(i,j,k) + \frac{\Delta t}{\varepsilon(i,j,k)}\left[\frac{H_y^{n+1/2}(i+1/2,j,k) - H_y^{n+1/2}(i-1/2,j,k)}{\Delta x} - \frac{H_x^{n+1/2}(i,j+1/2,k) - H_x^{n+1/2}(i,j-1/2,k)}{\Delta y}\right] \tag{1.23}$$

$$H_x^{n+1/2}(i,j,k) = H_x^{n-1/2}(i,j,k) + \frac{\Delta t}{\mu(i,j,k)}\left[\frac{E_y^n(i,j,k+1/2) - E_y^n(i,j,k-1/2)}{\Delta z} - \frac{E_z^n(i,j+1/2,k) - E_z^n(i,j-1/2,k)}{\Delta y}\right] \tag{1.24}$$

$$H_y^{n+1/2}(i,j,k) = H_y^{n-1/2}(i,j,k) + \frac{\Delta t}{\mu(i,j,k)}\left[\frac{E_z^n(i+1/2,j,k) - E_z^n(i-1/2,j,k)}{\Delta x} - \frac{E_x^n(i,j,k+1/2) - E_x^n(i,j,k-1/2)}{\Delta z}\right] \tag{1.25}$$

$$H_z^{n+1/2}(i,j,k) = H_z^{n-1/2}(i,j,k) + \frac{\Delta t}{\mu(i,j,k)}\left[\frac{E_x^n(i,j+1/2,k) - E_x^n(i,j-1/2,k)}{\Delta y} - \frac{E_y^n(i+1/2,j,k) - E_y^n(i-1/2,j,k)}{\Delta x}\right] \tag{1.26}$$

The parameter $E_x^n(i,j,k)$ is the x component of the electric field at the n time step in the *(i,j,k)* subcell.

Finite size systems can be easily modeled. This widely used technique [44,45] can be utilized to find either the steady-state or transient response of arbitrary systems containing dielectric or metallic components, as well as materials with nonlinear dielectric properties. In the perfect metallic code, the E field vanishes inside the metal. The FDTD can be used with a Gaussian pulse source. The fields are numerically integrated to obtain the fields at long times (>1000 time steps). The FT of the scattered and incident fields generates the frequency-dependent response of the system. Alternatively, the system may be subject to a source field with a single frequency ω, and one can determine the steady-state response of the system at that frequency. Such steady-state calculations may then be repeated at desired frequencies.

At the edges of the FDTD cell outer radiation, boundary conditions are frequently employed. Here the incident wave at the boundary is absorbed. Methods to transform the near fields to radiating far fields are then employed. This is particularly necessary for antenna problems where far-field radiation pattern are desired. The FDTD method is a very powerful design tool in simulating the EM response of systems, covering a broad range of frequencies.

We have used this method to calculate the radiation patterns of dipole antennas placed on PBG crystals (Fig. 1.14). The dipole antenna is driven by either a voltage pulse or a steady-state sinusoidal excitation and the radiated far fields are determined. The symmetry of our PBG crystal used a computational cell that is one-fourth the size of the actual system [46]. The calculations are in very good agreement with measurements [46]. It is also possible to calculate the currents flowing in the antenna and to calculate the gain of the system. We have driven a finite length dipole oscillator on the surface of the PBG crystal at a frequency of 13 GHz near the center of the band gap. The dipole is at the intersection of the first and second layers and the radiation patterns are calculated and measured at different heights z above the surface (Fig. 1.14). The agreement in both the E and H planes with measurements is very good.

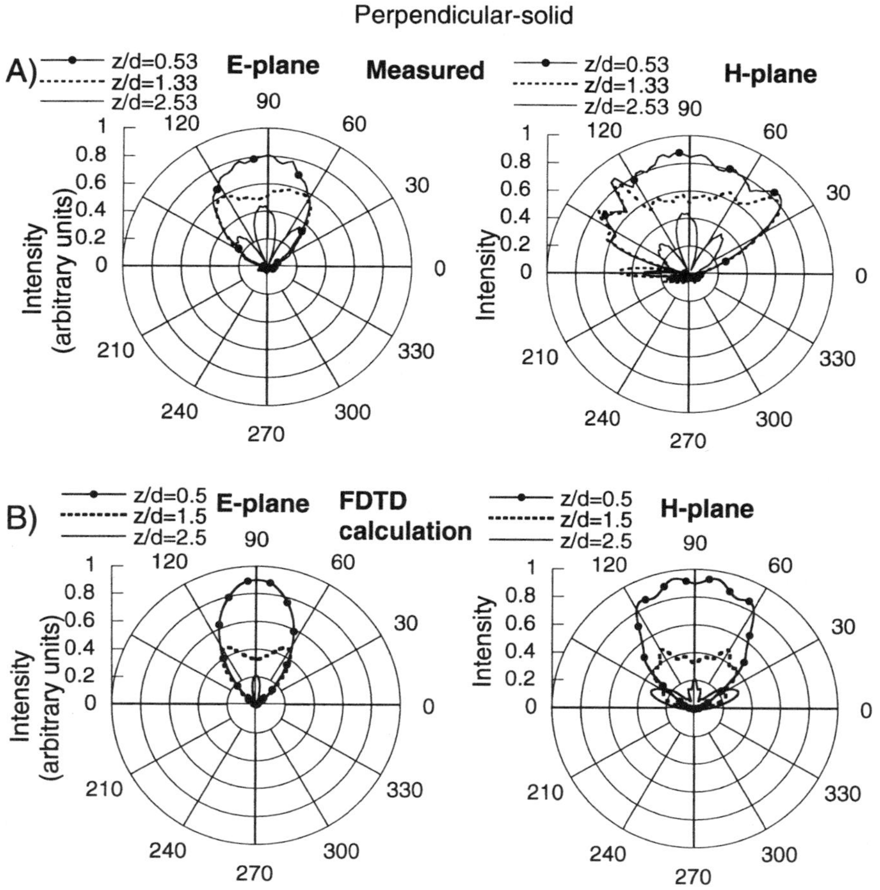

Figure 1.14. Measured (a) and FDTD calculations (b) of the antenna radiation in the E and H planes for a dipole antenna driven at 13 GHz at the center of the PBG. The three curves are for different heights z of the antenna above the surface, expressed as a ratio of z/d, where d is the diameter of the dielectric rod. Antenna is at the intersection of the first and second layers, perpendicular to the first layer.

1.3. RESULTS AND DISCUSSION

In this section, we study some of the recent achievements in the field and we point out some of the difficulties that may arise in the future especially for photonic crystals operating at the optical frequencies. We start with the defect cases.

1.3.1. Defects

In this section, we study 3D layer-by-layer photonic crystals [24–30]. The structure is made of layers of cylindrical alumina rods with a stacking sequence that repeats itself every four layers with repeat distance, $c = 1.272$ cm. Within each layer, the rods are arranged with their axes parallel and separated by a distance $a = 1.123$ cm. The orientations of the axes are rotated by 90° between adjacent layers. To obtain the periodicity of four layers in the direction of stacking, the rods of the second neighbor layers are shifted by a distance of $a/2$ in the direction perpendicular to the rods axes [24–30]. In order to simulate this structure with the TMM, we divide the unit cell into $7 \times 7 \times 8$ subcells assuming that the z axis is along the stacking direction.

Figure 1.15 shows the transmission of EM waves incident on a layer-by-layer photonic crystal with 4 unit cell thickness (16 layers of rods). The k vector of the incident wave is along the stacking direction (z axis). For the periodic case (dotted lines in Fig. 1.15), there is a gap between 11 and 15.7 GHz for both polarizations. We introduce a defect in this structure by removing every other rod in the eighth layer. A defect peak appears at 12.58 GHz. The width of the peak (0.016 GHz) is almost the same for both polarizations, and the transmission at the top of the peak (-3.4 and -29.7 dB for each polarization) is higher for the polarization where the incident electric field is parallel to the axis of the removed rods. In general, the transmission for the parallel polarized waves is more affected by the defect than the perpendicular polarized waves. The Q factor and the defect frequency are in very good agreement with measurements in the same configuration [30]. However, the measured transmission at the top of the peak is ~ 10 dB smaller than the calculated, most probably due to some small absorption of the alumina rods [30]. By increasing the thickness to eight unit cells (32 layers of rods), the width of the peak becomes 10^{-6} GHz, which correspond to $Q > 10^6$, while the defect frequency and the transmission at the top of the peak remain almost the same (12.61 GHz and -3.8 dB, respectively).

1.3.2. Effect of Absorption

In the optical wavelength region, the dielectric constant of most materials has an appreciable imaginary part. In order to study the effect of the absorption on the peak transmission due to defects, we calculated the transmission of a layer-by-layer structure with a defect (Fig. 1.16). The structure is similar with the one described in Fig. 1.15, in which we removed every second rod from the eighth layer (the system contains 16 layers of rods). We assume the dielectric constant is given by $\varepsilon = 9.61 + ix$. By increasing the imaginary part of the dielectric constant, the Q, as well as the transmission at the peak, decrease. In particular, $Q = 800$, 262,163, and 90 for $x = 0, 0.05, 0.1$, and 0.2, respectively, while the transmission at the peak is -3.4, -14.0, -18.7, and -24.0 dB (Fig. 1.16). The introduction of the absorption makes the peak wider while the transmission of the peak is smaller.

Even in periodic structures, the effect of the absorption could significantly change the transmission. Figure 1.17 show the transmission of a periodic layer-by-layer structure similar with the one described in Fig. 1.15 with a 3 unit cell thickness. The real part of the dielectric constant is 9.61. By increasing the imaginary part, the transmission decreases at all the frequencies. Especially at the upper edge of the gap, the transmission has dropped by almost 10 dB compared to the case with the zero imaginary part. In the nonabsorbing cases, it is commonly accepted that the photonic

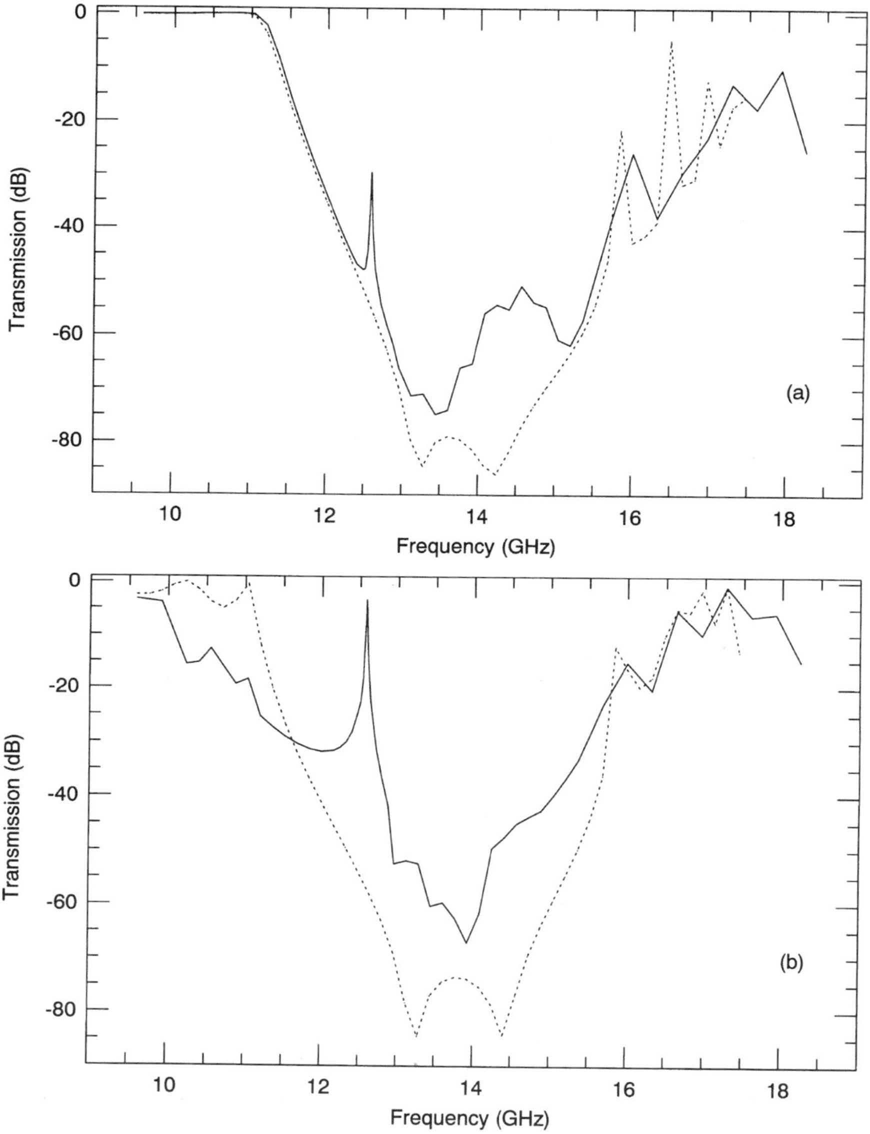

Figure 1.15. The transmission of EM waves propagating through a 3D layer-by-layer PBG consisting of 16 layers of rods. The dotted line correspond to the periodic case while the solid line correspond to the defect case in which every other rod from the eighth layer has been removed. Panels (a) and (b) correspond to the polarization with the electric field parallel and perpendicular to the first layer of rods.

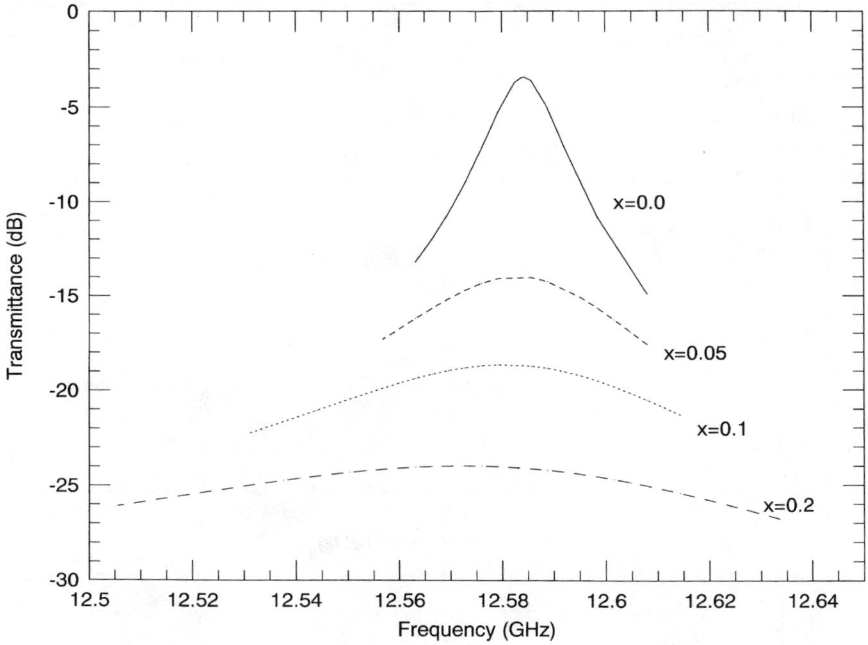

Figure 1.16. The transmission of EM waves propagating in a system similar to the one described in Fig. 1.15. The dielectric constant of the rods is $\varepsilon = 9.61 + ix$. The E field is parallel to the axis of the removed cylinders.

crystal must be as thick as possible, because the transmission inside the gap decreases as the thickness of the crystal increases. However, in photonic crystals constructed of materials with significant absorption, the transmission is thickness dependent at all frequencies. So, it is possible that we will not be able to measure the transmission at the upper edge of the gap, which is more affected by the absorption, if it is less than the noise level of our measurements.

1.3.3. Metallic Photonic Crystals

More recently, a different approach of creating photonic crystals using metallic components has been introduced. For the metal, we use the following frequency-dependent dielectric constant:

$$\varepsilon(\nu) = 1 - \frac{\nu_p^2}{\nu(\nu - i\gamma)} \tag{1.27}$$

where $\nu_p = 3600\,\text{THz}$, $\gamma = 340\,\text{THz}$ are the plasma frequency and the absorption coefficient, respectively. From Eq. (1.27), it turns out that the conductivity is [59] $\sigma = \nu_p^2\gamma/2(\gamma^2 + \nu^2)$. For frequencies $< 100\,\text{THz}$, σ can be practically assumed

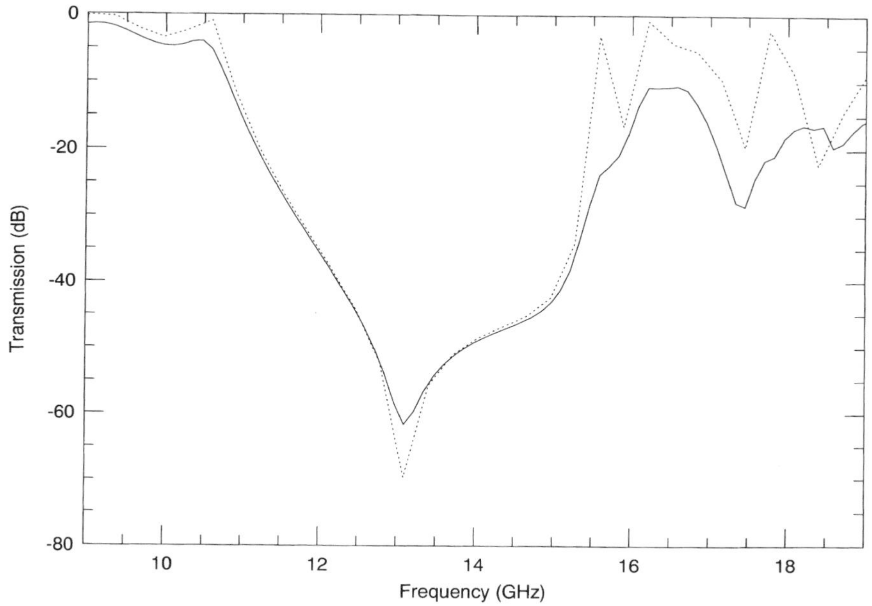

Figure 1.17. The transmission of EM waves propagating in a layer-by-layer system similar to the one in Fig. 1.4 with 3 unit cells thickness. The dielectric constant of the rods is $\varepsilon = 9.61 + ix$ where x is 0 and 0.2 (dotted and solid lines, respectively).

independent of frequency and equal to $0.22 \times 10^5 (\Omega\mathrm{cm})^{-1}$ which is very close with the σ of Ti. However, the conclusions are similar for other metals. The skin depth is $\delta = c(\mu\nu\sigma)^{-1/2}/2\pi$, where c is the velocity of light and μ is the magnetic permeability, which is 1 in our case [59]. Thus, for $\nu = 100$ and 10 THz, the skin depth is 0.035 and 0.11 μm, respectively.

First, we study 3D systems consisting of isolated metallic scatterers embedded in air (cermet topology [60,61]). Figure 1.18 shows the transmission and absorption of EM waves propagating in a simple cubic (s.c.) lattice consisting of metallic spheres with filling ratio $f = 0.03$. The system is infinite along the x and y directions, while its thickness along the z axis is $L = 4a$ and the incident waves with **k** along the z axis. The results for both polarizations are the same, due to the symmetry of the lattice. For the present as well for all the following cases, each unit cell is divided into $10 \times 10 \times 10$ cells, which gives a convergence of better than 5% for the periodic cases, and better than 15% for the defect cases. There are two drops in the transmission (Fig. 1.18); the first $\nu a/c = 0.45$ and the second (and sharpest one) ~ 0.85. The wavevector, **k**, parallel to the z axis corresponds to the $\Gamma - X$ direction in the k space; in this case, we expect the first gap to appear at the edge of the zone (in the X point) for $\nu a/c$ 0.5, which is slightly higher than the frequency where the first drop in the transmission appears in this direction (Fig. 1.18). Due to the small filling ratio, there is no full band gap since the gaps in different directions do not overlap. We find similar results for fcc, bcc, and diamond structures with isolated

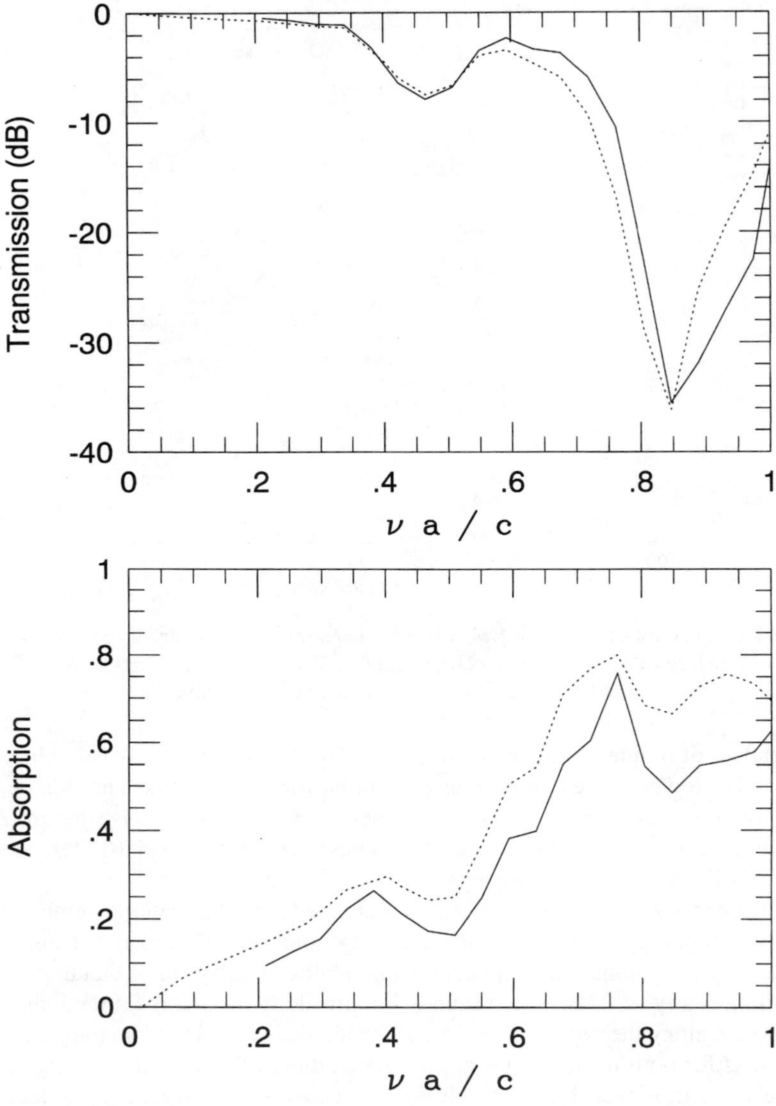

Figure 1.18. Transmission and absorption versus the dimensionless frequency $\nu a/c$ for EM waves propagating in a 3D s.c. lattice consisting of metallic spheres with $f = 0.03$; $L = 4a$ and $\theta = 0°$. Solid and dotted lines correspond to $a = 1.27$ and 12.7 μm, respectively.

metallic spheres or cubes. For the cases where the metal forms isolated scatterers, the results are similar to those of the dielectric PBG materials. The present results for the isolated metallic scatterers are in agreement with the results of a recent work [62] in which monolayers consisting of metallic spheres with a radius between 10 and 100 nm were studied.

From Fig. 1.18, we can also determine what happens as we scale the dimensions of the system and assume that the filling ratio of the system remains the same. By comparing the results for two lattice constants $a = 1.27$ and $12.7\,\mu m$ (dotted and solid lines in Fig. 1.18), we find the transmission is almost the same for both cases as long as $\nu a/c$ is $\lesssim 0.55$. For higher frequencies, the transmission of the $a = 12.7$-μm case is slightly higher than the one for the $a = 1.27$-μm case. The absorption, however, is always smaller for the $a = 12.7$-μm case. By increasing the lattice constant, the frequency decreases and the absolute value of the dielectric constant of the metal becomes larger; this means the metal reflects more (and absorbs less) of the power. For this reason, the absorption decreases as the lattice constant increases. Also, for any lattice constant, there is a drop in the absorption at frequencies where there is a drop in the transmission as a result of the fact that only the few first layers of the material contribute to the absorption.

We also study metallic photonic crystals with network topology. In that case, the metal forms a network throughout the crystal. We use a s.c. lattice consisting of metallic tetragonal rods connecting nearest neighbors with $f = 0.03$, $a = 1.27$ mm, and the surrounding medium is air. A supercell has been used with width $2a$ along the x and y axis and periodic boundary conditions are imposed at the edges of the supercell; the system is finite along the z axis with thickness $L = 3a$. For the periodic case (solid line in Fig. 1.19), there is a sharp drop in the transmission from zero up to a cut-off frequency, $\nu_c = 105$ THz. By increasing the incident angle ν_c increases. A defect can be introduced by removing part of the metal, which is included in a sphere of radius r_d centered at one of the crossing points of the rods in the second layer. Figure 1.19 shows the transmission and absorption for such a defect structure and incident waves with k parallel to the z axis and $a = 1.27$ mm. Once again the results are identical for both polarizations due to the symmetry of the structure. For $r_d/a = 0.15$, a small peak in the transmission appears at ~ 31 THz; the quality factor is very small ($Q = 3$) and the transmission at the top of the peak is also small (-37.2 dB). Apart from the frequency region around the defect, the transmissions of the defect and the periodic structures are almost the same (cf. dotted and solid lines in Fig. 1.19). For $r_d/a = 0.5$ (dashed line in Fig. 1.19), there is a peak in the transmission at higher frequency (60 THz), which is even wider and with higher transmission at the top of the peak. Thus, one can adjust the frequency of the defect inside the gap by just changing the volume of the removed metal; the higher the amount of the removed metal, the higher the frequency where the defect peak appears. Studies in dielectric PBG materials [32–34] have shown that a defect band emerges from the lower edge of the gap and approaches the upper edge of the gap as the volume of the removed dielectric material increases. The behavior of the defect band is similar in the present case, despite the fact that the lower edge of the gap is actually at zero frequency. The absorption for $r_d/a = 0.15$ is almost identical with the absorption of the periodic case except for a small maximum at the frequency where the defect peak appears (hardly noticed by comparing dotted and solid lines in Fig. 1.19). However, the differences in the absorption between $r_d/a = 0.5$ and the periodic cases (of dashed and solid lines in Fig. 1.19) are more obvious with a more prominent peak of the absorption in the $r_d/a = 0.5$ case at the frequency where a

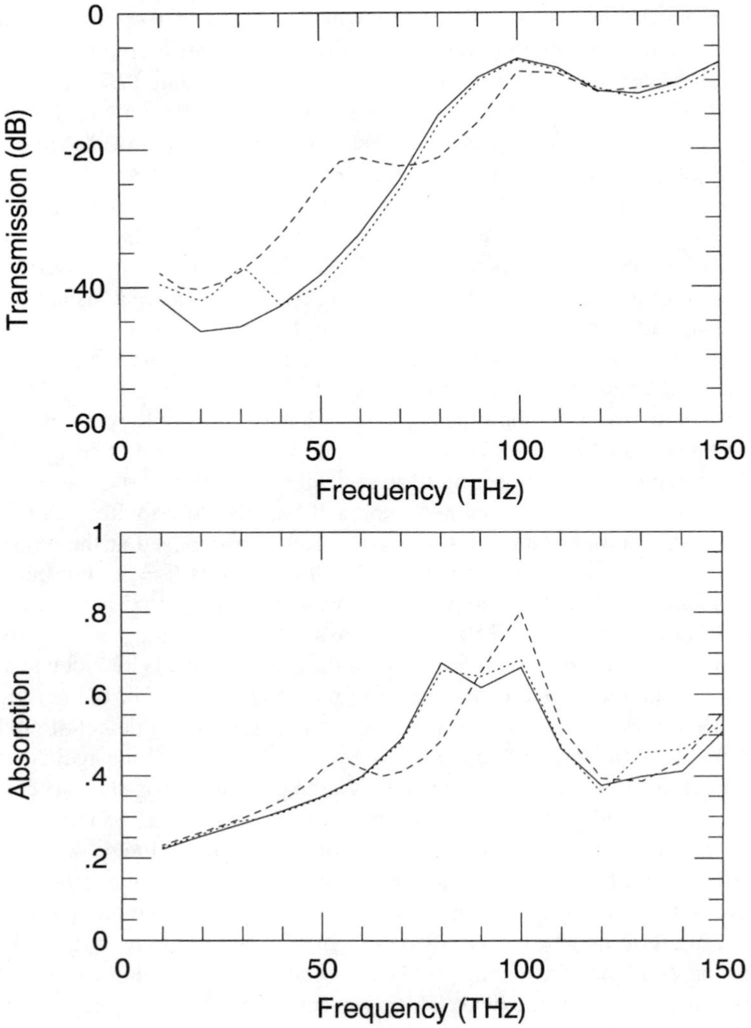

Figure 1.19. Transmission and absorption for EM waves propagating in a 3D s.c. lattice consisting of metallic tetragonal rods connecting nearest neighbors with $f = 0.03$, $L = 3a$, and incident angle $\theta = 0°$. A supercell of width $2a$ has been used with periodic boundary conditions at the edges of the supercell. Part of the metal, which is included in a sphere with a center in one of the crossing points of the rods at the second layer and radius r_d has been removed. Solid, dotted, and dashed lines correspond to $r_d/a = 0$, 0.15 and 0.5, respectively.

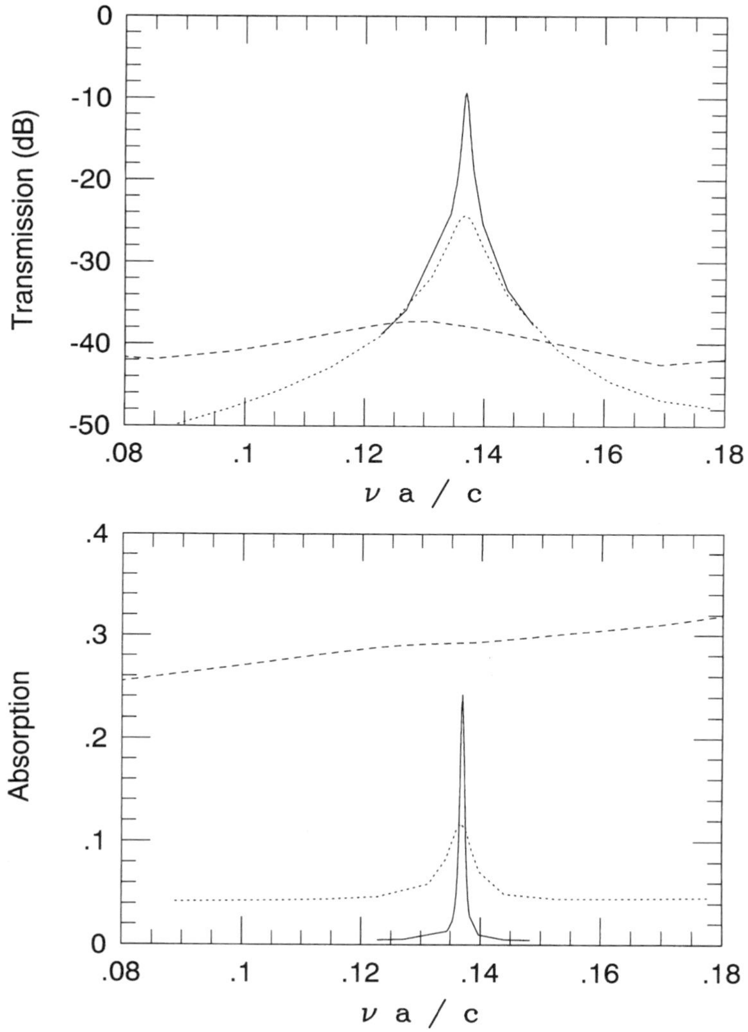

Figure 1.20. Transmission and absorption for EM waves propagating in a defect-structure similar with the one described in Fig. 1.19. with $r_d/a = 0.15$. Solid, dotted, and dashed lines correspond to $a = 1.27$, 12.7, and 127 μm, respectively.

defect appears (60 THz). Since the light is trapped around the defect region, one actually expects the absorption will be higher in this case.

Figure 1.20 shows transmission and absorption as functions of the dimensionless frequency $\nu a/c$ for a defect structure (similar to the one described in the previous paragraph) with $r_d/a = 0.15$. As we mentioned earlier, the results change as we scale the dimensions of the structure due to the frequency dependence of the dielectric constant. By increasing the lattice constant, the transmission at the top of the defect peak, T_d, and the Q factor increase by orders of magnitudes; in particular, $T_d = -37, -24$, and -9 dB, while $Q = 3$, 29, and 137 for $a = 1.27$, 12.7, and 127 μm. However, the dimensionless frequency of the defect, $\nu_d a/c$, increases slightly, reaching a constant value at high lattice constants ($\nu_d a/c = 0.1312, 0.1367$, and 0.1369 for $a = 1.27$, 12.7, and 127 μm). The absorption at $\nu_d a/c$, on the other hand, exhibits a more peculiar behavior. In general (see Fig. 1.13), there is a peak in the absorption exactly at $\nu_d a/c$ (although, this peak is hardly noticeable for the $a = 1.27$ μm), which becomes sharper as the lattice constant increases. As we mentioned earlier for the periodic cases, by increasing the lattice constant, the overall absorption decreases. Similarly, in the present case for frequencies well above or below the defect frequency, the overall absorption (we call it background absorption) decreases as the lattice constant increases. At the defect frequency, however, the wave becomes more localized as the lattice constant increases, because the metal becomes a better reflector, hence, the transmission and absorption peaks become sharper.

1.3.4. Waveguides

Recent studies of 2D photonic band-gap waveguides have shown encouraging results for the use of photonic crystals in order to improve waveguide efficiency [63,64]. In one of the studies [63], a 2D square lattice was used consisting of dielectric cylinders. A line of cylinders was removed in order to create the waveguide geometry. Numerical simulations using the FDTD method revealed complete transmission at certain frequencies, and very high transmission (>95%) over wide frequency ranges. High transmission is observed even for 90° bends with zero radius of curvature, with a maximum transmission of 98% as opposed to 30% for analogous conventional dielectric waveguides.

As an example, we theoretically study waveguide geometries in layer-by-layer photonic crystals consisting of alumina rods (dielectric constant of 9.61). The rectangular cross-section of the rods measures $w = 0.667$ mm by $c/4 = 0.7$ mm. The separation between the rods in each layer is $d = 2.5$ mm. The unit cell consists of 4 layers of rods with a total thickness of $c = 2.8$ mm. We study a system consisting of 14 rods within each layer and having 20 layers. We remove part of one rod in the tenth layer and part of the rod in the eleventh layer. The two rods are perpendicular to each other and form an L-shaped waveguide, which is an experimentally feasible design. We use a dipole parallel to the stacking direction to excite the waveguide modes. The dipole is located in the entrance of the waveguide.

For all the following cases, we calculate the integrated Poynting vector over two areas located two unit cells before and two unit cells after the bend. These areas are centered at the waveguide section. The cross-section of these areas is 6.67 mm within the layers and 7 mm along the stacking direction. We are using the FDTD method. The grid points used in our calculation were $464 \times 464 \times 200$ along the x,y, and z directions (z is the stacking direction). Each grid point is 0.083 mm and the time step is 0.163 ps. The numerical space is terminated with second-order Liao boundary conditions [65].

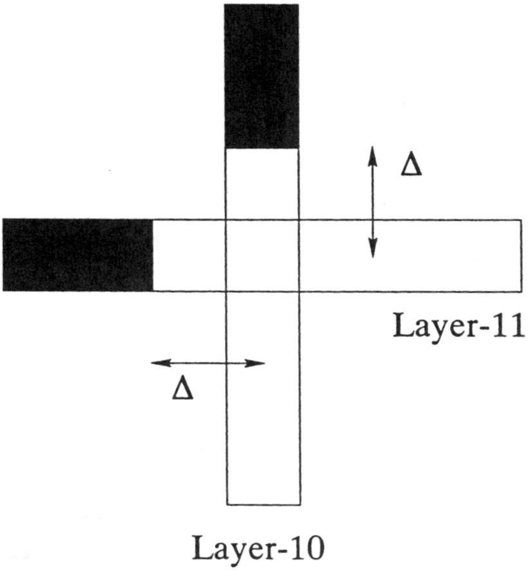

Figure 1.21. A diagram of the waveguide bend showing the remaining part of the rods (black area) and the part of the rods that have been removed (white area). Rods are in two different layers of the crystal.

For the waveguide bend, we remove part of one rod at the tenth layer and part of one rod at the eleventh layer (the rods are perpendicular to each other; see diagram in Fig. 1.21). In particular, we remove the rods up to the intersection of their axes plus an additional length Δ (Fig. 1.21), where Δ is the distance between the edge of the rod and the axis of the perpendicular rod. Both rods are handled in a symmetric way.

It has been shown that 100% transmission along the bend can be achieved for this particular structure at 49 GHz and Δ equal to the width of a rod w [66]. In this particular case, the power distribution at the middle of the defect layers is shown in Fig. 1.22. Close to the dipole the power has a maximum [see Fig. 1.22(a)]. The wave propagates first in the layer where the dipole is located [Fig. 1.22(a)] then turns around the bend and moves to the next layer [Fig. 1.22(b)]. By comparing the maximum of the power inside the guide, we can see that the maximum of the power just before the bend is equal with the maximum of the power just after the bend.

1.3.5. Resonant Cavity Antennas

In another interesting theoretical study [67], a thin slab of 2D photonic crystal was shown to drastically alter the radiation pattern of spontaneous emission. More specifically, by eliminating all guided modes at the transition frequencies, spontaneous emission can be coupled entirely to free space modes, resulting in a greatly enhanced extraction efficiency. Such structures might provide a solution to the long-standing problem of poor light extraction from high refractive index semiconductors in light-emitting diodes [67]. Extension of these studies into 3D photonic crystals will be very useful.

Resonant cavity enhanced detectors built around a 3D photonic band-gap crystal have been studied recently [68]. The measured power enhancement factor was as high as 3450 for planar cavity structures [68]. In a more recent study [69], photonic band-gap technology has been considered as a promising new solution to the problem of higher frequencies with conventional patch antenna designs. Problems such as narrow bandwidth, low gain, and surface wave losses can be improved using photonic crystals [69].

1.3.6. Nonlinearities

There are also works on nonlinear photonic band-gap materials that are basically focused on 1D photonic crystals [70–73]. Under certain circumstances, there may be nonlinear wave propagation within the photonic band gap. For a large-scale

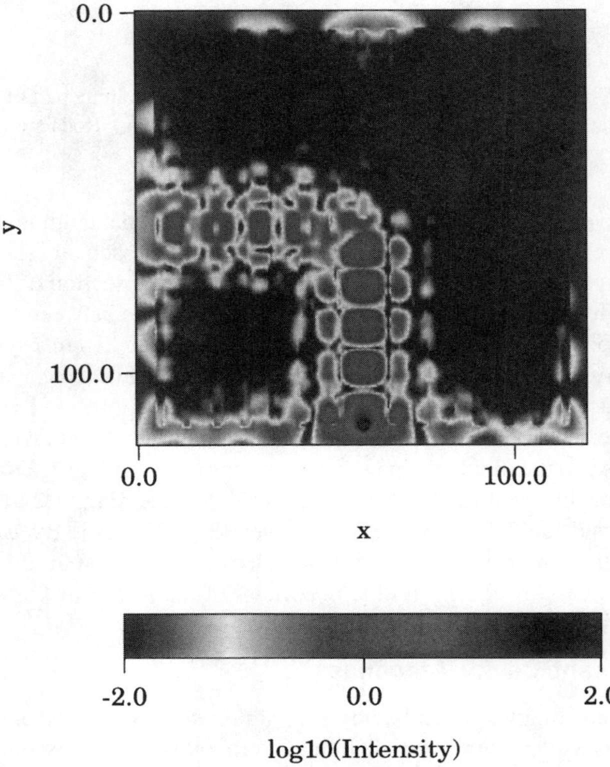

(a)

Figure 1.22. The power distribution for the case where the transmission along the bend is 100%.

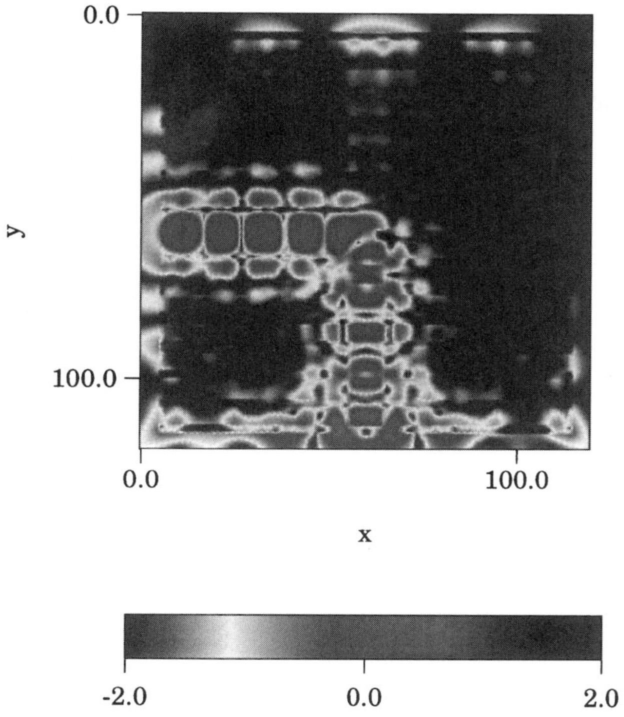

(b)

Figure 1.22. *(Continued)*

photonic band-gap material, the propagation of high-intensity, nonlinear solitary waves may provide a practical way of coupling large amounts of optical energy into, and out of, the otherwise impenetrable photonic band gap. A stationary solitary wave may be regarded as a self-localized state. In perfectly periodic material, the high light intensity itself creates a localized dielectric defect through the nonlinear Kerr coefficient. Unlike the localized state induced by static disorder, the localized dielectric defect is free to move with light intensity field. The result is a solitary wave that can move through the bulk photonic band-gap material with any velocity ranging from zero up to the average speed of light in the medium. Using a variational method, John and Akozbek [74] found a variety of different solitary wave solutions in two dimensions. Their work suggests that photonic band-gap materials in higher dimensions may have a variety of interesting bistable switching properties that go beyond the simple characteristic of 1D dielectrics. Since exact solution is no longer possible in two and three dimensions, numerical methods are required to solve this problem. The FDTD method described earlier with implementation of the formalism described in [75] is very promising for the solution of this problem.

1.3.7. Quantum Electrodynamics

There are also very interesting implications of photonic crystals on quantum electrodynamics. For a single excited atom with transition frequency ω_0 to the ground state that lies within the band gap, there is no true spontaneous emission of light. A photon that is emitted by the atom finds itself within the classically forbidden energy gap of the photonic crystal. The result is a coupled eigenstate of the electronic degrees of freedom of the atom and the electromagnetic modes of the photonic crystal. This photon–atom bound state [76–78] is the optical analog of an electron impurity level bound state in the gap of a semiconductor. When a collection of atoms is placed into the photonic crystal, a narrow photonic impurity band is formed within the larger photonic band gap. This may lead to new effects on nonlinear optics and laser physics [76].

ACKNOWLEDGMENTS

It is a pleasure to thank our colleagues G. Tuttle, K. Constant, C. T. Chan, E. Ozbay, W. Leung, and J. S. McCalmont for their insights and collaborations. We thank D. Crouch for providing FDTD simulations. This work was supported by the Director for Energy Research, Office of Basic Energy Sciences and Advanced Energy Projects. We also acknowledge support from the Department of Commerce through the Center for Advanced Technology (CATD). The Ames Laboratory is operated for the U.S. Department of Energy by Iowa State University under Contract No. W-7405-Eng-82.

REFERENCES

1. See, for example, the special features issue, C. M. Bowden, J. P. Dowling, and H. O. Everitt, Eds., "Development and applications of materials exhibiting photonic band gaps," *J. Opt. Soc. Am. B* 10 (1993).
2. For a review see, C. M. Soukoulis, Ed., *Photonic Bandgaps and Localization*, Plenum, New York, 1993.
3. J. D. Joannopoulos, R. D. Meade, and J. N. Minn, *Photonic Crystals*, Princeton University Press, Princeton, NJ, 1995; J. D. Joannopoulos, P. R. Villeneuve, and S. Fan, "Photonic Crystals," *Nature (London)* ***386***, 143 (1997).
4. For a review see, C. M. Soukoulis, Ed., *Photonic Band Gap Materials*, Kluwer, Dordrecht, The Netherlands, 1996.
5. E. Yablonovitch, "Inhibited spontaneous emission in solid state physics and electronics," *Phys. Rev. Lett.* ***58***, 2059 (1987).
6. J. Martorell and N. M. Lawandy, "Spontaneous emission in a disordered dielectric medium," *Phys. Rev. Lett.* ***66***, 887 (1991).
7. J. Martorell and N. M. Lawandy, "Observation of inhibited spontaneous emission in a periodic dielectric structure," *Phys. Rev. Lett.* ***65***, 1877 (1990).
8. B. Y. Tong, P. K. John, Y. Zhu, Y. S. Liu, S. K. Wong, and W. R. Ware, "Fluorescence-lifetime measurements in monodispersed suspensions of polystyrene particles," *J. Opt. Soc. Am. B* ***10***, 356 (1993).

9. S. John, "Strong localization of photons in certain disordered dielectric superlattices," *Phys. Rev. Lett.* ***58***, 2486 (1987); S. John, "The localization of light," *Comments Cond. Mater. Phys.* ***14***, 193 (1988).
10. A. Z. Genack and N. Garcia, "Electromagnetic localization and photonics," *J. Opt. Soc. Am. B* ***10***, 408 (1993).
11. N. Garcia and A. Z. Genack, "Anomalous photon diffusion at the threshold of the Anderson localization transition," *Phys. Rev. Lett.* ***66***, 1850 (1991).
12. S. John, "The localization of light," *Phys. Today* ***44***, 32 (1991).
13. L. M. Brekhovskikh, *Waves in Layered Media*, Academic, New York, 1960.
14. P. Yeh, *Optical Waves in Layered Media*, Wiley, New York, 1988.
15. A. Thelen, *Design of Optical Interference Coatings*, McGraw-Hill, New York, 1989.
16. C. M. Bowden, J. P. Dowling, and H. O. Everitt (Eds.), "Development and applications of materials exhibiting photonic band gaps," in *J. Opt. Soc. Am. B* ***10*** (1993).
17. See the special issue of *J. Mod. Opt.* ***41***, 209 (1994).
18. E. Yablonovitch and T. J. Gmitter, "Photonic band structure: the face centered cubic case," *Phys. Rev. Lett.* ***63***, 1950 (1989).
19. K. M. Leung and Y. F. Liu, "Full vector wave calculation of photonic band structures in face centered cubic dielectric media," *Phys. Rev. Lett.* ***65***, 2646 (1990).
20. Z. Zhang and S. Satpathy, "Electromagnetic wave propagation in periodic structures: Bloch wave solution of Maxwell's equations," *Phys. Rev. Lett.* ***65***, 2650 (1990).
21. K. M. Ho, C. T. Chan, and C. M. Soukoulis, "Existence of a photonic gap in periodic dielectric structures," *Phys. Rev. Lett.* ***65***, 3152 (1990).
22. C. T. Chan, K. M. Ho, and C. M. Soukoulis, "Photonic band gaps in experimentally realizable periodic structures," *Europhys. Lett.* ***16***, 563 (1991).
23. E. Yablonovitch, T. J. Gmitter, and K. M. Leung, "Photonic band structure: The face-centered-cubic case employing nonspherical atoms," *Phys. Rev. Lett.* ***67***, 2295 (1991).
24. K. M. Ho, C. T. Chan, C. M. Soukoulis, R. Biswas, and M. M. Sigalas, "Photonic band gaps in three dimensions: New layer by layer periodic structures," *Solid State Commun.* ***89***, 413 (1994).
25. E. Ozbay, A. Abeyta, G. Tuttle, M. Tringides, R. Biswas, C. M. Soukoulis, C. T. Chan, and K. M. Ho, "Measurement of a three-dimensional photonic band gap in a crystal structure made of dielectric rods," *Phys. Rev. B* ***50***, 1945 (1994).
26. E. Ozbay, G. Tuttle, R. Biswas, M. Sigalas, and K. M. Ho, "Micromachined millimeter-wave photonic bandgap crystals," *Appl. Phys. Lett.* ***64***, 2059 (1994).
27. E. Ozbay, E. Michel, G. Tuttle, R. Biswas, K. M. Ho, J. Bostak, and D. M. Bloom, "Terahertz spectroscopy of three-dimensional photonic band-gap crystals," *Opt. Lett.* ***19***, 1155 (1994).
28. E. Ozbay, G. Tuttle, M. Sigalas, R. Biswas, K. M. Ho, J. Bostak, and D. M. Bloom, "Double-etch geometry for millimeter-wave photonic band-gap crystals," *Appl. Phys. Lett.* ***65***, 1617 (1994).
29. E. Ozbay, G. Tuttle, J. S. McCalmont, M. Sigalas, R. Biswas, and K. M. Ho, "Laser-micromachined millimeter-wave photonic band gap cavity structures," *Appl. Phys. Lett.* ***67***, 1969 (1995).
30. E. Ozbay, G. Tuttle, M. Sigalas, C. M. Soukoulis, and K. M. Ho, "Defect structures in a layer-by-layer photonic band gap crystal," *Phys. Rev. B* ***51***, 13961 (1995).

31. M. C. Wanke, O. Lehmann, K. Muller, Q. Wen, and M. Stuke, "Laser rapid prototyping of photonic band-gap microstructures," *Science* ***275***, 1284 (1997); J. G. Fleming, and S. Y. Lin, "Three dimensional photonic crystal with a stop band from 1.35 to 1.95 microns," *Opt. Lett.* ***24***, 49 (1999); S. Y. Lin, J. G. Fleming, D. L. Hetherington, B. K. Smith, R. Biswas, K. M. Ho, M. M. Sigalas, W. Zubrzycki, S. R. Kurtz, and J. Bur, "A three dimensional photonic crystal operating at infrared wavelengths," *Nature (London)* ***394***, 251 (1998).
32. E. Yablonovitch, T. Gmitter, R. Meade, A. Rappe, K. Brommer, and J. Joannopoulos, "Donor and acceptor modes in photonic band structure," *Phys. Rev. Lett.* ***67***, 3380 (1991).
33. R. D. Meade, A. M. Rappe, K. D. Brommer, and J. D. Joannopoulos, "Photonic bound states in periodic dielectric materials," *Phys. Rev. B* ***44***, 13772 (1991).
34. R. D. Meade, A. M. Rappe, K. D. Brommer, J. D. Joannopoulos, and O. L. Alerhand, "Accurate theoretical analysis of photonic band-gap materials," *Phys. Rev. B* ***48***, 8434 (1993).
35. S. Satpathy, Z. Zhang, and M. R. Salehpour, "Theory of photon bands in three-dimensional periodic dielectric structures," *Phys. Rev. Lett.* ***64***, 1239 (1990).
36. R. Biswas, C. T. Chan, M. Sigalas, C. M. Soukoulis, and K. M. Ho, "Photonic band gap materials," in *Photonic Band Gap Matterials*, C. M. Soukoulis, Ed., Kluwer, Dordrecht, The Netherlands, 1996, p. 23.
37. J. B. Pendry and A. MacKinnon, "Calculation of photonic dispersion relations," *Phys. Rev. Lett.* ***69***, 2772 (1992); J. B. Pendry, "Photonic band structures," *J. Mod. Opt.* ***41***, 209 (1994).
38. M. M. Sigalas, C. M. Soukoulis, E. N. Economou, C. T. Chan, and K. M. Ho, "Photonic band gaps and defects in two dimensions: Studies of the transmission coefficient," *Phys. Rev. B* ***48***, 14121 (1993).
39. M. M. Sigalas, C. M. Soukoulis, C. T. Chan, and K. M. Ho, "Electromagnetic wave propagation through dispersive and absorptive photonic band gap materials," *Phys. Rev. B* ***49***, 11080 (1994).
40. M. M. Sigalas, C. T. Chan, K. M. Ho, and C. M. Soukoulis, "Metallic photonic band gap materials," *Phys. Rev. B* ***52***, 11744 (1995).
41. J. S. McCalmont, M. M. Sigalas, G. Tuttle, K. M. Ho, and C. M. Soukoulis, "A layer-by-layer metallic photonic band gap structure," *Appl. Phys. Lett.* ***68***, 2759 (1996).
42. M. M. Sigalas, J. S. McCalmont, K. M. Ho, and G. Tuttle, "Directional filters: theory and experiment," *Appl. Phys. Lett.* ***68***, 3525 (1996).
43. K. S. Yee, "Numerical solution of initial boundary value problems involving Maxwell's equations in isotropic media," *IEEE Trans. Antennas Propagation*, ***14***, 302 (1966).
44. K. Kunz and R. Luebbers, Eds., *The Finite Difference Time Domain Method for Electromagnetics*, CRC Press, Boca Raton, FL, 1993.
45. K. Umashankar and A. Taflove, Eds., *Computational Electromagnetics*, Artech House, Boston, MA, 1993.
46. M. M. Sigalas, R. Biswas, Q. Li, D. Crouch, W. Leung, R. Jacobs-Woodbury, B. Lough, S. Nielsen, S. McCalmont, G. Tuttle, and K.-M. Ho, "Dipole antennas on photonic band gap crystals—Experiment and simulations," *Microwave Opt. Technol. Lett.* ***15***, 153 (1997).
47. T. F. Krauss, R. M. De La Rue, and S. Brand, "Two dimensional photonic bandgap structures operating at near infrared wavelengths," *Nature (London)* ***383***, 699 (1997).

48. J. O'Brien, O. Painter, R. Lee, C. C. Cheng, A. Yariv, and A. Scherer, "Lasers incorporating 2D photonic bandgap mirrors," *Electron. Lett.* ***32***, 2243 (1996).

49. M. Kanskar, P. Paddon, V. Pacradouni, R. Morin, A. Busch, J. F. Young, S. R. Johnson, J. MacKenzie, and T. Tiedje, "Observation of leaky slab modes in an air-bridged semiconductor waveguide with a two-dimensional photonic lattice," *Appl. Phys. Lett.* ***70***, 1438 (1997).

50. V. Berger, O. Gauthier-Lafaye, and E. Costard, "Fabrication of a 2D photonic bandgap by a holographic method," *Electron. Lett.* ***33***, 425 (1997).

51. P. W. Evans, J. J. Wierer, and N. Holonyak, Jr., "Photopumped laser operation of an oxide post GaAs-A1As superlattice photonic lattice," *Appl. Phys. Lett.* ***70***, 1119 (1997).

52. U. Gruning and V. Lehmann, "Fabrication of 2D infrared photonic crystals in macroporous silicon," in C. M. Soukoulis, Ed., *Photonic Band Gap Materials*, Kluwer, Dordrecht, The Netherlands, 1996, p. 453.

53. C. Cheng and A. Scherer, "Fabrication of photonic band gap crystals," *J. Vacuum Sci. Tech. B* ***13***, 2696 (1995).

54. G. Feiertag, W. Ehrfeld, H. Freimuth, G. Kiriakidis, H. Lehr, T. Pedersen, M. Schmidt, C. Soukoulis, and R. Weiel, "Fabrication of three dimensional photonic band gap material by x ray lithography," in C. M. Soukoulis, Ed., *Photonic Band Gap Materials*, Kluwer, Dordrecht, The Netherlands, 1996, p. 63; G. Feiertag, W. Ehrfeld, H. Freimuth, H. Kolle, H. Lehr, M. Schmidt, M. M. Sigalas, C. M. Soukoulis, G. Kiriakidis, T. Pedersen, J. Kuhl, and W. Koenig, "Fabrication of photonic crystals by deep X-ray lithography," *Appl. Phys. Lett.* ***71***, 1441 (1997).

55. V. N. Astratov, Yu. A. Vlasov, O. Z. Karimov, A. A. Kaplyanskii, Yu. G. Musikhin, N. A. Bert, V. N. Bogomolov, and A. V. Prokofiev, "Photonic band gaps in 3D ordered fcc silica matrices," *Phys. Lett. A* ***222***, 349 (1996).

56. V. N. Bogomolov, S. V. Gaponenko, A. M. Kapitonov, A. V. Prokofiev, A. N. Ponyavina, N. I. Silvanovich, and S. M. Samoilovich, "Photonic band gap in the visible range in a three dimensional solid state lattice," *Appl. Phys. A* ***63***, 613 (1996).

57. R. D. Pradhan, J. A. Bloodgood, and G. H. Watson, "Photonic band structure of bcc colloidal crystals," *Phys. Rev. B* ***55***, 9503 (1997).

58. R. Mayoral, J. Requena, J. S. Moya, C. Lopez, A. Cintas, H. Miguez, F. Meseguer, L. Vasquez, M. Holgado, and A. Blanco, "3D Long-range ordering in an SiO_2 submicrometer-sphere sintered superstructure," *Adv. Mater.* ***9***, 257 (1997).

59. J. D. Jackson, *Classical Electrodynamics*, Wiley, New York, 1975.

60. W. Lamb, D. M. Wood, and N. W. Ashcroft, "Long wavelength electromagnetic propagation in heterogeneous media," *Phys. Rev. B* ***21***, 2248 (1980).

61. E. N. Economou and M. M. Sigalas, "Classical wave propagation in periodic structures: Cermet vs network topology," *Phys. Rev. B* ***48***, 13434 (1993).

62. N. Stefanou and A. Modinos, "Scattering of light from a two-dimensional array of spherical particles on a substrate," *J. Phys. Condens. Matter* ***3***, 8135 (1991).

63. A. Mekis, J. C. Chen, I. Kurland, S. Fan, P. R. Villeneuve, and J. D. Joannopoulos, "High transmission through sharp bends in photonic crystal waveguides," *Phys. Rev. Lett.* ***77***, 3787 (1996); S. Fan, P. R. Villeneuve, R. D. Meade, and J. D. Joannopoulos, "Design of three dimensional photonic crystals at submicron lengthscales," *Appl. Phys. Lett.* ***65***, 1466 (1994); S. Y. Lin, E. Chow, V. Hietala, P. R. Villeneuve, and J. D. Joannopoulos,

"Experimental demonstration of guiding and bending of electromagnetic waves in a photonic crystal," *Science* ***282***, 274 (1998).

64. J. G. Maloney, M. P. Kesler, B. L. Shirley, and G. S. Smith, "A simple description for waveguiding in photonic bandgap materials," *Microwave Opt. Technol. Lett.* ***14***, 261 (1997).
65. Z. P. Liao, H. L. Wong, B. P. Yang, and Y. F. Yuan, *Sci. Sinica* ***A27***, 1063 (1984).
66. M. M. Sigalas, R. Biswas, K. M. Ho, C. M. Soukoulis, D. Turner B. Vasiliu, S. C. Kothari, and S. Lin, "Waveguide bends in three dimensional layer by layer photonic band gap materials," submitted for publication.
67. S. Fan, P. R. Villeneuve, J. D. Joannopoulos, and E. F. Schubert, "High extraction of spontaneous emmission from slabs of photonic crystals," *Phys. Rev. Lett.* ***78***, 3294 (1997).
68. B. Temelkuran, E. Ozbay, J. P. Kavanaugh, G. Tuttle, and K. M. Ho, "Resonant cavity enhanced detectors embedded in photonic crystals," *Appl. Phys. Lett.* ***72***, 2376 (1998).
69. Y. Qian, R. Coccioli, D. Sievenpipper, V. Radisic, E. Yablonovitch, and T. Itoh, "A microstrip patch antenna using novel photonic band gap structures," *Microwave J.*, 66 (Jan. 1999).
70. M. D. Tocci, M. J. Bloemer, M. Scalora, J. P. Dowling, and C. M. Bowden, "Thin-film nonlinear optical diode," *Appl. Phys. Lett.* ***66***, 2324 (1995).
71. E. Lidorikis, Q. Li, and C. M. Soukoulis, "Wave propagation in nonlinear multilayer structures," *Phys. Rev. B* ***54***, 10249 (1996); E. Lidorikis, K. Busch, Q. Li, C. T. Chan, and C. M. Soukoulis, "Optical nonlinear response of a single nonlinear dielectric layer sandwiched between two linear dielectric structures," *Phys. Rev. B* ***56***, 15090 (1997).
72. T. Hattori, N. Tsurumachi, and H. Nakatsuka, "Analysis of optical nonlinearity by defect states in one-dimensional photonic crystals," *J. Opt. Soc. Am.* ***14***, 348 (1997).
73. R. Wang, J. Dong, and D. Y. Xing, "Dispersive optical bistability in one dimensional doped photonic band gap structures," *Phys. Rev. E* ***55***, 6301 (1997).
74. S. John and N. Akozbek, "Nonlinear optical solitary waves in a photonic band gap," *Phys. Rev. Lett.* ***71***, 1168 (1993).
75. R. W. Ziolkowski and J. B. Judkins, "Full wave vector Maxwell equation modeling of the self-focusing of ultrashort optical pulses in a nonlinear Kerr medium exhibiting a finite response time," *J. Opt. Soc. Am.* ***B10***, 186 (1993).
76. S. John and J. Wang, "Quantum optics of localized light in a photonic band gap," *Phys. Rev. B* ***43***, 12772 (1991); S. John, and T. Quang, "Spontaneous emission near the edge of a photonic band gap," *Phys. Rev. A* ***50***, 1764 (1994).
77. S. Bay, P. Lambropoulos, and K. Molmer, "Atom–atom interaction at the edge of a photonic band gap," *Opt. Commun.* ***132***, 257 (1996).
78. A. Kamli, M. Babiker, A. Al-Hary, and N. Enfati, "Dipole relaxation in dispersive photonic band gap structures," *Phys. Rev. A* ***55***, 1454 (1997).

CHAPTER TWO

"Holey" Silica Fibers

JONATHAN C. KNIGHT, TIMOTHY A. BIRKS, and PHILIP St. J. RUSSELL

Optoelectronics Group, Department of Physics, University of Bath,
Claverton Down, Bath BA2 7AY, UK

2.1. INTRODUCTION

The behavior of light in wavelength-scale structures is of growing interest, because continuing technological advances in microstructure fabrication are opening new vistas in optical engineering and design. The properties of *periodically* patterned structures with a periodicity or pitch of the same order of magnitude as the wavelength of interest, which we refer to as *photonic crystals*, are potentially of technological as well as of academic interest [1–3]. Such periodic structures have been fabricated in one, two, and even three dimensions (1D, 2D, and 3D) using a range of methods, many of them derived from those familiar from semiconductor and planar device work. Here, we describe a different way of using the remarkable properties of photonic crystals, by drawing the crystals in the form of fine fibers. Like this, we have fabricated 2D periodic structures that are effectively infinite in the 3D, and have used them to form new types of optical fiber waveguide. These fibers derive their novelty from the unusual optical response of the photonic crystal material that forms the fiber cladding.

In this chapter, we shall first describe the fabrication procedure for photonic crystal fiber (PCF, also referred to as "holey" fiber), setting the stage for the rest of the chapter and clarifying the technological constraints under which the work is being done. This description is necessary because the fabrication technology is new, and its limits are only now being established. We will then briefly describe some of the light patterns that can be observed in these fibers using an optical microscope, and our interpretation of these. Thereafter, we will proceed to present our work on "effective index" and "band gap" waveguiding in these structures.

Optics of Nanostructured Materials, Edited by Vadim A. Markel and Thomas F. George
ISBN 0-471-34968-2

2.2. FABRICATION OF PHOTONIC CRYSTAL FIBERS

The idea of using fiber drawing to reduce the transverse scale of composite glass objects is not a new one. Mosaic glass has been used for thousands of years to make decorative microstructured glass, culminating in the very intricate *millefiore* (thousands of flowers) patterns produced in nineteenth century Venice [4]. This process involves drawing down a composite of many different colored glasses to reduce the transverse dimensions. The drawn cane is then sliced up, each slice holding the reduced image of the original composite. A similar technology is used in the production of face plates, where a large number of macroscopic glass fiber waveguide preforms are stacked together, typically in a close-packed array, and are drawn down. The drawn stack is then typically stacked and drawn several more times, producing a large array of closely spaced waveguides.

In recent years, this process has been extended to allow for the formation of periodically microstructured glasses with the relatively large refractive index contrast required for the formation of photonic band gaps. Tonucci et al. [5] and Haus and co-workers [6] formed periodic microstructures using two different glasses that are repeatedly stacked together and drawn down, a process related to our own [7,8]. They have then sliced the composite material into thin slices (typically of the order of 1 mm) and used the differential etch rates of the two glasses to acid-etch one of the two phases of the material away, leaving a single glass shot through with a periodic array of air holes. They were then able to investigate the optical response of this material in the periodic plane, and compare the resulting reflection/transmission curves to those predicted by theory [9,10]. However, because their work resulted in only short microstructured samples with a significant index contrast, they were unable to investigate out-of-plane propagation or waveguiding effects in their samples.

The fabrication of photonic crystal fibers is in some ways very similar, and involves no new or high-technology processes. However, in contrast to all of the processes described above, our stacks are not composed of several solid materials, but of a single material incorporating purposeful voids. This means that we are able to draw high-contrast microstructured glass fibers directly, enabling us to form many meters—even kilometers—of continuous, uniform high-contrast photonic crystal material. As a result, not only are we able to investigate the out-of-plane propagation of optical waves through the material, but we have been able to produce some of the first practical and functional optical devices that exploit the unique properties of photonic crystals at optical frequencies.

Our work has been done using synthetic fused silica (SiO_2), which is the most commonly used material for fabricating optical fibers. The reason is that it has ideal optical properties (the lowest optical absorption of any known solid material) and also very attractive thermal and mechanical properties. It has a refractive index of ~ 1.46 at visible-light frequencies and is highly transparent at wavelengths from ~ 350 nm to nearly 2 μm. It has minima in optical loss at wavelengths of ~ 1.3 and 1.55 μm (the telecommunications windows), and a zero in the material group velocity dispersion at 1.29 μm. It has a softening point of 1600°C and a very slow variation of viscosity with temperature. Fiber drawing is typically carried out in the temperature range from 1650 to 2050°C. High-quality synthetic fused silica is readily available from various suppliers in the form of tubes or rods with differing cross-sections, which are ideally suited to our purposes.

Our fabrication process relies on our being able to form, in a morphologically stable way, the basic crystal structure of interest on a macroscopic scale, by stacking 1 mm diameter capillaries (canes) together to form a preform. This stack, which is typically $\sim$40 cm in length and perhaps 20 mm in diameter, is then held together while being fused and drawn into a fiber on a fiber-drawing tower. In order to stack a sufficient number of canes while keeping the overall preform diameter to a manageable scale, the diameter of the canes needs to be chosen with care—too small and the canes quickly become unmanageable: too large, and the overall preform diameter becomes too large to fit into the open furnace used for fiber drawing. We find that capillary diameters in the range of 0.5–1.0 mm are the most useful. Such canes can readily be drawn from a silica tube using a fiber-drawing tower and a cane-puller. (The capillaries are not sufficiently flexible to be wound onto a drum—the cane-puller is simply an alternative traction mechanism for such inflexible "fibers".) In order for the stack of canes to remain stable later in the drawing process (when the surface tension forces within the fiber become very strong) it is important that there be sufficient uniformity in the diameter and shapes of the canes. The stack of canes, which typically contain a few hundred capillaries and/or solid canes, is held together using tantalum wire at several points, enabling it to be held vertically. A long tube is fused onto one end to act as a handle, and the stack is then suspended by the handle while being drawn to fiber.

Drawing photonic crystal fiber from a preform is very similar to drawing conventional fiber, except that the drawing parameters are more critical for PCF. At higher temperatures the viscosity of the silica decreases, so that the air holes present in the structure tend to collapse under surface-tension forces. Consequently, the fiber must be drawn at a lower temperature than normal fiber. If the temperature becomes too low, the viscosity increases to the point where the strength of the fiber is exceeded and the fiber snaps during drawing. Also, vitreous silica is stable only under a limited range of temperature conditions. Devitrification results in the presence of crystalline SiO_2, the presence of which serves as an optical loss mechanism and also a source of weaknesses in the final fiber. One partial solution to these problems is to embed the PCF preform in a thick solid silica jacket, which serves to stabilize and to strengthen the final fiber. This embedding can be done by surrounding the crystalline stack with a few layers of solid silica canes, or by collapsing a thick-walled tube onto the stack. One is then able to draw the fiber at a higher temperature than would otherwise be possible. It is often easier to draw the preform down using two drawing stages—in the first, the stack is drawn down to a "subpreform" with a diameter in the range of 1–4 mm by using the cane-puller. The subpreform is cut to lengths of around 1 m, and the cross-section of these lengths is continually monitored under a microscope while being drawn to ensure that the furnace temperature is appropriate to properly fuse the different capillaries without allowing the collapse of the air holes. Each subpreform is then used as a preform for a third drawing stage in which the final fiber is drawn.

As with any fiber fabrication, contamination of the silica preform—especially at the surfaces—before drawing provides nucleation sites for devitrification and thus facilitates this process. Contamination is especially a difficulty when working with a large number of capillaries that in general hold an electrostatic charge. Consequently, it is essential to clean the preform at each stage of the fabrication process and to work in a clean environment.

The range of 2D microstructures that can be made using these techniques is in principle unlimited. In practice, the structures are constrained by the difficulty of working with capillaries $\ll 0.5$ mm in diameter, and by the need to keep the overall diameter of the preform within that of the furnace bore. The simplest preform to stack is a triangular array of capillaries—a close-packed stack of circular or hexagonal cylinders. Examples of fiber cross-sections resulting from such a preform are shown in Fig. 2.1. Both of these fibers were drawn without a silica jacket, and the differences between them illustrate two points.

1. The strong dependence of surface-tension forces on the hole diameter means that the maximum attainable ratio of the air-hole diameter d to the pitch Λ decreases as the pitch decreases. For fiber drawn from a typical unjacketed fiber subpreform with a diameter of perhaps 1.5 mm, the range of values of d/Λ is from $d/\Lambda \approx 0.1$ for a final pitch $\Lambda = 1\ \mu$m up to perhaps $d/\Lambda \approx 0.6$ for $\Lambda = 5\ \mu$m or above. At any given value of the pitch Λ, smaller values of d/Λ are obtainable by drawing the fiber at a higher temperature (lower tension). A wider range of parameters are accessible by using a silica jacket on the preform.
2. A range of actual structures is possible from a given preform as a result of the differential collapse rates of the different sized air holes. Thus, fiber drawn from a simple stack of circular capillaries changes from a relatively complicated triangular lattice with hexagonal large holes and smaller triangular interstitial holes, to a simple triangular array of small circular air holes when drawn at a higher temperature or to a smaller pitch.

Certain optical properties that are obtainable using a triangular lattice of air holes require large air holes but no interstitial holes. This can be achieved by starting the fabrication procedure with a silica tube having a hexagonal external shape. Such a tube can be prepared by grinding down the sides of a circular tube using a diamond milling machine, or alternatively can be bought from a commercial supplier of silica tubing. Yet more variations on the structure are made possible by the use of vacuum or gas pressure inside the structure before or during the drawing process.

The optical properties of these triangular-lattice structures turn out to be useful for designing new fiber waveguides, but other crystal structures are also of interest. Optical micrographs illustrating two further readily formed structures are shown in Fig. 2.2, which shows optical micrographs of subpreform for making square and a)

b)

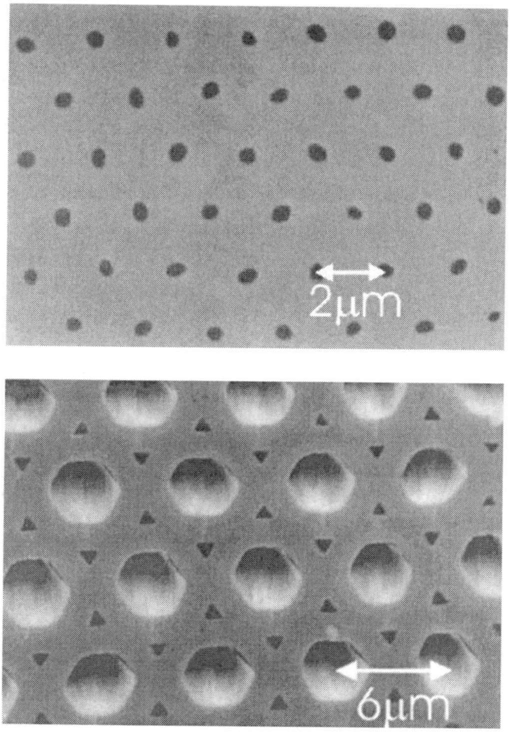

Figure 2.1. Electron micrographs of cleaved surfaces of triangular-lattice photonic crystal fiber. In both cases, the structure was formed by stacking together hundreds of circular silica capillaries, and then drawing the stack down to a fiber. The two structures shown are different because they have been drawn to different values of the interhole spacing or pitch Λ and at different temperatures. (a) The small scale and higher drawing temperature have resulted in a small air-filling fraction in the final fiber, which displays a simple triangular array of air holes. (b) The fiber shown was drawn at a lower temperature, so that the interstitial holes, which arise because we stacked circular capillaries into a triangular array, have survived in the final fiber. The optical properties of these two microstructured materials differ dramatically.

honeycomb PCF. The honeycomb structure [Fig. 2.2(a)] is formed by stacking similarly sized solid and hollow canes into a triangular array as illustrated. A cross-section of a fiber drawn from such a preform is shown in the electron micrograph in Fig. 2.3. This structure is highly stable when drawn (far more so than the triangular structure, perhaps because of the lattice of solid canes that form the skeleton of the structure) and is attractive because it enables photonic band gaps to be found for relatively small values of air-filling fraction [11]. The square lattice can be fabricated as shown in Fig. 2.2(b) by stacking together large capillaries with smaller solid canes to fill the interstitial gaps, as illustrated. This structure also draws down to a stable and highly periodic fiber.

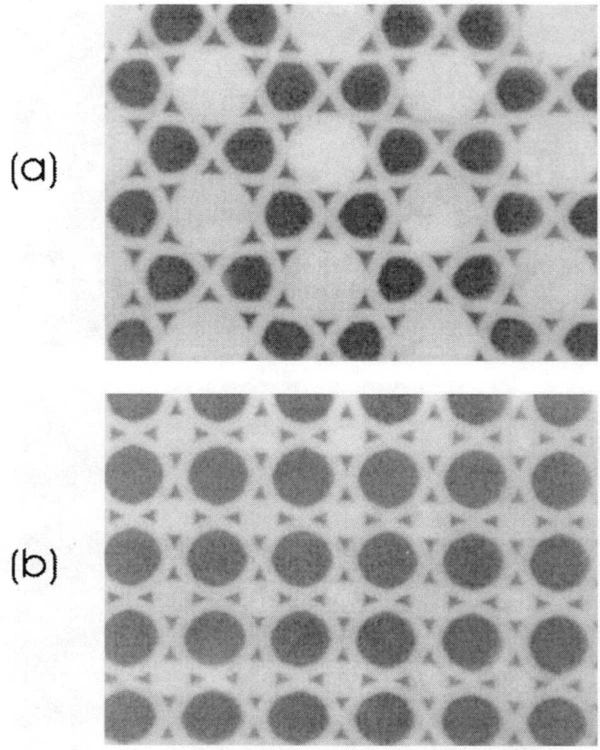

Figure 2.2. Optical micrographs of subpreform cane illustrating how one can form a honeycomb (a) or a square lattice (b) by stacking capillaries and solid canes. In both cases, the periodicity of the cane shown is $\sim 50\,\mu m$. The preform cane maintains the original stack, while the final fiber can look quite different (see Fig. 2.3).

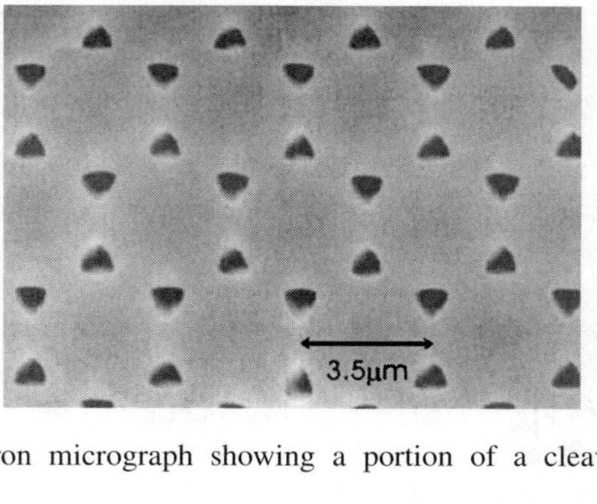

Figure 2.3. An electron micrograph showing a portion of a cleaved honeycomb fiber end-face.

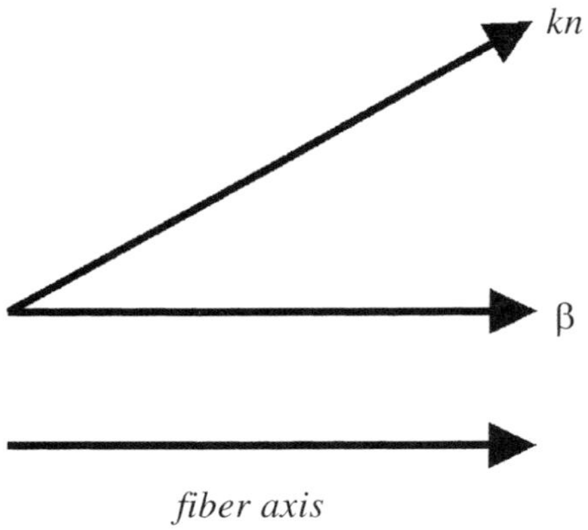

Figure 2.4. The propagation constant β of a mode of a fiber is the component of wavevector parallel to the fiber axis.

2.3. GUIDANCE OF LIGHT IN PHOTONIC CRYSTAL FIBERS

In order to understand how light is guided along a PCF, it is instructive to consider first how a conventional "step index" fiber works. Such a fiber comprises a central core surrounded by a cladding that has a lower refractive index [12]. (In a telecommunications fiber, both core and cladding are made from fused silica glass, but one or the other is doped to change the index in the required way.) Light is confined to the core by total internal reflection (TIR) if the angle of incidence at the core-cladding boundary exceeds the critical angle, as obtained from Snell's law.

An alternative view [13], is to consider that the wavevector of a plane wave in a medium of index n has magnitude kn, where $k = 2\pi/\lambda$ is the free space wave constant. The component of wavevector along the fiber's axis must therefore $\leq kn$. This component is termed the propagation constant β (see Fig. 2.4). Since the fiber is uniform along its axis, β is conserved during reflection and refraction, and so is a constant of the motion for a given mode at a given frequency. The condition for TIR, and hence confinement to the core, is then

$$kn_{\mathrm{co}} > \beta > kn_{\mathrm{cl}} \tag{2.1}$$

If this condition is satisfied, light can propagate in the core of index n_{co} but not in the cladding with the lower index n_{cl}. Otherwise, light can propagate in both core and cladding, though it can still be confined by TIR at the outer (cladding-air) boundary of the fiber as a "cladding mode."

The situation is illustrated in the "band diagrams" of Fig. 2.5(a) and (b). Ranges of allowed β for propagating waves in uniform core and cladding materials are shaded; there is a narrow range of β that is allowed in the core but forbidden in the cladding.

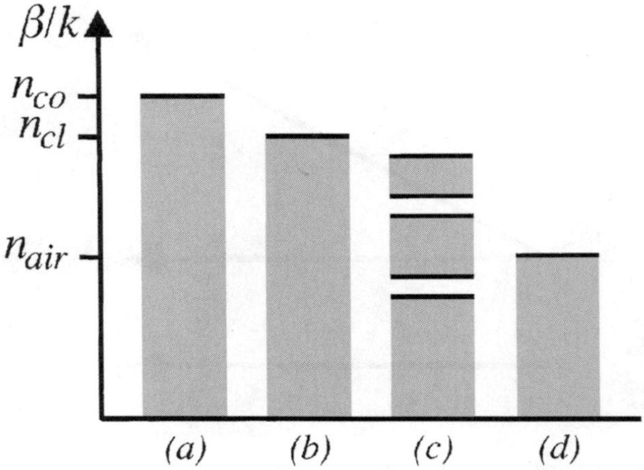

Figure 2.5. Schematic diagrams of the ranges of β allowed for propagating light waves in (a) the doped core of a step-index fiber, (b) pure fused silica (e.g., the cladding of a step-index fiber), (c) a photonic crystal comprising air holes in pure silica, and (d) air. The photonic crystal depicted in (c) exhibits photonic band gaps, though this is not always the case.

Although β is a constant of the motion, the magnitude of the transverse component of the wavevector k_T is not. It depends on the local refractive index as

$$k_T = \sqrt{k^2 n^2 - \beta^2} \tag{2.2}$$

It is k_T that determines the spatial variation of the intensity distribution within the medium. Thus a more rapidly varying intensity distribution indicates a greater value of k_T, and hence a smaller value of β through Eq. (2.2). Conversely, in some regions of low index β can exceed kn; k_T is then imaginary in such regions, and the light wave is evanescent there.

The range of β in Eq. (2.1) corresponds to a range of k_T in a given medium n using Eq. (2.2). This determines an (annular) area of transverse k space available to guided modes. Multiplying this with the area of the core gives the volume of a transverse phase space. From this, a dimensionless parameter called the V-value can be defined, which determines how many modes are guided in a particular fiber with a core of radius ρ:

$$V = \frac{2\pi\rho}{\lambda} \sqrt{n_{\mathrm{co}}^2 - n_{\mathrm{cl}}^2} \tag{2.3}$$

Only one spatial mode (with two states of polarization) is guided if V is less than a critical "cutoff" value V_{co}. Such a fiber is described as being "single mode". For standard step index fibers $V_{\mathrm{co}} = 2.405$, a number that arises from the theory of Bessel functions. Note that for a step index fiber V is essentially proportional to $1/\lambda$, because the wavelength dependence of the refractive indexes is very slight.

Even modes that are strictly guided in a step index fiber suffer some attenuation if the fiber is bent. Such bend loss is normally small if the radius of curvature of the bend exceeds a critical value:

$$R_c = \frac{8\pi^2 n^2 \rho^3}{W^3 \lambda^2} \tag{2.4}$$

where W is a parameter from the theory of waveguides that varies in such a way that R_c increases monotonically with wavelength. Thus fibers become ever more susceptible to bend loss as the wavelength increases.

Although the PCF is considerably more complicated than a step index fiber, many of its properties can be understood from similar concepts. Figure 2.5 (c) depicts a band diagram for a 2D photonic crystal with air holes in the same material as (b). There are two notable features. First, the maximum value of β in the photonic crystal is less than that in the uniform medium. Thus an extended region of the high-index material can act as the core of a fiber guiding light by TIR; there is again a range of β that is allowed in the core but not in the photonic crystal cladding. The photonic crystal can be assigned an effective cladding index, defined as the maximum value of β divided by k, from which an effective V value can be found from Eq. (2.3).

Second, in some circumstances additional ranges of β are forbidden, even though they are less than the maximum allowed value of β. These are the photonic band gaps. They give rise to the possibility of modes guided in cores of low refractive index. When they arise from Bragg scattering effects, they can lie at values of β even $< k$, making a hollow air-filled or evacuated core possible. For example, in Fig. 2.5 there is a range of β that is forbidden in the photonic crystal (c) but allowed in air (d).

2.4. OPTICAL MICROSCOPY AND MODE PATTERNS IN PHOTONIC CRYSTAL FIBER

One can learn a tremendous amount about the optical properties of PCF by exciting the modes of the fiber structure directly, by coupling in through a cleaved fiber end-face. This is especially useful in the search for confined or guided modes, as these can be observed directly at the end-face of the fiber. An optical micrograph showing the end-face of a short length of fiber viewed in reflection is shown in Fig. 2.6(a). The fiber appears as a reasonably uniform array of air holes with spacing 2.6 μm. The next two figures show the same fiber end-face, but are illuminated from below using a low-numerical aperture (NA) white-light source. Figure 2.6(b) shows the fiber being illuminated along its axis, so that the coupling is efficient into the lowest possible modes of the fiber structure. These modes have the smallest transverse k vector. This finding means that they are readily excluded from portions of the fiber cross-section, which have just a little more air than the rest: they do not have sufficient "transverse momentum" to enable them to propagate through these regions. Indeed, regions containing even miniscule additional air holes, or where the air holes are slightly larger than in the structure as a whole, are quite dark in the

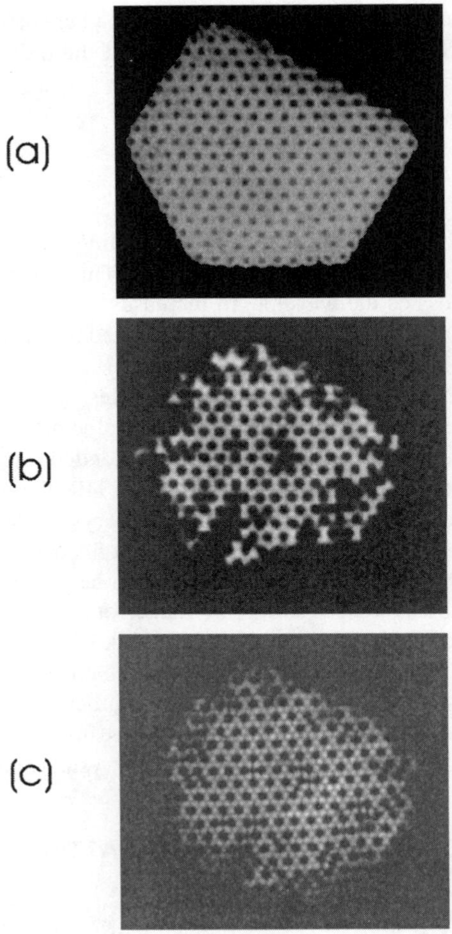

Figure 2.6. Optical micrographs of a fiber end-face. In (a), the fiber is illuminated from above with a white-light source (i.e., viewed in reflection), while (b) and (c) show the same fiber viewed in transmission. In (b) the white-light source has been adjusted so as to illuminate the fiber end-face normally, exciting mainly the lowest order modes of the microstructured fiber material. These modes show the least structure, and are the most sensitive to small fluctuations in the fiber cross-section because of their very small transverse *k* vector component. In (c), the source has been adjusted so as to excite higher order modes of the material.

transmitted-light image. In (c), the light is incident upon the bottom face of the fiber at an angle, exciting higher order modes of the periodic structure. These modes have a larger transverse *k* vector, sharper variations in intensity in the periodic plane, and are far less sensitive to variations in the fiber microstructure.

The limited number of modes available in the second and third cases might be expected to give a poorer resolution image of the fiber surface. Indeed, small air holes appear far larger when viewed in transmission than in reflection, where they are more accurately portrayed. On the other hand, the modes that make up the image in (b) and (c) are exactly those that match the structure (especially the pitch) and a result of this is that it is possible to "see" holes that cannot be resolved using free-space illumination. (One way of thinking about this is that the transmitted light is not resolving the holes themselves, but instead the silica bridges between the holes, which are considerably larger. The holes themselves remain unresolved in the Rayleigh sense.) It is thus apparent that a surprising amount of information about the modes supported by the PCF can be gained by studying optical images in transmission, whereas information on the morphological structure of the samples must be obtained using reflected light or electron microscopy. The usefulness of optical microscopy becomes even more obvious when waveguiding samples are being studied (see Sections 2.5 and 2.6). The guided modes manifest as localized high-intensity regions within the fiber (providing only that an objective lens with an appropriate numerical aperture is used to collect the light). Where the waveguiding process has a strong spectral dependence the high-intensity regions can be brilliantly colored. By using index-matching oil applied to the sides of the fiber, it is sometimes possible to completely strip off that light that is not confined to the center of the fiber, so that all portions of the image except those corresponding to guided modes appear dark when viewed in transmission. While many useful and informative images can be obtained using a conventional laboratory microscope set up to observe samples in transmission, for more selective excitation of specific modes a purpose-built optical configuration can be used. Many of the features to be discussed in this chapter can be seen in such images.

2.5. INDEX-GUIDING PCF

2.5.1. Introduction

Periodically, microstructured materials such as PCF have effective optical properties analogous to those of conventional homogenous materials, although these properties can vary sharply with the wavelength and details of the microstructure, as well as being anisotropic. One thus needs to exercise special caution in using such effective properties. However, as with any material, it is possible to find conditions under which light will be totally internally reflected from the surface of a sample. Clearly, in the case of air holes in silica, we would expect that the effective refractive index of the composite material will normally fall in a range between the indexes of the constituent materials—that is, the index will fall between those of silica and of air [14]. One can then expect to form an optical fiber waveguide using the microstructured material as the cladding. The natural way to form a waveguide in this structure using total internal reflection is to introduce into the structure a localized region with a higher value of the effective index [7,8,14,15]. This could be a missing air hole, for example, forming a pure silica region within the PCF structure. Alternatively, it could be a region with one or more smaller air holes. We have found that such waveguides have most unusual properties: Indeed, they have more in common with fibers studied in the late 1960s and early 1970s [16] than they do with modern commercial single-mode optical fiber, and require that some of the best-known rules of optical fiber performance be forgotten.

A simple index-guiding PCF structure is illustrated schematically in Fig. 2.7, while Fig. 2.8(a) shows a short length of such a fiber illuminated from below with an

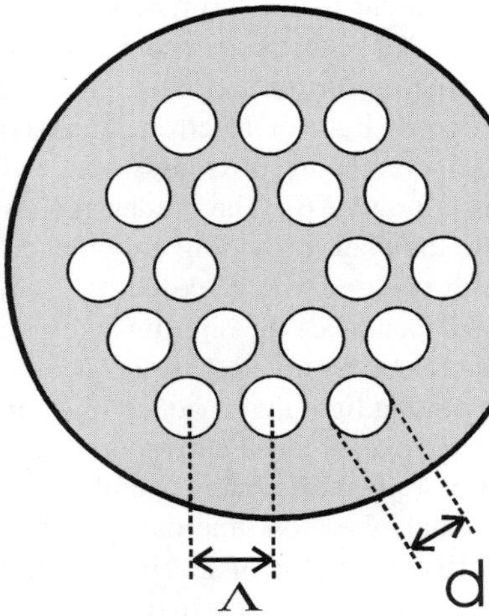

Figure 2.7. Schematic diagram showing a simple index-guiding photonic crystal fiber structure.

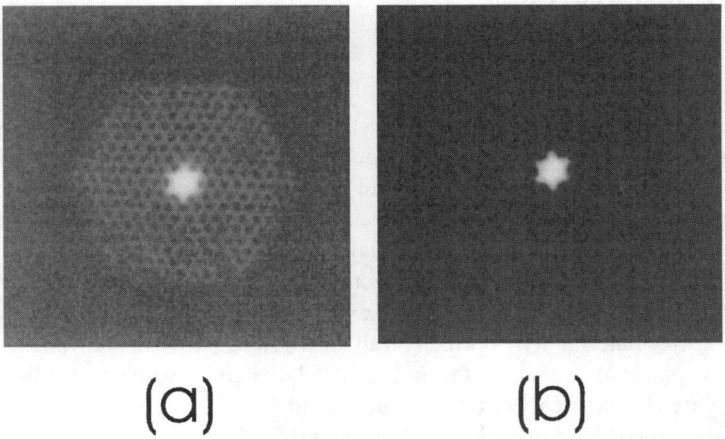

Figure 2.8. Optical micrograph showing an index-guiding PCF illuminated from below. The excitation of cladding modes and a confined core mode are very apparent in (a), while (b) shows the same fiber and excitation condition, but where index-matching liquid applied to the fiber surface has been used to strip off light in cladding modes. The light in the fiber core is unaffected.

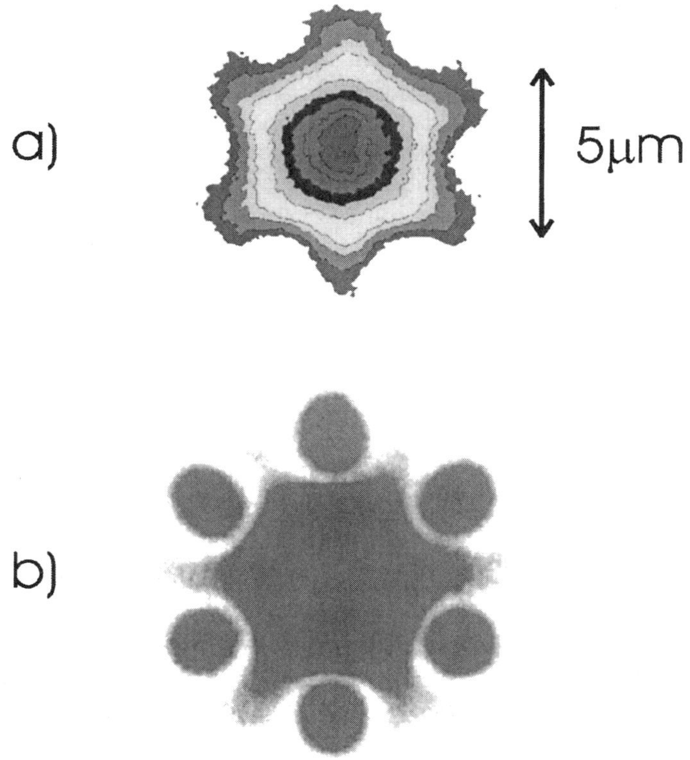

Figure 2.9. Near-field (a) and far-field (b) patterns observed using a laser source coupled into a single-mode triangular-lattice index-guiding PCF.

incoherent white-light source. The optical micrograph shows clearly that the region of higher effective index—the hole-free silica region or "high-index defect" in the crystal structure—supports guided modes. By applying index-matching oil to the external surface of the fiber, we are able to strip off cladding modes, which rely on reflection from this external surface of the fiber to remain trapped, and the only light left in the fiber is in the core [Fig. 2.8(b)]. By taking a similar fiber and coupling a laser source into one end using standard means, we find that light coupled into this "core" does indeed travel with low losses along long lengths of fiber [7,8]. The light leaving the far end of the fiber in a mode confined to the high-index defect [Fig. 2.9(a)] diffracts into free space, and forms the distinctive and very stable pattern shown in Fig. 2.9(b). The stability of this far-field pattern indicates that the fiber guides only a single mode. If there were more than a single guided mode, then the pattern would change as the coupling of the light at the input and the path of the fiber on the bench were varied. The pattern would also lack symmetry except by accident.

2.5.2. Effective Index and the Effective Normalized Frequency

Although the waveguiding mechanism in this fiber is total internal reflection (as in conventional fibers), the usual relationships between the wavelength of the radiation and the scale of the structure no longer apply in PCF [14]. In conventional fibers, a certain combination of core size and refractive index contrast between fiber core and cladding is required to make the fiber single mode, and this combination can only hold true for a relatively small range of wavelengths. Typically, a conventional fiber designed to operate single mode at a particular wavelength will not be single mode if the wavelength of operation is halved, because V in Eq. (2.3) becomes too large. On the other hand, if the wavelength is doubled, then the fiber will remain single moded, but will be very sensitive to bend loss because R_c in Eq. (2.4) becomes too large. In the case of PCF, the V value behaves quite differently: Instead of rising indefinitely as the wavelength decreases, it tends toward a finite upper bound, Fig. 2.10. This is because, for sufficiently short wavelengths, the effective cladding index is not constant, but approaches the core index in such a way as to keep V constant. This behavior can be understood by observing that light penetrates the air holes more as the wavelength is increased. The low index of the holes is therefore sampled to a greater extent, depressing the effective cladding index.

The value of the upper bound of V depends on the size of the air holes so that whatever value of V_{co} applies for PCFs, a fiber can be designed so that the upper bound is less than it. This makes the fiber "endlessly" monomode (i.e., monomode at all wavelengths, or monomode at a particular wavelength quite independently of the scale of the fiber). Numerical calculation show that V_{co} for a PCF is in fact near 4.1 if the core radius is taken to equal the pitch of the holes [17]. The general result explains our experimental observations, where we showed that a triangular-lattice index-guiding PCF can be formed so as to be monomode over the entire transmission window of silica. Conversely, we have shown that at a fixed wavelength a second

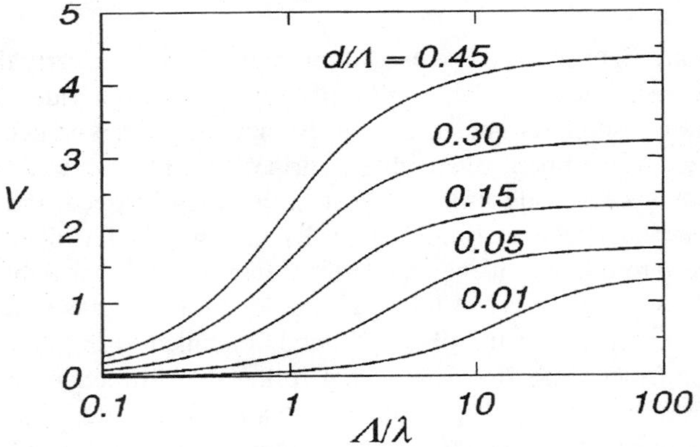

Figure 2.10. The calculated effective V value of a PCF comprising a hexagonal array of air holes of diameter d and separation Λ in silica, where one hole is absent. Note that the horizontal scale is logarithmic.

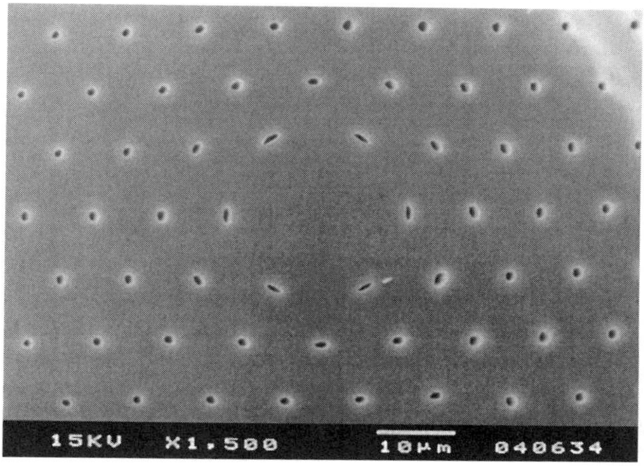

Figure 2.11. Electron micrograph of the core region of a large-mode-area PCF. The fiber shown has the parameters $\Lambda = 9.6\,\mu\text{m}$ and $d/\Lambda = 0.15$. The distance across the core is 24 µm and the fiber guides light single mode at a wavelength of 458 nm.

mode is still not guided even when the scale of the fiber—and of the core diameter—is increased by a factor $> 5\times$, provided only that the diameter of the air holes relative to the pitch remains constant. In a particular example [18,19], we have shown that a fiber with a core diameter of 24 µm guides light of wavelength $\lambda = 458$-nm single-mode (see Fig. 2.11). This is remarkable because to fabricate a conventional optical fiber to guide this wavelength single mode in such a core would require a very small refractive index contrast between the fiber core and the cladding ($< 10^{-4}$), and such small index differences are difficult to fabricate uniformly and repeatably.

The fact that the scale of the fiber can be increased endlessly without ever introducing a second guided mode suggests that it is the arrangement rather than the scale of the holes within the fiber that determines the number of guided modes, and there is strong analytical and numerical evidence to support this. It is thus essential that we are able to fabricate precisely the same structure over a range of absolute scales, which is not trivial because of the very strong dependence of the all-important surface-tension forces on scale—see Section 2.2. The importance of this is emphasized by the observation that we have never observed a single-moded fiber in which the waveguiding core is surrounded by interstitial as well as the primary holes. An example of such a fiber is shown in Fig. 2.12. Likewise, fibers created by omitting two adjacent air holes support at least two modes, for a broad range of air-filling fractions, pitches, and wavelengths. Theoretical work supports the observation that it is very difficult to form a monomode waveguide using such a structure. On the other hand, in the absence of interstitial holes the ratio d/Λ can be rather large without introducing extra guided modes.

Some understanding can be gained by imagining light in the core of a PCF to be surrounded by a "picket fence" of holes. The fundamental mode comprises a single

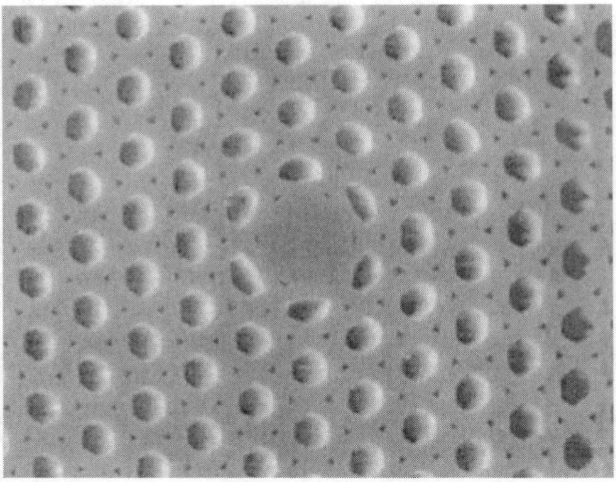

Figure 2.12. An index-guiding PCF that is "always multimode" can be formed using a triangular array of air holes with interstitial holes as well. The small extra holes play a quite disproportionate role in providing extra confinement to the high-index defect.

intensity lobe that fills the core, and cannot fit between the holes and escape. The second-order mode comprises two smaller lobes that can fit between the holes—unless additional interstitial holes are present. Thus PCFs without interstitial holes are far more likely to be single mode than PCFs with them. This intuitive picture makes contact with rigor if we remember that more rapid spatial variations in intensity correspond to lower values of β. A cladding with interstitial holes forces propagating waves to have a higher spatial frequency, thus depressing its effective index and increasing V. A similar picture helps to explain the unusual nature of the related fiber waveguides reported in [16].

While simple triangular-lattice PCF can be fabricated to be "endlessly single mode," it can also be made to support more than one mode, by increasing the relative size of the air holes in the fiber cladding. Much of the time, this is undesirable. The ratio of air hole size to pitch is the single most important parameter in determining whether a second mode will be guided in a particular structure. As mentioned above, if the ratio d/Λ is chosen appropriately, then the value of V_{eff} will approach an asymptotic value $< V_{co}$ for high frequencies. In designing an index-guiding PCF, it is important to be able to identify the parameters required for single-mode operation, and so it is important to be able to identify the value of V_{co}. Our early experimental work [15] suggested that the cut-off value of V_{eff} for second-mode guidance was $\sim V_{co} = 2.5$ (so that d/Λ should be less than about 0.2 for the fiber to remain endlessly single-mode). However, it soon became apparent that this was not in agreement with theoretical predictions, which suggested a rather larger value of V_{co} and of d/Λ. The apparent discrepancy has been resolved by the realization that the critical value for d/Λ is very sensitive to the relative size of the pure silica core. The assumption that the spacing between the nearest holes across the fiber core is exactly twice the spacing of the holes in the cladding region is not valid in some practical fibers, where the fact that the final fiber has smaller air holes than were present in the preform means that the entire cladding structure has collapsed somewhat during drawing. Naturally, the solid core is not able to collapse, so that the final "core diameter" is more than twice the inter-hole spacing. As one might anticipate from the arguments given in the preceding paragraph, this sharply reduces the maximum value of d/Λ for monomode operation—it also degrades the fiber performance in general, increasing bend loss problems for large fiber cores. In order to predict the fiber performance, the exact ratio of the core size (distance between nearest-neighbor air holes) to the cladding inter-hole spacing needs to be taken into account. It is worth emphasizing again that the presence of interstitial holes in the crystal lattice makes it very difficult to form a single-mode fiber of any size—there are simply too many air holes surrounding the core. Thus, while the index-guiding mechanism is very forgiving of variations in the scale of the structure, and in the air-filling fraction, it is uncompromising in the geometrical structure of the fiber.

2.5.3. Practical Matters: Bend Loss, Cladding Modes, Required Nature of the Cladding Material

Although the core of an index-guiding PCF can be made as large as we wish without introducing extra guided modes, the useful range of single-moded operation is limited by the appearance of bend loss. (This would also be the case in a "perfect" conventional fiber, if such a fiber could be made with arbitrarily low refractive index contrast.) Bend loss in PCF occurs at both the long- and the short-wavelength ends of the transmission window, but that at the short wavelength end is most likely to be a problem. (This is because of the scalability of the PCF properties: When we increase the fiber size at a fixed wavelength we move into the short wavelength regime. In the long-wavelength regime, the PCF acts very much like a conventional fiber.) A typical transmission spectrum through a bent length of triangular-lattice PCF is shown in Fig. 2.13, where the bend loss at the shorter wavelengths is apparent. (Similar behavior can be observed in other simple lattice geometries.) A natural question to ask is what is the largest *practical* fiber for operation at a particular wavelength. The trend in behavior can be determined from Eq. (2.4) by noting that the parameter W is a function of V only. Thus, in the short wavelength limit, it is a constant, and R_c varies as $\sim \Lambda^3/\lambda^2$. Thus bend loss rapidly becomes problematic as the scale of the fiber (represented by Λ) increases.

A related question is "what is the numerical aperture (or effective numerical aperture) of the guided mode in an index-guiding PCF?" Strict application of the usual definitions of numerical aperture is problematic, because the refractive index and the core size need to be defined. Also, the noncircular nature of the guided mode implies some variation in the NA with angle. However, the quasi-Gaussian nature of the modal pattern (for a wide range of fiber parameters) does enable workable definitions to be found. Experimental study [20] shows that in an endlessly single-mode fiber, the beam divergence is proportional to the wavelength at short

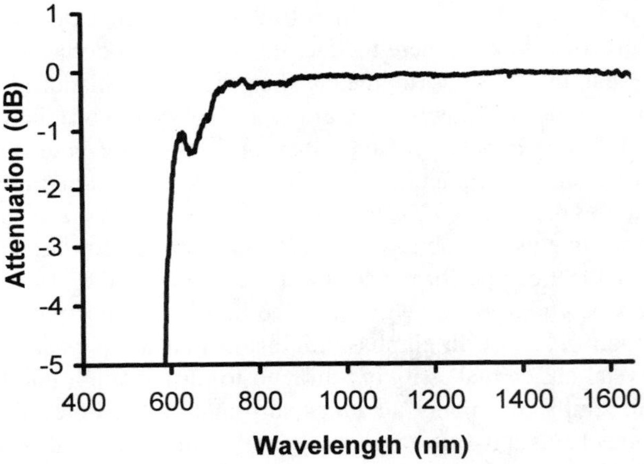

Figure 2.13. A transmission spectrum recorded from a bent PCF, using a white-light source, and showing the appearance of a bend-loss edge at the short-wavelength edge of the transmission band. The fiber is wound once around a mandrel of radius 4 mm.

wavelengths and flattens off at longer wavelengths. For fibers with much larger air holes, which might not be endlessly single mode but that might display other useful properties as a result of their large refractive index difference, the numerical aperture can become very large indeed.

A second practical question is the ease with which cladding modes can be suppressed. Cladding modes negate the advantages of a single-mode fiber if they are present, and in conventional fibers they are "stripped" away by a high index buffer coating around the fiber. This coating destroys TIR at the outer boundary and causes light in cladding modes to refract out of the fiber. However, a certain length of coated fiber is needed before this mechanism is fully effective. It is possible to show that the length required for the effective stripping of cladding modes in a PCF also varies as $\sim \Lambda^3/\lambda^2$. Thus large core PCFs can be dogged by problems of cladding mode stripping as well as bend loss. On the other hand, proper matching of the numerical aperture of the fiber mode minimizes the amount of light in cladding modes.

A further question that arises frequently about index-guiding PCF is How many periods of air hole are required to confine a mode to the core and prevent leakage to the external surface of the fiber? The question is analogous to asking "how thick does the cladding of a conventional fiber need to be?" The answer to that question depends on the refractive index difference and the core size of the fiber being discussed. For example, some of the fibers described in [16] have only two air holes. Clearly, to answer the question for PCF one needs to be quite specific about the fiber in question. One obvious way of restricting the fibers being discussed is to confine our attention to fiber structures that are single mode and that can be fabricated using the stack-and-draw procedure described in Section 2.2. The relevant structures can then be defined by a core radius that is roughly equal to the pitch of the periodic cladding, and by a relative air-hole size in the cladding region of up to perhaps

$d/\Lambda = 0.4$. We have fabricated a range of index-guiding fibers with more than one core in a single fiber, and have studied the coupling between these cores to determine the behavior of the evanescent part of the guided mode in the cladding region. Our primary interest in fabricating multicore fibers has been to use these in bend sensing applications [21–23]—the PCF fabrication technology lends itself very naturally to the formation of complex fiber cross-sections containing several cores that are difficult to form using conventional fiber technology. To predict the performance of such fibers, we formed a number of two-core fiber preforms in which the cores were spaced by varying amounts, and then drew each of these to fiber with a range of values of Λ and d/Λ. Each length of fiber was then incorporated into a coupling-length experiment, which was used to measure the strength of the coupling between the cores as a function of the core–core spacing, the air-hole size, and the pitch. One conclusion of this work was that typically four air holes (i.e., four periods) are required between adjacent guiding cores in a triangular-lattice fiber to ensure that there is no coupling between them over propagation distances of the order of meters. There is every reason to believe that at least as many periods are required between a core and the external surface of the fiber, for a similar range of the fiber parameters. This will become less important for larger values of d/Λ.

Another common question is How important is it that the air holes are arranged periodically in the cladding region? Certainly, the strict periodicity that is required for some of the band-gap effects that we have observed in PCF is not required for effective index guidance, and indeed, none of our fibers are strictly periodic because of imperfections in our fabrication process. However, in order to observe the interesting and technologically useful effects described in this section, one needs to use fiber that has holes that are a significant fraction of a wavelength in diameter, and that are spaced by a wavelength or more. In this regime, using a genuinely random arrangement of air holes will by definition lead to some regions of the cladding that effectively have a slightly higher or lower index than their surroundings. This variation is undesirable because high-index regions can act as individual unwanted cores, while low-index regions bordering on an intentional core, can cause that core to support higher order bound modes. On the other hand, systematic variations in the effective index of the structure (e.g., by changing the air-hole size in the cladding as a function of position) can be a useful additional tool in the design of high-performance PCF waveguides.

2.5.4. Dispersion

One of the most important properties of optical fiber is the group velocity dispersion (GVD). This is important in large part because the very nature of optical fibers results in their being used in long lengths, where the transmission of even moderately short optical pulses requires careful control of the dispersion. For example, one might wish to design a fiber that has zero GVD at a particular wavelength, to avoid distortion of short pulses during propagation. The group velocity dispersion of SiO_2 has a zero at 1.29 μm, and diverges away from zero at other wavelengths. In conventional single-mode optical fiber, the additional effect of waveguide dispersion means that this zero-dispersion point can be pulled to longer wavelengths by fiber design. For example, it is possible to design a conventional fiber that has zero dispersion at 1.55 μm, which is useful for telecommunications applications. To our knowledge, no-one has demonstrated a conventional single-mode fiber that has zero GVD at wavelengths much shorter than 1.29 μm (e.g., to transmit ultrashort pulses from a mode locked Ti:sapphire laser system), although a conventional multimode fiber can be formed in which the GVD of the fundamental mode is zero at these shorter wavelengths. However, PCF remains single moded even when a conventional fiber would support more than one mode, so that such a PCF can be designed to be single mode. We have studied the GVD of PCF structures both theoretically [24] and experimentally [25], and have found broad agreement between the calculated values of GVD and those found experimentally, observing an anomalous waveguide dispersion in a single mode fiber at a wavelength of 813 nm. The PCF can be designed to have zero dispersion over a very broad range of wavelengths compared to conventional fibers or alternatively can be made to have a very large negative dispersion, suggesting its use as a dispersion-compensating fiber element. Other applications of such specialized fibers are in soliton formation from ultrashort pulses from a Ti: sapphire laser at $\sim$800-nm wavelength and in the generation of super-broad-band continuum generation [26]. They might also be significant in any attempt to use PCF as a medium for enhancing nonlinear interactions between a guided mode and a liquid or gas in the air holes. Many of the useful dispersion properties arise from the very strong waveguide dispersion, which is attainable as a result of the large refractive index contrast attainable in PCF. In order to maximize this, one needs to form structures with a small pitch and a large air-filling fraction [27]. Much of our present effort in this area is directed toward this goal. Similar morphological parameters promise other uses, too, such as highly birefringent PCF, which is needed to make polarization-preserving fiber.

2.6. PHOTONIC BAND-GAP WAVEGUIDING IN PHOTONIC CRYSTAL FIBERS

Guiding light in an optical fiber using a photonic band gap as a confinement mechanism represents a radical departure from conventional optical fiber design. While the index-guiding mechanism described above displays some remarkable features that are quite unfamiliar from conventional fiber waveguide theory, truly astounding properties can be predicted (and are presently being investigated) for fibers that guide light by a photonic band gap. The design of the required structures is difficult, and the technological challenge in fabricating them is substantial, but photonic band-gap fibers promise waveguiding features that are truly impossible by other means. Perhaps the most striking of these is the possibility of guiding light down an air hole. Hollow waveguides are very familiar at microwave frequencies, but the high losses associated with using metals at optical frequencies forbid their use for optical waveguides. Microstructured dielectrics can display some of the properties of metals (especially their very high reflectance) under certain conditions, enabling us to form "quasimetallic" hollow waveguides, in which light can be guided down an air hole. Clearly, this represents a significant departure from conventional optical waveguides: potential technological applications range from ultrahigh-power laser transmission, or transmission of wavelengths that are difficult to guide at present because of absorption, to guiding of cold atoms, delivery of microscopic particles, and harmonic generation by guiding ultrashort pulses in a gas-filled air hole. However, less radical results are still surprising: Fundamentally, because the band-gap guidance mechanism enables the structure to support guided modes with an effective index less than the average index of the cladding.

Our work toward demonstrating band-gap guidance of light in PCF has revolved around two different crystal structures—a triangular and a honeycomb array of air holes. The first of these two structures is perhaps of greater fundamental interest for reasons outlined below, while the honeycomb array of holes is easier to fabricate to the specifications required to observe band-gap waveguiding. Forming a triangular array requires that only a single capillary be stacked to form each unit cell, so that for a given number of capillaries in the stack (fixed by the capillary diameter and the furnace bore) one can obtain the smallest pitch in a final fiber with a given diameter. On the other hand, theoretical studies of different material systems have shown that the ranges of parameters that give full 2D photonic band gaps are often wider in a honeycomb structure [28]. This is especially important for our fiber work, where the pitch and air-filling fraction are dictated by the fabrication procedure. In 1998, we observed band-gap waveguiding at a low-index defect site in a honeycomb PCF, and very recently we have observed strong evidence of air guidance in a triangular-lattice structure.

A working band-gap fiber has two crucial parts. First, the periodic "cladding" region needs to be designed and fabricated so as to exhibit a photonic band gap for the required wavelength range. This band gap corresponds to a range of β where there are no propagating modes in the cladding. Second, there needs to be a "core" embedded within the fiber. This core needs to be designed so as to support modes that fall within the band gap of the cladding. This chapter will describe some of the structures that we have fabricated, and how the different aspects of these structures have been characterized experimentally.

We showed several years ago that it is possible in theory to obtain full 2D photonic band gaps for out-of-plane propagation of light using a triangular array of air holes in silica [29]. This behavior is quite different from the situation for propagation in the periodic plane, where it is quite impossible to obtain full 2D band gaps for such a low refractive index contrast. One way of understanding this difference is to recognize that the reflection at a single silica–air interface is a sensitive function of the incident angle, increasing for larger angles of incidence and becoming unity under conditions of total internal reflection. This would suggest that photonic band gaps could be more easily obtained for larger angles of incidence, and indeed this is our general experience. However, because the reflection from each surface becomes so much more significant at larger angles, the effects of incoherent scattering and those of Fresnel reflections at the sample surface can become problematic from an experimental perspective. Furthermore, higher order band gaps, which we expect to observe for values of the pitch $\Lambda \gg \lambda$, are likely to be more sensitive to distortions and imperfections in the crystal structure. The situation is compounded by the difficulty (or slowness) of performing accurate numerical computations for the relatively large values of normalized frequency resulting from a large pitch. The optimum pitch for a full 2D band gap (given the fabrication limitations) is thus far from evident. However, the very strong dependence of the surface-tension forces on the air-hole size during the fabrication process—and the consequent strong dependence of the attainable air-filling fraction on the pitch—mean that large air-filling fraction samples have only been studied for relatively large values of the pitch.

One can think of the physical origins of the band-gap formation in the two structures as being rather different. In the *triangular* structure, band gaps can be though of as arising as a result of coherent scattering from a number of different unit cells of the structure. When such multiple scattering satisfies a phase condition, strong reflections occur, quite like in a 1D Bragg stack. We refer to such band gaps as Bragg band gaps. This "strong Bragg scattering" mechanism is important when the light is able to propagate in both the high- and the low-index materials. In the absence of strict and long-range (several periods) periodicity of the microstructured material, the band gap quickly breaks down because the band gap arises from constructive multiple reflections from several periods of the structure.

On the other hand, the *honeycomb array* of air holes can be viewed as a close-packed array of fine silica strands (see Fig. 2.3). Each of these strands will (if isolated) support waveguide modes at discrete values of the effective index. When a large number of such strands is stacked together, each waveguide mode of an individual strand couples to the neighboring strands, giving rise to a band of modes centered on each of the modes of the single strand (see Fig. 2.14.) Depending on the strength of the coupling, these bands might or might not overlap. The effect is of a series of "band windows" in a region of β where propagation through the composite material would normally be forbidden—propagation is only permitted due to tunneling between adjacent fiber strands. Thus this "frustrated tunneling" mechanism is significant when the light is able to propagate in the high-index phase but is evanescent in the low-index phase of the composite material. The band gap arises mainly from multiple scattering within each single strand, and therefore long-range periodicity is not nearly as important. Thus, in this case the band windows are the exception rather than the rule: Band gaps are therefore far easier to obtain and are more robust. A general description of some of these issues can be found in [13]. First, we will describe our work with honeycomb-lattice PCF before proceeding to describe the prospects for guiding in an air-core fiber by using triangular arrays of air holes.

2.6.1. Band-Gap Waveguiding Using the Honeycomb Structure

Complete 2D band gaps in general can be obtained more easily experimentally by using a honeycomb rather than a triangular array of air holes in silica [11]. These "frustrated tunneling" band gaps rely heavily on total internal reflection at the

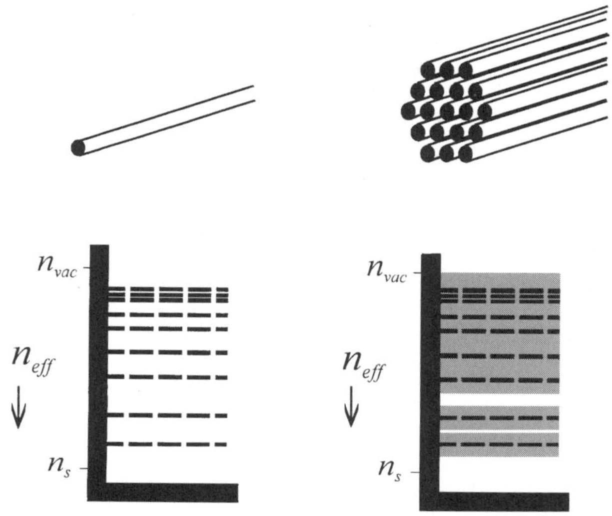

Figure 2.14. A material comprising a honeycomb array of air holes can be usefully considered as a close-packed array of simple waveguiding fibers (one of which is shown on the left, with associated guided modes). The "frustrated tunneling" band gaps (shown lower right) occur between the "bands" of modes which develop as a result of the coupling of the individual guided modes of the single strands.

silica–air interface for their existence: They thus appear in the regime where a single strand will support guided modes (i.e., where the effective index of the mode $n_{\text{eff}} = \beta/k$ lies between the indexes of silica and of air, $1 < n_{\text{eff}} < n_s$) so that the light is evanescent in air. They are thus not expected to enable us to achieve our goal of guiding an optical mode in an air hole. Nonetheless, they are genuine band gaps (in the sense referred to in this work) and are expected to exhibit unusual behavior in their optical response, as well as enable effective guided-mode indexes substantially less than the index of pure silica in an appropriate structure.

An obvious way to exploit such a band-gap material is to replace a single one of the waveguiding strands with a different type of strand. This different strand will support waveguide modes that are distinct from those of the surrounding strands, and that occur at different values of n_{eff}. It should then be possible to tune the parameters of the defect (while keeping those of the periodic material constant) so that one of the strand modes falls in the band gap of the surrounding material. The modes of the core strand are then uncoupled from those of the cladding. A natural choice for the different strand is to insert an extra air hole in the center of one of the silica strands. This is readily done in practice, by replacing a single solid cane with a hollow capillary during the fabrication process. An electron micrograph of such a fiber is shown in Fig. 2.15. The pitch of the structure shown is 3.3 μm, so that each individual silica strand is expected (if isolated in air) to support several waveguide modes. Such modes are illustrated in Fig. 2.16, where it is seen that the higher order modes in a microstructured fiber cladding are quite different from the corresponding

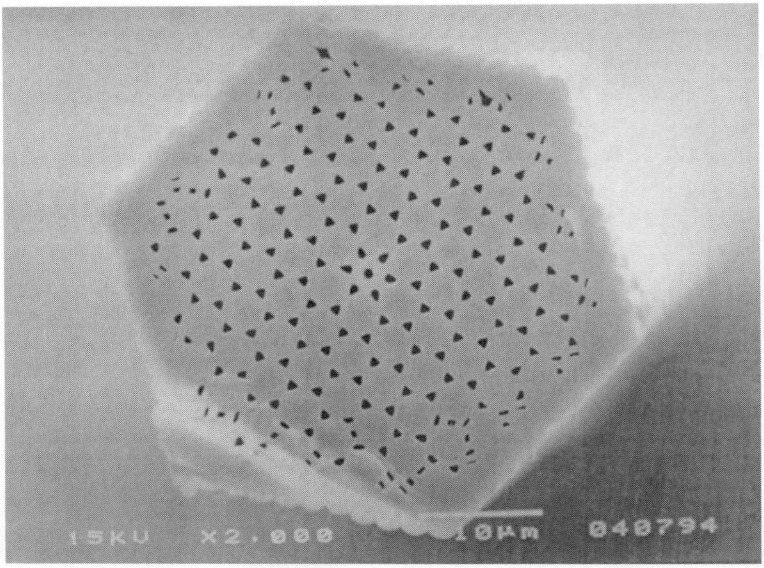

Figure 2.15. An electron micrograph of honeycomb photonic band-gap waveguide. The air filling fraction shown is $\sim 5.3\%$ and the nearest-air-hole spacing is 1.9 μm. The core region is defined by an extra air hole in the center of the fiber that has a diameter of 0.8 μm. This fiber guides green light.

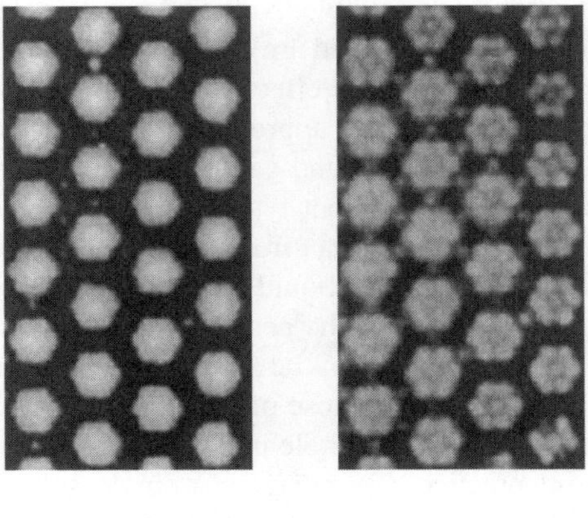

(a) (b)

Figure 2.16. (a) Lowest order and (b) higher order cladding mode patterns in a honeycomb PCF. The photographs were taken with a low-NA white-light source and an optical microscope.

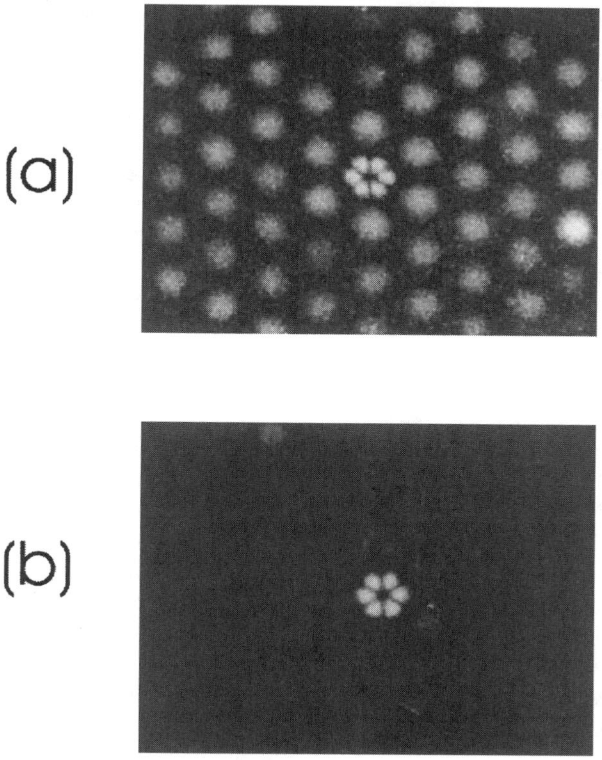

Figure 2.17. Optical micrographs showing a band-gap-guided mode in a honeycomb structure. (a) Shows the output face of the fiber when it is illuminated from below using a white-light source. The larger weaker spots represent the coupled guided modes of the fine close-packed fibers forming the cladding, while the smaller brighter spots in the middle, which are brightly colored, are due to the light trapped in the guide mode around the central air hole (see Fig. 2.14). This is more apparent in (b), where index-matching fluid applied to the fiber surface has been used to strip off much of the light which is not confined to the core.

modes in a conventional fiber. On the other hand, the defect in the photonic crystal caused by the extra air hole is not as easy to analyze. Clearly, one would expect that the highest β modes in such a region would be strongly confined to the silica phase of the microstructured material, while possible modes with $\beta^2 < k^2$ would be concentrated mainly in the air. Therefore, in a photonic crystal where the band gaps occur only for relatively large values of $\beta(\beta^2 > k^2)$, we should not expect to find defect modes trapped in the air.

The photographs in Fig. 2.17 show an optical image of the output face of a short lengths of fiber similar to that in Fig. 2.15, when the fiber is illuminated from below using a white-light source. In both photographs, it is apparent that some light is present in the center of the fiber, in the ring of silica surrounding the central "extra" air hole. To the eye, this localized light is strikingly colored (see Fig. 2.3 in [30]) despite only a white-light source being used for excitation. When index-matching fluid is applied to the side of the fiber [Fig. 2.17(b)] much of the light in the cladding is stripped off, while the light in the central core is unaffected—evidence of its being confined to the region of the low-index defect by the 2D band gap of the surrounding photonic crystal. The bright colors are not due to an insufficient numerical aperture in the objective lens used to collect the light, because the colors remain for a wide range of values of NA in the collection optics. Likewise, it is not a result of chromatic aberration in the optical system because the colors are different for different fibers but are virtually unchanged for different excitation and focusing conditions. Rather, colors result from the strong spectral dependence of the optical properties of the material. The modal intensity pattern consists of six bright lobes separated by deep minima, and distributed around the central air hole. The field is concentrated in the silica, consistent with the guided-mode index being above the refractive index of air.

By choosing a length of fiber that guides an appropriate color when excited with the white-light source, we can couple laser light into the guided mode. The rather unusual field pattern of the mode means that efficient coupling can only be obtained by use of a high-power/high-NA objective lens. The coupling efficiency is also increased by setting an angle between the optical axis of the fiber and the incident beam. The reason for that becomes apparent when one considers the far-field pattern of the guided mode, which is illustrated in Fig. 2.18. The six lobes surround a deep zero in the far-field intensity pattern, and the lobes diverge at an angle of $\sim 30^\circ$. Such a far-field pattern arises as a result of phase reversals between adjacent lobes in the near field. Consequently, in contrast to the fundamental mode of an optical fiber, the modal overlap with a plane wave incident along the fiber axis is strictly zero, and the mode can only be excited by tight focusing and/or off-axis incidence. In practice, the most efficient coupling has been obtained by using a $60\times$ objective lens with a numerical aperture of 0.6 and with an angle of $\sim 30^\circ$ between the incident optical axis and that of the fiber.

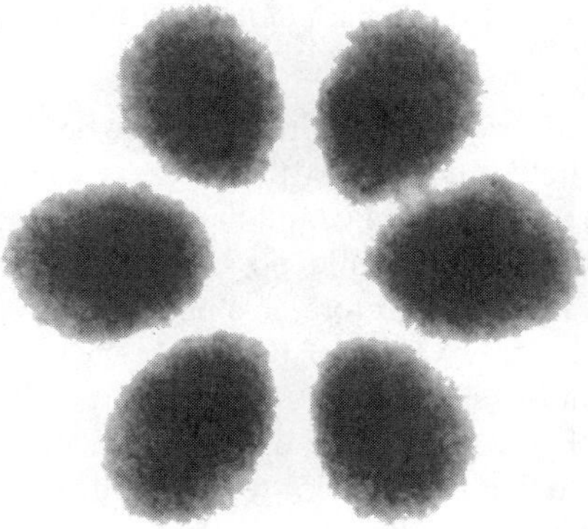

Figure 2.18. Far-field intensity pattern recorded using a laser source to excite the band-gap guided mode in a honeycomb PCF structure such as that in Fig. 2.14.

When the fiber mode is excited in this way, a significant fraction of the incident laser light can be coupled to the guided mode. We have studied both the near- and far-field patterns of the guided mode at the output end of the fiber, and have made the following observations.

1. Both the near- and the far-field patterns remain fixed as the input coupling is varied, except that the overall intensity changes, and there may or may not be significant light in the cladding region. The cladding light is easily distinguished in the far field by its very different divergence and by its highly multimode behavior.
2. The output patterns observed are highly sensitive to the wavelength being used, perhaps not surprising in light of the optical micrographs observed under white-light illumination. For certain wavelengths (corresponding roughly to the colors observed in the core when using white-light illumination) the core mode was tightly confined. For other wavelengths, identical excitation conditions would result in a deep minimum in the optical image of the output face, all the light being coupled to cladding modes. This remains true even for some relatively closely spaced wavelengths (e.g., a particular fiber has been found to have a strongly confined core mode at $\lambda = 488$ nm but no evidence of confinement at $\lambda = 528$ nm).

The guided wavelengths vary sharply with only minor changes in the fiber structure, so that it is extraordinarily difficult to fabricate long lengths of fiber (many meters or more) that are sufficiently uniform along their length to guide a single wavelength. Work on extending the range of measurements that have been done on this fiber is continuing in tandem with improvements in our fabrication procedure, but is hampered by the difficulty of dealing with a guided mode that has a very strong spectral dependence and—effectively—a very large numerical aperture.

Computations of the dispersion [31] of guided modes in such a structure suggest that this is highly engineerable and can differ markedly from that of conventional fibers—strong waveguide dispersion allows for large values of the overall dispersion, or for keeping the total dispersion low over an extended wavelength range. Experimental research into GVD in band-gap guiding photonic crystal fibers is continuing.

2.6.2. Band Gaps in a Triangular Structure—Progress Toward an Air-Guided Mode

One of the long-term goals of this research has been to demonstrate an optical fiber in which the guided mode is trapped within an air hole. The honeycomb band-gap fiber described in Section 2.6.1 does enable light to be trapped in a region of reduced refractive index when compared to the cladding, which effect is quite unattainable using conventional fiber waveguiding principles. This allows one to form guided modes in a silica/air fiber that have modal indexes well below that of silica. However, the (admittedly simple) model of band-gap formation in these honeycomb structures (the coupled resonator model of Section 2.6.1) suggests that the range of available modal indexes is limited between the index of air and that of silica. On the other hand, a guided mode *propagating* in air must have a modal index less than the index of air ($\beta/k < 1$). Hence, we are unlikely to observe such guided modes using a honeycomb lattice. On the other hand, band gaps in this range of β have been predicted in triangular structures [29], and our efforts to observe air guiding have therefore focused on that structure. Again, we need to consider separately the design and the optical properties of the fully periodic cladding and also the nature and details of the defect that will form the waveguiding core. It is evident that a relatively large air-filling fraction is required in a triangular lattice to observe full 2D band gaps. Modeling for relatively small values of normalized frequency $k\Lambda$ (as in [29]) suggests that at least 30% of air and preferably 40% or more is required. Fibers with such a large air-filling fraction are far easier to fabricate for large values of the pitch: Unfortunately, numerical modeling of the optical properties of such large structures is often prohibitively slow.

Fibers such as that shown in Fig. 2.1(b) have parameters for which theory predicts the existence of a complete band gap, although imperfections in the fiber structure are not taken into account in the computations. Part of the difficulty of analyzing such structures is that scattering [32] and fluorescence [33] studies provide only a limited view of their optical properties. Scattering measurements can probe only part of the band structure because of the presence of "uncoupled" modes. They are also complicated by the small transverse dimensions of the samples and by strong Fresnel reflections at the hexagonal external surfaces. The distribution of fluorescence from an Er^{3+}-doped core in the fiber is in good agreement with theoretical predictions, but covers only a limited range of wavelengths and angles. Given these difficulties, and recognizing that we were able to fabricate structures that (at least on paper) possessed a full 2D band gap, we have fabricated samples incorporating a purposeful air defect—an extra large air hole, in which it is hoped that we might see some confined modes. Optical microscopy is then arguably the most powerful tool in analyzing the structures and looking for evidence of waveguiding (at least on a quantitative level), as it enables excitation of fiber modes with a very broad range of optical frequencies and β values.

Only very recently have we observed strong evidence of photonic band-gap guidance of light in an air mode in one of these structures [34]. The band-gap airguide structure is shown in the electron micrograph in Fig. 2.19, and consists of a triangular array of thin-walled capillaries with a central air defect formed by omitting seven adjacent capillaries. We have observed air-guiding effects in a range of such structures with air-filling fractions from 30% up to around 48%. The very large air defect is needed because the relatively narrow band gaps that occur in such structures will in general not support a guided mode for smaller air defects. This can be understood in terms of a general phase-space argument detailed in [34]. An optical micrograph of this fiber illuminated from below with a white-light source is shown in Fig. 2.20, illustrating a concentration of light in the air core. This air mode is quite different to that observed in "capillary waveguides" consisting of a

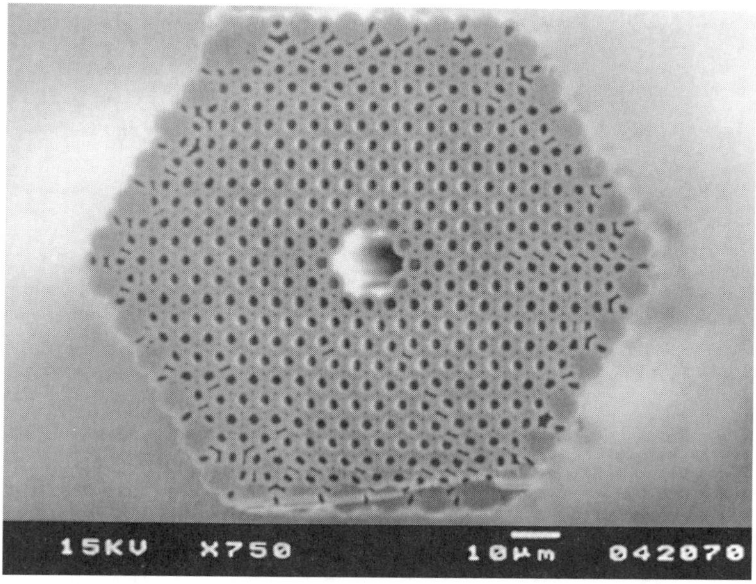

Figure 2.19. Scanning electron micrograph of an air-guiding fiber structure fabricated at the University of Bath. The fiber shown has an external diameter of 105 μm and a pitch of 4.9 μm, and guides green light.

fine air-filled capillary [35]. Simple air-filled capillaries are "leaky" waveguides, in which light is trapped in the air core by imperfect reflection at the air–silica interface. They have a wide range of applications despite their relatively poor performance. The fundamental mode in such structures can be shown to attenuate with a decal length given by $\alpha^{-1} = 2.6a^3/\lambda^2$, so that an air hole of diameter 14 μm in such a capillary will have an attenuation length of 1.3 mm. The total attenuation after propagating 30 mm (as in Fig. 2.20) would be > 70 dB. In contrast, we have recorded > 35% transmission of laser light through the air hole in such a length of fiber, including the input coupling losses.

As with the honeycomb band-gap fiber described above, the light transmitted through the core is brightly colored even when the excitation source is broad band, and we have measured the spectrum of the guided light using an optical spectrum analyzer [34]. The spectra show a number of transmitted bands that are typically a few tens of nanometers broad. These guided bands are seperated by regions of high loss—transmission through the air core is at least 20 dB down over fiber lengths of a few centimeters. (The measurement is limited by a low signal-to-noise ratio due to the white-light source used for the spectral measurements.) Combinations of several of these bands at widely separated wavelengths give rise to some very crisp colors at the output face. By using a fiber for which one of the transmitted wavelength bands includes a laser wavelength, we are able to guide laser light down the fiber in the guided mode. Light can be coupled in to the guided mode easily, and with a relatively high efficiency when compared to the band-gap guided mode described in

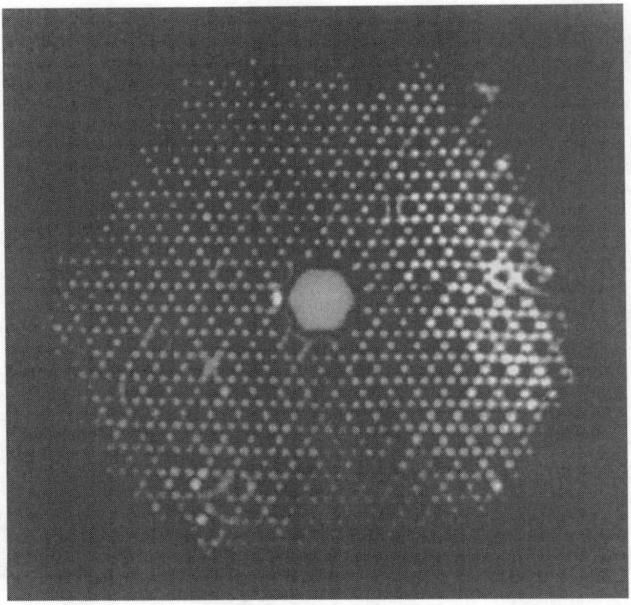

Figure 2.20. Light trapped in the air-guided mode, as observed under an optical microscope. The smooth lobe in the center of the fiber is brightly colored, despite the use of a white-light excitation source.

Section 2.6.1. Part of the reason for this is that the guided-mode index is very close to unity, so that there can be a vanishingly small refractive index mismatch at the input and output faces. As a result, we expect that it will be possible to avoid Fresnel reflections at the fiber surface. We have verified this by noting the complete absence of Fabry–Perot fringes on the light transmitted through a short length of fiber. On the other hand, we have incorporated such a short length of fiber into one arm of a Mach-Zehnder interferometer to verify that the light transmitted through the core in this way is transmitted in a single mode of propagation, so that the lack of fringes is not due to the excitation of a large number of modes. Work is currently in progress to measure the loss of the guided mode, the modal field profiles as a function of wavelength, the dispersion of the guided mode, and its effective index. We are also investigating a wide range of applications [36,37] for the air-guiding fiber, including particle delivery, Raman wavelength-shifting, self-phase modulation, higher harmonic generation, ultrahigh-power single-mode delivery, and short-wavelength transmission.

2.7. CONCLUSIONS

The photonic crystal fiber has proven to contain a wealth of new physics and not a few surprises. It has the capability to provide some waveguiding features that are superior to those available from conventional optical fiber technologies. Some of these (unusual dispersion properties, large-mode-area operation, high-numerical aperture guiding, etc.) are almost immediately realizable and open to commercial exploitation, while others (especially guiding at an air defect) hold tremendous promise for both technological and academic applications in the medium term. Further opportunities are offered by the possibility of using PCF technology to form single-mode optical fibers using non-silicate glasses: The fact that only a single material is required to form a fiber is significant for the exploitation of mid-infrared (IR) transmitting glasses for which no thermally compatible core–cladding combination of glasses can be found. Some of the observed effects have extended our understanding of the physics so as to enable us to design different structures with similar properties [38].

Much of the excitement of being involved in the work presented in this chapter has stemmed from the opportunities to turn conventional wisdom around, and to investigate and demonstrate counterintuitive effects. Optical fiber waveguides that cannot support a second mode at any wavelength or scale of structure, waveguides in which light is confined to a region of reduced refractive index, and even light trapped within an air hole are potentially important technological developments. These are only the first phase of the development of photonic crystal fiber, while being delightful demonstrations of exciting new optical effects using a very old technology—drawing fibers from glass.

ACKNOWLEDGMENTS

The work described in this chapter has benefited from contributions by many people, among them Brian Mangan, Bob Cregan, and Marcin Franczyk at the University of Bath; Terry Shepherd, John Roberts, and John Rarity at DERA Malvern; Jean-Phillipe de Sandro and Guillaume Vienne at the University of Southampton; Alan Greenaway and colleagues at DERA Malvern; Julian Jones and Roy McBride at Heriot-Watt University; Jes Broeng and co-workers at DTU Copenhagen; the group headed by Miguel Andres at the University of Valencia; and researchers at Corning, Inc. This work would not have been possible without the use of a fiber-drawing tower donated to us by the BT Research Laboratories, Ipswich, UK. The research has been supported by the Defence Evaluation and Research Agency, Malvern, UK, and by the Engineering and Physical Sciences Research Council, UK.

REFERENCES

1. See, for example, the special issues of C. M. Bowden, J. P. Dowling, and H. O. Everitt, Eds., (1993). *J. Opt. Soc. Am B* ***10***, 279–413, G. Kurizki and J. W. Haus, Eds., *J. Mod. Opt.* ***41***, 2 (1994).
2. P. St. J. Russell, "Photonic band gaps," *Phys. World* ***5***, 37–42 (1992).
3. J. D. Joannopoulos, R. D. Meade, and J. N. Winn, *Photonic Crystals: Molding the Flow of Light,* Princeton University Press, Princeton, NJ, 1995.

4. P. Phillips, *The Encyclopedia of Glass*, Heinemann, London, 1981.
5. R. J. Tonucci, B. L. Justus, A. J. Campillo, and C. E. Ford, "Nanochannel array glass," *Science* ***258***, 783 (1992).
6. K. Inoue, M. Wada, K. Sakoda, A. Yamanaka, M. Hayashi, and J. W. Haus, "Fabrication of 2-dimensional photonic band-structure with near-infrared band-gap," *Jpn. J. Appl. Phys. 2-Lett.* ***33***, L1463 (1994).
7. J. C. Knight, T. A. Birks, D. M. Atkin, and P. St. J. Russell, "Pure silica single-mode fiber with hexagonal photonic crystal cladding," OFC'96 Proceedings PD3-1, San Diego, CA, 1996.
8. J. C. Knight, T. A. Birks, P. St. J. Russell, and D. M. Atkin, "All-silica single-mode optical fiber with photonic crystal cladding," *Opt. Lett.* ***21***, 1547 (1996), *errata* ***22***, 484–485 (1997).
9. H. B. Lin, R. J. Tonucci, and A. J. Campillo, "Observation of two-dimensional photonic band behavior in the visible," *Appl. Phys. Lett.* ***88***, 2927 (1996).
10. M. Wada, Y. Doi, K. Inoue, and J. W. Haus, "Far-infrared transmittance and band-structure correspondence in two-dimensional air-rod photonic crystals," *Phys. Rev. B* ***55***, 10443 (1997).
11. J. Broeng, S. E. Barkou, A. Bjarklev, J. Knight, T. Birks, and P. St. J. Russell, "Highly increased photonic band gaps in silica/air strucutes," *Opt. Commun.* ***156***, 240 (1998).
12. J. M. Senior, *Optical Fiber Communications*, 2nd ed., Prentice Hall, New York, 1992.
13. P. St. J Russell, T. A. Birks, and F. D. Lloyd-Lucas, "Photonic Bloch waves and photonic band gaps," in E. Burstein and C. Weisbuch, Eds., *Confined Electrons and Photons*, Plenum, New York, 1995, pp. 585–633.
14. T. A. Birks, J. C. Knight, and P. St. J. Russell, "Endlessly single-mode photonic crystal fiber," *Opt. Lett.* ***22***, 961 (1997).
15. J. C. Knight, T. A. Birks, P. St. J. Russell, and J.-P. de Sandro, "Properties of photonic crystal fiber and the effective index model," *J. Opt. Soc. Am. A****15***, 748 (1998).
16. P. Kaiser and H. W. Astle, "Low-loss single-material fibers made from pure fused silica," *Bell Syst. Tech. J.* ***53***, 1021 (1974).
17. T. A. Birks, D. Mogilevtsev, J. C. Knight, P. St. J. Russell, J. Broeng, P. J. Roberts, J. A. West, D. C. Allan, and J. C. Fajardo, "The analogy between photonic crystal fibers and step-index fibers," OFC '99, paper FG4-1, 114, San Diego, CA, 1999.
18. T. A. Birks, J. C. Knight, R. F. Cregan, and P. St. J. Russell, "Single mode photonic crystal fiber with an indefinitely large core," CLEO '98, San Francisco, CA, 1998.
19. J. C. Knight, T. A. Birks, R. F. Cregan, P. St. J. Russell, and J.-P. de Sandro, "Large mode area photonic crystal fiber," *Electron. Lett.* ***34***, 1347 (1998).
20. M. J. Gander, R. McBride, J. D. C. Jones, T. A. Birks, J. C. Knight P. St. J Russell, P. M. Blanchard, J. G. Burnett, and A. H. Greenaway, "Measurement of the wavelength dependence of beam divergence for photonic crystal fibre," *Opt. Lett.* ***24***, 1017, (1999).
21. J. C. Knight, P. St. J. Russell, B. J. Mangan, T. A. Birks, G. C. Vienne, and J. P. de Sandro, "Multicore photonic crystal fibers," OFS '97, postdeadline paper, Williamsburg, VA, 1997.
22. B. J. Mangan, J. C. Knight, T. A. Birks, and P. S. J. Russell, "Dual-core photonic crystal fiber," paper JFB8, CLEO '99, Baltimore, MD, 1999.

23. P. M. Blanchard, J. G. Burnett, G. R. G. Erry, A. H. Greenaway, P. Harrison, B. J. Mangan, J. C. Knight, P. St. J. Russell, M. J. Gander, R. McBride, and J. D. C. Jones, "2-Dimensional bend sensing with a single multi-core optical fiber," *Smart Materials and Structures* ***9***, 132–140 (2000).
24. D. Mogilevtsev, T. A. Birks, P. St. J. Russell, "Group-velocity dispersion in photonic crystal fibers," *Opt. Lett.* ***23***, 1662 (1998).
25. M. J. Gander, R. McBride, J. D. C. Jones, D. Mogilevtsev, T. A. Birks, J. C. Knight, and P. St. J. Russell, "Experimental measurement of group velocity dispersion in photonic crystal fiber," *Electron. Lett.* ***35***, 63 (1999).
26. J. K. Ranka, R. S. Windeler, A. J. Stenz, "Efficient visible continuum generation in air-silica microstructure optical fiber with an anomolous dispersion at 800 nm," CLEO '99, paper CPD8, Baltimore, MD, 1999.
27. T. A. Birks, D. Mogilevtsev, J. C. Knight, and P. St. J. Russell, "Dispersion compensation using single-material fibers," *IEEE Photon. Technol. Lett.* ***11***, 674 (1999).
28. D. Cassagne, C. Jouanin, and D. Bertho, "Hexagonal photonic-band-gap structures," *Phys. Rev.* ***B53***, 7134 (1996).
29. T. A. Birks, P. J. Roberts, P. St. J. Russell, D. M. Atkin, and T. J. Shepherd, "Full 2-d photonic bandgaps in silica/air structures," *Electron. Lett.* ***31***, 1941 (1995).
30. J. C. Knight, J. Broeng, T. A. Birks, and P. St. J. Russell, "Photonic band gap guidance in optical fibers," *Science* ***282***, 1476 (1998).
31. S. E. Barkou, J. Broeng, and A. Bjarklev, "Dispersion properties of photonic bandgap guiding fibers," paper FG5, 117–119, OFC-99, San Diego, CA, 1999.
32. J. C. Knight, T. A. Birks, R. F. Cregan, J. Broeng, and P. St. J. Russell, "From scattering to waveguiding: photonic crystal fibers" to appear in *Scattering from Microstructures*, F. Moreno and F. Gonzalez, Ed., Springer-Verlag, New York, 2000.
33. R. F. Cregan, J. C. Knight, and P. St. J. Russell, "Distribution of spontaneous emission from an erbium-doped photonic crystal fiber," to be published.
34. R. F. Cregan, B. J. Mangan, J. C. Knight, T. A. Birks, P. St. J. Russell, P. J. Roberts, and D. G. Allan, "Photonic band gap guidance of light in air," to be published.
35. E. A. J. Marcatili and R. A. Schmeltzer, "Hollow metallic and dielectric waveguides for long distance optical transmission and lasers," *Bell Syst. Tech. J.* ***43***, 1783 (1964).
36. M. J. Renn and R. Pastel, "Particle manipulation and surface patterning by laser guidance," *J. Vac. Sci. Technol. B* ***16***, 3859 (1998).
37. K. Kim, Z. Chang, H. Wang, H. C. Kapteyn, and M. Murnane, "Spectral broadening in hollow-core fibers as a function of input chirp," CLEO '99, paper CWF20 (1999); Y. Tamaki, C. Nagura, J. Itatani, Y. Nagata, M. Obara, and K. Midorikawa, "Phase-matched high-order-harmonic generation by guided femtosecond pulses," CLEO '99, paper JWA3 (1999); C. Durfee, A. R. Rundquist, M. C. Muranane, and H. C. Kapteyn, "Phase-matched generation of coherent x-rays," CLEO '99, paper JWA1, Baltimore, MD, 1999.
38. E. Silvestre, P. St. J. Russell, T. A. Birks, and J. C. Knight, "Endlessly single-mode heat sink waveguide," CLEO '98, paper CTh059, San Francisco, CA, 1998.

CHAPTER THREE

Near-Field Optics of Nanostructured Surfaces

SERGEY I. BOZHEVOLNYI

Institute of Physics, Aalborg University, DK-9220 Aalborg Øst, Denmark

3.1. INTRODUCTION

Near-field optics is a relatively new branch of optics that deals with optical phenomena related to light scattering by subwavelength structured objects. Since an electromagnetic wave (EM) incident upon a subwavelength surface structure is partially diffracted into evanescent (i.e., nonpropagating) field components with corresponding spatial frequencies, the use of conventional optical techniques for studies of such a scattering is limited to indirect methods, for example, angular resolved measurements of scattered light intensity. Usage of indirect methods implies the necessity of dealing with an inverse problem, a procedure that often results in loss of details and poor accuracy. Only with the development of scanning near-field optical microscopy (SNOM) did it become possible (1) to *directly* probe scattered fields *near* illuminated surface structures, and (2) to image optical field distributions with *subwavelength* spatial resolutions. This technique has helped to significantly enrich our knowledge and enormously promote the understanding of optical phenomena occurring at the nanometer scale.

For a long time, it has been generally accepted that the spatial resolution of optical imaging systems is limited to half of the light wavelength used. The origin of this limit has been linked first to the phenomenon of light diffraction by Abbe [1] and further elucidated by Lord Rayleigh [2]. Therefore it is usually called the Rayleigh–Abbe diffraction limit. Until recently this wavelength limit was considered fundamental and often related to the Heisenberg uncertainty principle. However, as early as in 1928, Synge [3] proposed to leave "the beaten track of microscopic work" and to build a qualitatively new type of microscope, which would be capable of imaging with virtually unlimited resolution. The main idea put forward by Synge was to image a sample point by point by scanning an illuminated subwavelength

Optics of Nanostructured Materials, Edited by Vadim A. Markel and Thomas F. George
ISBN 0-471-34968-2

aperture along the sample surface while simultaneously detecting the transmitted light. He has convincingly argued that the spatial resolution in such a configuration should be related only to the aperture size and not to the light wavelength.

One might be tempted to proclaim that Synge has profoundly shattered and broken the magic spell of the wavelength limit. In fact, nothing drastic like that should be conceived provided that one would consistently use the conventional meaning of wavelength as a spatial period of field oscillations. Indeed, such a period is somewhat firmly related to the frequency of light only for propagating homogeneous plane waves that are infinitely extended in space, just like a purely monochromatic wave should be of infinite duration in time. Once an EM field is confined in a small region, for example, in an aperture, the very notion of wavelength as a spatial period connected with the light frequency (and the refractive index) becomes meaningless, since the spatial period is now determined by the aperture size. Therefore, generally speaking, the resolution limit is always set by the wavelength of EM field interrogating the sample. In the traditional description based on Fourier optics and adopted hereafter, subwavelength resolution is associated with the usage of evanescent waves whose spatial period is smaller than the wavelength. Finally, following the same line of reasoning one can easily establish that there is no contradiction between the Heisenberg uncertainty principle and the subwavelength resolution [4].

The issue of spatial resolution attainable in SNOM is intimately associated with one of the most fascinating phenomena in near-field optics of nanostructured surfaces, namely, the phenomenon of subwavelength light confinement. Actually, more than a century ago, Lord Rayleigh [2] directly connected the resolution limit and light confinement by relating the resolution of a telescope to the smallest spot size of light focused by a lens. Synge's idea of using a tiny aperture to achieve a subwavelength resolution can also be considered from the point of view of light confinement in such an aperture.

One can reason that EM fields are strongly enhanced near sharp corners and small scatterers [5], and thereby confined in their immediate vicinity by virtue of the inverse third power distance dependence of near-field components. The realization of this fact and understanding of its implications to SNOM have led to a rapid progress in high-resolution SNOM techniques based on the usage of homogeneous (apertureless) near-field optical probes [6]. Bearing in mind Fourier optics and the well-known perturbative description of light scattering [7], the subwavelength light confinement near nanostructured surfaces appeares as a natural consequence of the fact that a spatial Fourier spectrum of the scattered field contains spatial frequencies corresponding to those of the surface profile. Symbolically speaking, light scattering by *nanostructured surfaces* result in *nanostructured optical fields*. In this chapter, various phenomena connected with nanostructured optical fields and, in particular, the phenomenon of subwavelength light confinement are considered along with the appropriate SNOM techniques used for their studies. In the rest of this section new terms are introduced that are related to this phenomenon [8] and used hereafter.

The aforementioned configurations leading to the subwavelength light confinement, namely, light scattering by an aperture in a screen or by a scatterer in free space, are rather straightforward and complementary to each other if one bears in mind Babinet's principle [5]. In both cases, light confinement is bound to a certain (subwavelength-sized) geometrical structure and, therefore, can be referred to as the subwavelength *aperture-like* light confinement. It was already mentioned that light confinement is the manifestation of a specific spatial spectrum of an optical field. Taking into account that it is possible to transform virtually any given spatial spectrum into any desirable one by use of a proper combination of diffraction gratings (i.e., with a proper hologram), it is relatively straightforward to arrive at the conclusion that the subwavelength *apertureless* light confinement can be realized with near-field holography [9] or phase conjugation [10]. These approaches are more sophisticated than those considered above, and have the advantage of collecting the light from a hologram area or an area of phase-conjugating mirror, that can be much larger than the region of light confinement.

Recalling that the aperture-like light confinement is most often associated with the regime of *single* scattering of light, one should look for apertureless light confinement among the effects related to *multiple* scattering. Indeed, it turned out that multiple scattering of surface plasmon polaritons (SPPs) by surface roughness may lead to the SPP localization [11], a phenomenon that results in optical field enhancement within subwavelength-sized regions, that is, in subwavelength light confinement. Interference in multiple scattering of light by fractal structures (Chapter 8) is yet another mechanism resulting in subwavelength light confinement in the form of localized dipolar excitations that have been recently observed by direct near-field imaging [12,13]. Light confinement by multiple scattering is closely related to the phenomenon of strong (Anderson) localization of light (Chapter 5), which is one of the most intriguing and exciting phenomena in modern physics [14].

From the viewpoint of Fourier optics, it is clear that an optical field, which is confined within a subwavelength-sized region, is composed mainly of wave components with high spatial frequencies, that is, of evanescent waves. Thereby, the light intensity is exponentially decaying away from the plane, at which the optical field is confined within a subwavelength-sized spot. This means in turn that such a light confinement is always bound to the surface plane, which coincides with the (material) interface separating a half-space with the subwavelength light spot and a half-space containing propagating (to and from the interface) field components. Let us call the light spots, which appear near the surface due to the subwavelength apertureless light confinement, *surface light dots* emphasizing the circumstances that they are bound to the surface and are of subwavelength size, that is, no longer limited by diffraction.

In near-field phase conjugation and holography, surface light dots are formed (deliberately) at particular points of the surface plane. Randomly situated surface light dots occur (inadvertently) due to localization of surface plasmons or dipolar excitations. In both cases, it is difficult to indicate the ultimate limit for the size of surface light dots, and it is impossible to claim that arbitrarily small light dots can be realized. The same is true concerning the degree of field enhancement (with respect to the incident field) that can be achieved in surface light dots. Actually, the problem of light confinement on the subwavelength scale has many challenging aspects and is a subject of ongoing research [15]. In this chapter, only some physical effects related to specific configurations that may influence the size of and field enhancement in surface light dots are mentioned.

The organization of this chapter is as follows. In Section 3.2 the fundamentals of SNOM including a somewhat detailed description of a microscopic point-dipole approach and resonant interactions are presented. Important issues of optical resolution, topographical artifacts, and influence of a probe (when imaging near-field intensity distributions) are discussed. In Section 3.3, main experimental results and relevant numerical simulations concerning subwavelength light confinement by phase conjugation of optical near fields are considered. Different regimes of SPP scattering by surface features and localization phenomena resulting from interference in multiple SPP scattering are considered in Section 3.4, in which the latest results on near-field imaging of localized dipolar excitations are also incorporated. Finally, in Section 3.5, possible applications and further investigations of the phenomenon of subwavelength apertureless light confinement are outlined.

3.2. FUNDAMENTALS OF NEAR-FIELD OPTICAL MICROSCOPY

Underlying principles of near-field optical microscopy can be presented in a number of different ways. There exist a few review articles [16–18] that contain both instructive considerations of main principles and comprehensive discussions of different aspects of SNOM. In addition, a detailed description of SNOM instrumentation and various applications can be found in the first book on near-field optics [19]. Here, the arguments based on Fourier optics are used to introduce the main ideas of Sections 3.2.1–3.2.2. In Sections 3.2.3–3.2.6, a microscopic point-dipole description is outlined and subsequently employed to discuss general issues of multiple scattering, resonant interactions, topographical artifacts, and probe–sample interaction in SNOM.

3.2.1. Main Principles

The general idea of near-field microscopy can be formulated in terms of Fourier optics as follows: A propagating EM field incident upon a subwavelength surface structure is partially diffracted into an evanescent field with corresponding spatial frequencies, which is in turn (partially) converted by a subwavelength-sized probe into a field propagating toward a detector [20]. This idea is illustrated with the following simple example.

Let us consider imaging of a thin one-dimensional (1D) film (object), whose amplitude transmittance $t(x)$ contains an information that should be recovered by use of microscopy (Fig. 3.1). A plane monochromatic s-polarized wave (electric field is perpendicular to the plane of incidence, i.e., to the figure plane) with the amplitude A is sent on the transparency at normal incidence. The transmitted wave can be used to form an image with a conventional (*far-field*) arrangement, in which a lens situated *far* (as compared to the light wavelength) from the film projects the transmitted field

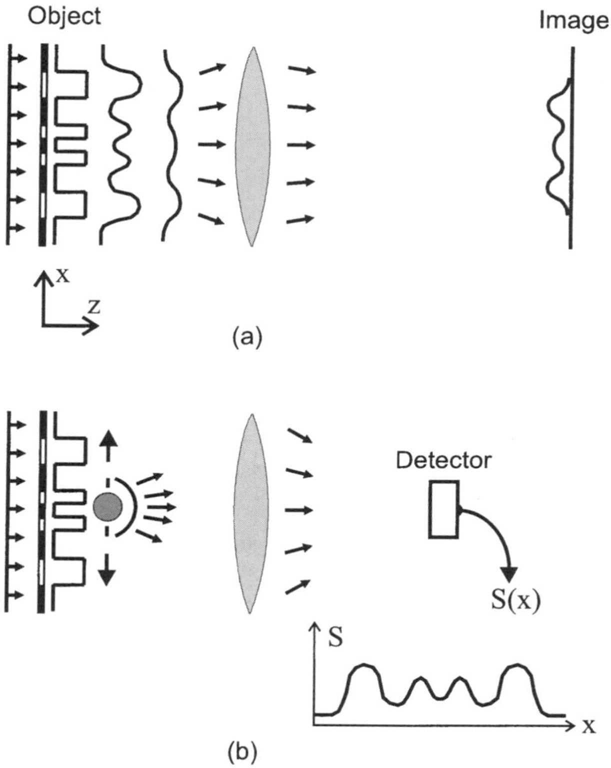

Figure 3.1. Schematic representation of (a) conventional far-field and (b) scanning near-field microscopy arrangements. In the far-field arrangement (a), an optical field intensity distribution in the image plane represents an image of the object (transparency). In the near-field arrangement (b), a dependence of the detected scattered power on the x coordinate of a probe produces an image. Qualitative distributions of EM amplitude at different cross-sections along the propagation direction are displayed illustrating the loss of resolution, which is especially pronounced for the far-field microscopy arrangement.

distribution (at the object plane) onto the image plane [Fig. 3.1(a)]. Alternatively, an image can be built point by point with the help of a small scatterer that is scanned *near* the film plane [Fig. 3.1(b)]. The operation principle of this (*near-field*) configuration is based on the circumstance that the scattered power depends on the incident field at the site of the scatterer. It is intuitively clear already from Fig. 3.1 that the resolving power is quite different for these configurations.

This difference can be easily quantified by tracking modifications of the transmitted wave during its propagation. For this purpose, let us represent the transmitted field $E(x, z \geq 0)$ by a superposition of plane waves [21]:

$$E(x,z) = \frac{A}{2\pi}\int a(p)\exp\left[i\left(px + \sqrt{k^2 - p^2}z\right)\right]dp \qquad (3.1)$$

with

$$a(p) = \int t(x)\exp(-ipx)dx \tag{3.2}$$

where $Aa(p)$ is the Fourier (spatial) spectrum of the transmitted field at the output plane $z=0$ (Fig. 3.1), and k is the wavenumber of light: $k = 2\pi/\lambda$, where λ is the wavelength of light. The information about the object is encoded in the transmitted wave in the form of the Fourier transform (FT) $a(p)$ of the film transmittance [Eq. (3.2)]. It follows from Eq. (3.1) that the spatial spectrum F of the transmitted field *changes* along the propagation direction:

$$F(z) = Aa(p)\begin{cases} \exp\left(i\sqrt{k^2 - p^2}z\right), & \text{for} \quad |p| \le k \\ \exp\left(-\sqrt{p^2 - k^2}z\right), & \text{for} \quad |p| > k. \end{cases} \tag{3.3}$$

so that the high spatial frequencies, namely, those with $|p| > k$, are progressively filtered out. These frequencies correspond to the so-called *evanescent* waves whose amplitudes are *exponentially decaying* away from the film surface [Eq. (3.1)]. Note that the characteristic spatial period of evanescent waves is related to p [not to k (!)], which corresponds to a magnitude of the wavevector projection on the plane along which evanescent waves do propagate ($z=0$ in our case). For sufficiently high spatial frequencies, this period is simply given by $\lambda_{ew} \cong 2\pi/p$.

A complete recovering of the information is clearly out of the question for both of the imaging arrangements (Fig. 3.1). In the far-field configuration, not only all spatial frequencies corresponding to the evanescent waves but also those corresponding to the waves propagating outside the acceptance angle ϑ of the imaging lens are irrevocably lost. The maximum spatial frequency of the wave that *can* reach the image plane is $p_{ff} \cong k\sin\vartheta$. Note that if a microscopic objective is used instead of the lens, the $\sin\vartheta$ should be changed on the numerical aperture of the objective. For the near-field configuration, the maximum spatial frequency is determined by the distance z_p between the scatterer (the generic term for it is "a near-field optical probe") and the film surface. If we use the usual criterion for cutting off exponential decay, we can obtain the required relation: $z_p(p_{nf}^2 - k^2)^{0.5} \cong 1$, which for sufficiently small distances leads to a simple formula: $p_{nf} \cong 1/z_p$. The loss of a high-frequency part of the spatial spectrum would inevitably result in the loss of details in the image obtained, that is, in a limited spatial resolution.

There are various approaches to the problems of definition and determination of the resolution of an imaging system [22]. The most widely used definition of resolution is as the smallest distance between two point-like objects that can still be resolved. In our configuration (Fig. 3.1), the appropriate transmittance should contain two δ functions: $t(x) = \delta(x - 0.5d) + \delta(x + 0.5d)$. The main difference between the corresponding spatial spectrum $a(p) = 4\pi\cos(pd/2)$ and the spectrum of an individual δ function $\{F[\delta(x)] = 2\pi\}$ is the presence of oscillations in the former.

One can then argue that, in the presence of the cutoff in the spectrum, the two δ sources will be well resolved if the first zero of $a(p)$ will be well within the detected spectrum: $p_{max}d/2 \gg \pi/2$. For the far-field configuration, this means that the following relation should be satisfied $d \gg \lambda/2\sin\vartheta$. Incidentally, the limiting case of this relation corresponds exactly to the classical Rayleigh–Abbe diffraction limit of resolution: $d = \lambda/2\sin\vartheta$. For the near-field configuration, the spatial resolution is no longer limited by one-half of the light wavelength but only by the probe–surface separation: $d \gg \pi z_p$. Since the probe–surface distance means actually the distance between the scattering *center* of a probe and the surface, it is plausible to state that *the SNOM resolution is determined by the size of a probe and its distance to the sample surface*. On the other hand, recalling that an evanescent wave can also be characterized with a spatial period ($\lambda_{ew} = 2\pi/p$) one can relate the SNOM resolution to the smallest period of evanescent waves that are efficiently scattered by the probe toward a detector ($d \sim \pi/p_{max} = \lambda_{min}/2$). The resolution limit in SNOM is therefore determined by *the smallest period of optical field participating in the probe–sample interaction*. The latter conclusion does not contradict the former one but rather generalizes it (this point has already been raised in the introduction).

3.2.2. Main Configurations

The above consideration, though being rather primitive, reveals the essential components that should be present in any SNOM configuration: a sample with subwavelength surface features illuminated by light, a subwavelength-sized probe that is scanned along the surface at a close distance, and a detection system that records the light scattered by the probe as a function of scanning coordinates. One can also use a subwavelength-sized source of the incident light that is scanned along the sample surface and detect the light scattered by the sample. The main physical phenomenon in SNOM is the conversion of the evanescent field components that contain the information about surface features into the waves propagating toward a detector. Since the important light interactions take place on a subwavelength scale (nanometer scale for light in visible), the technological requirements for SNOM components are rather high [16,17,19]. It is very important to fabricate a well-defined optical probe and to realize its positioning with very high precision. It is equally important to accurately detect the relevant scattered waves whose power is necessarily low because it rapidly decreases with the decrease of the probe size $a(\sim a^{-6})$. These quite challenging demands are hardly possible to satisfy for all possible applications, and, since the first experiments on SNOM in visible [23], many different configurations oriented toward specific areas of interest have been developed [24]. Next, the main SNOM configurations will be briefly outlined.

The near-field configuration shown in Fig. 3.1(b) contains a small isolated particle that acts as a near-field optical probe. Even though there are certain advantages in using such a probe (the probe particle can be put in a liquid environment) [25], it is very difficult to precisely control the scanning of an individual particle. Most of the near-field optical probes are *extended* bodies with

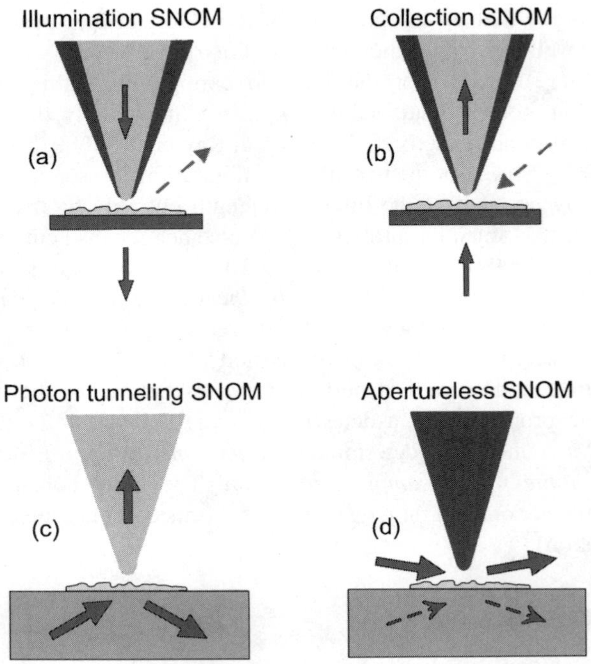

Figure 3.2. Different configurations employed in scanning near-field optical microscopy (SNOM).

pointed sharp tips that, by being inserted in an evanescent (nanostructured) optical field bound to the sample surface, scatter the light similarly to individual particles.

A metal-coated tapered optical fiber with a small opening at the very end [26] represents a logical development of the original idea of using a small aperture in an otherwise opaque screen [3]. Depending on the configuration used for illuminating the sample and detecting the scattered light, such a probe provides a well defined subwavelength-sized source or detector of light interrogating the sample. In the *illumination* configuration [Fig. 3.2(a)], the fiber serves to guide the radiation from a source to the aperture that illuminates the sample, and the light transmitted or reflected by the sample is detected. In the *collection* configuration [Fig. 3.2(b)], the sample is illuminated by a remote optical source, and the scattered light is collected by an aperture and guided toward a detector (light excitation and detection through an aperture are considered in Chapter 4). Despite many technological problems with the manufacturing of these probes in a controllable manner, the aperture probes are most widely used in SNOM because of a high degree of light confinement ensuring high resolution and a large contrast in near-field imaging [24,26].

An alternative approach is to use uncoated sharp fiber probes, whose fabrication process is less demanding. However the field distribution around an uncoated fiber probe is rather complicated and not limited to a subwavelength region, a circumstance that makes near-field imaging with these probes more cumbersome

[27]. One can get around the problem of confinement by using the confined illumination, for example, by a totally internally reflected wave incident on the surface from the side of the sample [28]. In this case [Fig. 3.2(c)], the background field is tightly bound to the sample surface, and a sharp pointed tip can be within certain approximations considered as a point-like probe of electric near-field intensity [29]. Due to the apparent analogy with the principle used in scanning tunneling microscopy (STM), this configuration is often referred to as the *photon tunneling* (SNOM) configuration [28]. Actually, in such a configuration, the usage of an uncoated probe (instead of metal-coated aperture probe) is preferable because the relatively low refractive index of optical fiber decreases the risk of perturbation of the measured field. This problem is very important in the context of mapping electromagnetic fields with SNOM, and is discussed in detail in Section 3.2.6.

The two probes considered above have the advantage of using a fiber to guide the light to or from the sample. However, these probes also have a certain drawback: light confinement in the best case of a metal-coated aperture probe is limited by the electric field skin depth of the metal coating (e.g., $\cong$ 12 nm for an Al coating at the light wavelength of 633 nm). This strict limit of light confinement (and the corresponding resolution limit) in the aperture SNOM along with the realization that the light is also confined near sharp material boundaries has stimulated interest in *apertureless* SNOM techniques [6]. The apertureless SNOM configuration [Fig. 3.2(d)] is similar to the configuration with a small spherical scatterer [Fig. 3.1(b)], with the only difference being that the scatterer is replaced with a sharp homogeneous tip made of a material with a high refractive index. Note that this technique is highly sensitive with respect to the polarization of the incident light. It has been demonstrated both theoretically [30] and experimentally [31] that in order to realize a significant field enhancement (near the tip) the incoming field should have a large component along the axial direction. For this reason, it is advantageous to use a p-polarized light beam (electric field is in the plane of incidence) incident under an oblique angle either from the side of the tip or from the side of the sample [Fig. 3.2(d)] for illumination. In addition, a vertical dithering of the tip along with lock-in detection is usually implemented to separate the near-field related signal from the background [6,31].

The resolution in SNOM is determined by the highest spatial frequencies of the optical field components that take part in the probe–sample interaction. This means that the probe in any SNOM configuration should be scanned very close to the surface inspected. Rapid progress in SNOM techniques and their applications during prior years has been largely stimulated by the development of shear-force feedback [32], which allows one to maintain the probe–surface distance during scanning. The shear-force regulation of the tip–surface distance can be implemented as follows [33]. The fiber probe tip is resonantly vibrated (3–30 kHz), and the vibration amplitude ($\sim$ 10 nm) is detected with the help of a laser beam passing through the fiber perpendicular to its axis (Fig. 3.3). A part of the resulting diffraction pattern is registered by a photodiode, whose signal at the frequency of fiber oscillations is measured with a lock-in amplifier. Near the surface, atomic forces between the tip and the sample drastically attenuate the vibration amplitude, a circumstance that

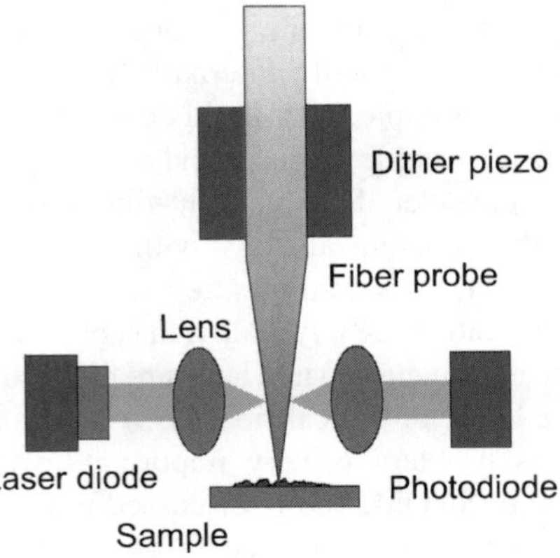

Figure 3.3. Schematic diagram showing the detection system used for shear-force feedback.

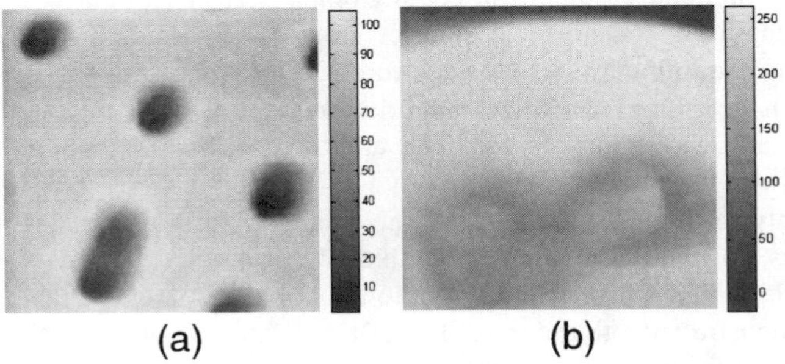

Figure 3.4. Gray-scale topographical $4.5 \times 4.5\text{-}\mu\text{m}^2$ images of (a) an uncoated compact disc and (b) a part of a leukocyte. Corresponding bar scales reflect the depth (in nm) of the images.

makes the photodiode signal suitable for a reliable feedback. Parameters of the optical detection part of the shear force setup can be optimized in order to maximize the detected signal for a given amplitude of fiber oscillations [34]. The range of tip–surface distances, in which the amplitude decreases from an initial value to nearly zero, is typically < 20 nm, implying that the tip–surface distance can be kept with nanometer accuracy. In general, stable operation of the feedback system is of great importance in SNOM not only because accurate imaging of surface topography is indispensable for identification of surface structures, but also for the reason that evanescent field components, whose detection is essential for subwavelength resolution, result in a strong dependence of the detected optical signal upon the tip–surface distance. In the considered system, high-quality topographical images with the spatial resolution of < 50 nm can be obtained for different surfaces. The images shown in Fig. 3.4 demonstrate the reliable operation of our shear-force feedback when scanning a plastic surface of a compact disc and a soft cell membrane [33]. One can even notice two nuclei of a polymorphonuclear leukocyte (white blood cell) on Fig. 3.4(b), indicating that the cell membrane was sufficiently pliable at the time of the experiment. Note that apart from the linear slope correction, the presented images are unprocessed. Finally, it should be mentioned that apertureless SNOM configurations are usually based on commercial atomic force microscopes (AFMs) [or scanning tunneling microscopes (STMs)], and the resolution in topographical imaging can be on the subnanometer scale [6].

3.2.3. Microscopic Point-Dipole Description

Theoretical description of near-field optical phenomena is a challenging problem. The small sizes of the interacting constituents (probe, surface features) and the small distances involved call for an approach that can deal with the problem of multiple scattering of light on the nanometer scale and in a rather sophisticated geometry. In near-field optics, the familiar concepts of refractive index and reflection factors become meaningless, and *light scattering* becomes the fundamental mechanism that governs interactions in the probe–sample system [18]. A number of theoretical models that can be used to deal adequately with this problem have been developed [35]. Here a microscopic point-dipole description of probe–sample interactions is introduced, emphasizing the importance of the self-consistent treatment of multiple scattering [36]. This approach allows one to deal in essentially the same manner with rather complicated extended structures and very simple systems whose behavior can be described by explicit analytical relations.

Let us represent a near-field optical probe and object structure by small spherical scatterers placed near the bulk (Fig. 3.5). The particle-field interaction can then be treated in the electric dipole approximation with the optical response (to an incident EM field) of an individual sphere being expressed in terms of the isotropic dipole polarizability. By assuming that the incident light is monochromatic with an angular frequency ω, the *total* electric field $\mathbf{E}(\mathbf{r}, \omega)$ at an arbitrary point above the bulk can

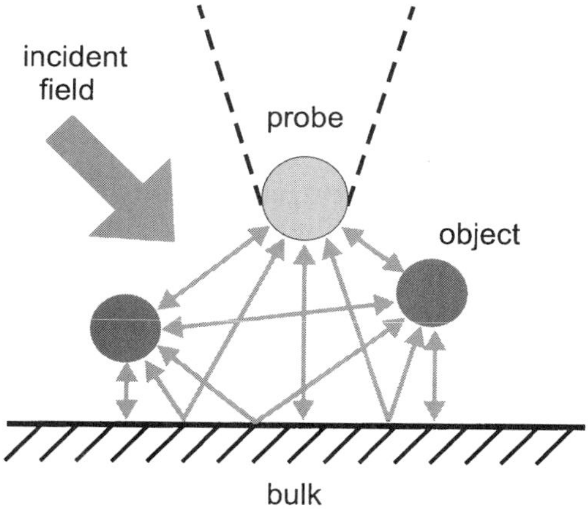

Figure 3.5. Schematic diagram showing the probe–sample system represented by spherical scatterers near the bulk and illuminated by light. The interactions between constituents of the system are indicated by arrows.

be written down as follows:

$$\mathbf{E}(\mathbf{r},\omega) = \mathbf{E}_{\text{in}}(\mathbf{r},\omega) - \mu_0\omega^2 \sum_{j=1}^{n} \alpha_j(\omega)\boldsymbol{G}(\mathbf{r},\mathbf{r}_j,\omega)\cdot\mathbf{E}(\mathbf{r}_j,\omega) \tag{3.4}$$

where $\mathbf{E}_{\text{in}}(\mathbf{r},\omega)$ is the incident field, that is, the field that would prevail in space if the scatterers were absent; $\boldsymbol{G}(\mathbf{r},\mathbf{r}_j,\omega)$ is the appropriate field propagator (or dyadic Green's function), which describes the field propagation from a source point $\mathbf{r}_j$ to an observation point $\mathbf{r}$; and $\mathbf{E}(\mathbf{r}_j,\omega)$ is the *self-consistent* field at the site of jth dipole scatterer with the polarizability $\alpha_j(\omega)$. In the long-wavelength electrostatic approximation, the polarizability of a small sphere can be related to the sphere's radius a_j and its dielectric constant $\varepsilon_j(\omega)$ by using the well-known expression

$$\alpha_j(\omega) = 4\pi\varepsilon_0 a_j^3 \frac{\varepsilon_j(\omega)-1}{\varepsilon_j(\omega)+2} \tag{3.5}$$

Note that the self-consistent field at the site of the scatterer is *different* from the incident field as it includes the fields scattered from *all* other scatterers (Fig. 3.5). The self-consistent field $\mathbf{E}(\mathbf{r}_j,\omega)$ at the site of jth dipole is determined by a set of self-consistent equations obtained from Eq. (3.4) by letting $\mathbf{r}$ coincide in turn with each of dipole positions:

$$\mathbf{E}(\mathbf{r}_j,\omega) = \mathbf{E}_{\text{in}}(\mathbf{r}_j,\omega) - \mu_0\omega^2 \sum_{i=1}^{N} \alpha_i(\omega)\boldsymbol{G}(\mathbf{r}_j,\mathbf{r}_i,\omega)\cdot\mathbf{E}(\mathbf{r}_i,\omega). \tag{3.6}$$

This system of self-consistent equations can be rigorously solved with standard procedures of linear algebra resulting in the local fields at the sites of the dipoles and allowing one to determine the total field at any point (e.g., at the position of a detector) with the help of Eq. (3.4).

Now, let us consider the field propagator $\boldsymbol{G}$. In principle, it consists of two parts, namely, the direct propagator $\boldsymbol{G}$ describing the dipole scattering in free space and the indirect propagator $\boldsymbol{I}$ accounting for the presence of an interface between free space and the bulk: $\boldsymbol{G} = \boldsymbol{D} + \boldsymbol{I}$. The complete retarded form of the direct propagator is given by

$$\boldsymbol{D}(\mathbf{r},\mathbf{r}_s,\omega) = \frac{1}{4\pi}\left[\left(-\frac{1}{R} - \frac{ic}{\omega R^2} + \frac{c^2}{\omega^2 R^3}\right)\boldsymbol{U} + \left(\frac{1}{R} + \frac{3ic}{\omega R^2} - \frac{3c^2}{\omega^2 R^3}\right)\mathbf{e}_R\mathbf{e}_R\right]\exp(ikR) \tag{3.7}$$

where $\mathbf{r}_s$ is the source point, $R = |\mathbf{r} - \mathbf{r}_s|$, $\mathbf{e}_R = (\mathbf{r} - \mathbf{r}_s)/R$, $\boldsymbol{U}$ the unit tensor, k is the wavenumber of light in free space: $k = \omega/c$. The direct propagator contains contributions with different distance dependencies. The *far-field* contribution is proportional to R^{-1} and it is the only one that survives at large distances from the source point, that is, when $kR \gg 1$, giving rise to the *propagating* components of the scattered field. In the opposite limit of very small distances, that is, when $kR \ll 1$, the leading contribution is proportional to R^{-3}, and this *near-field* contribution is mainly responsible for the *evanescent* field components [37]. It is these two limiting cases that should be borne in mind when one considers far- and near-field imaging configurations (Section 3.2.1).

If all distances involved are much smaller than the light wavelength, one can keep only the near-field contribution and neglect retardation effects, since $\exp(ikR) \to 1$, in Eq. (3.7). Thus, the near-field direct propagator takes the form

$$\boldsymbol{D}_{\text{nf}}(\mathbf{r}, \mathbf{r}_s, \omega) = -\frac{c^2}{4\pi\omega^2} \frac{3\mathbf{e}_R\mathbf{e}_R - \boldsymbol{U}}{R^3} \tag{3.8}$$

The indirect propagator has a rather complicated form and is usually expressed via its FT [38]. In the nonretarded and local limit of the bulk response, the indirect dipole–dipole interaction can be treated as a direct interaction between the dipole and its *mirror* image [36,39]. If the bulk surface coincides with the plane $z = 0$, the near-field indirect propagator can be expressed as follows:

$$\boldsymbol{I}_{\text{nf}}(\mathbf{r}, \mathbf{r}_s, \omega) = \boldsymbol{D}_{\text{nf}}(\mathbf{r}, \mathbf{r}_{\text{ms}}, \omega) \cdot \boldsymbol{M}(\omega)$$

where

$$\boldsymbol{M}(\omega) = \frac{\varepsilon_b(\omega) - 1}{\varepsilon_b(\omega) + 1} \begin{pmatrix} -1 & 0 & 0 \\ 0 & -1 & 0 \\ 0 & 0 & 1 \end{pmatrix} \tag{3.9}$$

$\boldsymbol{r}_{\text{ms}} = (x_s, y_s, -z_s)$ points to the position of the mirror image of the source, and $\varepsilon_b(\omega)$ is the dielectric constant of the bulk. The factor in front of the matrix in Eq. (3.9) is the nonretarded (electrostatic) reflection coefficient for *p*-polarized light: $r^p(\omega) = [\varepsilon_b(\omega) - 1]/[\varepsilon_b(\omega) + 1]$. Note that the nonretarded reflection coefficient for *s*-polarized light is zero [36] and that the bulk response is especially strong when $\varepsilon_b(\omega) \to -1$, that is, when the frequency is close to that of the SPP resonance. Finally, remember that, contrary to the indirect propagator, the direct propagator when written in the above form [Eqs. (3.7) and (3.8)] is not defined for $R = 0$. However, the appropriate self-consistent field contribution is already incorporated in the concept of polarizability [Eq. (3.5)] and, therefore, the direct propagator to the dipole itself should be omitted, for example, in the set of self-consistent equations [Eq. (3.6)]. More detailed discussion on different forms of the field propagators can be found in [18,35].

The structure of the propagator $\boldsymbol{G}$ determines the distance dependence and polarization characteristics of the field $\mathbf{E}_{\text{sc}}(\mathbf{r})$ scattered by an individual dipole:

$$\mathbf{E}_{\text{sc}}(\mathbf{r}, \omega) = -\mu_0\omega^2\alpha(\omega)\boldsymbol{G}(\mathbf{r}, \mathbf{r}_s) \cdot \mathbf{E}_{\text{in}}(\mathbf{r}_s) \tag{3.10}$$

where $\alpha(\omega)$ is the polarizability of the dipole located at $\mathbf{r}_s$. If the total propagator $\boldsymbol{G}$ can be approximated by the near-field propagators ($\boldsymbol{G} \cong \boldsymbol{G}_{\text{nf}} + \boldsymbol{I}_{\text{nf}}$), then the scattered field coincides with the field of an *electrostatic* dipole. It should have been expected since the non-retarded limit ($kR \to 0$) also means that the frequency of light vanishes: $\omega \to 0$. The properties of near-field direct and indirect contributions of the scattered field are of great importance for SNOM. Their detailed discussions can be found in the aforementioned reviews [16–18]. Here, it should be stressed that, even though the near-field region is a domain of electrostatics, there does exist the *phase* of near-field components that should not be left out of consideration [18,40].

In general, the following circumstances should be taken into account when considering the phase of scattered near-field components: the dipole polarizability is complex [Eq. (3.5)], the direct and indirect propagators exhibit different phases [Eq. (3.9)], and, finally, the phase of different near-field components is different. The latter fact is rather important for imaging properties in SNOM and, for example, it accounts for different optical contrasts obtained in photon tunneling SNOM for the same nanoparticle but with different light polarizations [41]. This feature can be illustrated by considering a dipole located at the coordinate origin in free space and subjected to a linearly polarized field $\mathbf{E}_{\text{in}} = (0, 0, E_{\text{in}})$. The scattered near-field components can then be expressed [Eqs. (3.8 and 3.10)] as follows: $\mathbf{E}_{\text{sc}}(x, y, z) \propto E_{\text{in}}(3xz, 3yz, 3z^2 - R^2)/R^5$. By comparing the scattered field amplitudes along the x and z axis, one can immediately see that these amplitudes are of different magnitudes and opposite phases: $\mathbf{E}_{\text{sc}}(x, 0, 0) \propto (0, 0, -E_{\text{in}})/R^3$, and $\mathbf{E}_{\text{sc}}(0, 0, z) \propto (0, 0, 2E_{\text{in}})/R^3$. The total field consists of the incident and scattered fields, and its intensity distribution in different planes exhibits thereby different (negative and positive) contrasts [18]. Note that the scattered *near* field is *stronger* along the *nonradiative* direction (a dipole does not radiate along its axis [5]). Actually, the latter feature constitutes the main physical reason for the aforementioned phenomenon (Section 3.2.2) of field enhancement near the tip illuminated with *p*-polarized light in apertureless SNOM [30,31]. These features of scattered near field, [viz., that the near field scattered in the radiative direction is out of phase with the incident field and considerably weaker than the field scattered in the nonradiative direction (the latter is also in phase with the incident field)] are rather fundamental and result in many phenomena in near-field optics. Some phenomena illustrating these properties of scattered near fields will be considered in Section 3.2.4, which deals with resonant interactions in the system of dipoles.

3.2.4. Resonant Interactions

The system of self-consistent equations [Eq. (3.6)] is often solved by using the *first* Born approximation, in which the field at the site of a dipole is assumed to be equal to the sum of the incident field and the primary scattered fields from other dipoles:

$$\mathbf{E}^1(\mathbf{r}_j, \omega) = \mathbf{E}_{\text{in}}(\mathbf{r}_j, \omega) - \mu_0\omega^2 \sum_{i=1}^{N} \alpha_i(\omega)\boldsymbol{G}(\mathbf{r}_j, \mathbf{r}_i, \omega) \cdot \mathbf{E}_{\text{in}}(\mathbf{r}_i, \omega) \tag{3.11}$$

This approximation corresponds to the regime of *single* scattering as far as the field at the dipoles' sites is concerned. However, if the field $\mathbf{E}^1$ is used in Eq. (3.4), the total field at the observation point not coinciding with the dipoles includes secondary scattered waves. The corresponding regime of *double* scattering is already close to the regime of strong interactions, that is, to the regime of multiple (recurrent) scattering, and, for a system of randomly located scatterers, can result in a remarkable phenomenon of *enhanced backscattering* of light. This phenomenon arises from a constructive interference (in the backscattering direction) between two waves scattered along the same path in the opposite directions [42], and is discussed in more detail in Section 3.4. If the scattering becomes stronger, one can invoke the *second* Born approximation, in which the field $\mathbf{E}^1$ found in the first approximation is substituted in Eq. (3.11) as the field incident on the dipoles, and so on.

Let us now consider the situation when the interactions between dipole scatterers are sufficiently strong, that is, when the regime of well-developed multiple scattering is realized. In such a case, the system of self-consistent equations [Eq. (3.6)] has to be solved exactly, that is, without resorting to Born approximations. In order to clarify this point, let us rewrite the system by using super-vectors [36]: $\mathbf{E} = \mathbf{E}_{\text{in}} + \boldsymbol{F} \cdot \mathbf{E}$. The exact solution can then be expressed in the form $\mathbf{E} = (\boldsymbol{U} - \boldsymbol{F})^{-1} \cdot \mathbf{E}_{\text{in}}$, while the Born series expansion becomes $\mathbf{E} = \mathbf{E}_{\text{in}} + \boldsymbol{F} \cdot \mathbf{E}_{\text{in}} + (\boldsymbol{F})^2 \cdot \mathbf{E}_{\text{in}} + \cdots$. This expansion is converging *only* if the inter-dipole interactions are small enough, that is, $\boldsymbol{U} > \boldsymbol{F}$, symbolically written. In the case of strong interactions, $\boldsymbol{F} \geq \boldsymbol{U}$, the Born approach cannot be used no matter how many terms in the expansion are taken into account. The exact solution leads to the resonance condition: $\det|\boldsymbol{U} - \boldsymbol{F}| = 0$. In general, this condition has complex solutions resulting in finite values of the self-consistent field at resonance. One can reach a very important conclusion from the general consideration presented above: *at resonance*, the dipole system supports *the self-consistent field* (eigenmode) whose *configuration is solely determined by the configuration of* strongly interacting *dipoles*. For this reason (illustrated later in this section), the resonances in the system of dipoles can be regarded as *configurational* resonances [36].

Multiple scattering of light manifests itself via various phenomena depending on the system configuration. For random scatterers, the aforementioned enhanced backscattering of light already appears in the regime of double scattering. As a result of enhanced backscattering, the forward propagating wave slows down, and, eventually, when the scattering becomes sufficiently strong, gets stopped or, in other words, captured in a "random" cavity. This phenomenon is called *strong* (Anderson) *localization* of light [14] and, is extensively dealt with in Chapter 5. In this chapter, strong localization of quasi 2D waves, such as SPPs, is considered in Section 3.4. Strong localization is not the only phenomenon associated with the regime of multiple scattering of light. Actually, it is extremely difficult to realize strong localization [14,43], and at least it requires a large (in comparison with the wavelength) volume filled with strong scatterers. The latter condition (along with the requirement of multiple scattering) is even more important for the realization of photonic band-gap structures, that is, infinitely periodic structures that do not transmit light in a certain range of wavelengths (Chapters 1 and 2). Strong multiple scattering of light in fractal nanostructures results in the self-consistent field configurations (dipolar eigenmodes) that can be localized at different length scales, from the size of the structure to the size of the smallest constituents (Chapter 8). This

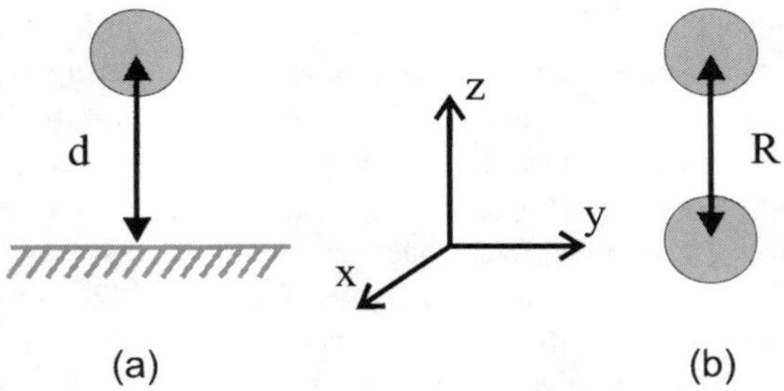

Figure 3.6. Schematic diagram showing the geometry of (a) the dipole–surface system and (b) the system of two dipoles.

phenomenon is usually referred to as *inhomogeneous* localization. Resonant interactions can also occur in the systems with only a few dipoles, or even in the system consisting of an individual dipole and the bulk. Multiple scattering in these simple configurations is considered below.

3.2.4.1. Dipole–Surface System. Electromagnetic interactions in the dipole–surface system are discussed at length in the review paper by Pohl [16]. By using the near-field approximation, the self-consistent field $\mathbf{E}(\mathbf{r}_d)$ at the site of the dipole placed near the bulk [Fig. 3.6(a)] can readily be obtained by combining Eqs. (3.6), (3.8), and (3.9) as follows [44]:

$$\mathbf{E}(\mathbf{r}_d, \omega) = \left[\frac{E_{\text{in}}^x(\mathbf{r}_d, \omega)}{1-\rho}, \frac{E_{\text{in}}^y(\mathbf{r}_d, \omega)}{1-\rho}, \frac{E_{\text{in}}^z(\mathbf{r}_d, \omega)}{1-2\rho}\right]$$

where

$$\rho = \frac{\alpha(\omega) r^p(\omega)}{4\pi\varepsilon_0 (2d)^3} \tag{3.12}$$

the field $\mathbf{E}_{\text{in}}(\mathbf{r}_d) = (E_{\text{in}}^x, E_{\text{in}}^y, E_{\text{in}}^z)$ is incident on the dipole with the polarizability $\alpha(\omega)$, d is the dipole–surface distance, and $r^p(\omega)$ is the nonretarded reflection coefficient for p-polarized light (Section 3.2.3): $r^p(\omega) = [\varepsilon_b(\omega) - 1]/[\varepsilon_b(\omega) + 1]$. It is seen from Eq. (3.12) that if the dipole–surface interaction is strong, then the self-consistent field can be appreciably enhanced. The corresponding resonance condition for this system is given by

$$\left[\frac{\alpha(\omega) r^p(\omega)}{4\pi\varepsilon_0 (2d)^3} - 1\right]^2 \left[\frac{2\alpha(\omega) r^p(\omega)}{4\pi\varepsilon_0 (2d)^3} - 1\right] = 0 \tag{3.13}$$

This equation shows the existence of two different resonance positions of the dipole with respect to the bulk, of which one is doubly degenerated. The degenerated resonance corresponds to the enhancement of field components parallel to the surface plane, and the nondegenerated resonance that occurs at a larger distance d is related to the perpendicular field component. The difference in resonance distances is connected with the aforementioned difference in the near-field magnitude for different directions (Section 3.2.3). The considered situation represents, in fact, a simple example of different eigenmodes (differently polarized resonant fields) that can be excited in different configurations (at different dipole–surface distances).

In order to observe the above resonances in a real system, the resonance distances should be larger than the radius of a spherical scatterer, a circumstance that imposes strict limitations on the values of the dipole polarizability $\alpha(\omega)$ and the reflection coefficient $r_p(\omega)$ [44]. Actually, the light frequency should be close to the frequency of single-particle plasmon resonance, so that $\varepsilon(\omega) \to -2$ [Eq. (3.5)], or to the frequency of SPP resonance, so that $\varepsilon_b(\omega) \to -1$ [Eq. (3.9)]. In near-field optics, this kind of resonance effect has been observed in experiments by Fisher and Pohl [45], who detected the light scattered from one of a number of probe gold-covered spheres approaching the surface. Sharp resonance peaks in distance dependencies of detected signal were seen only for p-polarized incident light, that is, for the field with a nonzero component perpendicular to the surface. This feature indicates that the resonance for parallel components (that can be realized at a shorter distance) could not have been excited. It should be mentioned that, for large ensembles of particles (protrusions or molecules), resonance excitations of particle plasmons and/or surface plasmon polaritons have been known to play an important role, for example, in surface-enhanced Raman scattering [46] and surface-induced modification of molecule's radiative lifetime [47].

3.2.4.2. System of Two Dipoles. Multiple scattering and resonance interactions in the system of two dipole scatterers have been considered in many papers, for example, in relation with light scattering by fractal clusters [48], as the main phenomenon behind so-called space-group resonances [49], and in relation to the resolution limit in near-field optics [50]. It is interesting to note, that the term "space-group resonances" was employed to emphasize the fact that the resonances existing in the system of interacting dipoles are associated with a certain spatial arrangement [49], that is, for exactly the same reason as the one used to introduce configurational resonances [36].

The system of two dipoles can be considered as the simplest model for SNOM: one dipole represents a probe and another one, an object [Fig. 3.6(b)]. Variations of the self-consistent field intensity at the site of the probe dipole when it is scanning near the object dipole may then be used to simulate a near-field optical image of the object. By using the near-field approximation for the direct propagator [Eq. (3.8)], the self-consistent field at the site of the probe $\mathbf{E}(\mathbf{r}_p)$ immediately can be written down in an explicit form [50], namely,

$$\mathbf{E}(\mathbf{r}_p, \omega) = \left[E_{\text{in}}^x(\mathbf{r}_p, \omega)\frac{1-\rho_2}{1-\rho_1\rho_2}, E_{\text{in}}^y(\mathbf{r}_p, \omega)\frac{1-\rho_2}{1-\rho_1\rho_2}, E_{\text{in}}^z(\mathbf{r}_p, \omega)\frac{1+2\rho_2}{1-4\rho_1\rho_2} \right] \tag{3.14}$$

where $\rho_{1,2} = \alpha_{1,2}(\omega)/(4\pi\varepsilon_0 R^3)$ and R is the separation between the probe dipole with polarizability $\alpha_1(\omega)$ and the object dipole with polarizability $\alpha_2(\omega)$. It is seen that, similar to the previous case, there are two resonant configurations determined by the conditions $(1 - \rho_1\rho_2)^2 = 0$ and $(1 - 4\rho_1\rho_2) = 0$ for the field components perpendicular and parallel to the line connecting two dipoles, respectively. In a sense, our system of two dipoles can be considered as a simple *near-field resonator* with the resonance condition being that the field scattered by one of the dipoles after reflection from the other dipole is equal to the driving field of the first dipole. Actually, the dipole–surface system treated in Section 3.2.4.1 represents a near-field resonator of the same kind with the field scattering occurring between the dipole and its mirror image.

Note that for both of the above configurations, the interaction operator $(\boldsymbol{U} - \boldsymbol{F})^{-1}$ is diagonal in the coordinate system that has one of the axes parallel to the line connecting the two interacting dipoles [Fig. 3.6 and Eqs. (3.12) and (3.14)]. This means that the field polarized along one of the axes of such a system can individually and, under the aforementioned conditions, resonantly excite the corresponding eigenmode, that is, polarized linearly in the direction parallel to the incident field polarization. For any other polarization, that is, when the field has nonzero projections on more than one axis, all three components of the self-consistent field may, in general, be excited even with a linearly polarized incident field. Finally, resonance conditions can be used as stated above only in the absence of damping. The presence of damping in the system changes the resonance conditions and leads to a finite height of the resonance peaks [44,50]. In the limit of small damping, only the real part of the expressions in the left-hand side of these conditions should be put to zero, for example, $\mathrm{Re}(1 - 4\rho_1\rho_2) \cong 0$ is the resonance condition for the z component of the field in the system of two dipoles [Fig. 3.6(b)].

The resonance conditions for the system of two dipoles are somewhat more demanding than the conditions for the dipole–surface system. The corresponding resonances can be realized only for the frequencies of light close enough to the frequencies of single–particle plasmon resonances [44]. Strong interaction between two dipole scatterers can result in a significant field enhancement and, for identical particles, splitting of the single-particle plasmon resonance. The latter phenomenon has been observed in the course of measurements of near-field transmission spectra of individual gold nanoparticles with the illumination SNOM configuration [51]. The dipole–dipole interaction can be enhanced by placing nanoparticles near the surface structure that supports propagating optical surface modes, for example, SPPs or optical waveguide modes. Electromagnetic fields scattered in the form of surface modes are quasi-two-dimensional (2D) waves and, thereby, have considerably softer decay than the scattered dipolar fields. Enhanced dipole–dipole interaction mediated by waveguide modes has been observed with random silver nanoparticle arrays fabricated onto waveguide structures [52]. Localization phenomena due to strong multiple scattering of SPPs are considered in Section 3.4.

Let us now discuss the influence of dipole–dipole interaction on the resolution that can be achieved in our simple model [Fig. 3.6(b)] of SNOM. If the interaction is weak, that is, $|\rho_1\rho_2| \ll 1$, the self-consistent field at the site of the probe replicates the total field existing in the absence of probe, that is, the probe acts as a passive detector (the influence of a probe is considered in Section 3.2.6). Then, it follows from the properties of scattered near fields (Section 3.2.3) that the intensity distribution of the field at the site of the probe dipole, which is scanned in the (x,y) plane near the object dipole, produces an object image in positive contrast, if the incident field is polarized along the z axis, or in negative contrast, if the incident field is polarized in the (x,y) plane. The same conclusions can be immediately obtained from the above expression for the self-consistent field [Eq. (3.14)]. In both cases, the contrast is weak, and the size of the image is approximately given by the dipole separation R, because the field perturbation ρ_2 caused by the object dipole is inversely proportional to R^3. If the dipole–dipole interaction is strong when the dipoles are at the smallest distance, that is, $|\rho_1\rho_2| \sim 1$, then the self-consistent field being strongly enhanced (at this position of the probe) decreases rapidly when the system is tuned away from resonance by moving the probe dipole away from the object one. It can be shown that the smallest size of the image (of the object dipole) is realized when the smallest dipole separation corresponds to the resonance distance, and that the resolution in such a case is limited only by the system damping [50]. In the limit of small damping, that is, when $\mathrm{Im}(\alpha_1\alpha_2) \ll \mathrm{Re}(\alpha_1\alpha_2)$, the resonance distance R_0 for the z component of the field, the field intensity enhancement Γ at resonance and the full width Δ of the image at half-maximum can be expressed as follows [50]:

$$R_0 = \left(\frac{|A|^2}{(2\pi\varepsilon_0)^2 \mathrm{Re}\,A}\right)^{1/6} \qquad \Gamma \cong \frac{|A|^2}{(\mathrm{Im}\,A)^2} \qquad \Delta \cong 2R_0\sqrt{\frac{|\mathrm{Im}\,A|}{3\mathrm{Re}\,A}} \tag{3.15}$$

where $A = \alpha_1(\omega)\alpha_2(\omega)$. It is thus clear that the resolution in SNOM can be significantly improved by employing the resonance interactions between the probe and surface features. In fact, remarkably high spatial resolutions have been demonstrated in the experiments with sharp metal tips being scanned along rough metal surfaces [53], so that the conditions close to resonance might be expected.

The resolution improvement related to the strong (close to resonant) dipole–dipole interaction is illustrated with the numerically simulated images shown in Fig. 3.7. Two identical spherical particles with the radius of 10 nm made of glass ($\varepsilon = 2.25$) in the first case and of silver ($\varepsilon = -3.77 + 0.67i$) in the second one are compared from the point of view of near-field imaging. The intensity distributions of the total field were calculated for different polarizations of the incident field at a wavelength of 400 nm with one particle being scanned along the plane $z = 20$ nm and another one being located at the origin [Fig. 3.6(b)]. The light wavelength was chosen deliberately close to the resonant one for single-particle plasmon resonance of silver sphere [54]. Note that, in both cases, the image contrast is opposite for the two considered polarizations and is in complete agreement with the preceding discussion. The expected field enhancement and resolution improvement are indeed clearly seen for the silver spheres and especially pronounced for the incident light polarized parallel to the z axis indicating that *the dipole–dipole interaction is*

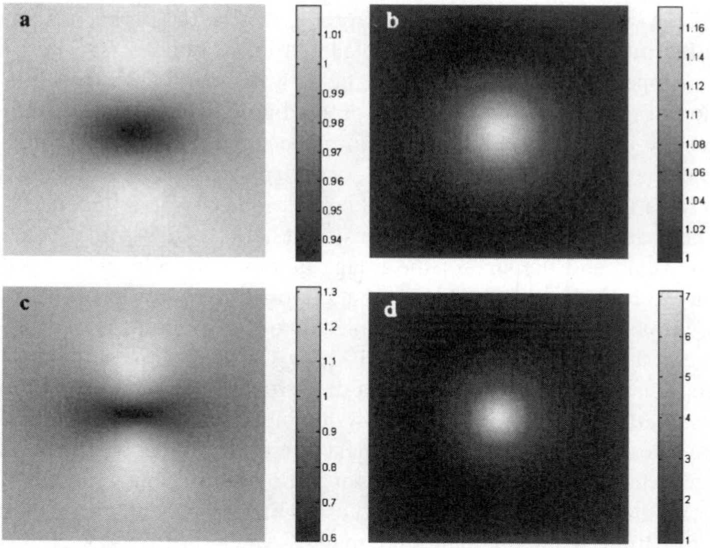

Figure 3.7. Gray-scale representations ($80 \times 80\,\text{nm}^2$) of the intensity distributions of the self-consistent field at the site of the probe sphere, which is scanning in the (x,y) plane at $z = 20\,\text{nm}$ [Fig. 3.6(b)], for the incident field (a,c) $\mathbf{E} = (1,0,0)$ and (b,d) $\mathbf{E} = (0,0,1)$. Probe and object 10-nm-radius spheres made of (a,b) glass and (c,d) silver and illuminated by light at the wavelength of 400 nm. The direction of the x axis is vertical.

stronger for the field component parallel to the line connecting the dipoles. The reason is the same as the one used to explain the field enhancement near the tip illuminated with p-polarized light and the difference in resonance distances for different field components, namely, that the scattered near field is stronger along the nonradiative direction (Section 3.2.3).

The total field intensity distributions displayed in Fig. 3.7 mainly reflect the field component parallel to the polarization of the incident field. The other two components though being weaker are also different from zero for the probe dipole being moved away from the z axis (Fig. 3.8). The intensity distributions of these components are dictated by nondiagonal elements of the near-field direct propagator [Eq. (3.8)], and thereby produce complicated patterns. On the other hand, the corresponding images have the advantage of achieving an ultimate contrast of 100% (because of canceling out the incident field), a circumstance that can be advantageously exploited in SNOM especially when the dipole–dipole interaction is relatively weak [Fig. 3.7(a)]. If, for example, total noise of the detection process would amount to a few percent of the detected signal, then the image obtained with the dielectric probe and surface features for the incident field polarized in the plane of scanning can become completely blurred. At the same time, an image obtained in the cross-polarization configuration would hardly be affected. The cross-polarized detection has been successfully used, for example, in the reflection SNOM configuration with an uncoated fiber tip not only to increase the image contrast but

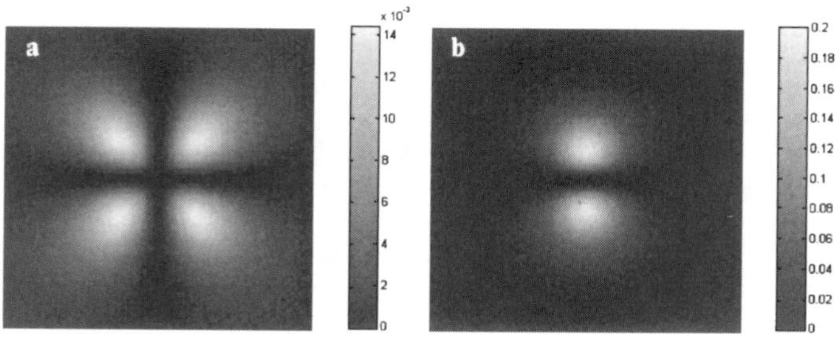

Figure 3.8. Gray-scale representations ($80 \times 80\,\text{nm}^2$) of the square magnitude of the self-consistent field (a) y and (b) z components at the site of the scanning silver sphere for the incident field $\mathbf{E} = (1,0,0)$. All else is as in Fig. 3.7.

also to cancel propagating field components that have the same polarization as the incident field [55].

3.2.4.3. System of Three Dipoles. A characteristic feature of multiple scattering is that the self-consistent field established at a given point in the process of multiple scattering is determined not only by the local surroundings, but also by the large-scale geometrical structure. This remarkable feature put a clear distinction between the regimes of single and multiple scattering and can be qualitatively explained as follows. In the regime of single scattering, the total field consists of the incident field and the primary scattered waves. The amplitudes of these waves decrease rapidly with the distance from the scatterers (as R^3) and, therefore, only the nearest scatterers contribute to the total field. In the regime of multiple scattering, the total field can be expanded by using the eigenmodes of the self-consistent problem in question (Chapter 8 and [13]). Any variation in the structure geometry leads to the modifications of eigenmodes and thereby changes the self-consistent field distribution. An immediate consequence is that the field intensity distribution observed (in the presence of multiple scattering) near a surface structure does not correlate with the local surface topography.

This feature is illustrated with the self-consistent field intensity distributions calculated for two different systems composed of three identical silver spherical scatterers (Fig. 3.9). The parameters of spheres and the wavelength of light used are the same as in the preceding numerical examples (Figs. 3.7 and 3.8). It is clearly seen that the self-consistent field intensity distributions [Fig. 3.10(a) and (b)] calculated for a linear chain of spheres [Fig. 3.9(a)] are rather different from those one would have expected for the regime of single scattering. Indeed, in the regime of single scattering, the intensity distributions for different polarizations of the incident light should look as if they are composed of the appropriate distributions produced by an individual scatterer (Fig. 3.7). In our case, the field enhancement is significantly larger than the one obtained for an individual sphere, and the field intensity is distributed nonuniformly along the chain (i.e., there is little correlation between these distributions and the dipole structure). Note that the enhancement is stronger for the incident field polarized along the chain of dipoles, a feature that is again

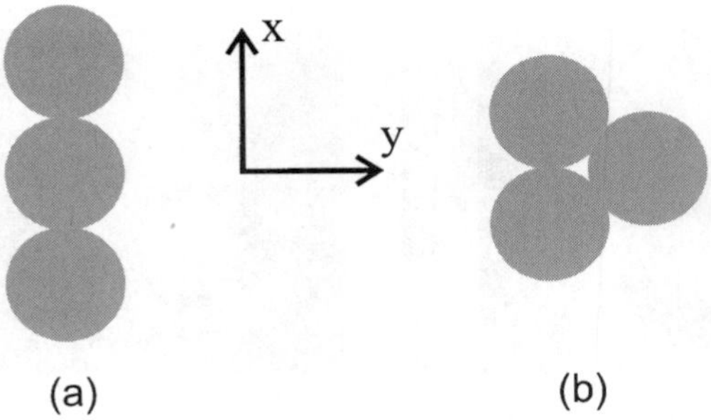

Figure 3.9. Schematic diagram (top view) showing the geometry of (a) chain and (b) triangular arrangements of three spheres.

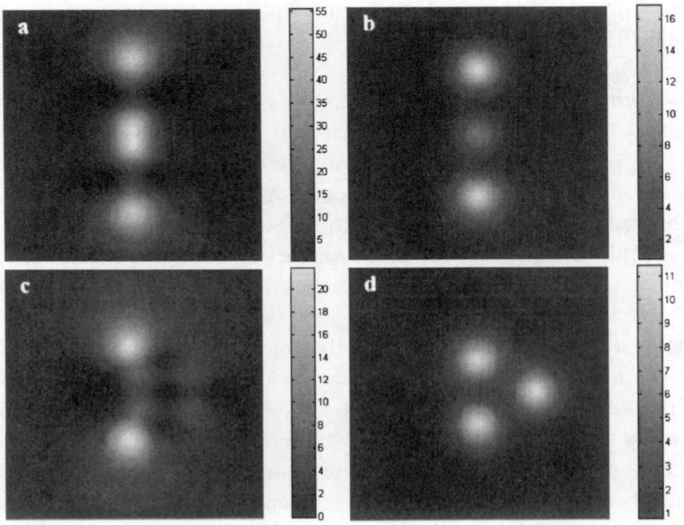

Figure 3.10. Gray-scale representations ($80 \times 80\,\text{nm}^2$) of the intensity distributions of the self-consistent field in the (x,y) plane at the distance of 11 nm from the (a,b) chain and (c,d) triangular silver sphere systems (Fig. 3.9) for the incident field (a,c) $\mathbf{E} = (1,0,0)$ and (b,d) $\mathbf{E} = (0,0,1)$ at the wavelength of 400 nm. The direction of the x axis is vertical.

related to the fact that the scattered near-field components are stronger along the dipole direction. For this polarization, the inter-dipole interaction is so strong that nearly the same intensity distribution was calculated for the case with the incident field irradiating only one of the dipoles. This circumstance is consistent with the discussion of resonant interactions in the beginning of this section and with the recently reported simulations of the light energy transport via linear chain of nanoparticles [56].

The intensity distributions calculated for the triangular configuration of spheres [Fig. 3.9(b)] exhibit essentially the same features as those discussed above. However, the difference between the distributions obtained for the two polarizations of the incident field is even more striking in this case. The intensity distribution for the field polarized along the z axis faithfully reproduces the geometry of dipoles [cf. Figs. 3.9(b) and 3.10(d)], whereas that for the field polarized along the x axis [Fig. 3.10(c)] shows only two bright spots (for three scatterers!). The first feature can be accounted for by the symmetry of the configuration as a whole (i.e., the system geometry together with the polarization of the incident field). The second observation demonstrates that, even in a system with only a few strongly interacting scatterers, the self-consistent field intensity distribution can exhibit spatially localized and quite significant intensity enhancement in the form of bright spots whose positions are not correlated with the locations of adjacent scatterers. The simulated distributions (Fig. 3.10) demonstrate thereby not only the aforementioned characteristic feature of multiple scattering but also the phenomenon of subwavelength apertureless light confinement (and provide the examples of surface light dots). One can, of course, increase the number of scatterers and obtain more complicated intensity distributions. In Sections 3.3 and 3.4, particular multiple scattering phenomena leading to light confinement, namely, phase conjugation of optical near fields and SPP localization caused by surface roughness, are discussed in detail.

3.2.5. Topographical Artifacts

It has already been emphasized that in order to obtain optical images with subwavelength resolution, the probe should be scanned along and sufficiently close to the sample surface. This problem is usually solved by incorporating the appropriate feedback system based on nonoptical short-range probe–sample interactions, of which so-called shear forces [32] are the most widely used (Section 3.2.2). Such feedback is implemented into the most modern SNOM configurations [24], allowing one to image a surface topography while simultaneously recording a near-field optical image. However, SNOM operation in constant (probe–surface) distance mode brings up a serious problem of coupling between the topographical and the near-field optical images [57]. This coupling severely hinders access to pure optical information (i.e., not related to surface topography), a circumstance that is of crucial importance for near-field imaging of nanostructured fields (e.g., surface light dots) established in the course of multiple scattering of light. Topographically induced features in the optical images (topographical artifacts) also complicate the determination of pure optical resolution, since high spatial frequencies in the optical images may be dictated solely by these artifacts. Both phenomenological [58] and rigorous [59,60] descriptions of the artifacts have been developed and many important issues have been clarified. In the following paragraphs, the topographical artifacts are considered within the framework of the microscopic approach (Section 3.2.3).

Topographical artifacts can be defined as the features that appear in a near-field optical image obtained in the *absence* of an object when a probe follows (during the scanning) a surface profile of the object. Let us consider the optical field that reaches a remote detector in the SNOM configuration depicted in Fig. 3.11. In the absence of

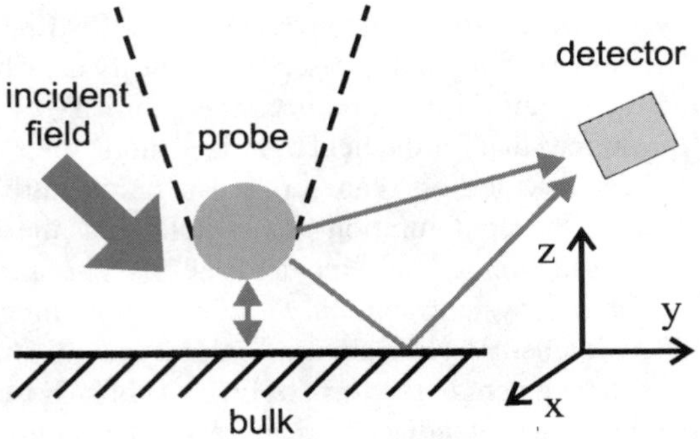

Figure 3.11. Schematic diagram showing the geometry of the probe–surface system illuminated by light and a remote detector. The directions of scattered waves are indicated by arrows.

an object, the detected optical field $\mathbf{E}(\mathbf{r}_d)$ is, in general, composed of the incident field and the field scattered by a probe [Eq. (3.4)]:

$$\mathbf{E}(\mathbf{r}_d, \omega) = \mathbf{E}_{\text{in}}(\mathbf{r}_d, \omega) - \mu_0\omega^2\alpha_p(\omega)[\boldsymbol{D}(\mathbf{r}_d, \mathbf{r}_p, \omega) + \boldsymbol{I}(\boldsymbol{r}_d, \boldsymbol{r}_p, \omega)] \cdot \mathbf{E}(\mathbf{r}_p, \omega) \quad (3.16)$$

where $\alpha_p(\omega)$ is the polarizability of the probe dipole located at $\mathbf{r}_p$ and the self-consistent field $\mathbf{E}(\mathbf{r}_p)$ at the site of the probe is given by Eq. (3.12). The field $\mathbf{E}(\mathbf{r}_p)$ is thereby determined by the incident field at the site of the probe and the probe–surface coupling and can be expressed in the following form: $\mathbf{E}(\mathbf{r}_p) = \Gamma(\alpha_p, \varepsilon_b, z_p)\cdot \mathbf{E}_{\text{in}}(\mathbf{r}_p)$ with ε_b and z_p being the bulk dielectric constant and the probe–surface distance, respectively. When the probe follows the surface profile S of the object $z_p(x, y) = z_0 + S(x, y)$, the detected optical field $\mathbf{E}(\mathbf{r}_d)$ changes accordingly resulting in an optical image induced by the z motion of the probe, that is, in a topographical artifact [57–60]. This artifact represents, in the first approximation, an exact replica of the topographical image of the object. Depending on the sign of the signal derivative with respect to the probe–surface distance, the topography can be reproduced in positive or negative contrast. Equation (3.16) allows us to analyze different contributions to the topographical artifacts and evaluate their magnitudes for a particular SNOM configuration.

The artifact can be caused by the probe z motion in the incident field since $\mathbf{E}_{\text{in}}(\mathbf{r}_p) = \mathbf{E}_{\text{in}}(z_p)$. If the incoming field is incident from the side of the sample and totally internally reflected (e.g., in photon tunneling SNOM), the incident field is evanescent above the sample surface, and any increase in the probe–surface distance would produce a decrease in the field at the site of the probe. If the incoming field is incident from above, as shown in Fig. 3.11, the resulting interference pattern between the incident and the reflected (by the sample) fields would again result in a similar artifact. In both cases, variations of the detected signal can be up to a few percent per nanometer of the probe z motion [58]. Finally, illumination by a strongly focused light beam would introduce an additional contribution to this *artifact by illumination*. The artifact by illumination is so far the only contribution (also called liftoff contribution) that has been taken into consideration in experimental studies [61].

The z-motion artifact can also be connected with the modification of the self-consistent field at the site of the probe via the probe–surface interaction (Section 3.2.4.1). This modification decreases drastically when the probe–surface distance increases ($\sim z_p^{-3}$), and its contribution can be significant only if the field is sufficiently enhanced when the probe is very close to the surface. By applying Eq. (3.12) to the case of the probe sphere that is in contact with the surface, one can evaluate the modification of, for example, the z component of the field:

$$E_z(z_p = a_p, \omega) = E_{\text{in}}^z(z_p, \omega)\left[1 - \frac{1}{4}\,\frac{\varepsilon_p(\omega) - 1}{\varepsilon_p(\omega) + 2}\,\frac{\varepsilon_b(\omega) - 1}{\varepsilon_b(\omega) + 1}\right]^{-1} \tag{3.17}$$

where a_p and ε_p are the radius of the probe sphere and its dielectric constant. For a glass probe and substrate ($\varepsilon_p = \varepsilon_b = 2.25$), the field enhancement is rather small $\Gamma \sim 3\%$ (for the intensity, it is accordingly $\sim 6\%$), but for a (nonresonant) metal probe and substrate ($\varepsilon_p, \varepsilon_b \to -\infty$) it becomes appreciable: $\Gamma \sim 30\%$. Finally, in the case of resonant probe–surface interactions, the enhancement can amount to several orders of magnitude (Section 3.2.4.1). Therefore, the *artifact by probe–surface coupling* can be noticeable and should be taken into account, especially when overall signal variations constitute only a fraction of the maximum signal.

The third artifact contribution is related to the interference between the field directly scattered (by the probe) toward a detector and the scattered field reflected by the surface (Fig. 3.11). The corresponding artifact can be evaluated by calculating the direct and indirect field propagators in Eq. (3.16). This *artifact by interference of scattered waves* depends strongly on the position of the detector. If the direction to the detector is close to the surface normal and the surface reflectivity is relatively high, the variations in the detected signal can be on the same level as for the artifact by illumination, that is, a few percent per nanometer of the probe z motion. For the detector being close to the surface plane, the contrast of the interference pattern is always close to 100%, since the surface reflectivity for oblique angles of incidence is close to -1 for any bulk material. However, the variations in the detected signal due to the probe z motion can be rather slow, especially for the detector placed far away from the probe, and, for example, would amount to only a few tens of a percent per nanometer for the probe–detector distance of ~ 1 mm [62].

It is clear that the only way to completely avoid any topographical artifacts is to scan a probe in the constant plane mode, that is, to keep the same z coordinate of the probe during the scanning [58]. On the other hand, the near-field imaging in the constant distance mode allows us to realize the best optical resolution, which deteriorates with the increase of the probe–surface distance, and to simultaneously image the surface topography. The latter circumstance is of special relevance for the situation when successive images of the same area are being taken, and one has to accurately account for the eventual drift of the sample with respect to the probe. These reasons make the problem of dealing correctly with the topographical artifacts rather important.

One might be tempted to subtract the artifact contribution from the detected signal and to clear the near-field optical image of the topographical artifact [61]. However, such a subtraction is far from being a trivial issue, and extreme care should be taken when implementing it. The most complicated problem is related to the fact that the detected signal in conventional detection schemes is proportional to the *intensity* of the field incident on the detector. This means that not only the aforementioned three contributions would interfere with each other but also the overall artifact contribution would interfere with the object contribution (and not simply adding up). Consequently, the detected signal *after* subtraction of the topographically induced signal, which is calculated from the results of the signal measurement in the absence of an object [61], would still contain the *cross-talk* signal. Only in the case of both the artifact and object contributions being significantly weak in comparison with the (coherent) background field [e.g., the incident field at the site of the detector, Eq. (3.16)], can this subtraction be correctly (to the first order) implemented. Remember, for *weak* signals, the issue of noise becomes very important. Clearly, the noise contribution should be notably smaller than the difference between the signals detected with and without the object. Finally, remember that the reference surface plane (used to determine the artifacts) is in practice the result of averaging of a sample surface within the corresponding interaction area, that is, within the sensitivity window [60]. The size of this area is, in general, different for different SNOM configurations and probes used. Therefore, in order to subtract the artifact contribution, especially when dealing with heterogeneous samples, one should measure the probe–surface distance dependent signal at (nearly) the same place of the sample as that to be imaged.

The aforementioned contributions to the topographical artifact can be regarded as *linear* artifacts because the above treatment is valid only if the *probe–object coupling* is negligibly small. The self-consistent field at the site of the probe, which is scanned along an object surface, can then be decomposed into the field existing in the absence of the object and the field induced by the object. It is mapping of the latter field intensity distribution that provides pure optical information about the object. However, in the case of strong probe–object coupling, the self-consistent object field can be significantly different from the field existing in the absence of the probe. The resulted optical image, even after clearing out all topographical artifacts, would then represent rather a map of the strength of the probe–sample interaction than the intensity distribution of the object field existing in the absence of the probe. This interaction thereby constitutes a source of the *nonlinear artifact by probe–object coupling*. Such an artifact can hardly be subtracted under any circumstances, but to a certain extent it can be controlled and even avoided. This issue is addressed in Section 3.2.6.

3.2.6. Influence of a Probe

The detection and measurement of confined and/or nanostructured EM fields is a conceptually new application of SNOM that has appeared in addition to mainstream applications related to the extension of far-field optical techniques (microscopy, spectroscopy, lithography, etc.) into the subwavelength region. It has been convincingly demonstrated that the use of SNOM can provide invaluable information on multiple scattering of light by nanostructured surfaces and corresponding localization phenomena [10–13]. One can envisage the use of SNOM in studies of other multiple scattering phenomena, for example, the formation of a full photonic band gap due to the scattering of SPPs on a periodically textured surface [63] or enhanced dipole–dipole interaction via surface modes [52]. Therefore, the problem of influence of a probe on the resulting field intensity distribution becomes of crucial importance for interpretation and understanding of the near-field optical images obtained. It is usually accepted that a sharp uncoated fiber tip used as a probe in the photon tunneling SNOM configuration can be considered to be a passive detector of near-field components [29,64]. Indeed, the probe–sample coupling has been numerically investigated for an individual (dielectric or metallic) surface defect resulting in the conclusion that the coupling can be disregarded [65]. However, it is not *a priori* given that this conclusion would remain valid for surface structures with enhanced multiple scattering. Actually, one could in fact suggest the opposite, as it is known that the resonantly excited systems are easily influenced by an external perturbation.

3.2.6.1. Qualitative Consideration. The key point to be understood is that a near-field optical probe can be considered to be passive and, consequently, near-field imaging can be described using the transfer function only if there is no multiple scattering (coupling) between the probe and a sample [18,60]. Clearly, if the probe–sample coupling cannot be neglected, the problem of image interpretation is practically hopeless, since it can be tackled only with the help of rigorous numerical simulations. Another important circumstance is that the existence of multiple scattering is determined solely by the configuration of the probe–sample system and is not affected by the illumination and detection methods [60]. This allows one to deal with the problem of probe–sample coupling in a rather general way [66].

Let us again consider the probe–sample system as being composed of electric dipoles represented by small spheres (Fig. 3.5) whose responses to the EM field are described by isotropic dipole polarizabilities (Section 3.2.3). A near-field optical image is produced by detecting the scattered fields as a function of the position of a probe. Variations in the detected field are dictated by variations in the self-consistent field $\mathbf{E}(\mathbf{r}_p)$ at the site of the probe (Section 3.2.5). This field is in turn determined by the self-consistent fields at the sites of the object dipoles [Eq. (3.6)]. Therefore, if the self-consistent field at the object dipoles in the presence of the probe is (nearly) the same as that in its absence, the probe–sample coupling can be neglected, and the field at the probe would reproduce the field existing in the absence of the probe. The resulted optical image, when corrected for topographical artifacts, would then represent the *unperturbed* near-field intensity distribution. To proceed further, let us introduce the difference $\mathbf{E}^1(\mathbf{r}_j)$ between the self-consistent fields at the site of the jth object dipole in the presence $\mathbf{E}(\mathbf{r}_j)$ and in the absence $\mathbf{E}^0(\mathbf{r}_j)$ of the probe: $\mathbf{E}^1(\mathbf{r}_j) = \mathbf{E}(\mathbf{r}_j) - \mathbf{E}^0(\mathbf{r}_j)$. It is straightforward to show that the difference field satisfies the following system [66]:

$$\begin{aligned}\mathbf{E}^1(\mathbf{r}_j,\omega) &= -\mu_0\omega^2\alpha_p \boldsymbol{G}(\mathbf{r}_j,\mathbf{r}_p,\omega)\cdot\mathbf{E}(\mathbf{r}_p,\omega)\\ &\quad - \mu_0\omega^2\sum_{i=1}^{N-1}\alpha_i \boldsymbol{G}(\mathbf{r}_j,\mathbf{r}_i,\omega)\cdot\mathbf{E}^1(\mathbf{r}_i,\omega)\end{aligned} \tag{3.18}$$

where the probe dipole is denoted as the Nth dipole.

The structure of this self-consistent system is the same as that of the general system [Eq. (3.6)], and the only difference is in the origin of the incident field [cf. Eqs. (3.6) and (3.18)]. Consequently, if there is a resonant enhancement of the self-consistent field in the absence of the probe, then the difference field $\mathbf{E}^1(\mathbf{r}_j)$ will also be resonantly enhanced and can become comparable with $\mathbf{E}^0(\mathbf{r}_j)$. By introducing the field enhancement (at the site of the probe) factor Γ and using the near-field part of the direct field propagator [Eq. (3.8)] together with Eq. (3.5) for the probe polarizability, one can obtain an estimate of the enhancement that would result in a strong perturbation (when $|\mathbf{E}^0| \sim |\mathbf{E}^1|$):

$$\Gamma\frac{\mu_0\alpha_p c^2}{4\pi R^3} \propto 1 \qquad \Rightarrow \Gamma \propto \left(\frac{R}{a_p}\right)^3 \frac{\varepsilon_p + 2}{\varepsilon_p - 1} \tag{3.19}$$

where R is the probe–sample distance, or, more precisely, the smallest distance between the probe dipole and one of the object dipoles. For example, by considering a metal probe ($|\varepsilon_p| \gg 1$) being in contact with an object dipole, that is, $R \approx 2a_p$, one obtains from Eq. (3.19) the field enhancement value $\Gamma \approx 8$, which is similar to that observed experimentally in SPP scattering by surface roughness [11] and localized dipolar excitations on fractal clusters [12,13].

The above analysis demonstrates that the problem of field perturbation by a probe becomes rather important in configurations with well-developed multiple scattering by surface features, especially for metal probes. It is also clear that the probe influence can be strongly suppressed if the probe–sample distance were to be increased [Eq. (3.19)], although probably at the expense of the resolution obtained. Therefore, by comparing near-field optical images taken at different probe–surface distances, one should be able to deduce the extent of field perturbation introduced by the probe. The idea is that *the images at different distances should be similar to each other when the probe–sample coupling becomes negligibly small.* This idea is based on the fact that the dipole–dipole coupling scales as R^{-6} [R is the dipole–dipole separation, Eq. (3.14)], whereas the field distributions at different distances from the sample differ only because of relatively soft filtering out evanescent field components [Eq. (3.3)]. Because near-field optical images are usually obtained simultaneously with topographical ones by using a (shear-force) feedback system, the recommended procedure of imaging at different probe–sample distances would also help to distinguish topographical artifacts (Section 3.2.5).

3.2.6.2. Numerical Example. A top view of the probe–sample system used in the numerical simulations [66] is shown on Fig. 3.12 together with the system parameters. The dielectric constant ($\varepsilon = -3$) of object spheres was chosen in order to ensure strong interactions and, therefore, large field enhancement ($Q \sim 100$) in the absence of the probe. All numerical results presented here correspond to exact solutions of the self-consistent problem with the complete retarded (direct) field propagator [Eqs. (3.4–3.7)] and the incident field in the form $\mathbf{E} = (1, 0, 1)/\sqrt{2}$. The field intensity distributions calculated without the probe (for different distances from the object plane) are compared in Fig. 3.13 with the intensity of

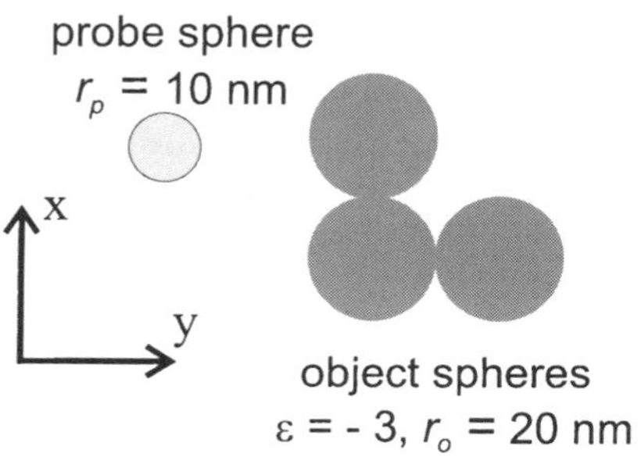

Figure 3.12. Schematic diagram (top view) showing the object and probe spheres and indicating their parameters used in the numerical simulations.

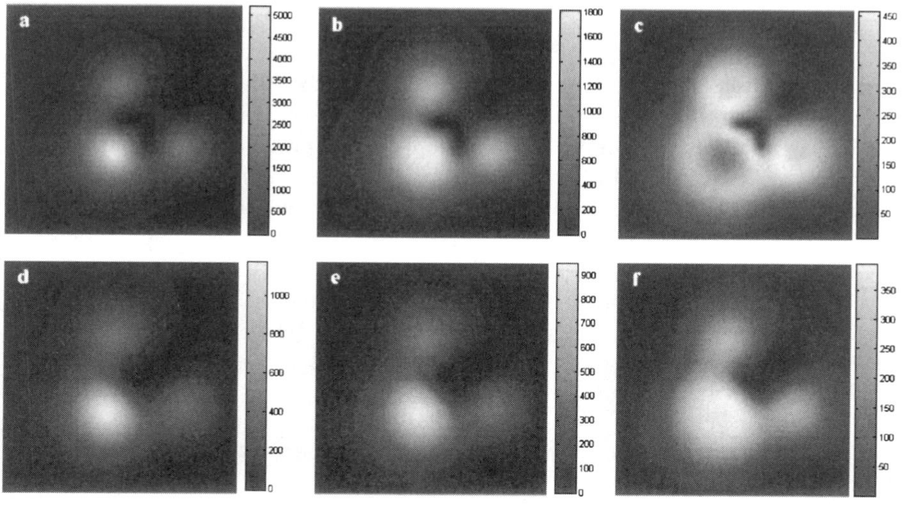

Figure 3.13. Gray-scale representations ($200 \times 200\,\text{nm}^2$) of the intensity distributions of the self-consistent field calculated (a,d) without the probe and at the site of the scanning probe made of (b,e) glass ($\varepsilon_p = 2.25$) and (c,f) metal ($\varepsilon_p = -10$) for different distances from the object plane: (a,b,c) 30 nm; (d,e,f) 40 nm. The incident field is in the form $\mathbf{E} = (1,0,1)/\sqrt{2}$ and at the light wavelength of 633 nm. The direction of the x axis is vertical. (Adapted from [66].)

the self-consistent field at the site of the scanning probe (for different probe materials). Note that the incident field is polarized in the (x,z) plane, but the field intensity distributions have the same symmetry as the system of object spheres. This is a fingerprint of resonant field enhancement, since the self-consistent field at resonance is self-sustained and its distribution is determined by the system configuration (Section 3.2.4). Different probe materials were used to simulate different experimental configurations, that is, uncoated ($\varepsilon_p = 2.25$) and metal-coated fiber probes ($\varepsilon = -10$). Comparing the field intensity distributions calculated (at the smallest distance from the object plane) in the absence [Fig. 3.13(a)] and in the presence [Fig. 3.13(b) and (c)] of the probe, it is apparent that the field perturbation by the probe is stronger the larger the polarizability of the probe. Note that the perturbation not only decreases the field intensity (probe quenches the resonance) but also changes its distribution and quite dramatically for the metal probe. As the probe–sample distance increases, the intensity distributions calculated for the two probes become similar to each other and to the unperturbed intensity distribution [cf. Fig. 3.13(d, e, and f)]. Also notice that, in the case of the relatively weak-field perturbation by the glass probe, the difference between the images obtained at different probe–object distances is rather small and clearly related to the filtering of high-frequency (evanescent) field components [cf. Fig. 3.13(b and e)]. These features are in complete agreement with the above qualitative consideration.

The results presented in this section indicate that, in any configuration with well-developed multiple scattering, near-field imaging of the field intensity distribution should be conducted with great care and consideration of the possibility of field perturbation by the probe used. The probe–sample coupling is determined by the polarizability of the probe (strictly speaking, it can be introduced only for a small particle used as a probe) and the probe–surface distance. If the near-field imaging is carried out in constant probe–surface distance mode, for example, by using shear-force feedback, then this coupling depends only on the probe material. One can then formulate a simple rule of thumb: If the ratio $(\varepsilon - 1)/(\varepsilon + 2)$ is much smaller for the probe than for surface defects, then the probe–sample coupling can be disregarded. Even in this case, however, it is recommended also to monitor the coupling (and the presence of topographical artifacts) by imaging the field distribution in the constant plane mode (just a few nanometers away from the sample surface) and comparing it with the image obtained in the constant distance mode. The experimental examples of the described procedure are presented in Sections 3.4.3 and 3.4.4.

3.3. PHASE CONJUGATION OF OPTICAL NEAR FIELDS

Interactions in the system of small scatterers placed near a phase-conjugating mirror (PCM) are of great interest in both fundamental and applied areas of physics. It was Agarwal [67] who first brought up this subject and considered the interaction of atoms with the PCM, and later, together with Wolf [68], developed a theory of distortion correction by a PCM. Until recently, phase conjugation of scattered (by atoms or small scatterers) radiation has been treated in the far-field approximation, in which only propagating components of the optical field were taken into account [69]. Such an approximation was used for obvious reasons: It was not clear whether phase conjugation of evanescent field components was possible, and, even if it were possible, whether the (already strong) interaction would be seriously affected. By using degenerate four-wave mixing in a photorefractive crystal, phase conjugation of light emitted by a near-field optical fiber probe was realized, and subwavelength-sized phase-conjugated light spots formed near the crystal surface were observed [10]. This experiment can be regarded as the first direct demonstration of phase conjugation of near-field (evanescent) optical wave components.

The above experiment stimulated investigations of various fascinating phenomena related to phase conjugation of evanescent waves. Agarwal and Gupta [70] considered the interaction of an electric dipole with the PCM with evanescent waves being taken into account and they predicted several interesting consequences of near-field coupling, such as the possibility of thresholdless lasing in resonators with PCMs and spontaneous excitation. Actually, the latter effect reformulated for small scatterers is very closely related to the experimentally observed effect of scattered light enhancement near the PCM that was accounted for by multiple-phase conjugation of scattered evanescent field components [71]. As an example of further development of the experiments on phase conjugation of light emitted by a fiber probe, one can mention the demonstration of the possibility of forming two phase-conjugated light spots only 240-nm apart [72]. This experiment suggests the potential possibility of using the SNOM–PCM combination for high-density optical data storage. The experimental results obtained were successfully modeled with a macroscopic self-consistent model for near-field microscopy [73]. This modeling was complemented by consideration of the phase conjugation process from the point of view of holography [9] leading to the formulation of the concept of near-field optical holography [74]. In this section, main experimental results on near-field phase conjugation and relevant numerical simulations are presented.

3.3.1. Subwavelength Light Confinement by Phase Conjugation

Apertureless light confinement can be achieved by focusing light with an optical element, for example, lens, mirror, or diffractive optical element (Fresnel or Bragg lens). The latter case is similar to phase conjugation of a divergent spherical wave incident on the PCM, since phase conjugation can be viewed as diffraction of a pump wave on the grating formed due to interaction between a signal wave and another (counterpropagating) pump wave [75]. In either configuration, such a light confinement is diffraction limited if evanescent field components can be disregarded [76]. This is usually the case because, in conventional (far-field) optics, all distances involved are much larger than the wavelength of light used. In near-field optics, optical interactions between nanostructured surfaces and a probe occur on the subwavelength scale. It is natural to expect that, once the distance between the source of outgoing spherical waves and the PCM becomes less than the light wavelength, the confinement of phase-conjugated light would no longer be diffraction limited.

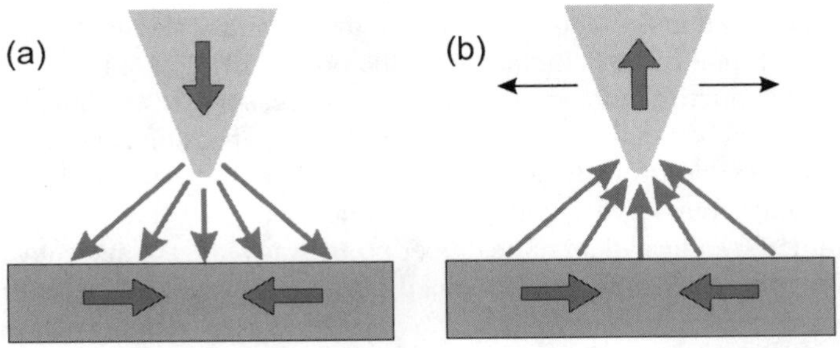

Figure 3.14. Schematic representation of light interactions during the recording (a) and imaging (b) of a phase-conjugated surface light dot.

3.3.1.1. Experimental Results. In the experiments on near-field phase conjugation, degenerate four-wave mixing in a photorefractive crystal was used to realize optical phase conjugation, and the SNOM arrangement was designed to provide a signal wave and to detect a phase-conjugated wave [10]. The idea behind these experiments is displayed schematically in Fig. 3.14. First, degenerate four-wave mixing is realized in a photorefractive crystal through the use of counter-propagating (near the crystal surface) waves as the pump beams and the use of light radiated from an uncoated fiber probe as the signal wave [Fig. 3.14(a)]. This process gives rise to a wave that is phase conjugated with respect to the signal wave. After some exposure time the signal wave is switched off, and the phase-conjugated light collected by the probe is detected when the probe is scanned along the crystal surface [Fig. 3.14(b)]. One can observe the phase-conjugated light in the absence of the signal wave if recording/erasure-time constants of the crystal used are sufficiently large. If the light radiated by the probe contains evanescent field components and these components are efficiently phase conjugated, the phase-conjugated wave should converge (toward a source point) and become localized within a subwavelength-sized region, resulting thereby in a surface light dot. The radiation collected by the (scanning near the surface) probe would then produce a near-field optical image of the phase-conjugated field intensity distribution.

Technical details of the conducted experiments are described elsewhere [10,72]. The radiation of a 10-mW He–Ne laser ($\lambda = 633$ nm) was used to form both the signal and pump waves, the latter being totally internally reflected at the surface of an Fe:$LiNbO_3$ 0.04-wt.% photorefractive crystal. An uncoated fiber probe used as a near-field optical probe in our SNOM could be scanned along the surface at a constant distance ($\sim$5 nm) by means of shear-force feedback [33]. Typical topographical and near-field optical images obtained after a few minutes of exposure, during which the fiber tip was kept near the center of the imaging area, are shown in Fig. 3.15. The fact that the spot size is $\sim$180 nm [10] indicates unequivocally that evanescent field components do take part in the phase conjugation process. This experiment can be regarded as the first demonstration of surface light dots and the phenomenon of subwavelength apertureless light confinement.

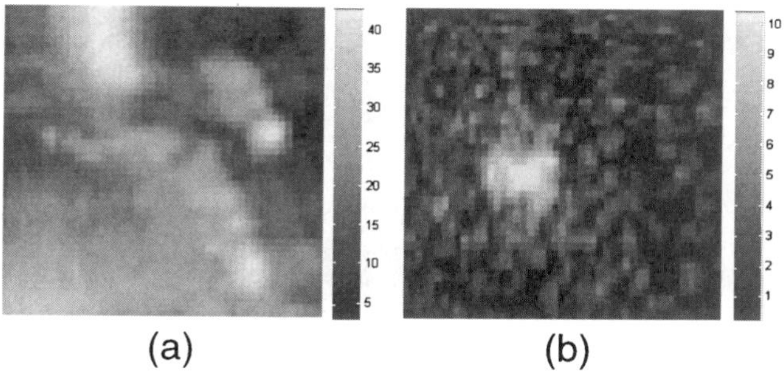

Figure 3.15. Gray-scale (a) topographical and (b) near-field optical images ($2 \times 2\,\mu m^2$) of the crystal surface immediately after the exposure, when a fiber probe was positioned near a center of the image area. Corresponding bar scales reflect (a) the depth (in nm) and (b) the detected phase-conjugated light power (in pW). (Adapted from [10].)

Remember, the actual size of the phase-conjugated surface light dot observed (Fig. 3.15) can be considerably smaller than the image size due to the limited resolution of the SNOM with an uncoated fiber probe [27,73].

Phase conjugation of optical near fields has immediate and very important implications to the problem of interaction of atoms (small scatterers) with the PCM. An already strong interaction between a dipole radiator and the PCM (since the latter focuses back the dipole radiation) [67,69] becomes progressively stronger with the decrease of the dipole–PCM distance due to near-field coupling in the system [70]. Qualitatively, this effect can be explained by the circumstance that, for small dipole–PCM distances, the interaction via propagating waves is complemented by the interaction in which the evanescent field components are involved. The strength of the latter interaction increases exponentially with the decrease of the dipole–PCM distance. The dipole–PCM system can thereby form a very efficient near-field *resonator* (Section 3.2.4) leading eventually to the thresholdless lasing and spontaneous excitation [70].

Strong enhancement of light scattered by small defects on the PCM surface has been experimentally observed with the PCM–SNOM arrangement [71] similar to that used to create phase-conjugated surface light dots. In this case, an uncoated fiber probe was positioned at a distance of 7–10 μm from the surface of the photorefractive crystal exposed to counterpropagating pump waves. The crystal surface was first exposed to the light radiated from the fiber probe during ~20 min, and the phase-conjugated light was then collected by the probe scanned along the surface by using shear-force feedback. The detected near-field optical signal was rather small (~1 pW on average), but it was found to be strongly enhanced (by up to 10 times) near surface features, especially near individual subwavelength-sized surface bumps (Fig. 3.16). It has also been found that the optical images showing the enhancement of phase-conjugated light at surface scatterers deteriorated in time much slower than one would have expected from the measured value of the erasure-time constant. Similar experiments were carried out recently with latex spheres (diameter ~200 nm) placed on the crystal surface in order to investigate the time evolution

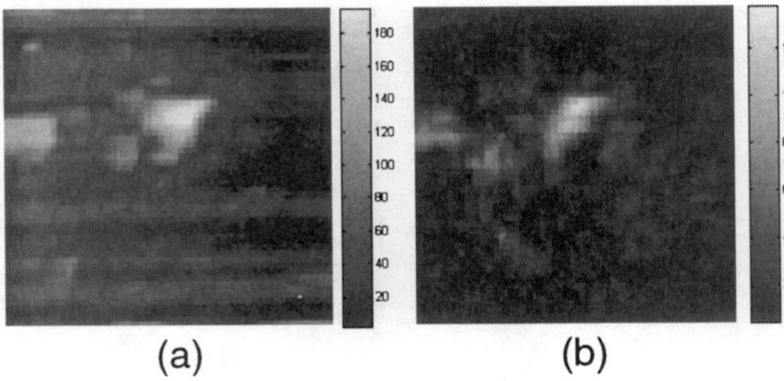

Figure 3.16. Gray-scale (a) topographical and (b) near-field optical images ($4 \times 4\ \mu m^2$) of the crystal surface with a scatterer after the exposure with the probe–surface distance of $\sim 8\ \mu m$. Corresponding bar scales reflect (a) the depth (in nm) and (b) the detected phase-conjugated light power (in pW). (Adapted from [71].)

of scattered light enhancement near the PCM by using well-defined scatterers [77]. It appeared that some bright spots could sustain over the whole period of the experiment (>1 h) without visible degradation, a circumstance that points directly to the *resonant* character of light interactions in the scatterer–PCM system.

3.3.1.2. Qualitative Consideration. Phase conjugation of light scattered by a subwavelength-sized object (probe or surface scatterer) near a PCM is a complicated problem to treat [73], but qualitative considerations may already help to explain the experimental results described above. Let us consider the optical field scattered by a subwavelength object placed near the PCM as a superposition of propagating and evanescent plane waves (Section 3.2.1). By using k_p to denote the numerical magnitude of the wavevector projection along the surface plane, one can express the conditions for the propagating and evanescent components, respectively, $k_p < \omega/c$ and $k_p > \omega/c$. However, an appreciable part of the *evanescent* wave components, namely, those satisfying the criterion $\omega n/c > k_p > \omega/c$ (n is the refractive index of the PCM), gives rise to the waves *propagating* inside the PCM. These waves certainly will experience phase conjugation with the same efficiency as the propagating waves having $k_p < \omega/c$. A phase-conjugated surface light dot composed only of the propagating (inside the PCM) wave components is not expected to be smaller than the diffraction limited spot, that is, smaller than $\lambda/2n \cong 140$ nm [71]. Taking into account the limited resolution of the SNOM used [27] this value agrees resonably well with the spot size (~ 180 nm) in the near-field optical image (Fig. 3.15).

In order to explain the scattered light enhancement near the PCM, let us recall that the radiation of a dipole increases drastically when it approaches the surface of a denser medium and that this increase is due to those evanescent waves, which are transformed into propagating ones in the dense medium [78]. It is clear that in the case of a finite sized scatterer, these wave components (with $\omega n/c > k_p > \omega/c$) can be noticeable in the scattered light only if the scatterer is of subwavelength size.

Therefore, a subwavelength surface bump scatters the incident light producing in the PCM strong wave components with $\omega n/c > k_p > \omega/c$, which in turn give rise to the strong phase-conjugated reflected light. This light is focused back on the scatterer (because it is phase conjugated with respect to the scattered light) and scattered by it again, thereby producing secondary scattered waves. During a sufficiently long exposure, the secondary waves can also experience phase conjugation, and so on. This process leads to the formation of a self-consistent scattered field, which can be substantially stronger than the primarily scattered field. In such a case, switching off the original signal wave from the fiber probe should not drastically change the situation since most of the scattered light power comes from the pump beams. The experimental fact that the optical images deteriorate very slowly (if at all) [71,77], indicates that an enhanced self-consistent scattered field is established during the aforementioned process of multiple scattering $\leftrightarrow$ phase conjugation (especially, for wave components being evanescent in air but propagating in the PCM).

The situation with phase conjugation of waves having $k_p > \omega n/c$, which are evanescent inside the PCM as well as in air, is entirely different. These evanescent waves are bound to the PCM surface and propagate along it. Phase conjugation of such a wave can occur if there is a sufficient overlap between the fields of counter-propagating pump waves and that of the evanescent signal wave. For the considered configuration, this overlap should be different from zero, because the pump waves upon reflecting by the PCM surface create a nonzero field near the surface. In this case, phase conjugation of the evanescent waves is somewhat similar to the 2D phase conjugation of surface plasmons [79] and can be viewed as Bragg reflection of surface waves. The most important problem in this context is to relate the efficiency of phase conjugation of the evanescent waves to that of propagating (inside the PCM) waves. This is a challenging question in its own right, and one possible approach to its solution will be considered in Section 3.3.2.

3.3.2. Phase Conjugation by a Surface Hologram

The size of phase-conjugated light dots and the scattered light enhancement near the PCM depend ultimately on the highest spatial frequency of the signal wave that can still be phase conjugated efficiently. The evanescent waves decaying exponentially in the PCM penetrate into the PCM on the decay length d_p that is inversely proportional to the wave's spatial frequency: $d_p = (k_p^2 - \omega^2 n^2/c^2)^{-0.5}$. For this reason, the efficiency of phase conjugation is expected to vanish in the limit of high frequencies. Another problem appears due to the circumstance that the evanescent wave, which is phase conjugated with respect to an individual evanescent harmonic, should have an amplitude growing *along* the surface. It means that phase conjugation of an individual evanescent wave results in the field represented by a *spectrum* of evanescent waves (not by an individual harmonic). These specific problems with phase conjugation of evanescent (inside the PCM) waves have been left out of the first theoretical considerations concerning near-field phase conjugation [70,73]. A theoretical model based on the concept of near-field holography allows one to evaluate, essentially in the same manner, contributions of the propagating and

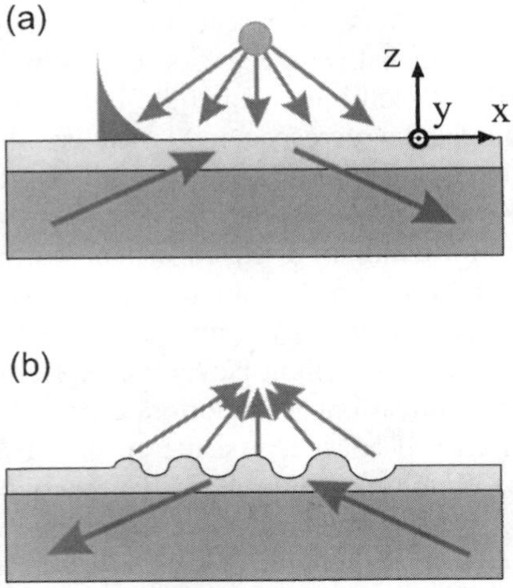

Figure 3.17. Schematic representation of light interactions during the recording (a) of a near-field hologram of a small object and the reconstruction (b) of the phase-conjugated scattered object field.

evanescent waves in the phase-conjugated field, and to introduce a realistic cutoff for high spatial frequencies of the phase-conjugated field [9].

The idea is to consider phase conjugation as a result of reconstruction by a surface hologram with an optical wave that is phase conjugated with respect to the reference wave used during the recording process. The realization of this idea is displayed schematically in Fig. 3.17 for near-field holography with a small (point-like) object. First, a thin photosensitive layer is exposed to the intensity distribution created due to interference between the electric fields of totally internally reflected (reference) plane wave and outgoing spherical wave scattered off the object [Fig. 3.17(a)]. Here, it is assumed that the scatterer is sufficiently close to the surface and, thereby, immersed in the evanescent field of the reference wave. Development of the exposed layer results in a surface profile representing a hologram of the scattered object field. Note that such an arrangement greatly reduces requirements on the coherence length of the light used (down to a few microns). The surface hologram is then illuminated with another plane wave propagating in the direction opposite to that of the reference wave [Fig. 3.17(b)]. Diffraction of this wave gives rise to a wave, which can be under certain approximations regarded as phase conjugated with respect to the scattered object field.

3.3.2.1. Theoretical Framework. The appropriate formulas can be readily written down [9] by treating a small scatterer (object) in the electric dipole approximation (Section 3.2.3) and by using a perturbative approach [7,18] to describe light scattering by the surface hologram. For a given object dipole, the scattered dipole field [Eq. (3.10)] can be expressed via its Fourier transform $\mathbf{F}_0$ at the sample surface by using a 2D analog of Eq. (3.1) for all three field components.

Assuming that the substrate and the layer have the same refractive index, the Fourier spectrum **F** of the object dipole field transmitted through the air–sample interface can be found by using the transfer matrix $\boldsymbol{T}_{12}$ of Fresnel coefficients: $\mathbf{F} = \boldsymbol{T}_{12} \cdot \mathbf{F}_0$ (the expression of the transfer matrix can be found, e.g., in [80]). The object field $\mathbf{E}_\mathrm{d}$ is then determined everywhere inside the sample via the FT *just inside* the sample, that is, at $z = -0$ (Fig. 3.17):

$$\mathbf{E}_d(\mathbf{r}_p, z) = \frac{1}{4\pi^2}\int\int \mathbf{F}(\mathbf{k}_p)\exp\left[i(\mathbf{k}_p \cdot \mathbf{r} - wz)\right]d\mathbf{k}_p \tag{3.20}$$

where $\mathbf{r}_p = (x, y), \mathbf{k}_p = (u, v), w = (n^2\omega^2/c^2 - k_p^2)^{0.5}$, and n is the refractive index of the sample. Inside a thin photosensitive layer (with the thickness $t \ll c/n\omega$), the dipole field $\mathbf{E}_d$ interferes with the reference field $\mathbf{E}_r = \mathbf{E}_r^0\exp(iu_r x)$, where $u_r = n\omega\sin\vartheta/c$, with ϑ being the angle of incidence. Considering the object dipole field to be sufficiently weak: $|\mathbf{E}_d| \ll |\mathbf{E}_r|$, the leading term in the intensity distribution is given by $2\mathrm{Re}\{\mathbf{E}_d \cdot \mathbf{E}_r^*\}$. Assuming development of the film to be linear with respect to the field intensity, the depth of the resulting surface profile is given by

$$\begin{aligned} S(x, y) = \frac{\gamma}{4\pi^2}\int\int &\{\mathbf{F}^*(-\mathbf{k}_p) \cdot \mathbf{E}_r^0\exp[i(u + u_r)x + ivy] \\ &+ \mathbf{F}(\mathbf{k}_p) \cdot \mathbf{E}_r^{0^*}\exp[i(u - u_r)x + ivy]\}g(k_p)d\mathbf{k}_p \end{aligned} \tag{3.21}$$

where γ is the scaling factor, and $g(k_p)$ is the weighting function, which takes into account the decay of the evanescent waves $(k_p > n\omega/c)$ inside the photosensitive layer:

$$\begin{aligned} g(k_p) &= 1 && \text{if} \quad k_p \le n\omega/c \\ &= \frac{1 - \exp(-|w|t)}{|w|t} && \text{if} \quad k_p \ge n\omega/c \end{aligned} \tag{3.22}$$

The reconstructed field, that is created upon diffraction of the wave $\mathbf{E}_i = \mathbf{E}_r^{0^*}\exp[-i(u_r x - wz)]$ by the hologram, can be readily determined by using a perturbative approach [7,18]. The spectrum $\mathbf{F}_d$ of the diffracted field in the surface plane *just above* the hologram is related to the spectrum F_s of the surface profile via the Fresnel transfer matrix $\boldsymbol{T}_{21}$ and the diffraction matrix $\boldsymbol{D}$ [80]:

$$\mathbf{F}_d(\mathbf{k}_p) = i(n^2 - 1)\boldsymbol{D}(\mathbf{k}_p) \cdot F_s(u + u_r, v)[\boldsymbol{T}_{21}(\mathbf{k}_p) \cdot \mathbf{E}_r^{0^*}] \tag{3.23}$$

The reconstructed field can then be found everywhere above the hologram [Eq. (3.20)]:

$$\mathbf{E}_{rf}(\mathbf{r}_p, z) = \frac{1}{4\pi^2}\int\int \mathbf{F}_d(\mathbf{k}_p)\exp[i(\mathbf{k}_p \cdot r + wz)]d\mathbf{k}_p \tag{3.24}$$

where $w_0 = (\omega^2/c^2 - k_p^2)^{0.5}$. Here, as well as in Eq. (3.20), it is implicitly assumed that $\mathrm{Im}(w_0) > 0$, a condition ensuring that evanescent waves would indeed be exponentially decaying in the considered half-space [21]. Finally, one should bear in mind that the perturbative approach can be used only for shallow surface profiles with the maximum height $h \leq \lambda/20$ and with spatial frequencies $< 1/h$ [18].

In order to gain physical insight into phase conjugation in the described configuration, let us qualitatively compare the reconstructed field with the object field at the plane $z = d$, with d being the dipole–sample distance. As the object dipole field propagates toward the sample, its spatial spectrum loses high-frequency components corresponding to the evanescent waves [Eq. (3.3)]. When penetrating into the film, the field spectrum is modified further upon transmission through the sample surface. As a result of the recording process, the spectrum $\mathbf{F}$ of the transmitted field is encoded in the hologram surface profile, whose spatial spectrum consists of two parts shifted with respect to the origin by $\pm u_r$ [Eq. (3.21)]. These parts are related to the spectra of the transmitted field and its phase-conjugated replica. Note that neither of these spectra is perfectly encoded into the surface profile. The scalar products in Eq. (3.21) reflect the circumstance that the transmitted field is *projected* on the reference field in the recording process [81]. In addition, since the hologram topography is less influenced by evanescent waves with shorter penetration depth, very high spatial frequencies (i.e., frequencies of waves with penetration depths much smaller than the film thickness) are effectively cut off [Eq. (3.22)]. Illumination of the hologram with the wave phase conjugated with respect to the reference wave results in the reconstructed field whose spectrum is shifted by u_r with respect to the hologram spectrum [Eq. (3.23)]. The reconstructed spectrum thereby consists of the two parts that can ultimately be related to the following spectra: $\mathbf{F}^*(-\mathbf{k})$ and $\mathbf{F}(\mathbf{k} + 2u_r)$, that is, the spectrum of the *phase-conjugated* (with respect to the object dipole field in the sample) field and the *shifted* spectrum of the dipole field. The latter spectrum contains mostly high spatial frequencies and represents the diffracted field that is relatively small and noticeable only near the sample surface. Note that the components of both diffraction and Fresnel matrixes are rapidly decreasing with the increase of the wave's spatial frequency, especially for evanescent fields [80]. To summarize, the reconstructed field at the plane $z = d$ is dominated by the phase-conjugated field, whose spectrum (in comparison with that of the object field) is (1) filtered with respect to the evanescent waves due to propagation over the distance $2d$; (2) modified twice upon transmission through the air–sample interface; (3) transformed during the recording (polarization effects) and diffraction processes; and (4) weighted with respect to the field components evanescent in the sample.

3.3.2.2. Numerical Results. The qualitative consideration of field spectrum changes during phase conjugation can be illustrated with the following simulations. Let us consider the reconstructed field intensity distributions at the dipole plane $z = d$ as a function of the dipole–surface distance, when the object dipole is induced (in the direction perpendicular to the plane of incidence) by the s-polarized plane wave [9]. The following parameters were used in the calculations: the light

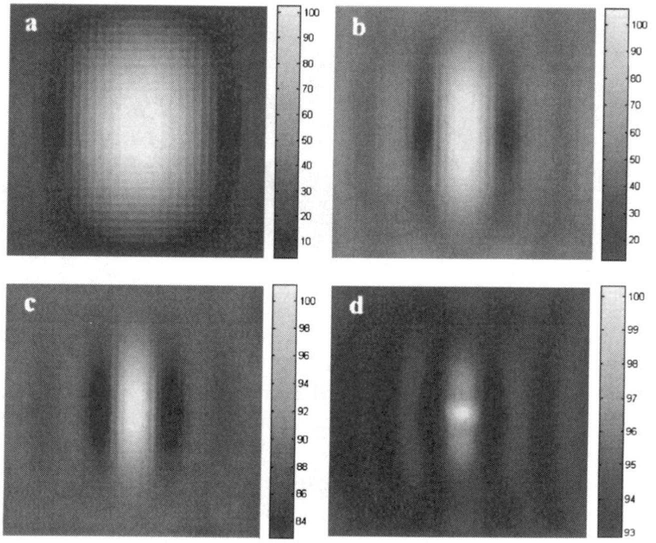

Figure 3.18. Gray-scale representations ($1 \times 1\ \mu m^2$) of the magnitude distributions of the reconstructed field in the plane $z = d$ (the x axis is horizontal) for different distances between the scatterer and the hologram surface: $d = 1\ \mu m$ (a), 220 nm (b), 100 nm (c), and 50 nm (d). (Adapted from [9].)

wavelength $\lambda = 633$ nm, the angle of reference wave incidence $\vartheta = 60°$, the refractive index $n = 2.2$, and the film thickness $t = 10$ nm. When calculating the surface profile of the hologram, the constant exposure level related to the intensity of the reference wave was neglected and only the interference contribution to the exposure was taken into account. By using photosensitive materials with highly nonlinear response (especially for low exposures), it is feasible to adjust the fabrication conditions in the appropriate manner. The scaling factor γ was chosen so that the maximum depth of the surface profile was always equal to the layer thickness. For large distances ($d > \lambda$), the intensity distribution of the reconstructed field was found independent on d representing a diffraction limited spot [Fig. 3.18(a)]. For subwavelength distances d, the phase-conjugated light spot decreases rapidly in size [Fig. 3.18(b)–(d)] because the amount of evanescent field components participating in phase conjugation exponentially increases. At the same time, the background level due to the evanescent field of the reconstructing wave is also increasing, resulting in the decrease of the contrast. It is also seen that the phase-conjugated light spot formed by the propagating field components is elliptical, with the long axis parallel to the dipole [Fig. 3.18(a)]. This elongation of the spot along the dipole is due to the fact that a dipole does not irradiate (propagating waves) along its axis. Therefore, the effective aperture of the converging phase-conjugated wave is about two times less in the cross-section parallel to the dipole direction than in the one perpendicular to it [cf. Fig. 3.19(a) and (b)].

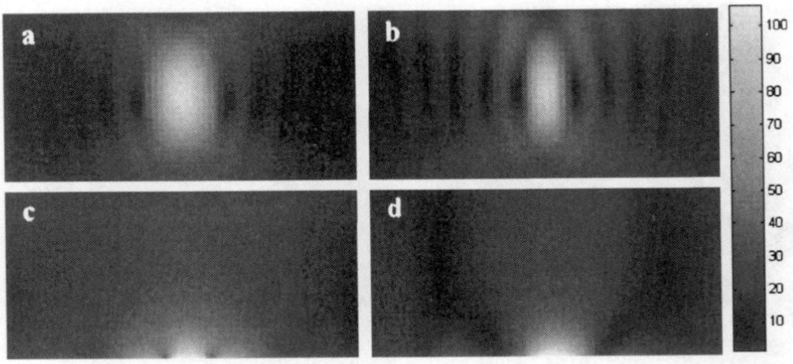

Figure 3.19. Gray-scale representations of the magnitude distributions of the phase-conjugated field in the cross-sections $x = 0$ (a,c) and $y = 0$ (b,d) for $d = 1\,\mu m$ (a,b) and 50 nm (c,d). Note that the image sizes are different: $4 \times 2\,\mu m^2$ (a,b) and $1 \times 0.5\,\mu m^2$ (c,d). The lower boundary of images corresponds to the hologram plane $z = 0$. (Adapted from [9].)

The distributions of phase-conjugated light in the planes perpendicular to the hologram surface are also rather different for large and small distances d (Fig. 3.19). The reconstructed field intensity distribution for the distance of 1 μm [Figs. 3.18(a) and 3.19(a) and, (b)] is a good example of diffraction limited apertureless light confinement realized, in this case, with the help of the hologram acting as a diffractive lens. Spatial dimensions of this confinement when being related to the wavelength ($\sim 0.5 \times 1 \times 1.5\lambda^3$) are nearly the smallest that are possible to be achieved with the propagating waves. For small distances, evanescent field components are dominating, and the reconstructed field is confined within a phase-conjugated surface light dot [Figs. 3.18(a) and 3.19(c) and (d)]. Note that (for subwavelength distances) the field maximum is at the hologram surface and not at the site of the object dipole. This remarkable feature is connected with the following important symmetry relation between evanescent components of conjugate fields: The decay direction of the evanescent components of a field, which is generated from a signal field by phase conjugation followed by reflection, is *opposite* to the direction of evanescent decay of the signal field components [82]. In the considered configuration, the dipole (signal) field components, which are evanescent toward the surface, give rise to the phase-conjugated field components, which are evanescent toward the dipole location. This feature has been illustrated and discussed in detail with simulations based on the macroscopic (phenomenological) model for phase conjugation of optical near fields [73].

The holographic approach to near-field phase conjugation is not limited to a single object dipole but the same principle can be extended to treat more complicated objects, for instance, by dividing them into a set of elementary scatterers, leading to the concept of near-field holography [74]. Furthermore, computer generation of the near-field holograms providing surface light dots with special intensity distributions might prove advantageous in the context of high-density data storage. This technique can easily be adapted to various configurations: The surface profile can be changed to an equivalent profile of index variation, the hologram can be considered as being recorded and read out with two counterpropagating waves, and combining the above modifications will allow one to consider phase conjugation by wave mixing in nonlinear thin films. Finally, the hologram can be recorded with an additional wave illuminating the object and/or in a bulk material, for example, photorefractive crystal. In the latter case, the phase-conjugated field can be rather strong, even much stronger than the evanescent field of the reconstructing wave. Note, that the developed description [Eqs. (3.20)–(3.24)] cannot be applied to bulk holograms, for which the relation between the efficiencies of phase conjugation for the evanescent and propagating field components is not yet clear [73].

3.3.2.3. Experimental Example. The experiments on phase conjugation of light scattered by latex spheres that were placed on the surface of a Fe:$LiNbO_3$ crystal subjected to two counterpropagating pump beams can be regarded as the first experiments utilizing the principle of near-field holography [74,77]. In this configuration, the hologram of latex spheres is continuously recorded (and read out) with two counterpropagating waves whose scattering by spheres results in a signal (object) wave. After a few minutes of exposure, the phase-conjugated light spots appear at the appropriate places as a result of diffraction of the pumping waves on volume holograms, which are created in the crystal via the photorefractive effect. The appropriate near-field optical image related to the phase-conjugated field is produced by scanning an uncoated fiber tip along the crystal surface (with shear-force feedback) and detecting the optical power trapped by the fiber. In contrast with the two-stage holographic procedure described above (Fig. 3.17), the reconstructed optical field is observed together with the object spheres (Fig. 3.20). For reasons yet to be established, not all latex spheres show up in the optical image, indicating that the field *directly* scattered by the spheres is considerably weaker than the (self-sustained) phase-conjugated field enhanced due to the process of multiple scattering ↔ phase conjugation (Section 3.3.1.2).

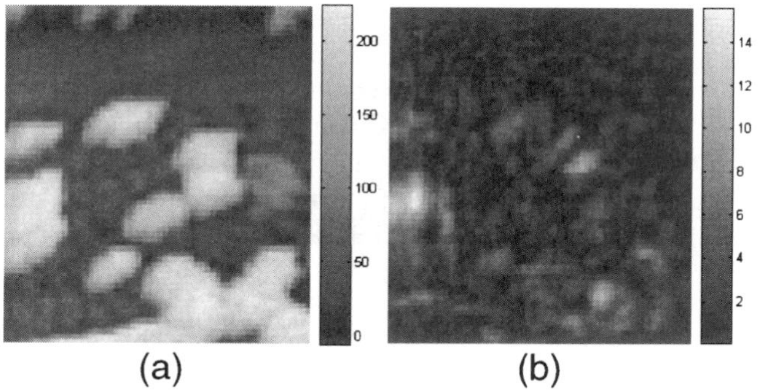

Figure 3.20. Gray-scale (a) topographical and (b) near-field optical images ($4.3 \times 4.9\ \mu m^2$) of latex spheres with the diameter of 200 nm placed on the photorefractive crystal surface continuously illuminated by two counterpropagating laser beams in the total internal reflection configuration. Corresponding bar scales reflect (a) the depth (in nm) and (b) the detected light power (in pW). (Adapted from [74].)

The absence of spheres in the near-optical images can be related to the specific properties of the crystal used (e.g., limited capacity for multiplexed holograms and/or competition between multiple scattering processes involving different spheres). Note that the light spots that can be seen on the optical image [Fig. 3.20(b)], are rather small, a circumstance that demonstrates the possibility of achieving the subwavelength resolution in near-field holography.

3.3.2.4. Size Limitations. Finally, let us briefly discuss the factors that can limit the size of phase-conjugated surface light dots. It is clear that the size decreases with the decrease of the distance between the PCM surface and the scattering center and with the increase of the cutoff (spatial) frequency for the evanescent waves being phase conjugated [9,73]. Using small (illuminated) scatterers placed on the PCM surface, one can approach the size limit imposed by physical effects responsible for phase conjugation of evanescent waves. In the case of bulk nonlinear media, one of the limiting factors is related to the circumstance that the overlap integral for interacting waves [79] decreases rapidly with the increase of the spatial frequency of mixing evanescent waves. This decrease sets the cutoff frequency and, consequently, the size limit at a fraction of the light wavelength in the medium. For sufficiently thin nonlinear films, the size limit could then be related to the film thickness instead of the wavelength in the film [72]. Ultimately, one can employ phase conjugation in a quantum well though probably at the expense of the phase-conjugated light power [83].

The approach based on the idea of light focusing by a thin-film hologram leads again to the limit related, in the first place, to the hologram thickness. It relies also on the regime of *linear* recording (of the interference pattern) and, therefore, as the distance d decreases, the hologram profile reduces to a single pit (bump) that ensures the light confinement [81]. This means, in turn, that the confinement effect becomes more aperture-like and loses the advantage of *collecting* the light from the area larger in size than the region of light confinement. Consequently, the confined field becomes progressively weak in comparison with the evanescent field of the reconstructing wave (Fig. 3.18). Overall, the thin-film holography technique seems most effective for subwavelength light confinement with sizes just beyond the diffraction limit ($\sim \lambda/6$). The size of corresponding surface light dots can be further reduced if a specific *nonlinear* regime of development is used to enhance the contribution from the evanescent waves. Such an enhancement can also be realized by using thin nonlinear films as discussed above.

3.4. SURFACE PLASMON POLARITON LOCALIZATION

Localization of light is an essentially interference phenomenon related to multiple elastic scattering in random media [14]. When a wave propagates through a strong-scattering and nonabsorbing random medium, the mean free path gets reduced due to interference in multiple scattering. Strong (Anderson) localization implies that the mean free path vanishes and propagation no longer exists—the wave is captured in a "random" cavity (Chapter 5). Localization of light is very similar to, yet quite different from, the electron localization, which is a well-studied phenomenon. Like electrons, the localization of optical waves is an interference phenomenon that is, unlike electrons, purely a result of multiple scattering of noninteracting waves (i.e., without complications arising from electron–electron interactions). Localization is expected once the (elastic) scattering mean free path decreases below the light wavelength. Unlike electrons, this criterion is extremely difficult to satisfy, because the scattering potential is frequency dependent for EM waves resulting in the divergence of the mean free path in the limit of both low and high frequencies [84]. Direct experimental evidence of light localization in three dimensions has only recently been reported in strong scattering media of semiconductor powders [43]. Another interference phenomenon that has been extensively studied is enhanced backscattering (also referred to as weak localization), which is already present in lower orders of multiple scattering and considered to be a precursor of strong localization [42]. The weak localization effect arises from a constructive interference (in the backscattering direction) between two waves scattered along the same path in the opposite directions, and, therefore, shows up already in double scattering.

The situation with localization is completely different in two dimensions: light, as well as electrons, are localized with any degree of disorder, at least in the absence of absorption [84]. Qualitatively, it can be explained by the fact that a random walk is recurrent in two dimensions, implying that the effective cross-section of a single scatterer tends to infinity, drastically reducing the effective mean free path [85]. Surface plasmon polaritons propagating along a plane interface represent (quasi) 2D waves, and, therefore, should exhibit localization effects caused by surface roughness [86]. It should be stressed that, here, the consideration is concerned with the (elastic) SPP scattering in the surface plane. The SPP scattering into a free space is an unwanted process leading to the additional (radiative) losses experienced by the SPP. These losses together with the internal absorption limit the length of the SPP propagation path and can eventually destroy the SPP localization [86,87]. Note that the coupling between SPPs and propagating (in air) field components due to surface roughness is responsible for the remarkable phenomenon of backscattering enhancement in the diffusely scattered (out of the surface plane) light [88].

Localization phenomena related to elastic SPP scattering are difficult to observe because of the spatial confinement of the SPP field in the direction perpendicular to the surface plane. Traditionally, interaction of SPPs with surface roughness has been investigated by using far-field measurements of light scattered into a free space [89]. Thus, weak localization of SPPs has been observed by detecting a sharp peak in the angular dependence of the efficiency of second harmonic generation (SHG) in the direction perpendicular to the sample surface [90]. Such a peak is a manifestation of the enhanced backscattering of SPPs, since SHG in the normal direction is related to the nonlinear interaction between counterpropagating (i.e., between the excited and backscattered) SPPs at the fundamental frequency [91]. Only with the advent of scanning probe techniques did it become possible to probe (with high spatial resolution) the SPP field directly and locally, thus opening additional possibilities for studying the SPP scattering in the surface plane. In this section, main results of near-field optical experiments concerned with localization phenomena in elastic SPP scattering are presented along with relevant numerical simulations.

3.4.1. Qualitative Consideration

According to the scaling theory of localization [84,92] 2D waves are localized in the absence of absorption with any degree of disorder. The situation with SPPs is different for several reasons. First and foremost, SPPs, being confined in the direction perpendicular to the surface plane, are quasi-2D waves. Thus, in general, any surface inhomogeneity leads not only to the elastic SPP scattering, which is a cause of localization, but also to the (inelastic) SPP scattering out of the surface plane. The latter process gives rise to radiative losses, thereby, decreasing the length of SPP propagation path. However, even for a perfectly flat interface between air and a homogeneous metal, the SPP propagation length L is finite due to internal damping. In the case of semiinfinite media on both sides of the air–metal interface, the SPP propagation length (or the lateral decay length) is given by [89]

$$L = \lambda \left[4\pi \,\mathrm{Im} \left(\frac{\varepsilon}{\varepsilon + 1} \right)^{1/2} \right]^{-1} \tag{3.25}$$

where ε is the dielectric constant of metal. Usually, SPPs are excited at the air–metal interface of a thin metal layer placed on the surface of a glass (or silica) prism by using the light beam incident from the side of the prism, that is, in the Kretschmann configuration [89]. In such a case, the SPP propagation length L is smaller than the one expressed by Eq. (3.25) due to the coupling between the SPP and the field components propagating in the prism, which results in the resonant reradiation of the SPP. For a realistic interface, L is further reduced due to the aforementioned inelastic scattering. However, for a standard vacuum deposited metal layer, L is mainly determined by the first two processes, and one can use Eq. (3.25) for the estimation of L keeping in mind the reduction of L due to the finite-layer thickness [93].

It is clear that the regime of multiple scattering of light associated with localization can be realized only if light paths are sufficiently long with respect to the elastic mean free path. In the case of weak localization of light, deterioration of the coherent backscattering peak by cutting off long light paths with absorption or confined geometry has been experimentally demonstrated [94]. The condition of weak dissipation of SPPs means accordingly that the SPP propagation length should be much larger than the elastic scattering mean free path: $L \gg l$ [86]. However, this may not be enough to ensure strong localization of SPPs, especially with weak disorder (i.e., when $l \gg \lambda_{sp}$, with λ_{sp} being the SPP wavelength).

In the limit of weak scattering, strong localization of light in two dimensions can be viewed as a consequence of the fact that, for a classical random walker, the total sojourn time in a small finite region around the origin is infinite [14,85]. In the presence of losses, the sojourn time T is apparently finite, and can be evaluated in the wavelength-sized region as follows:

$$T = \frac{\lambda^2}{4D\pi}\int_{l/c}^{L/c}\frac{dt}{t} = \frac{\lambda^2}{4D\pi}\ln\frac{L}{l} \tag{3.26}$$

where D is the classical diffusion coefficient ($D \approx lc/3$), c is the speed of light in a medium, and L is the inelastic mean free path (the SPP propagation length in the case of SPPs). It seems reasonable to suggest [87] that this time should be sufficiently large in comparison with the time l/c of free propagation in order for strong localization to occur. This condition leads to the following relation for the inelastic mean free path:

$$L \gg l\exp\left(\frac{4\pi l^2}{3\lambda^2}\right) \tag{3.27}$$

Exponential divergence of the above condition with respect to the ratio l/λ means that the strong localization of light (in two dimensions) by weak disorder is a very problematic issue in the presence of losses.

In the case of SPPs, the SPP propagation length L in the visible and near-infrared (IR) region can be at best hundreds of microns, implying that the localization condition [Eq. (3.27)] cannot be satisfied at weak disorder ($l \gg \lambda_{sp}$). Therefore, strong localization of SPPs can be realized only at strong disorder ($l \sim \lambda_{sp}$). One can expect to find sufficiently strong disorder in the case of a film with an island structure similar to that used for the first observation of strong SPP localization [11], for example, produced by evaporating a thin film on the surface covered with a sublayer of colloid gold particles [95]. Consequently, at weak disorder, SPPs can exhibit only weak localization, that is, of course, if L is large enough to ensure at least double scattering: $L > l$. Such a condition is relatively easy to satisfy with noble metals (e.g., silver films have been used to observe weak localization effects in the SPP scattering) [90,96]. Evaluation of the elastic mean free path l for SPPs is a challenging problem that is of interest in its own right. However, it can be roughly estimated by using the circumstance that standard metal films have smooth surfaces with rarely spaced micron-sized bumps [11,96]. Introducing the average bump size a and the average separation R between surface bumps, one can estimate l in the limit of geometrical optics ($a \gg \lambda_{sp}$) as $l \sim R^2/a$. Such an evaluation is convenient to use in near-field experiments, in which the surface topography is measured simultaneously with the SPP intensity distribution. As the internal damping increases, the regime of multiple scattering (at weak disorder) changes to the regime of single scattering, when the following relation is valid: $l > L \gg \lambda_{sp}$. Finally, if $L \sim \lambda_{sp}$, then the SPP propagation and scattering in the surface plane become meaningless, even though the SPP excitation (in the Kretschmann configuration) can still be quite pronounced in the angular dependence of the reflected light power.

There are several specific features of near-field experiments with SPPs, that should be taken into account when looking for the localization effects in the elastic

SPP scattering. First, it is very important that the SPP scattering into propagating (in air) field components should be sufficiently weak. This scattering decreases L and contributes to the detected optical signal, thus distorting recorded images of the SPP intensity distribution. Second, the surface area that can be imaged is usually much smaller than the size of an incident laser beam used for the SPP excitation. It means that the image area should be inside of the spot of the incident beam. Therefore, it seems hardly possible to measure transmission or reflection of the SPP through a corrugated region, as could have been suggested by extrapolating the ideas of experiments carried out in three dimensions. On the other hand, the usage of SNOM techniques provides a unique opportunity to observe interference phenomena in the elastic SPP scattering directly by imaging the appropriate interference pattern. In the case of strong localization of SPPs, the near-field optical images are expected to exhibit bright spots [11]. Weak localization of SPPs is related to the formation of the backscattered SPP, which upon interference with the excited SPP should produce interference fringes perpendicular to the propagation direction of the excited SPP [96]. However, it is often very difficult to unambiguously interpret the near-field optical images, because the SPP intensity distribution within an image area is determined by excitation and scattering of SPPs in the *surrounding* area within the SPP propagation length. It turned out that, in this case, an analysis of the spatial Fourier spectra of optical images can help to evaluate the scattering order [95,96]. This issue will be addressed in Section 3.4.2, which is concerned with modeling of SPP scattering.

3.4.2. Modeling of Elastic Surface Polariton Scattering

Surface polariton scattering by surface inhomogeneities has been considered in many theoretical papers (a good overview can be found in a paper by Pincemin et al. [97]). In general, this problem is very complicated, and even a fairly simple case of the SPP scattering, that is, scattering of a homogeneous SPP by a circularly symmetric defect on a metal surface, requires elaborate numerical simulations [98]. The approximation of point-like dipolar scatterers has been recently used to describe local excitation and scattering of SPPs out of the surface plane [99] as well as to simulate localization effects [100]. In the latter work, the indirect coupling between dipoles via the excitation of SPPs has been neglected, thus leaving the elastic SPP scattering out of the consideration and making the results obtained of limited use. As far as the elastic SPP scattering is concerned, one should bear in mind that the scattering observed experimentally [11,101] was close to isotropic. Isotropic scatterers have also been pointed out as a limiting case of small scatterers in a rigorous consideration of SPP scattering [98] and used in rather simplified simulations of SPP excitation with individual surface defects [102]. By taking this into account, one can approximate a scattered SPP by a cylindrical SPP, which is described by the Hankel function with the lowest angular number ($m = 0$) and with the wavenumber determined by the same dispersion relation as for a plane SPP [103]. Such an approach allows one to circumvent the very complicated problem of SPP scattering by surface inhomogeneities and concentrate the efforts on modeling of various phenomena [104] that were experimentally observed in (multiple) elastic SPP scattering.

3.4.2.1. Theoretical Framework. The model used in this section is based on two assumptions, namely, (1) the elastic scattering of SPPs is dominant with respect to the SPP scattering out of the surface plane, and (2) the SPP scattered by an individual surface defect represents an isotropic cylindrical SPP. The first assumption is justified by the experimental observations showing the appropriate interference patterns, whose periods corresponded to the interference between the incident and scattered SPPs [11,87,96,101], and signal (probe–surface) distance dependencies, that exhibited exponential behavior related to the evanescent decay of the SPP field [95]. In addition, recent theoretical results for an individual surface defect indicated that the elastic SPP scattering becomes dominant over the scattering out of the surface plane in the case of resonant SPP scattering [98]. The second assumption is primarily based on the circumstance that well-pronounced parabolic interference fringes were observed with SPP scattering by individual defects [11,101]. Such a pattern corresponds to the interference between the excited SPP with a plane phase front and the isotropically scattered SPP with a cylindrical phase front. Furthermore, it has been theoretically shown that the scattering by a circularly symmetric defect becomes isotropic for small scatterers [98].

In the regime of linear scattering, the field amplitude of a scattered SPP is proportional to the field of an incident SPP at the site of the scatterer. Surface defects can thus be viewed as point-like scatterers characterized by their effective polarizabilities that phenomenologically relate the scattered field amplitudes to the incident ones. Let us now evoke the circumstance that the geometry of elastic SPP scattering is essentially 2D. This means that, at any surface coordinate $\mathbf{r}$, the total SPP field being a superposition of cylindrical SPPs with the same wavenumber β (equal to that of a plane SPP [103]), exhibits the same spatial dependence along the direction perpendicular to the surface plane. Therefore, the well-known exponential decays of the SPP field into the neighboring media [89] can be omitted. Furthermore, it is reasonable to assume that the SPP field component parallel to the surface plane is negligibly small in comparison with the perpendicular field component [89]. In the following paragraph, the total SPP field is represented by the magnitude $E(\mathbf{r})$ of its normal component at the surface plane.

By taking into account the aforementioned considerations, the main relation for the total SPP field at an arbitrary surface point $\mathbf{r}$, which does not coincide with the position of scatterers, can be written down as follows [104]

$$E(\mathbf{r}) = E_{\text{in}}(\mathbf{r}) + \sum_{j=1}^{N} \alpha_j E(\mathbf{r}_j) G(\mathbf{r}, \mathbf{r}_j) \qquad \text{with} \qquad G(\mathbf{r}, \mathbf{r}_j) = \frac{i}{4} H_0^{(1)}(\beta|\mathbf{r} - \mathbf{r}_j|) \tag{3.28}$$

where E_{in} in is the incident SPP field; α_j is the effective polarizability of the jth scatterer located at the surface coordinate $\mathbf{r}_j$; N is the number of scatterers; H is the

Hankel function; and β is the SPP propagation constant, which is determined by the dielectric constants of media that are adjacent to the interface. For semiinfinite media on both sides of the air–metal interface, the SPP propagation constant is given by [89]: $\beta = (2\pi/\lambda)[\varepsilon/(\varepsilon + 1)]^{0.5}$. The effective polarizability α is dimensionless, and corresponds to the scatterer's strength as far as the elastic SPP scattering is concerned. By using the far-field approximation for the Hankel function [105], it is relatively straightforward to express the total (elastic) cross-section σ of the scatterer [98] via the effective polarizability: $\sigma = \alpha^2/[4\mathrm{Re}(\beta)]$. Calculation of the polarizability or the cross-section from the actual surface profile of the scatterer is rather complicated [98], and, in the following numerical simulations, the estimate of α based on the experimental near-field optical images is used [104].

The general relation for the total SPP field [Eq. (3.28)] is a self-consistent equation of multiple scattering, and can be used only after the self-consistent field at the sites of the scatterers $E(\mathbf{r}_j)$ has been determined (Section 3.2.4). Depending on the regime of SPP scattering in question, different approaches can be used to deal with the self-consistent problem. Now, let us recall (Section 3.4.1) that the regime of SPP scattering is determined by the ratio between the SPP propagation length L [Eq. (3.25)] and the elastic scattering mean free path $l \sim R^2/\sigma$. If $l > L$, the regime of multiple scattering reduces to the regime of single scattering [Fig. 3.21(a)], and, consequently, the zeroth-order Born approximation can be used to evaluate the

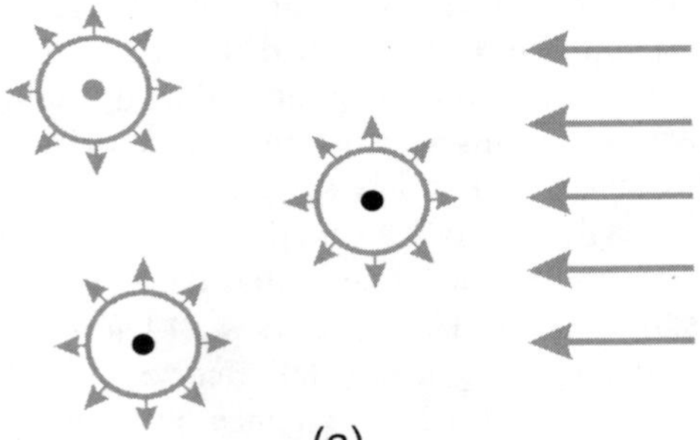

(a)

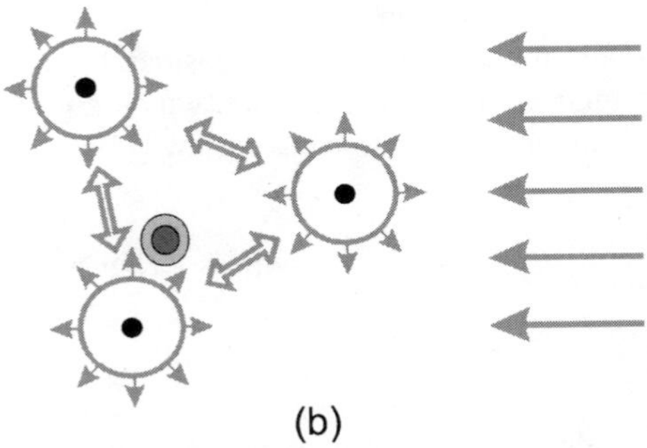

(b)

Figure 3.21. Schematic representation of the regimes of (a) single and (b) multiple scattering.

self-consistent field at the sites of the scatterers: $E^0(\mathbf{r}_j) = E_{\text{in}}(\mathbf{r}_j)$. As the SPP propagation length increases, the regime of single scattering changes to the regime of double scattering (when $2l > L > l$), and the first Born approximation should be invoked

$$E^1(\mathbf{r}_j) = E_{\text{in}}(\mathbf{r}_j) + \sum_{l=1,\, l\neq j}^{N} \alpha_l E_{\text{in}}(\mathbf{r}_l) G(\mathbf{r}_j, \mathbf{r}_l) \tag{3.29}$$

Note that for randomly situated scatterers, the effect of weak SPP localization would already have become pronounced in double scattering [42,87,106].

Finally, for $L \gg l$, the regime of multiple SPP scattering prevails [Fig. 3.21(b)], and the successive Born iterations should be used to calculate the self-consistent field at the scatterers' sites:

$$E^n(\mathbf{r}_j) = E_{\text{in}}(\mathbf{r}_j) + \sum_{l=1,\, l\neq j}^{N} \alpha_l E^{n-1}(\mathbf{r}_l) G(\mathbf{r}_j, \mathbf{r}_l) \tag{3.30}$$

Multiple scattering may eventually result in strong SPP localization, but only in the case of sufficiently strong disorder ($l \sim \lambda_{\text{sp}}$), that is, when the SPP propagation length satisfies the condition expressed by Eq. (3.27). Remember, in the case of sufficiently strong (resonance) interaction, the Born iterations become divergent, and the exact solution of the self-consistent equation has to be employed [36]. However, in the conducted numerical simulations, the Born series expansion converged rapidly, and a few iterations [typically, $n < 10$ in Eq. (3.30)] were sufficient to obtain stable values of the self-consistent field at the sites of the scatterers.

3.4.2.2. Numerical Results. In the presented simulations [104], 25 equivalent scatterers with the effective polarizability $\alpha = 3$ are placed randomly in the area $5 \times 5 \mu\text{m}^2$ and illuminated by a plane wave with the wavelength $\lambda = 633\,\text{nm}$ propagating from the right side toward the left side in the horizontal direction. The propagation constant β is calculated for semiinfinite media on both sides of the air–metal interface with the metal dielectric constant $\varepsilon = -16 + i$, which is a typical value for silver films [95,101]. Then, the total (elastic) cross-section of an individual scatterer can be evaluated as described in Section 3.4.2.1: $\sigma \approx 0.22\,\mu\text{m}$. Note that for a symmetric surface defect treated rigorously [98], the same cross-section would correspond (in the first Born approximation) to the scatterer with a height of $0.1\,\mu\text{m}$ and a radius of $0.7\,\mu\text{m}$, which are parameters that are similar to those of experimentally observed scatterers [11,101]. In the considered configuration, the propagation length ($L \sim 23\,\mu\text{m}$) calculated with Eq. (3.25) is sufficiently large in comparison with the elastic scattering mean free path $l \sim R^2/\sigma \approx 4.5\,\mu\text{m}$. Therefore, by using the Born approximations with different orders, one can subsequently simulate the regimes of single, double, and multiple scattering [104]. Experimentally, this would correspond to the usage of different wavelengths or/and different metal films to change the relation between the SPP propagation length and the elastic mean free path [87]. Finally, remember that the Fourier spectrum $F(\mathbf{k})$ of an intensity distribution, which is a real function of spatial variables, has Hermitian symmetry and, therefore, its magnitude distribution is symmetric with respect to the origin, that is, $|F(-\mathbf{k})| = |F(\mathbf{k})|$. For example, the intensity interference pattern for two (unequal in amplitude) plane waves with wavevectors $\mathbf{k}_1$ and $\mathbf{k}_2$ would result in the Fourier spectrum, whose magnitude (besides being nonzero at $\mathbf{k} = 0$) has the same value for $\mathbf{k} = \pm(\mathbf{k}_2 - \mathbf{k}_1)$. Hereafter, the spectrum magnitude at the origin (that corresponds to the average value of intensity) is not shown on gray-scale representations of Fourier spectra to emphasize weak scattered waves.

In the regime of single scattering, that is, when the zeroth-order Born approximation is applied, the intensity distribution reflects the interference between the incident plane wave and the scattered cylindrical waves [Fig. 3.22(a)]. Note that the field intensity distribution in the immediate vicinity of a scatterer is averaged over the area of $150 \times 150\,\text{nm}^2$ centered at the scatterer's location (to avoid anomalously large and physically meaningless values of the field intensity). The calculated intensity distribution [Fig. 3.22(a)] is actually quite complicated and shows some bright spots that are somewhat similar to the spots expected for SPP localization [11]. The corresponding spatial Fourier spectrum represents a pair of open circles with the radius equal to the SPP propagation constant β [Fig. 3.22(c)]. This spectrum

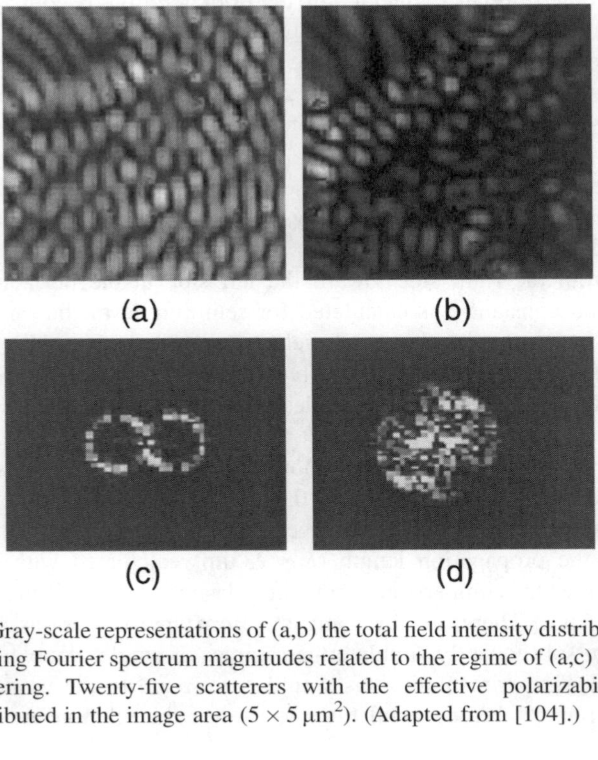

Figure 3.22. Gray-scale representations of (a,b) the total field intensity distributions and (c,d) the corresponding Fourier spectrum magnitudes related to the regime of (a,c) single and (b,d) multiple scattering. Twenty-five scatterers with the effective polarizability $\alpha = 3$ are randomly distributed in the image area ($5 \times 5\,\mu\text{m}^2$). (Adapted from [104].)

immediately reveals that the single-scattered waves are relatively weak, and only the interference between the incident and the scattered waves contribute to the resulting intensity distribution. In the regime of multiple scattering [Eq. (3.30)], the scattered waves are comparable in amplitude with the incident wave resulting in the interference pattern that shows very strong (and spatially localized) field enhancement [Fig. 3.22(b)]. Consequently, the interference between the multiple-scattered waves determines the intensity distribution and its spatial Fourier spectrum that represents a nearly filled circle with the radius that is twice the propagation constant β [Fig. 3.22(d)]. These features of spatial spectra are in complete agreement with the experimental results obtained with SPP scattering on smooth and rough surfaces [95], and shall be illustrated in section 3.4.3. It should be emphasized that the positions of bright spots in Fig. 3.22(b) are not related to the positions of the nearest scatterers but determined by the overall scattering configuration. It was found that a displacement of any scatterer changes the locations of bright spots as should have been expected for the regime of multiple scattering (Section 3.2.4).

Let us now consider the influence of the wavelength λ on the intensity distribution established in the regime of multiple scattering. For the propagation length L being smaller than the lateral extension of the scatterers' area, the corresponding interference pattern should become different when the wavelength is changed by $\delta\lambda \sim \lambda^2/L \sim 15$ nm. In our case, the area dimension is noticeably smaller than the propagation length, and variations in the intensity distribution (bright spots in particular) were found pronounced for the wavelength variation of 30 nm (Fig. 3.23). Such a wavelength dependence is yet another characteristic feature related to interference in multiple scattering. The corresponding spatial Fourier spectra are similar to each other and the one shown in Fig. 3.22(d). Note that the overall appearance of bright spots (Fig. 3.23) is quite similar to the bright spots that were observed in experimental studies of SPP scattering by rough metal films and attributed to strong SPP localization [11,87,95]. Being the result of interference of evanescent waves

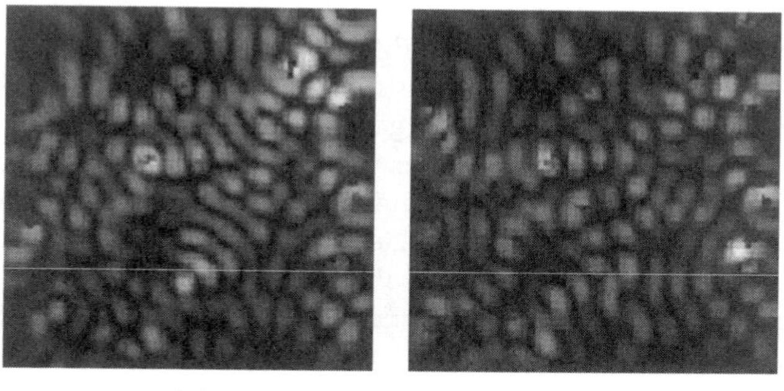

(a) (b)

Figure 3.23. Gray-scale representations of the total field intensity distributions in the area of $5 \times 5\ \mu m^2$ calculated in the regime of multiple scattering (by 25 scatterers with the effective polarizability $\alpha = 3$ randomly distributed in the image area) for different wavelengths of light: $\lambda = 570$ (a), and 600 (b) nm. (Adapted from [104].)

(SPPs), the bright spots represent *per definition* surface light dots and constitute an example of subwavelength apertureless light confinement by multiple scattering in random media. However, even though the simulated bright spots are undoubtedly the result of interference in multiple scattering, they should not be directly related to strong localization as the size ($d = 5\,\mu m$) of the scatterers' area used in our calculations is too small in comparison with the propagation length ($L \sim 23\,\mu m$). Actually, the opposite relation should be satisfied: $d \gg L$, as well as the localization condition [Eq. (3.27)]. Complying with these demands means an increase in the number of scatterers to be considered by at least two orders of magnitude, and such extensive (and time consuming) calculations are yet to be carried out.

3.4.3. Near-Field Imaging of Surface Polariton Scattering

Different near-field techniques have been employed to investigate various SPP properties (overviews can be found in papers by Bozhevolnyi et al. [11] and Smolyaninov et al. [102]). Photon tunneling–SNOM (PT–SNOM) [28], in which an uncoated fiber tip is used to probe an evanescent field of the light being totally internally reflected at the sample surface, seems to be the most suitable technique for local and unobtrusive probing of the SPP field. It has already been mentioned (Section 3.2.6) that, due to the relatively low refractive index of optical fiber, such a tip can be considered within certain approximations as a passive probe of the electric field intensity [65,66]. Detailed experimental investigations have also demonstrated that, for the SPP being resonantly excited at a relatively smooth surface, near-field optical images (obtained with an uncoated fiber probe) can be directly related to the intensity distributions of the total SPP field, that is, the field of the excited and scattered SPPs [95].

The PT–SNOM configuration used to obtain experimental results presented in this section consists of a stand alone near-field microscope combined with a shear-force based feedback system and an arrangement for SPP excitation in the usual Kretschmann configuration. This configuration is described in detail elsewhere [11,104]. Schematic representation of SPP excitation and scattering, and near-field imaging of the resulting SPP intensity distribution is shown in Fig. 3.24. The SPP is excited at a surface of thin metal (silver, gold) film by *p*-polarized (electrical field is parallel to the plane of incidence) laser radiation directed on the surface at the resonant angle ϑ, which provides the minimum in the angular dependence of the reflected light power. The resonant angle is determined by the phase matching condition: $\beta = (2\pi/\lambda)n \sin\vartheta$, where β is the SPP propagation constant and n is the prism refractive index [89]. A film surface can be rough due to the fabrication process [11], or it can be made rough if, for example, the film is evaporated on a sublayer of colloid gold particles [95,104]. Any surface feature would scatter the excited SPP both in the surface plane (into a cylindrical SPP) and out of it (into field components propagating away from the surface). The first process is often dominating (see the discussion in Section 3.4.2.1), and results in an SPP interference pattern, which is imaged with an uncoated fiber probe scanning along the surface (Fig. 3.24).

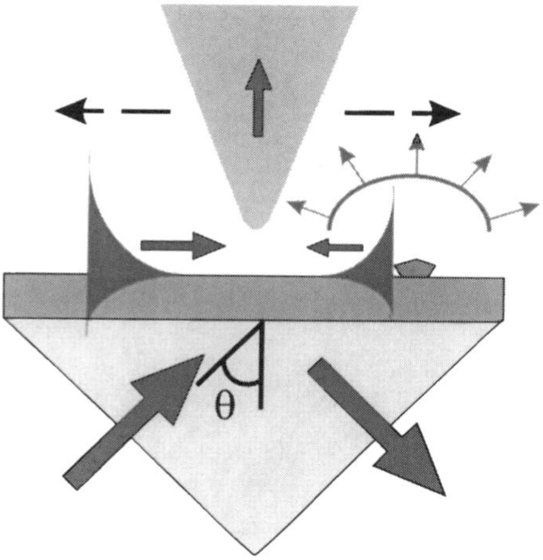

Figure 3.24. Schematic representation of excitation and scattering of surface plasmon polaritons and near-field imaging of the resulting field intensity distribution along the sample surface.

The average optical signal was usually found to decrease exponentially with the increase of the probe–surface distance in accord with the evanescent decay of the SPP field [95]. Such a distance dependence of the signal indicates the possibility of inducing topographical artifacts in the near-field optical images obtained in the constant probe–surface distance mode (Section 3.2.5). In the case of strong multiple scattering of SPPs, it should also be borne in mind that the field distribution can be perturbed by a probe (Section 3.2.6). Recording optical images in constant distance and constant height modes allows one to control both the artifact contribution and the field perturbation. Typical images obtained with a relatively rough silver film [104] are shown in Fig. 3.25. The only difference found between the images in constant distance and constant height modes is related to the displacement of the fiber tip when getting out of contact with the sample surface. This means that the field perturbation by an uncoated fiber probe is negligibly small, and that the contrast observed in near-field images is pure optical, that is, not induced by topographical variations. The latter fact can be explained by rather strong and rapid variations of the near-field intensity in the surface plane. This explanation is supported by the circumstance that the detected optical signal decreased noticeably when the probe–surface distance was increased by $\sim 15\,\text{nm}$ [cf. Fig. 3.25(b) and (c)]. In Sections 3.4.3.1 and 3.4.3.2, the optical images obtained with shear-force feedback (in constant distance mode) are shown along with the simultaneously recorded topographical images. All images are oriented in the way that the excited SPP propagates from the right side toward the left in the horizontal direction.

3.4.3.1. Single and Double Scattering. Different relatively smooth silver and gold films prepared by thermal evaporation on the base of a glass prism in vacuum ($\sim 10^{-5}$ Torr) were used in the experiments on SPP scattering [87,95,96,101,104].

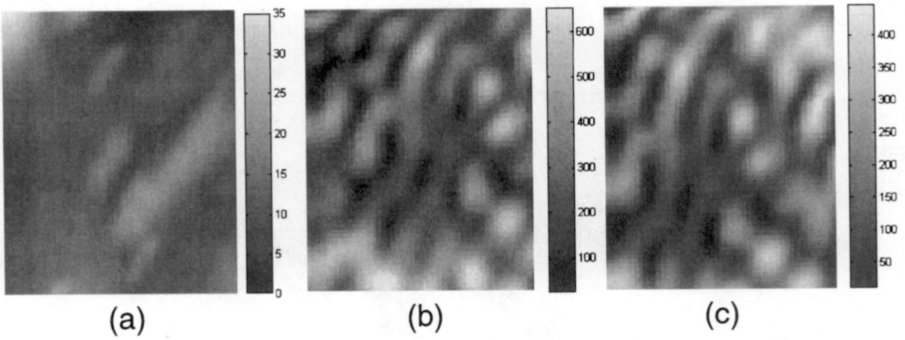

Figure 3.25. Gray-scale (a) topographical and (b,c) near-field optical images ($2 \times 2.5\,\mu m^2$) obtained with the 45-nm-thick silver film and the SPP being resonantly excited at the light wavelength of 594 nm, when operating the microscope with shear-force feedback (b) and in constant height mode (c) with the probe–surface distance of ~ 15 nm. Corresponding bar scales reflect (a) the depth (in nm) and (b,c) the detected light power (in pW). (Adapted from [104].)

Topographical images of these films showed a smooth surface with rarely spaced (several micrometers apart) micron-sized bumps. The SPP propagation length at $\lambda = 633\,\mu m$ can be estimated by using Eq. (3.25) and the available optical constants [54] as ~ 25 and $\sim 8\,\mu m$ for silver and gold films, respectively. Depending on the spatial distribution of surface scatterers in a particular area, the conditions for both single and double scattering, that is, $L < l$ and $L > l$, could be realized. Typical near-field optical and topographical images corresponding to the regime of single scattering are shown in Fig. 3.26. In the vicinity of a scatterer, well-pronounced parabolic interference fringes were observed due to the interference of the excited SPP (with a plane phase front) and the scattered SPP (with a cylindrical phase front) [Fig. 3.26(b)]. Such a pattern can be used for determination of the effective polarizability of the scatterer or its elastic cross-section, a procedure that was actually used to obtain the value $\alpha = 3$ used in the above numerical simulations [104]. If the imaging area is located away from scatterers but within the SPP propagation length, the interference pattern becomes more complicated revealing the interference of several scattered SPPs [Fig. 3.26(d)]. Note that, in the latter case, the image contrast is considerably smaller due to the decrease in amplitudes of scattered SPPs with the increase of distances from the scatterers [cf. Fig. 3.26(b) and (d)].

It is seen that, even in the simplest case of single scattering, the resulting interference pattern can be quite complicated. Proper interpretation based on direct inspection of the near-field optical image obtained may thus turn out to be quite difficult if possible at all. Therefore, it is always recommended to consider the corresponding spatial Fourier spectrum of the optical image as described in Section 3.4.2. For the image in question [Fig. 3.26(d)], the spatial spectrum represents a pair of open circles with the radius corresponding to the SPP propagation constant (Fig. 3.27), a spectrum that is expected for the regime of single SPP scattering

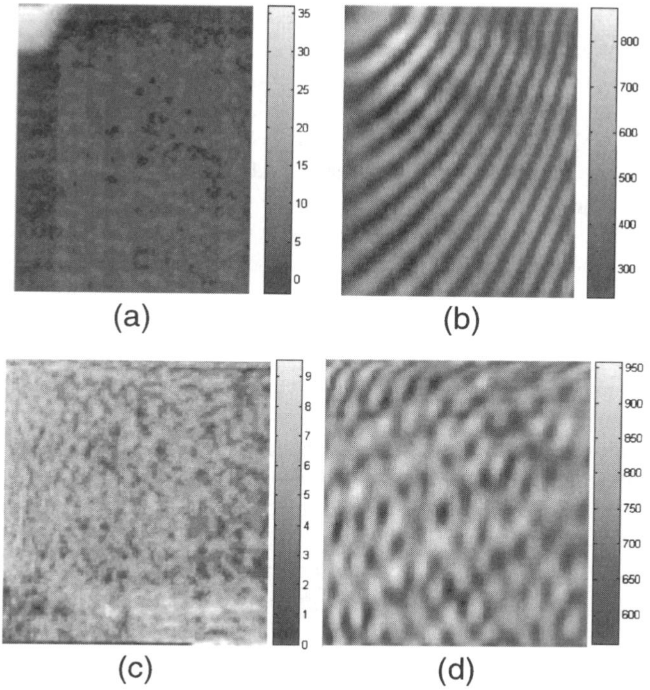

Figure 3.26. Gray-scale (a,c) topographical and (b,d) near-field optical images obtained with the 45-nm-thick silver film and the SPP being resonantly excited at the light wavelength of 633 nm. The sizes of the imaged surface regions are (a,b) $3 \times 4\,\mu m^2$ and (c,d) $3.5 \times 4\,\mu m^2$. Corresponding bar scales reflect (a,c) the depth (in nm) and (b,d) the detected light power (in pW). (Images (a) and (b) adapted from [104].)

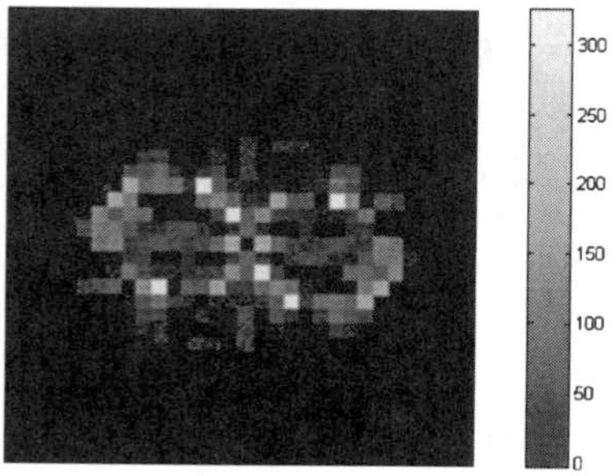

Figure 3.27. Gray-scale representation of Fourier spectrum magnitude corresponding to the near-field optical image shown in Fig. 3.26(d).

[Fig. 3.22(c)]. A slight tilt of the line connecting circle centers is most probably caused by misalignment of the excited SPP with respect to the scanning axes. When comparing experimental and theoretical spectra, remember that the number of experimental data available is usually limited, (e.g., the experimental spectrum shown in Fig. 3.27 was calculated from a 64×64 data set of the optical image) and some spatial frequencies can be accentuated. Also remember that it is not only the information on the scattering regime that can be sought by use of Fourier analysis of optical images. The presence of nonzero spectral components *inside* the circles indicates that the interference between the excited SPP and the inelastically scattered (out of the surface plane) field components does contribute to the optical image obtained. Since this contribution also decreases with the increase of the probe–surface distance, it might be difficult to properly evaluate its magnitude only from the measurements of distance dependencies of the detected optical signal [95].

If the distances between scatterers to satisfy the condition $L > l$ are sufficiently small, the regime of double scattering can be realized and eventually lead to the effect of enhanced SPP backscattering (weak SPP localization) [87,96]. The corresponding near-field optical image should then exhibit interference fringes, that are formed by interference between the excited and backsacttered SPPs and, therefore, have the spacing of $\lambda_{sp}/2$ and oriented perpendicular to the propagation direction of the excited SPP. From the preceding discussion, it is clear that the desirable interference pattern can be substantially influenced by the presence of other interference patterns, especially those related to single scattering [106]. In three dimensions, measurements of backscattered light are carried out *outside* a medium with randomly distributed scatterers, and special (e.g., polarization) techniques are usually used to cut off the single-scattered light [42]. Similarly, the formation of the backscattered SPP should be most pronounced relatively far away from individual scatterers. For this reason, it is quite cumbersome to find surface regions that would result in the near-field optical images showing properly oriented interference fringes

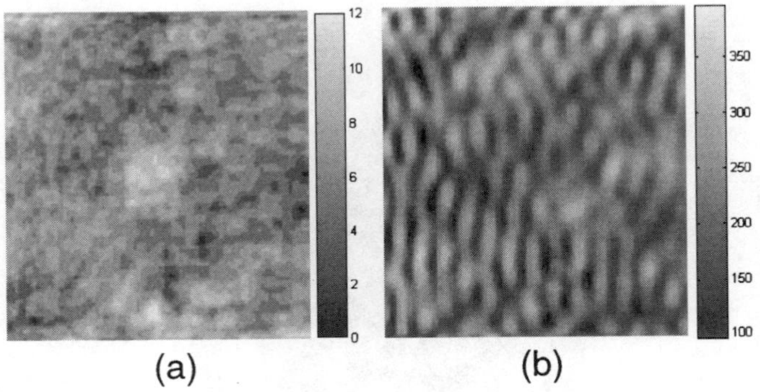

Figure 3.28. Gray-scale (a) topographical and (b) near-field optical images ($4 \times 4.4\,\mu m^2$) obtained with the 53-nm-thick gold film and the SPP being resonantly excited at the light wavelength of 633 nm. Corresponding bar scales reflect (a) the depth (in nm) and (b) the detected light power (in pW). (Adapted from [95].)

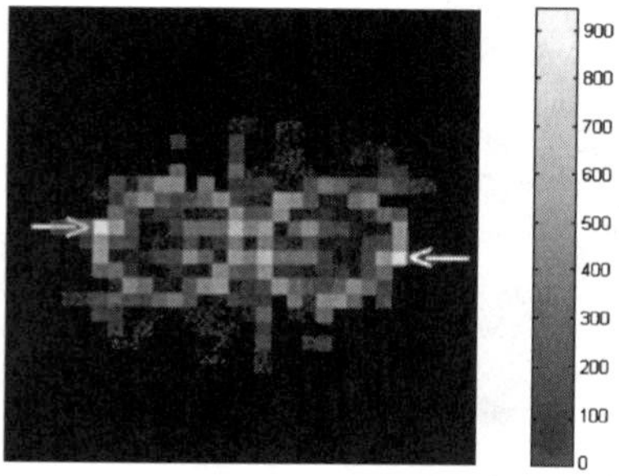

Figure 3.29. Gray-scale representation of Fourier spectrum magnitude corresponding to the near-field optical image shown in Fig. 3.28(b). Arrows indicate the bright spots corresponding to the spectral frequencies related to the interference between the excited and backscattered SPPs.

[96]. Typical topographical and optical images exhibiting the presence of the backscattered SPP, that is, the effect of weak SPP localization, are shown in Fig. 3.28. The overall appearance of the optical image is quite similar to that of the image obtained in the single scattering regime [cf. Figs. 3.26(d) and 3.28(b)], but vertically oriented interference fringes with the proper spacing are also visible. Consequently, the spatial spectrum of the former image (Fig. 3.29) shows two very bright spots corresponding to these fringes, a feature that immediately reveals the effect of enhanced SPP backscattering. The presence of single-scattered waves that result in the double-circle spectrum is also clearly seen [cf. Figs. 3.27 and 3.29].

3.4.3.2. Multiple Scattering. The regime of multiple SPP scattering can be realized only if the separation between scatterers is considerably smaller than the SPP propagation length. For the wavelength of light in the visible, this means that the metal films have to be fabricated under special conditions. The first (80-nm-thick gold) film, that appeared suitable for observations of multiple SPP scattering [11], has been evaporated in a relatively poor vacuum (~ 10–5 Torr) on the cold (room temperature) substrate for a very short time interval (~ 0.1 s). Topographical images of the film surface showed the typical island structure consisting of bumps with various heights (5–100 nm) and sizes (50–1000 nm) in the surface plane [Fig. 3.30(a)]. Since the surface scatterers are practically adjacent to each other, it is plausible to assume that $R \sim \sigma \sim \lambda$ and, therefore, $L \gg l \sim \lambda$ (Section 3.4.1). In such a case, strong SPP localization leading to spatially localized SPP field enhancement should be expected. Indeed, near-field optical images exhibited well-pronounced bright spots related to spatially localized (within 150–250 nm) signal enhancement by up to seven times (Fig. 3.30).

It has been pointed out (Section 3.2.4) that one of the main characteristic features of the localization spot related to multiple scattering is independence of their positions on local topography. This feature is illustrated with the optical images recorded at the same place for different angles ϑ of incidence of the excited SPP [Figs. 3.30(b)–(d)]. It is seen that angular variation of $\delta\vartheta \sim 2^\circ$ is enough to

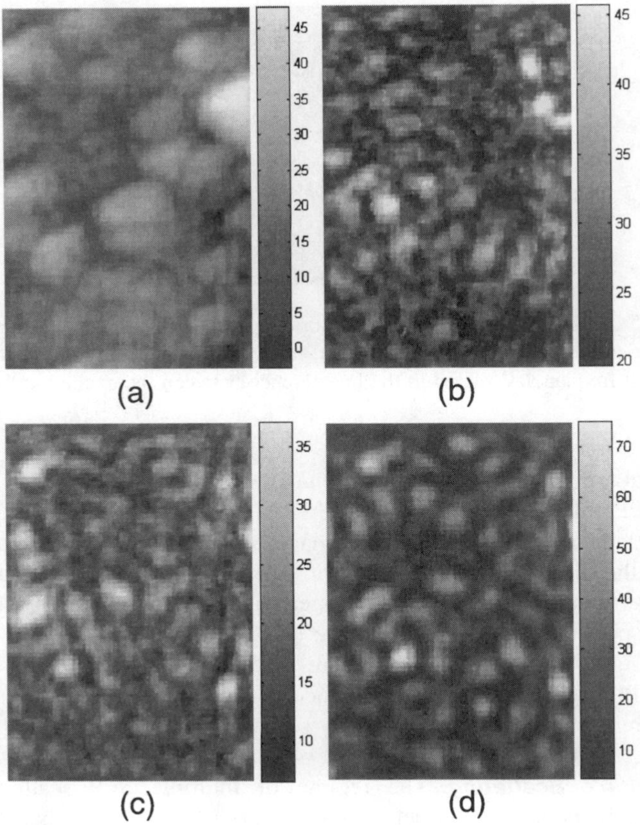

Figure 3.30. Gray-scale (a) topographical and (b–d) near-field optical images ($2.5 \times 3.8\,\mu m^2$) obtained with the 80-nm-thick rough gold film and the SPP excited (at the light wavelength of 633 nm) at different angles of the exciting beam incidence (Fig. 3.24): $\vartheta \cong$ (b) 47°, (c) 50°, and (d) 52°. Corresponding bar scales reflect (a) the depth (in nm) and (b–d) the detected light power (in pW). (Adapted from [8].)

extinguish or to lighten up a bright (localization) spot. Taking into account that the bright SPP spots are the result of interference of scattered SPPs, one can argue that, instead of the phase matching condition for the SPP excitation at a flat surface, a similar one exists for the excitation of a localized plasmon field at a rough surface. The observed angular width (within which a bright spot is pronounced) can be used to estimate the size S of the surface area within which SPP scattering contributes to the formation of the particular bright spot. In the considered experiment [11], $n = 1.56$ and $S \sim \lambda/2\delta\vartheta n \cos\vartheta \sim 8\,\mu m$, a value that is actually correspondent to the SPP propagation length L for smooth gold films. In the case of well-developed multiple SPP scattering, the typical Fourier spatial spectrum represents a nearly filled circle with the radius of 2β [cf. Fig. 3.31 and 3.22(d)] showing that the scattered SPPs are comparable in amplitude to the excited SPP. One can visualize

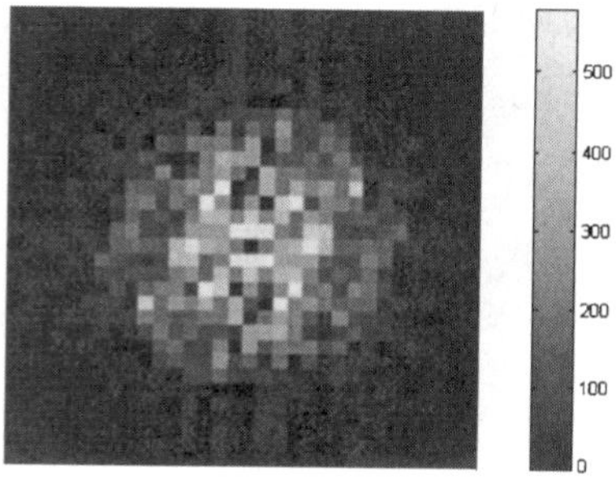

Figure 3.31. Typical gray-scale representation of Fourier spectrum magnitude corresponding to the near-field optical images obtained with rough gold films. (Adapted from [95].)

transformation of the spectrum corresponding to the single scattering regime (Fig. 3.27) into that of multiple scattering (Fig. 3.31) as rotation of the pair of circles the around the origin, because in the latter regime any scattered SPP can be viewed as the incident (excited) SPP.

Strong localization of SPPs has been also observed with artificially roughened gold and silver films [95,104]. Bases of glass prisms were first covered with thin layers of colloid gold particles (diameter $\approx 40\,\text{nm}$) dried up in atmosphere, and silver or gold films were then evaporated on these bases. The corresponding topographical images showed a rough surface revealing the sublayer of randomly located gold particles (or particle clusters) [Fig. 3.32(a)]. The elongated appearance of the particles is probably induced by an asymmetrical shape of the fiber tip (taking the relatively large cone-angle into account) and/or a tilt of the fiber axis with respect to the normal direction (to the film surface). Such an asymmetry in topographical imaging has also been observed in the experiments with latex spheres placed on a crystal surface, which have been imaged with a similar fiber tip [77]. The corresponding near-field optical images of the SPP intensity distributions exhibited rather bright spots that are almost round in shape and located differently for different wavelengths [cf. Fig. 3.32 (c) and (e)]. The wavelength dependence of the locations of bright spots is typical for the regime of multiple scattering (Fig. 3.23) that can eventually result in strong SPP localization. In general, an interference pattern produced by multiple scattering is very sensitive to variations in any parameter (wavelength, phase distribution, positions of scatterers, etc.). The sensitivity with respect to scatterers' locations was directly observed in our experiments [104]. The two sets of images shown in Fig. 3.32 have been taken at the same place of the gold film, whose topography has been modified (after obtaining the left set but before recording the right set) by pressing the fiber tip against the film surface. It is clearly seen that, after such a modification, the near-field optical images changed drastically (and differently) for both wavelengths.

The above experimental results demonstrate the effect of spatially localized enhancement of the SPP field intensity at rough metal surfaces. The enhancement ratio, subwavelength sizes of the observed bright spots, the spectral content of the near-field optical images, together with the fact that the positions of the bright spots

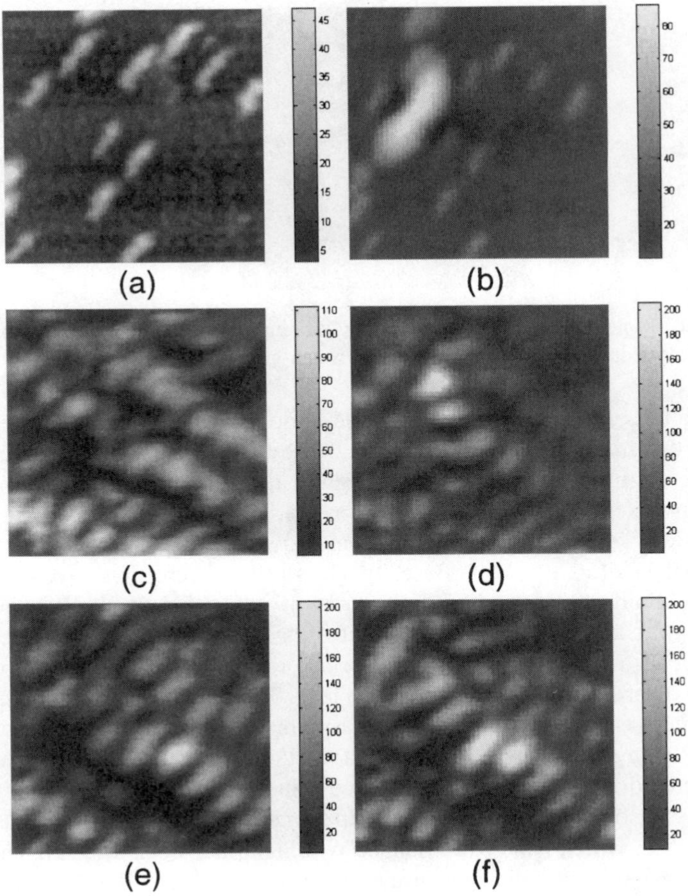

Figure 3.32. Gray-scale (a,b) topographical and (c–f) near-field optical images ($3 \times 3\ \mu m^2$) obtained with the 40-nm-thick gold film (evaporated on a sublayer of gold 40-nm-diameter particles) before (a,c,e) and after (b,d,f) surface modification (the tip was pressed against the surface to the left from the center of the scan area) and the SPP being resonantly excited in turn at two light wavelengths: (c,d) 594 nm and (e,f) 633 nm. Corresponding bar scales reflect (a,b) the depth (in nm) and (c–f) the detected light power (in pW). (Adapted from [104].)

do not correlate with the local surface topography but depend on the incident angle and the wavelength of the exciting beam, altogether provide conclusive evidence of strong SPP localization caused by interference in multiple SPP scattering. Thereby, in the regime of multiple SPP scattering, one can also expect to observe the phenomenon of subwavelength apertureless light confinement in the form of randomly located SPP dots. Even though these surface light dots are somewhat similar to those obtained via phase conjugation with a surface hologram (Section 3.3.2), there is a principal difference between these two phenomena. Surface light dots by localization are the product of *multiple* scattering of SPPs, whereas dots by phase conjugation in thin-film holograms are related to single scattering of the incident wave. Consequently, positions and brightness of random SPP dots are angular and wavelength dependent, whereas those of holographic light dots are practically independent on the angle of incidence and the light wavelength [81]. The difference in the underlying physical principles leads to the circumstance that the factors influencing the size of surface light dots are also different for SPP and phase-conjugated light dots.

3.4.3.3. Size Limitations. The size of surface light dots associated with SPP localization is related to the localization length, which is of the order of the elastic scattering mean free path l at strong disorder (when $l \sim \lambda$) [84]. The mean free path is, in turn, determined by sizes (more precisely, elastic scattering cross-sections) of scatterers and their separations. Subwavelength light confinement requires that the surface should exhibit roughness on a subwavelength size scale (to ensure near-field light interactions). The problem is that, in the limit of small scatterers, the scattering cross-section decreases drastically with the decrease of scatterer's dimensions (e.g., $\sigma \sim d(d/\lambda)^3(h/\lambda)^2$, where d and h are the diameter and height of a circularly symmetric surface defect [98]). On the other hand, when the separations between scatterers are smaller than the wavelength, resonance effects in near-field interactions may help to substantially increase the scattering cross-section and, consequently, to decrease the mean free path [50,107]. In fact, strong SPP localization has been first observed with a gold island film at the light wavelength of 633 nm [11], and, in a similar arrangement, single-particle plasmons have been resonantly excited [45]. Furthermore, one should not underestimate the importance of wavelength-sized scatterers, whose presence is also essential for SPP localization. Large surface features result in strong multiple scattering of propagating SPPs, whose interference makes the SPP propagation virtually impossible (leading instead to localization). Apparently, a film surface should exhibit roughness in a sufficiently large range of sizes, a condition that can be formulated in a more quantitative way by using, for example, fractal surface characterization [108].

As far as the size limit for surface light dots is concerned, let us note that SPP scattering in the limit of very small random scatterers is similar to light scattering in fractal aggregation of metal colloid particles. The latter phenomenon has been theoretically studied (Chapter 8), and it was established that such a scattering may exhibit resonant dipolar excitations (with the field enhancement being several orders of magnitude) localized within the minimum roughness scale [109]. Actually, the experimental results on SPP localization are consistent with this conclusion if one takes into account the finite spatial resolution of SNOM with an uncoated fiber tip [73,95]. It has been shown that the rough gold film surface that resulted in strong SPP localization has a fractal structure in the spatial range 80–640 nm [108]. At the same time, the smallest size ($\sim$ 150 nm) and the strongest signal enhancement ($\sim$ 7 times) observed [e.g., the bright spot in Fig. 3.30(d)] indicate that the actual size of the corresponding SPP dot is probably even smaller ($\sim$ 100 nm) and the intensity enhancement even stronger (10–20 times). Overall, there are sufficient reasons to believe that random surface light dots associated with multiple SPP scattering can be extremely bright and only a few nanometers in size (for the light wavelength in the visible).

3.4.4. Direct Observation of Localized Dipolar Excitations

Multiple scattering of light by rough nanostructured films and surfaces (typically, fractal) is yet another interference phenomenon leading to subwavelength apertureless light confinement. Resonant dipolar excitations in fractal structures with sufficiently small roughness scale (e.g., composed of nanoparticles) can be localized in subwavelength-sized regions and exhibit strong frequency and polarization dependence of their locations [109]. This also means that the spatial locations of light-induced dipole excitations are determined not only by the local surface topography, but also by the large-scale geometrical structure. Although there have been conducted extensive theoretical investigations (Chapter 8), direct experimental evidences of existence of the localized (dipolar) excitations in fractal structures are still scarce. Usage of near-field techniques is probably inevitable for direct imaging of these excitations, and interesting experimental observations have been reported for fractal metal colloid clusters [110] and self-affine fractal interfaces [12]. Here the experimental results concerning direct near-field imaging of localized dipolar excitations in silver colloid fractal structures [13] are presented.

The experimental setup was similar to that used in investigations of SPP scattering [11,13]. The polarization of the illuminating light beam that can be directed from any of the two He–Ne lasers ($\lambda_1 = 633$ nm and $\lambda_2 = 594$ nm), is controlled with a $\lambda/4$ plate and a linear polarizer. The *p*- or *s*-polarized light beam (electric field is parallel or perpendicular to the incidence plane, respectively) at one of the two wavelengths is used to illuminate a sample (placed on the base of a prism) under an angle that exceeds the angle of total internal reflection. Fractal aggregates of silver colloid particles ($\sim$7 nm in radius) were prepared originally in solution, then deposited onto a glass prism and the water was soaked out. The density of the aggregates in the solution was relatively low so that different clusters could be imaged individually. Images obtained with commercial atomic force and scanning electron microscopes showed a typical fractal structure of aggregates of silver particles [13]. For all (polarization and wavelength) configurations and differently structured surface areas, the near-field optical images exhibited a spatially localized (within 150–250 nm) intensity enhancement of up to 20 times compared to the incident field intensity. Positions and brightness of the observed well-defined and subwavelength-sized bright spots were found to be strongly dependent on the wavelength and polarization of the incident light beam (Fig. 3.33). These features are typical for interference phenomena in multiple scattering and consistent with the theoretical predictions of localized optical excitations in fractal structures [109]. One can conclude that the bright spots observed in the near-field optical images constitute an example of fractal surface light dots related to the localized dipolar

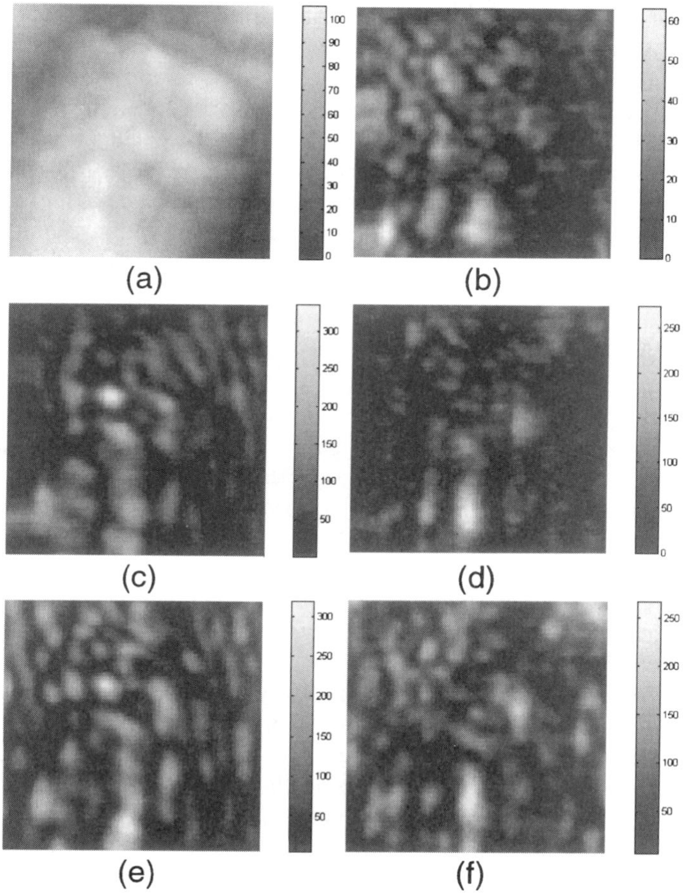

Figure 3.33. Gray-scale (a) topographical and (b–f) near-field optical images ($3.5 \times 3.5\,\mu m^2$) obtained with a silver colloid surface aggregate at two light wavelengths: (b) 594 nm and (c–f) 633 nm, for different polarizations and incident angles (measured inside the orism) of the illuminating beam: (b,d) 48° and p wave, (c) 48° and s wave, (e) 54° and s wave, and (f) 54° and p wave. Corresponding bar scales reflect (a) the depth (in nm) and (b–f) the detected light power (in pW).

excitations. Note that the influence of field polarization on the self-consistent field established (via multiple scattering) is also discussed and demonstrated with simple numerical simulations in Section 3.2.4.3.

The observed spectral and polarization dependence of the spatial positions of the bright spots is an anticipated experimental result. It has also been found that these bright spots are (relatively weakly) influenced by the angle of light beam incidence [cf. Fig. 3.33(c), (e), (d), and (f)]. The origin of this new phenomenon is so far not as clear as that of the spectral and polarization dependence. It is most probably related to the fact that, due to the finite spectral width of the exciting laser beam, several dipolar eigenmodes (with different localization lengths) can be excited simultaneously at the illuminated surface region. If these modes are inhomogeneous and overlap with each other, a change in the angle of incidence can result in redistribution of the total field composed of the excited mode fields [13]. Therefore, the angular resolved near-field imaging provides certain information about the localization of eigenmodes. This feature should be definitely investigated further, as it might be important for practical applications associated, for example, with persistent holes induced by laser radiation in the spectra of fractal structures [111].

Finally, let us compare the elastic SPP scattering by rough surfaces and multiple light scattering in fractal surface structures. The main difference is related to the fact that the SPP scattering involves propagating 2D waves, whereas dipolar eigenmodes are related to multiple scattering in three dimensions with the considerable contribution of near-field (nonpropagating) components. There are also certain similarities between these phenomena. It is well known that a small scatterer placed near a denser medium scatters light primarily in the form of waves propagating in the dense medium nearly parallel to the surface (and evanescent outside the medium) [78]. It is also clear that, when the distances between surface scatterers become smaller than the SPP wavelength, one can hardly separate the interactions mediated via the elastic SPP scattering from the total near-field interactions, because the very definition of elastic scattering relies on the *propagation* concept. In fact, one could probably treat the problem of SPP localization in the same way as the problem of light scattering in fractal structures but incorporating the additional interaction channel (i.e., the interaction via the elastic SPP scattering). Note that the near-field optical images obtained for these configurations also bear close resemblance to each other (cf. Figs. 3.30 and 3.33). These circumstances make scattering of light by rough nanostructured (fractal) surfaces rather similar to the SPP scattering considered above, so that most of the consideration regarding the size limit of random SPP dots (Section 3.4.3.3) can be straightforwardly applied to light scattering by fractal surface structures.

3.5. OUTLOOK

In this chapter, the basic ideas of near-field microscopy and the main experimental results concerning the phenomenon of subwavelength apertureless light confinement along with relevant numerical simulations were reviewed. As far as resonant interactions (or more broadly multiple scattering) and the issues closely related to the appropriate experiments (i.e., topographical artifacts and influence of a probe) are concerned, the results of the latest developments in near-field optics were tried to be presented. Given the rate of progress in this exciting field, it is hardly surprising that this was not accomplished, and many very interesting results did not find their way in to the above material. Even the results on the microoptics of SPPs [101,102, 104], a subject that is closely related to the phenomenon of SPP elastic scattering, were left out of the consideration. The experimental observation of near-field optical effects with a linear chain of gold nanoparticles [112] and a rather unexpected turn in the understanding of image formation in the collection SNOM [113] are two of the most recent results that should be mentioned. One should also be aware that near-field optics of semiconductor nanostructures is rather thoroughly covered in Chapter 4. As far as surface light dots by phase conjugation, holography, and multiple scattering at random surfaces are concerned, it should be stressed that there are many avenues left unexplored that may lead to both exciting developments of fundamental issues and important practical applications.

The use of a shallow *surface* hologram for subwavelength apertureless light confinement is limited from the outset with respect to the efficiency and the extent of light confinement, since a diffracted (phase-conjugated) wave is weaker than an incident one, at least within the approximations used. Light diffraction by deep surface structures requires a self-consistent treatment, and it remains to be seen whether light confinement can be improved by using deep holograms. The situation is different in the configuration with *bulk* phase conjugation, in which a phase-conjugated wave can be much stronger than evanescent fields of pumping waves. However, phase conjugation of *evanescent* waves is a very complicated problem (especially for the waves with the penetration depth on the nanometer scale) that has yet to be explored [83]. As far as experimental issues are concerned, advances in time-resolved near-field spectroscopy of single molecules [114] suggest an exciting avenue for lifetime measurements with molecules placed on the PCM. These kind of experiments would undoubtedly fuel theoretical considerations that have continued since the pioneering work by Agarwal [67]. Computer-generated near-field surface holograms may turn out to be a better alternative for high-density data storage than the direct (point-to-point) near-field techniques. The effect of scattered light enhancement near the PCM might be purposefully used to lighten up small surface features, and thereby to improve the optical contrast when imaging with SNOM [77].

Random surface light dots appear due to strong SPP localization or light-induced dipolar excitations resulting from interference in multiple elastic scattering by nanostructured surfaces. Since *bulk* phase conjugation also involves *multiple* scattering, there is a certain similarity between these phenomena, that is, in both cases, surface light dots can be very bright. However, the physics behind the phenomenon of localization of light is very complicated with many issues, such as the influence of dissipation and wave–wave interaction, which are still not clarified [14]. Some of these issues (e.g., influence of inelastic scattering) can be experimentally investigated by using near-field microscopy of SPPs [87]. Recent observation of a full photonic band gap for SPPs excited along a periodically textured silver surface [63] suggests new possibilities for localization studies, since localization should be realized even with a relatively weak disorder near an edge of the band gap [14]. Taking into account that the field near a metal surface is already enhanced in the case of resonant SPP excitation, one can conclude that the total field enhancement at an SPP light dot may easily reach several orders of magnitude. Similar enhancement is expected at random surface light dots due to localization of resonant dipolar excitations in fractal structures [109]. Both effects can be advantageously exploited, for example, in near-field (spatially resolved) linear and nonlinear spectroscopy, surface chemical and biosensing. A final remark would be that it is more than likely that a number of new exciting issues related to subwavelength apertureless light confinement will appear in the near future.

ACKNOWLEDGMENTS

I gratefully acknowledge all my co-workers and colleagues whose fruitful contributions and critical comments during our collaboration made writing this chapter possible. In addition, I thank my very special co-worker, Elena Bozhevolnaya, for help with numerical simulations of multiple scattering in dipole systems.

REFERENCES

1. E. Abbe, *Arch Mikrosk*. ***9***, 413 (1873).
2. Lord Rayleigh, *Philos. Mag*. ***8***, 261 (1879).
3. E. H. Synge, *Philos. Mag*. ***6***, 356 (1928).
4. J.-M. Vigoureux and D. Courjon, *Appl. Opt*. ***31***, 3170 (1992).
5. J. D. Jackson, *Classical Electrodynamics*, Wiley, New York, 1975.
6. F. Zenhausern, M. P. O'Boyle, and H. K. Wickramasinghe, *Appl. Phys. Lett*. ***65***, 1623 (1994); N. H. P. Moers, R. G. Tack, N. F. van Hulst, and B. Bölger, *J. Appl. Phys*. ***75***, 1254 (1994); Y. Inouye and S. Kawata, *Opt. Lett*. ***19***, 159 (1994); U. C. Fisher, J. Koglin, and H. Fuchs, *J. Microsc*. ***176***, 231 (1994); R. Bachelot, P. Gleyzes, and A. C. Boccara, *Opt. Lett*. ***20***, 1924 (1995).
7. A. Maradudin and D. Mills, *Phys. Rev. B* ***11***, 1392 (1975); J. M. Elson, *Phys. Rev. B* ***12***, 2541 (1975); G. Agarwal, *Phys. Rev. B* ***14***, 846 (1976).
8. S. I. Bozhevolnyi, *Opt. Memory Neural Networks* ***7***, 267 (1998).
9. S. I. Bozhevolnyi and B. Vohnsen, *Opt. Commun*. ***135***, 19 (1997).
10. S. I. Bozhevolnyi, O. Keller, and I. I Smolyaninov, *Opt. Lett*. ***19***, 1601 (1994).
11. S. I. Bozhevolnyi, I. I. Smolyaninov, and A. V. Zayats, *Phys. Rev. B* ***51***, 17916 (1995); S. I. Bozhevolnyi, B. Vohnsen, I. I. Smolyaninov, and A. V. Zayats, *Opt. Commun*. ***117***, 417 (1995).
12. P. Zhang, T. L. Haslett, C. Douketis, and M. Moskovits, *Phys. Rev. B* ***57***, 15513 (1998).
13. S. I. Bozhevolnyi, V. A. Markel, V. Coello, W. Kim, and V. M. Shalaev, *Phys. Rev. B* ***58***, 11441 (1998).
14. S. John, in P. Sheng, Ed., *Scattering and Localization of Classical Waves in Random Media*, World Scientific, Singapore, 1990, p. 1.
15. O. Keller, *J. Nonlinear Opt. Phys. Mater*. ***5***, 109 (1996); *Mater. Sci. Eng. B* ***48***, 175 (1997).
16. D. W. Pohl, in T. Mulvey and C. J. R. Sheppard, Eds., *Advances in Optical and Electron Microscopy*, Academic, London, 1991, p. 243.

17. D. Courjon and C. Bainier, *Rep. Prog. Phys.* ***57***, 989 (1993).
18. J.-J. Greffet and R. Carminati, *Prog. Surf. Sci.* ***56***, 133 (1997).
19. M. A. Paesler and P.J. Moyer, *Near-Field Optics*, J Wiley, New York, 1996.
20. J. M. Vigoureux, C. Girard, and D. Courjon, *Opt. Lett.* ***14***, 1039 (1989).
21. J. W. Goodman, *Introduction to Fourier Optics*, McGraw-Hill, New York, 1968.
22. A. J. den Dekker, and A. van den Bos, *J. Opt. Soc. Am. A* ***14***, 547 (1997).
23. D. W. Pohl, W. Denk, and M. Lanz, *Appl. Phys. Lett.* ***44***, 651 (1984); A. Lewis, M. Isaacson, A. Harootunian, and A. Muray, *Ultramicroscopy* ***13***, 227 (1984).
24. D. W. Pohl and D. Courjon, Eds., *Near Field Optics*, Kluwer, Dordrecht, The Netherlands, 1993; O. Marti and R. Möller, Eds., *Photons and Local Probes*, Kluwer, Dordrecht, The Netherlands, 1995; M. Nieto-Vesperinas and N. García, Eds., *Optics at the Nanometer Scale*, Kluwer, Dordrecht, The Netherlands, 1996; *Ultramicroscopy* (special issues) ***57*** (**2–3**), (1995); ***61*** (**1–4**), (1995); ***71*** (**1–4**), (1998).
25. T. Sugiura, T. Okada, Y. Inouye, O. Nakamura, and S. Kawata, *Opt. Lett.* ***22***, 1663 (1997); M. Gu and P. C. Ke, *Opt. Lett.* ***24***, 74 (1999).
26. E. Betzig and J. Trautman, *Science* ***257***, 189 (1992).
27. S. I. Bozhevolnyi and B. Vohnsen, *J. Opt. Soc. Am. B* ***14***, 1656 (1997).
28. D. Courjon, K. Sarayeddine, and M. Spajer, *Opt. Commun.* ***71***, 23 (1989); R. C. Reddick, R. J. Warmack, and T. L. Ferrel, *Phys. Rev. B* ***39***, 767 (1989); F. de Fornel, J. P. Goudonnet, L. Salomon, and E. Lesniewska, *Proc. SPIE* ***1139***, 77 (1989).
29. D. Van Labeke and D. Barchiesi, *J. Opt. Soc. Am. A* ***10***, 2193 (1993).
30. W. Denk and D. W. Pohl, *J. Vac. Sci. Technol. B* ***9***, 510 (1991); O. J. F. Martin and C. Girard, *Appl. Phys. Lett.* ***70***, 705 (1997); L. Novotny, R. X. Bian, and X. S. Xie, *Phys. Rev. Lett.* ***79***, 645 (1997).
31. L. Aigouy, A. Lahrech, S. Grésillon, H. Cory, A. C. Boccara, and J. C. Rivoal, *Opt. Lett.* ***24***, 187 (1999).
32. E. Betzig, P. L. Finn, and J. S. Weiner, *Appl. Phys. Lett.* ***60***, 2484 (1992); R. Toledo-Crow, P. C. Yang, Y. Chen, and M. Vaez-Iravani, *Appl. Phys. Lett.* ***60***, 2957 (1992); A. Shchemelin, M. Rudman, K. Lieberman, and A. Lewis, *Rev. Sci. Instrum.* ***64***, 3538 (1993).
33. S. I. Bozhevolnyi, I. I. Smolyaninov, and O. Keller, *Appl. Opt.* ***34***, 3793 (1995).
34. B. Vohnsen, S. I. Bozhevolnyi, and R. Olesen, *Ultramicroscopy* ***61***, 207 (1995).
35. C. Girard and A. Dereux, *Rep. Prog. Phys.* ***59***, 657 (1996).
36. O. Keller, M. Xiao, and S. Bozhevolnyi, *Surf. Sci.* ***280***, 217 (1993).
37. T. Setälä, M. Kaivola, and A. T. Friberg, *Phys. Rev. E* ***59***, 1200 (1999).
38. G. S. Agarwal, *Phys. Rev. A* ***11***, 230 (1975); C. Girard, A. Dereux, O. J. F. Martin, and M. Devel, *Phys. Rev. B* ***52***, 2889 (1995).
39. J. E. Sipe, J. F. Young, J. S. Preston, and H. M. van Driel, *Phys. Rev. B* ***27***, 1141 (1983).
40. R. Carminati, *Phys. Rev. E* ***55***, R4901 (1997).
41. C. Girard, A. Dereux, and J.-C. Weeber, *Phys. Rev. E* ***58***, 1081 (1998).
42. M. P. van Albada, M. B. van der Mark, and A. Lagendijk, in P. Sheng, Ed., *Scattering and Localization of Classical Waves in Random Media*, World Scientific, Singapore, 1990, p. 97; Yu. N. Barabanenkov, Yu. A. Kravtsov, V. D. Ozrin, and A. I. Saichev, in E. Wolf, Ed., *Progress in Optics*, Vol. 29, Elsevier, New York, 1991, p. 65.

43. D. S. Wiersma, P. Bartolini, A. Lagendijk, and R. Righini, *Nature (London)* ***390***, 671 (1997).
44. M. Xiao, S. Bozhevolnyi, and O. Keller, *Appl. Phys. A* ***62***, 115 (1996).
45. U. Ch. Fischer and D. W. Pohl, *Phys. Rev. Lett.* ***62***, 458 (1989).
46. R. K. Chang and T. P. Furtak, Eds., *Surface-Enhanced Raman Scattering*, Plenum, New York, 1982.
47. K. H. Drexhage, in E. Wolf, Ed., *Progress in Optics*, Vol. 12, North-Holland, Amsterdam, The Netherlands, 1974, p. 163.
48. V. A. Markel, *J. Mod. Opt.* ***39***, 853 (1992).
49. Yu. N. Barabanenkov and V. V. Shlyapin, *Phys. Lett. A* ***170***, 239 (1992).
50. O. Keller, S. Bozhevolnyi, and M. Xiao, in D. W. Pohl and D. Courjon, Eds., *Near Field Optics*, Kluwer, Dordrecht, The Netherlands, 1993, p. 229.
51. T. Klar, M. Perner, S. Grosse, G. von Plessen, W. Spirkl, and J. Feldman, *Phys. Rev. Lett.* ***80***, 4249 (1998).
52. H. R. Stuart and D. G. Hall, *Phys. Rev. Lett.* ***80***, 5663 (1998).
53. M. Specht, J. D. Pedarnig, W. M. Heckl, and T. W. Hänsch, *Phys. Rev. Lett.* ***68***, 476 (1992); J. Koglin, U. C. Fischer, and H. Fuchs, *Phys. Rev. B* ***55***, 7977 (1997).
54. E. D. Palik, *Handbook of Optical Constants of Solids*, Academic, New York, 1985.
55. S. I. Bozhevolnyi, M. Xiao, and O. Keller, *Appl. Opt.* ***33***, 876 (1994); S. Madsen, S. I. Bozhevolnyi, and J. M. Hvam, *Opt. Commun.* ***146***, 277 (1998).
56. M. Quinten, A. Leitner, J. R. Krenn, and F. R. Aussenegg, *Opt. Lett.* ***23***, 1331 (1998).
57. E. Betzig, in D. W. Pohl and D. Courjon, Eds., *Near Field Optics*, Kluwer, Dordrecht, The Netherlands, 1993, p. 7.
58. B. Hecht, H. Bielefeldt, Y. Inouye, D.W. Pohl, and L. Novotny, *J. Appl. Phys.* ***81***, 2492 (1997).
59. R. Carminati, A. Madrazo, M. Nieto-Vesperinas, and J.-J. Greffet, *J. Appl. Phys.* ***82***, 501 (1997); C. Girard and D. Courjon, *Surf. Sci.* ***382***, 9 (1997).
60. S. I. Bozhevolnyi, *J. Opt. Soc. Am. B* ***14***, 2254 (1997).
61. Y. Martin, F. Zenhausern, and H. K. Wickramasinghe, *Appl. Phys. Lett.* ***68***, 2475 (1996); H. F. Hamann, A. Gallagher, and D. J. Nesbitt, *Appl. Phys. Lett.* ***73***, 1469 (1998).
62. S. I. Bozhevolnyi, O. Keller, and M. Xiao, *Appl. Opt.* ***32***, 4864 (1993).
63. S. C. Kitson, W. L. Barnes, and J. R. Sambles, *Phys. Rev. Lett.* ***77***, 2670 (1996).
64. R. Carminati and J.-J. Greffet, *Opt. Commun.* ***116***, 316 (1995).
65. J. C. Weeber, F. de Fornel, and J. P. Goudonnet, *Opt. Commun.* ***126***, 285 (1996).
66. S. I. Bozhevolnyi, *J. Microsc.* ***194***, 561 (1999).
67. G. S. Agarwal, *Opt. Commun.* ***42***, 205 (1982).
68. G. S. Agarwal and E. Wolf, *J. Opt. Soc. Am.* ***72***, 321 (1982); G. S. Agarwal, A. T. Friberg, and E. Wolf, *J. Opt. Soc. Am.* ***73***, 529 (1983).
69. R. J. Cook and P. W. Milonni, *IEEE J. Quantum Electron.* ***QE-24***, 1383 (1988); H. F. Arnoldus and T. F. George, *Phys. Rev. A* ***43***, 3675 (1991); E. J. Bochove, *J. Opt. Soc. Am. B* ***9***, 266 (1992); H. F. Arnoldus and T. F. George, *Phys. Rev. A* ***48***, 3910 (1993), and references cited therein.
70. G. S. Agarwal and S. D. Gupta, *Opt. Commun.* ***119***, 591 (1995).

71. S. I. Bozhevolnyi, O. Keller, and I. I. Smolyaninov, *Opt. Commun.* ***115***, 115 (1995).
72. S. I. Bozhevolnyi and I. I. Smolyaninov, *J. Opt. Soc. Am. B* ***12***, 1617 (1995).
73. S. I. Bozhevolnyi, E. A. Bozhevolnaya, and S. Berntsen, *J. Opt. Soc. Am. A* ***12***, 2645 (1995); S. I. Bozhevolnyi, B. Vohnsen, E. A. Bozhevolnaya, and S. Berntsen, *J. Opt. Soc. Am. A* ***13***, 2381 (1996).
74. S. I. Bozhevolnyi and B. Vohnsen, *Phys. Rev. Lett.* ***77***, 3351 (1996).
75. M. C. Gower and D. Proch, *Optical Phase Conjugation*, Springer, Berlin, 1994.
76. M. Nieto-Vesperinas, *Scattering and Diffraction in Physical Optics*, Wiley, New York, 1991.
77. B. Vohnsen and S.I. Bozhevolnyi, *Opt. Commun.* ***148***, 331 (1998).
78. W. Lukosz and R. E. Kunz, *J. Opt. Soc. Am.* ***67***, 1607 (1977).
79. K. Ujihara, *Opt. Commun.* ***42***, 1 (1982); *J. Opt. Soc. Am.* ***73***, 610 (1983).
80. D. Van Labeke, D. Barchiesi, and F. Baida, *J. Opt. Soc. Am. A* ***12***, 695 (1995).
81. B. Vohnsen and S. I. Bozhevolnyi, *J. Opt. Soc. Am. A* ***14***, 1491 (1997).
82. M. Nieto-Vesperinas and E. Wolf, *J. Opt. Soc. Am. A* ***2***, 1429 (1985).
83. T. Andersen and O. Keller, *Phys. Rev. B* ***57***, 14793 (1998).
84. S. John, H. Sompolinsky, and M. J. Stephen, *Phys. Rev. B* ***27***, 5592 (1983); S. John, *Phys. Rev. Lett.* ***53***, 2169 (1984).
85. B. Souillard, in J. Souletie, J. Vannimenus, and R. Stora, Eds., *Chance and Matter*, North-Holland, Amsterdam, The Netherlands, 1987, p. 305.
86. K. Arya, Z. B. Su, and J. L. Birman, *Phys. Rev. Lett.* ***54***, 1559 (1985).
87. S. I. Bozhevolnyi, *Phys. Rev. B* ***54***, 8177 (1996).
88. A. R. McGurn, A. A. Maradudin, and V. Celli, *Phys. Rev. B* ***31***, 4866 (1985); A. R. McGurn and A. A. Maradudin, *J. Opt. Soc. Am. B* ***4***, 910 (1987); J. A. Sanchez-Gil and M. Nieto-Vesperinas, *Phys. Rev. B* ***45***, 8623 (1992); C. S. West and K. A. O'Donnel, *J. Opt. Soc. Am. A* ***12***, 390 (1995), and references cited therein.
89. H. Raether, *Surface Plasmons*, Springer Tracts in Modern Physics Vol. **111**, Springer, Berlin, 1988.
90. O. A. Aktsipetrov, V. N. Golovkina, O. I. Kapusta, T. A. Leskova, and N. N. Novikova, *Phys. Lett. A* ***170***, 231 (1992); S. I. Bozhevolnyi and K. Pedersen, *Surf. Sci.* ***377–379***, 384 (1997).
91. A. R. McGurn, T. A. Leskova, and V. M. Agranovich, *Phys. Rev. B* ***44***, 11441 (1991).
92. E. Abrahams, P. W. Anderson, D. C. Licciardello, and T. V. Ramakrishnan, *Phys. Rev. Lett.* ***42***, 673 (1979).
93. N. Kroo, W. Krieger, Z. Lenkefi, Z. Szentirmany, J. P. Thost, and H. Walther, *Surf. Sci.* ***331–333***, 1305 (1995).
94. S. Etemad, R. Thompson, M. J. Andrejco, S. John, and F. C. MacKintosh, *Phys. Rev. Lett.* ***59***, 1420 (1987).
95. V. Coello, S. I. Bozhevolnyi, and F. A. Pudonin, *Proc. SPIE* ***3098***, 536 (1997).
96. S. I. Bozhevolnyi, A. V. Zayats, and B. Vohnsen, in M. Nieto-Vesperinas and N. Garcia, Eds., *Optics at the Nanometer Scale*, Kluwer, Dordrecht, The Netherlands, 1996, p. 163.
97. F. Pincemin, A. A. Maradudin, A. D. Boardman, and J.-J. Greffet, *Phys. Rev. B* ***50***, 15261 (1994).
98. A. V. Shchegrov, I. V. Novikov, and A. A. Maradudin, *Phys. Rev. Lett.* ***78***, 4269 (1997).

99. L. Novotny, B. Hecht, and D. W. Pohl, *J. Appl. Phys.* ***81***, 1798 (1997).
100. M. Xiao, A. Zayats, and J. Siqueiros, *Phys. Rev. B* ***55***, 1824 (1997).
101. S. I. Bozhevolnyi and F. A. Pudonin, *Phys. Rev. Lett.* ***78***, 2823 (1997).
102. I. I. Smolyaninov, D. L. Mazzoni, J. Mait, and C. C. Davis, *Phys. Rev. B* ***56***, 1601 (1997).
103. V. A. Kosobukin, *Phys. Solid State* ***35***, 457 (1993); P. J. Valle, E. M. Ortiz, and J. M. Saiz, *Opt. Commun.* ***137***, 334 (1997).
104. S. I. Bozhevolnyi and V. Coello, *Phys. Rev. B* ***58***, 10899 (1998).
105. M. Abramowitz and I. A. Stegun, *Handbook of Mathematical Functions*, 9th ed. Dover, New York, 1972.
106. A. V. Shchegrov, *Phys. Rev. B* ***57***, 4132 (1998).
107. K. Arya, Z. B. Su, and J. L. Birman, *Phys. Rev. Lett.* ***57***, 2725 (1986).
108. S. I. Bozhevolnyi, B. Vohnsen, A. V. Zayats, and I. I. Smolyaninov, *Surf. Sci.* ***356***, 268 (1996).
109. V. M. Shalaev, *Phys. Rep.* ***272***, 61 (1996).
110. D. P. Tsai, J. Kovacs, Z. Wang, M. Moskovits, V. M. Shalaev, J. S. Suh, and R. Botet, *Phys. Rev. Lett.* ***72***, 4149 (1994).
111. V. P. Safonov, V. M. Shalaev, V. Markel, Y. E. Danilova, N. N. Lepeshkin, W. Kim, S. G. Rautian, and R. L. Armstrong, *Phys. Rev. Lett.* ***80***, 1102 (1988).
112. J. R. Krenn, A. Dereux, J. C. Weeber, E. Bourillot, Y. Lacroute, J. P. Goudonnet, G. Schider, W. Gotschy, A. Leitner, F.R. Aussenegg, and C. Girard, *Phys. Rev. Lett.* ***82***, 2590 (1999).
113. S. I. Bozhevolnyi and E. A. Bozhevolnaya, *Opt. Lett.* ***24***, 747 (1999).
114. X. S. Xie and J. K. Trautman, *Annu. Rev. Phys. Chem.* ***49***, 441 (1998).

CHAPTER FOUR

Near-Field Optics of Nanostructured Semiconductor Materials

BERND HANEWINKEL, ANDREAS KNORR, PETER THOMAS, and STEPHEN W. KOCH

Department of Physics and Material Sciences Center, Philipps-Universität, Renthof 5, D-35032 Marburg, Germany

4.1. INTRODUCTION

4.1.1. Principles of Near-Field Optics

For a conventional optical microscope, the diameter of the minimal illuminated spot size is determined by the Rayleigh criterion

$$\Delta = \frac{1.22\lambda}{N_A} \tag{4.1}$$

where λ denotes the wavelength of the incident light and N_A is the numerical aperture of the microscope. The parameter Δ is a measure for minimum distance of two objects, which can be resolved with the microscope. Practically, this leads to a maximum resolution of the order of $\lambda/2$, which is approximately one-half of a micrometer at a wavelength corresponding to the fundamental band edge of gallium-arsenid (GaAs). Generally, the Rayleigh criterion applies in the far-field limit, where the observation distance is larger than the wavelength of light. However, placing a test object (often called near-field probe) into the subwavelength vicinity of the sample, modifies the field distribution on a shorter scale. Techniques, exploiting this, are referred to as near-field optics, and in combination with scanning the probe over the sample as scanning near-field optical microscopy (SNOM or NSOM). This concept was first applied in the microwave region [1], where a resolution of $\lambda/60$ was reported. Since the early 1980s this technique has been applied in the optical region [2,3], where a resolution of 20 nm has been achieved [4]. The experimental realization of a low temperature environment is especially important for applications to semiconductor optics [5].

Optics of Nanostructured Materials, Edited by Vadim A. Markel and Thomas F. George
ISBN 0-471-34968-2

In general, one can distinguish between techniques where the probe is scanned over the sample (SNOM) and experimental setups in which a probe is stationary deposited in the vicinity of the sample. The versatility of scanning near-field techniques is higher, however, the method is experimentally much more demanding than the stationary setup. Stationary apertures can be realized with diameters down to 50 nm in a metal coating on top of the sample [6]. Another possibility is the deposition of small scatterers, like metal clusters or even moleculesclose to the sample [7].

In SNOM experiments, the commonly used probes are optical fiber tips [5] and atomic force microscopy (AFM) cantilevers [8]. With the latter information, synchronously SNOM and AFM data can be recorded, yielding simultaneous information on optical and topographic properties of the sample. By using optical fibers, etching [9], or pulling and heating [5] yields tips with an apex curvature in the range of 50 nm. In most cases, the fiber tips are additionally metal coated in a way that a small aperture is left in the apex. The metal guides the light in the fiber through the subwavelength hole, but unfortunately, it also leads to strong reflection and absorption such that the transmitted intensity is rather weak [10].

In principle, the probe can either be used to locally illuminate the sample in combination with detection in the far field or to locally collect the radiation, after far-field excitation. The illumination mode is used if one wants to excite only the material locally. In the collection mode, the sample is excited homogeneously and only the probe selects the local response (e.g., in a photon tunneling setup).

The combination of illumination and collection modes with a metal coated fiber tip suffers from a large intensity loss resulting from the need to pass the aperture twice. Nevertheless, this idea has been used in combination with uncoated fiber tips (internal reflection SNOM), where subwavelength resolution has been realized experimentally [11,12] and modeled theoretically [13]. It turns out, however, that the influence of the topography has to be considered carefully [14] and that subwavelength resolution occurs only for operation with polarized light if excitation and detection of the signal are perpendicularly polarized. Moreover, the signal is mostly sensitive to changes in the optical material properties, such as edges of domains with different refractive indexes.

From this brief discussion, it should already be clear that the analysis of near-field optical experiments requires thorough theoretical modeling. In general, in contrast to far-field spectroscopy, the interaction of probe and sample (e.g., due to multiple scattering processes in near-field spectroscopy) is important and cannot be neglected.

4.1.2. Near-Field Optics of Semiconductors

Near-field optics is a very promising technique for the study of manmade semiconductor structures. Although different techniques and therefore different possible resolutions have been reported, it can be said that most commonly used near-field optical devices provide a resolution down to some tens to hundreds of nanometers.

In the last decade, artificial structuring of semiconductors on mesoscopic scales (i.e., on length scales above the lattice constant but below or comparable to the wavelength of resonant light), has lead to numerous new applications and improvements of semiconductor devices. In particular, structures of reduced dimensionality like quantum wells (QWs), quantum wires (QWIs), and quantum dots (QDs) have widely been investigated and implemented (e.g., in optoelectronic semiconductor devices [15]). These structures have features that are determined by a large ensemble of atoms resembling the crystalline band structure of the material, as well as effective additional mesoscopic potentials, which modify the material properties and lead to quantum confinement effects on the mesoscopic length scale. In particular, the interplay of mesoscopic structure and Coulomb interaction between excited carriers introduces new features such as a strong enhancement of the optical response of low-dimensional structures compared to the bulk material [16].

For its high spatial resolution power, near-field optics is in some sense the spatial analogy to ultrafast spectroscopy. Here the availability of ultrashort pulses and corresponding measurement techniques (down to the order of 10 fs), make it possible to study the ultrafast dynamics of semiconductors. In particular the combination of near-field techniques and time-resolved measurements promises to yield valuable insights into the spatio-temporal dynamics of excitations in semiconductor structures [17,18]. Probing light emitting semiconductor structures, such as lasers or diodes, with near-field optical techniques directly yields information about their optical properties. This is an advantage over other scanning probe techniques, for example, atomic force microscopy, which offer higher resolution but can be used only to probestructural features. Indeed, SNOM has been applied in *in situ* measurements of optoelectronic devices like diodes, waveguides [19], and lasers [20]. Here, the local mode structure or spatially resolved emission characteristics were determined.

Basic investigations focus on the optical properties of lower dimensional structures like QWs, QWIs, and QDs. In particular, QDs have been the subject of near-field optical investigations [6,21–23]. Due to the three-dimensional (3D) carrier confinement, ideal QDs are expected to have atomic-like optical spectra consisting of discrete sharp lines [24]. In contrast to atomic systems, where ensembles of identical particles only differ in their spectral properties, which are due to different velocities or to their interaction with the environment, the individual dots in aQD ensembles may vary in size, shape, and composition. Since optical spectra only give information averaged over the spot size of the excitation, very small spot sizes are necessary to observe single QDs. Successful examples are the so-called μPL [25] and near-field optical experiments [6]. In this context, line shape, fine structure, and even hyperfine structures [6], as well as effects of magnetic fields [26] have been studied.

The investigations on structures with one- and two-dimensional (1D and 2D) confinement (QWs, QWIs) can in general be divided into studies of the inherent optical properties, for example, transport dynamics in the directions of translational invariance [17,27], and into local investigations of deviations from the ideal quantum confinement. Here, potential fluctuations in QWs due to monolayer fluctuations or local modifications of the alloy composition have been studied. Knowledge of the local optical properties makes it possible to draw conclusions about the underlying mesoscopic material structure that are often not available via surface sensitive microscopic techniques (e.g., in a QW structure where the layers of interest are buried inside the sample).

Most near-field optical experiments of semiconductors so far have been photoluminescence (PL) or PL excitation (PLE) measurements in which the near-field probe either excites the sample or collects the photoluminescence. Additionally, this technique has been combined with nonlinear excitation, time-resolved PL or magnetic fields.

To illustrate differences between near- and far-field excitation of a sample, we briefly analyze the field distribution of a simple radiating source: a point dipole. Its field is discussed in Section 3.3.2.3 in detail. In the far field, where the distance R is large compared to the wavelength λ, the amplitude of the field components decrease with $1/R$, whereas in the near field the electric field diverges with $1/R^3$. This result leads to a high localization of the field intensity in the vicinity of the dipole source, which is determined by the distance R and can be small compared to the wavelength. Hence, by assuming that the dipole radiation excites a sample, we can expect subwavelength resolution in the subwavelength vicinity of the dipole. This finding can be exploited to test inhomogeneous samples and will be discussed for the case of excitons that are spatially localized (cf. Section 4.3.4). For homogeneous samples, it can also be used to exploit nonlocal material properties, which are related to optical transport (cf. Section 4.3.7).

In connection with the pronounced field localizations, strong gradients occur in the field distribution since the field varies on a subwavelength scale. This can be important if the interaction of the dipole radiation with a sample is expanded in a multipole series. Here one should expect higher order moments to become more important than in the case of far-field excitation. This can be the case for dipole forbidden interband transitions in semiconductors. (cf. Sections 4.3.4.5 and 4.3.6).

In Chapter 3, the vector structure of the field distribution of the dipole in the near-field is also discussed [Eq. (3.8)]. Clearly, a dipole source scatters the incoming field and creates new vector components. For anisotropic materials, which are sensitive on the polarization direction of the excitation, this can lead to modifications in the response. In semiconductors, such an anisotropy can be caused by the mesoscopic structure, which is discussed in Section 4.3.5 for a QW.

4.2. THEORY OF SEMICONDUCTOR NEAR-FIELD INTERACTION

In this section, we present a microscopic theory to describe the response of semiconductor structures and the near-field excitation. We treat the configuration where an external electromagnetic (EM) field is incident on the near-field probe and excites the semiconductor structure. The induced polarization of probe and semiconductor are sources for the EM fields. Material equations for the semiconductor, which describe the response of the semiconductor to the field, as well as Maxwell's equations, which describe the creation of the EM fields by the material sources have to be solved self-consistently. The linear and local parts of the material response, like the probe and the semiconductor background polarization, are described by a dielectric function $\varepsilon(r)$. Quantum mechanical equations of motions are derived for the resonant response of the semiconductor structure in the spectral region around the fundamental band edge. These equations reflect the special conditions of the near-field excitation, such as highly localized excitation, high-field gradients, and strong polarization mixing. Thus a highly nonlocal, anisotropic response is expected. Moreover, due to the strong-field gradients in the near-field higher multipole, contributions should be important for the description of the field–matter interaction.

In general, the near-field/semiconductor system is characterized by its Hamiltonian. The dynamics of the observables can then be deduced from the Heisenberg equation of motion for the corresponding operators. In Section 4.2.1, the microscopic Hamiltonian is derived and the treatment of structures of reduced dimensionality is discussed.

4.2.1. Microscopic Hamiltonian

The dynamics of the semiconductor as a system of charged particles is determined by the minimal coupling Hamiltonian [28]. It is convenient to divide the Hamiltonian into the following contributions:

$$\mathscr{H} = \mathscr{H}_0 + \mathscr{H}_{\text{conf}} + \mathscr{H}_f + \mathscr{H}_C + \mathscr{H}_I \tag{4.2}$$

$\mathscr{H}_0$ describes the motion of the electrons in the lattice potential of the semiconductor crystal, $\mathscr{H}_{\text{conf}}$ contains an additional confinement potential, which stems from changes in the alloy composition. The Hamiltonian of the free electromagnetic field is denoted by $\mathscr{H}_f$, and $\mathscr{H}_C$ and $\mathscr{H}_I$ describe the interaction via the EM field, which is usually split into the interaction via the longitudinal part $\mathscr{H}_C$ and the transverse part $\mathscr{H}_I$. In the formalism of second quantization, $\mathscr{H}$ is expanded in Fermionic material operators $a_{\lambda\mathbf{k}}, a^{\dagger}_{\lambda\mathbf{k}}$, which annihilate and create electrons in the single-particle Bloch states$|\lambda\mathbf{k}>= (1/\sqrt{V})\, u_{\lambda\mathbf{k}}(r)e^{i\mathbf{k}\cdot\mathbf{r}}$ in an infinitely extended crystal lattice.

The Bloch states are eigenfunctions of the unperturbed Hamiltonian: $\mathbf{p}^2/2m_0 + V_0$. Here, V_0 is the lattice potential, $\mathbf{p}$, the momentum operator and m_0 is the free electron mass. The eigenenergies are denoted by $\varepsilon_{\lambda\mathbf{k}}$, labeled by wavenumber $\mathbf{k}$ and band index λ.

As $\mathscr{H}_0$ describes the free motion of electrons, it is diagonal:

$$\mathscr{H}_0 = \sum_{\lambda\mathbf{k}} \varepsilon_{\lambda\mathbf{k}} a^{\dagger}_{\lambda\mathbf{k}} a_{\lambda\mathbf{k}} \tag{4.3}$$

The Hamiltonian describing the confinement potential in the low-dimensional semiconductor structures, $\mathscr{H}_{\text{conf}}$ will be discussed in Section 4.2.3.1.

The three remaining parts of the total Hamiltonian include the interaction with the EM field and describe the optical response of the system. This response can be divided into contributions from the dielectric environment and the resonantly excited polarization, that is, into the respective off-resonant and resonant sources. The off-resonant parts are modeled by a dielectric function $\varepsilon(\mathbf{r})$, which represents the background refractive index of the semiconductor material as well as the dielectric or metallic material of the near-field probe. The dynamics of the EM fields in the background material can then be included in the Hamiltonian of the free field [29]. It is convenient to choose the gauge freedom for the vector potential $\mathbf{A}$ and scalar potential ϕ of the EM field in a way that $\mathbf{A}$ satisfies the generalized Coulomb condition [29]:

$$\nabla \cdot (\varepsilon \mathbf{A}) = 0 \tag{4.4}$$

Then the scalar potential ϕ is a function of the charge density only:

$$-\nabla \cdot (\varepsilon \nabla \phi) = 4\pi\rho \tag{4.5}$$

In the case of a homogeneous medium $\varepsilon(\mathbf{r}) = \varepsilon_0$, this equation reduces to the Laplace equation with the potential given by the Coulomb integral:

$$\phi(\mathbf{r}) = \frac{1}{\varepsilon_0} \int \frac{\rho(\mathbf{r}')}{|\mathbf{r} - \mathbf{r}'|}\, d^3r' \tag{4.6}$$

More generally, the potential can be represented as a functional of the charge density,

$$\phi(\mathbf{r}) = \int V_C(\mathbf{r}, \mathbf{r}')\rho(\mathbf{r}')\, d^3r' \tag{4.7}$$

where $V_C(\mathbf{r}, \mathbf{r}')$ is the potential at $\mathbf{r}$ of a point-like charge at $\mathbf{r}'$. Due to the presence of boundaries, we may include image charges to account for the boundary conditions at the interfaces. Typically, these deviations from the homogeneous case are important only for electrons and holes in close proximity to a surface. A measure for the characteristic distance is the excitonic Bohr radius of the semiconductor.

The Coulomb interaction part of the Hamiltonian in second quantization is written as

$$\mathscr{H}_C = \tfrac{1}{2} \sum V^{\lambda_1\lambda_2\lambda_3\lambda_4}_{\mathbf{k}_1\mathbf{k}_2\mathbf{k}_3\mathbf{k}_4} a^\dagger_{\lambda_1\mathbf{k}_1} a^\dagger_{\lambda_2\mathbf{k}_2} a_{\lambda_4\mathbf{k}_4} a_{\lambda_3\mathbf{k}_3} \tag{4.8}$$

where the Coulomb matrix elements are given by

$$V^{\lambda_1\lambda_2\lambda_3\lambda_4}_{\mathbf{k}_1\mathbf{k}_2\mathbf{k}_3\mathbf{k}_4} = \frac{1}{V^2} \int d^3r d^3r'\; u^\star_{\lambda_1\mathbf{k}_1}(\mathbf{r}) u^\star_{\lambda_2\mathbf{k}_2}(\mathbf{r}') V_C(\mathbf{r}, \mathbf{r}') u_{\lambda_3\mathbf{k}_3}(\mathbf{r}) u_{\lambda_4\mathbf{k}_4}(\mathbf{r}') \times e^{-i(\mathbf{k}_1-\mathbf{k}_3)\cdot\mathbf{r}} e^{-i(\mathbf{k}_2-\mathbf{k}_4)\cdot\mathbf{r}'} \tag{4.9}$$

The interaction with the EM field is given by the coupling to the vector potential in the generalized Coulomb gauge, Eq. (4.4):

$$H_I = \frac{1}{V}\sum \int d^3r\, u^{\star}_{\lambda_1 \mathbf{k}_1}(\mathbf{r}) e^{-i\mathbf{k}_1\cdot\mathbf{r}} \left(-\frac{e}{m_0 c}\mathbf{A}\cdot\mathbf{p} + \frac{e_2}{2m_0c^2}\mathbf{A}^2 \right) u_{\lambda_2\mathbf{k}_2}(\mathbf{r}) e^{i\mathbf{k}_2\cdot\mathbf{r}} \times a^{\dagger}_{\lambda_1\mathbf{k}_1} a_{\lambda_2\mathbf{k}_2} \tag{4.10}$$

The dynamics of the free field including the dielectric background material is defined by

$$\mathscr{H}_f = \int \frac{d^3r}{8\pi\varepsilon}\left((\nabla\times\mathbf{A})^2 + \frac{\varepsilon}{c^2}(\dot{\mathbf{A}})^2 \right) \tag{4.11}$$

In Eq. (4.12), the EM fields will be treated classically, because our main interest in this chapter is the description of semiclassical and coherent phenomena. The usual Hamiltonian equations lead to the equation of motion for the vector potential

$$-\nabla\times\nabla\times - \frac{\varepsilon(\mathbf{r})}{c^2}\frac{\partial^2}{\partial t^2}\mathbf{A} = -\frac{4\pi}{c}\langle \mathbf{j}_t \rangle \tag{4.12}$$

where $\langle \mathbf{j}_t \rangle$ is the transverse part of the current, which is specified by

$$\begin{aligned} \mathbf{j}_t &= -c\left(\frac{\delta H}{\delta \mathbf{A}}\right)_t \\ &= -\left(\sum u^{\star}_{\lambda_1\mathbf{k}_1}(\mathbf{r}) e^{-i\mathbf{k}_1\cdot\mathbf{r}} \left(-\frac{e}{m_0}\mathbf{p} + \frac{e^2}{m_0 c}\mathbf{A} \right) u_{\lambda_2\mathbf{k}_2}(\mathbf{r}) e^{i\mathbf{k}_2\cdot\mathbf{r}}\, a^{\dagger}_{\lambda_1\mathbf{k}_1} a_{\lambda_2\mathbf{k}_2} \right)_t \end{aligned} \tag{4.13}$$

With the definition of the scalar potential, the gauge invariant expressions for the dynamics of the electric and magnetic fields are obtained as

$$\begin{aligned} \nabla\times\mathbf{E}(\mathbf{r},t) &= -\frac{1}{c_0}\partial_t \mathbf{B}(\mathbf{r},t) \\ \nabla\times\mathbf{B}(\mathbf{r},t) &= \frac{\varepsilon(\mathbf{r})}{c_0}\partial_t \mathbf{E}(\mathbf{r},t) + \frac{4\pi}{c_0}\mathbf{j}(\mathbf{r},t) \end{aligned} \tag{4.14}$$

These equations determine the motion of the full (longitudinal and transversal) EM fields (**E**, **B**) excited by the resonant currents **j** including the nonresonant background material (ε).

4.2.2. Averaged Hamiltonian

The Hamiltonian, Eq. (4.2), describes the system on a fully microscopic length scale characterized by the dimensions of the elementary cell, that is, a few angstroms. The spatially localized excitation by the near-field probe, however, varies on a mesoscopic scale that is at least an order of magnitude larger. To simplify the analysis further, we can therefore assume that it is a very good approximation to average the electromagnetic fields over one elementary cell, which then makes it possible to use a multipole expansion of the interaction over the elementary cell. This approach is sensible, especially in systems with Wannier-like excitons, whose relative-motion wave functions extend over many elementary cells. Under these conditions, the potential ϕ (and analogously also $\mathbf{A}$) can be approximated by

$$\phi(\mathbf{r}) = \sum_{\mathbf{k}} \phi(\mathbf{k}) e^{-i\mathbf{kx}} \tag{4.15}$$

where the summation is only over the wavevectors of the first Brioullin zone (BZ). Higher Fourier coefficients are neglected. In real-space, this treatment implies that any field can be represented by a macroscopic field, which is averaged over the elementary cell. Note that in this case the Fourier transform (FT) fulfills the relation,

$$\phi(\mathbf{k}) = \frac{1}{V} \int e^{i\mathbf{kr}} \phi(\mathbf{r})\, d^3r = \frac{\Omega_0}{V} \sum_{\mathbf{R}} \phi(\mathbf{R}) \mathbf{e}^{\mathbf{ikR}} \tag{4.16}$$

where $\mathbf{R}$ denotes lattice points only and Ω_0 is the volume of the elementary cell. With this treatment, the averaged material Hamiltonian is obtained as

$$\begin{aligned}
\mathscr{H} &= \sum_{\mathbf{k}\lambda} \varepsilon_{\lambda\mathbf{k}} a^{\dagger}_{\lambda\mathbf{k}} a_{\lambda\mathbf{k}} \\
&+ \frac{1}{2} \sum_{\lambda\mathbf{k}} \frac{4\pi e^2}{V|\mathbf{k}_1 - \mathbf{k}_3|^2} \langle \lambda_1 \mathbf{k}_1 | \lambda_3 \mathbf{k}_3 \rangle \, \langle \lambda_2 \mathbf{k}_2 | \lambda_4 \mathbf{k}_4 \rangle \\
&\times a^{\dagger}_{\lambda_1\mathbf{k}_1} a^{\dagger}_{\lambda_2\mathbf{k}_2} a_{\lambda_4\mathbf{k}_4} a_{\lambda_3\mathbf{k}_3} \\
&- \frac{1}{c_0} \frac{e}{m_0} \sum_{\lambda\mathbf{k}} (\langle \lambda_1 \mathbf{k}_1 | \mathbf{p} + \mathbf{q} | \lambda_2 \mathbf{k}_2 \rangle \mathbf{A}_{\mathbf{k}_2 - \mathbf{k}_1} \\
&- \frac{e}{2c_0} \langle \lambda_1 \mathbf{k}_1 | \lambda_2 \mathbf{k}_2 \rangle \mathbf{A}^2_{\mathbf{k}_2 - \mathbf{k}_1}) a^{\dagger}_{\lambda_1\mathbf{k}_1} a_{\lambda_2\mathbf{k}_2}
\end{aligned} \tag{4.17}$$

with the brackets indicating the integral over the elementary cell

$$\langle \lambda_1, \mathbf{k}_1 | \lambda_2, \mathbf{k}_2 \rangle = \frac{1}{\Omega_0} \int_{\Omega_0} u^{\star}_{\lambda_1\mathbf{k}_1}(\mathbf{r}) u_{\lambda_2\mathbf{k}_2}(\mathbf{r})\, d^3r \tag{4.18}$$

The averaged charge and current densities are analogously derived from the microscopic expressions, Eq. (4.13),

$$\rho(\mathbf{R}) = \frac{e}{V} \sum_{\lambda_1,\lambda_2,\mathbf{k}_1,\mathbf{k}_2} e^{-i(\mathbf{k}_1 - \mathbf{k}_2)\mathbf{R}} \langle \lambda_1 \mathbf{k}_1 | \lambda_2 \mathbf{k}_2 \rangle a^{\dagger}_{\lambda_1\mathbf{k}_1} a_{\lambda_2\mathbf{k}_2}$$

$$\text{J}(\mathbf{R}) = \frac{e}{Vm_0} \sum_{\lambda_1,\lambda_2,\mathbf{k}_1,\mathbf{k}_2} e^{-i(\mathbf{k}_1-\mathbf{k}_2)\mathbf{R}}(\langle\lambda_1\mathbf{k}_1|\mathbf{p}|\lambda_2\mathbf{k}_2\rangle$$
$$+ \langle\lambda_1\mathbf{k}_1|\mathbf{k}_2 - \frac{e}{c_0}\mathbf{A}(\mathbf{R})|\lambda_2\mathbf{k}_2\rangle)a^{\dagger}_{\lambda_1\mathbf{k}_1}a_{\lambda_2\mathbf{k}_2} \tag{4.19}$$

It can be shown that these relations satisfy the continuity equation for currents and charges.

We still need to evaluate the unit cell integrals. The expansion of the Bloch functions around the Γ point in a direct gap semiconductor leads to a multipole expansion of the interaction. By assuming that the simple model system with a nondegenerate state at $\mathbf{k} = 0$, we can use $\mathbf{k}\cdot\mathbf{p}$ theory [16] to obtain the expansion up to first order in the wavevector as

$$|\lambda\mathbf{k}\rangle = |\lambda 0\rangle + \frac{\hbar}{m_0}\sum_{\eta\neq\lambda}\frac{|\eta\rangle\mathbf{k}\cdot\langle\eta|\mathbf{p}|\lambda\rangle}{E_\lambda - E_\eta} \tag{4.20}$$

4.2.3. Refined Treatment of the Hamiltonian

In Section 4.2.2, we presented the general outline of the basic theory. In the following sections, we now refine this discussion by including the effects of the electron spin and the quantum confinement.

4.2.3.1. Treatment of Confined Structures. By changing the alloy composition of the semiconductor, the band structure is influenced. In many cases, it can be assumed that this effect can be modeled by additional potentials $V_\lambda(R)$, which describe the space-dependent energy shift of band λ [52]. The parameter $\mathcal{H}_{\text{conf}}$ is thus given by

$$\mathcal{H}_{\text{conf}} = \sum_{\lambda\mathbf{k}_1\mathbf{k}_2}\int V_\lambda(\mathbf{R})e^{-i(\mathbf{k}_1-\mathbf{k}_2)\mathbf{R}}a^{\dagger}_{\lambda\mathbf{k}_1}a_{\lambda\mathbf{k}_2}\, d^3R \tag{4.21}$$

4.2.3.2. Inclusion of Electron Spin. By lumping the spin index together with the band index, $\lambda = (\tilde{\lambda}, s)$, and defining

$$|\lambda\mathbf{k}\rangle = |\tilde{\lambda}, \mathbf{k}\rangle|s\rangle \tag{4.22}$$

and

$$\langle\lambda_1, \mathbf{k}_1|\lambda_2, \mathbf{k}_2\rangle = \frac{1}{\Omega_0}\int_{\Omega_0} u^{\star}_{\tilde{\lambda}_1\mathbf{k}_1}(\mathbf{r})u_{\tilde{\lambda}_2,\mathbf{k}_2}\, d^3r\langle s_1|s_2\rangle \tag{4.23}$$

the spin can be included, and the results presented above remain valid. However, for a realistic description of the semiconductor band structure spin–orbit coupling has to be taken into account. This leads to a classification of bands corresponding to the total angular momentum. In the case of a semiconductor with a zinc blende crystal structure (e.g., GaAs), the valence bands at $\mathbf{k} = 0$ are p-like, which means they result from electrons with angular momentum $l = 1$. Together with the spin, these band electrons have total angular momentum $j = \frac{3}{2}$ or $j = \frac{1}{2}$. While the $j = \frac{1}{2}$ bands are split off to lower energy due to spin–orbit interaction, the $j = \frac{3}{2}$ bands are degenerate at the center of the BZ in the bulk material. However, for $\mathbf{k} \neq 0$ the degeneracy is lifted, resulting in different effective masses of the $j = \frac{3}{2}$ (heavy hole) and the $j = \frac{1}{2}$ (light hole band). The inclusion of spin–orbit interaction modifies the results for the expansion of the Bloch functions (Eq. 4.20). The perturbation is given by a generalized momentum Π, which includes effects from the spin–orbit coupling and can be written as:

$$\Pi = \mathbf{p} + \frac{\hbar}{4m_0c^2}(\sigma \times \nabla V(\mathbf{r})) \tag{4.24}$$

For the $\mathbf{k} \cdot \mathbf{p}$ expansion of the Bloch functions, degenerate perturbation theory has to be applied. However, the perturbation does not mix the degenerate states and the first-order correction is only due to remote bands [30]:

$$|\lambda\mathbf{k}\rangle = |\lambda 0\rangle + \frac{\hbar}{m_0} \sum_{E_\eta \neq E_\lambda} \frac{|\eta\rangle \mathbf{k} \cdot \langle \eta|\Pi|\lambda\rangle}{E_\lambda - E_\eta} \tag{4.25}$$

Adding an additional confinement potential lifts the degeneracy at the Γ point and leads to modified (generally nonisotropic) effective masses (cf. Fig. 4.12). Anti-crossings of the bands also occur for higher $\mathbf{k}$ values, whose description involves contributions beyond the simple parabolic effective mass approximation.

4.2.4. Linear Optics in the Dipole and Rotating Wave Approximation

4.2.4.1. Dynamics of the Polarization. We restrict the general discussion to a simple two band model and treat the optical response in the rotating wave approximation (RWA) [16]. Only the leading terms in $\mathbf{k}$ are taken into account in the expansion of Bloch functions, Eq. (4.20), corresponding to a dipole approximation on the elementary cell. In this case, the interaction Hamiltonian can be written as

$$\mathcal{H}_I = -\frac{1}{c_0}\frac{e}{m_0} \sum_{\mathbf{k}_1\mathbf{k}_2} e^{-i(\mathbf{k}_1-\mathbf{k}_2)\cdot\mathbf{R}} \mathbf{p}_{vc} \cdot \mathbf{A}(\mathbf{k}_2 - \mathbf{k}_1) a^\dagger_{v\mathbf{k}_1} a_{c\mathbf{k}_2} + h.c. \tag{4.26}$$

In the same approximation, current and charge density are given by

$$\rho(\mathbf{R}) = \frac{e}{V}\sum_{\mathbf{k}_1,\mathbf{k}_2} i(\mathbf{k}_1 - \mathbf{k}_2)\cdot \mathbf{d}_{vc} e^{-i(\mathbf{k}_1-\mathbf{k}_2)\mathbf{R}} a^{\dagger}_{v\mathbf{k}_1} a_{c\mathbf{k}_2} + h.c.$$
$$\mathbf{j}(\mathbf{R}) = \frac{e}{Vm_0}\sum_{\mathbf{k}_1,\mathbf{k}_2} \mathbf{p}_{vc} e^{-i(\mathbf{k}_1-\mathbf{k}_2)\mathbf{R}} a^{\dagger}_{v\mathbf{k}_1} a_{c\mathbf{k}_2} + h.c. \tag{4.27}$$

where $\mathbf{d}_{vc} = \langle v|e\mathbf{r}|c\rangle$ and $\mathbf{p}_{vc} = \langle v|\mathbf{p}|c\rangle$ denote the dipole and momentum matrix elements, taken at the center of the BZ. In this limit, we obtain the Coulomb matrix elements as

$$V^{\lambda_1\lambda_2\lambda_3\lambda_4}_{\mathbf{k}_1\mathbf{k}_2\mathbf{k}_3\mathbf{k}_4} = V_{\mathbf{k}_1-\mathbf{k}_3}\delta_{\mathbf{k}_1+\mathbf{k}_2,\mathbf{k}_3+\mathbf{k}_4} \times \left(e^2\delta_{\lambda_1\lambda_3}\delta_{\lambda_2\lambda_4} + (\mathbf{k}_1-\mathbf{k}_3)\cdot\mathbf{d}_{\lambda_1\lambda_3}(\mathbf{k}_1-\mathbf{k}_3)\cdot\mathbf{d}_{\lambda_2\lambda_4}\right) \tag{4.28}$$

The first term is the usual monopole–monopole interaction, which is responsible for the formation of the exciton. For the sake of consistency, however, we also considered terms up to dipole–dipole interaction in the following equations.

Now that we have defined the complete Hamiltonian, we can obtain equations of motion for the operators O with the help of the Heisenberg, equation of motion

$$\dot{O} = i/\hbar[\mathcal{H}, O] \tag{4.29}$$

The current Eq. (4.27), which acts as a source term in Maxwell's equations, is determined by the off-diagonal elements of the single particle density matrix:

$$P^{vc}_{\mathbf{k}_1\mathbf{k}_2} = \langle a^{\dagger}_{v\mathbf{k}_1} a_{c\mathbf{k}_2}\rangle \tag{4.30}$$

As they determine the polarization, these elements are referred to as interband polarization or as interband coherence in order to distinguish them from the macroscopic polarization. However, the equation of motion for $P^{vc}_{\mathbf{k}_1\mathbf{k}_2}$ leads to the common hierarchy problem of coupling to higher order correlations. Closing the equations of motion by the dynamic Hartree–Fock approximation [16], which is exact to first order in the field, yields

$$-i\dot{P}^{vc}_{\mathbf{k}_1\mathbf{k}_2} = (\varepsilon_{v\mathbf{k}_1} - \varepsilon_{c\mathbf{k}_2})P^{vc}_{\mathbf{k}_1\mathbf{k}_2} + \sum_{\mathbf{q}} V_{\mathbf{q}} P^{vc}_{\mathbf{k}_1-\mathbf{q}\mathbf{k}_2-\mathbf{q}} + \mathbf{d}_{cv}\cdot\mathbf{E}(\mathbf{k}_1-\mathbf{k}_2) + \sum_{\mathbf{q}}\left(V_v(\mathbf{k}_1-\mathbf{q})P^{vc}_{\mathbf{q}\mathbf{k}_2} - V_c(\mathbf{q}-\mathbf{k}_2)P^{vc}_{\mathbf{k}_1\mathbf{q}}\right) \tag{4.31}$$

Note that the dynamics of the interband polarization is coupled to the total electric field $\mathbf{E} = \mathbf{E}_l + \mathbf{E}_t$. This coupling to the total field, not the transverse field alone, results from the contributions by the transverse field [$\mathbf{E}_t \approx -i\omega\mathbf{A}$ in slowly variing envelope approximation (SVEA)] in H_I and the longitudinal field E_l created by the polarization. This longitudinal field stems from the dipole–dipole interaction included in H_C.

In the next step, we now separate relative and center-of-mass (COM) motion. Therefore the COM momentum $\mathbf{Q}$ and the momentum of the relative motion $\mathbf{q}$ are

$$\mathbf{Q} = \mathbf{k}_1 - \mathbf{k}_2 \tag{4.32}$$

$$\mathbf{q} = \frac{m_e}{M}\mathbf{k}_1 + \frac{m_h}{M}\mathbf{k}_2 \tag{4.33}$$

where m_e and m_h denote electron and hole mass and $M = m_e + m_h$ is the exciton mass. With these definitions, the equation of motion takes the form:

$$\begin{aligned} -i\dot{P}^{vc}(\mathbf{q},\mathbf{Q}) &= (-\varepsilon_{\text{gap}} - \varepsilon_{\mathbf{q}} - \varepsilon_{\mathbf{Q}})P^{vc}(\mathbf{q},\mathbf{Q}) + \sum_{\mathbf{q}'} V_{\mathbf{q}'}P^{vc}(\mathbf{q}-\mathbf{q}',\mathbf{Q}) \\ &+ \mathbf{d}_{cv}\cdot\mathbf{E}(\mathbf{Q}) + \sum_{\mathbf{Q}'}\Big(V_v(\mathbf{Q}')P^{vc}\Big(\mathbf{q} - \frac{m_e}{M}\mathbf{Q}',\mathbf{Q}-\mathbf{Q}'\Big) \\ &- V_c(\mathbf{Q}')P^{vc}\Big(\mathbf{q} + \frac{m_h}{M}\mathbf{Q}',\mathbf{Q}-\mathbf{Q}'\Big) \end{aligned} \tag{4.34}$$

where ε_{gap} denotes the gap energy and ε_q and ε_Q are the energies corresponding to relative motion and COM motion. Equation (4.34) shows that the electric field couples to the COM motion of the excitons. In Eq. (4.35), we analyze Eq. (4.32) to describe excitons in a bulk system and in a QW.

For the description of bound electron–hole states, it is useful to expand Eq. (4.32) into solutions φ_α of the Wannier equation:

$$P^{vc}(\mathbf{q},\mathbf{Q}) = \sum P_\alpha(\mathbf{Q})\varphi_\alpha(\mathbf{q}) \tag{4.35}$$

If we focus only on the energetically lowest exciton, neglect all additional couplings, and transform into real-space, we arrive at

$$\begin{aligned} -i\hbar\partial_t P_{1s}^{vc}(\mathbf{R},t) &= \left(-\varepsilon_{1s} + \frac{\hbar^2\triangle_R}{2M} - V(\mathbf{R})\right)P_{1s}^{vc}(\mathbf{R},t) \\ &+ \varphi_{1s}^*(\mathbf{r}=0)\mathbf{d}_{cv}\cdot\mathbf{E}(\mathbf{R},t) \end{aligned} \tag{4.36}$$

where the potential $V(\mathbf{R})$ is given by [31]

$$V(\mathbf{R}) = \int d^3r\ \varphi_{1s}^2(\mathbf{r})\left(U^c\left(\mathbf{R} - \frac{m_v}{M}\mathbf{r}\right) + U^v\left(\mathbf{R} - \frac{m_c}{M}\mathbf{r}\right)\right) \tag{4.37}$$

Equation (4.36) describes the coherent, linear dynamics of excitons in a 3D semiconductor with additional static potentials U^v and U^c. However, we should keep in mind that this representation of the polarization as a product of excitonic wave function and COM motion is valid only if the contribution of other exciton eigenfunctions can be neglected.

4.2.4.2. Dynamics of the Density. In Section 4.2.4.1, equations of motion for the interband polarization in the linear regime were derived. The occupation numbers for electrons and holes, which are related to the diagonal elements of the single-particle density matrix,

$$f^e_{\mathbf{k}_1\mathbf{k}_2} = \langle a^\dagger_{c\mathbf{k}_1} a_{c\mathbf{k}_2}\rangle \tag{4.38}$$

$$f^h_{\mathbf{k}_1\mathbf{k}_2} = 1 - \langle a^\dagger_{v\mathbf{k}_1} a_{v\mathbf{k}_2}\rangle \tag{4.39}$$

do not contribute in linear order to the optical field. However, by evaluating the equations of motion up to second order in the electric field (χ^2) the following conservation law can be obtained [32]:

$$\frac{d}{dt} f^e_{\mathbf{k}_1\mathbf{k}_2} = \frac{d}{dt}\sum_{\mathbf{k}} P^{cv}_{\mathbf{k}_1\mathbf{k}} P^{vc}_{\mathbf{k}\mathbf{k}_2} \tag{4.40}$$

$$\frac{d}{dt} f^h_{\mathbf{k}_1\mathbf{k}_2} = \frac{d}{dt}\sum_{\mathbf{k}} P^{vc}_{\mathbf{k}_1\mathbf{k}} P^{cv}_{\mathbf{k}\mathbf{k}_2} \tag{4.41}$$

If we assume that for early times (before the optical excitation) no polarization or occupation density exists, these relations do not only hold for the time derivatives, but also for the populations and polarizations themselves.

In the case of a pure $1s$ excitonic response, if we factor analogous to Eq. (4.35) and transformation into real-space, the spatial occupation function of electron and hole states is obtained as

$$f^{e/h}(\mathbf{R}) = \frac{M^2}{2m^2_{h/e}} \int d^3r \left| P^{vc}_{1s}(\mathbf{R}+\mathbf{r})\varphi\left(\mp\frac{M}{m_{h/e}}\mathbf{r}\right)\right|^2 \tag{4.42}$$

The occupation is hence determined by the square of the interband polarization folded with the square of the excitonic wave function. The distribution of electrons and holes corresponds to their distribution within the exciton, which is determined by their different masses. Equation (4.42) suggests the interpretation of

$$n_X(\mathbf{R}) = \left|P^{vc}_{1s}(\mathbf{R})\right|^2 \tag{4.43}$$

as the excitonic density.

4.2.4.3. Inclusion of Strong Confinement. In this section, we now extend the discussion to mesoscopic structures in which the carriers are strongly quantum confined.

In this case, it is convenient to transform Eq. (4.31) into real-space. Conduction and valence band are approximated by parabolic band shapes with effective masses m_e and m_h, respectively. We define Φ as the FT of P^{vc}, $\Phi(r_h, r_e) = \mathrm{FT}\{P^{vc}(\mathbf{k}_1, -\mathbf{k}_2)\}$, such that Φ fulfills a two-particle Schroedinger equation:

$$\begin{aligned}
-i\hbar\partial_t\Phi(\mathbf{r}_h, \mathbf{r}_e, t) = & \left(-\varepsilon_{\mathrm{gap}} + \frac{\hbar^2\Delta_h}{2m_h} + \frac{\hbar^2\Delta_e}{2m_e}\right)\Phi(\mathbf{r}_h, \mathbf{r}_e, t) \\
& + V_{Cb}(r_e, r_h)\Phi(\mathbf{r}_h, \mathbf{r}_e, t) - (V_c(\mathbf{r}_e) + V_h(\mathbf{r}_h))\Phi(\mathbf{r}_h, \mathbf{r}_e, t) \\
& + \mathbf{d}^{vc}\cdot\mathbf{E}(\mathbf{r}_e, t)\delta(\mathbf{r}_e - \mathbf{r}_h). \qquad (4.44)
\end{aligned}$$

From this equation, we obtain Eq. (4.36) as an approximation only if the energy variation of the two-particle eigenstates is small compared to the excitonic binding energy. This may be the case if either the potential is weak or if it varies only on long scales.

In semiconductor structures of reduced dimensionality, one can distinguish between the strong confinement situation that effectively reduces the dimensionality of the electronic motion, in contrast to additional static potentials that can possibly be treated in the limit of Eq. (4.36). To describe QW excitons we assume a strong confinement potential, which restricts the motion of carriers in the z direction. If the width of the well is small compared to the Bohr radius of the system, the z dependence in the Coulomb potential can be neglected such that we are left with a 2D Coulomb interaction. In this case, the in-plane motion and the motion in the growth direction can be separated as well, which leads to different subbands belonging to different electron and hole states in the z-potential $V_\lambda(z)$. The resulting equation of motion for the QW excitons differs from Eq. (4.36) only in that the variable $\mathbf{R}$ is restricted to the QW plane. The parameter φ^*_{1s} denotes the 2D excitons wave function [16] and coupling to the field is given by $\mathbf{d}_{vc}\cdot\mathbf{E}(\mathbf{R}, t)|\xi(z)|^2$, where $\xi(z)$ is the confinement function for electrons and holes. As the EM field varies on the scale of the wavelength, $|\xi(z)|^2$ can be approximated by a δ function.

4.2.5. Linear Optics Beyond the Dipole Approximation

Due to the strong spatial field gradient below a near-field tip, higher order multipole moments on the length scale of the elementary cell become increasingly important. In this section, we therefore extend our analysis of the semiconductor-light coupling beyond the dipole approximation on the elementary cell level. However, note that in the treatment of mesoscopic structures there is not one but two important scales: (1) the dimension of the elementary cell and (2) the dimension of the mesoscopic structure. Here, we treat the multipole expansion of the fields within the scale of the elementary cell. On the next level, the response of the entire mesoscopic structure to its total response may again be classified in terms of multipoles. Thus, in a mesoscopic structure there may be transitions that are dipole forbidden even though they belong to interband transitions with nonvanishing dipole matrix elements. An example is treated in more detail in Section 4.3.4.5.

We obtain the higher order contributions on the scale of the elementary cell by including the expansion of the Bloch functions up to first order [see Eq. (4.20) for nondegenerate bands]. By assuming resonant exciton excitation, the current Eq. (4.19) is determined by

$$\mathbf{j}(\mathbf{R},t) = \frac{\partial}{\partial t}\sum_{\lambda\gamma}\Big[\mathbf{d}_{\lambda_1\lambda_2} + \Big(\nabla_{\mathbf{r}} + \xi_{\lambda_2}^{\lambda_1\lambda_2}\nabla_{\mathbf{R}}\Big)\cdot \boldsymbol{B}_{\lambda_1\lambda_2}^{\lambda_1}$$
$$+ \boldsymbol{B}_{\lambda_1\lambda_2}^{\lambda_2}\cdot\Big(\nabla_{\mathbf{r}} - \xi_{\lambda_1}^{\lambda_1\lambda_2}\nabla_{\mathbf{R}}\Big)\Big]\,\varphi_{\gamma\mathbf{r}=0}^{\lambda_1\lambda_2}\,P_{\gamma}^{\lambda_1\lambda_2}(\mathbf{R},t) \qquad (4.45)$$

Here, we introduced the following definitions [33]:

$$\xi_{\lambda_1}^{\lambda_1\lambda_2} = \frac{m_{\lambda_1}}{m_{\lambda_1}+m_{\lambda_2}} \quad \text{and} \quad \xi_{\lambda_2}^{\lambda_1\lambda_2} = \frac{m_{\lambda_2}}{m_{\lambda_1}+m_{\lambda_2}} \qquad (4.46)$$

and the matrices $\boldsymbol{B}_{\lambda_1\lambda_2}^{\lambda}$ with elements

$$B_{\lambda_1\lambda_2}^{\alpha\beta\lambda} = e\sum_{v}\frac{\langle\lambda_1|p^{\alpha}|v\rangle|p^{\beta}|\lambda_2\rangle}{m_0^2(\varepsilon_{\lambda_1}-\varepsilon_{\lambda_2})(\varepsilon_{\lambda}-\varepsilon_{v})} \qquad (4.47)$$

Equation (4.45) contains distributions from the dipole allowed transitions described by the dipole matrix element $\mathbf{d}_{\lambda_1\lambda_2}$ and higher order terms that are related to the two photon absorption coefficients $\boldsymbol{B}_{\lambda_1\lambda_2}^{\lambda_1}$ and $\boldsymbol{B}_{\lambda_1\lambda_2}^{\lambda_2}$ [34], where can be viewed as quadrupole or magnetic dipole contributions. The latter have two contributions. The first one is given by the interband coherence and the spatial gradient of the exciton wave function at $\varphi(\mathbf{r}=0)$, which is nonzero only for excitons with a p-like wave function. The second contribution is given by the derivative of the interband polarization and the $\varphi(\mathbf{r}=0)$ value of the exciton wave function, which is nonzero only for s-like excitons.

The equation of motion for the interband coherence derived within the same approximations as in Section 4.2.4.3 (RWA, SVEA, linear optics), but here also including the higher order coupling, is obtained as

$$-i\partial_t P_{\gamma}^{vc}(\mathbf{R},t) = \Big(-\varepsilon_{\gamma} - \frac{\triangle_R}{2M^{cv}} + V(\mathbf{R})\Big)P_{\gamma}^{vc}(\mathbf{R},t) + \varphi_{\gamma\mathbf{r}=0}^{*}\mathbf{d}_{vc}\cdot\mathbf{E}(\mathbf{R},t)$$
$$+\Big[\Big(\nabla_{\mathbf{r}} - \xi_{\lambda_1}^{\lambda_1\lambda_2}\nabla_{\mathbf{R}}\Big)\varphi_{\gamma\mathbf{r}=0}^{*}\cdot\boldsymbol{B}_{cv}^{c} + \boldsymbol{B}_{cv}^{v}\cdot\Big(\nabla_{\mathbf{r}} + \xi_{\lambda_2}^{\lambda_1\lambda_2}\nabla_{\mathbf{R}}\Big)\varphi_{\gamma\mathbf{r}=0}^{*}\Big]\mathbf{E}(\mathbf{R},t) \qquad (4.48)$$

This equation shows that field coupling occurs not only via the dipole moment but also in connection with the two-photon absorption coefficient. This contribution consists of terms coupling to both the electric field and the gradient of the field. The structure of this term is similar to the expression for the current: the p excitons couple to the electric field, the s excitons couple to the gradient of the field. Therefore their relative contribution is dependent on the ratio of the field to its derivatives. In Section 4.3.6, an example for the modification of quadrupole-like transitions in the near field is discussed.

4.3. APPLICATIONS

In this section, we apply the theory to study coherent near-field excitation of semiconductor structures, that is, we focus on purely radiative phenomena where the expectation value of the EM field is finite (classical field). We assume as our standard situation the configuration where the semiconductor structure under investigation is situated below a subwavelength aperture in a metallic screen. In order to characterize the field distributions, we analyze first the optical properties of the aperture configuration alone. Details of the geometry are described in Section 4.3.1 and the near-field distribution created by the external excitation is computed in Section 4.3.2. As an additional preparatory analysis, we discuss the detection process, that is, the transformation of the signals from the near to the far field (Section 4.3.3). The results of these studies are then utilized to analyze the excitation of localized excitons (Section 4.3.4), to derive selection rules (Sections 4.3.5 and 4.3.6), and to study the excitation of free excitons, including the formation of excitonic wavepackets (Section 4.3.7).

4.3.1. Model Configuration

Most near-field probes used in experiments consist of a metallized fiber tip [35,36] or a thin metallic aperture [6,25] on top of the sample, which in our case is a semiconductor structure. To model such a situation, the near field is assumed to be generated by the transmission of a plane wave, or a short optical pulse, through a circular aperture in a perfectly conducting screen. This aperture forms the near-field probe situated within subwavelength distance from the semiconductor sample (cf. Fig. 4.1). As a representative case we study the situation where the diameter of the aperture is $d_a = 60\,\text{nm}$. This diameter is much smaller than the incident wavelength of $\lambda = 826\,\text{nm}$ if we discuss resonant excitation near the fundamental band edge of GaAs. The distance between aperture and dielectric material is chosen to be $d_m = 10\,\text{nm}$ (or $d_m = 0\,\text{nm}$ if we discuss the direct metal coating of a sample). The QW is assumed to be situated at $d_{qw} = 30\,\text{nm}$ below the aperture, and 20 nm under the surface of the dielectric material (30 nm in the case of a direct coating of the sample). Besides this 3D configuration, additional results are presented for a 2D model (Section 4.3.7), where the aperture has the shape of a slit infinitely extended in the x direction, and has a width $d_a = 60\,\text{nm}$ in the y direction.

To investigate the described situation, we solve the macroscopic Maxwell equations (4.14) for the electric and magnetic fields, $\mathbf{E}(\mathbf{r}, t)$ and $\mathbf{B}(\mathbf{r}, t)$, respectively.

Equation (4.14) includes the aperture geometry, the resonantly excited semiconductor material (excitonic response close to the fundamental band edge) and the nonresonant dielectric host material (semiconductor barrier). Aperture and barrier material are described by the local, frequency independent dielectric function $\varepsilon(\mathbf{r})$. The current $\mathbf{j}(\mathbf{r}, t)$, Eq. (4.19), occurring in Maxwell's equations results from the resonantly excited excitons and is the relevant quantity linking Maxwell's equations with the semiconductor material equations. In contrast to far-field excitation, the near field cannot simply be modeled as a purely external field. Instead, the aperture has to be included in the dynamical description because the sample generated fields influence the near-field probe and vice versa [37,38]. This situation is very different from the conventional scenario in far-field optics, where it can be assumed that the infinitely distant sources of the external fields are completely decoupled from the sample.

In the following list, we split the discussion into three parts:

1. The excitation through the near-field probe without any resonant sources. This is the problem of transforming the incoming field into the near field of the probe.
2. The transformation of the sample near field into the far field. This is the detection problem.
3. The dynamics of the resonantly excited excitons in the near-field setup. Here we focus on excitons in localized states, the modification of selection rules in the case of heavy-hole (hh) and light-hole (lh) transitions in GaAs, on dipole forbidden quadrupole transitions in Cu_2O, and on the nonlocal excitonic response in a GaAs quantum well.

To numerically solve the coupled Maxwell and polarization equations, a finite difference time domain method (FDTD) is applied [33,39] (cf. the appendix). The equations are discretized in time with $\Delta t \simeq 0.01$ fs and in space with $\Delta r \simeq 10$ nm. The numerical solution for a monochromatic incident wave yields the stationary field distribution, whereas pulsed excitation yields the time-resolved information about the field and the polarization current.

4.3.2. Excitation through an Aperture

4.3.2.1. Ideal Aperture in Vacuum. Here we consider the diffraction of a plane wave with amplitude E_0, traveling in the z direction and polarized in the x direction by a small circular aperture with radius $a = d_a/2$ in an ideal metallic screen. For this case, analytic results are available for the field distribution within the aperture [40]. Special attention has to be given to the boundary conditions at the rim where the EM field diverges due to the ideal metallic properties. For apertures that are small in comparison to the wavelength ($a \ll \lambda$), the field distribution within the aperture ($z = 0$) can be expanded in orders of ak, where $k = 2\pi/\lambda$ denotes the wavevector of the incident light. The leading term is given by

$$\frac{E_x}{E_0} = -\frac{8ik}{3\pi}\left(\sqrt{a^2-\rho^2} + \frac{\rho^2}{2\sqrt{a^2-\rho^2}}\cos^2\varphi\right)$$
$$\frac{E_y}{E_0} = -\frac{8ik}{3\pi}\frac{\rho^2}{2\sqrt{a^2-\rho^2}}\cos\varphi\,\sin\varphi$$
$$\frac{E_z}{E_0} = 0 \tag{4.49}$$

Here, ρ and φ denote cylindrical coordinates with their origin at the center of the aperture. It can be seen that the electric field diverges at the rim of the aperture in the polarization direction of the incoming light (x polarized, thus $\varphi = 0$). The electric field strength at the center is given by $-8ika/3\pi$, which is approximately a factor of ka smaller than the field of the incoming beam. This indicates that the transmission through a small aperture is not only limited by the geometrical hole size, but for a radius much smaller than the wavelength it is also strongly suppressed due to the boundary conditions of the metal interface. This is in contrast to Kirchhoff diffraction theory where generally the field in the aperture is set equal to the incident field [41]. The magnetic field in the aperture is given by [40]

$$\frac{H_x}{H_0} = 0$$
$$\frac{H_y}{H_0} = 1$$
$$\frac{H_z}{H_0} = -\frac{4}{\pi}\frac{\rho}{\sqrt{a^2-\rho^2}}\sin\varphi \tag{4.50}$$

At the center of the aperture, the ratio of magnetic-to-electric field is enhanced by $H_y/E_x \approx ka$, which indicates that the aperture may have an impact on magnetic spectroscopy.

The fields in front of and behind of the aperture are uniquely defined by the distribution in the aperture [41]:

$$\mathbf{E}_{\rm ap}(\mathbf{r}) = \pm\frac{1}{2\pi}\nabla\times\int_{\rm ap}(\mathbf{e}_z\times\mathbf{E}(\mathbf{r}'))\frac{e^{ik|\mathbf{r}-\mathbf{r}'|}}{|\mathbf{r}-\mathbf{r}'|}\,d^2r' \tag{4.51}$$

where the integration only runs over the aperture. Behind the screen ($z > 0$), the upper sign holds and the total field is given by Eq. (4.51). In front of the screen, the total field is the sum of incident field, the field reflected by the screen with the aperture closed, and the electric field given by Eq. (4.51) with the lower sign. While Eq. (4.51) represents an exact expression, it can be expanded in multipoles. Within the expansion $|\mathbf{r}-\mathbf{r}'| \approx \mathbf{r} - \mathbf{r}'\cdot\mathbf{r}/|\mathbf{r}|$, the fields created by a small aperture are determined by an electrical dipole moment $\mathbf{p}$ and a magnetic dipole moment $\mathbf{m}$

$$\mathbf{p} = \frac{\mathbf{e}_z}{4\pi} \int_{\text{ap}} (\mathbf{r}' \cdot \mathbf{E}_{\tan}(\mathbf{r}'))\, d^2r' \tag{4.52}$$

$$\mathbf{m} = \frac{1}{2\pi ik} \int_{\text{ap}} (\mathbf{e}_z \times \mathbf{E}_{\tan}(\mathbf{r}'))\, d^2r' \tag{4.53}$$

where $\mathbf{E}_{\tan}$ is the electric field vector tangential in the plane of the metal screen. Thus, the aperture can be viewed in a way that the incident field creates an electric dipole that is directed perpendicular to the screen, and a magnetic dipole that is oriented in the plane of the screen. The corresponding fields behind the aperture can be written as

$$\mathbf{E}_{\text{dip}}(\mathbf{r}) = k^2 \left((\mathbf{n} \times \mathbf{p}) \times \mathbf{n} + (3\mathbf{n}(\mathbf{n} \cdot \mathbf{p}) - \mathbf{p}) \left(\frac{1}{r^2} - \frac{ik}{r} \right) \right) \frac{e^{ikr}}{r} \tag{4.54}$$

$$\mathbf{E}_{\text{mag}}(\mathbf{r}) = -k^2 (\mathbf{n} \times \mathbf{m}) \frac{e^{ikr}}{r} \left(1 - \frac{1}{ikr} \right) \tag{4.55}$$

with $\mathbf{n} = \mathbf{r}/r$. Note that the field in the transmission region is the field excited by the sources $\mathbf{m}$ and $\mathbf{p}$, whereas in the reflection region the fields are those emitted by $-\mathbf{m}$ and $-\mathbf{p}$.

For the discussion of the field distribution, one can distinguish between three spatial regions [42]: (1) the vicinity of the aperture ($r < d_a$) in which the multipole approximation does not hold, (2) an intermediate region in which the fields are described by the multipoles defined above, and (3) a far-field region ($r \gg \lambda$) in which only the radiation parts of Eqs. (4.54) and (4.55) proportional to $1/r$ are important. For the excitation with a plane wave, with a field distribution given by Eq. (4.49), the electric dipole moment, Eq. (4.52), vanishes and the magnetic dipole is given by

$$\mathbf{m} = -\frac{4}{3\pi} a^3 \mathbf{e}_y \tag{4.56}$$

In this situation, the far field of the aperture acts like a magnetic dipole source with the total radiated power

$$s = \frac{c}{8\pi} \int \boldsymbol{E} \times \boldsymbol{H}^* = \frac{c}{8\pi} \frac{64}{27\pi} k^4 a^6 \tag{4.57}$$

This is approximately a factor of $(ka)^4$ smaller than the power flux of the incident wave through the area of the aperture.

As pointed out earlier, the combination of near-field optical techniques and short pulse excitation is a very promising concept. For the analysis, the influence of the aperture on the shape of a short light pulse has to be examined. Equation (4.49) shows that the field distribution in the aperture (and also in the vicinity) has a linear dependence on the frequency of the incoming light. In the time domain, this means that the shape of the pulse is the original shape plus an additional term proportional to the derivative of the pulse envelope. Thus the relative change is approximately proportional to the pulse duration/optical cycle ratio. For pulses with typical durations of 100 fs, this leads only to a small modification in the range of a few percent. However, this effect is only that small since we assumed an infinitely thin screen. If the screen is thicker or comparable to the wavelength, we can think of the aperture as a small waveguide with a certain cut off frequency, which leads to a strong change of the transmission in the range of the cutoff. Effects of this kind have, for example, been observed for teraherz radiation through an aperture [43].

The results presented above describe the excitation through an aperture in a homogeneous background without any additional current sources or spatially inhomogeneous dielectrics (Fig. 4.1). Inclusion of the background material leads to modifications. However, as we assume no inhomogeneity on the side of the incident field of the screen, Eq. (4.51) is applicable to obtain the reflected field from the field distribution in the aperture.

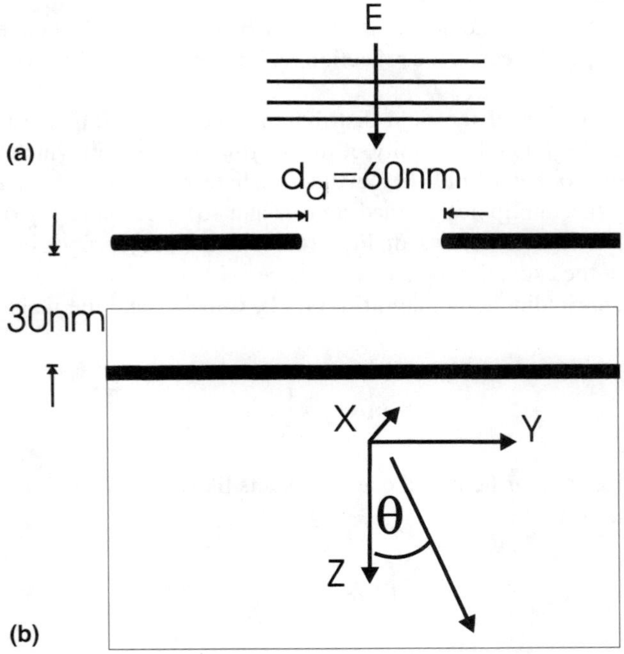

Figure 4.1. Schematics of the configuration investigated in this section: An aperture in an ideal metal film is placed above a semiconductor sample. The semiconductor is optically excited through the aperture by a plane wave, with propagation direction parallel to the z axis. Either a circular aperture with diameter $d = 60\,\mathrm{nm}$ or an infinitely extended slit in the x direction of width $d = 60\,\mathrm{nm}$ is discussed.

4.3.2.2. Aperture above a Semiconductor Substrate. In this section, we discuss the circular components of the field distribution since this allows a convenient projection on the dipole moment vectors for transitions in the semi-conductor material. The results are obtained numerically from FDTD calculations. Assuming a circularly polarized σ^+ monochromatic plane wave $E = E_0 e^{-i(\omega t - k_z z)}$ with the photon energy $\hbar\omega = 1.5\,\text{eV}$ (excitonic transition energy of GaAs) as the incident field [cf. Fig. 4.1(a)], we calculate the stationary field distribution for the vector components of the electric field $|E^j(\mathbf{R}, z)|$ in the x, y plane for different values of z below the aperture in an ideal conducting metal. The index $j = +, -, z$ denotes the vector components of the electric field that we choose as $\sigma^\pm = 1/\sqrt{2}(1, \pm i, 0)$ and $\sigma^z = (0, 0, 1)$, where σ^z is the vector component in the propagation direction of the field and $\sigma^\pm$ are the two circular polarization components in the plane of the aperture. Here, no resonant sources due to semiconductor structures are taken into account. Figure 4.2 shows the computed normalized field distribution in the plane of the aperture at a distance of $z = 30\,\text{nm}$ below the aperture including the influence of the background material. At this relatively small distance, the peak ratio between the incident field and the transmitted σ^+ component $|E^+(\mathbf{R} = \mathbf{0})|/|E_{\text{inc}}|$ is already below 0.06. Furthermore, the transmitted field does not only contain the incident σ^+ component but also the σ^- and σ^z components. These new vector contributions are generated by light scattering at

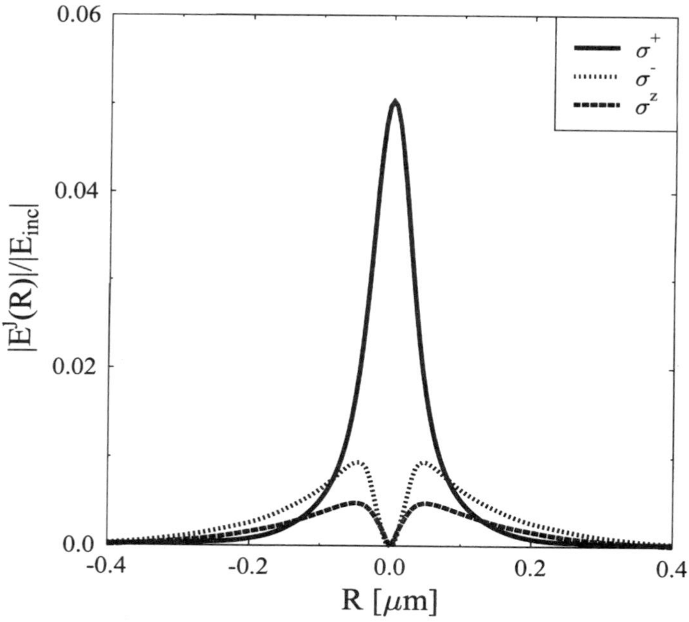

Figure 4.2. Field distribution in a plane of the dielectric barrier material (dielectric constant $\varepsilon = 12$) at a distance of $z = 30\,\text{nm}$ below the aperture (distance between material and aperture $d_m = 10\,\text{nm}$).

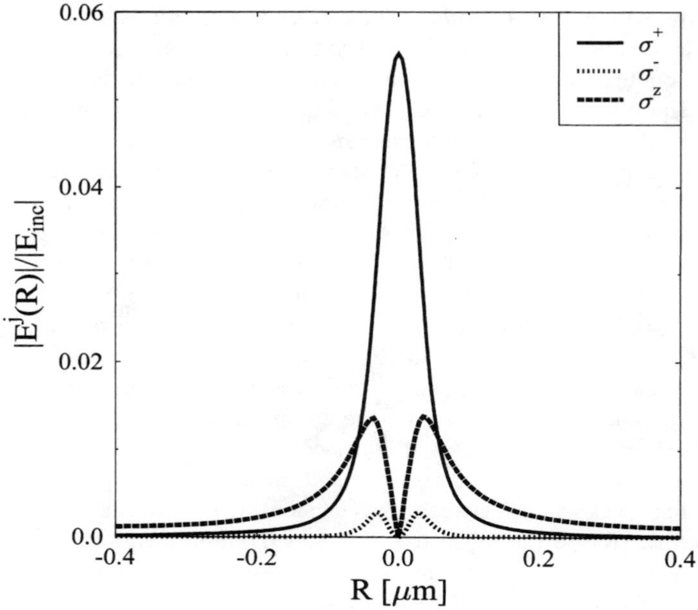

Figure 4.3. Stationary electric field distribution in the (x, y) plane of the aperture at a distance of $z = 30\,\text{nm}$ in vacuum. Additional vector components emerge from the aperture.

the subwavelength aperture. Whereas the σ^+ component of the electric field is peaked directly below the aperture, the two aperture generated components have their maxima below the edges of the metal cladding. Below the rim of the aperture, the σ^z component reaches the same order of magnitude as the σ^+ component transmitted from the incident field. Besides, this aperture induced polarization mixing, Fig. 4.2, shows that strong field gradients also appear in the x, y plane.

To estimate the influence of the background medium on the field distribution, Fig. 4.3 shows the field distribution without a background medium. Even though the distributions have qualitatively the same shape, there are some important differences. The σ^+ component is only weakly effected apart from a slightly smaller amplitude. If we compare this with the field in the center of the aperture that is given by $30\,\text{nm} * nk_0 = 0.25$, according to Eq. (4.49) we see that there is a substantial spatial decay due to the evanescent character of the field (cf. the discussion of evanescent waves in Chapter 3). For σ^- and σ^z components, there is an inversed trend in the field magnitude in comparison to the presence of the dielectric background. The σ^- is strongly enhanced within the background material. This happens because the more strongly localized modes are allowed to propagate more effectively in the high refractive index material than in a vacuum. The modes are evanescent for transverse confinement higher than the corresponding wavelength, which is in the background material a factor of $n = \sqrt{\varepsilon}$ higher. As the σ^- component is strongly localized under the rim of the aperture, it benefits from this enhancement more than the σ^+ mode.

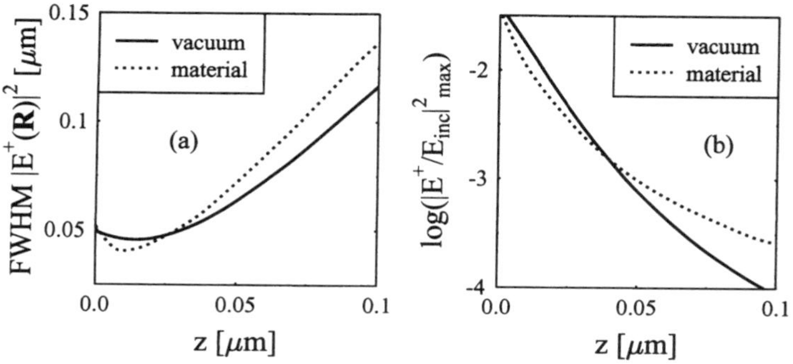

Figure 4.4. Full width at half-maximum of the in-plane intensity distribution and maximum of the intensity $|E^+(\mathbf{R})|^2$ of the σ^+ component of the field as a function of the distance z from the aperture for vacuum and a dielectric material with $\varepsilon = 12$. (a) fwhm of the σ^+-intensity and (b) maximum of the σ^+ field intensity. The fwhm can be viewed as a measure of the optical resolution and remains constant close to the aperture.

This argument should also apply to the σ^z component, but there is another, in this case even more important compensating effect. In contrast to σ^+ and σ^-, which are continuous at the surface of the medium, the normal component of the electric field σ^z suffers a decrease by a factor ε, which reduces the z component in the semiconductor material. Hence, we conclude that the background material has a very important influence on the vectorial field distribution and can in general not be neglected.

To investigate the optical resolution achieved by excitation through the aperture, we discuss the full width of half-maximum (fwhm) of the transmitted σ^+ intensity as a function of the distance z from the aperture [see Fig. 4.4(a)]. In the vacuum case (solid line), close to the aperture ($z \leq 30\,\text{nm}$) the width of the field distribution remains basically constant, indicating that the full resolution power of the SNOM is available up to a distance of roughly the aperture radius. For larger distances, the spot size increases almost linearly. For our parameter values at a distance of $z = 100\,\text{nm}$, the optical resolution is less than one-half of that available in the vicinity of the aperture. Concerning the influence of the dielectric material, we find no additional broadening in the vicinity of the aperture ($z \leq 30\,\text{nm}$). Due to interference effects the spot size may even be slightly reduced. However, at larger distances, the transmitted light is less focused than in a vacuum. A decrease of the resolution at $z = 100\,\text{nm}$ of $\sim 10 - 20\%$ can be expected. At the same time, the maximum of the field intensity decreases rapidly with increasing distance from the aperture. It can be recognized in Fig. 4.4(b) that during the first 50 nm below the aperture the maximum intensity is reduced by more than one order of magnitude. Inside the dielectric material the decrease is slightly weaker than in vacuum due to photon tunneling. Here, at $z = 100\,\text{nm}$, the peak ratio between incident and transmitted σ^+ intensity is rather small. In the present case it is only 0.00035 in the dielectric material.

4.3.3. Detection through an Aperture

So far, the externally excited field distributions close to the near-field probe have been investigated. This section is devoted to the analysis of the detection of the emitted radiation, that is, to calculation of observable signals that are emitted from a source below the aperture. In the far field of region I (cf. Fig. 4.1), the emitted radiation can be attributed to magnetic and electric dipole moments induced in the aperture by the resonant sources.

In order to discuss the excitation of the magnetic and electric dipole moments by a polarization density in the semiconductor, a transmission function t^{ij} is defined. As the relation between sources and induced moments is linear, the following ansatz can be made

$$m^i_{\text{ind}}(\omega) = \sum_j \int t^{ij}(\mathbf{r}, \omega) P^j(\mathbf{r}, \omega)\, d^3r \tag{4.58}$$

where the indexes refer to components corresponding to a circular basis $i, j = +, -, z$. To simplify the notation, let m^i, $i = 1 \cdots 3$, denote the triple (m^+, m^-, p^z). The induced moment m^i_{ind} by a point dipole $\mathbf{P}(\mathbf{r}') = \delta(\mathbf{r}' - \mathbf{r})\mathbf{e}^j p_0$, which is located at $\mathbf{r}$ and is oriented in direction j, is then given by $t^{ij}(r, \omega)p_0$. Thus t^{ij} describes the ratio of the induced to the exciting dipole p_0. Next, we discuss the numerically obtained dependence of the functions $t^{ij}(x, z)$ on the distance x to the center of the aperture for a fixed distance to the screen $z = -30$ nm. Figure 4.5(a) shows the transmission t for the vacuum background. The excited electric and magnetic dipole moments in the aperture, normalized to the strength of the source dipole, reach maximum values of ~ 0.05. This is in the same range as the ratio of excited to transmitted fields discussed in Fig. 4.3. The magnetic moment m^+ has a single peak at the center of the aperture with a fwhm of 80 nm, whereas the electric dipole excitation is double peaked, but reaches the same magnitude. The m^-

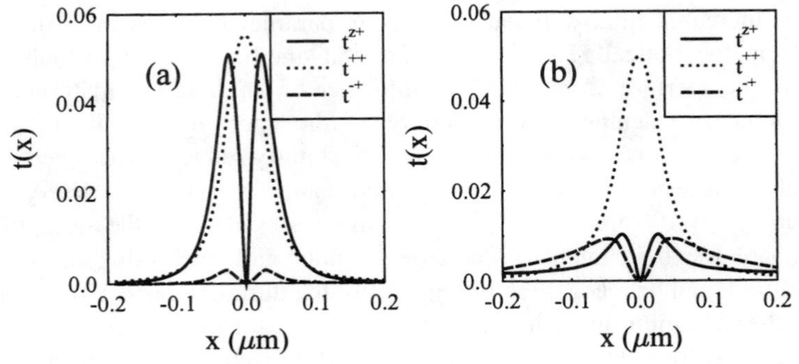

Figure 4.5. Transmission function t, which describes the excitation of moments in the aperture by a dipole source located $z = 30$ nm below the aperture. The dipole is σ^+ polarized. (a) For a vacuum backgound and (b) for a semiconductor background.

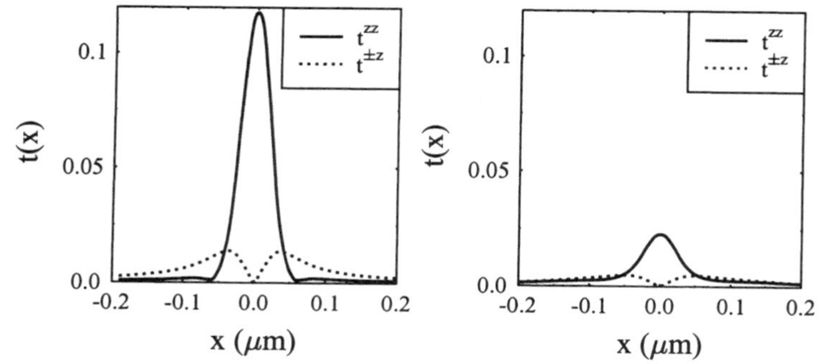

Figure 4.6. Transmission function for an exciting dipole oriented in the z direction. (a) For a vacuum background and (b) for a semiconductor background.

contribution is about an order of magnitude smaller and also shows a double peak. In comparison to the field distribution under the aperture for external σ^+ excitation, the shapes of the excitation curves are very similar if the pairs (m^+, E^+), (m^-, E^-), and (p^z, E^z) are compared. This can be explained in the following way: The field distribution under the aperture is similar to the electric field distribution of a magnetic dipole. It is known [41] that this is equal to the magnetic field distribution of an electric dipole. The reason for the similar shape of the curves is the assumption that the aperture merely couples to the magnetic field of the electric source dipole. Figure 4.5(b) shows the modification by the dielectric background material. The magnetic moment m^+ is only affected a little in comparison to the vacuum case, however, the p_z component is strongly suppressed and the m^- moment is enhanced. The effect of the background material on the excited dipole moments is very similar to the effect on the transmitted field distribution, (cf. Figs 4.2 and 4.3). Figures 4.6(a) and (b) shows the excited moments for a source polarization oriented in the z direction, with and without background material. In both cases, the excitation of the electric dipole moment in the aperture is dominant. The maximum value is approximately twice the maximum value of the m^+ excitation by a σ^+ polarized source. The magnetic moment is double peaked and about one order of magnitude smaller. Due to symmetry, both moments (left and right circular) are excited with the same strength. Taking into account the background material, Fig. 4.6(b), the excitation of moments in the aperture is strongly screened in comparison to the vacuum case. The p^z excitation is reduced by a factor of 5, whereas the m^+ and m^- are about three times smaller.

The observed intensity of the reflected light is related to the dipole moments in the far-field approximation of Eqs. (4.52) and (4.53). The two moments have different radiation characteristics. Both electric and magnetic dipoles do not radiate in the direction of their orientation. The electric dipole, oriented along the z axis, mainly radiates into the in-plane direction whereas the magnetic moments, lying in the plane of the screen, also radiate into the z direction. Thus the contribution of moments to the total signal can be changed by varying the position of the detector. In the z direction, the magnetic moments are dominant, whereas in the x, y plane both magnetic and electric moments contribute.

Let us discuss the case of a detector placed under a small angle with respect to the z axis in such a way that it only gathers light scattered by the aperture but not the reflected light of the screen. In this case, a reflectivity can be defined that describes the relative change of the intensity due to the sources below the aperture. Because the intensity is proportional to the square of the moments [Eq. (4.57)] the reflection $\mathcal{R}$ is given by

$$\mathcal{R} = \left(\frac{|\mathbf{m}_0 + \mathbf{m}_{\text{ind}}|}{|\mathbf{m}_0|} \right)^2 \tag{4.59}$$

where $\mathbf{m}_0$ denotes the moment induced by the external excitation of the aperture and $\mathbf{m}_{\text{ind}}$ is the moment induced by the resonant current sources. For excitation with σ^+ polarized light, m_0 has only a nonvanishing m_0^+ component, which is valid in the presence of the background semiconductor material as well. If it is further assumed that the induced moment $\mathbf{m}_{\text{ind}}$ is small in comparison to $\mathbf{m}_0$ Eq. (4.59) reduces to

$$\mathcal{R} = 1 + \left(\frac{m_{\text{ind}}^+}{m_0^+} + \text{c.c.} \right) \tag{4.60}$$

From Eq. (4.60), it can be recognized that the modification of the aperture reflection is proportional to the induced magnetic moment.

To discuss the excitation with a metal coated fiber tip, we assume the same basic geometry as discussed above. In this case, only light can be guided by the fiber that is totally reflected at the surface between core and cladding. For a single mode fiber, this is only the case for light incident under a very small angle to the surface as a consequence of the very small step in the refractive index. As the induced electric dipole moment of the aperture mainly radiates in the plane of the screen it can be assumed that radiation of the magnetic moment is coupled into the fiber more effectively, and the signal is dominated by the magnetic dipole radiation.

After the characterization of the near-field probe and the discussion of the calculation of observables, we now focus on the application of SNOM to semiconductors.

4.3.4. Near-Field Optics of Localized Excitons

Localized excitons, for example, in ensembles of quantum dots or in disordered heterostructures, are often studied using conventional optical techniques, such as pump–probe spectroscopy, nonlinear wave mixing, and so on. However, such experiments typically yield information only on the averaged system properties because of the spatially broad excitation spot. Since real samples usually contain distributions of dot sizes, shapes, or composition [24,44], the signals are often dominated by inhomogeneous broadening, masking many of the signatures of the individual QDs. In addition to the attempt to prepare the ultimate homogeneous samples, experimental information on the optical properties of individual dots can be obtained only by using optical techniques with subwavelength spatial resolution [6,25]. Such idealized experiments are analyzed in this section. First, we present a general discussion of the description of localized excitons. Then (Section 4.3.4.1), the exciton line shape in a spatially homogeneous background material is analyzed. In Sections 4.3.4.2 and 4.3.4.3, the investigations are extended to the inhomogeneous environment, that is, sample boundaries, including the near-field probe. Measurements of the influence of localized states on the reflection geometry are treated in three steps: First, for a mesoscopically dipole allowed transition (Section 4.3.4.4); second for a quadrupole transition (Section 4.3.4.5) with a comparison of both in Section 4.3.4.6; and third, results for the most general case including a mixture of localized and spatially extended states are presented (Section 4.3.4.7).

To model a situation with substantial inhomogeneous broadening, excitons in statically disordered bulk material or quantum wells are investigated. If the typical variation of the potential is small compared to the exciton binding energy, the motion of excitons is described by Eq. (4.36), with the additional averaged potential given by Eq. (4.37). Note that for a system where the hole mass is much larger than the electron mass, only the electron potential is averaged over the entire exciton extension, whereas the effective disorder potential for the hole is averaged over a shorter space scale. However, the lighter particle has a higher localization energy, therefore the localization potential U^c has a stronger dependence on interface roughness.

We concentrate on situations where the COM motion of $1s$ excitons is localized in individual minima of long-range disorder potentials of sufficient strength (i.e., the case of Anderson localization is not envisaged). For disorder with a correlation length on the order of the dimension of the near-field probe one then expects that the near-field response is determined by a small number of localized states in effective QDs. An expansion of the interband polarization in the stationary solutions of Eq. (4.36) is useful,

$$P_{1s}^{vc}(\mathbf{R}, \mathbf{t}) = \sum_{\mu} a_{\mu}(t)\psi_{\mu}(\mathbf{R}) \tag{4.61}$$

Here, we assume that the coordinate **R** is not restricted to the QW plane. In this case, the equation of motion for QW and bulk excitons that are weakly confined in 3D structures become formally identical. The only differences between 2D and 3D treatment are that φ in Eq. (4.36) denotes the 2D or the 3D exciton wave function, respectively, and ψ in the 3D case is a solution of the 3D Schrödinger equation normalized to the volume, and in the 2D case is the product of a solution of the 2D Schrödinger equation (normalized to the area) and the confinement function $|\xi(z)|^2$.

In both cases, the expansion coefficients fulfill the equation of motion

$$\dot{a}_{\mu}(t) = -i\omega_{\mu}a_{\mu} - \Gamma^{\text{int}}a_{\mu} + i\varphi(\mathbf{r}=0)\mathbf{d}_{vc} \cdot \int \psi_{\mu}^{*}(\mathbf{R})\mathbf{E}(\mathbf{R}, t)\, d^3R \tag{4.62}$$

where Γ^{int} describes intrinsic dephasing of the polarization (e.g., due to phonons). In the frequency domain, Eq. (4.62) yields

$$a_\mu(\omega) = -\frac{\varphi(\mathbf{r}=0)\mathbf{d}_{vc}\cdot\int\psi_\mu^*(\mathbf{R})\mathbf{E}(\mathbf{R},t)\,d^3R}{\omega-\omega_\mu+i\Gamma^{\text{int}}} \tag{4.63}$$

The occupation number of the state ψ_μ is according to Eq. (4.43) given by

$$n_\mu = a_\mu^*(t)a_\mu(t) \tag{4.64}$$

and the macroscopic polarization $\mathcal{P}$ is

$$\begin{aligned}\mathcal{P}(\mathbf{R}) &= \sum_\mu \mathbf{P}_\mu(\mathbf{R})\\ &= |\varphi_{1s}(\mathbf{r}=0)|^2\mathbf{d}_{vc}\otimes\mathbf{d}_{vc}\left(\frac{\psi_\mu(\mathbf{R})\int\psi_\mu^*(\mathbf{R}')\mathbf{E}(\mathbf{R}')\,d^3R'}{\omega-\omega_\mu+i\Gamma^{\text{int}}}+\text{c.c.}(-\omega)\right)\end{aligned} \tag{4.65}$$

where c.c. is the complex conjugate. Equation (4.63) shows that spatial and spectral overlap with the field are necessary to excite a state ψ_μ. The corresponding susceptibility is nonlocal.

If the electric field does not vary over the size of the state, Eq. (4.63) leads to the common far-field selection rules stating, for example, that in a symmetric potential no state with odd parity is excited. In the case of near-field excitation, however, the electric field may vary considerably over the extension of the state, which modifies the selection rules and may enhance forbidden transitions for localized excitons or QDs.

Due to the strong field localization, near-field optics makes it possible to study single (or a few) QDs such that their spectral properties are not masked by inhomogeneous broadening. On the other hand, the presence of the probe may have an influence on the line shapes and resonance frequencies of optical transitions. Therefore, in the following equation the line shape and its modification due to the interaction of the near-field probe with localized exciton are discussed. For this purpose, the EM field of a localized current source can be described by Green's function (field propagator) $\boldsymbol{G}(\mathbf{r},\mathbf{r}')$ of the wave equation [45] defined by

$$(\nabla^2-\nabla\nabla\cdot+k^2\varepsilon(\mathbf{r}))\boldsymbol{G}(\mathbf{r},\mathbf{r}')=\boldsymbol{\delta}(\mathbf{r}-\mathbf{r}') \tag{4.66}$$

(see also the discussion in Section 3.2.3) Here the dyadic delta function is given by $\boldsymbol{\delta}(\mathbf{r})=\delta(\mathbf{r})\sum\delta_{ij}\mathbf{e}_i\otimes\mathbf{e}_j$. Due to the vectorial character of the EM field, Green's function is a 3×3 dyade [45]. The component $G^{ij}(\mathbf{r},\mathbf{r}')$ is the ith component of the electric field at point $\mathbf{r}$ from a dipole source at $\mathbf{r}'$ oriented in direction j. In the homogeneous case ($\varepsilon=$ const), Green's function is

$$G_{ij}=-\left(\delta_{ij}+\frac{1}{n^2k^2}\partial_i\partial_j\right)\frac{\exp(ink|r-r'|)}{4\pi|r-r'|} \tag{4.67}$$

By using $\boldsymbol{G}(\mathbf{r}, \mathbf{r}')$, we can express the electric field radiated by an excited state ψ_μ as

$$\mathbf{E}_\mu(\mathbf{r}) = -4\pi k^2 \int \boldsymbol{G}(\mathbf{r}, \mathbf{r}')(\varphi_{1s}(\mathbf{r} = 0)\mathbf{d}_{vc} a_\mu \psi_\mu(\mathbf{r}') + \text{c.c.})\, d^3 r' \tag{4.68}$$

Inclusion of the interaction via the EM field into the equation of motion leads to a modified equation for the expansion coefficients a_μ:

$$\sum_\nu \left(\delta_{\mu\nu} + \frac{\Sigma_{\mu\nu}}{\omega - \omega_\mu + i\Gamma^{\text{int}}} \right) a_\nu = -\frac{\varphi_{1s}(0)\mathbf{d}_{vc} \int \psi^*(\mathbf{R})\mathbf{E}_{ap}(\mathbf{R})\, d^3R}{\omega - \omega_\mu + i\Gamma^{\text{int}}} \tag{4.69}$$

where $\mathbf{E}_{ap}$ denotes the external field only, that is, a homogeneous solution of Maxwell's equation without the resonantly excited excitons. The radiative coupling between state μ and ν is described by the self energy

$$\Sigma_{\mu\nu} = -4\pi k^2 |\varphi|^2 \int\int \psi_\mu(\mathbf{r})^* \mathbf{d}_{vc} \cdot \boldsymbol{G}(\mathbf{r}, \mathbf{r}') \cdot \mathbf{d}_{vc} \psi_v(\mathbf{r}')\, d^3r\, d^3r' \tag{4.70}$$

Taking into account the diagonal part of the coupling only, one obtains a closed equation for the amplitude

$$a_\mu = -\frac{\varphi \mathbf{d}_{vc} \cdot \int \psi_\mu^*(\mathbf{R})\mathbf{E}(\mathbf{R})\, d^3R}{\omega - \omega_\mu + i\Gamma^{\text{int}} + \Sigma_{\mu\mu}} \tag{4.71}$$

If we assume that the intrinsic line width is small, the spectral properties of Green's function do not vary over the range of the exciton line. This assumption is good as long as special design of the background dielectric ε (e.g., in a cavity), does not introduce narrow resonances of optical modes. For such a Green's function, the solution of Eq. (4.71) has a Lorentzian line shape with a modified width and resonance energy. The real part ($\Re$) of Green's function is connected to an energetic shift, which is given by

$$\delta\omega_\mu = -\Re(\Sigma_{\mu\mu}) \tag{4.72}$$

For a state that has an extension that is small compared to the optical wavelength, this shift is dominantly induced by the interaction with the longitudinal component of the EM field, as the longitudinal field diverges most strongly for small distances. In general, for a nonspherical eigenstate ψ_μ the energy shift depends on the orientation of the induced polarization relative to the state. Thus $\delta\omega_\mu$ has been discussed as the origin for line shifts observed in PL spectra for excitation in different polarization directions [6]. The imaginary part ($\Im$) of Green's function is connected to the radiative damping

$$\Gamma_{\mu}^{\text{rad}} = \Im(\Sigma_{\mu\mu}) \tag{4.73}$$

The nondiagonal parts of Eq. (4.69) describe the coupling of the QDs via the EM field. Effects like superradiance, which have been reported for atomic systems and for radiatively coupled QWs [46] have their origin in this interaction. For an efficient radiative coupling, the dot resonances need to have spectral overlap. If an ensemble of QDs is mainly radiatively damped, most QDs will have a very small line width in comparison to their energetic splittings resulting from the variations in size, composition and shape. The resulting shifts of the individual transition resonances make the realization of radiative coupling very difficult. Taking this into consideration, one can see that for efficient coupling, energy relaxation has to be possible, which means that another quasi-particle (like a phonon in the Förster process [47]) has to balance the energy conservation rule. Therefore, the description of the coherent dynamics is in Section 4.3.4.1 restricted to the diagonal part of the self-interaction.

4.3.4.1. Homogeneous Environment. In this section, the line width of excitonic states in a homogeneous environment ($\varepsilon = \text{const}$) is discussed, before we analyze the influence of additional boundaries and the near-field probe in Section 4.3.4.2.

For small quantum dots ($nkr \ll 1$ over the dot extension), Green's function for the homogeneous case, Eq. (4.67), can be expanded, and the lifetime is given by [48]

$$\Gamma^{\text{rad}} = \frac{2}{3} n k^3 d^2 |\varphi|^2 \left| \int \psi_{\mu}(\mathbf{r})\, d^3 r \right|^2 \tag{4.74}$$

This result is similar to that obtained from quantum optical calculations for atomic systems [49], when we identify $d^2|\varphi|^2| \int \psi_{\mu}(\mathbf{r})\, d\mathbf{r}|^2$ with the square of the dipole moment of the atomic transition. Thus with respect to the radiative coupling, small dots behave similar to atomic systems. For QW excitons and GaAs parameters, this leads to lifetimes in the range of 200 ps for states of 20 nm localization length [48]. Equation (4.74) is of course only applicable if the localization length of the state is larger than the Bohr radius a_0 of the material, because this is the only case where the limitation to the $1s$ exciton is valid.

If we compare the lifetime of a 2D exciton in a QW and a 3D exciton in a 3D potential we find

$$\frac{\Gamma_2}{\Gamma_3} = \frac{1}{8a_0} \frac{|\int \psi_2(x,y)\, dxdy|^2}{|\int \psi_3(x,y,z)\, dxdydz|^2} \tag{4.75}$$

If the extension of the 3D state is l_z and the 3D and 2D states both have similar transverse shapes one can approximate,

$$\frac{\Gamma_2}{\Gamma_3} = \frac{l_z}{8a_0} \tag{4.76}$$

Thus for small dots ($l_z < 8a_0$) the lifetime of three dimensionally localized excitons is higher than that of QW excitons. This can be interpreted as a consequence of the higher probability of finding an electron and a hole in the same lattice unit cell in a 2D geometry.

It is instructive to compare two circularly symmetric states, ψ_1, ψ_2, with similar shapes but different in-plane extensions in a QW:

$$\psi_1 = \psi(r) \quad \text{and} \quad \psi_2 = \alpha\psi(\alpha r) \tag{4.77}$$

Both states are assumed to be normalized. Equation (4.74) shows that

$$\Gamma_2^{\text{rad}} = \frac{1}{\alpha^2}\Gamma_1^{\text{rad}} \tag{4.78}$$

Thus, for small dots, the radiative damping is proportional to the volume of the state.

Figure 4.7 shows the computed radiative damping for a localized QW exciton in a circular potential hole with radius R. Results are also included for configurations with more extended states for which the relation $nkr < 1$ does not hold. For small R, the quadratic increase of the radiative damping with the radius can be seen [inset to Fig. 4.7]. For larger radii, the damping saturates at $\sim 1/10$ ps for radii in the range of 100 nm. The phase factor $\exp(ink|r - r'|)$ in Green's function leads to small

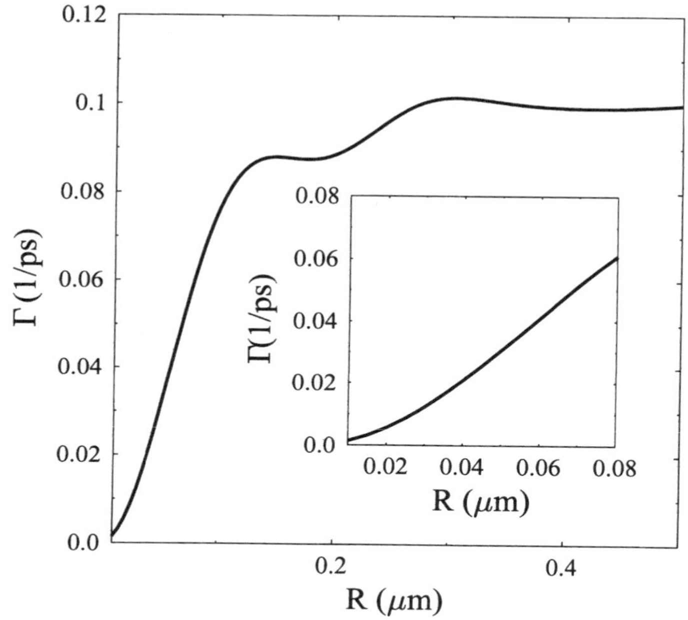

Figure 4.7. Radiative decay rate of a weakly confined QW exciton in a circular potential depression with radius R.

oscillations of the self interaction if the extension of the state varies on the scale of the wavelength.

Experimental studies of QD excitons are often performed by PL or PLE measurements. Here, lifetimes in the range of picoseconds up to nanoseconds have been reported. These lifetimes describe the decay of the exciton density. In the coherent limit, however, the damping of the density is twice the polarization damping that has been calculated here. Thus our estimates are in rough agreement with current experimental results [6].

4.3.4.2. Inhomogeneous Environment: Surfaces. Whereas the discussion of Section 4.3.4.1 focuses on localized excitons in a homogeneous environment, we now investigate which modifications are caused under near-field excitation conditions. On the one hand, there is the additional presence of the probe (e.g., a metal coated fiber tip), which is in close proximity to the confined exciton. Furthermore, the semiconductor structure of interest has to be close to the surface of the embedding background material in order to be able to use the resolution advantages of near-field spectroscopy. Thus, next we will discuss the influence of additional surfaces on the excitonic polarization. This is done by using Green's function for the EM field in the presence of boundaries. It can be decomposed into a direct part, belonging to the homogeneous environment [Eq. (4.67)] and a part reflected by the surface (cf. Section 3.2.3). In the case of an ideal metal coating of the sample, the reflected fields can be accounted for by introducing appropriate image charges. This leads to a mirror dipole at the opposite side of the metal. The mirror dipole moment depends on the polarization direction of the optically excited exciton. For polarization in (perpendicular to) the plane of the surface, the mirror dipole moment has the opposite (same) sign as the dipole moment of the exciton in the sample (QD). The total electric field at the position of the exciton is then the sum of field created in a homogeneous environment plus the field of the image dipole. If the exciton is close to the surface ($d \ll \lambda$), the fields add up destructively (in plane polarization) or constructively (polarization perpendicular to surface). Therefore for in-plane polarization the backaction of the EM field onto the polarization is decreased, which leads to a smaller radiative damping. On the other hand for polarization perpendicular to the surface the enhanced interaction with the EM field leads to a faster radiative decay.

For a nonideal metal, absorption in the lossy material is also important and must be taken into account by a realistic dielectric function for the metal. This can be included consistently by a mode expansion of Green's function [45] and taking into account a mode-dependent reflection coefficient. Here, the evanescent modes also have to be considered. This leads to integral representations, equivalent to the results first obtained by Sommerfeld [50], for the radiation of an antenna in the presence of a flat earth. For the following results, the imaginary part of Σ has been obtained by numerical integration of the integral representation of Green's function. Figure 4.8 shows the lifetime in the presence of a surface for a metal–GaAs interface (a,b) and a vacuum–GaAs interface (c,d). Figure 4.8 (a and c) are for perpendicular (to the surface) and (b and d) for parallel (to the surface) orientation of the polarization. The

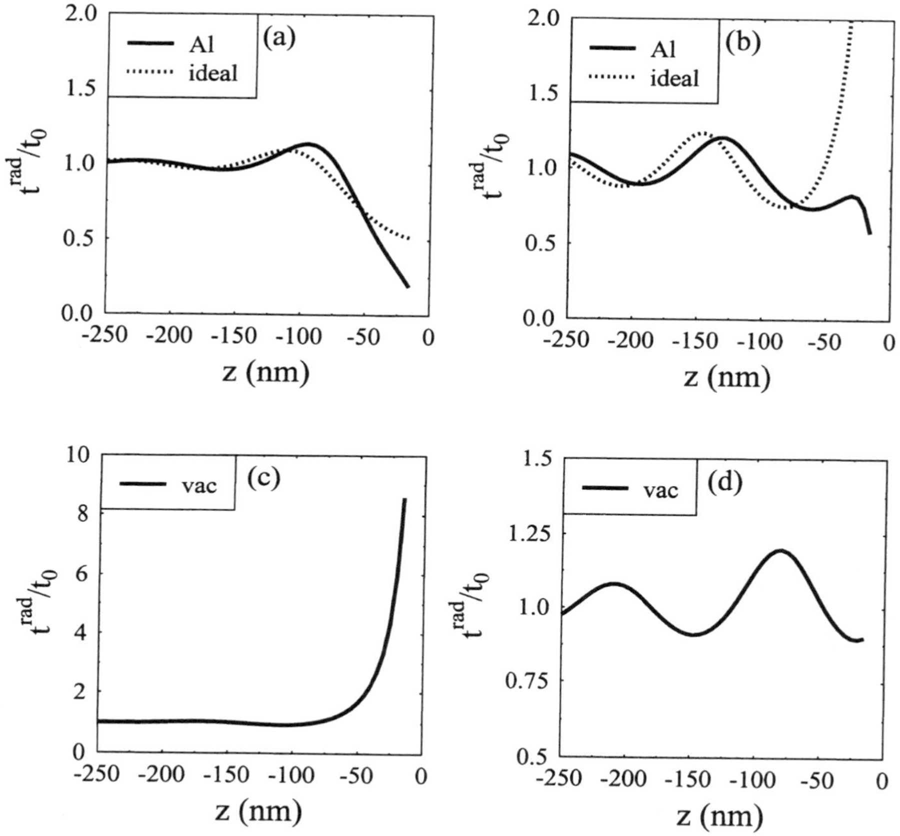

Figure 4.8. Radiative damping for an oscillating dipole embedded in GaAs in front of a surface as a function of the distance to the surface. In (a) and (b), a metal surface is considered and in (c) and (d) a vacuum boundary is considered. The dipole is oriented perpendicular [(a) and (c)] or parallel [(b) and (d)] to the surface. The damping is normalized to the radiative damping in the homogeneous GaAs environment.

lifetime is normalized to the lifetime t_0 in a homogeneous GaAs environment. Aluminum is described by a dielectric constant $\varepsilon = (2.8 + 8.45i)^2$ [51]. For distances >50 nm, the presence of the surface causes only a small deviation from t_0. It shows spatial oscillations corresponding to the half-wavelength representing the phase interference of the original and the mirror dipole. The oscillations are pronounced for polarization parallel to the surface [Figs. 4.8(b) and (d)], but only weak for perpendicular polarization. This can be understood by the far-field radiation pattern of an electric dipole, which does not radiate into its orientation direction. Therefore for polarization perpendicular to the surface, the influence of the surface on the lifetime is small. In the near field, however, this argument does not hold and for both orientations there are deviations from the lifetime t_0. For an aluminum surface, there is a decrease of the lifetime to $\sim \frac{1}{2}$ at a distance of 20 nm for

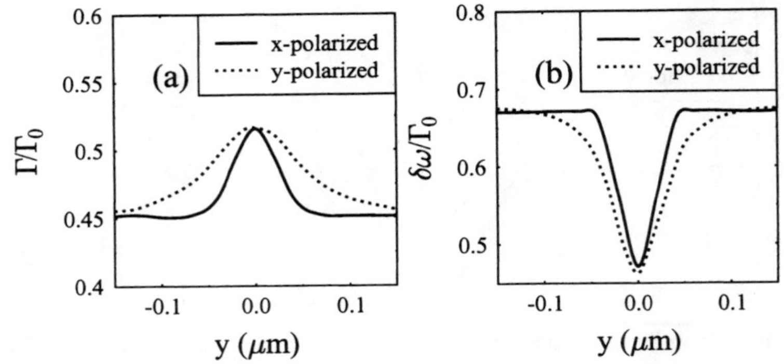

Figure 4.9. Radiative line width (a) and shift of the resonance frequency (b) due to the presence of the aperture. The curves are normalized to the line width Γ_0 in homogeneous GaAs background material. The shift refers to the deviation from the resonance frequency in the homogeneous case. The dipole is located at $z = 30\,\text{nm}$ below the aperture, transversally it is scanned along the y axis, and the polarization is oriented in the x and y direction, respectively.

both polarization directions. Only calculations for an ideal metal predict an increase of lifetime in the case of polarization parallel to the surface. In the case of a vacuum interface there are only small deviations from t_0 in the case of parallel polarization, while in the case of perpendicular polarization, however, there is a dramatic increase of lifetime.

4.3.4.3. Inhomogeneous Environment: Apertures. In general, the EM field radiated by a point dipole in front of a screen with an aperture can be divided into two parts. First, the field radiated in the presence of the screen with the aperture closed and second, the field scattered by the aperture. The contribution of the first part to the self-interaction has been discussed in Section 4.3.4.2, here the modifications by the aperture are analyzed [38]. Figure 4.9(a) and (b) show the change in radiative damping and the shift of the resonance frequency due to the presence of the aperture for a dipole source polarized in x and y direction. The position of the dipole is scanned along the y axis. The radiative damping is normalized to the damping in homogeneous background material. For a large in-plane distance between dipole and aperture ($|y| > 100\,\text{nm}$) the influence of the aperture is small. Therefore the radiative damping is that of a dipole in front of a homogeneous screen, which is a factor of 0.45 smaller than in a homogeneous environment (cf. the discussion of an ideal metal surface in Section 4.3.4.2). Directly under the aperture, however, the damping is $\sim 10\%$ higher. This can be explained by a reduced influence of the metallic screen.

Figure 4.9(b) shows the shift $\delta\omega$ of the resonance as compared to the resonance in the homogeneous environment. It is normalized to the radiative damping, that is, the line width in the homogeneous environment. Due to the surface, the resonance is shifted by approximately two thirds of the line widths. The aperture reduces this shift to one-half of the line width for a source located centrally below aperture.

For a source located at the center of the aperture due to symmetry, the influence of the aperture on the radiative damping and the resonance frequency is not dependent on the polarization direction. Therefore the same results also apply for circular polarization at $y = 0$. However, for $y \neq 0$, depending on the polarization direction, the spatial influence is more peaked if the dipole is oriented perpendicular to the scanning direction.

4.3.4.4. Observable Near-Field Signals from a QD Dipole. Generally, the polarization induced by the excitation of an excitonic state, Eq. (4.65), is nonlocal and anisotropic. However, when discussing the excitation of a state with an extension small compared to the wavelength ($nkr < 1$), its response can be expanded into multipoles. This multipole expansion is performed on the scale of the mesoscopic structure and has to be distinguished from the multipole expansion on the scale of the elementary cell discussed in Section 4.2.5. In leading order, the total polarization couples to the field at the center of the state $\mathbf{r} = \mathbf{r}_0$. In this case, an effective susceptibility can be defined that relates the total dipole $\mathbf{p}$ to the exciting field. Including the self-interaction, this effective susceptibility can be defined by the relation

$$\mathbf{p} = \chi_{\text{eff}} \mathbf{E}_{\text{ap}}(\mathbf{r}_0) \tag{4.79}$$

According to Eq. (4.65) the susceptibility χ_{eff} is then obtained as

$$\chi_{\text{eff}} = |\varphi_{1s}(\mathbf{r} = 0)|^2 \mathbf{d}_{vc} \otimes \mathbf{d}_{vc} \left(\frac{|\int \psi_\mu(\mathbf{r}')\, d\mathbf{r}'|^2}{\omega - \omega_\mu + i\Gamma^{\text{int}} + \Sigma_{\mu\mu}} + \text{c.c.}(-\omega) \right)$$

The notion effective susceptibility is related to the polarizability as introduced in Section 3.3.2.3., which also contains the self-interaction of a small dipole source via the EM field. Note, however, that $\Sigma_{\mu\mu}$ as defined here contains the self-interaction in the presence of the (inhomogeneous) environment, that is, it also includes the fields reflected from a surface.

To calculate the reflected far-field signals, the dipole moment of the aperture has to be determined. The total moment in the aperture can be split into the moment $\mathbf{m}_0$ excited by the incident beam plus the induced moment $\mathbf{m}_{\text{ind}}$ by the exciton polarization. With the excitation field given by $\mathbf{E}_{ap}$, the excited polarization (4.79), and the induced moments defined by Eq. (4.58), the reflection in Eq. (4.60) is given by

$$\mathcal{R} = 1 + \frac{(\boldsymbol{t}(\mathbf{r}_0)\hat{\chi}_{\text{eff}} \mathbf{E}^{ap}(\mathbf{r}_0))^+}{m_0^+} + \text{c.c.} \tag{4.80}$$

Due to the dyadic character of $\hat{t}$, all polarization directions can contribute to the signal, depending on the nondiagonal elements of $\hat{t}$.

In the case of an excitation of a heavy hole band, the polarization has no z component. Moreover, the contribution from σ^- is weak as $\hat{t}^{+-}$ and $\mathbf{E}_{ap}^-$ are comparably weak (cf. Figs. 4.2 and 4.58). In this case, the reflection is dominated by

$$\mathcal{R} = 1 + \frac{t^{++}(\mathbf{r}_0)\chi_{\text{eff}}^{++} E_{ap}^{+}(\mathbf{r}_0)}{m_0^+} + \text{c.c.} \tag{4.81}$$

The relative strength of the response is determined by the product of the excitation field E^+ and the excitation function t^{++}. It is strongly localized under the aperture [cf. Figs. 4.2 and 4.5(b)]. Depending on the phase of t^{++}, m_0^+, and E^+, however, the spectrum can either be dominated by the real part of the susceptibility or by the imaginary part. For $z = 30\,\text{nm}$, we find from numerical calculations that under the aperture the phase of $t^{++}E^+/m_0^+$ has nearly no transversal spatial dependence and $t^{++}E^+/m_0^+ +$ is negative and real. Thus, under the aperture the signal is given by

$$\mathcal{R} = 1 - 2\left|\frac{t^{++}(\mathbf{r}_0)E_{ap}^{+}(\mathbf{r}_0)}{m_0^+}\right| \Re(\chi_{\text{eff}}^{+}) \tag{4.82}$$

Spectrally, the reflection resembles the real part of the effective susceptibility χ_{eff}^+. Depending on the relative position of the exciton and aperture, it includes small shifts of the resonance and modifications of the line width as discussed in Section 4.3.4.3. The strength of the reflected signal is given by the product of the transmission function and the field distribution under the aperture.

4.3.4.5. Observable Signals from a QD Quadrupole. If $\int \psi_\mu(\mathbf{r}')\, d\mathbf{r}'$ vanishes, a state ψ_μ does not contribute to the response in dipole approximation. The next higher order contribution becomes especially important in this case, yielding a coupling to the gradient of the electric field. In general, the gradient of the electric field can be divided into its symmetric and its antisymmetric part. The antisymmetric part can be expressed by $\nabla \times \mathbf{E}$ and, due to Maxwell's equations, is related to the magnetic field. This leads to the definition of a magnetic susceptibility, which describes the response to the magnetic field. Similarly, a quadrupole susceptibility can be introduced to describe the response to the symmetric part of $\nabla\mathbf{E}$. Here we do not distinguish between these two contributions, because formally they can be treated in the same way. In this sense, a generalized quadrupole moment is defined by

$$\boldsymbol{Q} = \int d\mathbf{R}\mathbf{R} \otimes \mathcal{P}(\mathbf{R}) \tag{4.83}$$

The effective susceptibility $\hat{\hat{\chi}}_{\text{eff}}$ relates $\boldsymbol{Q}$ to the gradient of the electric field,

$$\boldsymbol{Q} = \hat{\hat{\chi}}_{\text{eff}} \nabla\mathbf{E}_{ap} \tag{4.84}$$

According to Eq. (4.65), the susceptibility $\hat{\hat{\chi}}_{\text{eff}}$ is given by

$$\hat{\hat{\chi}}_{\text{eff}} = |\varphi_{1s}(\mathbf{r}=0)|^2 \int \psi_\mu(\mathbf{R})\mathbf{R}\, d\mathbf{R} \otimes \mathbf{d}_{vc} \otimes \mathbf{d}_{vc} \otimes \int \psi_\mu(\mathbf{R})\mathbf{R}\, d\mathbf{R} \times \left(\frac{1}{\omega - \omega_\mu + i\Sigma_{\mu\mu}} + \text{c.c.}(-\omega) \right) \tag{4.85}$$

In analogy to Section 4.3.4.4, the reflected signal can be expressed by the transmission function and the exciting field,

$$\mathcal{R} = 1 + \frac{(\nabla t(\mathbf{r}_0)\hat{\hat{\chi}}_{\text{eff}} \nabla \mathbf{E}_{ap}(\mathbf{r}_0))^+}{m_0^+} + \text{c.c.} \tag{4.86}$$

It can be seen that in comparison to Eq. (4.80), which describes the excitation of a dipole source, the signal here is determined by the derivatives of the EM field and the derivatives of the transmission function.

In Section 4.3.4.6, a simple example is discussed to illustrate the implications of Eqs. (4.80 and 4.86).

4.3.4.6. Comparison of Dipole and Quadrupole Excitation. As shown in Section 4.3.4.5, a dipole source is driven by the electric field at the position of the dipole, whereas a quadrupole source couples to its derivatives. As a typical example we now discuss the excitation and detection of an excited dipole and quadrupole state in a QW-potential fluctuation. The potential is modeled as a box with infinitely high barriers. In this case, the ground state for an exciton is given by

$$\psi_0 = \frac{2}{l} \cos\left(\frac{\pi}{l}x\right) \cos\left(\frac{\pi}{l}y\right) \tag{4.87}$$

and one of the first excited states is

$$\psi_1 = \frac{2}{l} \sin\left(\frac{2\pi}{l}x\right) \cos\left(\frac{\pi}{l}y\right) \tag{4.88}$$

The state ψ_1 has a vanishing dipole moment and, in the case of plane wave excitation from the z direction, it would neither be excited in a dipole nor in a quadrupole approximation. However, in the case of near-field excitation both states contribute to the reflected signal. For excitation with σ^+ polarized light, with frequencies corresponding to the heavy hole interband transition, the relative strength of the reflected signal (subtracting the background from the aperture) is given by

$$\frac{\mathcal{R}_1 - 1}{\mathcal{R}_2 - 1} = \frac{(\partial_x t^{++})(\partial_x E_{ap}^+)}{t^{++} E_{ap}^+} \frac{\int \psi_1(\mathbf{r}) x d\mathbf{r}}{\int \psi_0(\mathbf{r}) d\mathbf{r}} \tag{4.89}$$

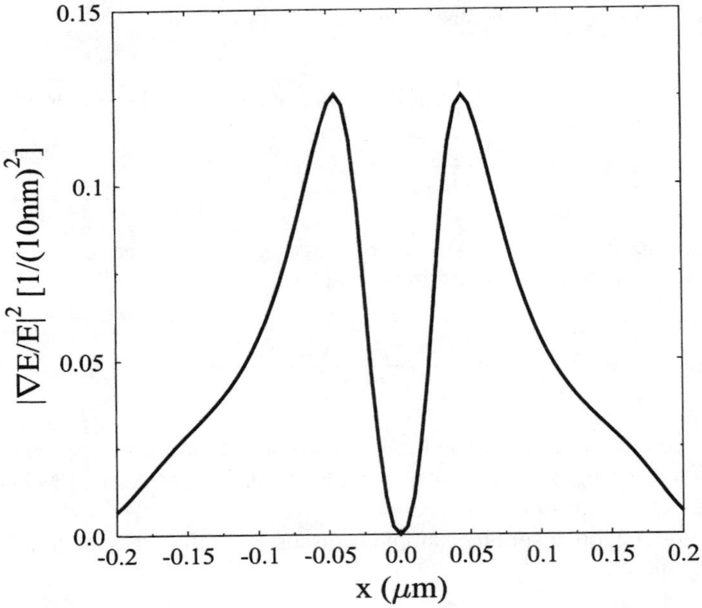

Figure 4.10. Ratio of the field derivatives to the field for the σ^+ component of the electric field distribution below the aperture. As a quadrupole couples to the derivative of the field and a dipole couples to the field itself, the ratio determines the contribution of dipole forbidden transitions to the polarization.

[cf. Eqs. (4.80) and (4.86)]. The first factor on the right-hand side (rhs) of Eq. (4.89) is determined by the geometry of the aperture. It describes the relative changes of exciting field and transmission function below the aperture. Figure 4.10 shows this factor as a function of distance on the x axis below the aperture at $z = 30\,\text{nm}$. The considered ratio has a maxima at the edges of the aperture around $x = 50\,\text{nm}$. At the center, it vanishes due to the symmetry. Thus the relative strength of the response from the excited state is maximal below the rim of the aperture.

The second factor on the (rhs) of Eq. (4.89) only depends on the states ψ_0 and ψ_1 and is roughly connected to the square of the extension of the states. With the definitions [Eqs. (4.87) and (4.88)] it is given by $l^2/16$. For $l = 30\,\text{nm}$, this leads to a maximum reflected signal from the dipole forbidden state on the order of one-tenth of the allowed transition. Due to the quadratic dependence in R, the quadrupole becomes important only if the exciting field varies considerably on the scale of the ratio of quadrupole moment to dipole moment. This is in the range of the extension of the individual system. For an atom and an aperture of 60 nm diameter, we thus expect no considerable quadrupole excitation compared to the dipole excitation. In mesocopic semiconductor structures, however, the length scale of the field variation and the length scale of the material structure can be of the same order, thus allowing for considerable modifications of the optical selection rules.

4.3.4.7. Extended and Localized States. In general, not only localized states contribute to the signal, but also extended excitonic states have to be considered, such as in a QW with a weak static disorder. For these extended states, the expansion into dipole and quadrupole terms is not useful.

As an illustration we discuss the excitation of a quantum wire embedded within a QW for the 2D configuration (cf. Section 4.3.1). The wire is oriented parallel to the slit in the metal film, the aperture is excited in TE-mode. The heavy hole polarization spectrum $|P(y, \omega)|^2$ is registered as a function of the distance y between wire and near-field source. Figure 4.11 shows the computed spectra at different positions with respect to the quantum wire. The spectra show two peaks and the relative strength of the peaks is changed by spatially scanning in the y direction over the wire. This result can be understood by remembering that spatial overlap has to be provided by the optical excitation to generate excitons in a certain quantum state. Therefore, by starting the spatial scan close to the wire ($\Delta y = 0$), the spectrum should be dominated by the first bound state in the quantum wire only ($\omega \approx \omega_{\text{wire}}$), where ω_{wire} is the resonance of the bound state. Note that slight deviations from the ideal quantum wire response occur due to the interaction of the metal and the excitonic response, yielding energetic shifts. By increasing the distance Δy of the near-field source with respect to the wire, the excitonic continuum states in the quantum well

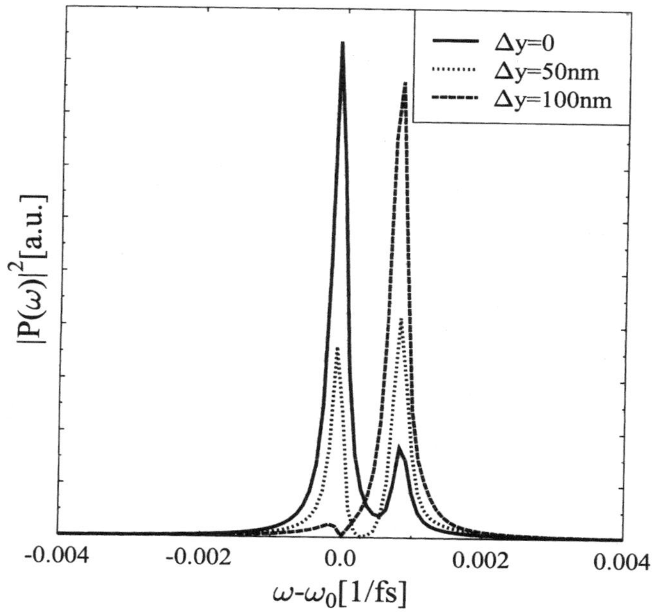

Figure 4.11. Near-field excitation of a quantum wire, embedded in a quantum well. The parameter Δy denotes the distance between wire and aperture, at $\Delta y = 0$ the wire is centered under the aperture.

become increasingly important ($\omega \approx \omega_0$). For large distances, the excitonic quantum well spectrum is recovered.

4.3.5. Excitation of Heavy- and Light-Hole Excitons in a Quantum Well

Generally, semiconductor nanostructures give rise to a spatially anisotropic response, due to the reduced symmetry induced by the confinement potential. As the near-field probe mixes the polarization and induces new components, it is well suited to test this anisotropy. We discuss a typical QW as a representative example.

The carriers are strongly confined in the growth direction of the QW. The excitonic transitions result from the interband Coulomb attraction between electrons and holes from the p-like valence and s-like conduction bands. The states at the extrema of the conduction and valence bands (Γ point of the Brillouin zone) can be classified by the quantum number of the total angular momentum j, (cf. Section 4.2.3.2). As a consequence of the reduced symmetry the degeneracy of the $|j = \frac{3}{2}, m_j = \pm\frac{1}{2}\rangle$ and the $|j = \frac{3}{2}, m_j = \pm\frac{3}{2}\rangle$ hole states is lifted in the QW and the corresponding heavy- and light-hole excitonic transitions are separated in the optical spectra. In the dipole approximation on the elementary cell, the selection rules for optical transitions are determined by the dipole matrix elements

$$\mathbf{d}_{m_j^v m_j^c} = \frac{1}{\Omega_0} \int dV u^*_{m_j^v}(\mathbf{r}) e\mathbf{r} u_{m_j^c}(\mathbf{r}) \tag{4.90}$$

where $u_{m_j^v}$ and $u_{m_j^c}$ are the Bloch functions of valence and conduction bands at $k = 0$ ($m_j^v = \pm\frac{1}{2}, \pm\frac{3}{2}, m_j^c = \pm\frac{1}{2}$) and the integral is taken over the elementary cell. For the following discussion, the resulting dipole transitions are schematically shown in Fig. 4.12. The dipole matrix elements are labeled hh and lh for transitions from the heavy-hole and light-hole valence band, respectively. The transitions with $\Delta m_j = \pm 1$

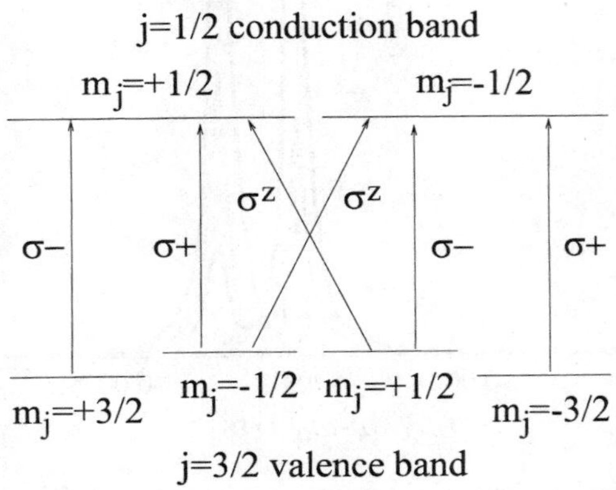

Figure 4.12. Sketch of the possible dipole allowed transitions in a GaAs quantum well with a p-like valence band and an s-like conduction band.

are excited by $\sigma^{\pm}$ polarized light, respectively, whereas the transition with $\Delta m_j = 0$ is excited by z polarized light. With these notations the only nonvanishing dipole moments are $d^{\pm}_{\text{hh}} = d_{cv}\sigma^{\pm}$, $d^{\pm}_{\text{lh}} = 1/\sqrt{3}d_{cv}\sigma^{\pm}$, and $d^{z}_{\text{lh}} = \sqrt{2/3}d_{cv}\sigma^{z}$ [52]. These matrix elements show that heavy-holes can be excited only by $\sigma^{\pm}$ polarized light, whereas the light-holes can be excited by all polarization directions. Because both resonances are energetically non-degenerate, they form an ideal system to study highly polarization dependent near-field distributions. For a similar study in the case of QDs see Ref. [53].

In the following calculations, the excitons are assumed to be generated by a plane wave, σ^{+} polarized, temporally Gaussian shaped pulse ($\simeq 10\,\text{fs}$ pulse duration) incident on the aperture, (cf. Fig. 4.1). The short pulse excitation is assumed to reduce the numerical integration time but it is not relevant for the results discussed below. To determine an experimentally observable quantity we calculate the occupation number of excitons in the state ν, which is determined by $N_{\nu} \equiv \int d^2\mathbf{R}_{\|}|P^{vc}_{\nu}(\mathbf{R}_{\|})|^2$, where the integral in the x, y plane at the position of the quantum well is taken over a square of $1 \times 1\,\mu\text{m}$. This averaged exciton number can be taken as a rough measure of the strength of the respective absorption.

For a quantum well embedded in the dielectric barrier material (Fig. 4.1, $\varepsilon = 12$), the computed time evolution of $N(t)$ is plotted in Fig. 4.13. For far-field excitation, where only a σ^{+} component of the field exists, the number of excited heavy holes is three times the number of excited light holes, because the corresponding dipole moments scale as $\sqrt{3} : 1$. We see in Fig. 4.13(a) that this behavior is not changed for the σ^{+} component of the near field. However, due to the polarization mixing below the aperture a strong σ^{z} component also exists that excites only light-hole transitions. If we add up all heavy- and light-hole transitions, we find a ratio of $\sim 2 : 1$ under near-field excitation compared with the far-field excitation ratio 3:1

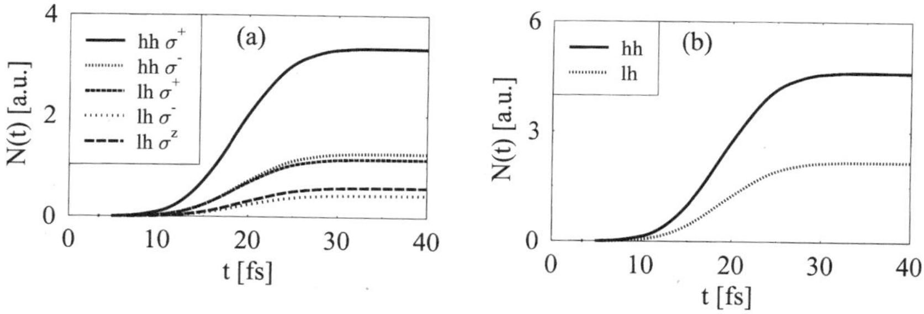

Figure 4.13. Excitonic light and heavy-hole density as a function of time in a GaAs quantum well embedded in a dielectric barrier material (distance between aperture and barrier material $d_m = 10\,\text{nm}$, dielectric constant $\varepsilon = 12$) situated at $z = 30\,\text{nm}$ below the near-field aperture by excitation with a 10 fs, σ^{+} polarized Gaussian pulse: (a) excitonic heavy and light-hole density for the different polarizations; (b) total heavy-hole density compared with the total light-hole density. Due to the near-field distribution, different transitions than in far-field optics are excited.

[Fig. 4.13(b)]. The same calculation for a free-standing quantum well (without background medium) yields that the excitation ratio is even more increased.

4.3.6. Excitation of Second Order Dipole Transitions and Quadrupole Enhancement

In this section, we study the optical excitation of a thin layer of bulk semiconductor material at a transition with vanishing dipole matrix elements. An example is the spectral region around the $1s/2p$ exciton transition in cuprous oxide, Cu_2O. Since there is no first-order dipole contribution to the polarization, Eq. (4.45), it is necessary to take into account higher order terms in the multipole series, such as the second-order dipole and quadrupole transitions [55]. The corresponding equation of motion of the excitonic polarization has been derived in Eq. (4.48). The parameter P_ν^{vc} depends on the electric field $\mathbf{E}(\mathbf{R})|_{z=z_0}$ at the position of the semiconductor layer and its spatial derivative $\nabla_\mathbf{R}\mathbf{E}(\mathbf{R})|_{z=z_0}$. The term proportional $\mathbf{E}$ multiplied with the spatial derivative of the excitonic wave function yields a non vanishing contribution for p like excitons, with $\nabla_\mathbf{r}\varphi_\nu(\mathbf{r}=0)$. This term therefore describes the second order dipole transitions, resulting in p-excitons. The next term in Eq. (4.48) depends on the gradient of the electric field and the value of the wave function at $\mathbf{r}=0$. This corresponds to a quadrupole transition of s-like excitons. In far-field excitation, this contribution is usually weak in comparison to the second-order dipole transitions of p-like excitons. Their relative contribution is determined by the scale on which the exciton wave function changes compared to the scale on which the electric field varies. For far-field excitation, this is the ratio of Bohr radius to wavelength, which is generally small. However, in the near-field case close to the aperture the field varies on a scale given by the diameter of the aperture d_a, which can be much smaller than the wavelength of the incident light. Therefore the response of s-excitons should be enhanced in comparison to p excitons in the near field.

Since the $1s$ exciton and the $2p$ exciton resonance are energetically nondegenerate, the near-field enhancement of the quadrupole transition is directly observable by comparison of the strength of the optical response at the two resonances. In the corresponding numerical calculations, we solve Eq. (4.48) for a thin layer of semiconductor material with dipole forbidden transitions, such as Cu_2O (dielectric constant $\varepsilon \simeq 7$). We assume that this material is situated at $z=z_0$ below the aperture, (cf. Fig. 4.1), and compute the response for resonant stationary excitation through the aperture. To discuss a measure for the quadrupole enhancement, we focus on the ratio between the number of excited p and s excitons in the material, N_p and N_s [56]. The second term in Eq. (4.48) describes the strength of the p exciton, whereas the third term determines the strength of the quadrupole transition of s excitons. To roughly estimate the ratio between N_p and N_s we assume all matrix contributions in the quadrupole-like moments $\boldsymbol{B}_{cv}^{\lambda}$ to be equal. Then, N_p/N_s in the material can be estimated by

$$\frac{N_p}{N_s}=\frac{|P_{2p}^{vc}|^2}{|P_{1s}^{vc}|^2}\simeq\frac{\int d^2\mathbf{R}_\parallel\,|\sum_{j=1}^3 \partial_{x_j}\varphi_{2p}(\mathbf{r})|_{\mathbf{r}=0}^2\cdot|\sum_{j=1}^3 E^j(\mathbf{R})|_{z=z_0}^2}{\int d^2\mathbf{R}_\parallel\,|\varphi_{1s}(\mathbf{r})|_{\mathbf{r}=0}^2\cdot|\nabla_\mathbf{R}\mathbf{E}(\mathbf{R})|_{z=z_0}^2} \tag{4.91}$$

As the derivative of the $2p$ wave functions is

$$[\nabla_r \varphi_{2p}(\mathbf{r})]_{\mathbf{r}=0} = \frac{1-\sqrt{2}i}{\sqrt{2^5 a_B^2}} \varphi_{1s}(0)$$

Eq. (4.91) reads

$$\frac{N_p}{N_s} = \frac{3}{2^5 a_B^2} \frac{\int d^2\mathbf{R}_{\|} |\sum_{j=1}^{3} E^j(\mathbf{R})|^2_{z=z_0}}{\int d^2\mathbf{R}_{\|} |\nabla_{\mathbf{R}} \mathbf{E}(\mathbf{R})|^2_{z=z_0}} \tag{4.92}$$

Here, a_B is the Bohr radius of an exciton in Cu_2O, $a_B \simeq 1$ nm. The integration area is again a square of $1 \times 1\,\mu$m in the (x, y) plane.

Figure 4.14 shows the ratio of excited p and s excitons for different distances between aperture and sample. Two cases are compared: a free standing sample of Cu_2O without any dielectric background medium, and a thin layer of Cu_2O embedded in a nonresonant dielectric bulk material ($\varepsilon = 7$). In the latter case the resonant semiconductor layer is situated at 10 nm below the surface of the dielectric barrier material, (cf. Fig 4.1). It can be recognized in Fig. 4.14 that with decreasing

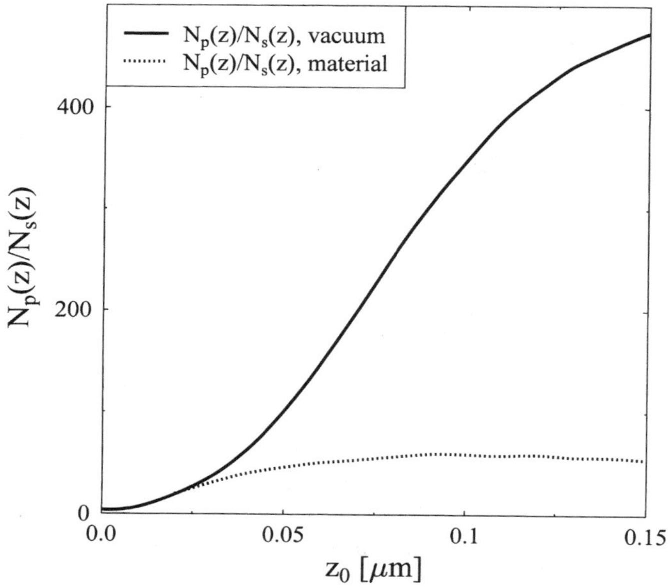

Figure 4.14. Ratio between steady-state second-order dipole p-exciton and quadrupole s-exciton density as a function of distance z_0 between aperture and sample, for a freestanding Cu_2O layer (vacuum) and a Cu_2O layer embedded in a dielectric bulk material with dielectric constant $\varepsilon = 7$ (material). With decreasing distance the quadrupole transition shows a stronger enhancement than the second-order dipole transition.

distance between aperture and sample the strength of the quadrupole transition increases stronger than the second order dipole transition, resulting in a decreasing ratio N_p/N_s. Close to the aperture the quadrupole transition reaches the same order of magnitude as the second order dipole transition ($N_p/N_s \approx 1$). In the case of a free-standing semiconductor layer without any dielectric background material, the ratio between the number of excited p and s excitons is ≈ 1 close to the aperture and changes to $N_p/N_s \approx 500$ at larger distances ($z \approx 150\,\text{nm}$) from the aperture. In far-field excitation, (e.g., for a plane wave), this ratio should become $N_p/N_s = (3/2^5 a_B^2 k_0^2) \approx 700$, where k_0 is the wavenumber in vacuum. This shows that in the numerical calculation, the far-field limit is not yet reached. However, the ratio of excited p- and s-excitons N_p/N_s in a semiconductor layer embedded in a dielectric barrier material is smaller than in vacuum. Close to the aperture the ratio is similar to the vacuum case, but at larger distances from the aperture it changes only to ≈ 60. Here N_p/N_s is expected to be $3/2^5 a_B^2 k^2 \approx 100$ for excitation with a plane wave (k is the wave number in the material, $k = \sqrt{\varepsilon} k_0$). However, in general, due to the large field gradients in an optical near field the quadrupole transition is strongly enhanced in comparison to far-field excitation.

4.3.7. Time-Dependent Transport Phenomena and Wavepacket Dynamics

In this section, the use of optical near-field techniques for the study of optical transport phenomena is discussed. Formally, transport is connected to a nonlocal material response described by susceptibility $\chi(\mathbf{r}, \mathbf{r}', t - t')$. In this case, the polarization is given by $P(\mathbf{r}, t) = \int \chi(\mathbf{r}, \mathbf{r}', t - t') E(\mathbf{r}', t')\, d\mathbf{r}' dt'$, describing polarization induced also at points with vanishing excitation.

Microscopically, the nonlocal response is induced by the ultrafast spatio-temporal dynamics of the electron–hole pairs. Since an optical pulse that is short compared to typical scattering times prepares a coherent superposition of different states (i.e., a spatio-temporal wavepacket [32]), one expects that on short space and timescales ballistic (free) motion should be observable. For longer times, however, scattering mechanisms (e.g., by phonons or static disorder) lead to a subsequent transition to diffusive or even sub-diffusive motion.

There is a distinct difference between dynamics of bound and free electron-hole pairs. For bound pairs [excitons as described in Eq. (4.36)], the relative motion is determined by the Wannier equation and the EM field couples to the COM motion. The induced motion is thus given by the momentum transfer from photons to COM motion of the exciton. For excitation in the band the free electron hole pairs are excited, their center of mass motion is again determined by the photon momentum, but their relative motion is mostly determined by the energy transfer to the pair, which depends on the spectral position of the excitation in comparison to the band edge [32].

First, experimental studies in the interband excitation configuration report the observation of diffusive motion on a nanosecond timescale [17]. However, deviations from diffusive motion of the electron–hole continuum have been predicted [58] and observed [59] on a picosecond scale using diffraction limited optics with micrometer resolution.

According to theoretical predictions [32], the spatial picosecond dynamics of bound electron–hole pairs at low temperatures cannot be resolved within conventional optics, in agreement with the corresponding experimental conclusions [59]. The reason for this is that the momentum transfer of the photons to the COM motion of the exciton is small, as for a standard optical excitation with a microscope lens, the spot size is limited to one-half of the wave length of the incident light, which is about a micrometer at the band edge of a typical semiconductor like GaAs. The velocity of the excited excitons with mass M can roughly be estimated as $\hbar/Ma$, with a being the extension of the wavepacket and $\hbar/a$ the exciton momentum, as long as no heating by phonons occurs [60]. For $a = 1\,\mu\text{m}$ and $M = 0.5m_0$ the resulting velocity of 0.2μ m/ns indicates dynamics on a nanosecond scale, where scattering dominates the motion.

An order of magnitude smaller spot size is available with the help of near-field excitation. For $a = 50$ nm, which can be reached by todays SNOMs [61], the velocity is 4 nm/ps. Hence, the excitonic density evolves significantly on a picosecond timescale [58] such that coherent and radiative optical effects become observable for high quality samples [62]. However, in general the influence of scattering mechanisms has to be considered. Especially static disorder, which may result from surface roughness or alloy disorder is supposed to be of considerable importance [63], because even for near-field excitation the kinetic energy $\hbar^2/2Ma^2 \approx 0.03$ meV of the excitons is very small. Thus even small potential perturbations may have a strong impact on the excitonic motion.

In this section, we study the coherent-pulse near-field excitation of excitons in the local semiconductor environment. Because a pulsed near-field contains a broad frequency and wavenumber spectrum, the generated excitons propagate in the form of wavepackets. In order to be close to realistic situations, we additionally include the influence of structural disorder.

The configuration is chosen as in Fig. 4.1, but here we discuss the case of excitation through a slit (width 60 nm), which leads to a 2D configuration. A vanishing distance between metal layer and semiconductor surface is chosen $d_l = 0$, corresponding to a metal coating of the surface. This geometry allows for the investigation of two independent linear polarizations [39], one in the direction of the slit (x polarized, TE) and the other one perpendicular to the slit (y polarized, TM), respectively. Due to the translational invariance in the x direction, scattered TE fields remain x polarized, whereas the scattered electric field for the TM case also has a z component.

4.3.7.1. Polariton Modes. In this section, we investigate the properties of exciton polaritons [64] and formulate the in plane dispersion $\omega(Q)$ relation for the propagation of the polarization $P(Q, \omega)$ relevant for our study of exciton wavepackets. Formally, the discussion can be connected to the equations given in Section 4.3.4, where the excitation of eigenstates ψ_μ of the equation of motion (4.36) for the interband $P^{vc}_{1s}(\mathbf{R}, t)$ polarization has been discussed. In an ideal

QW, ψ_μ are plane waves labeled by the in-plane momentum $\mu = \mathbf{Q}$. In analogy to Section 4.3.4, the equation of motion for the polarization and the coupling to the EM field lead to a self-interaction Σ, which changes the lifetime and resonance frequency of the excitons. However, in contrast to the case of localized states, for a layered structure that is homogeneous in the transverse directions, the coupling between different eigenstates by the EM field vanishes. This makes the concept of coupled modes of field and polarization, the 2D-exciton polariton, very useful. The self-interaction Σ, as introduced in Section 4.3.4, can in this sense be interpreted as the deviation of the polariton dispersion relation from the pure exciton dispersion relation.

The equation of motion of the polarization in momentum $\mathbf{Q}$ representation has already been given in Eq. (4.34). For the excitation of the in-plane FT of the electric field $\mathbf{E}(\mathbf{Q}, z, \omega)$, Maxwell's equations yield the following wave equation:

$$(\partial_z^2 + \kappa^2)\mathbf{E} - (i\mathbf{Q} + \mathbf{e}_z\partial_z)(i\mathbf{Q} + \mathbf{e}_z\partial_z) \cdot \mathbf{E} = -4\pi k_0^2 \mathbf{P} \tag{4.93}$$

Here $k_0 = \omega/c_0$ denotes the vacuum wavevector, $\mathbf{Q}$ is the in-plane component, and $\kappa = \sqrt{n^2 k_0^2 - Q^2}$ is the z component of the wavevector, respectively. The parameter $n = \sqrt{\varepsilon}$ is the refractive index of the background accounting for the nonresonant polarization of the semiconductor, where we have assumed that the background refractive index of the barrier equals that of the QW.

Considering the TE and TM polarized modes introduced above, we note for the TE case that the polarization and the electric field are always perpendicular to the field gradients. Thus these fields are divergence free and the second term on the left hand side (lhs) of Eq. (4.93) vanishes:

$$(\partial_z^2 + \kappa^2)\mathbf{E}^{TE} = -4\pi k_0^2 \mathbf{P} \tag{4.94}$$

For TM excitation, we have from Coulomb's law for a constant background refractive index

$$\nabla \cdot \mathbf{E}^{\mathrm{TM}} = -\frac{4\pi}{n^2} \nabla \cdot \mathbf{P} \tag{4.95}$$

Inserting Eq. (4.95) into Eq. (4.93) yields for the TM modes

$$\begin{aligned}(\partial_z^2 + \kappa^2)\mathbf{E}^{\mathrm{TM}} &= -\frac{4\pi}{n^2}(\kappa^2 \mathbf{P} + \partial_z^2(\mathbf{P} \cdot \mathbf{e}_z)\mathbf{e}_z \\ &\quad + i(\mathbf{Q} \cdot \partial_z \mathbf{P})\mathbf{e}_z + i\mathbf{Q}\partial_z(\mathbf{P} \cdot \mathbf{e}_z))\end{aligned} \tag{4.96}$$

Let us look in paticular at a case where the dipole matrix element has no z component, such as the heavy-hole exciton in a GaAs quantum cell. By considering only in-plane polarization ($\mathbf{P} \cdot \mathbf{e}_z = 0$), the components of the electric fields are given by the formal solution of the respective wave equations (4.94) and (4.96)

$$\mathbf{E}(\mathbf{Q}, z, \omega) = \frac{1}{2i\kappa} \int dz' e^{-i\kappa(z-z')} \mathbf{f}(z') \tag{4.97}$$

with the inhomogeneity $\mathbf{f}(z)$. The solutions for the in-plane component of the TE and the TM modes are

$$E^{\mathrm{TE}}(Q, z, \omega) = 2\pi i \frac{k_0^2}{\kappa} e^{i\kappa|z|} P(Q, \omega) \tag{4.98}$$

$$E^{\mathrm{TM}}(Q, z, \omega) = 2\pi i \frac{\kappa}{n^2} e^{i\kappa|z|} P(Q, \omega) \tag{4.99}$$

For $Q > nk_0$ ($\kappa^2 < 0$), the z component of the wavenumber κ becomes imaginary and thus E^{TE} and E^{TM} decay exponentially in the z direction. These contributions are evanescent fields that do not transport energy perpendicular to the QW. Equations (4.98) and (4.99) show that the $\mathbf{Q}$ component of the polarization only couples to the corresponding electric field component. This is a direct consequence of the translational invariance of the system in the transverse direction: Excitons with in-plane momentum $\mathbf{Q}$ only couple to photons with the same in-plane momentum.

Inserting the expressions for the electric fields (4.98) and (4.99) at the quantum well ($z = 0$) into Eq. (4.36) yields a dispersion relation for the combined transversal motion of polarization and EM field: the 2D-exciton polariton [64]. Neglecting the nonresonant contributions in P^{vc} the polariton dispersion relations are

$$\omega^{\mathrm{TE}} = \omega_{1s} + \frac{\hbar Q^2}{2M} - i\Gamma \frac{nk_0}{\kappa} \tag{4.100}$$

$$\omega^{\mathrm{TM}} = \omega_{1s} + \frac{\hbar Q^2}{2M} - i\Gamma \frac{\kappa}{nk_0} \tag{4.101}$$

Here, $\omega_{1s} = \varepsilon_{1s}/\hbar$ and $\Gamma = 2\pi k_0^2 |d_{vc}\varphi(0)|^2/n$ denotes the radiative line width of the QW. Figure 4.15 shows the real and the imaginary parts (i.e., energy and damping of the resulting polariton modes). For free excitons, the real part is just the parabolic energy of the excitonic COM motion. The imaginary part vanishes in this case since there is no decay mechanism. By including the interaction, the modes are damped for $Q < nk_0$, but for higher Q values the modes do not decay. Excitons with higher momentum than nk_0, only couple to evanescent modes of the electric field. These modes are not radiative but are guided by the QW acting as inhomogeneity in the bulk medium. For $Q < nk_0$ radiative decay leads to a damping of the modes, for $Q = 0$ we find the radiative broadening for plane wave excitation, which is $\Gamma \approx 1/10\,\mathrm{ps}$ for the parameters given in Section 4.3.1. The dispersion curve for the TE cases is rather smooth, while for the TM case there is a peak at $Q \approx nk_0$. The gradient of the dispersion curve gives the group velocity of a wavepacket thus indicating a fast traveling mode with large decay.

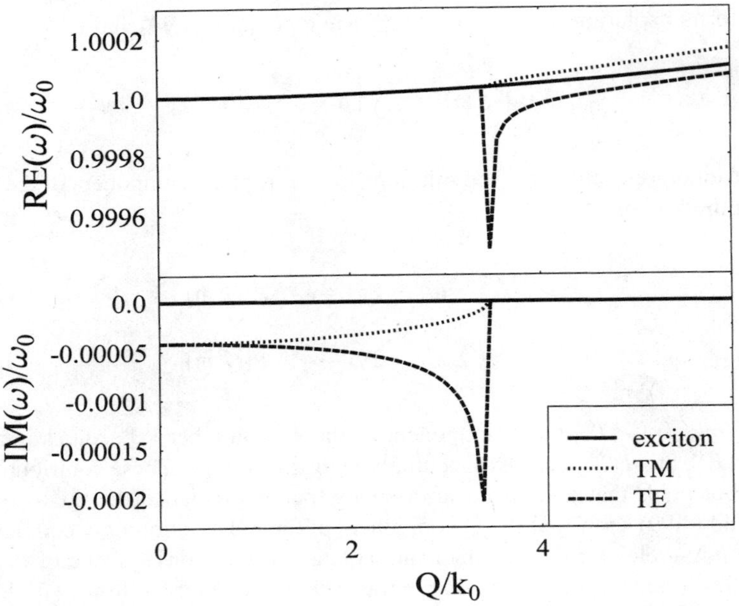

Figure 4.15. Real and imaginary part of the dispersion relation of the exciton polariton in infinitely extended background mateiral.

4.3.7.2. Quantum Well between Finite Barriers. In this section, we investigate the influence of the boundary at the sample–metal transition (Fig. 4.1), where the EM field is reflected acting back on the quantum well. The electric field E_B at the QW is the sum of the field E without boundary, given by Eq. (4.98) for $z = 0$, plus the reflected field,

$$E_B = E(1 + re^{i\kappa(2\Delta z)}) \tag{4.102}$$

Here Δz is the distance between the boundary and the quantum well and r denotes the reflection coefficient at the boundary. This reflection depends on the in-plane wavevector Q and on the polarization direction and is given by Fresnel's coefficients [65],

$$r^{\text{TE}} = \frac{\kappa - \kappa_b}{\kappa + \kappa_b} \tag{4.103}$$

$$r^{\text{TM}} = \frac{n_b\kappa - n\kappa_b}{n_b\kappa + n\kappa_b} \tag{4.104}$$

In these equations, n and n_b are the refractive indexes of the semiconductor and the boundary, respectively, and κ and $\kappa_b = \sqrt{n_b^2 k_0^2 - Q^2}$ are the corresponding z components of the wavevector. Figure 4.16 shows the TM dispersion relations computed for a semiconductor–vacuum interface ($n_b = 1$) and for a metal coating ($\Delta z = 30$ nm). Here, the metal has been modeled as ideal ($n_b = \infty$) in contrast, for

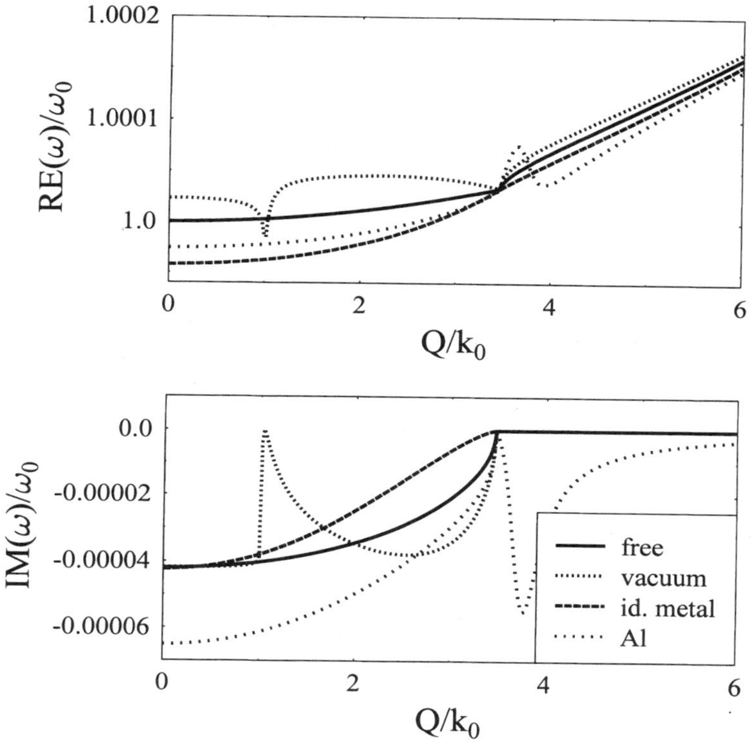

Figure 4.16. Real and imaginary part of the dispersion relation of the exciton polariton in the presence of an additional surface (TM mode).

example, to an aluminum coating ($n_b = 2.5 + 8.5i$). For the semiconductor–vacuum interface, we find a traveling undamped mode at $Q \geq k_0$. This mode is a consequence of the total reflection at the boundary for $Q > k_0$. For $r\exp(i\kappa 2\Delta z) = -1$, the interference of the reflected and unreflected field is destructive and the exciton polarization is not damped at all.

For an ideal metallic layer on the surface ($r = -1$) the damping is similar to the free case, however, quantitatively the damping is smaller for almost all modes (apart from $Q \approx 0$). This behavior is typical for an ideal metal forcing $E = 0$ at the boundary, which effectively reduces the coupling of the bare exciton and field. For an aluminum coating (i.e., a lossy material), almost all modes, even the nonradiating ones with $Q > nk_0$ are damped.

For the real part of the dispersion relation we note an important difference in comparison to the boundary free case of Fig. 4.15, where for $Q < nk_0$ the dispersion with EM interaction is always identical to the parabolic dispersion relation of the exciton, as long as the background material is nonabsorbing. In the presence of the boundary, the free dispersion is also renormalized by a shift of the excitonic resonance for $Q < nk_0$. These momentum states are observable in far-field optics.

For $Q = 0$, the shift relates to the imaginary part of the phase of the reflected wave$-\Gamma(\Im(r\exp(i\kappa(2\Delta z))$, which is on the order of the radiative line width $\hbar\Gamma = 50\,\mu\text{eV}$. Note that the dispersion relations are also strongly dependent on the distance of the quantum well to the boundary, therefore the curves presented here only serve as an example.

4.3.7.3. Spatio-Temporal Dynamics of an Excitonic Wavepacket Excited through a Subwavelength Aperture. In this section, the spatio-temporal dynamics of a nonstationary excitonic wavepacket is investigated. The wavepacket represents the superposition of different states $(\mathbf{Q}, \omega)$ determined by the excitation conditions realized in a short pulse SNOM. To treat the problem numerically, we used finite discretization of the coupled partial differential equations. For Maxwell's equation, this can be done effectively with the FDTD formalism [39] and for the material equations we use a predictor–corrector method [66].

Next, we treat the case of the QW in the geometry of Fig. 4.1 excited by a short laser pulse (100 fs,TM polarized) spectrally centered at the hh–exciton resonance. Figure 4.17 shows the numerically computed spatio-temporal dynamics of the exciton density $n_X = |P_{1s}^{vc}|^2$ [67] as a function of the in-plane coordinate at different times after the arrival of the pulse. The initially excited exciton distribution

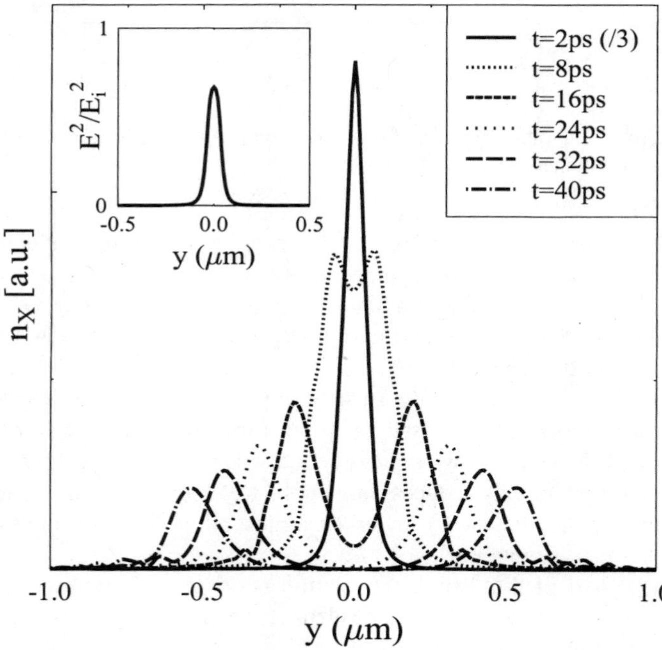

Figure 4.17. Spatio-temporal dynamics of the exciton density after a short-pulse excitation through the aperture. The inset shows the field intensity distribution under the aperture for resonant continuous wave (CW) excitation.

resembles the field shape ($t = 2$ ps) shown in the inset of Fig. 4.17 at 30 nm under the aperture for CW excitation.

The computed wavepacket dynamics can be understood qualitatively on the basis of the polarization dispersion, Fig. 4.16. The aperture excitation creates a momentum distribution that is roughly bounded by the inverse slit width. Therefore, damped and undamped modes around nk_0 are excited, propagating in the form of an excitonic wavepacket. This wavepacket spreads in both directions, propagating out of the focus of the aperture. For $Q > nk_0$, the momentum distribution of the excitons remains close to that of the initial contribution and vanishes for $Q < nk_0$ due to the radiation decay (cf. Fig. 4.16). Therefore, at late times ($t > \Gamma^{-1}$) the exciton distribution is peaked around $Q \approx nk_0$. The group velocity can be estimated by $v_g \approx \hbar n k_0 / M = 6$ nm/ps.

4.3.7.4. Influence of Disorder. Real QW samples usually contain an intrinsic amount of composition fluctuation and/or interface roughness. To analyze the influence of these static structural imperfections, we now discuss the influence of disorder on the formation and propagation of a wavepacket. Due to interface roughness or alloy disorder, the exciton polaritons are not the true eigenstates of such a system, but will be scattered between different momentum states. Significant disorder induced modifications of the space-time dynamics are expected if the kinetic energy of the excitons is on the order of the potential fluctuations or smaller. As the dispersion relation of the exciton is rather flat, this can be expected to be an important effect even for high quality samples. For a typical exciton with a momentum of $\hbar n k_0$, the kinetic energy is $50\,\mu$eV. This is on the order of the radiative line width of the QW. Such a purely qualitative argument already shows that the free-motion results presented in Section 4.3.7.4 can only be realized in QWs where the average disorder potential fluctuations are lesser than the radiative energy $\hbar\Gamma$.

By following the discussion of Section 4.2.4, the disorder is described by an effective potential $V(\mathbf{R})$, which is the average over microscopic potential fluctuations due to the spatial extension of the internal exciton wave function, 4.37. For simplicity, and in order to preserve the translational invariance in the x direction (cf. Fig. 4.1) the disorder is assumed here to be restricted to the y direction. For strained samples, this can be a reasonable approximation [6]. As usual, we assume Gaussian distributed disorder with a correlation length of 8 nm. The deviation $\hbar\Delta$ of the distribution has been chosen in the range of the typical kinetic energy of the exciton COM motion. Note that by solving Eq. (4.36) with the additional potential given by Eq. (4.37), disorder is treated exactly within the limits of the numerical approach, that is, no perturbation theory is involved and possible localization is described correctly.

Figure 4.18 shows the shape of the density distribution 20 ps after the initial excitation for different strengths of disorder. We see that even for relatively weak disorder $\Delta = 3\Gamma$, the shape of the distribution is significantly modified by the local potential, but we still notice spreading of the wavepacket. For stronger disorder, $\Delta = 10\Gamma$ (0.5 meV) the spreading is reduced. A more quantitative picture of the

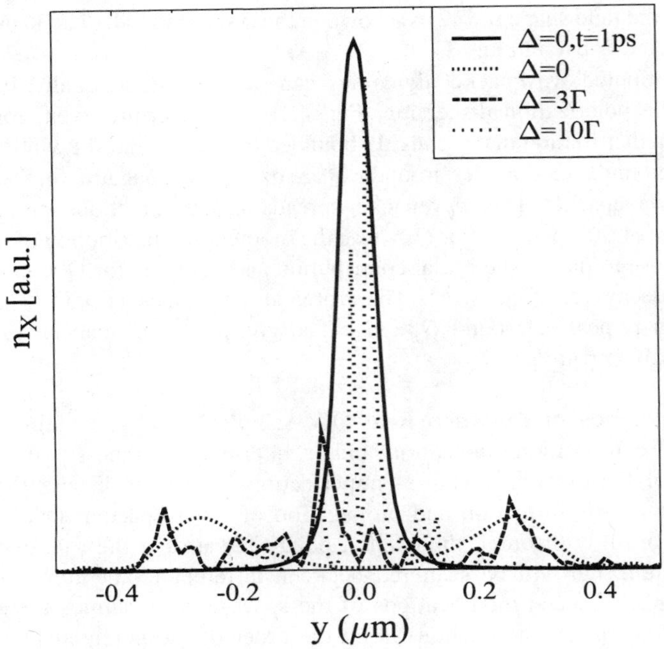

Figure 4.18. Spatial shape of the excitonic wavepacket 20 ps after short-pulse excitation through the aperture for different static disorder Δ in the QW. The dotted curve shows a reference distribution for the ordered QW shortly (1 ps) after the excitation.

spreading can be obtained from the mean square displacement (MSD)

$$\text{MSD} = \frac{1}{N(t)} \int dx \, n(x,t)x^2 \qquad (4.105)$$

For free ballistic motion, the MSD has a parabolic time dependence, whereas for diffusive motion the MSD varies linearly in time. In a 1D system, any wavepacket becomes localized for sufficient large times due to the enhanced backscattering probability (Anderson localization) [68]. A numerical analysis of the curves shown in Fig. 4.18 has the following results: For $\Delta = 0$ and $\Delta = 3\Gamma$, the motion is ballistic on a timescale of 10 ps, with only small deviations in the latter case. However, for $\Delta = 10\Gamma$ after 4 ps already the time dependence of the MSD becomes linear and the motion becomes diffusive. For later times (10 ps), the motion is slower than in the diffusive regime, here localization effects begin to show up. For $\Delta = 30\Gamma$, the wavepacket is localized on a picosecond timescale. Due to the random character of the potential there are fluctuations from ideal behavior.

4.3.7.5. Optical Detection of Wavepackets. We now discuss to which degree the temporal development of the excitonic density can be detected in an optical far-field signal. The reflected signal is determined by the excitation of the

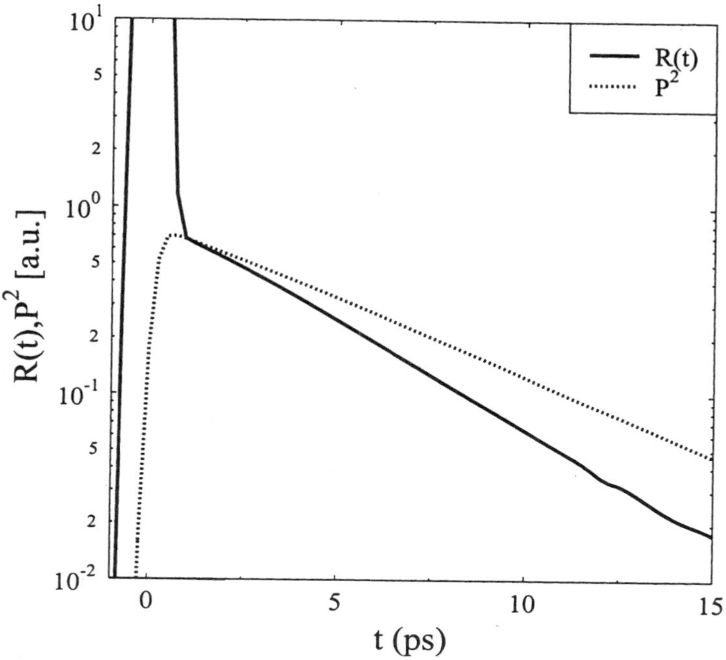

Figure 4.19. Time resolved reflected signal of the QW under the aperature. For comparison the total polarization $P(Q = 0)^2$ is shown.

magnetic dipole in the aperture, which is connected to the polarization in the QW via the transmission function, Eq. (4.58). Since this transmission function is centrally peaked under the aperture, the reflected signal is mainly determined by the polarization centrally under the aperture. Figure 4.19 shows the square of the magnetic moment excited in the aperure as a function of time. The square of the magnetic moment is proportional to the reflected signal, as discussed in Section 4.33. For comparison, the square of the integral of the polarization is shown, which in the Fourier domain is given by $P(Q = 0)^2$. If the influence of the aperture on the lifetime is neglected, $P(Q = 0)$ decays with the radiative lifetime in the presence of the metallic screen. For the parameters discussed here, $P(Q = 0)^2$ decays with a damping time of 5 ps. However, the reflected signal decays faster than that, the curve can be fitted with an exponential decay with lifetime of 3.8 ps. Thus in the reflected signal there is evidence for the transport of the polarization out of the focus of the aperture.

4.4. SUMMARY

We have given an overview of near-field excitation of semiconductor heterostructures. Important differences between near and far-field excitation have been pointed out. Besides the desired high field localization these differences include strong field gradients, a complex vector structure of the EM field and the interaction between probe and sample. The relevance of these effects to the spectroscopy of semiconductor heterostructures has been discussed. Here especially the possibility to investigate single localized states and to study optical transport phenomena has been demonstrated. Selection rules of radiative decay rates can be significantly modified in comparison to far-field excitation. For the description of the dynamics of the electrodynamic field and its coupling to the semiconductor sources in the highly inhomogeneous setup of a near-field experiment, the FDTD method is applicable and promises to become an important tool. For further investigations, an inclusion of incoherent processes would be desirable. The influence of phonons on transport mechanisms and on the interaction of localized states is especially of interest here. In addition, the extension to nonlinear dynamics (e.g., to describe pump–probe experiments), is important. These experiments give additional information on the carrier population dynamics in locally excited semiconductor structures. As pointed out, theoretical modeling plays an important role in the description of near-field experiments. We have discussed a simple model to study the mean effects. However, when modeling a certain experimental setup a refined treatment of the tip geometry is desirable. A more detailed treatment of the optical properties of the metal may be important.

ACKNOWLEDGMENTS

We thank Anna v. d. Heydt, Frank Steininger, and Harald Giessen for useful discussions. This work is supported by the Deutsche Forschungsgemeinschaft (DFG) through the Sonderforschungsbereich 383, the Quantenkohärenz Schwerpunkt, the Leibniz prize, and by the HLRZ Jülich through grants for extended CPU time on their supercomputer systems.

APPENDIX: NUMERICAL METHODS

Different numerical approaches have been used to model near-field optical experiments: especially boundary element methods (BEM) [70], calculations based on Green's function approaches [71], and FDTD simulations [39]. In BEM, which was used for the calculation of the field propagation in a fiber tip [36], spatial regions are split into homogeneous parts and the sources are accounted for by fictious currents in the boundaries, that have to be solved for. This generally requires the solution of a coupled system of N equations, which takes N^3 steps. In comparison to BEM, Green's function approaches have the advantage that the problem can be split into a reference system, for which a Green's function is available, and in spatial regions that deviate from the reference. Only in these deviating regions do the fields have to be computed numerically. This solution also requires N^3 steps. However, for the description of semiconductor structures nonlinear and nonlocal features have to be incorporated. This is in general difficult with *MM* and BEM.

The numerical results presented in this chapter have been obtained with the FDTD method [39]. Here Maxwell's equations are discretized in the space and time domains. To obtain an efficient algorithm, the first-order partial differential equations are split into central-difference second-order accurate difference equations. This can be achieved if the fields are discretized on a special temporal and spatial grid (Yee-grid). In the time domain, this leads to a explicit discretization scheme (leapfrog arrangement), which is most stable. The polarization equation, which has the formal structure of the Schrödinger equation, $i\hbar\partial_t P = HP$, has been treated analogously. A common discretization procedure of the Schrödinger equation in space and time is the Crank–Nicholson scheme [72]

$$P^{n+1} = \frac{1-(i/2\hbar)H}{1+(i/2\hbar)H}P^n \tag{4.A.1}$$

where P^{n+1} is the polarization at the next time step. If H is Hermitian, this induces a unitary transformation. Thus the conservation of particles is incorporated in this discretization, which leads to the very good stability quality of the Crank–Nicholson method. The system of linear equations is tridiagonal.In our calculations, we used a combination with a predictor–corrector scheme, splitting each time step into $j = 1 \cdots f$ substeps

$$P^{n+1,j+1} = P^n - \frac{i}{\hbar}H(P^{n+1,j} + P^n) \tag{4.A.2}$$

In a first step, $(j = 1)$ $P^{n+1,1}$ on the rhs is set to P^n. With the resulting value for $P^{n+1,2}$ the equation is iterated. The scheme converges very fast ($f \approx 5$ steps). After convergence, $(P^{n+1,j+1} = P^{n+1,j})$ and Eq. (4.A.2) give the same result obtained with a Crank–Nicholson time step. Each time step requires fN steps, where f is the number of steps for the predictor–corrector scheme to converge. With this method the solution of large 2D and 3D grids (up to the order of 10^6) is possible. However, at the position of the quantum well, not only the polarization is updated with the scheme describe above, but also the electric field

$$\begin{aligned} P^{n+1,j+1} &= F(P^n, P^{n+1,j}, E^n, E^{n+1,j}) \\ E^{n+1,j+1} &= F(P^n, P^{n+1,j}, E^n, E^{n+1,j}) \end{aligned}$$

Previously, a scheme like this had been used to describe nonlinear properties of materials that are modeled by the optical Bloch equations [67].

REFERENCES

1. E. A. Ash and G. Nichols, *Nature (London)* ***237***, 510 (1972).
2. D. W. Pohl, W. Denk, and M. Lanz, *Appl. Phys. Lett.* ***44***, 651 (1984).

3. A. Lewis, M. Isaacson, A. Murray, and A. Harootunian, *Biophys. J.* ***41***, 405a (1983).
4. U. Düring, D. W. Pohl, and F. Rohner, *J. Appl. Phys.* ***59***, 3318 (1986).
5. R. D. Grober, T. D. Harris, J. K. Trautman, E. Betzig, W. Wegscheider, L. Pfeiffer, and K. West, *Appl. Phys. Lett.* ***64***, 1421 (1994).
6. D. Gammon, E. S. Snow, B. V. Shanabrook, D. S. Katzer, and D. Park, *Phys. Rev. Lett.* ***76***, 3005 (1996).
7. D. W. Pohl, *Adv. Opt. Electron Micros.* ***12***, 243 (1991).
8. E. Oesterschulze, O. Rudow, C. Mihalcea, W. Scholz, and S. Werner, *Ultramicroscopy* ***71***, 85, (1998)
9. S. Jiang, N. Tomita, H. Oshwa, and M. Ohtsu, *Jpn. J. Appl. Phys.* ***30***, 2107 (1991).
10. G. A. Valaskovic, M. Holton, and G. H. Morrison, *Appl. Opt.* ***34***, 1215 (1995).
11. W. Langbein, J. M. Hvam, S. Madsen, M. Hetterich, and C. Klingshirn, *Phys. Stat. Sol. A*, ***164***, 541 (1997).
12. Ch. Adelmann, J. Hetzler, G. Scheiber, Th. Schimmel, M. Wegener, H. B. Weber, and H.v. Löhneysen, *Appl. Phys. Lett.* ***74***, 179 (1999).
13. G.v. Freymann, Th. Schimmel, M. Wegener, B. Hanewinkel, A. Knorr, and S. W. Koch, *Appl. Phys. Lett.* ***73***, 1170 (1998).
14. V. Sandoghar, S. Wegscheider, G. Krausch, and J. Mlynek, *J. Appl. Phys.* ***81***, 2499 (1997).
15. S. L. Chuang, *Physics of Optoelectronic Devices*, Wiley, New York, 1997.
16. H. Haug and S. W. Koch, *Quantum Theory of the Optical and Electronic Properties of Semiconductors*, 3rd ed., World Scientific, Singapore 1994.
17. A. Richter, G. Behme, M. Stüpitz, Ch. Lienau, Th. Elsässer, M. Ramsteiner, R. Nötzel, and K. H. Ploog, *Phys. Rev. Lett.* ***79***, 2145 (1997).
18. B. A. Nechay, U. Siegner, F. Morier-Genoud, A. Schertel, and U. Keller, *Appl. Phys. Lett.* ***74***, 61 (1999).
19. N. Van Hulst, M. Moers, and E. Bogonjen, in O. Marti and R. Möller, Eds., *Photons and Local Probes*, Kluwer Academic Publishers, Dordrecht, 1996.
20. Ch. Lienau, A. Richter, and T. Elsaesser, *Appl. Phys. Lett.* ***69***, 325 (1996).
21. K. Brunner, G. Abstreiter, G. Böhm, G. Tränkle, and G. Weimann, *Phys. Rev. Lett.* ***73***, 1183 (1994).
22. M. J. Gregor, P. G. Blome, R. G. Ulbrich, P. Grossmann, S. Grosse, J. Feldmann, W. Stolz, E. O. Göbel, D. J. Arent, M. Bode, K. A. Bertness, and J. M. Olson, *Appl. Phys. Lett.* ***67*** (24), 3572 (1995).
23. B. Hanewinkel, A. Knorr, P. Thomas, and S. W. Koch, *Phys. Rev. B* ***55***, 13715 (1997).
24. L. Banyai and S. W. Koch, "Semiconductor quantum dots," *World Series On Atomic, Molecular and Optical Physics*, Vol. 2, World Scientific, Singapore, 1993.
25. A. Zrenner, L. V. Butov, M. Hagn, G. Abstreiter, G. Böhm, and G. Weinmann, *Phys. Rev. Lett.* ***72***, 3382 (1994).
26. Y. Toda, S. Shinomori, K. Suzuki, and Y. Arakawa, *Phys. Rev. B* ***58*** (1998).
27. A. Vertikov, I. Ozden, and A. Nurmikko, *Appl. Phys. Lett.* ***74*** (1999).
28. G. D. Mahan, *Many-Particle Physics*, Plenum, New York, 1990.
29. W. Vogel, *Lectures on Quantum Optics*, 1st. ed., Akademische Verlag, Berlin, 1994.
30. W. W. Chow, S. W. Koch, and M. Sargent, *Semiconductor-Laser Physics*, Springer-Verlag, Berlin, 1994.

31. S. D. Baranovskii and A. L. Efros, *Fiz. Tekh. Poluprovodh.* ***12***, 2233 (1978). [*Sov. Phys. Semicond.* ***12***, 1328 (1978)].
32. F. Steininger, A. Knorr, T. Stroucken, P. Thomas, and S. W. Koch, *Phys. Rev. Lett.* ***77***, 550 (1996).
33. A. Knorr, B. Hanewinkel, H. Giessen, and S. W. Koch, *Advances in Solid Sate Physics* ***38***, Vieweg Verlag, Wiesbaden, Germany, 1998, p. 311.
34. D. Fröhlich, in O. Madelung, Ed., *Advances in Solid State Physics* ***10***, Vieweg, Braunschweig, Germany, 1970, p. 227.
35. E. Betzig, J. K. Trautman, T. D. Harris, J. S. Weiner, and R. L. Kostelak, *Science* ***251***, 1468 (1991).
36. L. Novotny, D. W. Pohl, and B. Hecht, *Opt. Lett.* ***20***, 970 (1995).
37. C. Girard, O. J. F. Martin, and A. Dereux, *Phys. Rev. Lett.* ***75***, 3098 (1995).
38. R. X. Bian, R. C. Dunn, S. Xie, and P. T. Leung, *Phys. Rev. Lett.* ***75***, 4772 (1995).
39. A. Taflove, *Computational Electrodynamics—The Finite-Difference Time-Domain Method.*, Artech House Inc., Norwood, 1995.
40. C. J. Bouwkamp, *Philips Res. Rep.* ***5***, 401 (1950).
41. J. D. Jackson, *Classical Electrodynamics* 2nd. ed., Wiley, New York, 1975.
42. Y. Leviatan, *J. Appl. Phys.* ***60***, 1577 (1986).
43. J. Bromage, S. Radic, G. P. Agrawal, C. R. Stroud, P. M. Fauchet, and R. Sobolewski, *Opt. Lett.* ***22***, 627 (1997).
44. U. Woggon, "Optical properties of semiconductor quantum dots," *Springer Tracts in Modern Physics*, ***136***, Springer, 1996.
45. C. Tai, *Dyadic Green Functions in Electromagnetic Theory*, 2nd ed., IEEE Press, 1994.
46. M. Hübner, J. Kuhl, T. Stroucken, A. Knorr, S. W. Koch, R. Hey, and K. Ploog, *Phys. Rev. Lett.* ***76***, 4199 (1996).
47. D. L. Dexter, *J. Chem. Phys.* ***21***, 836 (1953).
48. U. Bockelmann, *Phys. Rev. B* ***48***, 17637 (1994).
49. P. Meystre and M. Sargent, *Elements of Quantum Optics*, 2nd ed., Springer Verlag, Berlin, 1991.
50. A. Sommerfeld, *Partielle Differentialgleichungen der Physik.*, Akademische Verlagsgessell Schaft, Leipzig, 1962.
51. M. Bass, *Handbook of Optics*, 2nd ed., McGraw-Hill, New York, 1995.
52. G. Bastard, *Wave Mechanics Applied to Semiconductor Heterostructurs*, Les Editions de Physique, Paris, 1988.
53. G. W. Bryant, *Appl. Phys. Lett.* ***72***, 768 (1998).
54. A. Otto, *Z. Physik* ***216***, 398 (1964).
55. P. Y. Yu and M. Cardona, *Fundamentals of Semiconductors*, Springer, Berlin, 1996.
56. R. J. Elliott, *Phys. Rev.* ***124***, 340 (1961).
57. D. W. Pohl, *Advances in Optical and Electron Microscopy*, ***12*** (1991); E. Betzig, J. K. Trautman, T. D. Harris, J. S. Weiner, and R. L. Kostelak, *Science* ***251***, 1468 (1991).
58. A. Knorr, F. Steininger, B. Hanewinkel, S. Kuckenburg, P. Thomas, and S. W. Koch, *Phys. Status Solidi B* ***206***, 139 (1998).
59. M. Vollmer, H. Giessen, W. Stolz, and W. Rühle, submitted for publication.

60. S. Grosse, R. Arnold, G. von Plessen, M. Koch, J. Feldmann, V. M. Axt, T. Kuhn, R. Rettig, and W. Stolz, *Phys. Status Solidi B* ***204***, 147 (1997).
61. O. Marti and R. Möller, Eds., *Photons and Local Probes*, Kluwer Academic Publishers, Dordrecht, 1996.
62. M. Hübner, J. Kuhl, T. Stroucken, A. Knorr, S. W. Koch, R. Hey, and K. Ploog, *Phys. Rev. Lett.* ***76***, 4199 (1996).
63. S. D. Baranovskii, U. Doerr, P. Thomas, A. Naumov, and W. Gebhardt, *Phys. Rev. B* ***48***, 17149 (1993); E. Runge and R. Zimmermann, *J. Lumin.* ***60***, 320 (1994).
64. V. M. Agranovitch and O. A. Dubovskii, *JETP Lett.* ***3***, 223 (1966).
65. J. D. Jackson, *Classical Electrodynamics*, Wiley, New York, 1962.
66. R. W. Ziolkowski, *IEEE Trans. Ant. Prop.* ***45***, 375 (1997).
67. A. Knorr, A. v. d. Heydt, B. Hanewinkel, and S. W. Koch, to be published.
68. P. A. Lee and T. V. Ramakrishnan, *Rev. Mod. Phys.* ***57***, 287 (1985).
69. D. Fröhlich et al., *Phys. Rev. Lett.* ***67***, 2343 (1991).
70. Ch. Hafner, *The Generalized Multiple Multipole Technique for Computational Electrodynamics*, Artech, Boston, MA, 1990.
71. O. J. F. Martin, C. Girard, and A. Dereux, *Phys. Rev. Lett.* ***74***, 526 (1995).
72. H. William, *Numerical Recipes*, Cambridge University Press, Cambridge, 1986.

CHAPTER FIVE

Localization of Light in Three-Dimensional Disordered Dielectrics

MARIAN RUSEK* and ARKADIUSZ ORŁOWSKI

Instytut Fizyki, Polska Akademia Nauk, Aleja Lotników 32/46, 02-668 Warszawa, Poland

5.1. INTRODUCTION

Investigations of the electron transport in disordered solids, usually semiconductors, led to the concept of localization of the electron wave functions. This phenomenon, known now as the Anderson localization, became a prominent part of contemporary condensed matter physics and is still a vivid subject of theoretical and experimental research. As shown by Anderson [1], in a sufficiently disordered infinite material, an entire *band* of electronic states can be spatially localized. In fact, the Anderson localization may be viewed as a transition from particle-like behavior described by the diffusion equation to wave-like behavior, which results in localization by interference. Indeed, the most plausible explanation of the Anderson localization is based on the interference effects in multiple elastic scattering of electrons on the material impurities [2].

As interference is the common property of all wave phenomena, the quest for some analogs of electron localization for other types of waves has been undertaken and many generalizations of electron localization exist, especially in the realm of electromagnetic (EM) waves [3–6]. So-called weak localization of EM waves manifesting itself as enhanced coherent backscattering is presently relatively well understood theoretically [7–9] and established experimentally [10–12]. The question is whether interference effects in three-dimensional (3D) random dielectric media can reduce the diffusion constant to zero, leading to strong localization. The crucial parameter is the mean free path l, which should be rather short [13–15].

*Presently on leave at Commissariat à l'Energie Atomique, DSM/DRECAM/SPAM, Centre d'Etudes de Saclay, 91191 Gif-sur-Yvette, France.

Optics of Nanostructured Materials, Edited by Vadim A. Markel and Thomas F. George
ISBN 0-471-34968-2

Of course, apart from remarkable similarities between scattering of electrons and light waves, there are also striking differences. The long-wavelength limit of elastic scattering, for example, is very different. For electrons, we mainly have s-wave scattering, which is spatially isotropic and wavelength independent. For light, we observe p-wave scattering. In this case, there is forward-backward symmetry but scattering is nonisotropic. In inelastic scattering, electrons change their energy but their total number is conserved. For light, we have strong absorption and the intensity decreases. Moreover, electrons are described by scalar wave functions (or two-component spinors if the spin is included). To describe scattering and localization of EM waves correctly, we need to consider, in general, 3D vector fields.

The Anderson localization of EM waves could be observed experimentally in the scaling properties of the transmission T. Imagine a slab of thickness L containing randomly distributed nonabsorptive scatterers. Usually, propagation of EM waves in weakly scattering random media can be described adequately by a diffusion process [16,17]. Thus the equivalent of Ohm's law holds and the transmission decreases linearly with the thickness of the sample, that is, $T \propto L^{-1}$ (for sufficiently large L). However, when the fluctuations of the dielectric constant become large enough, the EM field ceases to diffuse and becomes localized due to interference. Anderson localization occurs when this happens. In such a case, the material behaves as an optical equivalent of an insulator and the transmission decreases exponentially with the size of the system $T \propto e^{-L/\xi}$ [4,18].

A convincing experimental demonstration that Anderson localization is indeed possible in 3D disordered dielectric structures has been given recently [19]. The strongly scattering medium has been provided by semiconductor powders with a very large refractive index. By decreasing the average particle size, it was possible to observe a clear transition from linear scaling of transmission ($T \propto L^{-1}$) to an exponential decay ($T \propto e^{-L/\xi}$). Some localization effects also have been reported in previous experiments on microwave localization in copper tubes filled with metallic and dielectric spheres [20]. However, the latter experiments were plagued by large absorption, which makes the interpretation of the data quite complicated.

Thus, from the experimental point of view, there are indeed some reasonable indications that strong localization could be possible in 3D random dielectric structures. On the other hand, it would be desirable to have a reasonably simple yet realistic theoretical model providing deeper insight into the localization of light. It especially concerns those problems where the polarization effects have to be taken into account. Such considerations should assume the vector character of EM fields from the very beginning. To achieve this goal in a consistent way, they should be based directly on the Maxwell equations. On the other hand, they should be simple enough to provide calculations without very many crude approximations. In this chapter, we explicitly construct such a model for the 3D localization of EM waves. The resulting model is thoroughly analyzed and its major consequences are elaborated.

The main advantage of the presented approach is that we do not need to perform any averaging over the disorder. Generally speaking, there is a temptation to apply averaging procedures as soon as "disorder" is introduced into the model. Averaging of the scattered intensity over some random variable leads to a transport theory of localization [21–23]. But there is a very important and fundamental truth about random systems we must always keep in mind: No real atom is an average atom, nor is an experiment done on an ensemble of samples [24]. What we really need to properly understand the existing experimental results are probability distributions, not averages. Indeed, to perform any mathematically meaningful averaging procedure the assumption of infinite medium is needed. On the other hand, in all experiments we can study finite media only. Within our approach we can see how localization "sets in" for an increasing number of scatterers by studying the probability densities of eigenvalues of some random matrices.

This chapter is organized as follows. In Section 5.2 we introduce the Lippmann-Schwinger integral equations as a very convenient and effective tool for studying scattering of light by bounded dielectric media. By using this formalism in Section 5.3, we present general considerations dealing with elastic scattering of light waves. They will be used in Section 5.5 to derive the explicit form of coupling between a point-like scatterer and the electric field of the wave incident on it. In Section 5.4, a definition of light waves localized in dielectric media is proposed and its consequences are elaborated. It is shown that a way of dealing with localized states in the formalism of Lippmann-Schwinger equations is to solve them as a homogeneous system of equations, that is, for the incoming wave equal to zero. In Section 5.5, we recall the point-scatterer approximation and analyze the basic ideas behind it. A representation for the scatterer that rigorously fulfills the optical theorem and conserves energy in the scattering processes is derived. In Section 5.6, we arrive at the system of linear equations determining the polarization of the medium for a given incident wave. Eigenvalues of the random Green matrix corresponding to this set of equations are studied. The Breit–Wigner type model of the single scatterer allows us to give a clear physical interpretation of the obtained results. In this particular case, the real and imaginary parts of the eigenvalues of the Green matrix can be considered as first-order approximations to the relative widths and positions of the resonances. In Section 5.7, a numerical scattering experiment is performed. The positions and widths of the resonances are compared with the values obtained in Section 5.6. In Section 5.8, self-averaging of the real parts of the eigenvalues emerging in the limit of an infinite medium is discovered numerically. This phenomenon is illustrated graphically and observed features are compared with one-dimensional (1D) results. Note that in 1D the possibility of self-averaging can be proved analytically. In Section 5.9, a sound physical interpretation of the obtained results is proposed. Self-averaging of the real parts of the eigenvalues is considered as the signature of the appearance of the band of localized EM waves, emerging in the limit of infinite system. It can be understood as a counterpart of Anderson localization in solid-state physics. We finish with some comments and conclusions in Section 5.10.

5.2. BASIC ASSUMPTIONS

Here, we restrict ourselves to the study of properties of stationary solutions

$$\mathbf{E}(\mathbf{r},t) = \mathrm{Re}\{\mathcal{E}(\mathbf{r})e^{-i\omega t}\} \qquad \mathbf{H}(\mathbf{r},t) = \mathrm{Re}\{\mathcal{H}(\mathbf{r})e^{-i\omega t}\} \tag{5.1}$$

of the Maxwell equations. Consequently, the polarization of the medium is considered to be the oscillatory function of time

$$\mathbf{P}(\mathbf{r},t) = \mathrm{Re}\{\mathcal{P}(\mathbf{r})e^{-i\omega t}\} \tag{5.2}$$

The polarization of the isotropic linear and lossless dielectric medium described by the real dielectric constant $\epsilon(\mathbf{r})$ is related to the electric field by the well-known relation [25]:

$$\mathcal{P}(\mathbf{r}) = \frac{\varepsilon(\mathbf{r}) - 1}{4\pi}\mathcal{E}(\mathbf{r}) \tag{5.3}$$

Let us stress that in all experiments we can investigate only systems confined to certain finite regions of space. It is therefore reasonable to restrict our analysis to bounded media consisting of a finite number of dielectric particles. In this case, the polarization (5.3) satisfies the condition

$$\mathcal{P}(\mathbf{r}) = 0 \qquad \text{for} \qquad |\mathbf{r}| > R \tag{5.4}$$

Note that in Eq. (5.4) we have explicitly introduced the characteristic length-scale R, which will be used later on. If we denote by $k = \omega/c$ the wavenumber in vacuum and introduce the complex Hertz vector

$$\mathcal{Z}(\mathbf{r}) = \int d^3r'\,\mathcal{P}(\mathbf{r}')\,\frac{e^{ik|\mathbf{r}-\mathbf{r}'|}}{|\mathbf{r}-\mathbf{r}'|} \tag{5.5}$$

the field scattered by the finite dielectric medium (5.4) can be written in the following form [26]:

$$\mathcal{E}^{(1)}(\mathbf{r}) = \nabla\times\nabla\times\mathcal{Z}(\mathbf{r}) \qquad \mathcal{H}^{(1)}(\mathbf{r}) = -ik\,\nabla\times\mathcal{Z}(\mathbf{r}) \tag{5.6}$$

The total field may now be considered as the sum of the scattered field (5.6) and the free field $\mathcal{E}^{(0)}(\mathbf{r}), \mathcal{H}^{(0)}(\mathbf{r})$

$$\mathcal{E}(\mathbf{r}) = \mathcal{E}^{(0)}(\mathbf{r}) + \mathcal{E}^{(1)}(\mathbf{r}) \qquad \mathcal{H}(\mathbf{r}) = \mathcal{H}^{(0)}(\mathbf{r}) + \mathcal{H}^{(1)}(\mathbf{r}) \tag{5.7}$$

The system of equations (5.3) and (5.5)–(5.7) fully determines the EM field $\mathcal{E}(\mathbf{r})$, $\mathcal{H}(\mathbf{r})$ everywhere in space for a given field of the free wave $\mathcal{E}^{(0)}(\mathbf{r}), \mathcal{H}^{(0)}(\mathbf{r})$ incident on the system. Analogous relationships between the stationary outgoing wave and the stationary incoming wave are known in general scattering theory as the Lippmann–Schwinger equations [27].

Let us observe that far from the medium (i.e., for $|\mathbf{r}| \gg R$) the Hertz vector (5.5) can be approximated by

$$\mathcal{Z}(\mathbf{r}) \approx \mathcal{P}(k\,\mathbf{n})\,\frac{e^{ik|\mathbf{r}|}}{|\mathbf{r}|} \tag{5.8}$$

where

$$\mathcal{P}(\mathbf{k}) = \int d^3r\,\mathcal{P}(\mathbf{r})\,e^{-i\,\mathbf{k}\cdot\mathbf{r}} \tag{5.9}$$

is the spatial Fourier transform (FT) of the polarization and

$$\mathbf{n} = \frac{\mathbf{r}}{|\mathbf{r}|} \tag{5.10}$$

is the versor pointing to the direction of observation. We see from Eq. (5.8) that for each direction of observation $\mathbf{n}$ the far field scattered by the localized dielectric medium (5.4) looks like the field radiated by a certain Hertz dipole described by the polarization

$$\mathcal{P}(\mathbf{r}) = \mathcal{P}(k\,\mathbf{n})\,\delta(\mathbf{r}) \tag{5.11}$$

5.3. ELASTIC SCATTERING

We are interested in lossless media only, where localization is due to interference effects in elastic scattering of light by various parts of the medium. Obviously, if the bounded dielectric medium (5.4) is lossless, then the time-averaged field energy flux integrated over a surface surrounding it should vanish:

$$\int d\mathbf{s}\cdot\mathcal{S}(\mathbf{r}) = \frac{c}{4\pi}\frac{1}{2}\mathrm{Re}\int d\mathbf{s}\cdot\{\mathcal{E}(\mathbf{r})\times\mathcal{H}^*(\mathbf{r})\} = 0 \tag{5.12}$$

It turns out that the energy conservation condition (5.12) leads to restrictions imposed on the polarization $\mathcal{P}(\mathbf{r})$. As we will see in this section, the requirement that the time-averaged total-field energy flux integrated over a surface surrounding the considered bounded and lossless medium should vanish for the arbitrary incident wave $\mathcal{E}^{(0)}(\mathbf{r}), \mathcal{H}^{(0)}(\mathbf{r})$ uniquely determines the form of the dependence of the polarization of the medium on the field. The results of those general considerations, which deal with elastic scattering by bounded dielectric media, will be used in

Section 5.5 to derive of the explicit form of the coupling between the point-like scatterer and the electric field of the light wave incident on it.

After inserting formula (5.7) into the expression (5.12), it may be split into three terms. The first term

$$\int d\mathbf{s} \cdot \mathcal{S}^{(1)}(\mathbf{r}) = \frac{c}{4\pi}\frac{1}{2}\mathrm{Re}\int d\mathbf{s} \cdot \{\mathcal{E}^{(1)}(\mathbf{r}) \times \mathcal{H}^{(1)*}(\mathbf{r})\} \tag{5.13}$$

corresponds to the time-averaged energy radiated by the medium per unit time. To obtain its explicit form, we use the formula for the Poynting vector of the field radiated by the Hertz dipole (5.11) in the far-field limit (see, e.g., [26]):

$$\mathcal{S}^{(1)}(\mathbf{r}) = \mathbf{n}\,\frac{1}{8\pi}\frac{ck^4}{|\mathbf{r}|^2}|\mathcal{P}_T(k\,\mathbf{n})|^2 \tag{5.14}$$

where

$$\mathcal{P}_T(\mathbf{k}) = \mathcal{P}(\mathbf{k}) - \frac{\mathbf{k}(\mathbf{k}\cdot\mathcal{P}(\mathbf{k}))}{|\mathbf{k}|^2} \tag{5.15}$$

denotes the transverse part of the FT of the polarization. By integrating Eq. (5.14) over a sphere with radius $|\mathbf{r}|$ surrounding all sources, we get the following expression for the energy radiated on average by the dielectric medium (5.4):

$$\int d\mathbf{s} \cdot \mathcal{S}^{(1)}(\mathbf{r}) = \frac{ck^4}{8\pi}\int d\Omega|\mathcal{P}_T(k\,\mathbf{n})|^2 \tag{5.16}$$

The second term describes the total time-averaged energy flux integrated over a closed surface for the free field and thus vanishes

$$\frac{c}{4\pi}\frac{1}{2}\mathrm{Re}\int d\mathbf{s} \cdot \{\mathcal{E}^{(0)}(\mathbf{r}) \times \mathcal{H}^{(0)*}(\mathbf{r})\} = 0 \tag{5.17}$$

To calculate the last (third) interference term

$$\frac{c}{4\pi}\frac{1}{2}\mathrm{Re}\int d\mathbf{s} \cdot \{\mathcal{E}_0(\mathbf{r}) \times \mathcal{H}^{(1)*}(\mathbf{r}) + \mathcal{E}^{(1)}(\mathbf{r}) \times \mathcal{H}^{(0)*}(\mathbf{r})\} \tag{5.18}$$

we use the following identity (Lorentz theorem)

$$\nabla\cdot \{\mathcal{E}^{(0)}(\mathbf{r}) \times \mathcal{H}^{(1)*}(\mathbf{r}) + \mathcal{E}^{(1)*}(\mathbf{r}) \times \mathcal{H}^{(0)}(\mathbf{r})\} = -4\pi\, ik\, \mathcal{P}^*(\mathbf{r}) \cdot \mathcal{E}^{(0)}(\mathbf{r}) \tag{5.19}$$

which follows directly from the Maxwell equations. If we integrate (5.19) over a volume containing the isolated part of the medium under consideration and calculate the real part, we see that the Eq. (5.12) may be written in an equivalent form

$$\int d\mathbf{s}\cdot \mathcal{S}^{(1)}(\mathbf{r}) = \frac{1}{2}\, ck\, \mathrm{Re} \int d^3r\, \{i\mathcal{P}^*(\mathbf{r})\cdot \mathcal{E}^{(0)}(\mathbf{r})\} \tag{5.20}$$

Thus, on average the energy radiated by the medium must be equal to the energy given to the medium by the incident wave. Condition (5.20) together with Eq. (5.16) determines the relation between polarization and the electric field of the incident wave. It may be viewed as a generalized version of the optical theorem.

5.4. LOCALIZED WAVES

The standard approach to localized EM waves [3, 5, 28–33] is based on the similarities between the Helmholtz equation for the electric field amplitude in an isotropic lossless dielectric

$$\nabla \times \nabla \times \mathcal{E}(\mathbf{r}) + k^2[1 - \varepsilon(\mathbf{r})]\mathcal{E}(\mathbf{r}) = k^2\, \mathcal{E}(\mathbf{r}) \tag{5.21}$$

and the time-independent Schrödinger equation

$$\left\{-\frac{\hbar^2}{2m}\nabla^2 + V(\mathbf{r})\right\}\psi(\mathbf{r}) = E\,\psi(\mathbf{r}) \tag{5.22}$$

The term $k^2[1 - \varepsilon(\mathbf{r})]$ corresponds to the potential $V(\mathbf{r})$ providing localization of the electron wave function and the squared wavenumber in vacuum $k^2 = \omega^2/c^2$ plays the role analogous to the energy eigenvalue E. By definition, an EM wave is localized in a certain region of space if its magnitude is (at least) exponentially decaying in any direction from this region. We will show now that EM waves localized in the finite dielectric medium (5.4) correspond to nonzero solutions $\mathcal{E}(\mathbf{r}) \neq 0$ of Eqs. (5.3) and (5.5)–(5.7) for the incoming wave equal to zero, that is, $\mathcal{E}^{(0)}(\mathbf{r}) \equiv 0$.

Indeed, let us suppose that the field is exponentially localized in the vicinity of the bounded dielectric medium (5.4). First, let us observe that due to Eqs. (5.6) and (5.8), the scattered field $\mathcal{E}^{(1)}(\mathbf{r})$, $\mathcal{H}^{(1)}(\mathbf{r})$ tends to zero if $|\mathbf{r}| \to \infty$. Thus, if the total field (5.7) is exponentially localized, then the free-field $\mathcal{E}^{(0)}(\mathbf{r}), \mathcal{H}^{(0)}(\mathbf{r})$ must also tend to zero in this limit. But it is known from the vector form of the Kirchhoff integral formula [25] that if the free-field vanishes on a closed surface, then it is zero everywhere inside this surface.

The proof works also the other way round. Suppose that $\mathcal{E}(\mathbf{r})$ is a solution of Eqs. (5.3) and (5.5)–(5.7) for $\mathcal{E}^{(0)}(\mathbf{r}) \equiv 0$. For $z > R$ (the choice of the z axis is arbitrary), the scattered field $\mathcal{E}^{(1)}(\mathbf{r})$ can be expanded into the plane waves propagating into the positive z direction and the evanescent plane waves:

$$\mathcal{E}^{(1)}(\mathbf{r}) = \int \frac{d^2q}{(2\pi)^2}\, \mathbf{A}(\mathbf{q})\, e^{i\mathbf{k}\cdot\mathbf{r}} \qquad \text{where} \qquad \mathbf{k} = \left[q_x, q_y, \sqrt{k^2 - |\mathbf{q}|^2}\right] \tag{5.23}$$

If we use the Lorentz theorem, Eq. (5.19), and perform the straightforward but lengthly calculations (see, e.g., [25]) we easily arrive at the following expressions determining the coefficients $\mathbf{A}(\mathbf{q})$ that correspond to the propagating waves

$$\mathbf{A}(\mathbf{q}) = 2\pi\, ik\, \mathcal{P}_T(\mathbf{k}) \qquad \text{where} \qquad |\mathbf{q}| < k \tag{5.24}$$

As the considered medium is nondissipative, the time average energy stream integrated over a closed surface surrounding it must vanish:

$$\int d\mathbf{s} \cdot \mathcal{S}^{(1)}(\mathbf{r}) = 0 \tag{5.25}$$

By inserting (5.16) into condition (5.25), we see that the transverse part of the polarization of the medium vanishes on the light cone

$$\mathcal{P}_T(\mathbf{k}) = 0 \qquad \text{for} \qquad |\mathbf{k}| = \mathbf{k} \tag{5.26}$$

This means that there are no propagating plane waves in the scattered field (which in the case $\mathcal{E}^{(0)}(\mathbf{r}) \equiv 0$ is equal to the total field). Therefore, the field consists only of evanescent plane waves and thus is exponentially localized.

5.5. POINT SCATTERERS

Usually, localization of light is studied experimentally in microstructures consisting of dielectric spheres with diameters and mutual distances being comparable to the wavelength [15]. It is well known that the theory of multiple scattering of light by dielectric particles is tremendously simplified in the limit of point scatterers. In principle, this approximation is justified only when the size of the scattering particles is much smaller than the wavelength. In practical calculations, however, many multiple scattering effects can be obtained qualitatively for coupled electrical dipoles. Examples are universal conductance fluctuations [34], enhanced backscattering [35], dependent scattering [36], and strong localization in two [37,38], and three dimensions [39,40]. We believe that what really counts for localization is the scattering cross-section and not the geometrical shape and real size of the scatterer. Therefore, we will represent the dielectric particles located at the points $\mathbf{r}_a$ by *single* electric dipoles

$$\mathcal{P}(\mathbf{r}) = \sum_a \mathbf{p}_a\, \delta(\mathbf{r} - \mathbf{r}_a) \tag{5.27}$$

with properly adjusted scattering properties.

Let us mention that in practice any dielectric medium may be modeled by a set of discrete electric dipoles. This so-called coupled-dipole approximation was successfully used to study light scattering by a dielectric sphere [41] and more recently to obtain the scattering coefficients of arbitrarily shaped particles [42]. This method works well only if there are many dipoles in a volume whose dimensions are of the order of the wavelength [26]. In numerical calculations performed on supercomputers, a single small dielectric particle is built out of 10^6 dipoles (see, e.g., [43]). This difference between the coupled-dipole approximation and our qualitative approach is important. In our case, a single dielectric particle with a diameter comparable to the wavelength, is modeled only by *one* dipole with properly adjusted scattering properties.

It is known that several mathematical problems emerge in the formulation of interactions of point-like dielectric particles with EM waves [36,44,45]. Instead of applying several complicated regularization procedures, we will show that it is possible to analyze light scattering by point-like dielectric particles as a special case of general considerations dealing with elastic scattering of EM waves presented in Section 5.3. Previous results corresponding to the two-dimensional (2D) case were based on the Kirchhoff integral formula for scalar waves. We are interested in lossless media only, where localization is due to interference effects in elastic scattering of light by various dielectric particles. Therefore, the time-averaged Poynting vector integrated over a surface surrounding each scatterer must vanish for the arbitrary incident wave. As was shown in Section 5.3, this condition will be fulfilled if the dipole moments $\mathbf{p}_a$ depend on the electric field of the wave incident on them.

To obtain an explicit form of the coupling, let us recall the formula for the energy radiated on average by the Hertz dipole [26]:

$$\int d\mathbf{s} \cdot \mathcal{S}^{(1)}(\mathbf{r}) = \tfrac{1}{3} c k^4 |\mathbf{p}|^2 \tag{5.28}$$

which can be derived by substituting the polarization of a point-scatterer into Eq. (5.16) and performing the straightforward integration. By inserting the above expression and Eq. (5.27), we may rewrite condition (5.20) in the following form

$$\left| ik^3\, \mathbf{p}_a + \tfrac{3}{4}\, \mathcal{E}'(\mathbf{r}_a) \right|^2 = \left| \tfrac{3}{4}\, \mathcal{E}'(\mathbf{r}_a) \right|^2 \tag{5.29}$$

where the field acting on the ath scatterer

$$\mathcal{E}'(\mathbf{r}_a) = \mathcal{E}^{(0)}(\mathbf{r}_a) + \sum_{b \neq a} \mathcal{E}_b(\mathbf{r}_a) \tag{5.30}$$

is the sum of the free-field and waves scattered by all other particles

$$\mathcal{E}_a(\mathbf{r}) = \tfrac{2}{3}\, ik^3\, \mathbf{g}(\mathbf{r} - \mathbf{r}_a) \cdot \mathbf{p}_a \tag{5.31}$$

which are expressed by the Green tensor (see, e.g., [26]):

$$\mathbf{g}(\mathbf{r}) = \frac{3}{2} \frac{e^{ik|\mathbf{r}|}}{ik|\mathbf{r}|} \left\{ \left(\frac{3}{(k|\mathbf{r}|)^2} - i\frac{3}{k|\mathbf{r}|} - 1 \right) \frac{\mathbf{r}\,\mathbf{r}}{|\mathbf{r}|^2} - \left(\frac{1}{(k|\mathbf{r}|)^2} - i\frac{1}{k|\mathbf{r}|} - 1 \right) \right\} \tag{5.32}$$

Assuming that the vector on the left-hand side (lhs) of Eq. (5.29) is a function of the vector on the right-hand side (rhs) and that the dielectric particles modeled by the dipoles are spherically symmetrical we get

$$\tfrac{2}{3}\, ik^3\, \mathbf{p}_a = \tfrac{1}{2}(e^{2i\phi} - 1)\mathcal{E}'(\mathbf{r}_a) \tag{5.33}$$

Thus, to provide conservation of energy, the dipole moments are coupled to the electric field of the incident wave by complex "polarizability" $(e^{2i\phi} - 1)/2$, which can take values from a circle on the complex plane.

If we divide Eq. (5.28) by the intensity of a plane wave given by [26]

$$I = \frac{c}{8\pi}|\mathcal{E}^{(0)}(\mathbf{r}_a)|^2 \tag{5.34}$$

and insert Eq. (5.33), we obtain the explicit formula for the total scattering cross-section σ of an individual scatterer:

$$k^2\,\sigma = 6\pi \sin^2\phi \tag{5.35}$$

The scatterers necessarily have an internal structure. Thus, in general the phase shift ϕ from Eq. (5.33) should be regarded as a function of frequency ω. For example, to model a simple scatterer with one internal Breit–Wigner type resonance one can write

$$\cot\phi = -\frac{\omega - \omega_0}{\gamma_0} \tag{5.36}$$

The total scattering cross-section σ takes then the familiar Lorentzian form

$$k^2\,\sigma = \frac{6\pi\,\gamma_0^2}{(\omega - \omega_0)^2 + \gamma_0^2} \tag{5.37}$$

5.6. EIGENPROBLEM

Now, inserting Eq. (5.31) into (5.30), and using (5.33), it is easy to obtain the system of linear equations determining the field acting on each dipole $\mathcal{E}'(\mathbf{r}_a)$ for a given incoming wave $\mathcal{E}^{(0)}(\mathbf{r}_a)$

$$\mathcal{E}'(\mathbf{r}_a) = \mathcal{E}^{(0)}(\mathbf{r}_a) + \frac{e^{2i\phi} - 1}{2}\sum_b \boldsymbol{G}_{ab}\cdot\mathcal{E}'(\mathbf{r}_b) \tag{5.38}$$

The elements of the $\boldsymbol{G}$ matrix from Eq. (5.38) are equal to the Green function calculated for the differences between the positions of the scatterers

$$\boldsymbol{G}_{ab} = \begin{cases} \mathbf{g}(\mathbf{r}_a - \mathbf{r}_b) & \text{for } a \neq b \\ 0 & \text{for } a = b \end{cases} \tag{5.39}$$

If we solve Eqs. (5.38) and use again Eq. (5.33) to find $\mathbf{p}_a$, then we are able to find the EM field everywhere in space.

A way of dealing with *resonances* in this formalism is to look for resonance poles in the complex ω plane. Resonance poles are frequencies, ω, for which it is possible to solve Eqs. (5.38) as homogeneous equations (i.e., for the incoming wave $\mathcal{E}^{(0)}$ equal to zero). The real and imaginary parts of the resonance frequencies determine the positions and widths of the resonances. This method has been applied recently to the analysis of resonances in a system of $N = 2$ s-wave scatterers [46,47]. An example of such a system is the case of fixed frequency sound incident on small identical air bubbles in water [48–50]. Electron or phonon scattering from defects or impurities in crystal lattices give another example. It turned out that very interesting phenomena can arise for a pair of identical scatterers placed very close together, well within one wavelength. An extremely narrow p-wave *proximity* resonance develops from a broad s-wave resonance of individual scatterers. A new s-wave resonance of the pair also appears [46]. A similar problem of scattering of light from two spherical particles has been discussed by Markel [51].

It is seen from Eqs. (5.38) that for $\mathcal{E}^{(0)} \equiv 0$ the latter system of equations is equivalent to the *eigenproblem* for the $\boldsymbol{G}$ matrix

$$\sum_{b=1}^{N} \boldsymbol{G}_{ab} \cdot \mathcal{E}'(\mathbf{r}_b) = \lambda\, \mathcal{E}'(\mathbf{r}_a) \qquad a = 1, \cdots, N \tag{5.40}$$

where

$$\lambda = -1 - i \cot \phi \tag{5.41}$$

The Green matrix defined by Eq. (5.39) depends only on the scaled distances between all pairs of the scatterers $k|\mathbf{r}_a - \mathbf{r}_b|$. Therefore for fixed positions of the scatters $\mathbf{r}_a$, its eigenvalues still remain functions of frequency ω. By using an explicit model of the scattering phase shift $\phi(\omega)$ and by solving Eq. (5.41) in a complex ω plane, it is possible to determine the positions and widths of the resonances. In the particular case of the Breit–Wigner type scatterers [Eq. (5.36)], the real and imaginary parts of the eigenvalues of the $\boldsymbol{G}$ matrix have a nice physical interpretation: They are equal to the relative widths $(\gamma - \gamma_0)/\gamma_0$ and positions $(\omega - \omega_0)/\gamma_0$ of the resonances. Indeed, by using the explicit form of the complex frequency $\omega \rightarrow \omega - i\gamma$ and by substituting the Breit–Wigner model of the scattering into Eq. (5.41) we get

$$\omega - i\gamma = \omega_0 - i\gamma_0[1 + \lambda(\omega - i\gamma)] \tag{5.42}$$

This system of two coupled nonlinear equations determines the values of the resonance poles $\omega - i\gamma$. In many physically interesting cases, Eqs. (5.42) can be solved numerically by iteration. For instance, in solving it up to the first order in

γ_0/ω_0 one substitutes $\lambda(\omega_0)$ for $\lambda(\omega - i\gamma)$ getting

$$\operatorname{Re}\lambda(\omega_0) \simeq \frac{\gamma - \gamma_0}{\gamma_0} \qquad \operatorname{Im}\lambda(\omega_0) \simeq \frac{\omega - \omega_0}{\gamma_0} \tag{5.43}$$

Let us start with a simplest possible example of a system of $N = 2$ scatterers separated by a distance d. In this case, the $\boldsymbol{G}$ matrix from Eq. (5.39) has four eigenvalues: $\lambda_{\pm}^{(T)} = \mp\frac{3}{2}\, e^{ikd}/ikd([1/(kd)^2] - i(1/kd) - 1)$ corresponding to the transverse oscillations of the dipoles, and $\lambda_{\pm}^{(L)} = \pm\frac{3}{2}\, e^{ikd}/ikd([2/(kd)^2] - i(2/kd))$ corresponding to the longitudinal oscillations of the dipoles. The upper sign (+) describes oscillations in phase, whereas the lower sign (−) corresponds to the oscillations in antiphase. For $k = k_0 = \omega_0/c$, the eigenvalues $\lambda_{\pm}^{(T)}$ and $\lambda_{\pm}^{(L)}$ may be considered as an approximate solution of Eq. (5.42) up to the first order in γ_0/ω_0.

The eigenvalues $\lambda_{\pm}^{(T)}$ and $\lambda_{\pm}^{(L)}$ are depicted in Fig. 5.1. They form a characteristic four-arms spiral. We see that for scatterers very close to each other ($d \to 0$) two arms of this spiral corresponding to the oscillations in antiphase approach the axis $\operatorname{Re}\lambda = -1$ ($\gamma = 0$). They are related to the very narrow "antisymmetric" resonances. On the other hand, in this limit the remaining two arms corresponding to the oscillations in phase tend asymptotically to the axis $\operatorname{Re}\lambda = 1$ ($\gamma = 2\gamma_0$). These arms are related to the "symmetric" resonances of the pair that are about twice as broad as the resonance of the single scatterer. For $d \to \infty$, both arms meet in the point $\lambda = 0$ ($\omega = \omega_0, \gamma = \gamma_0$) reproducing the results of the single scattering.

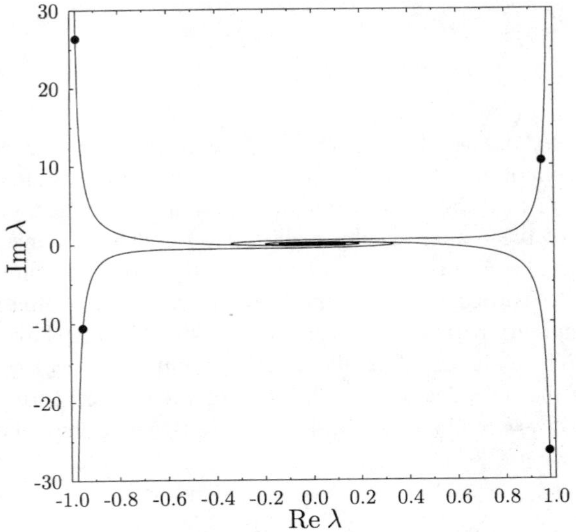

Figure 5.1. Eigenvalues λ of a $\boldsymbol{G}$ matrix corresponding to a system of $N = 2$ point-like scatterers. All four types of resonances are clearly visible. The four black dots correspond to the eigenvalues calculated for a certain specific value of the distance between the scatterers kd.

As a second example, let us consider systems of $N = 100$ and $N = 1000$ scatterers placed randomly inside a sphere, with the uniform density $n = 1$ scatterer per wavelength cubed l_0^3. The chosen wavelength, l_0, corresponds to the resonance frequency ω_0 (i.e., $l_0 = 2\pi/k_0$, where $k_0 = \omega_0/c$). In Figs. 5.2 and 5.3, we plot the eigenvalues λ obtained after numerically diagonalizing the corresponding $\boldsymbol{G}$ matrices. In the case of the Breit–Wigner type scatterers, these eigenvalues can be considered as the approximate positions of the resonance poles up to the first order in γ_0/ω_0. By comparing Figs. 5.2 and 5.3 with Fig. 5.1, we see that the tails corresponding to "antisymmetric" proximity resonances still persists for a larger number of scatterers. On the other hand, the remaining two arms of the spiral from Fig. 5.1 corresponding to the "symmetric" resonances between pairs of scatterers completely disappeared in the case of $N = 1000$. It follows also from inspection of Figs. 5.2 and 5.3 that for an increasing number of scatterers new collective effects start to appear. They are visible especially for $\mathrm{Im}\,\lambda \simeq 0$, that is, for frequencies that are close to the resonance frequency of a single scatterer ($\omega \simeq \omega_0$). For instance, in this range of frequencies quite a lot of eigenvalues are located near the $\mathrm{Re}\,\lambda = -1$ ($\gamma = 0$) axis. They correspond to narrow resonances with width $\gamma \simeq 0.25\,\gamma_0$. As we will show shortly, the width of these resonances γ decreases with an increasing number of scatterers N (while keeping the density constant), and in the limit of the infinite medium $N \to \infty$, they become localized states. It is also seen from Fig. 5.3 that a few new broad (i.e., $\gamma \simeq 2.5\,\gamma_0$) resonances appear for $\omega \simeq \omega_0$ in this limit.

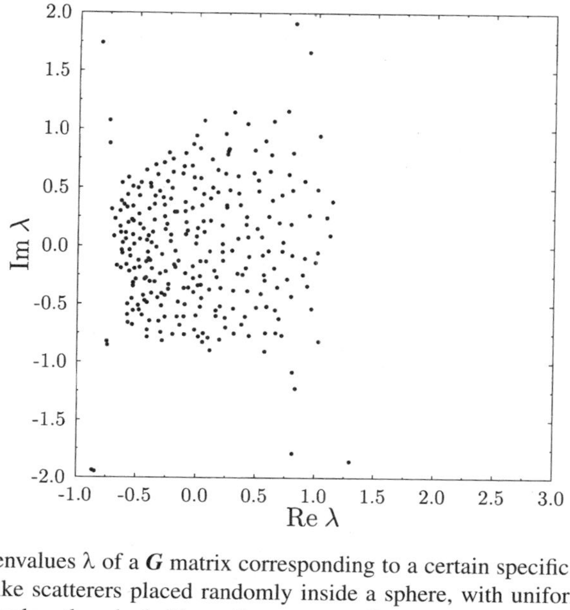

Figure 5.2. Eigenvalues λ of a $\boldsymbol{G}$ matrix corresponding to a certain specific configuration of $N = 100$ point-like scatterers placed randomly inside a sphere, with uniform density $n = 1$ scatterer per wavelength cubed. The tails corresponding to "symmetric" and "antisymmetric" proximity resonances from Fig. 5.1 still persist.

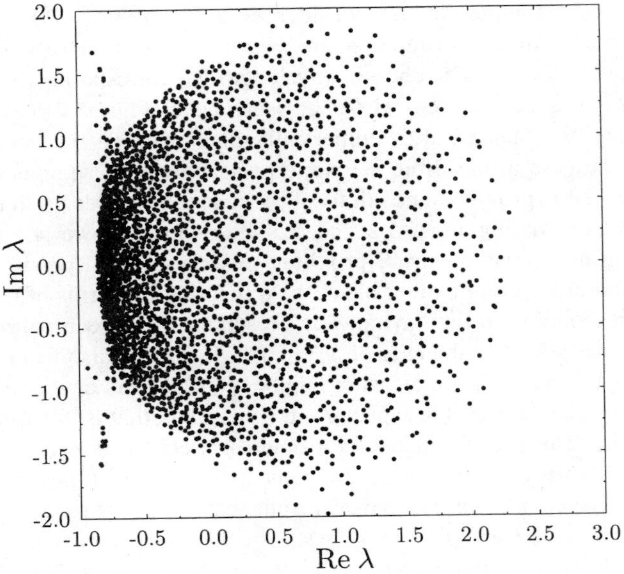

Figure 5.3. Eigenvalues λ of a $\boldsymbol{G}$ matrix corresponding to a certain specific configuration of $N = 1000$ point-like scatterers placed randomly inside a sphere, with uniform density $n = 1$ scatterer per wavelength cubed. Quite a lot of eigenvalues are located near the $\mathrm{Re}\,\lambda = -1$ axis.

Looking for resonance poles in a complex energy plane turns out to be an enormous numerical problem for a large number of scatterers. Nevertheless, as we have seen in the case of $N = 2$ Breit–Wigner type scatterers, sometimes it is possible to extract some qualitative information about the resonances just from the spectrum of the $\boldsymbol{G}$ matrix calculated for real values of energy. Moreover, as will be shown in Section 5.8, a striking phase-transition-like behavior appears in the spectra of such Green matrices when the number of scatterers increases. This transition may be interpreted as an appearance of the band of localized states emerging in the limit of the infinite medium. It is an interesting analog of the Anderson localization in noncrystalline solids, such as amorphous semiconductors or disordered metals.

5.7. SCATTERING EXPERIMENT

The actual properties of physical systems have to be observed experimentally and it is not enough just to know the properties of the stationary solutions of the Maxwell equations. These are only theoretical tools. Experiments deal rather with measurable quantities and, for many practical problems, a natural quantity to look for is the total scattering cross-section of a finite system σ and its dependence on the frequency ω. In this section, we would like to compare the positions and widths of the resonances of the total scattering cross-section of a system of point-like scatterers with the approximate positions of the complex resonance poles calculated in Section 5.6.

By inserting the free field in the form of a plane wave into Eq. (5.20):

$$\mathcal{E}^{(0)}(\mathbf{r}) = \mathcal{E}^{(0)}(0)\, e^{i\mathbf{k}\cdot\mathbf{r}} \tag{5.44}$$

and by substituting Eq. (5.27) for Eq. (5.33), then dividing the resulting equation by the intensity of the incident wave Eq. (5.34), and assuming $|\mathcal{E}^{(0)}(0)|^2 = 1$, we arrive at the explicit formula for the total scattering cross-section σ of a system of N identical point-like scatterers

$$k^2\,\sigma = -6\pi\,\mathrm{Re}\left\{\frac{e^{2i\phi}-1}{2}\sum_{a=1}^{N}\mathcal{E}'(\mathbf{r}_a)\cdot\mathcal{E}^{(0)*}(\mathbf{r}_a)\right\} \tag{5.45}$$

By substituting $\mathcal{E}^{(0)}(\mathbf{r}_a)$ from Eq. (5.30) after lengthly but straightforward calculations it is possible to rewrite Eq. (5.45) in an equivalent way

$$k^2\,\sigma = 6\pi\sin^2\phi\sum_{a=1}^{N}\sum_{b=1}^{N}\mathcal{E}'(\mathbf{r}_a)\cdot\mathrm{Re}\{\mathbf{g}(\mathbf{r}_a-\mathbf{r}_b)\}\cdot\mathcal{E}'^{*}(\mathbf{r}_b) \tag{5.46}$$

Equation (5.46) can also be derived by substituting the polarization of a system of point scatterers (5.27) into Eq. (5.16), using Eq. (5.33), and explicitly performing the integration over the solid angle.

Let us now suppose that $\mathcal{E}'(\mathbf{r}_a)$ is an eigenvector of the $\boldsymbol{G}$ matrix corresponding to the eigenvalue λ. In this case, Eq. (5.30) reduces to

$$\left\{1-\frac{e^{2i\phi}-1}{2}\lambda\right\}\mathcal{E}'(\mathbf{r}_a) = \mathcal{E}^{(0)}(\mathbf{r}_a), \qquad a = 1,\ldots,N \tag{5.47}$$

and Eq. (5.46) takes the following form:

$$k^2\,\sigma = \frac{6\pi}{(\cot\phi+\mathrm{Im}\,\lambda)^2+(1+\mathrm{Re}\,\lambda)^2} \times\sum_{a=1}^{N}\sum_{b=1}^{N}\mathcal{E}^{(0)}(\mathbf{r}_a)\cdot\mathrm{Re}\{\mathbf{g}(\mathbf{r}_a-\mathbf{r}_b)\}\cdot\mathcal{E}^{(0)*}(\mathbf{r}_b) \tag{5.48}$$

Therefore, the scattering cross-section σ considered as a function of $\cot\phi$ for a constant value of the frequency ω has a form of a Lorentzian curve of width $\mathrm{Re}\,\lambda + 1$ centered at $\mathrm{Im}\,\lambda = -\cot\phi$. Note that for Breit–Wigner type scatterers [Eq. (5.36)] we have$(\omega-\omega_0)/\gamma_0 = -\cot\phi$. Thus, in this particular case, the plot of σ as a function of $-\cot\phi$ computed for $\omega = \omega_0$ may be considered as an approximation of the plot of σ as a function of $(\omega-\omega_0)/\gamma_0$ up to the first order in γ_0/ω_0. Therefore, Eq. (5.48) is consistent with results of Section 5.6, where we have shown that in the case of Breit–Wigner type scatterers the first-order approximations to the relative

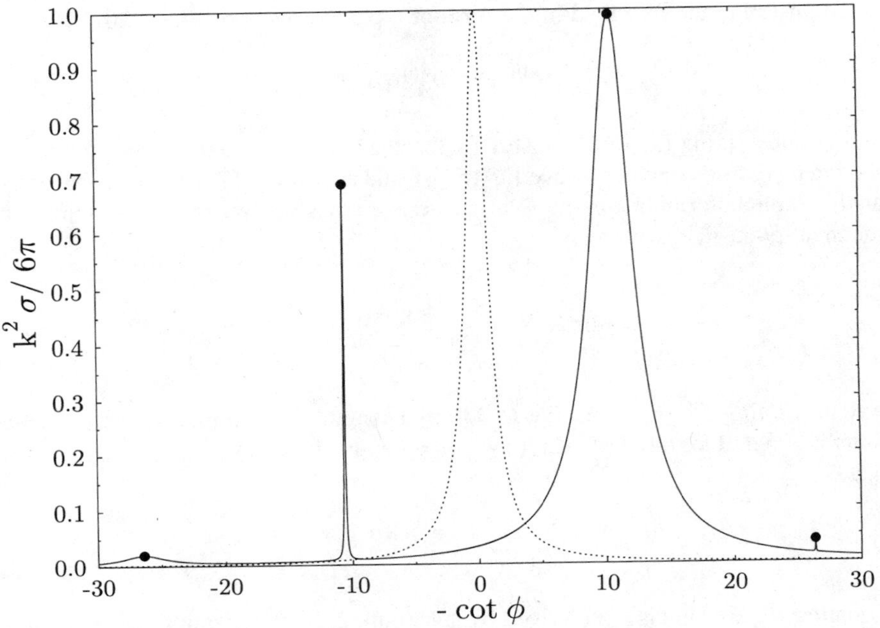

Figure 5.4. Scattering cross-section σ of a certain specific configuration of $N = 2$ point-like scatterers from Fig. 5.1 ploted as a function of $-\cot\phi$. The dashed line corresponds to the scattering cross-section of an individual scatterer. All four types of resonances are clearly visible.

widths $(\gamma - \gamma_0)/\gamma_0$ and positions $(\omega - \omega_0)/\gamma_0$ of the resonances are given by the real and imaginary parts of the eigenvalues of the $\boldsymbol{G}$ matrix calculated for $\omega = \omega_0$.

In Fig. 5.4, we have the scattering cross-section σ of a system of $N = 2$ point-like scatterers ploted as a function of $-\cot\phi$. The system was illuminated by a linearly polarized plane wave. This plot corresponds to the certain specific configuration of scatterers from Fig. 5.1. The dashed line shows the scattering cross-section of an individual scatterer given by Eq. (5.35). We see from inspection of Fig. 5.4 that all four types of resonances are clearly visible. From left to right, we have maxima of the scattering cross-section corresponding to the resonances related to the longitudinal oscillations in phase, transverse oscillations in antiphase, transverse oscillations in-phase, and longitudinal oscillations in antiphase. The difference in strength between the transverse and longitudinal oscillations follows from the orientation of the system with respect to the polarization of the incident wave. The black dots in Fig. 5.4 mark the positions of the resonances calculated from the eigenvalues depicted as black dots in Fig. 5.1 by using the formula $\mathrm{Im}\,\lambda = -\cot\phi$.

In Fig. 5.5, we present the scattering cross-section σ/N of systems of $N = 100$ and 1000 point-like scatterers from Figs. 5.2 and 5.3 ploted as a function of $-\cot\phi$. For comparison, we also include a plot of the scattering cross-section of an individual scatterer. The narrow collective resonances from Figs. 5.2 and 5.3 are not

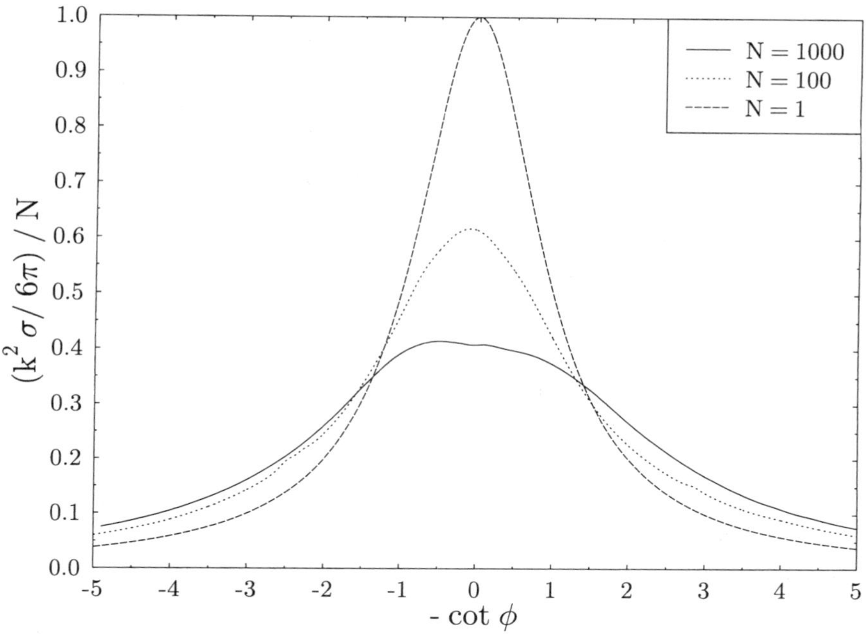

Figure 5.5. Scattering cross-section σ of systems of $N = 100$ and 1000 point-like scatterers from Figs. 5.2 and 5.3 ploted as a function of $-\cot\phi$. The dashed line corresponds to the scattering cross-section of an individual scatterer. A very broad collective resonance appears for an increasing number of scatters. It can be considered as a precursor of the Anderson localization.

visible. A possible reason for this is that the corresponding eigenvectors of the $\boldsymbol{G}$ matrix are orthogonal to the vector formed by the values of incident field calculated at the positions of the scatterers. Nevertheless, the appearance of a very broad collective resonance is readily seen. Note that in the case of a system of N scatterers scattering incoherently, the total scattering cross-section of the system should scale as N (i.e, $\sigma/N = \text{const}$). It is seen from Fig. 5.5 that for $|\cot\phi| < 1$ (or $|\omega - \omega_0| < \gamma_0$ in the case of Breit–Wigner scatterers) the scattering cross-section of a system is below this incoherent limit. On the other hand, for $|\cot\phi| > 1$ (or $|\omega - \omega_0| > \gamma_0$ in the case of Breit–Wigner scatterers) we observe enhanced coherent scattering. It may be considered as a precursor of the Anderson localization.

5.8. SELF-AVERAGING

To illustrate the appearance of the band of localized EM waves, emerging in the limit of infinite system, we have to study the properties of *finite* systems for an increasing number of dipoles N (while keeping the density constant). For each distribution of the dipoles $\mathbf{r}_a$ placed randomly inside a sphere with the uniform scaled density $n = 1$

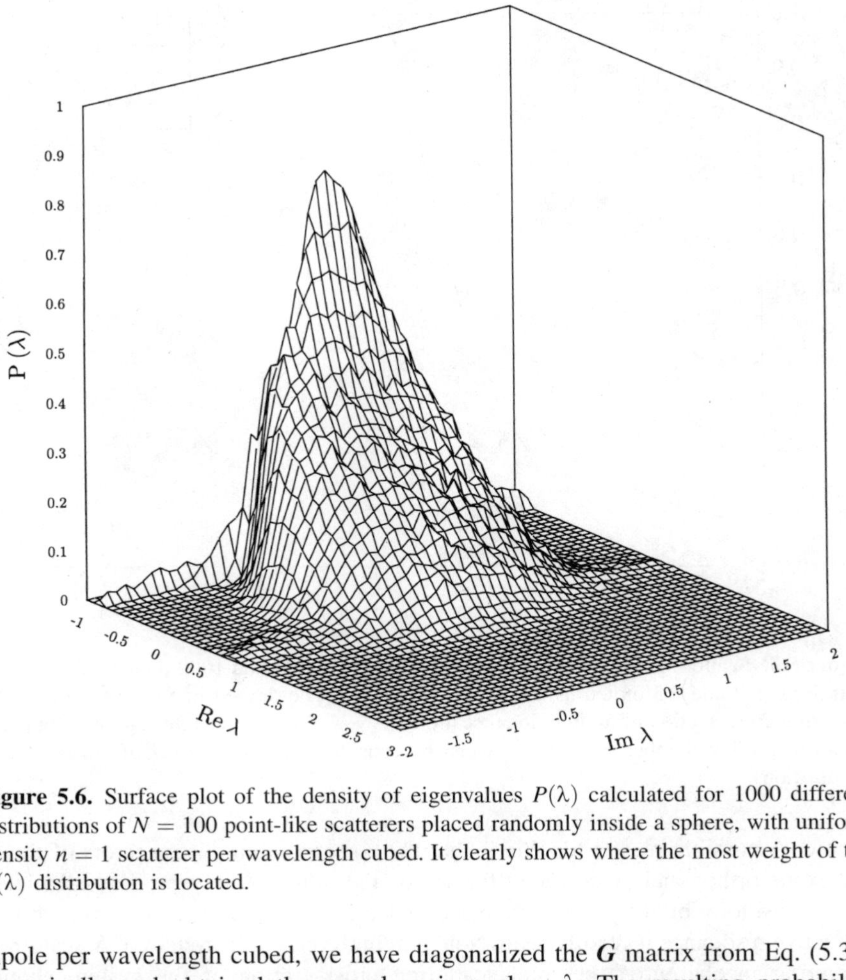

Figure 5.6. Surface plot of the density of eigenvalues $P(\lambda)$ calculated for 1000 different distributions of $N = 100$ point-like scatterers placed randomly inside a sphere, with uniform density $n = 1$ scatterer per wavelength cubed. It clearly shows where the most weight of the $P(\lambda)$ distribution is located.

dipole per wavelength cubed, we have diagonalized the $\boldsymbol{G}$ matrix from Eq. (5.39) numerically and obtained the complex eigenvalues λ. The resulting probability distribution $P(\lambda)$, calculated from several different distributions of N dipoles is normalized in the standard way $\int d^2\lambda\, P(\lambda) = 1$. Let us now compare the surface plots of $P(\lambda)$ (treated as a function of two variables $\mathrm{Re}\,\lambda$ and $\mathrm{Im}\,\lambda$) calculated for systems consisting of $N = 100$ and 1000 dipoles. They are presented in Figs. 5.6 and 5.7, respectively. It is seen from inspection of these plots that, for increasing size of the system (in our case it increased $\sqrt[3]{10} \simeq 2$ times), at some $\mathrm{Im}\,\lambda$ the probability distribution $P(\lambda)$ apparently moves toward the $\mathrm{Re}\,\lambda = -1$ axis and its variance simultaneously decreases. This tendency is easily seen, for example, for values of $|\mathrm{Im}\,\lambda|$ that are close to 0. Our numerical investigations indicate that in the limit of an infinite medium, the probability distribution $P(\lambda)$ tends to the delta function in $\mathrm{Re}\,\lambda$:

$$\lim_{N\to\infty} P(\lambda) = \delta(\mathrm{Re}\,\lambda + 1) f(\mathrm{Im}\,\lambda) \tag{5.49}$$

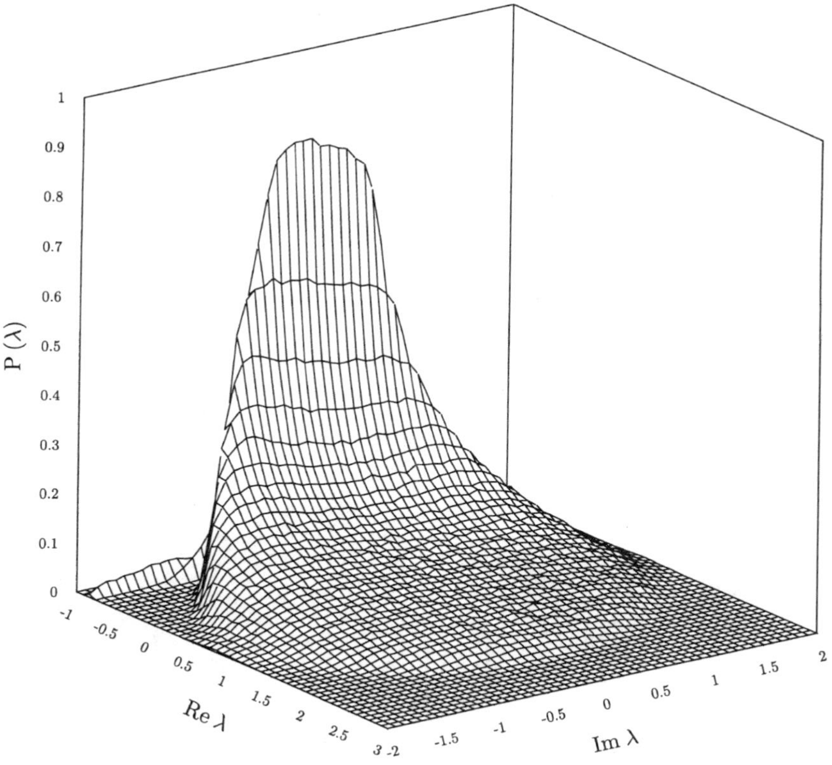

Figure 5.7. Surface plot of the density of eigenvalues $P(\lambda)$ calculated for 500 different distributions of $N = 1000$ point-like scatterers placed randomly inside a sphere, with uniform density $n = 1$ scatterer per wavelength cubed. For increasing values of N, the probability distribution $P(\lambda)$ apparently moves toward the $\mathrm{Re}\,\lambda = -1$ axis and, simultaneously, its variance along the $\mathrm{Im}\,\lambda = \mathrm{const}$ axes decreases.

We have some numerical evidence that this fact is a general property of $\boldsymbol{G}$ matrices, not restricted to the considered case of one dipole per wavelength cubed $n = 1$. Of course, we could justify Eq. (5.49) by a more orthodox approach based on a version of the finite size scaling analysis that leads, however, to an analogous conclusion.

It follows from Eq. (5.49) that in the limit $N \to \infty$ the distribution function $P(\lambda)$ has only one value of $\mathrm{Re}\,\lambda$ for which it is nonzero. The quantity $\mathrm{Re}\,\lambda$ is then "self-averaging" and the random process has in fact become deterministic. Knowledge of the average then provides knowledge about "almost every" individual realization of the random system. This property implies that the average value applies to *every single* realization of the system except for a few special ones (with measure zero). This means that for almost any random distribution of the dipoles $\mathbf{r}_a$, the equation $\mathrm{Re}\,\lambda = -1$ holds. Therefore, Eq. (5.41) can be fulfilled if the phase shift of the scatterers satisfies: $\mathrm{Im}\,\lambda = -\cot\phi$. In this case, the corresponding eigenvector

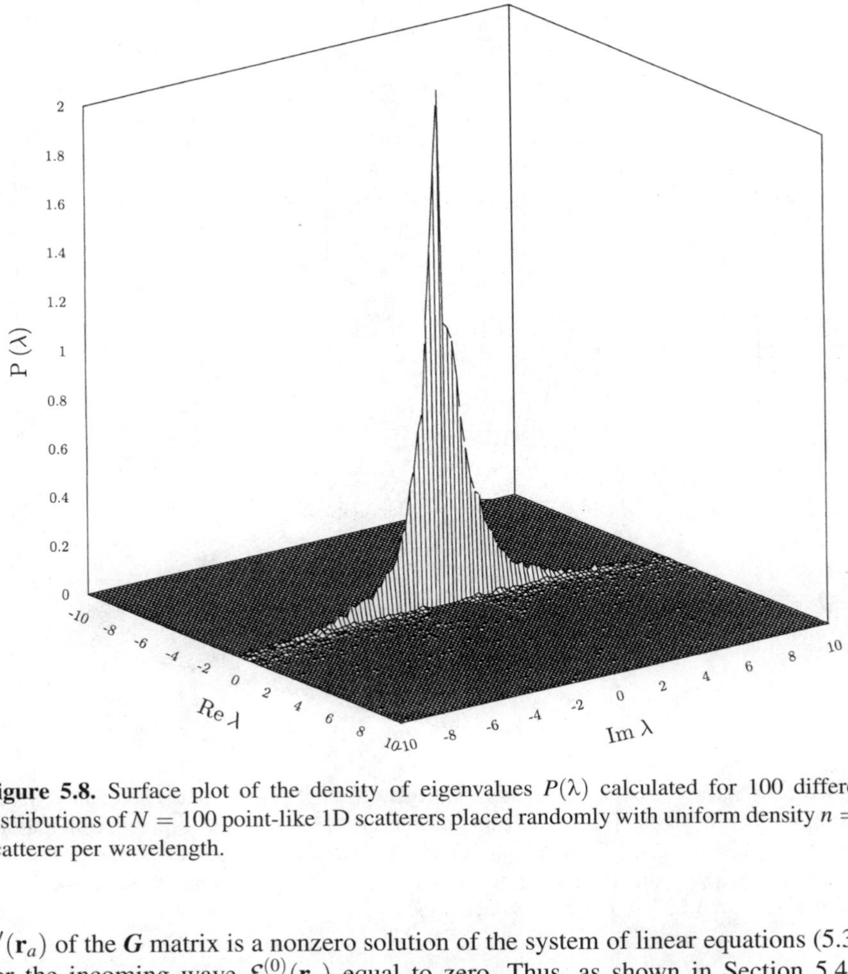

Figure 5.8. Surface plot of the density of eigenvalues $P(\lambda)$ calculated for 100 different distributions of $N = 100$ point-like 1D scatterers placed randomly with uniform density $n = 1$ scatterer per wavelength.

$\mathcal{E}'(\mathbf{r}_a)$ of the $\boldsymbol{G}$ matrix is a nonzero solution of the system of linear equations (5.38) for the incoming wave $\mathcal{E}^{(0)}(\mathbf{r}_a)$ equal to zero. Thus, as shown in Section 5.4, a localized wave exists.

In three dimensions, proofs of self-averaging are rare and in most cases quantities are not self-averaging [52]. For waves propagating in 1D random systems (meaning that two out of three dimensions are translationally invariant and only the third is random) self-averaging can be demonstrated mathematically. For 1D systems, it was shown that for "almost any" energy or frequency an eigenfunction decays exponentially in space for "almost any" realization of the disorder [53,54]. This fact is also unambiguously confirmed within the 1D version of our model. In Fig. 5.8, we present a 1D counterpart of Figs. 5.6 and 5.7. It is easily seen from inspection of these figures that Eq. (5.49) is also satisfied for systems consisting of 1D point-like scatterers. Moreover, this mathematical property of random Green matrices seems to be fulfilled in the 2D free-space case as well [38] and in the case of a 2D system with nontrivial boundary conditions [55]. Thus it appears to be truly universal.

5.9. ANDERSON LOCALIZATION

Electronic states in solids are usually either extended, by analogy with the Bloch picture for crystalline media, or are localized around *isolated* spatial regions such as surfaces and impurities. However, in the case of a sufficiently disordered system a countable set of discrete energies corresponding to localized states becomes dense in some finite interval. But, physically speaking, it is impossible to distinguish between the allowed energies, which may be arbitrarily close to each other; the spectrum is always a coarse-grained object. Therefore an entire continuous band of spatially localized electronic states appears. Anderson localization occurs when this happens [56]. Similarly, it is reasonable to expect that in the case of an infinite and random collection of dielectric particles there can exist a band of localized EM waves corresponding to a region of frequencies ω. This analogy allows us to elaborate on a physical interpretation of the results obtained with the point-scatterer model used.

Let us now apply our model to a system of identical dielectric spheres with given radii R and dielectric constants ε located randomly with uniform physical density η. In this case, the parameter ϕ from Eq. (5.33) remains a function of the frequency, that is, $\phi = \phi(\omega)$. It follows from Eq. (5.49), for almost any random distribution of the scatterers $\mathbf{r}_a$ (except maybe for a few special ones with measure zero), that an infinite number of eigenvalues λ of the $\boldsymbol{G}$ matrix satisfies the condition

$$\operatorname{Re} \lambda_j(\omega) = -1 \tag{5.50}$$

Note that we added an index j, which labels the localized waves. As pointed out before, this occurs not only for $n = 1$ but for a whole range of n and, therefore, for fixed physical density ρ, for a range of frequencies ω. Thus, at real values of frequency ω_j determined by the equation

$$\operatorname{Im} \lambda_j(\omega_j) = -\cot \phi(\omega_j) \tag{5.51}$$

these eigenvalues are solutions of Eq. (5.41). Now the corresponding eigenvectors $\mathcal{E}'_j(\mathbf{r}_a)$ of the $\boldsymbol{G}$ matrix describe localized states that exist at *discrete* frequencies ω_j. Note that this result does not depend on the particular model of the scatterer used.

Let us observe that the $\boldsymbol{G}$ matrix from Eq. (5.39) is a *traceless* matrix, (i.e., $\sum_i \lambda_i = 0$). This means that it is impossible for all eigenvalues to fulfill the localization condition Eq. (5.50). Inspection of Figs. 5.2 and 5.3 suggest that these eigenvalues may be first approximations to very broad resonances responsible for enhanced coherent backscattering from a random medium. This so-called weak localization is usually considered as a precursor of the strong localization (Anderson localization). The property $\sum_i \lambda_i = 0$ serves also as a good test for accuracy of our numerical procedures.

It seems reasonable to expect that the function f from Eq. (5.49) has compact support

$$f(\operatorname{Im} \lambda) = 0 \quad \text{for} \qquad |\operatorname{Im} \lambda| > \operatorname{Im} \lambda_{cr} \tag{5.52}$$

(the values of Im λ_{cr} should be regarded as functions of ω). According to Eq. (5.51), this means that localized waves of frequency ω can exist only if

$$|\phi(\omega)| \geq \phi_{cr}(\omega) \tag{5.53}$$

where

$$\cot \phi_{cr}(\omega) = \operatorname{Im} \lambda_{cr}(\omega) \tag{5.54}$$

We see from Eqs. (5.35) and (5.53) that the total scattering cross-section of individual particles σ must exceed some critical value $\sigma_{cr} = \sigma(\phi_{cr})$ before localization will take place in the limit $N \to \infty$. This fact is in perfect agreement with the scaling theory of localization [18]: In 3D random media, a certain critical degree of disorder is needed for localization.

Moreover, our preliminary calculations indicate that the value of $k^2 \sigma_{cr}$ may decrease with n but *slower* than n^{-2}. If we use the Rayleigh expression for the total scattering cross-section σ of a dielectric sphere with radius R and dielectric constant ε [25]

$$k^2 \sigma = \frac{8}{3\pi} (kR)^6 \left| \frac{\varepsilon - 1}{\varepsilon + 1} \right|^2 \tag{5.55}$$

we conclude that in the long-wave limit the system of dielectric spheres distributed with constant density $\eta = k^3 n/(2\pi)^3$ will be out of the localization regime. On the other hand, in the limit of small wavelengths, the propagation of light is ruled by the laws of geometrical optics and the point-scatterer approximation we use becomes invalid. Therefore, our results seem to agree with the common believe (see, e.g., [14,15]) that in 3D media localization of light is possible only in a certain frequency window:

$$\omega_{min} \leq \omega \leq \omega_{max} \tag{5.56}$$

In the limit $N \to \infty$, a countable set of discrete frequencies ω_j corresponding to localized waves becomes dense in this finite interval given by Eq. (5.56). Thus an entire *band* of spatially localized waves appears in the limit of an infinite medium. Physically speaking this means that different realizations of a sufficiently large *disordered* system are practically (i.e., by a transmission experiment) indistinguishable from each other. Similarly, as pointed out by Anderson, in a sufficiently disordered solid an entire band of spatially localized electronic states can be formed [1,24].

By analogy with the electron case, the phenomenon of Anderson localization of EM waves should manifest itself as an inhibition of the transmission in a spatially random dielectric medium. We already have some numerical evidence that it is actually true in the case of a 2D system consisting of randomly distributed dielectric cylinders [55]. The validity of this connection in the considered 3D model would attribute sound interpretation and clear physical meaning to the continuous region of frequencies corresponding to localized waves. We expect that for each point ω from this region, incident waves with frequency ω will be totally reflected by almost any random distribution of the spheres $\mathbf{r}_a$ with scattering properties $\phi(\omega)$.

5.10. BRIEF SUMMARY

We have presented a quite realistic point-scatterer model describing scattering of EM waves by a disordered dielectric medium. Its relative simplicity allowed us to discover some new features of the Anderson localization of EM waves in 3D dielectric media without using any averaging procedures. Within our theoretical approach, one can easily see how localization "sets in" for increasing size of the system. Very striking universal properties of the spectra of random matrices describing the scattering from a collection of randomly distributed point-like scatterers have been observed. Self-averaging of the real parts of the eigenvalues emerging in the limit of an infinite medium has been discovered numerically. For the first time (to our knowledge), the appearance of the band of localized EM waves in 3D was demonstrated. A connection between this phenomenon and a dramatic inhibition of the propagation of EM waves in a spatially random dielectric medium has been sketched. It can be understood as a counterpart of Anderson localization in solid-state physics.

REFERENCES

1. P. W. Anderson, "Absence of diffusion in certain random lattices," *Phys. Rev.* ***109***, 1492 (1958).
2. M. Kaveh, "What to expect from similarities between the Schrödinger and Maxwell equations," in W. van Haeringen and D. Lenstra, Eds., *Analogies in Optics and Micro Electronics*, Kluwer, Dordrecht, The Netherlands, 1990, p. 21.
3. S. John, "Electromagnetic absorption in a disordered medium near a photon mobility edge," *Phys. Rev. Lett.* ***53***, 2169 (1984).
4. P. W. Anderson, "The question of classical localization. A theory of white paint"? *Philos. Mag. B* ***52***, 505 (1985).
5. S. John, "Strong localization of photons in certain disordered dielectric superlattices," *Phys. Rev. Lett.* ***58***, 2486 (1987).
6. C. M. Soukoulis, Ed. *Photonic Band Gaps and Localization*, New York, NATO ASI Series, Plenum, New York, 1993.
7. E. Akkermans, P. E. Wolf, and R. Maynard, "Coherent backscattering of light by disordered media: Analysis of the peak line shape," *Phys. Rev. Lett.* ***56***, 1471 (1986).
8. M. J. Stephen and G. Cwillich, "Rayleigh scattering and weak localization: Effects of polarization," *Phys. Rev. B* ***34***, 7564 (1986).
9. F. C. MacKintosh and S. John, "Coherent backscattering of light in the presence of time-reversal-noninvariant and parity-nonconserving media," *Phys. Rev. B* ***37***, 1884 (1988).

10. Y. Kuga and A. Ishimaru, "Retroreflectance from a dense distribution of spherical particles," *J. Opt. Soc. Am. A* ***1***, 831 (1984).
11. M. P. van Albada and A. Lagendijk, "Observation of weak localization of light in a random medium," *Phys. Rev. Lett.* ***55***, 2692 (1985).
12. P.-E. Wolf and G. Maret, "Weak localization and coherent backscattering of photons in disordered media," *Phys. Rev. Lett.* ***55***, 2696 (1985).
13. S. John, "Localization and absorption of waves in a weakly dissipative disordered medium," *Phys. Rev. B* ***31***, 304 (1985).
14. S. John, "Photon localization: the inhibition of electromagnetism in certain dielectrics," in W. van Haeringen and D. Lenstra, Eds., *Analogies in Optics and Micro Electronics*, Kluwer, Dordrecht, The Netherlands, 1990, p. 105.
15. S. John, "Localization of light," *Phys. Today* ***44***(5), 32 (1991).
16. A. Ishimaru, *Wave Propagation and Scattering in Random Media*, Academic, New York, 1978.
17. A. Kerker, *The Scattering of Light and Other Electromagnetic Radiation*, Academic, New York, 1969.
18. E. Abrahams, P. W. Anderson, D. C. Licciardello, and T. V. Ramakrishnan, "Scaling theory of localization: Absence of quantum diffusion in two dimensions," *Phys. Rev. Lett.* ***42***, 673 (1979).
19. D. S. Wiersma, P. Bartolini, A. Lagendijk, and R. Righini, "Localization of light in a disordered media," *Nature (London)* ***390***, 671 (1997).
20. A. Z. Genack and N. Garcia, "Observation of photon localization in a three-dimensional disordered system," *Phys. Rev. Lett.* ***66***, 2064 (1991).
21. W. Götze, "A theory for the conductivity of a fermion gas moving in a strong three-dimensional random potential," *J. Phys. C* ***12***, 1279 (1979).
22. W. Götze, "The mobility of a quantum particle in a three-dimensional random potential," *Philos. Mag. B* ***43***, 219 (1981).
23. D. Vollhardt and P. Wölfle, "Diagramatic, self-consistent treatment of the Anderson localization problem in $d \leq 2$ dimensions," *Phys. Rev. B* ***22***, 4666 (1980).
24. P. W. Anderson, "Localized moments and localized states," *Rev. Mod. Phys.* ***50***, 191 (1978).
25. J. D. Jackson, *Classical Electrodynamics*, Wiley, New York, 1962.
26. M. Born and E. Wolf, *Principles of Optics*, Pergamon Press, Oxford–London, 1965.
27. B. A. Lippmann and J. Schwinger, "Variational principles for scattering processes," *Phys. Rev.* ***79***, 469 (1950).
28. M. P. van Albada, B. A. van Tiggelen, A. Lagendijk, and A. Tip, "Speed of propagation of classical waves in strongly scattering media," *Phys. Rev. Lett.* ***66***, 3132 (1991).
29. Yu. N. Barabanenkov and V. D. Ozrin, "Problem of light diffusion in strongly scattering media," *Phys. Rev. Lett.* ***69***, 1364 (1992).
30. B. A. van Tiggelen, A. Lagendijk, M. P. van Albada, and A. Tip, "Speed of light in random media," *Phys. Rev. B* ***45***, 12233 (1992).
31. J. Kroha, C. M. Sokoulis, and P. Wölfle, "Localization of classical waves in a random medium: A self-consistent theory," *Phys. Rev. B* ***47***, 11093 (1993).
32. B. A. van Tiggelen and E. Kogan, "Analogies between light and electrons: Density of states and Friedel's identity," *Phys. Rev. A* ***49***, 708 (1994).

33. C. M. Soukoulis, S. Datta, and E. N. Economou, "Propagation of classical waves in random media," *Phys. Rev. B* ***49***, 3800 (1994).

34. P. A. Lee and A. D. Stone, "Universal conductance fluctuations in metals," *Phys. Rev. Lett.* ***55***, 1622 (1985).

35. M. B. van der Mark, M. P. van Albada, and A. Lagendijk, "Light scattering in strongly scattering media: Multiple scattering and weak localization," *Phys. Rev. B* ***37***, 3575 (1988).

36. B. A. van Tiggelen, A. Lagendijk, and A. Tip, "Multiple-scattering effects for the propagation of light in 3d slabs," *J. Phys. C* ***2***(37), 7653 (1990).

37. M. Rusek and A. Orłowski, "Analytical approach to localization of electromagnetic waves in two-dimensional random media," *Phys. Rev. E* ***51***(4), R2763 (1995).

38. M. Rusek, A. Orłowski, and J. Mostowski, "Band of localized electromagnetic waves in random arrays of dielectric cylinders," *Phys. Rev. E* ***56***(4), 4892 (1997).

39. M. Rusek, A. Orłowski, and J. Mostowski, "Localization of light in three-dimensional random dielectric media," *Phys. Rev. E* ***53***(4), 4122 (1996).

40. M. Rusek and A. Orłowski, "Example of self-averaging in three dimensions: Anderson localization of electromagnetic waves in random distributions of pointlike scatterers," *Phys. Rev. E* ***56***(5B), 6090 (1997).

41. E. M. Purcell and C. R. Pennypacker, "Scattering and absorption of light by nonspherical dielectric grains," *Astrophys. J.* ***186***(186), 705 (1973).

42. C. F. Bohren and D. R. Huffman, *Absorption and Scattering of Light by Small Particles*, Wiley, New York, 1983.

43. B. T. Draine and P. J. Flatau, "Discrete-dipole approximation for scattering calculations," *J. Opt. Soc. Am. A* ***11***, 1491 (1994).

44. B. A. van Tiggelen, Ph.D. Thesis, University of Amsterdam, Amsterdam, The Netherlands, 1992.

45. T. M. Nieuwenhuizen, A. Lagendijk, and B. A. van Tiggelen, "Resonant point scatterers in multiple scattering of classical waves," *Phys. Lett. A* ***169***, 191 (1992).

46. E. J. Heller, "Quantum proximity resonances," *Phys. Rev. Lett.* ***77***(20), 4122 (1996).

47. J. S. Hersch and E. J. Heller, "Observation of proximity resonances in a parallel-plate waveguide," *Phys. Rev. Lett.* ***81***(15), 3059 (1998).

48. I. Tolstoy, "Superresonant systems of scatterers," *J. Acoust. Soc. Am.* ***80***(1), 282 (1986).

49. I. Tolstoy and A. Tolstoy, "Superresonant systems of scatterers. ii," *J. Acoust. Soc. Am.* ***83***(6), 2086 (1988).

50. C. Feuillade, "Scattering from collective modes of air bubbles in water and the physical mechanism of superresonances," *J. Acoust. Soc. Am.* ***98***(2), 1178 (1995).

51. V. A. Markel, "Scattering of light from two interacting spherical particles," *J. Mod. Opt.* ***39***, 853 (1992).

52. A. Lagendijk and B. A. van Tiggelen, "Resonat multiple scattering of light," *Phys. Rep.* ***270***(3), 143 (1996).

53. H. Furstenberg, "Noncommuting random matrices," *Trans. Am. Math. Soc.* ***108***, 377 (1963).

54. F. Deylon, H. Kunz, and B. Souillard, "One-dimensional wave equations in disordered media," *J. Phys. A* ***16***, 25 (1983).

55. M. Rusek and A. Orłowski, "Anderson localization of electromagnetic waves in confined dielectric media," *Phys. Rev. E* ***59***(3), 3655 (1999).

56. B. Souillard, "Waves and electrons in inhomogeneous media," in J. Souletie, J. Vannimenus, and R. Stora, Eds., *Chance and Matter*, North-Holland, Amsterdam, The Netherlands, 1987, p. 305.

CHAPTER SIX

Field Distribution, Anderson Localization, and Optical Phenomena in Random Metal–Dielectric Films

ANDREY K. SARYCHEV

Center for Applied Problems of Electrodynamics, 127412 Moscow, Russia

VLADIMIR M. SHALAEV

Department of Physics, New Mexico State University, Las Cruces, NM 88003

This chapter presents a theory of optical, infrared (IR), and microwave response of metal–dielectric inhomogeneous films. First, we describe the generalized Ohm's law approximation formulated for the case when the inhomogeneity length scale is much smaller than the wavelength, but is not smaller than the skin (penetration) depth for metal grains. In this approach, electric and magnetic fields *outside* a film are related to the currents *inside* the film. The computer simulations, based on the generalized Ohm's law approximation, reproduce the prominent absorption band near the percolation threshold and show that local electric and magnetic fields experience giant spatial fluctuations, which were detected in recent experiments. The fields are localized in small spatially separated peaks: electric and magnetic hot spots. A scaling theory, which is discussed in detail, predicts that the hot spots represent localized surface plasmons. The localization maps the Anderson transition problem, which is described by the random Hamiltonian with diagonal and off-diagonal disorder. The local fields exceed the applied field by several orders of magnitude, resulting in the enormous enhancement of various optical phenomena (Raman and hyper-Raman scattering, Kerr refraction, four-wave mixing, etc.). At percolation, a dip in the dependence of optical processes on the metal concentration is predicted. It is also shown that transmittance of a regular array of holes in a metal film is much enhanced when the incident wave is in resonance with one of the internal modes in the film.

Optics of Nanostructured Materials, Edited by Vadim A. Markel and Thomas F. George
ISBN 0-471-34968-2

6.1. INTRODUCTION

Random metal–dielectric films, also known as semicontinuous metal films, are usually produced by thermal evaporation or spattering of metal onto an insulating substrate. In the growing process, first, small clusters of metal grains are formed and, eventually, at a percolation threshold, a continuous conducting path appears between the ends of the sample, indicating a metal–insulator transition in the system. At high surface coverage, the film is mostly metallic with voids of irregular shape and, finally, the film becomes uniform. Over the past three decades, the electric transport properties of the semicontinuous metal films have been a topic of active experimental and theoretical study. The classical percolation theory had been employed to describe an anomalous behavior of the conductivity and other transport properties near the percolation threshold [1–4]. Recently, it was shown that quantum effects such as tunneling between metal clusters and electron localization become important at the percolation even at room temperature (see [5–8] and references cited therein). Thus low-frequency divergence of the dielectric constant was predicted theoretically [7,8] and then obtained experimentally [9].

In this chapter, we will consider the optical response of metal–insulator thin films that have been intensively studied both experimentally and theoretically along with transport phenomena (see, e.g., [3,4,10–22]). A two-dimensional (2D) nonhomogeneous film is a thin layer over which the local physical properties are not uniform. The response of such a layer to an incident wave depends crucially on the inhomogeneity length scale compared to the wavelength and also on the angle of incidence. Usually, when the wavelength is smaller than the inhomogeneity scale, the incident wave is scattered in various directions. The total field that is scattered in a certain direction is the sum of the elementary waves scattered in that direction by each elementary scatterer on the surface. As each elementary wave is given not only by its amplitude, but also by its phase, this sum will be a vector sum. The scattered wave is then distributed in various directions, though certain privileged directions may receive more energy than others. By contrast, when the inhomogeneity length scale is much smaller than the wavelength, the resolution of the wave is too small to "see" the irregularities, therefore the wave is then reflected specularly and transmitted in a well-defined direction, as if the film were a homogeneous layer with bulk effective physical properties (conductivity, permittivity, and permeability) that are uniform. The wave is coupled to the inhomogeneities in such a way that irregular currents are excited on the surface of the layer. Strong distortions of the field then appear near the surface; however, they decay exponentially so that far enough from the surface the wave resumes its plane wave character.

The question of scattering from a nonhomogeneous surface has attracted attention since the time of Lord Rayleigh [23]. Due to the wide range of potential applications in, for example, radiowave and radar techniques, most efforts have been concentrated in the regime where the scale of inhomogeneity is larger than the wavelength [24]. In the last decade, a problem of localization of surface polaritons [25] and other "internal modes" due to their interaction with surface roughness attracted a lot of attention. This localization was found to contribute a maximum to the angular dependence of the intensity of the nonspecularly reflected light in the antispecular direction [26] and other "resonance directions" [27,28] as well. The development of near-field scanning optical microscopy has opened the way to probe the surface polariton field above the surface and visualize its distribution. Vast progress in the near-field optics of various rough surfaces and metal grain structures is discussed in Chapter 3. The subject of this chapter belongs to another regime, where the inhomogeneity length scale is much smaller than the wavelength, but can be of the order or even larger than skin depth. In other words, coupling of a metal grain with an *electromagnetic* (EM) field is supposed to be strong in spite of its subwavelength size. In particular, we focus on the high-frequency response (optical, IR, and microwave) of thin, metal-dielectric random films.

The optical properties of metal–dielectric films show anomalous phenomena that are absent for bulk metal and dielectric components. For example, the anomalous absorption in the near-IR spectral range leads to unusual behavior of transmittance and reflectance. Typically, the transmittance is much higher than that of continuous metal films, whereas the reflectance is much lower (see, e.g., [3,4,10,11,16–18]). Near and well below the percolation threshold, the anomalous absorbance can be as high as 50% [12–16,20]. A number of theories were proposed for calculation of the optical properties of semicontinuous random films, including the effective-medium approaches [29,30], their various modifications [3,16,17,31–35], and the renormalization group method (see e.g., [4,36,37]). In most of these theories, the semicontinuous metal–dielectric film is considered as a fully 2D system and quasistatic approximation is invoked. However, usage of that approximation implies that both the electric and magnetic fields in the film are assumed to be 2D and curl-free. That assumption ceases to be valid when the fields are changed considerably in the physical film and in its close neighborhood, which is usually the case in a semicontinuous metal thin film, especially in the strong skin effect regime.

In an attempt to expand the theoretical treatment beyond the quasistatic approximation, an approach was recently proposed that was based on the full set of Maxwell's equations [18–20]. This approach does not use the quasistatic approximation because the fields are not assumed to be curl-free inside the physical film. Although that theory was proposed with metal–insulator thin films in mind, it is in fact quite general and can be applied to any kind of inhomogeneous film under appropriate conditions. For reasons that will be explained below, this theory is called the "generalized Ohm's law" (GOL). We present this new theory here.

Below, we restrict ourselves to the case where all the external fields are parallel to the plane of the film. This means that an incident wave, as well as the reflected and transmitted waves, are traveling in the direction perpendicular to the film plane. We focus our attention on the electric and magnetic field magnitudes at certain distances *away* from the film and relate them to the currents *inside* the film. We assume that inhomogeneities on a film are much smaller in size than the wavelength λ (but not necessarily smaller than the skin depth), so that the fields away from the film are curl-free and can be expressed as gradients of potential fields. The electric and magnetic induction currents averaged over the film thickness obey the usual 2D continuity equations. Therefore, the equations for the fields (e.g., $\nabla \times \mathbf{E} = 0$)

and the equations for the currents (e.g., $\nabla \cdot \mathbf{j} = 0$) are the *same* as in the quasistatic case. The only difference is that the fields and the averaged currents are now related by new constitutive equations and that there are magnetic as well as electric currents.

To determine these new constitutive equations, we find the electric and magnetic field distributions inside the conductive and dielectric regions of the film. The boundary conditions completely determine solutions of Maxwell's equations for the fields inside a grain when the frequency is fixed. Therefore the internal fields, which change very rapidly with position in the direction perpendicular to the film, depend linearly on the electric and magnetic field away from the film. The currents inside the film are linear functions of the local internal fields given by the usual local constitutive equations. Therefore, the currents flowing *inside* the film also depend linearly on the electric and magnetic fields *outside* the film. However, the electric current averaged over the film thickness now depends not only on the external electric field, but also on the external magnetic field. The same is true for the average magnetic induction current. Thus we have two linear equations that connect the two types of average internal currents to the external fields. These equations can be considered as a generalization of Ohm's law to the nonquasistatic case and they dub as GOL [19]. The GOL forms the basis of a new approach to calculate the EM properties of inhomogeneous films.

The continuity equations for the electric and magnetic currents use the GOL and take into account the potential character of the electric and magnetic fields outside the film. This allows us to determine the field and current distribution over a metal–dielectric film in the computer experiment, and, finally, to calculate the optical properties of the film. Computer simulations show that the local electric and magnetic fields both fluctuate strongly over the film at the metal concentration close to the percolation threshold. The fields are localized in small spatially separated peaks: electric and magnetic hot spots. When the skin effect in the metal grains is strong, the magnetic fluctuations are as large as fluctuations of the local electric field. The amplitude of local field fluctuations in the case of a strong skin effect is large regardless of losses in the metal. It is also important to note that giant magnetic field fluctuations is a purely nonquasistatic effect that cannot be obtained within the traditional approach used earlier.

We present a scaling theory of local field fluctuations in the random semicontinuous metal films [38–46]. The theory is based on the fact that the problem of optical excitations in semicontinuous metal films mathematically maps the Anderson transition problem. This allowed us to predict localization of surface plasmons in the films and to describe in detail the localization pattern. It is shown that the surface plasmons eigenstates are localized on a scale much smaller than the wavelength of the incident light. The surface plasmons eigenstates, with eigenvalues close to zero (resonant modes), are excited most efficiently by the external field. Since the eigenstates are localized and only a small portion of them are excited by the incident wave, the overlapping of the eigenstates can typically be neglected, which significantly simplifies theoretical consideration and allows one to obtain relatively simple expressions for the local field fluctuations. It is important to stress that the surface plasmon localization length is much smaller than the wavelength; in that sense, the predicted subwavelength localization of the surface plasmons quite differs from the long-time discussed localization of light due to strong scattering in a random homogeneous medium [6,47].

The developed scaling theory of local field fluctuations in semicontinuous metal films opens up new means to study the classical Anderson problem by taking advantage of unique characteristics of laser radiation, namely, its coherence and high intensity. For example, this theory predicts that at percolation there is a *minimum* in nonlinear optical responses of metal–dielectric composites, a fact that follows from the Anderson localization of surface plasmon modes and can be studied and verified in laser experiments.

The rest of this chapter is organized as follows: the GOL for semicontinuous metal films is derived in Section 6.2. In Section 6.3, it is shown how the optical properties: (reflectance, transmittance, and absorbance) are found in the GOL approximation. The original computer method and calculation of the local electric and magnetic field are presented in Section 6.4. In Section 6.5 and 6.6, analytical theory is developed for the giant local field fluctuations. In Section 6.7, the theoretical results are implemented to find equations for the spatial moments of the local fields. In Section 6.8, we consider optical properties of a metal film that is perforated with an array of subwavelength holes; we show here that the transmittance through such a film can be strongly enhanced in agreement with recent experimental observations. Section 6.9 summarizes and concludes this chapter.

6.2. GOL AND BASIC EQUATIONS

We base the following presentation on the results of [18–20]. In contrast to the traditional consideration, it is not assumed that the electric and magnetic fields inside a semicontinuous metal film are curl-free and z independent, where the z coordinate is perpendicular to the film plane.

First, let us consider a homogeneous conducting film with a uniform conductivity σ_m and thickness d, and assume constant values of the electric field $\mathbf{E}_1$ and magnetic field $\mathbf{H}_1$ at some reference plane $z = -d/2 - \ell_0$ behind the film, as shown in Fig. 6.1. Under these conditions the fields depend only on the z coordinate, and Maxwell's equations for a monochromatic field can be written in the following form:

$$\frac{d}{dz}\,\mathbf{E}(z) = -\frac{i\omega}{c}\,\mu(z)[\mathbf{n}\times\mathbf{H}(z)] \tag{6.1}$$

$$\frac{d}{dz}\,\mathbf{H}(z) = -\frac{4\pi}{c}\,\sigma(z)[\mathbf{n}\times\mathbf{E}(z)] \tag{6.2}$$

with boundary conditions

$$\mathbf{E}(z = -d/2 - l_0) = \mathbf{E}_1 \qquad \mathbf{H}(z = -d/2 - l_0) = \mathbf{H}_1 \tag{6.3}$$

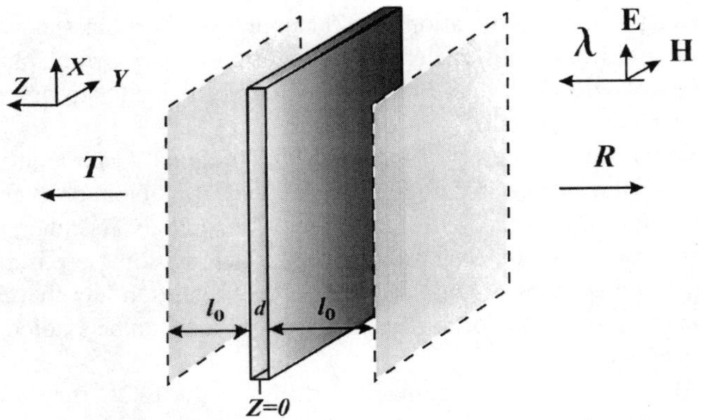

Figure 6.1. The scheme used in a theoretical model. Electromagnetic wave of wavelength λ is incident on a thin-metal–insulator film with thickness d. It is partially reflected and absorbed, and the remainder is transmitted through the film. The amplitudes of the electric and magnetic fields, which are averaged over the plane $z = -d/2 - l_0$ behind the film, are equal to each other.

where $\mathbf{E}_1$ and $\mathbf{H}_1$ are parallel to the film plane. Here, the conductivity $\sigma(z)$ is equal to the metal conductivity σ_m inside the film ($-d/2 < z < d/2$) and to $\sigma_d = -i\omega/4\pi$ outside the film ($z < -d/2$ and $z > d/2$), and similarly, the magnetic permeability $\mu(z)$ is equal to the film permeability μ_m inside the film and to one outside the film; the unit vector $\mathbf{n} = (0, 0, 1)$ is perpendicular to the film plane. When solving Eqs. (6.1) and (6.2), it is taken into account that the electric and magnetic fields are continuous at the film boundaries. In this way, the fields $\mathbf{E}(z)$ and $\mathbf{H}(z)$ are determined everywhere. Then, electric $\mathbf{j}_E$ and magnetic $\mathbf{j}_H$ currents flowing in between the two planes at $z = -d/2 - l_0$ and at $z = d/2 + l_0$ are calculated as

$$\mathbf{j}_E = -\frac{i\omega}{4\pi}\left[\int_{-d/2-l_0}^{-d/2} \mathbf{E}(z)dz + \int_{-d/2}^{d/2} \varepsilon_m \mathbf{E}(z)dz + \int_{d/2}^{d/2+l_0} \mathbf{E}(z)dz\right] \tag{6.4}$$

$$\mathbf{j}_H = \frac{i\omega}{4\pi}\left[\int_{-d/2-l_0}^{-d/2} \mathbf{H}(z)dz + \int_{-d/2}^{d/2} \mu_m \mathbf{H}(z)dz + \int_{d/2}^{d/2+l_0} \mathbf{H}(z)dz\right] \tag{6.5}$$

where $\varepsilon_m = 4i\pi\sigma_m/\omega$ is the metal dielectric constant. In what follows, it is assumed, for simplicity, that the magnetic permeability $\mu_m = 1$. Since the Maxwell equations are linear, the local fields $\mathbf{E}(z)$ and $\mathbf{H}(z)$ are linear functions of the boundary values $\mathbf{E}_1$ and $\mathbf{H}_1$ defined at the plane $z = -d/2 - l_0$

$$\mathbf{E}(z) = a(z)\mathbf{E}_1 + c(z)[\mathbf{n} \times \mathbf{H}_1] \tag{6.6}$$

$$\mathbf{H}(z) = b(z)\mathbf{H}_1 + d(z)[\mathbf{n} \times \mathbf{E}_1] \tag{6.7}$$

Note that $\mathbf{n}$ is the single constant vector in the problem, which let us to build polar $[\mathbf{n} \times \mathbf{H}_1]$ and axial $[\mathbf{n} \times \mathbf{E}_1]$ vectors in Eqs. (6.6) and (6.7). By substituting Eq. (6.6) for $\mathbf{E}(z)$ and Eq. (6.7) for $\mathbf{H}(z)$ in Eqs. (6.4) and (6.5), we express the currents $\mathbf{j}_E$ and $\mathbf{j}_H$ in terms of the boundary (surface) fields $\mathbf{E}_1$ and $\mathbf{H}_1$ as

$$\mathbf{j}_E = s\mathbf{E}_1 + g_1\,[\mathbf{n} \times \mathbf{H}_1] \tag{6.8}$$

$$\mathbf{j}_H = m\mathbf{H}_1 + g_2\,[\mathbf{n} \times \mathbf{E}_1] \tag{6.9}$$

In contrast to the usual constitutive equations, the planar electric current $\mathbf{j}_E$, which flows between the planes $z = -d/2 - l_0$ and $z = d/2 + l_0$, depends not only on the external electric field $\mathbf{E}_1$ but also on the external magnetic field $\mathbf{H}_1$. The same is true for the magnetic induction current $\mathbf{j}_H$. These equations constitute the GOL. The Ohmic parameters s, m, g_1, and g_2 have the dimension of surface conductivity (cm/s) and depend on the frequency ω, the metal dielectric constant ε_m, the film thickness d, and the distance l_0 between the film and the reference plane $z = -d/2 - l_0$. Below, the films are supposed to be invariant under reflection through the plane $z = 0$. In this case, $g_1 = g_2 = g$ as it is shown in [18], and the Ohmic parameter g can be expressed in terms of parameters s and m as

$$g = -\frac{c}{4\pi} + \sqrt{\left(\frac{c}{4\pi}\right)^2 - ms} \tag{6.10}$$

Then, the GOL equations (6.8) and (6.9) take the following form:

$$\mathbf{j}_E = s\mathbf{E}_1 + g\,[\mathbf{n} \times \mathbf{H}_1] \tag{6.11}$$

$$\mathbf{j}_H = m\mathbf{H}_1 + g\,[\mathbf{n} \times \mathbf{E}_1] \tag{6.12}$$

where the Ohmic parameter g is given by Eq. (6.10). The Ohmic parameters s and m can be expressed in terms of the film refractive index $n = \sqrt{\varepsilon_m}$ and film thickness d in the following way:

$$s = \frac{c}{8n\pi}\,[e^{-idkn}(n\cos(adk) - i\sin(adk))^2 - e^{idkn}(n\cos(adk) + i\sin(adk))^2] \tag{6.13}$$

$$m = \frac{c}{8n\pi}\,[e^{-idkn}(i\cos(adk) + n\sin(adk))^2 - e^{idkn}(-i\cos(adk) + n\sin(adk))^2] \tag{6.14}$$

where we still assume, for simplicity, that $\varepsilon = 1$ outside the film ($z < -d/2$, $z > d/2$) and introduce the wave vector $k = \omega/c$ and dimensionless parameter $a \equiv l_0/d$ (see [18,19]). The skin (penetration) depth δ is equal to $\delta = 1/k\,\mathrm{Im}\,n$ in these notations. In the microwave spectral range, metal conductivity is real and the dielectric constant ε_m is purely imaginary and the skin depth $\delta = c/\sqrt{2\pi\sigma_m\omega}$. On the other hand, the dielectric constant is negative for a typical metal in the optical and IR spectral ranges, therefore $\delta \cong 1/k\sqrt{|\varepsilon_m|}$ in this case

We now turn to the case of laterally inhomogeneous films. Then, the currents $\mathbf{j}_E$ and $\mathbf{j}_H$ defined by Eqs. (6.4) and (6.5), as well as the fields $\mathbf{E}_1$ and $\mathbf{H}_1$, are functions of the 2D vector $\mathbf{r} = \{x, y\}$. From Maxwell's equations, it follows that the fields and currents are connected by linear relations

$$\mathbf{j}_E(\mathbf{r}) = s\mathbf{E}_1 + g\,[\mathbf{n} \times \mathbf{H}_1] \tag{6.15}$$

$$\mathbf{j}_H(\mathbf{r}) = m\mathbf{H}_1 + g\,[\mathbf{n} \times \mathbf{E}_1] \tag{6.16}$$

where s, m, and g are some integral operators now. The metal islands in semicontinuous films usually have an oblate shape so that the grain diameter D is much larger than the film thickness d (see, e.g., [11]). When the thickness of a conducting grain d (or skin depth δ) and distance l_0 are much smaller than the grain diameter D, the relation of the fields $\mathbf{E}_1$ and $\mathbf{H}_1$ to the currents becomes fully local in Eqs. (6.15) and (6.16). The local Ohmic parameters $s = s(\mathbf{r})$, $m = m(\mathbf{r})$, and $g = g(\mathbf{r})$, given by Eqs. (6.10), (6.13), and (6.14), are determined by the local refractive index $n(\mathbf{r}) = \sqrt{\varepsilon(\mathbf{r})}$, where $\varepsilon(\mathbf{r})$ is a local dielectric constant. Equations (6.15) and (6.16) are the local GOL for semicontinuous films. For binary metal–dielectric semicontinuous films, the local dielectric constant is equal to either ε_m or ε_d. The electric $\mathbf{j}_E$ and magnetic $\mathbf{j}_H$ currents given by Eqs. (6.15) and (6.16) lie in between the planes $z = -d/2 - l_0$ and $z = d/2 + l_0$. These currents satisfy the 2D continuity equations

$$\nabla \cdot \mathbf{j}_E(\mathbf{r}) = 0 \qquad \nabla \cdot \mathbf{j}_H(\mathbf{r}) = 0 \tag{6.17}$$

which follow from the three-dimensional (3D) continuity equations when the z components of $\mathbf{E}_1$ and $\mathbf{H}_1$ are neglected at the planes $z = \pm(d/2 + l_0)$. This is possible because these components are small, in accordance with the fact that the average fields $\langle \mathbf{E}_1 \rangle$ and $\langle \mathbf{H}_1 \rangle$ are parallel to the film plane. Since we consider semicontinuous films with an inhomogeneity scale much smaller than the wavelength λ, the fields $\mathbf{E}_1(\mathbf{r})$ and $\mathbf{H}_1(\mathbf{r})$ are still the gradients of potential fields when considered as functions of x and y in the fixed reference plane $z = -d/2 - l_0$, that is,

$$\mathbf{E}_1(\mathbf{r}) = -\nabla\varphi_1(\mathbf{r}) \qquad \mathbf{H}_1(\mathbf{r}) = -\nabla\psi_1(\mathbf{r}) \tag{6.18}$$

By substituting these expressions in the continuity Eq. (6.17) and taking into account the GOL Eqs. (6.15) and (6.16), the system of two basic equations for the electric φ_1 and magnetic ψ_1 potentials are obtained

$$\nabla \cdot (s\nabla\varphi_1 + g[\mathbf{n} \times \nabla\psi_1]) = 0 \qquad \nabla \cdot (m\nabla\psi_1 + g[\mathbf{n} \times \nabla\varphi_1]) = 0 \tag{6.19}$$

where all variables are functions of the coordinates x and y in the reference plain. The above equations must be solved under the following conditions:

$$\langle \nabla\varphi_1 \rangle = \langle \mathbf{E}_1 \rangle \qquad \langle \nabla\psi_1 \rangle = \langle \mathbf{H}_1 \rangle \tag{6.20}$$

where the constant fields $\langle \mathbf{E}_1 \rangle$ and $\langle \mathbf{H}_1 \rangle$ are external (given) fields. Here and below $\langle \cdots \rangle$ denotes an average over coordinates "x" and "y".

The essence of the GOL can be summarized as follows. The entire physics of a 3D inhomogeneous layer, which is described by the full set of Maxwell's equations, has been reduced to a set of quasistatic equations in a (2D) reference plane. Part of the price for this achievement is the introduction of coupled electric and magnetic fields, currents, and dependence on one adjustable parameter, namely, the distance l_0 to the reference plane. Comparison of the numerical calculation and the GOL approximation for the metal film with periodic corrugation [19] shows that GOL results are generally not sensitive to the distance l_0. The original choice $l_0 = 0.25\,D$ [18] [i.e., parameter $a = D/4d$ in Eqs. (6.13) and (6.14)] allows us to reproduce most of the computer simulations except those where a surface polariton is excited in the corrugated film.

6.3. DECOUPLING GOL EQUATIONS AND EFFECTIVE MEDIUM APPROXIMATION FOR TRANSMITTANCE, REFLECTANCE, AND ABSORBANCE

To simplify the system of the basic equations (6.19) the electric and magnetic fields on both sides of the film are considered [19,20]. Namely, the electric and magnetic fields are considered at the distance l_0 behind the film $\mathbf{E}_1(\mathbf{r}) = \mathbf{E}(\mathbf{r},\ -d/2 - l_0)$, $\mathbf{H}_1(\mathbf{r}) = \mathbf{H}(\mathbf{r},\ -d/2 - l_0)$, and at the distance l_0 in front of the film $\mathbf{E}_2(\mathbf{r}) = \mathbf{E}(\mathbf{r},\ d/2 + l_0)$, $\mathbf{H}_2(\mathbf{r}) = \mathbf{H}(\mathbf{r},\ d/2 + l_0)$. Remember that $\mathbf{r} = \{x,\ y\}$ is a 2D vector in a plane perpendicular to the "z" axis. The components of the fields aligned with z are still neglected. Then, the second Maxwell's equation curl $\mathbf{H} = (4\pi/c)\mathbf{j}$ can be written as $\oint \mathbf{H} d\mathbf{l} = (4\pi/c)\mathbf{j}_E \Delta$, where the integration is over the contour $ABCD$ shown in Fig. 6.2, while the current $\mathbf{j}_E$ is given by the GOL Eq. (6.15). When $\Delta \to 0$ this equation takes the following form

$$\mathbf{H}_2 - \mathbf{H}_1 = -\frac{4\pi}{c}\,[\mathbf{n} \times \mathbf{j}_E] = -\frac{4\pi}{c}\,(s[\mathbf{n} \times \mathbf{E}_1] - g\mathbf{H}_1) \tag{6.21}$$

When the same procedure is applied to the first Maxwell equation, curl $\mathbf{E} = ik\mathbf{H}$, it gives

$$\mathbf{E}_2 - \mathbf{E}_1 = -\frac{4\pi}{c}\,[\mathbf{n} \times \mathbf{j}_H] = -\frac{4\pi}{c}\,(m[\mathbf{n} \times \mathbf{H}_1] - g\mathbf{E}_1) \tag{6.22}$$

where the GOL equation (6.16) has been substituted for the electric current $\mathbf{j}_H$ in Eq. (6.22). Then, electric field $\mathbf{E}_1$ can be expressed from Eq. (6.21) in terms of the magnetic fields $\mathbf{H}_1$ and $\mathbf{H}_2$ as

$$[\mathbf{n} \times \mathbf{E}_1] = \frac{g}{s}\,\mathbf{H}_1 - \frac{c}{4\pi s}\,(\mathbf{H}_2 - \mathbf{H}_1) \tag{6.23}$$

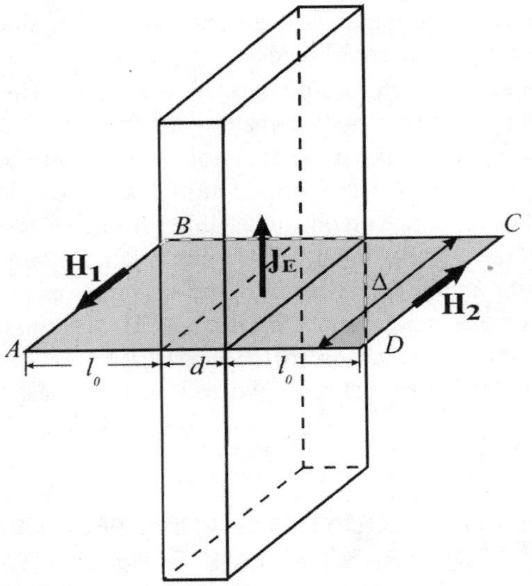

Figure 6.2. The left-hand side of the Maxwell's equation $\oint \mathbf{H}\, d\mathbf{l} = (4\pi/c)\mathbf{j}_E\Delta$ is integrated over the contour ABCD to obtain Eq. (6.21).

and the magnetic field $\mathbf{H}_1$ can be expressed from Eq. (6.22) in terms of the electric fields $\mathbf{E}_1$ and $\mathbf{E}_2$ as

$$[\mathbf{n} \times \mathbf{H}_1] = \frac{g}{m}\,\mathbf{E}_1 - \frac{c}{4\pi m}\,(\mathbf{E}_2 - \mathbf{E}_1) \tag{6.24}$$

Substitution the right-hand side (rhs) of Eq. (6.23) in the GOL Eq. (6.16) and substitution (6.24) in the GOL Eq. (6.15) results in

$$\mathbf{j}_E = s\mathbf{E}_1 + g\left(\frac{g}{m}\,\mathbf{E}_1 - \frac{c}{4\pi m}\,(\mathbf{E}_2 - \mathbf{E}_1)\right) \tag{6.25}$$

$$\mathbf{j}_H = m\mathbf{H}_1 + g\left(\frac{g}{s}\,\mathbf{H}_1 - \frac{c}{4\pi s}\,(\mathbf{H}_2 - \mathbf{H}_1)\right) \tag{6.26}$$

Finally, the relation (6.10) between the Ohmic parameters s, m, and g allows us to rewrite the above equations as

$$\mathbf{j}_E = u\mathbf{E} \qquad \mathbf{j}_H = w\mathbf{H} \tag{6.27}$$

where $\mathbf{E} = (\mathbf{E}_1 + \mathbf{E}_2)/2$, $\mathbf{H} = (\mathbf{H}_1 + \mathbf{H}_2)/2$ and parameters u and w are given by the following equations:

$$u = -\frac{c}{2\pi}\,\frac{g}{m} \qquad w = -\frac{c}{2\pi}\,\frac{g}{s} \tag{6.28}$$

Thus the GOL is diagonalized by introducing new fields **E** and **H** so that Eqs. (6.27) have the same form as constitutive equations of the macroscopic electrodynamics. The only difference is that the local conductivity σ is replaced by parameter u, and magnetic permeability μ is replaced by parameter $-i4\pi w/\omega$.

It follows from Eqs. (6.28) and (6.10), (6.13), and (6.14) that the new Ohmic parameters u and v are expressed in terms of the local refractive index $n = \sqrt{\varepsilon(\mathbf{r})}$ as

$$u = -i\,\frac{c}{2\pi}\,\frac{\tan(Dk/4) + n\,\tan(dkn/2)}{1 - n\,\tan(Dk/4)\,\tan(dkn/2)} \tag{6.29}$$

$$w = i\,\frac{c}{2\pi}\,\frac{n\,\tan(Dk/4) + \tan(dkn/2)}{n - \tan(Dk/4)\,\tan(dkn/2)} \tag{6.30}$$

where the parameter $a = D/4d$ is substituted as it is discussed at the end of Section 6.2. The refractive index n in the above equations takes values $n_m = \sqrt{\varepsilon_m}$ and $n_d = \sqrt{\varepsilon_d}$ for metal and dielectric regions of the film, respectively. In the quasistatic limit, when the optical thickness of metal grains is small, $dk|n_m| \ll 1$, while the metal dielectric constant is large in magnitude, $|\varepsilon_m| \gg 1$, the following estimates hold:

$$u_m \simeq -i\,\frac{\omega\varepsilon_m}{4\pi}d \qquad w_m \simeq i\,\frac{\omega}{4\pi}(d + D/2) \qquad (d/\delta \ll 1) \tag{6.31}$$

for the metal grains. In the opposite case of strong skin effect, when the skin depth (penetration depth) $\delta = 1/k\,\mathrm{Im}\,n_m$ is much smaller than the grain thickness d and the electromagnetic field does not penetrate in metal grains, the parameters u_m and w_m take values

$$u_m = i\,\frac{2c^2}{\pi D\omega} \qquad w_m = i\,\frac{\omega D}{8\pi} \qquad (d/\delta \gg 1) \tag{6.32}$$

For the dielectric region, when the film is thin enough so that $dkn_d \ll 1$ and $\varepsilon_d \sim 1$, Eqs. (6.29) and (6.30) give

$$u_d = -i\,\frac{\omega\varepsilon'_d}{8\pi}D \qquad w_d = i\,\frac{\omega}{4\pi}(d + D/2) \tag{6.33}$$

where the reduced dielectric constant $\varepsilon'_d = 1 + 2\varepsilon_d d/D$ is introduced. Note that in the limit of the strong skin effect the Ohmic parameters u_m and w_m are purely imaginary and the parameter u_m is of inductive character, that is, it has the sign opposite to the dielectric parameter u_d. In contrast, the Ohmic parameter w remains essentially the same $w \sim iD\omega/8\pi$ for dielectric and for metal regions regardless of the value of the skin effect.

Potentials for the fields $\mathbf{E}_2(\mathbf{r})$ and $\mathbf{H}_2(\mathbf{r})$ can be introduced for the same reason as potentials for the fields $\mathbf{E}_1(\mathbf{r})$ and $\mathbf{H}_1(\mathbf{r})$ [see the discussion accompanying

Eq. (6.18)]. Therefore the fields $\mathbf{E}(\mathbf{r})$ and $\mathbf{H}(\mathbf{r})$ in Eqs. (6.27) can in turn be represented as gradients of some potentials:

$$\mathbf{E} = -\nabla\phi' \qquad \mathbf{H} = -\nabla\psi' \tag{6.34}$$

By substituting these expressions into Eq. (6.27) and then into the continuity Eq. (6.17), we obtain the following equations:

$$\nabla \cdot [u(\mathbf{r})\nabla\varphi'(\mathbf{r})] = 0 \tag{6.35}$$

$$\nabla \cdot [w(\mathbf{r})\nabla\psi(\mathbf{r})] = 0 \tag{6.36}$$

which can be solved independently for the potentials φ' and ψ'. Eqs. (6.35) and (6.36) are solved under the following conditions:

$$\langle\nabla\varphi_1'\rangle = \langle\mathbf{E}\rangle \equiv \mathbf{E}_0 \qquad \langle\nabla\psi_1'\rangle = \langle\mathbf{H}_1\rangle \equiv \mathbf{H}_0 \tag{6.37}$$

where the constant fields $\mathbf{E}_0$ and $\mathbf{H}_0$ are external (given) fields that are determined by the incident wave. When the fields $\mathbf{E}$, $\mathbf{H}$ and currents $\mathbf{j}_E$, $\mathbf{j}_H$ are found from the solution of Eqs. (6.35)–(6.37), the local electric and magnetic fields in the plane $z = -l_0 - d/2$ are given by the equations

$$\mathbf{E}_1 = \mathbf{E} + \frac{2\pi}{c}[\mathbf{n} \times \mathbf{j}_H] \qquad \mathbf{H}_1 = \mathbf{H} + \frac{2\pi}{c}[\mathbf{n} \times \mathbf{j}_E] \tag{6.38}$$

which follow from Eqs. (6.21) and (6.22) and definitions of the fields $\mathbf{E}$ and $\mathbf{H}$. Note that the field $\mathbf{E}_1(\mathbf{r})$ usually is measured in a typical near-field experiment (see Chapter 3). The effective parameters u_e and w_e are defined in a usual way

$$\langle\mathbf{j}_E\rangle = u_e\mathbf{E}_0 \equiv u_e(\langle\mathbf{E}_1\rangle + \langle\mathbf{E}_2\rangle)/2 \tag{6.39}$$

$$\langle\mathbf{j}_H\rangle = w_e\mathbf{H}_0 \equiv w_e(\langle\mathbf{H}_1\rangle + \langle\mathbf{H}_2\rangle)/2 \tag{6.40}$$

These expressions are substituted into Eqs. (6.21) and (6.22), which are averaged over the $\{x, y\}$ coordinates to obtain equations

$$[\mathbf{n} \times (\langle\mathbf{H}_2\rangle - \langle\mathbf{H}_1\rangle)] = \frac{2\pi}{c}\, u_e(\langle\mathbf{E}_1\rangle + \langle\mathbf{E}_2\rangle) \tag{6.41}$$

$$[\mathbf{n} \times (\langle\mathbf{E}_2\rangle - \langle\mathbf{E}_1\rangle)] = \frac{2\pi}{c}\, w_e(\langle\mathbf{H}_1\rangle + \langle\mathbf{H}_2\rangle) \tag{6.42}$$

for the averaged fields that determine the optical response of an inhomogeneous film.

Let us suppose that the wave enters the film from the right half-space (see Fig. 6.1), so that its amplitude is proportional to e^{-ikz}. The incident wave is partially reflected and partially transmitted through the film. The electric field amplitude in the right half-space, away from the film, can be written as $\mathbf{e}[e^{-ikz} + re^{ikz}]$, where r is the reflection coefficient and $\mathbf{e}$ is the polarization vector. Then, the electric component of the EM wave well behind the film acquires the form $\mathbf{e}\ te^{-ikz}$, where t is the transmission coefficient. It is supposed for simplicity that the film has no optical activity, therefore wave polarization $\mathbf{e}$ remains the same before and after the film. At the planes $z = d/2 + l_0$ and $z = -d/2 - l_0$, the average electric field equals to $\langle \mathbf{E}_2 \rangle$ and $\langle \mathbf{E}_1 \rangle$, respectively. Now, the wave away from the film is matched by the average fields in the planes $z = d/2 + l_0$ and $z = -d/2 - l_0$, that is, $\langle \mathbf{E}_2 \rangle = \mathbf{e}[e^{-ik(d/2+l_0)} + re^{ik(d/2+l_0)}]$ and $\langle \mathbf{E}_1 \rangle = \mathbf{e}\ te^{ik(d/2+l_0)}$. The same matching, but with magnetic fields, gives $\langle \mathbf{H}_2 \rangle = [\mathbf{n} \times \mathbf{e}][-e^{-ik(d/2+l_0)} + re^{ik(d/2+l_0)}]$ and $\langle \mathbf{H}_1 \rangle = -[\mathbf{n} \times \mathbf{e}]te^{ik(d/2+l_0)}$ in the planes $z = d/2 + l_0$ and $z = -d/2 - l_0$, respectively. Substitution of these expressions for the fields $\langle \mathbf{E}_1 \rangle$, $\langle \mathbf{E}_2 \rangle$, $\langle \mathbf{H}_1 \rangle$, and $\langle \mathbf{H}_2 \rangle$ in Eqs. (6.41) and (6.42) gives two scalar, linear equations for reflection r and transmission t coefficients. Solution to these equations gives the reflectance and transmittance

$$R \equiv |r|^2 = \left| \frac{\frac{2\pi}{c}(u_e + w_e)}{\left(1 + \frac{2\pi}{c} u_e\right)\left(1 - \frac{2\pi}{c} w_e\right)} \right|^2 \tag{6.43}$$

$$T \equiv |t|^2 = \left| \frac{1 + \left(\frac{2\pi}{c}\right)^2 u_e w_e}{\left(1 + \frac{2\pi}{c} u_e\right)\left(1 - \frac{2\pi}{c} w_e\right)} \right|^2 \tag{6.44}$$

and absorbance

$$A = 1 - T - R \tag{6.45}$$

of the film. Therefore the effective Ohmic parameters u_e and w_e completely determine the optical properties of inhomogeneous films.

Thus the problem of the field distribution and optical properties of the metal–dielectric films reduces to uncoupled quasistatic conductivity problems Eqs. (6.35) and (6.36) to which extensive theory already exist. Thus numerous methods developed in the percolation theory can be used to find the effective parameters u_e and w_e of the film (see Section 6.4).

Now, let us consider the case of the strong skin effect in metal grains and trace the evolution of the optical properties of a semicontinuous metal film when the surface density p of metal is increasing. When $p = 0$, the film is purely dielectric and the effective parameters u_e and w_e coincide with the dielectric Ohmic parameters given by Eq. (6.33). By substituting $u_e = u_d$ and $w_e = w_d$ into Eqs. (6.43)–(6.45) and assuming that the dielectric film has no losses and is optically thin $(dk\varepsilon_d \ll 1)$, we obtain the reflectance $R = d^2(\varepsilon_d - 1)^2 k^2/4$, transmittance $T = 1 - d^2(\varepsilon_d - 1)^2 k^2/4$, and the absorbance $A = 0$ that coincides with the well-known results for a thin dielectric film [48,49].

It is not surprising that a film without losses has zero absorbance. The losses are also absent in the limit of full coverage, when the metal concentration $p = 1$, since the strong skin effect is considered when penetration length (skin depth) $\delta = 1/k \operatorname{Im} n_m$ is negligible in comparison with the film thickness d. In this case, the film is a perfect metal mirror. Indeed, by substituting the Ohmic parameters $u_e = u_m$ and $w_e = w_m$ from Eq. (6.32) into Eqs. (6.43)–(6.45) we obtain for the reflectance $R = 1$, while the transmittance T and absorbance A are both equal to zero. Note that the optical properties of the film do not depend on the particle size D for the metal concentration $p = 0$ and $p = 1$, since properties of the dielectric and continuous metal films do not depend on the shape of the metal grains.

Now, we consider the film at the percolation threshold $p = p_c$ with $p_c = \frac{1}{2}$ for a self-dual system [3,4]. A semicontinuous metal film may be thought of as a mirror, which is broken into small pieces with typical size D much smaller than the wavelength λ. At the percolation threshold, the exact Dykhne formulas $u_e = \sqrt{u_d u_m}$ and $w_e = \sqrt{w_d w_m}$ hold [50]. Thus the following equations for the effective Ohmic parameters are obtained from Eqs. (6.33) and (6.32)

$$\frac{2\pi}{c}\, u_e(p_c) = \sqrt{\varepsilon'_d} \qquad \frac{2\pi}{c}\, w_e(p_c) = i\,\frac{Dk}{4}\sqrt{1 + \frac{2d}{D}} \tag{6.46}$$

From these equations it follows that $|w_e/u_e| \sim Dk \ll 1$ and the effective Ohmic parameter w_e can be neglected in comparison with u_e. By substituting the effective Ohmic parameter $u_e(p_c)$ given by Eq. (6.46) in Eqs. (6.43)–(6.45), the optical properties at the percolation are obtained

$$R(p_c) = \frac{\varepsilon'_d}{(1 + \sqrt{\varepsilon'_d})^2} \tag{6.47}$$

$$T(p_c) = \frac{1}{(1 + \sqrt{\varepsilon'_d})^2} \tag{6.48}$$

$$A(p_c) = \frac{2\sqrt{\varepsilon'_d}}{(1 + \sqrt{\varepsilon'_d})^2} \tag{6.49}$$

Remember that reduced dielectric function $\varepsilon'_d = 1 + 2\varepsilon_d d/D$. When metal grains are oblate enough so that $\varepsilon_d d/D \ll 1$ and $\varepsilon'_d \to 1$, the above expressions simplify to the universal result

$$R = T = \tfrac{1}{4} \qquad A = \tfrac{1}{2} \tag{6.50}$$

Thus, there is effective adsorption in semicontinuous metal films even for the case when neither dielectric nor metal grains absorb light energy. The mirror broken into small pieces effectively absorbs energy from the EM field. The effective absorption in a loss-free film means that the EM energy is stored in the system and that the amplitudes of the local EM field increase up to infinity. In any real semicontinuous metal film, the local field saturates due to nonzero losses, but significant field fluctuations take place over the film when losses are small, as discussed below.

To find the optical properties of semicontinuous films for arbitrary metal concentration p, the effective medium theory can be implemented, which was originally developed to provide a semiquantitative description of the transport properties of percolating composites [3]. The effective medium theory, when being applied to Eqs. (6.35), (6.39), (6.36), and (6.40), results in the following equations for the effective parameters:

$$u_e^2 - \Delta p u_e(u_m - u_d) - u_d u_m = 0 \tag{6.51}$$

$$w_e^2 - \Delta p w_e(w_m - w_d) - w_d w_m = 0 \tag{6.52}$$

where the reduced concentration $\Delta p = (p - p_c)/p_c$, $\left(p_c = \frac{1}{2}\right)$ is introduced. It follows from Eq. (6.52) that for the considered case of a strong skin effect, when the Ohmic parameters w_m and w_d are given by Eqs. (6.32) and (6.33), the effective Ohmic parameter $|w_e| \ll c$ for all metal concentrations p. Therefore, the parameter w_e is negligible in Eqs. (6.43) and (6.44). For further simplification, the Ohmic parameter u_d can be neglected in comparison with u_m in the second term of Eq. (6.51) [cf. Eqs. (6.32) and (6.33)]. Then, introduction of the dimensionless Ohmic parameter $u_e' = (2\pi/c)u_e$ allows us to rewrite Eq. (6.51) as

$$u_e'^2 - 2i\,\frac{\lambda \Delta p}{\pi D}\,u_e' - \varepsilon_d' = 0 \tag{6.53}$$

Right at the percolation threshold $p = p_c = \frac{1}{2}$, when the reduced concentration $\Delta p = 0$, Eq. (6.53) gives the effective Ohmic parameter $u_e'(p_c) = \sqrt{\varepsilon_d'}$, which coincides exactly with Eq. (6.46) and results in reflectance, transmittance, and absorbance given by Eqs. (6.47)–(6.49), respectively. For concentrations different than the percolation threshold, Eq. (6.53) gives

$$u_e' = i\,\frac{\lambda \Delta p}{\pi D} + \sqrt{-\left(\frac{\lambda \Delta p}{\pi D}\right)^2 + \varepsilon_d'} \tag{6.54}$$

which becomes purely imaginary for $|\Delta p| > \pi D\sqrt{\varepsilon_d'}/\lambda$. Then, Eqs. (6.47)–(6.49) result in zero absorbance, $A = 1 - R - T = 1 - |u_e'|^2/|1 + u_e'|^2 - 1/|1 + u_e'|^2 = 0$ (recall that the effective Ohmic parameter w_e is neglected). In the vicinity of a percolation threshold, namely, for

$$|\Delta p| < \frac{\pi D}{\lambda}\sqrt{\varepsilon_d'} \tag{6.55}$$

the effective Ohmic parameter u_e' has a nonvanishing real part and, therefore, the absorbance

$$A = \frac{2\sqrt{-(\lambda\Delta p/\pi D)^2 + \varepsilon_d'}}{1 + \varepsilon_d' + 2\sqrt{-(\lambda\Delta p/\pi D)^2 + \varepsilon_d'}} \tag{6.56}$$

is nonzero and has a well-defined maximum at the percolation threshold; the width of the maximum is inversely proportional to the wavelength. The effective absorption in almost loss-free semicontinuous metal film means that local EM fields strongly fluctuate in the system, as was speculated above. The spectral width for the strong fluctuations should be the same as the width of the absorption maximum, that is, it is given by Eq. (6.55).

Note that the effective parameters u_e and w_e can be determined experimentally by measuring the amplitude and phase of the transmitted and reflected waves using, for example, a waveguide technique (see, [51] and references cited therein), or by measuring the film reflectance as a function of the fields $\mathbf{E}_1$ and $\mathbf{H}_1$. In this case, a metal screen placed behind the film can be used to control the values of these fields [52,53].

6.4. COMPUTER SIMULATIONS OF LOCAL ELECTRIC AND MAGNETIC FIELDS

6.4.1. Kirchhoff's Equations

To find the local electric $\mathbf{E}(\mathbf{r})$ and magnetic $\mathbf{H}(\mathbf{r})$ fields, Eqs. (6.35) and (6.36) should be solved. First, consider Eq. (6.35), which is convenient to rewrite in terms of the dimensionless "dielectric constant"

$$\tilde{\varepsilon} = 4\pi i u(\mathbf{r})/\omega d \tag{6.57}$$

as follows:

$$\nabla \cdot [\tilde{\varepsilon}(\mathbf{r})\nabla\phi(\mathbf{r})] = \mathcal{E} \tag{6.58}$$

where $\phi(\mathbf{r})$ is the fluctuating part of the potential $\phi'(\mathbf{r})$ so that $\nabla\phi'(\mathbf{r}) = \nabla\phi(\mathbf{r}) - \mathbf{E}_0$, $\langle\phi(\mathbf{r})\rangle = 0$, and $\mathcal{E} = \nabla \cdot [\tilde{\varepsilon}(\mathbf{r})\mathbf{E}_0]$. Remember that the "external" field $\mathbf{E}_0$ is defined by Eq. (6.37). For the metal–dielectric films considered here, local dielectric constant $\tilde{\varepsilon}(\mathbf{r})$ equals $\tilde{\varepsilon}_m = 4\pi i u_m/\omega d$ and $\tilde{\varepsilon}_d = \varepsilon_d' D/2d$ for the metal and dielectric regions, respectively. The external field $\mathbf{E}_0$ in Eq. (6.58) can be chosen real, while the local potential $\phi(\mathbf{r})$ takes complex values since the metal dielectric constant $\tilde{\varepsilon}_m$ is complex $\tilde{\varepsilon}_m = \tilde{\varepsilon}_m' + i\tilde{\varepsilon}_m''$. In the quasistatic limit, when the skin depth δ is much larger than the film thickness d, the dielectric constant $\tilde{\varepsilon}_m$ coincides with the metal dielectric constant ε_m as it follows from Eq. (6.31).

To get insight into the high-frequency properties of metals, a simple model known as a Drude metal is considered, which semiquantitaively reproduces the basic optical properties of a metal [54]. In this approach, the dielectric constant of metal grains can be approximated by the Drude formula

$$\varepsilon_m(\omega) = \varepsilon_b - (\omega_p/\omega)^2/(1 + i\omega_\tau/\omega) \tag{6.59}$$

where ε_b is contribution to ε_m due to the interband transitions: ω_p is the plasma frequency, and $\omega_\tau = 1/\tau \ll \omega_p$ is the relaxation rate. In the high-frequency range considered here, losses in metal grains are relatively small, $\omega_\tau \ll \omega$. Therefore, the real part ε'_m of the metal dielectric function ε_m is much larger (in modulus) than the imaginary part ε''_m $(|\varepsilon'_m|/\varepsilon''_m \cong \omega/\omega_\tau \gg 1)$, and ε'_m is negative for the frequencies ω less than the renormalized plasma frequency,

$$\tilde{\omega}_p = \omega_p/\sqrt{\varepsilon_b} \tag{6.60}$$

Thus, the metal conductivity $\sigma_m = -i\omega\varepsilon_m/4\pi \cong (\varepsilon_b\tilde{\omega}_p^2/4\pi\omega)[i(1 - \omega^2/\tilde{\omega}_p^2) + \omega_\tau/\omega]$ is characterized by the dominant imaginary part for $\tilde{\omega}_p > \omega \gg \omega_\tau$, that is, it is of inductive character. The same is true for the Ohmic parameter u_m in the quasistatic limit since it is just proportional to the metal conductivity in this limit. In the opposite case of the strong skin effect, the Ohmic parameter u_m is inductive according to Eq. (6.32) for all spectral ranges regardless of the metal properties. Therefore, the metal grains can be modeled as inductances L while the dielectric gaps can be represented by capacitances C. This model works for the optic and IR regardless of the metal grain size and holds for *all* spectral ranges when the skin effect is strong in the metal grains. Then, the percolation metal–dielectric film represents a set of randomly distributed L and C elements. The collective surface plasmons excited by the external field, can be thought of as resonances in different $L - C$ circuits, and the excited surface plasmon eigenstates are seen as giant fluctuations of the local field [50].

Note that Ohmic parameter w takes the same sign and rather close absolute values for metal and dielectric grains according to Eqs. (6.31)–(6.33). A film can be thought of as a collection of C elements in "w" space. Therefore the resonance phenomena are absent in a solution of Eq. (6.36). The fluctuations of the potential ψ' can be indeed neglected in comparison to the φ' fluctuations. For this reason, we concentrate attention on the properties of the "electric" field $\mathbf{E}(\mathbf{r}) = -\nabla\phi'(\mathbf{r}) = -\nabla\phi(\mathbf{r}) + \mathbf{E}_0$ that can be found from solution of Eq. (6.58).

Because of difficulties in finding a solution to the Poisson Eq. (6.35) or (6.58), a great deal of use is made of the tight binding model in which metal and dielectric particles are represented by metal and dielectric bonds of a square lattice. After such discretization, Eq. (6.58) acquires the form of the Kirchhoff's equations defined on a square lattice [3]. The Kirchhoff's equations can be written in terms of the local dielectric constant and it is assumed that the external electric field $\mathbf{E}_0$ is directed along the "x" axis. Thus, the following set of equations are obtained

$$\sum_j \tilde{\varepsilon}_{ij}(\phi_j - \phi_i) = \sum_j \tilde{\varepsilon}_{ij} E_{ij} \tag{6.61}$$

where ϕ_i and ϕ_j are the electric potentials determined at the sites of the square lattice and the summation is over the nearest neighbors of the site i. The electromotive force (emf) E_{ij} takes the value $E_0 a_0$ for the bond $\langle ij \rangle$ in the positive x direction (where a_0 is the spatial period of the square lattice) and $-E_0 a_0$, for the bond $\langle ij \rangle$ in the $-x$ direction; $E_{kj} = 0$ for the other four bonds at the site i. Thus the composite is modeled by a capacitor–inductor–resistor network represented by Kirchhoff's equations (6.61). The emf forces E_{ij} represent the external electric field applied to the system. In transition from the continuous medium described by Eq. (6.58) to the random network described by Eq. (6.61), it is usually supposed [1–4] that bond permittivities $\tilde{\varepsilon}_{ij}$ are statistically independent and a_0 is set to be equal to the metal grain size, $a_0 = a$. In the considered case of a two component metal–dielectric random film, the permittivities $\tilde{\varepsilon}_{ij}$ take the values $\tilde{\varepsilon}_m$ and $\tilde{\varepsilon}_d$, with probabilities p and $1 - p$, respectively. The assumption that the bond permittivities $\tilde{\varepsilon}_{ij}$ in Eq. (6.61) are statistically independent considerably simplify computer simulations as well as analytical consideration of local optical fields in the film. Note that important critical properties are universal, that is, they are independent of details of a model (e.g., possible correlation of permittivities $\tilde{\varepsilon}_{ij}$ in different bonds).

6.4.2. Numerical Model

Now, there exist very efficient numerical methods for calculating the effective conductivity of composite materials (see [1–4,8,55,56]), but they typically do not allow calculations of the field distributions. To calculate local field distribution, a new original method had been developed. It is based on the real-space renormalization group (RSRG) method that was suggested for percolation by Reynolds, et al. [57] and Sarychev [58], and then extended to study the conductivity [59] and the permeability of oil reservoirs [60]. By following the approach used by Aharony, the RSRG method was adopted to finding the field distributions in the following way [38,39,61]. First, a square lattice of L-R (metal) and C (dielectric) bonds was generated using a random number generator. As seen in Fig. 6.3, such a lattice can be considered as a set of "corner" elements. One of such elements is labeled as (ABCDEFGH) in Fig. 6.3. In the first stage of the RSRG procedure, each of these elements is replaced by the two Wheatstone bridges, as shown in Fig. 6.3. After this transformation, the initial square lattice is converted to another square lattice, with the distance between the sites twice larger and with each bond between the two nearest-neighboring sites being the Wheatstone bridge. Note that there is a 1:1 correspondence between the x bonds in the initial lattice and the x bonds in the x directed bridges of the transformed lattice, as seen in Fig. 6.3. The same 1:1 correspondence also exists between the y bonds. The transformed lattice is also a

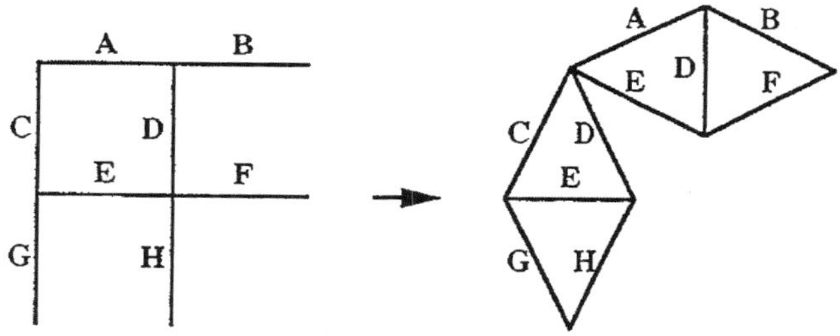

Figure 6.3. The real-space renormalization scheme.

square lattice, and we can again apply to it the RSRG transformation. We continue this procedure until the size $\mathcal{L}$ of the system is reached. As a result, instead of the initial lattice we have two large Wheatstone bridges in the x and y directions. Each of them has a hierarchical structure consisting of bridges with the sizes from 2 to $\mathcal{L}$. Because the 1:1 correspondence is preserved at each step of the transformation, the correspondence also exists between the elementary bonds of the transformed lattice and the bonds of the initial lattice. After using the RSRG transformation, the periodic boundary conditions (see, e.g., [8]) are applied and the Kirchhoff's equations (6.61) are solved to determine the fields and the currents in all the bonds of the transformed lattice. Due to the hierarchical structure of the transformed lattice, these equations can be solved exactly. Then, the 1:1 correspondence between the elementary bonds of the transformed lattice and the bonds of the initial square lattice is used to find the field distributions in the initial lattice as well as its effective dielectric constant. The number of operations to get the full distributions of the local fields is proportional to $\mathcal{L}^2$ and is much less than the $\mathcal{L}^7$ operations needed in the transform-matrix method [3,62] and the $\mathcal{L}^3$ operations needed in the well-known Frank–Lobb algorithm [55], which does not provide the field distributions but the effective conductivity only. The RSRG procedure is certainly not exact since the effective connectivity of the transformed system does not repeat the connectivity of the initial square lattice exactly. To check the accuracy of the RSRG, the $2d$ percolation problem was solved using this method [39]. Namely, the effective parameters were calculated for a two-component composite with the real metallic conductivity σ_m much larger than the real conductivity σ_d of the dielectric component $\sigma_m \gg \sigma_d$. It obtained the percolation threshold $p_c = 0.5$ and the effective conductivity at the percolation threshold that is very close to $\sigma(p_c) = \sqrt{\sigma_m \sigma_d}$. These results coincide with the exact ones for $2d$ composites [50]. This is not surprising since the RSRG procedure preserves the self-duality of the initial system. The critical exponents obtained by the RSRG are also in agreement with known values of the exponents from the percolation theory [1,3]. Thus the ratio of the critical exponent s for the static dielectric constant and the exponent ν for the percolation correlation length is equal to $s/\nu \simeq 0.94$, the ratio of the critical exponent t for the static conductivity and the exponent ν is equal to $t/\nu \simeq 0.82$. These results should be compared with $s/\nu \simeq t/\nu \simeq 1$ that follow from the percolation theory for $2d$ composites. Therefore, there are good reasons to believe that the numerical method describes at least qualitatively the field distributions on semicontinuous metal films. Below the RSRG exponents $s/\nu \simeq 0.94$ and $t/\nu \simeq 0.8$ are used when the computer results obtained by RSRG are compared with the scaling theory.

6.4.3. Giant Fluctuations of the Local Fields

The real-space renormalization method described above was employed to solve Eq. (6.61) and to calculate the potentials ϕ_i in the lattice. Then, we find the local field $\mathbf{E}(\mathbf{r})$ and electric current $\mathbf{j}_E(\mathbf{r})$ in terms of the average field $\mathbf{E}_0$. The effective Ohmic parameter u_e is determined by Eq. (6.39), which can be written as $\langle \mathbf{j}_E \rangle = u_e \mathbf{E}_0$. The effective dielectric constant $\tilde{\varepsilon}_e$ is equal to $4\pi i u_e/\omega d$. In the same manner, the field $\mathbf{H}(\mathbf{r})$, the magnetic current $\mathbf{j}_H(\mathbf{r})$, and the effective parameter w_e can be found from Eq. (6.36) and its lattice discretization. Note that the *same* lattice should be used to determined the fields $\mathbf{E}(\mathbf{r})$ and $\mathbf{H}(\mathbf{r})$. The directions of the external fields $\mathbf{E}_0$ and $\mathbf{H}_0$ may be chosen arbitraryly when the effective parameters u_e and w_e are calculated since the effective parameters do not depend on the direction of the field for a film, which is isotropic as a whole.

Though the effective parameters do not depend on the external field, the local electric $\mathbf{E}_1(\mathbf{r})$ and magnetic and $\mathbf{H}_1(\mathbf{r})$ fields do depend on the incident wave. The local fields $\mathbf{E}_1(\mathbf{r})$ and $\mathbf{H}_1(\mathbf{r})$ are define in the reference plane $z = -d/2 - l_0$ (see Fig. 6.1). Note that the field $\mathbf{E}_1(\mathbf{r})$ is observed in a typical near-field experiment (see Chapter 3). For the calculations below the electric and magnetic fields of the incident EM wave have been chosen in the form $\langle \mathbf{E}_1 \rangle = \{1, 0, 0\}$ and $\langle \mathbf{H}_1 \rangle = \{0, -1, 0\}$ in the plane $z = -l_0 - d/2$. This choice corresponds to the wavevector of the incident wave as $\mathbf{k} = (0, 0, -k)$, that is, there is only a transmitted wave behind the film (see Fig. 6.1). It follows from the average of Eq. (6.38), which can be written as $\langle \mathbf{E}_1 \rangle = \mathbf{E}_0 + (2\pi/c) w_e [\mathbf{n} \times \mathbf{H}_0]$ and $\langle \mathbf{H}_1 \rangle = \mathbf{H}_0 + (2\pi/c) u_e [\mathbf{n} \times \mathbf{E}_0]$ that the fields $\mathbf{E}_0$ and $\mathbf{H}_0$ are given by

$$\mathbf{E}_0 = \frac{\langle \mathbf{E}_1 \rangle - (2\pi/c) w_e [\mathbf{n} \times \langle \mathbf{H}_1 \rangle]}{1 + (2\pi/c)^2 u_e w_e} \qquad \mathbf{H}_0 = \frac{\langle \mathbf{H}_1 \rangle - (2\pi/c) u_e [\mathbf{n} \times \langle \mathbf{E}_1 \rangle]}{1 + (2\pi/c)^2 u_e w_e}. \qquad (6.62)$$

These values of the fields $\mathbf{E}_0$ and $\mathbf{H}_0$ are used to calculate the local fields $\mathbf{E}(\mathbf{r})$ and $\mathbf{H}(\mathbf{r})$. The local electric $\mathbf{E}_1(\mathbf{r})$ and magnetic $\mathbf{H}_1(\mathbf{r})$ fields are restored then from the fields $\mathbf{E}(\mathbf{r})$ and $\mathbf{H}(\mathbf{r})$ by using Eq. (6.38).

The local electric and magnetic fields have been calculated in silver-on-glass semicontinuous film as functions of the surface concentration p of silver grains. The typical glass dielectric constant ε_d is $\varepsilon_d = 2.2$. The dielectric function for silver was chosen in the Drude form (6.59); the following parameters were also used in Eq. (6.59): the interband-transition contribution $\varepsilon_b = 5$, the plasma frequency $\omega_p = 9.1$ eV, and the relaxation frequency $\omega_\tau = 0.021$ eV [63]. The metal grains are supposed to be oblate. The ratio of the grain thickness d (film thickness) and the grain diameter D has been chosen as $D/d = 3$, the same as used in [18]. To consider the skin effect of different strengths (i.e., different interactions between the electric and magnetic fields), we vary the size d of silver particles in a wide range, $d = 1 \div 100\,\text{nm}$. The size of metal grains in semicontinuous metal films is usually on the order of a few nanometers, but it can be increased significantly by using the proper method of preparation [64]. For microwave experiments [20] the films were prepared by the lithography method, so that the size of the metal particle could vary over a large range.

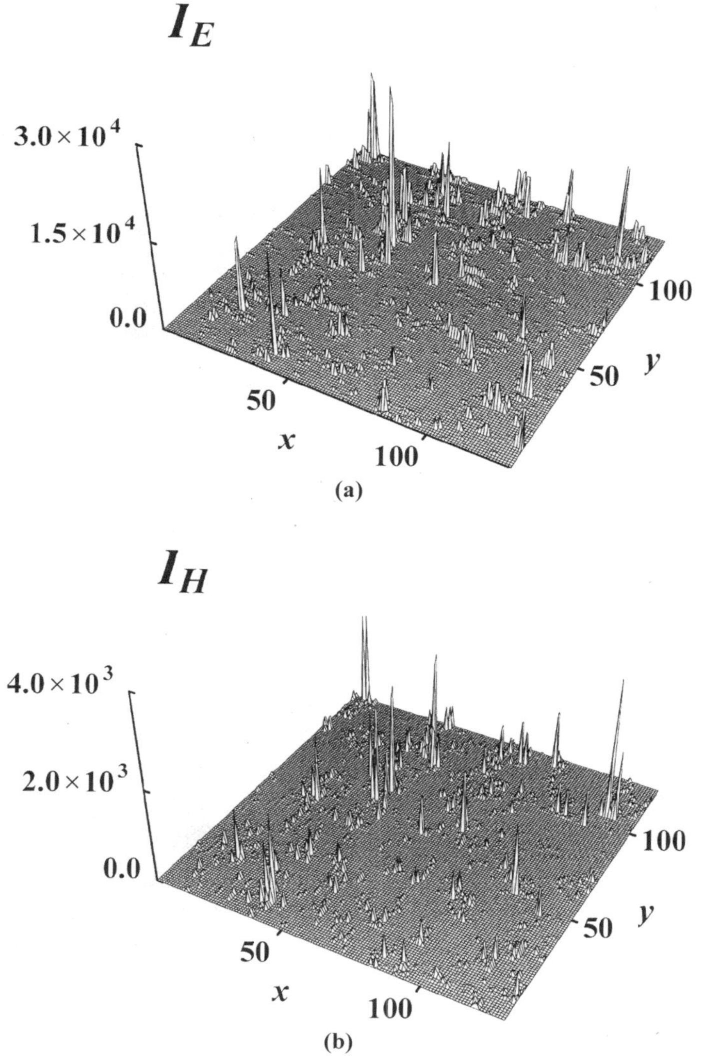

Figure 6.4. Distribution of local EM field intensities (a) $I_E = |\mathbf{E}_1(\mathbf{r})|^2/|\langle\mathbf{E}_1\rangle|^2$ and (b) $I_H = |\mathbf{H}_1(\mathbf{r})|^2/|\langle\mathbf{H}_1\rangle|^2$ in a semicontinuous silver film at the percolation threshold for $\lambda = 1\,\mu\text{m}$ and $\delta/d = 4.5$, where δ is the skin depth and d is the thickness of the film.

The space distribution of the electric and magnetic fields was calculated at two sets of parameters, as illustrated in Figs. 6.4–6.7. In Figs. 6.4 and 6.5, we show the electric and magnetic field distributions for $\lambda = 1\ \mu\text{m}$ and two different thicknesses d of the film, $d = 5$ and $d = 50\,\text{nm}$. The first thickness (Fig. 6.4) corresponds to a weak skin effect since the dimensionless thickness is small, $\Delta \equiv d/\delta = 0.2$, where $\delta = 1/k\,(\text{Im}\, n_m)$ is the skin depth. In this case, we observe the giant field fluctua-

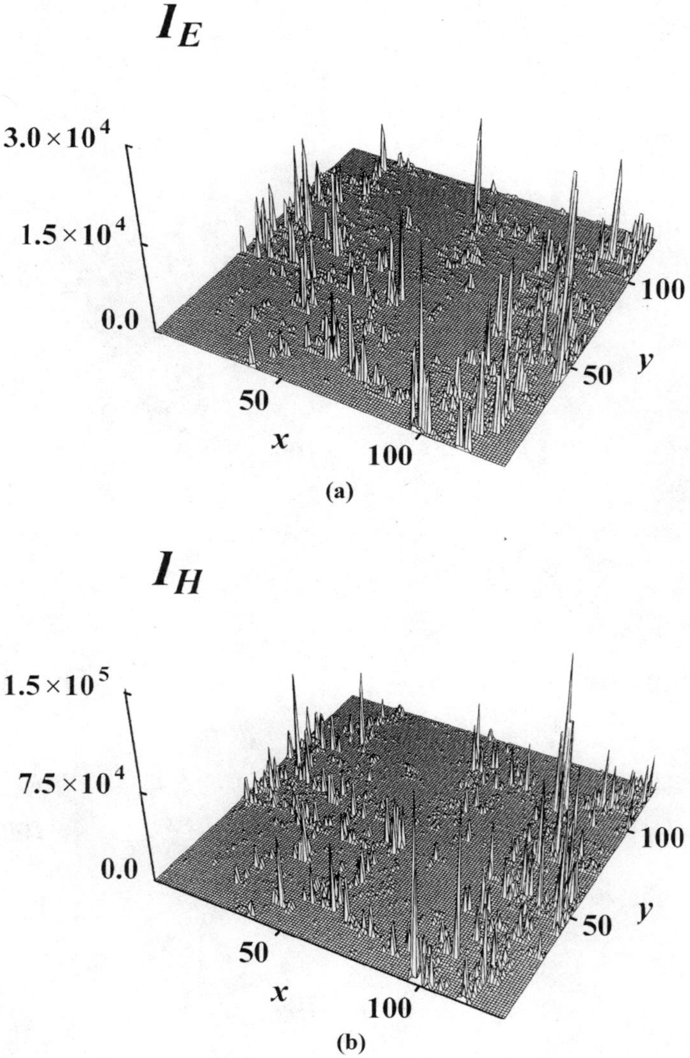

Figure 6.5. Distribution of local EM field intensities (a) $I_E = |\mathbf{E}_1(\mathbf{r})|^2/|\langle \mathbf{E}_1 \rangle|^2$ and (b) $I_H = |\mathbf{H}_1(\mathbf{r})|^2/|\langle \mathbf{H}_1 \rangle|^2$ in a semicontinuous silver film at the percolation threshold for $\lambda = 1\ \mu\text{m}$ and $\delta/d = 0.45$.

tions of the local electric field; the magnetic field also strongly fluctuates over the film but the field peaks are small compared to the electric field. The reason is that the film itself is not magnetic, $\mu_d = \mu_m = 1$, and the interaction of the magnetic field with the electric field through the skin effect is relatively small.

In Fig. 6.5, we show results for a significant skin effect, when the film thickness $d = 50\,\text{nm}$ and the dimensionless thickness exceeds 1, $\Delta = 2.2$. It is interesting to

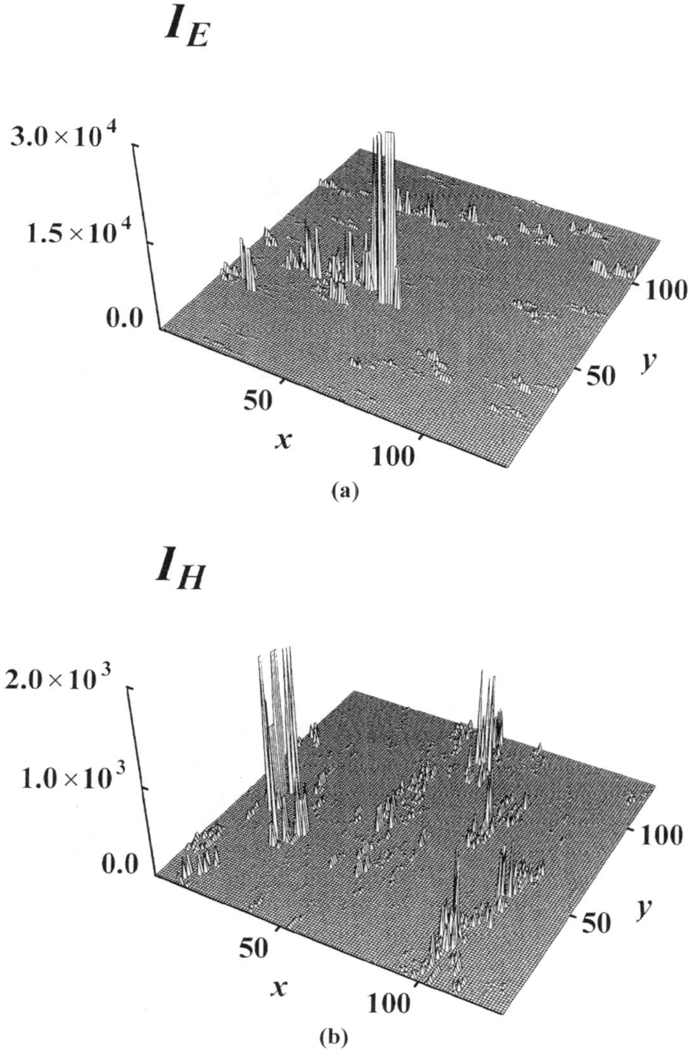

Figure 6.6. Distribution of local EM field intensities (a) $I_E = |\mathbf{E}_1(\mathbf{r})|^2/|\langle\mathbf{E}_1\rangle|^2$ and (b) $I_H = |\mathbf{H}_1(\mathbf{r})|^2/|\langle\mathbf{H}_1\rangle|^2$ in a semicontinuous silver film at the percolation threshold for $\lambda = 10\,\mu\text{m}$ and $\delta/d = 4.5$.

note that the amplitude of the electric field is roughly the same as in Fig. 6.4(a), despite the fact that the parameter Δ increased by one order of magnitude. In contrast, the local magnetic field (Fig. 6.5(b)) is strongly increased in this case so that the amplitude of the magnetic field in peaks is of the same order of magnitude as the electric field maxima. This behavior can be understood by considering the spatial moments of the local magnetic field as shown in Section 6.5.

In Figs. 6.6 and 6.7, we show results of the calculations for the local electric and magnetic fields at $\lambda = 10\,\mu\text{m}$, when the metal dielectric constant $|\varepsilon_m| \sim 10^4$. We see

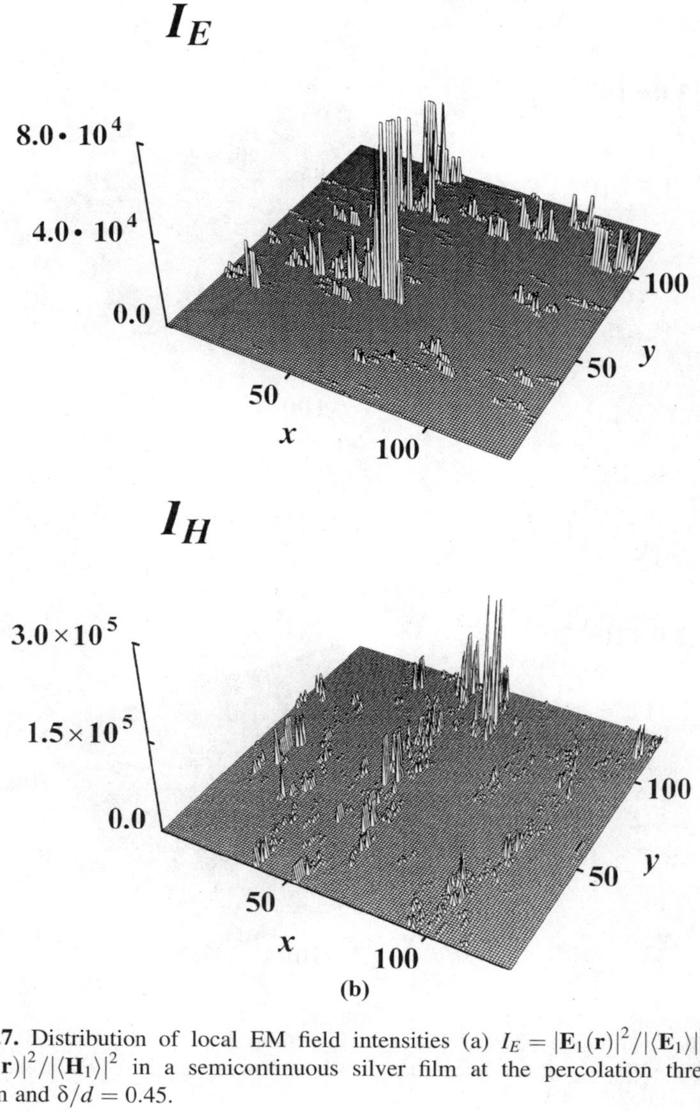

(b)

Figure 6.7. Distribution of local EM field intensities (a) $I_E = |\mathbf{E}_1(\mathbf{r})|^2/|\langle \mathbf{E}_1 \rangle|^2$ and (b) $I_H = |\mathbf{H}_1(\mathbf{r})|^2/|\langle \mathbf{H}_1 \rangle|^2$ in a semicontinuous silver film at the percolation threshold for $\lambda = 10\,\mu\text{m}$ and $\delta/d = 0.45$.

that in this case the local magnetic field can even exceed the electric field. Systematic computer simulation of the local electric field for a different wavelength and metal concentration a reader can find in [39,41–45]. The giant local field fluctuations were observe first in the microwave experiment [20] and then in the optic near-field experiments [21,22].

Being given the local fields, the effective parameters u_e and w_e can be found as well as the effective optical properties of the film. In Figs. 6.8 and 6.9, we show the

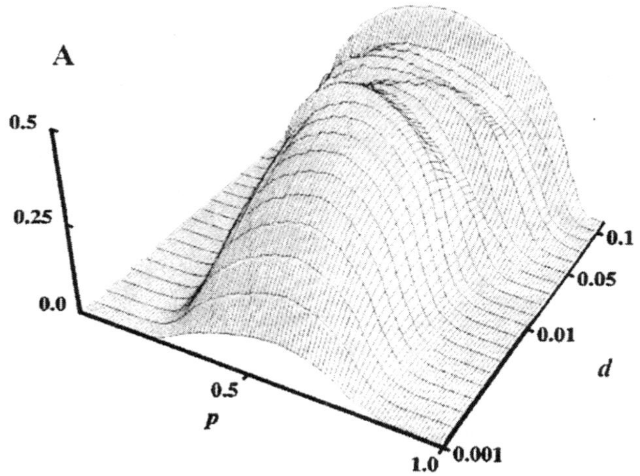

(a)

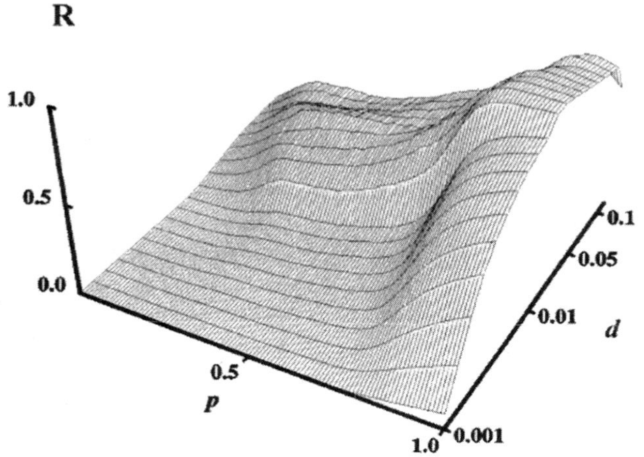

(b)

Figure 6.8. Computer simulation of (a) absorptance A, (b) reflectance R, and (c) transmittance T for a silver-on-glass semicontinous film as functions of metal concentration p and film thickness d (μm) at $\lambda = 1$ μm.

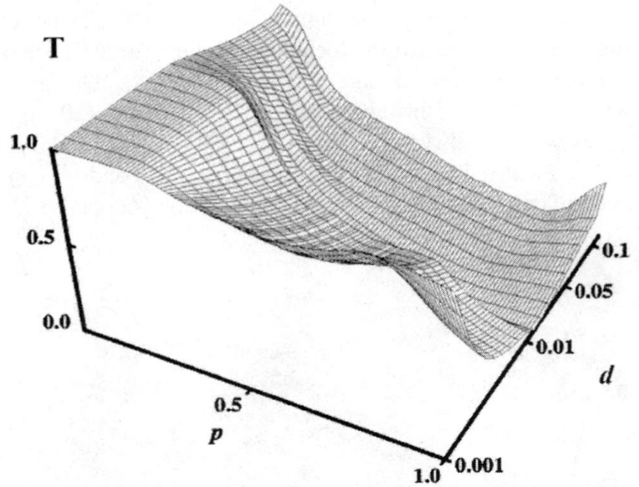

(c)

Figure 6.8. (*Continued*)

reflectance, transmittance, and absorbance as functions of silver concentration p, for wavelengths $\lambda = 1$ and $10\,\mu$m, respectively. The absorbance in these figures has an anomalous maximum in the vicinity of the percolation threshold that corresponds to the behavior predicted by Eq. (6.56). This maximum was first detected in the experiments [13–16]. The maximum in the absorption corresponds to strong fluctuations of the local fields. In Eq. (6.55), we have estimated the concentration range Δp

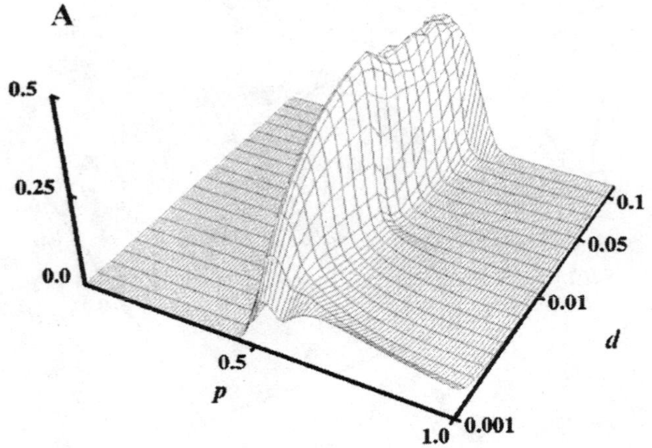

(a)

Figure 6.9. Computer simulation of (a) absorptance A, (b) reflectance R, and (c) transmittance T for a silver-on-glass semicontinous film as functions of metal concentration p and film thickness d (μm) at $\lambda = 10\,\mu$m.

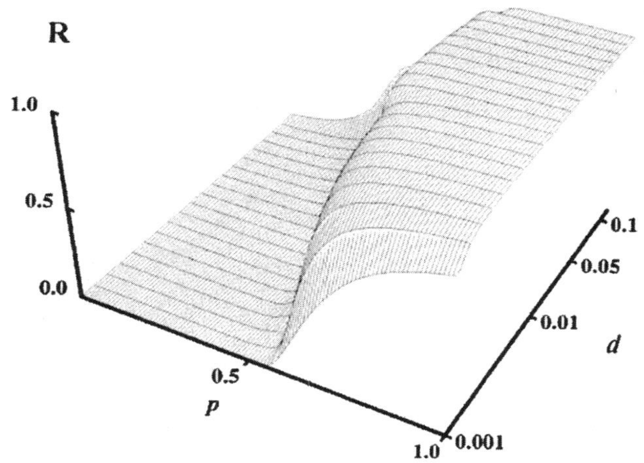

(b)

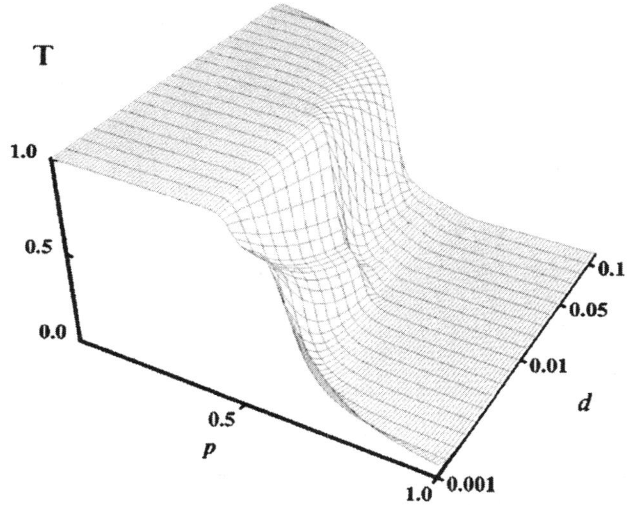

(c)

Figure 6.9. (*Continued*)

centered at the percolation threshold p_c (where the giant local field fluctuations occur) as $\Delta p \propto 1/\lambda$. Indeed, the absorbance shrinks at transition from Fig. 6.8 to Fig. 6.9, when the wavelength λ increases 10 times. In Figs. 6.10 and 6.11, we compare results of numerical simulations for the optical properties of silver semicontinuous films with calculations based on the effective medium approach [Eqs. (6.51) and (6.52)] represented in terms of the new Ohm's parameters u and v. Results of such a "dynamic" effective-medium theory are in good agreement with our numerical simulations for arbitrary skin effects.

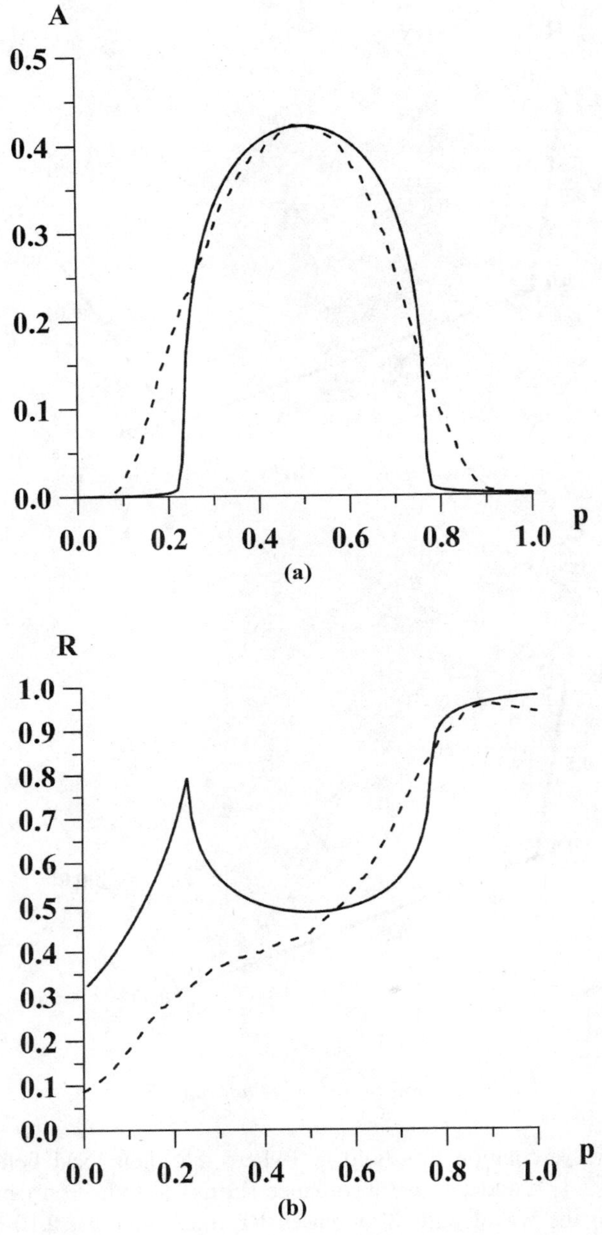

Figure 6.10. Results of computer simulations (dashed line) and the dynamic effective–medium theory (solid line) for (a) absorptance A, (b) reflectance R, and (c) transmittance T of a silver-on-glass semicontinuous film as a function of the metal concentration p at wavelengths $\lambda = 1\,\mu\text{m}$ and thickness $d = 50\,\text{nm}$.

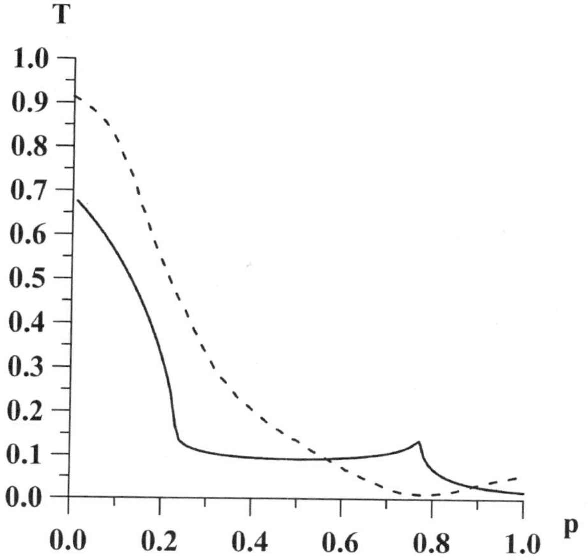

(c)

Figure 6.10. (*Continued*)

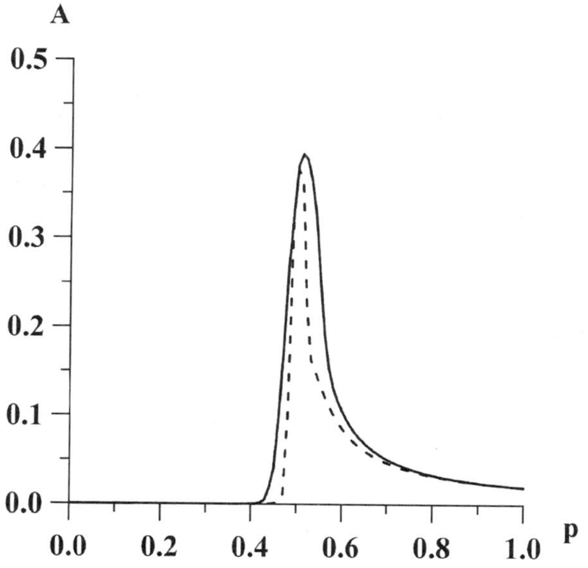

(a)

Figure 6.11. Results of computer simulations (dashed line) and the dynamic effective-medium theory (solid line) for (a) absorptance A, (b) reflectance R, and (c) transmittance T of a silver-on-glass semicontinuous film as a function of the metal concentration p at wavelengths $\lambda = 10\,\mu\text{m}$ and thickness $d = 5\,\text{nm}$.

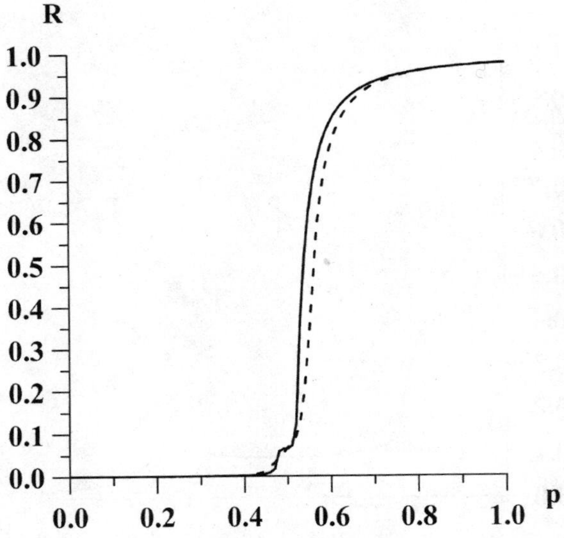

(b)

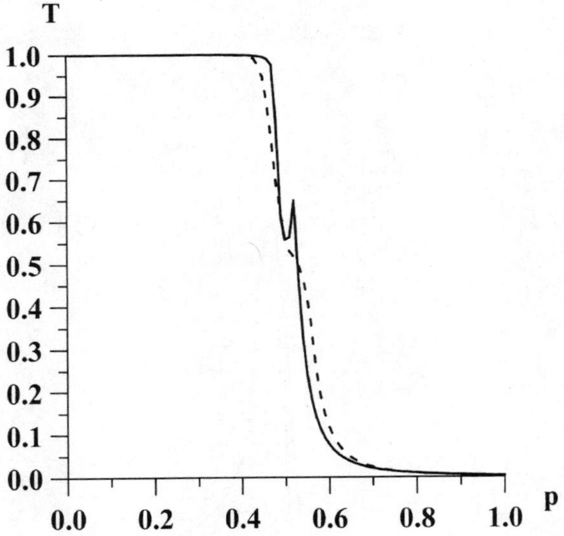

(c)

Figure 6.11. (*Continued*)

6.5. ANDERSON LOCALIZATION OF SURFACE PLASMONS

In this section, we follow the approach developed in recent papers [21,45,46]. To estimate the local field distribution analytically it is convenient to start with Kirchhoff's equations (6.61). For further consideration, it is assumed that the square lattice has a very large but finite number of sites N and rewrite Eq. (6.61) in matrix form with the "Hamiltonian" $\mathcal{H}$ defined in terms of the local dielectric constants,

$$\mathcal{H}\phi = \mathcal{E} \tag{6.63}$$

where ϕ is a vector of the local potentials $\phi = \{\phi_1, \phi_2, \ldots, \phi_N\}$ determined in all N sites of the lattice, vector $\mathcal{E}$ equals to $\mathcal{E}_i = \sum_j \tilde{\varepsilon}_{ij} E_{ij}$, as it follows from Eq. (6.61). The Hamiltonian $\mathcal{H}$ is $N \times N$ matrix that has off-diagonal elements $\mathrm{H}_{ij} = -\tilde{\varepsilon}_{ij}$ and diagonal elements defined as $\mathrm{H}_{ii} = \sum_j \tilde{\varepsilon}_{ij}$, where j refers to nearest neighbors of site i. The off-diagonal elements H_{ij} take values $\tilde{\varepsilon}_d > 0$ and $\tilde{\varepsilon}_m$, with probability p and $1 - p$, respectively. The diagonal elements H_{ii} are Eq. (6.61) distributed between $2d\tilde{\varepsilon}_m$ and $2d\tilde{\varepsilon}_d$, where d is the dimensionality of the space ($2d$ is the number of the nearest neighbors in d dimensional square lattice).

The dielectric constant $\tilde{\varepsilon}_m$ is negative in the visible and IR spectral ranges for a typical metal as it was discussed after Eq. (6.59). Therefore, $\tilde{\varepsilon}_m$ can be written as $\tilde{\varepsilon}_m = |\tilde{\varepsilon}'_m|(-1 + i\kappa)$, where the loss factor $\kappa = \tilde{\varepsilon}''_m/|\tilde{\varepsilon}'_m|$ is small, $\kappa \ll 1$. This equation for $\tilde{\varepsilon}_m$ holds in all spectral ranges if the skin effect is strong in the metal grains. It is shown below that the fluctuations of the local fields are significant when $\tilde{\varepsilon}'_m$ is negative and the losses are small. It is supposed in this and Sections 6.6 and 6.7 that this condition is fulfilled, that is, $\tilde{\varepsilon}'_m < 0$ and $\kappa \ll 1$.

It is convenient to represent the Hamiltonian $\mathcal{H}$ as a sum of two Hermitian Hamiltonians $\mathcal{H} = \mathcal{H}' + i\kappa\mathcal{H}''$, where the term $i\kappa\mathcal{H}''$ ($\kappa \ll 1$) represents losses in the system. The Hamiltonian $\mathcal{H}'$ formally coincides with the Hamiltonian of the problem of metal–insulator transition (Anderson transition) in quantum systems [65–68]. More specifically, the Hamiltonian $\mathcal{H}'$ maps the quantum mechanical Hamiltonian for the Anderson transition problem with both on- and off-diagonal correlated disorder. Since the off-diagonal matrix elements in $\mathcal{H}'$ have different signs, the Hamiltonian is similar to the so-called gauge-invariant model. This model, in turn, is a simple version of the random flux model, which represents a quantum system with random magnetic field [65] (see also recent numerical studies [69–71]). Hereafter, we refer to operator $\mathcal{H}'$ as to Kirchhoff's Hamiltonian (KH).

Thus, the problem of the field distribution in the system, that is, the problem of finding a solution to Kirchhoff's Eq. (6.61), becomes an eigenfunction problem for the KH, $\mathcal{H}'\Psi_n = \Lambda_n\Psi_n$, whereas the losses can be treated as perturbations. Since the real part ε'_m of the metal dielectric function $\tilde{\varepsilon}_m$ is negative ($\tilde{\varepsilon}'_m < 0$) and the permittivity of the dielectric host is positive ($\tilde{\varepsilon}_d > 0$), the manifold of the KH eigenvalues Λ_n contains eigenvalues that have the real parts equal (or close) to zero. Then, eigenstates Ψ_n that correspond to eigenvalues $|\Lambda_n/\tilde{\varepsilon}_m| \ll 1$ are strongly excited by the external field and seen as giant field fluctuations representing the resonant surface plasmon modes. If we assume that the eigenstates excited by the external field are localized, they should look like local-field peaks. The average distance between the field peaks can be estimated as $a(N/n)^{1/d}$, where n is the number of the KH eigenstates excited by the external field and N is the total number of the eigenstates.

Now, we consider in more detail behavior of the eigenfunctions Ψ_n of the KH $\mathcal{H}'$, in the special case when $\tilde{\varepsilon}'_m = -\tilde{\varepsilon}_d$, corresponding to the plasmon resonance of individual particles in a $2d$ system. Since a solution to Eq. (6.61) does not change when multiplying $\tilde{\varepsilon}_m$ and $\tilde{\varepsilon}_d$ by the same factor, we can normalize the system and set $\tilde{\varepsilon}_d = -\tilde{\varepsilon}_m = 1$.

According to the one-parameter scaling theory, the eigenstates Ψ_n are all localized for the $2d$ case (see, however, the discussion in [68,72]. On the other hand, it was shown in computer simulations [73] that there is a transition from chaotic [74,75] to localized eigenstates for the $2d$ Anderson problem [73], with an intermediate cross over region. First, we consider the case when metal concentration p is equal to the percolation threshold $p_c = \frac{1}{2}$ for the $2d$ bond percolation problem. Then, the on-diagonal disorder in the KH $\mathcal{H}'$ is characterized by $\langle \mathrm{H}'_{ii} \rangle = 0$, $\langle \mathrm{H}'^2_{ii} \rangle = 4$, which corresponds to the chaos-localization transition [73]. The KH also has a strong off-diagonal disorder, $\langle \mathrm{H}'_{ij} \rangle = 0, \langle \mathrm{H}'^2_{ij} \rangle = 1 (i \neq j)$, which favors localization [69,70]. Our conjecture is that eigenstates Ψ_n are localized for all Λ_n in the $2d$ system. (We cannot rule out a possibility of inhomogeneous localization, similar to that obtained for fractals [76], or the power-law localization [65,77]; note, however, that these possibilities are in strong disagreement with the one-parameter scaling theory of the Anderson localization [65].)

In the considered case of $\tilde{\varepsilon}_d = -\tilde{\varepsilon}_m = 1$ and $p = \frac{1}{2}$, all parameters in the KH $\mathcal{H}'$ are of the order of unity and its properties do not change under the transformation $\tilde{\varepsilon}_d \Longleftrightarrow \tilde{\varepsilon}_m$. Therefore, the real eigenvalues, Λ_n, are distributed symmetrically with respect to zero, in an interval on the order of 1. The eigenstates with $\Lambda_n \approx 0$ are effectively excited by the external field and represent the giant local-field fluctuations. When metal concentration p decreases (increases), the eigenstates with $\Lambda_n \approx 0$ are shifted from the center of the distribution toward its lower (upper) edge, which typically favors localization. Because of this, we assume that the eigenstates, or at least those with $\Lambda_n \approx 0$, are localized, for nearly all metal concentrations p in the $2d$ case.

Suppose we found all eigenvalues Λ_n and eigenfunctions Ψ_n of $\mathcal{H}'$. Then, we can express the potential ϕ in Eq. (6.63) in terms of the eigenfunctions as $\phi = \sum_n A_n \Psi_n$ and substitute it in Eq. (6.63). Thus we obtain the following equation for coefficients A_n:

$$(i\kappa b_n + \Lambda_n)A_n + i\kappa \sum_{m \neq n} (\Psi_n | \mathcal{H}'' | \Psi_m) A_m = \mathcal{E}_n \tag{6.64}$$

where $b_n = (\Psi_n | \mathcal{H}'' | \Psi_n)$, and $\mathcal{E}_n = (\Psi_n | \mathcal{E})$ is a projection of the external field on eigenstate Ψ_n. [The product of two vectors, e.g., Ψ_n and $\mathcal{E}$, is defined here in a usual way, as $\mathcal{E}_n = (\Psi_n | \mathcal{E}) \equiv \sum_i \Psi^*_{n,i} \mathcal{E}_i$, where the sum is over all lattice sites.] Since all parameters in the real Hamiltonian $\mathcal{H}''$ are on the order of 1, the matrix elements b_n are also on the order of 1. For a qualitative consideration they can be approximated by some constant b, which is about unity. The eigenstates Ψ_n are localized within spatial domains $\xi_A(\Lambda)$, where $\xi_A(\Lambda)$ is the Anderson localization length, which depends on Λ. Then, the sum in Eq. (6.64) converges and can be treated as a small perturbation. In the zeroth approximation,

$$A_n^{(0)} = \mathcal{E}_n/(\Lambda_n + i\kappa b) \tag{6.65}$$

The first-order correction to A_n is equal to

$$A_n^{(1)} = -i\kappa \sum_{m\neq n} (\Psi_n|\mathcal{H}''|\Psi_m)\mathcal{E}_m/(\Lambda_m + i\kappa b) \tag{6.66}$$

For $\kappa \to 0$, most important eigenstates in this sum are those with $|\Lambda_m| \leq b\kappa$. Since the eigenstates Λ_n are distributed in the interval of the order of unity, the spatial density of the eigenmodes with $|\Lambda_m| \leq b\kappa$ vanishes as $a^{-d}\kappa \to 0$ at $\kappa \to 0$. Therefore, $A_n^{(1)}$ is exponentially small $|A_n^{(1)}| \sim |\sum_{m\neq n}(\Psi_n|\mathcal{H}''|\Psi_m)\mathcal{E}_m| \propto \exp\{-\xi_e/\xi_A(0)\}$, where $\xi_e \sim a\kappa^{-1/d}$ is a typical distance between resonance eigenstates (with $|\Lambda_m| \leq b\kappa$) in real-space. Since $|A_n^{(1)}|$ is exponentially small it can be neglected when the losses in the system are small ($\kappa \ll 1$) and the field correlation length $\xi_e \gg \xi_A(0)$. Then, the local potential ϕ is equal to $\phi(\mathbf{r}) = \sum_n A_n^{(0)}\Psi_n = \sum_n \mathcal{E}_n\ \Psi_n(r)/(\Lambda_n + i\kappa b)$ (see Eq. (6.65)] and the fluctuating part of the local field $\mathbf{E}_f = -\nabla\phi(\mathbf{r})$ is given by

$$\mathbf{E}_f(\mathbf{r}) = -\sum_n \mathcal{E}_n \nabla\Psi_n(\mathbf{r})/(\Lambda_n + i\kappa b) \tag{6.67}$$

The average field intensity is as follows

$$\langle |E|^2 \rangle = \langle |\mathbf{E}_f + \mathbf{E}_0|^2 \rangle = E_0^2 + \left\langle \sum_{n,m} \frac{\mathcal{E}_n\mathcal{E}_m^*(\nabla\Psi_n(\mathbf{r}) \cdot \nabla\Psi_m^*(\mathbf{r}))}{(\Lambda_n + i\kappa b)(\Lambda_m - i\kappa b)} \right\rangle \tag{6.68}$$

where equality $\langle \mathbf{E}_f \rangle = \langle \mathbf{E}_f^* \rangle = 0$ is taken into account. Now, let us consider the eigenstates Ψ_n with eigenvalues Λ_n within a small interval $|\Lambda_n - \Lambda| \leq \Delta\Lambda \ll \kappa$ centered at some Λ. These states are denoted as $\Psi_n(\Lambda, \mathbf{r})$. Recall that the eigenstates are assumed to be localized so that eigenfunctions $\Psi_n(\Lambda, \mathbf{r})$ are well separated in space. The average distance between them, l, can be estimated as $l(\Delta\Lambda) \sim a[\rho(\Lambda)\Delta\Lambda]^{-1/d}$, where

$$\rho(\Lambda) = a^d \sum_n \delta(\Lambda - \Lambda_n)/V \tag{6.69}$$

is the dimensionless density of states for the KH $\mathcal{H}'$ and V is the volume of the system. We assume here that the metal concentration p is about one-half so that all quantities in the KH $\mathcal{H}'$ are about unity and, therefore, the density of states $\rho(\Lambda)$ is also about unity at the center of the spectrum, that is, at $\Lambda = 0$. Then, the distance $l(\Delta\Lambda)$ can be arbitrarly large for $\Delta\Lambda \to 0$; it is assumed, of course, that $l(\Delta\Lambda)$ is still much smaller than the system size, and the total number of eigenstates $\Psi_n(\Lambda, \mathbf{r})$

within interval $\Delta\Lambda$ is macroscopically large. When the interstate distance $l(\Delta\Lambda)$ is much larger than the localization length $\xi_A(\Lambda)$, the localized eigenfunctions $\Psi_n(\Lambda, \mathbf{r})$ can be characterized by the spatial positions of their "centers" $\mathbf{r}_n$ so that $\Psi_n(\Lambda, \mathbf{r}) = \Psi(\Lambda, \mathbf{r} - \mathbf{r}_n)$ and Eq. (6.68) acquires the following form:

$$\langle |E|^2 \rangle = E_0^2 + \sum_{\Lambda_1, \Lambda_2} \frac{\langle \sum_{n,m} \mathcal{E}_n \mathcal{E}_m^* [\nabla\Psi(\Lambda_1, \mathbf{r} - \mathbf{r}_n) \cdot \nabla\Psi^*(\Lambda_2, \mathbf{r} - \mathbf{r}_m)] \rangle}{(\Lambda_1 + i\kappa b)(\Lambda_2 - i\kappa b)} \tag{6.70}$$

where the first sum is over positions of the intervals $|\Lambda_n - \Lambda_1|$ and $|\Lambda_m - \Lambda_2|$ in the Λ space, whereas the sum in the numerator is over spatial positions $\mathbf{r}_n$ and $\mathbf{r}_m$ of the eigenfunctions. For each realization of a macroscopically homogeneous random film, the positions $\mathbf{r}_n$ of eigenfunctions $\Psi(\Lambda, \mathbf{r} - \mathbf{r}_n)$ take new values that do not correlate with the value of Λ. Therefore, the numerator in the second term of Eq. (6.70) can be independently averaged over positions $\mathbf{r}_n$ and $\mathbf{r}_m$ of the eigenstates Ψ_n and Ψ_m. By taking into account that $\langle \nabla\Psi_n(\mathbf{r}) \rangle = 0$, we obtain

$$\langle \mathcal{E}_n \mathcal{E}_m^* [\nabla\Psi(\Lambda_1, \mathbf{r} - \mathbf{r}_n) \cdot \nabla\Psi^*(\Lambda_2, \mathbf{r} - \mathbf{r}_m)] \rangle \simeq \langle |\mathcal{E}_n|^2 |\nabla\Psi(\Lambda_1, \mathbf{r} - \mathbf{r}_n)|^2 \rangle \delta_{\Lambda_1 \Lambda_2} \delta_{nm} \tag{6.71}$$

where we neglected possible correlations of eigenfunctions from different intervals Λ_1 and Λ_2, since the spatial density of the eigenfunctions excited effectively by the external field is estimated as $a^{-d}\rho(\Lambda)\kappa$, that is, it vanishes for $\kappa \to 0$. Substitution of Eq. (6.71) into Eq. (6.68) results in

$$\langle |E|^2 \rangle = E_0^2 + \sum_{\Lambda} \frac{\sum_n |\mathcal{E}_n|^2 \langle |\nabla\Psi_n(\Lambda, \mathbf{r})|^2 \rangle}{\Lambda^2 + (b\kappa)^2} \tag{6.72}$$

The localized eigenstates are not, in general, degenerate, so that the eigenfunctions Ψ_n can be chosen as real. Then, the term $|\mathcal{E}_n|^2 = |(\Psi_n|\mathcal{E})|^2 = |\sum_{i=1}^N \Psi_{n,i}\mathcal{E}_i|^2$ in Eq. (6.72) can be estimated by replacing the sum over all N sites of the system with integration over the system volume V, which gives $|\mathcal{E}_n|^2 \sim a^{-2d} |\int \Psi_n \mathcal{E} d\mathbf{r}|^2$. By using Eqs. (6.61) and (6.58), we find

$$|\mathcal{E}_n|^2 \sim a^{4-2d} \left| \int \Psi_n (\mathbf{E}_0 \cdot \nabla\tilde{\varepsilon}) d\mathbf{r} \right|^2 = a^{4-2d} \left| \int \tilde{\varepsilon} (\mathbf{E}_0 \cdot \nabla\Psi_n) d\mathbf{r} \right|^2 \tag{6.73}$$

where, to obtain the last relation, we integrated by parts and took into account that the eigenstates Ψ_n are all localized. Therefore, $\int (\nabla \cdot \mathbf{E}_0 \Psi_n \tilde{\varepsilon}) d\mathbf{r} = 0$. The local dielectric constant $|\tilde{\varepsilon}|$ is on the order of 1, $|\tilde{\varepsilon}| \sim 1$, the spatial derivative $\nabla\Psi_n$ is estimated as $\Psi_n/\xi_A(\Lambda)$, and Eq. (6.73) is estimated as

$$|\mathcal{E}_n|^2 \sim \frac{E_0^2 a^4}{a^{2d} \xi_A^2(\Lambda)} \left| \int \Psi_n(\mathbf{r}) d\mathbf{r} \right|^2 \sim \frac{E_0^2 a^4}{\xi_A^2(\Lambda)} \left| \sum_{i=1}^N \Psi_{n,i} \right|^2 \tag{6.74}$$

where summation over sites of the tight-binding model is restored. Since the eigenfunctions Ψ_n are normalized to unity, that is, $\langle\Psi_n|\Psi_n\rangle = \sum_{i=1}^{N}|\Psi_{n,i}|^2 = 1$ and localized within $\xi_A(\Lambda)$, they are estimated as $\Psi_{n,i} \sim [\xi_A(\Lambda)/a]^{-d/2}$ in the localization domain. Substitution of this estimate in Eq. (6.74) gives the final result for matrix element $\mathcal{E}_n$ of the external field, namely,

$$|\mathcal{E}_n|^2 \sim E_0^2 a^2 [\xi_A(\Lambda)/a]^{d-2} \tag{6.75}$$

The average spatial derivative in the numerator of Eq. (6.72) can be estimated in a similar way

$$\langle|\nabla\Psi_n(\Lambda,\mathbf{r})|^2\rangle \sim \xi_A^{-2}(\Lambda)\langle|\Psi_n(\Lambda,\mathbf{r})|^2\rangle \sim \xi_A^{-2}(\Lambda)N^{-1}\sum_{i=1}^{N}|\Psi_{n,i}|^2 \sim \xi_A^{-2}(\Lambda)/N \tag{6.76}$$

where $N = V/a^d$ is the total number of sites. Now the estimates (6.75) and (6.76) allow us to rewrite the numerator of Eq. (6.72) as

$$\sum_n |\mathcal{E}_n|^2\langle|\nabla\Psi_n(\Lambda,\mathbf{r})|^2\rangle \sim \frac{1}{N}\sum_n E_0^2[\xi_A(\Lambda)/a]^{d-4} \sim E_0^2[\xi_A(\Lambda)/a]^{d-4}\rho(\Lambda)\Delta\Lambda \tag{6.77}$$

where we took into account that the total number of the eigenstates within interval $\Delta\Lambda$ is equal to $N\rho(\Lambda)\,\Delta\Lambda$. By substituting Eq. (6.77) into Eq. (6.72) and replacing the summation by integration over Λ, the following estimate for the field intensity is obtained

$$\langle|E|^2\rangle \sim E_0^2 + E_0^2\int\frac{\rho(\Lambda)[a/\xi_A(\Lambda)]^{4-d}}{\Lambda^2 + (b\kappa)^2}d\Lambda \tag{6.78}$$

Since all matrix elements in KH $\mathcal{H}'$ are of the order of unity (in fact, the off-diagonal elements are ± 1), the density of states $\rho(\Lambda)$ and localization length $\xi_A(\Lambda)$ change significantly within an interval on an order of 1. In contrast, the denominator in Eq. (6.72) has an essential singularity at $\Lambda = \pm ib\kappa$. Then, the second moment of the local electric field $M_2^* \equiv M_{2,0}^* = \langle|E|^2\rangle/E_0^2$ is estimated as

$$M_2^* \sim 1 + \rho(a/\xi_A)^{4-d}\int\frac{1}{\Lambda^2 + (b\kappa)^2}d\Lambda \sim \rho(a/\xi_A)^{4-d}\kappa^{-1} \gg 1 \tag{6.79}$$

provided that $\kappa \ll \rho(a/\xi_A)^{4-d}$ [we set $\xi_A(\Lambda = 0) \equiv \xi_A$, $\rho(\Lambda = 0) \equiv \rho$ and approximated the constant b by 1].

Above, we defined the field correlation length

$$\xi_e^* \sim a(\rho\kappa b)^{-1/d} \sim a(\rho\kappa)^{-1/d} \tag{6.80}$$

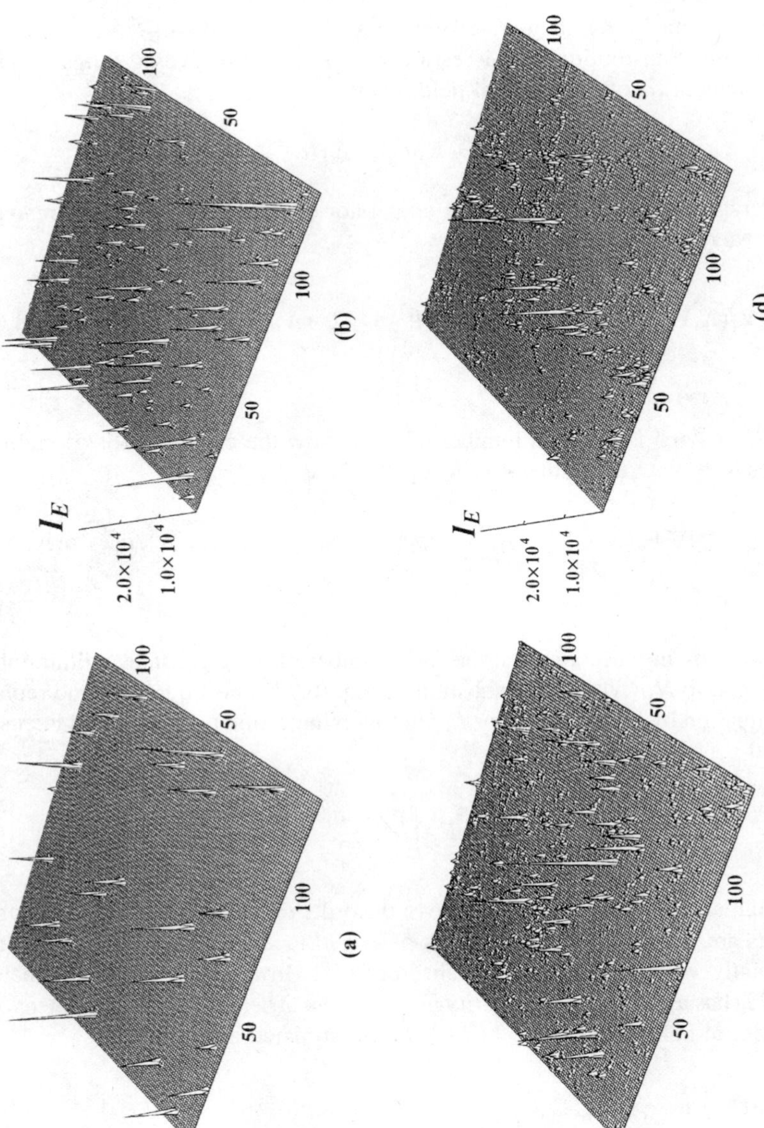

I_E
2.0×10^4
1.0×10^4
50
100
(a)
(b)
2.0• 10^4
1.0• 10^4
(c)
(d)

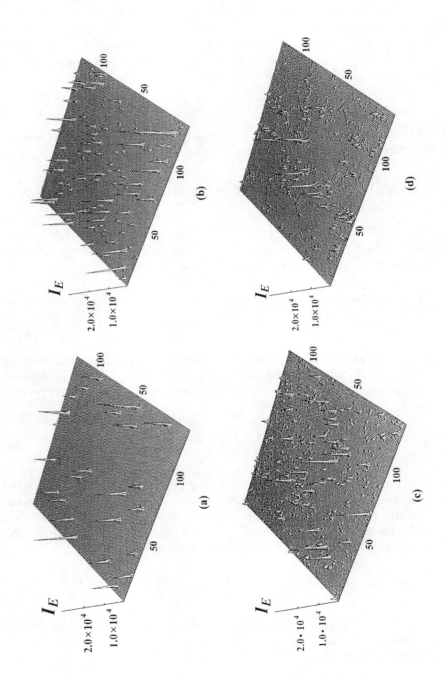

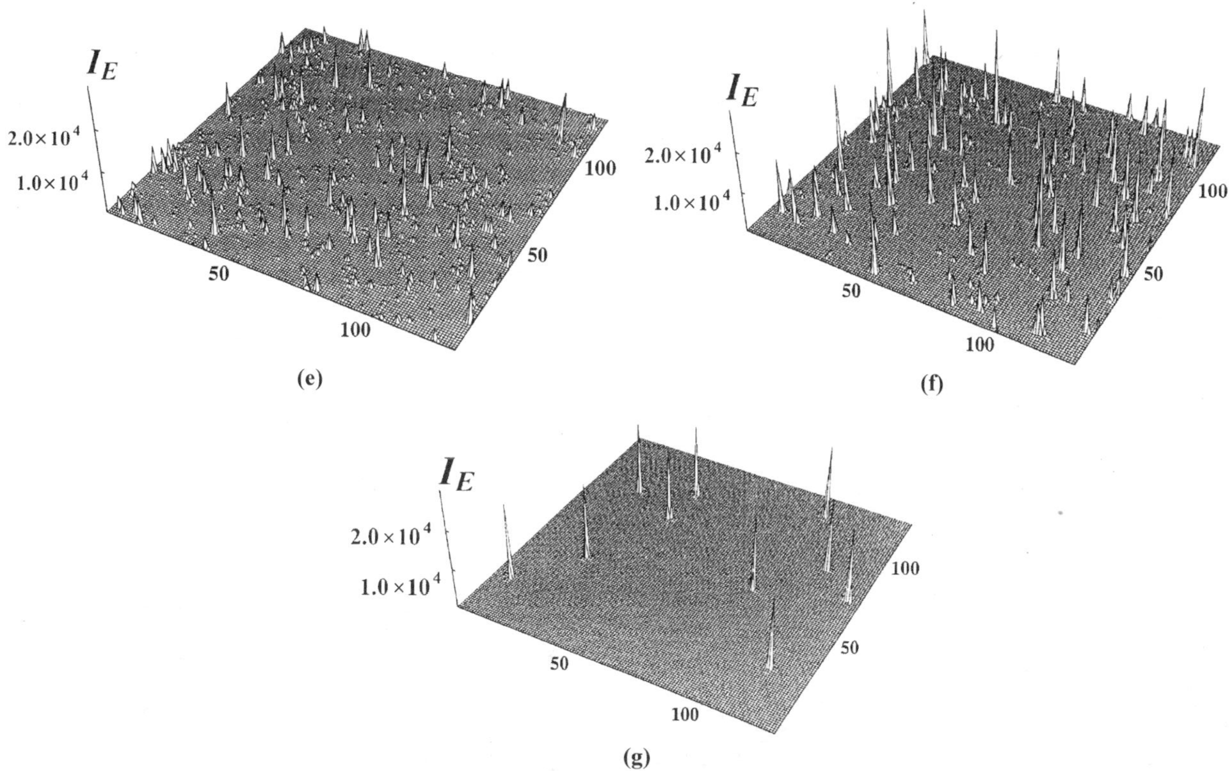

Figure 6.12. Distribution of the local field intensities $I_E = |\mathbf{E}_1(\mathbf{r})|^2/|\langle\mathbf{E}_1\rangle|^2$ in a metal–silver semicontinuous film for $\tilde{\varepsilon}'_m = -\tilde{\varepsilon}_d$ ($\lambda \approx 480$ nm, $d = 5$ nm, $D/d = 3$) at different metal concentrations, p. (a) $p = 0.001$, (b) $p = 0.01$, (c) $p = 0.1$, (d) $p = 0.5$, (e) $p = 0.9$, (f) $p = 0.99$, and (g) $p = 0.999$.

which is the typical spatial distance between the eigenstates KH Ψ_n excited effectively by the external field. Thus the field distribution can be thought of as a set of the peaks localized within ξ_A and separated in a distance by the field correlation length ξ_e^*. Then, it follows from Eq. (6.79) that the field peaks have the amplitudes

$$E_m^* \sim E_0 \kappa^{-1} (a/\xi_A)^2 \tag{6.81}$$

for the considered $d = 2$ case. The field maximum goes to infinity when losses vanish in the film. All of the above speculations leading to Eqs. (6.79) and (6.80) hold when the field correlation length ξ_e^* is much larger than the Anderson localization length, that is, $\xi_e^* \gg \xi_A$. This condition is fulfilled in the limit of small losses when $\kappa \to 0$.

Note that hereafter we mark the quantities, which are given for the special case $-\tilde{\varepsilon}_m = \tilde{\varepsilon}_d = 1$ by the superscript * (this sign, of course, should not be confused with the complex conjugation also denoted by an asterisk). Using the scale renormalization described in the Section 6.6, we will see how these quantities are transformed when $|\tilde{\varepsilon}_m/\tilde{\varepsilon}_d| \gg 1$, that is, in the long wavelength part of the spectrum. Note also that for ξ_A and ρ we omit the * sign in order to avoid complicated notations; it is implied that their values are always taken at $-\tilde{\varepsilon}_m = \tilde{\varepsilon}_d = 1$, even if the case of $|\tilde{\varepsilon}_m/\tilde{\varepsilon}_d| \gg 1$ is considered.

In the above estimates, the localization length ξ_A was to be proportional to the eigenstate "size". Generally, this assumption might not be exact for the Anderson system (e.g., see the discussion in [65]), but it is well confirmed by the numerical calculations shown in Figs. 6.4–6.7 and in Fig. 6.12. The local electric field in Fig. 6.12 had been obtained in the quasistatic limit.

Equations (6.79)–(6.81) are in good agreement with comprehensive numerical calculations performed in [38,43,61] for a $2d$ system with $\tilde{\varepsilon}_m/\tilde{\varepsilon}_d \approx -1$ and $p = p_c = \frac{1}{2}$. It was shown there that the average intensity of the local-field fluctuations, that is, the second moment M_2^* is estimated as $M_2^* \sim \kappa^{-\gamma}$, where the critical exponent $\gamma \approx 1.0$. The authors also found that the correlation length ξ_e^* of the field fluctuations diverge as $\xi_e^* \sim \kappa^{-\nu_e}$ at $\kappa \to 0$, where the critical exponent $\nu_e \approx 0.5$. For $d = 2$, these values of γ and v_e are very close to $\gamma = 1$ and $\nu_e = \frac{1}{2}$ found from the hypothesis of Anderson localization of the surface plasmons.

Above, the metal concentration p was about one-half, which corresponds to the percolation threshold for $2d$ self-dual semicontinuous metal films. The derivation of Eqs. (6.78) and (6.79) was based on the assumption that the density of states $\rho(\Lambda)$ is finite and about unity for $\Lambda = 0$. This assumption, however, is violated for a small metal concentration p, when the eigenvalue distribution shifts to the positive side of Λ, so that the eigenstates with $\Lambda \approx 0$ are shifted to the lower edge of the distribution. Then, the density of states ρ in Eq. (6.79) becomes a function of the metal concentration p. In the limit of $p \to 0$, the number of states effectively excited by the external field is proportional to the number of metal particles. Then, the function $\rho(p)$ can be estimated as $\rho(p) \sim p$, for $p \to 0$. The same consideration holds in the other limit, when a small portion of holes in otherwise continuous metal film resonate with the external field and the density of states can be estimated as $\rho(p) \sim 1 - p$, for

$p \to 1$. When the density of states decreases, localization becomes stronger and we estimate the localization length ξ_A as $\xi_A(\Lambda = 0,\ p \to 0) \sim \xi_A(\Lambda = 0,\ p \to 1) \sim a$. Then, it follows from Eq. (6.79) that strong field fluctuations ($M_2 > 1$) exist in a metal–dielectric composite with $\tilde{\varepsilon}_d = -\tilde{\varepsilon}'_m$ in a wide concentration range

$$\kappa < p < 1 - \kappa \qquad \kappa \ll 1 \tag{6.82}$$

This behavior of the local field fluctuations is easy to trace in Fig. 6.12, where the local electric field is shown for the silver-on-glass semicontinuous film with $\tilde{\varepsilon}_d \simeq -\tilde{\varepsilon}'_m$. When metal concentration p decreases (increases) from the percolation threshold p_c the field maxima become needle-like (Anderson localization length ξ_A decreases) and their amplitude increases according to Eq. (6.81).

Although in Eq. (6.82) we estimated local fields for the special case of $\tilde{\varepsilon}_d = -\tilde{\varepsilon}'_m$, all of the above speculations, which are based on the assumption that the eigenstates of KH are localized, hold in a more general case, when the real part of the metal dielectric constant $\tilde{\varepsilon}'_m$ is negative and its absolute value is of the order of $\tilde{\varepsilon}_d$. Section 6.6 will consider the important case of the large contrast when $|\tilde{\varepsilon}_m| \gg \tilde{\varepsilon}_d$.

Note that the above speculations, which lead to the prediction of giant field fluctuations described by Eqs. (6.79)–(6.81), do not require long-range spatial correlations (e.g., in fractal structures) in particle positions. The large field fluctuations have been seen in computer simulations, in particular, for the so-called random gas of a metal particle [78,79], that is, for metal particles randomly distributed in space. This, however, is not true when the contrast is large $|\tilde{\varepsilon}_m| \gg \tilde{\varepsilon}_d$; in Section 6.6 we show that in this case the internal structure of a composite becomes crucial.

6.6. SCALING THEORY OF THE FIELD FLUCTUATIONS IN SEMICONTINUOUS METAL FILMS

The scaling approach was developed to estimate the local-field fluctuations in percolation composites for the large contrast, $|\tilde{\varepsilon}_m|/\tilde{\varepsilon}_d \gg 1$, [39,41–43]. Here, we briefly recapitulate the main points of the scaling renormalization. First, consider a semicontinuous metal film where the metal concentration p is equal to the percolation threshold $p = p_c$. We divide a system into squares of size l and consider each square as a new renormalized element. All such squares can be classified into two types. A square is considered as a "conducting" element if the square contains a continuous path of metallic particles that spans in the x or y direction. A square without such an "infinite" cluster is considered as a nonconducting, "dielectric," element [57,58]. According to finite size scaling [1,3] the effective dielectric constant of the "conducting" square $\tilde{\varepsilon}_m(l)$ decreases with increasing its size l as $\tilde{\varepsilon}_m(l) \simeq (l/a)^{-t/\nu}\tilde{\varepsilon}_m$, whereas the effective dielectric constant of the dielectric square $\tilde{\varepsilon}_d(l)$ increases with l as $\tilde{\varepsilon}_d(l) \simeq (l/a)^{s/\nu}\tilde{\varepsilon}_d$ (t, s, and ν are the percolation critical exponents for the static conductivity, dielectric constant, and percolation correlation length, respectively; for the $2d$ case, $t \cong s \cong \nu \cong \frac{4}{3}$. Now, we set the square size l to be equal to

$$l = l_r = a(|\tilde{\varepsilon}_m|/\tilde{\varepsilon}_d)^{\nu/(t+s)} \simeq a\sqrt{|\tilde{\varepsilon}_m|/\tilde{\varepsilon}_d} \tag{6.83}$$

Then, in the renormalized system, where each square of size l_r is considered as a single element, the dielectric constant of these new elements takes either value $\tilde{\varepsilon}_m(l_r) = \tilde{\varepsilon}_d^{t/(t+s)}|\tilde{\varepsilon}_m|^{s/(t+s)}(\tilde{\varepsilon}_m/|\tilde{\varepsilon}_m|)$, for the element renormalized from the conducting square, or $\tilde{\varepsilon}_d(l_r) = \tilde{\varepsilon}_d^{t/(t+s)}|\tilde{\varepsilon}_m|^{s/(t+s)}$, for the element renormalized from the dielectric square. The ratio of the dielectric constants of these new elements is equal to $\tilde{\varepsilon}_m(l_r)/\tilde{\varepsilon}_d(l_r) = \tilde{\varepsilon}_m/|\tilde{\varepsilon}_m| \cong -1 + i\kappa$, where the loss factor $\kappa = \tilde{\varepsilon}''_m/|\tilde{\varepsilon}_m| \ll 1$ is the same as in the original system. According to the basic ideas of the renormalization group transformation [1,57,58], the concentration of conducting and dielectric elements does not change under the above transformation, provided that $p = p_c$. The field distribution in a two component system depends on the ratio of the dielectric permittivities of the components. Thus after the renormalization, the problem becomes equivalent to the considered above field distribution for the case $\tilde{\varepsilon}_d = -\tilde{\varepsilon}'_m = 1$. If we take into account that the electric field renormalizes as $E_0^* = E_0(l_r/a)$, we obtain from Eq. (6.81) that the field peaks in the renormalized system are

$$E_m \simeq E_m^*(l_r/a) \simeq E_0(a/\xi_A)^2(l_r/a)\kappa^{-1} \simeq E_0(a/\xi_A)^2\left(\frac{|\tilde{\varepsilon}_m|}{\tilde{\varepsilon}_d}\right)^{\nu/(t+s)}\left(\frac{|\tilde{\varepsilon}_m|}{\tilde{\varepsilon}''_m}\right) \tag{6.84}$$

where $\xi_A = \xi_A(p_c)$ is the Anderson localization length in the renormalized system. In the original system, each field maximum of the renormalized system locates in a dielectric gap in the "dielectric" square of the l_r size or in between two "conducting" squares of size l_r that are not necessarily connected to each other [58] (see Fig. 6.13). There is no a characteristic length in the original system that is smaller than l_r, except the microscopical length in the problem, which is a grain size a. Therefore it is plausible to suggest that the width of a local-field peak in the original system is $\sim a$. Then, the values of the field maxima E_m do not change when returning from the renormalized system to the original one. Therefore, Eq. (6.84) gives the values of the field maxima in the original system. Note that value E_m of the field maxima is different from the previously obtained estimate (6.81) due to the renormalization of the applied field E_0.

To get insight into a value of the local field, let us consider the case of small frequencies $\omega \ll \omega_p$ when the dielectric constant ε_m for a Drude metal [see Eq. (6.59)] takes the form

$$\varepsilon_m \cong -\left(\frac{\omega_p}{\omega}\right)^2\left(1 - i\frac{\omega_\tau}{\omega}\right) \tag{6.85}$$

where we suppose that $\omega \gg \omega_\tau$. For example, the silver plasma frequency ω_p estimates as $\omega_p \approx 10$ eV, the relaxation rate $\omega_\tau \approx 0.02$ eV [63], therefore the

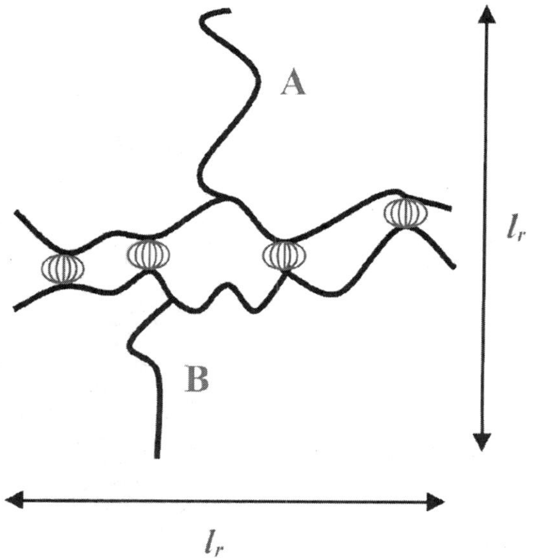

Figure 6.13. Typical configuration of the conducting clusters A and B that resonate at the frequency $\omega \ll \omega_p$. Local electric field concentrates in the chain of peaks.

condition $\omega_p \gg \omega \gg \omega_\tau$ holds for silver grains from the optical to far-IR spectral ranges. We also suppose, for simplicity, that skin depth δ is much larger than the grain size and the quasistatic approximation $\tilde{\varepsilon}_m = \varepsilon_m$ can be used [see Eqs. (6.31) and (6.57)]. Then, the expression (6.85) can be substituted into Eq. (6.84) for the dielectric constant $\tilde{\varepsilon}_m$ to obtain the following estimate:

$$E_m(\omega \ll \omega_p) \simeq E_0 \left(\frac{a}{\xi_A}\right)^2 \left(\frac{\omega_p}{\sqrt{\tilde{\varepsilon}_d}\,\omega}\right)^{2\nu/(t+s)} \left(\frac{\omega}{\omega_\tau}\right) \simeq E_0 \left(\frac{a}{\xi_A}\right)^2 \frac{\omega_p}{\omega_\tau} \tag{6.86}$$

where we omit the dielectric function $\tilde{\varepsilon}_d \sim 1$ and take into account that the critical exponents are equal to $\nu \cong t \cong s \cong \frac{4}{3}$. Thus we obtain an important result that the local-field peaks in metal semicontinuous films remain almost the same in the wide frequency range $\omega_p \gg \omega \gg \omega_\tau$. The computer simulations at the percolation threshold and different frequencies, shown on Fig. 6.14, are in good agreement with this theoretical prediction.

Since we know the peak amplitudes for the local electric field $\mathbf{E}$ we can now estimate the spatial moments $M_{n,m} = \langle |E|^n E^m \rangle / |E_0|^n E_0^m$ of the local electric field, where the amplitude $E^2 \equiv (\mathbf{E} \cdot \mathbf{E})$. The average distance ξ_e^* between the field maxima (i.e., eigenstates Ψ_n exited by the external field) in the renormalized system is given by Eq. (6.80), where the loss-factor $\kappa = \tilde{\varepsilon}''_m(l_r)/\left|\tilde{\varepsilon}'_m(l_r)\right| = \tilde{\varepsilon}''_m/|\tilde{\varepsilon}_m|$ is substituted. Then, the average distance ξ_e between the field maxima in the original

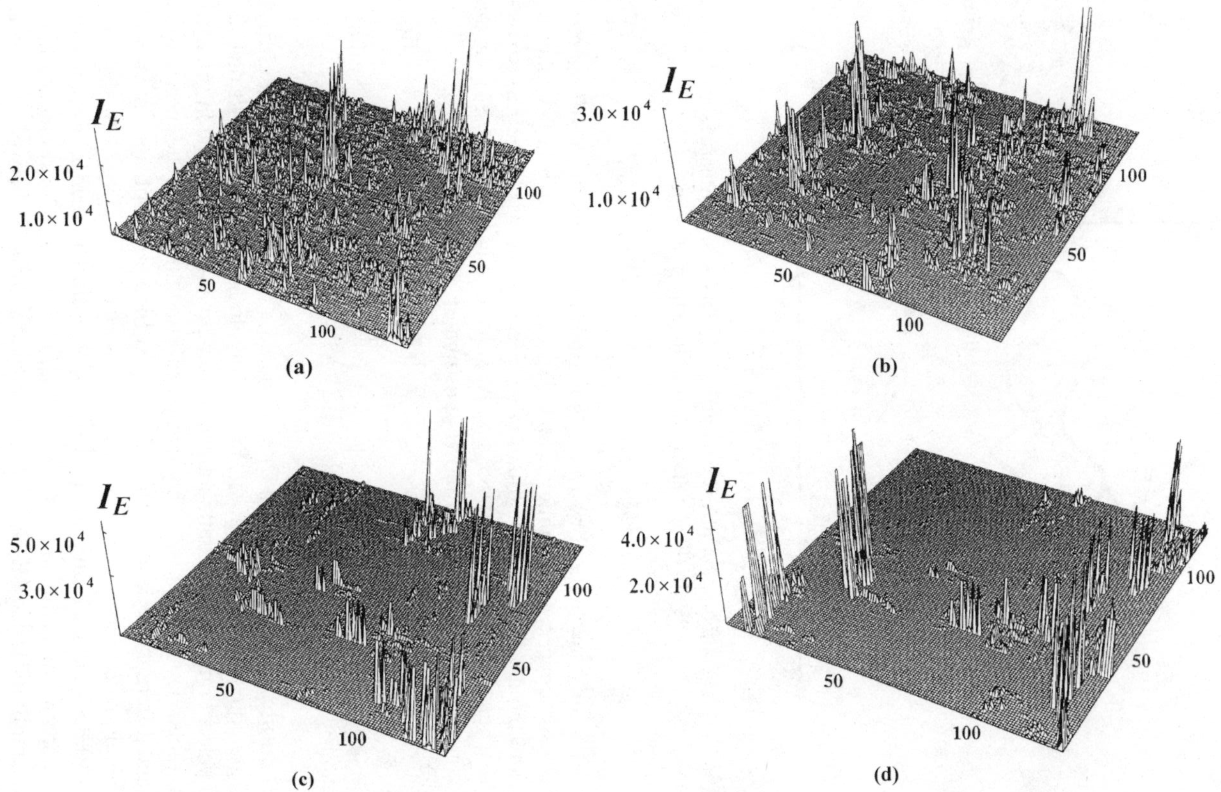

Figure 6.14. Distribution of the local field intensities $I_E = |\mathbf{E}_1(\mathbf{r})|^2/|\langle \mathbf{E}_1 \rangle|^2$ in a semicontinuous film at the percolation threshold for different wavelengths. (a) $\lambda = 0.6\,\mu\text{m}$, (b) $\lambda = 1.5\,\mu\text{m}$, (c) $\lambda = 10\,\mu\text{m}$, and (d) $\lambda = 20\,\mu\text{m}$; ($d = 5$ nm, $D/d = 3$).

system (provided that $\rho \sim 1$) is equal to

$$\xi_e \sim \left(\frac{l_r}{a}\right)\xi_e^* \sim \frac{l_r}{\kappa^{1/d}} \sim a\left(\frac{|\tilde{\varepsilon}_m|}{\tilde{\varepsilon}_d}\right)^{\nu/(t+s)}\left(\frac{|\tilde{\varepsilon}_m|}{\tilde{\varepsilon}''_m}\right)^{1/d} \sim a\frac{|\tilde{\varepsilon}_m|}{\sqrt{\tilde{\varepsilon}_d\tilde{\varepsilon}''_m}} \tag{6.87}$$

where the estimate $\nu \cong t \cong s \cong \frac{4}{3}$ for $d = 2$ is used again. In the renormalized system, a typical area of a field peak corresponds to the Anderson localization length squared ξ_A^2. Therefore, in the original system each field maximum is stretched over $(\xi_A/a)^2$ clusters of the size l_r. In each of these clusters, the field maximum splits into $n(l_r)$ peaks of the E_m amplitude located along a dielectric gap in the "dielectric" square of the l_r size as it is shown in Fig. 6.13. The gap "area" scales as the capacitance of the dielectric square, so does the number of peaks

$$n(l_r) \propto (l_r/a)^{d-2+s/\nu} \sim \sqrt{\frac{|\tilde{\varepsilon}_m|}{\tilde{\varepsilon}_d}} \tag{6.88}$$

By multiplying the amplitude of the field peaks E_m raised to the proper power by the number of peaks in one group $(\xi_A/a)^2 n(l_r)$ and by normalizing to the distance between the groups ξ_e, we obtain the following estimate for the local-field moments

$$M_{n,m} \sim a^2\frac{(E_m/E_0)^{n+m}(\xi_A/a)^2 n(l_r)}{\xi_e^2} \sim \rho\left[\frac{|\tilde{\varepsilon}_m|^{3/2}}{(\xi_A/a)^2\sqrt{\tilde{\varepsilon}_d\tilde{\varepsilon}''_m}}\right]^{n+m-1} \tag{6.89}$$

that hold for $n + m > 1$ and $n > 0$. Since $|\tilde{\varepsilon}_m| \gg \tilde{\varepsilon}_d$ and the ratio $|\tilde{\varepsilon}_m|/\tilde{\varepsilon}''_m \gg 1$, the moments of the local field are very large, $M_{n,m} \gg 1$, in the visible and IR spectral ranges. Note that the first moment $M_{0,1} \sim 1$ corresponds to the equation $\langle \mathbf{E}(\mathbf{r})\rangle = \mathbf{E}_0$. Again, we stress that the localization length ξ_A in Eq. (6.89) corresponds to the renormalized system with $\tilde{\varepsilon}_d = -\tilde{\varepsilon}'_m = 1$. The localization length in the original system, that is, a typical size of the eigenfunction, is estimated as $(l_r/a)\xi_A \gg a$. In other words, the eigenstates become macroscopically large in the limit of large contrast $|\tilde{\varepsilon}_m|/\tilde{\varepsilon}_d \gg 1$ and consist of sharp peaks scattered in space by a distance much larger than a. The predicted behavior of the eigenstates of KH $\mathcal{H}$ is in agreement with the computer simulation shown in Fig. 6.14.

Now, we consider the moments $M_{n,m}$ for $n = 0$, which correspond to the volume average of the mth power of the complex amplitude $E(\mathbf{r})$, namely, $M_{0,m} = \langle E^m(\mathbf{r})\rangle/E_0^m$. In the renormalized system, where $|\tilde{\varepsilon}_m(l_r)| = |\tilde{\varepsilon}_d(l_r)|$ and $\tilde{\varepsilon}_m(l_r)/\tilde{\varepsilon}_d(l_r) \cong -1 + i\kappa$, the field distribution coincides with the field distribution in the system with $\tilde{\varepsilon}_d \simeq -\tilde{\varepsilon}'_m \sim 1$. In the system with $\tilde{\varepsilon}_d \simeq -\tilde{\varepsilon}'_m \sim 1$, the field peaks E_m^* are different in phase and because of the destructive interference, the moment $M_{0,m}^* \sim O(1)$ in contrast to Eq. (6.81) as it is discussed in [45,46]. In transition to the original system, the peaks increase by the factor l_r/a, leading to the corresponding increase of the moment $M_{0,m}$. It is supposed that within a single "dielectric" square the field peaks are in phase, that is, the field maxima form chains of aligned peaks that are stretched out in a dielectric square. Then, the moment can be estimated as follows:

$$M_{0,m} \sim \frac{M_{0,m}^{*}(l_r/a)^m n(l_r)}{(\xi_e/a)^d} \sim \frac{\tilde{\varepsilon}_m''|\tilde{\varepsilon}_m|^{(m-3)/2}}{\tilde{\varepsilon}_d^{(m-1)/2}}. \tag{6.90}$$

which holds when $M_{0,m}$ given by this equation is >1.

Earlier in this section, for the sake of simplicity, we assumed that $p = p_c$. Now we estimate the concentration range $\Delta p = p - p_c$, where the above estimates for the local-field moments are valid [38,39]. We note that the above expressions for the local field and average-field moments $M_{n,m}$ hold in almost all concentration ranges given by Eq. (6.82) when $\tilde{\varepsilon}_m \simeq -\tilde{\varepsilon}_d$. The metal concentration range Δp, where the local electric field is strongly enhanced, shrinks, however, when $\tilde{\varepsilon}_m' \ll 0$. The above speculations are based on the finite-size scaling arguments, which hold provided the scale l_r of the renormalized squares is smaller than the percolation correlation length $\xi \cong a(|p-p_c|/p_c)^{-\nu}$ [1]. At the percolation threshold, where the correlation length ξ diverges, our estimates are valid for all frequencies ω when $\tilde{\varepsilon}_m'(\omega) \ll 0$ and losses are small $|\tilde{\varepsilon}_m'(\omega)|/\tilde{\varepsilon}_m''(\omega) \ll 0$. This frequency range includes all spectral ranges if skin effect is strong. For any particular frequency, we estimate the concentration range Δp, where the giant field fluctuations occur by equating the values of l_r and ξ, which results in the inequality

$$|\Delta p| \leq (\tilde{\varepsilon}_d/|\tilde{\varepsilon}_m|)^{1/(t+s)} \tag{6.91}$$

In the case of the strong skin effect, we substitute Eqs. (6.32), (6.33) and (6.57) into Eq. (6.91) and obtain

$$|\Delta p| \leq (\pi D\, \varepsilon_d'/\lambda)^{2/(t+s)} \tag{6.92}$$

where the reduced dielectric constant $\varepsilon_d' = 1 + 2\varepsilon_d d/D$, d is the thickness of the film and D is typical size (diameter) of a grain $(d < D)$. Note that in Section 6.3 the estimate (6.55) was obtained for the concentration band where a semicontinuous metal film has anomalous absorption. Equation (6.55) was obtained in the effective medium approximation for which the critical exponents $t = s = 1$ [3]. If the exponents $t = s = 1$ are substituted in Eq. (6.92) instead of $t \simeq s \simeq \frac{4}{3}$, this equation becomes identical to Eq. (6.55). Therefore the anomalous absorption and giant field fluctuation indeed take place in the same vicinity Δ_p to the percolation threshold. It is not surprising since anomalous absorption arises due to the giant field fluctuations as it was discussed after Eq. (6.50).

6.7. SPATIAL HIGH-ORDER MOMENTS OF LOCAL ELECTRIC AND MAGNETIC FIELDS

The results obtained in Section 6.6 allow us to find spatial moments of the local electric field $\mathbf{E}_1$ distributed in the reference plane $z = -d/2 - l_0$ (see Fig. 6.1) behind the film. The electric field $\mathbf{E}_1$ is expressed in terms of the fields $\mathbf{E}$ and $\mathbf{H}$ by means of Eq. (6.38). The fluctuations of the local magnetic current $\mathbf{j}_H(\mathbf{r}) = w(\mathbf{r})\mathbf{H}(\mathbf{r})$ can be neglected in the first Eq. (6.38) as it is discussed after Eq. (6.60). Therefore, the moment $M_{n,m}^E = \langle |\mathbf{E}_1(\mathbf{r})|^n \mathbf{E}_1^m(\mathbf{r})\rangle / (|\langle \mathbf{E}_1 \rangle|^n \langle \mathbf{E}_1 \rangle^m)$, where $\mathbf{E}_1^m \equiv (\mathbf{E}_1 \cdot \mathbf{E}_1)^{m/2}$, are approximately equal to the moments $M_{n,m}$ of the field $\mathbf{E}(\mathbf{r})$ estimated in Eq. (6.89).

In Fig. 6.15, the results of numerical and theoretical calculations are compared for the field moments $M_{n,m}^E$ obtained in the quasistatic limit for $2d$ silver

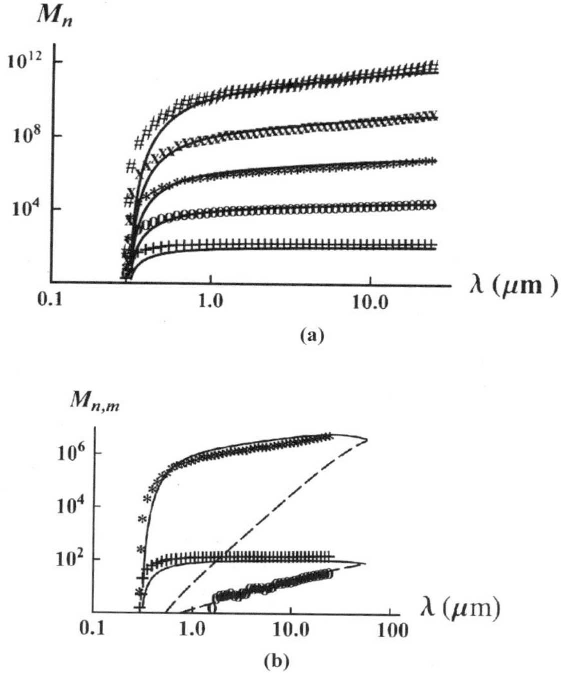

Figure 6.15. High-order field moments of a local electric field in semicontinuous silver films as a function of the wavelength λ at $p = p_c$. (a) Results of numerical calculations of the moments $M_n \equiv M_{n,0} = \langle |\mathbf{E}_1(\mathbf{r})|^n \rangle / |\langle \mathbf{E}_1 \rangle|^n$ for $n = 2$–6 are represented by $+, 0, *, \times$, and #, respectively. The solid lines describe M_n found from the scaling formula (6.89). (b) Comparison of the moment $M_{4,0} = \langle |\mathbf{E}_1(\mathbf{r})|^4 \rangle / |\langle \mathbf{E}_1 \rangle|^4$ [upper solid line — scaling formula (6.89), $*$ — numerical simulations] and moment $M_{0,4} = |\langle \mathbf{E}_1^4(\mathbf{r}) \rangle| / |\langle \mathbf{E}_1 \rangle|^4$ [upper dashed line — scaling formula (6.90)]. The moment $M_{2,0} = \langle |\mathbf{E}_1(\mathbf{r})|^2 \rangle / |\langle \mathbf{E}_1 \rangle|^2$ [lower solid line — scaling formula (89), $+$ — numerical simulations] vs. moment $M_{0,2} = |\langle \mathbf{E}_1(\mathbf{r})^2 \rangle| / |\langle \mathbf{E}_1 \rangle|^2$ [lower dashed line — scaling formula (6.90), 0 — numerical simulations]. In all presented analytical calculations, we set $\xi_A = 2a$ and $\rho = 1$ in Eqs. (6.89) and (6.90).

semicontinuous films on a glass substrate. We see that there is excellent agreement between the scaling theory [formulas (6.89) and (6.90)] and numerical simulations. To fit the data, the Anderson localization length $\xi_A \approx 2a$ was used. (Results of numerical simulations for $M_{0,4}$ are not shown in Fig. 6.15 since it was not possible to achieve reliable results in the simulations because of large fluctuations in values of this moment.) The small value of ξ_A indicates strong localization of surface plasmons in percolation composites, at least for the $2d$ case. As seen in Fig. 6.15 (b), the spectral dependence of enhancement $M_{n,m}$ differs strongly for processes with $(n \neq 0)$ and without $(n = 0)$ subtraction of photons.

Now, consider the moments $M^E_{n,m}$ for arbitrary strong skin effect assuming again that the metal dielectric constant ε_m is negative, large in absolute value, and can be approximated by Drude formula (6.85). The Drude formula, Eq. (6.85), is substituted into Eq. (6.29) to obtain the Ohmic parameter u_m in the limit $\omega \ll \omega_p$ and $\omega \gg \omega_\tau$; then, the Ohmic parameter u_m is substituted in Eq. (6.57) to obtain $\tilde{\varepsilon}_m$, and finally the moment $M^E_{n,m}$ is obtained from Eq. (6.89) as

$$M^E_{n,m} \sim \rho \left[\frac{\omega_p}{\omega_\tau} \left(\frac{a}{\xi_A} \right)^2 \sqrt{f_0} \right]^{n+m-1} \tag{6.93}$$

$$f_0 = \frac{2d}{D} \frac{4 \tanh^3 x \left(1 + \frac{D}{4\delta} \tanh x\right)}{x[\tanh x + x(1 - \tanh^2 x)]^2} \tag{6.94}$$

where $x = d/2\delta \simeq d\omega_p/2c$ is the ratio of the film thickness d to the skin depth $\delta \simeq c/\omega_p$. It follows from these equations that the moments of the local electric field are independent of the frequency in the wide frequency band $\omega_\tau < \omega \ll \omega_p$, which include (e.g., in the case of the silver semicontinuous films) optical and IR spectral ranges [cf. Fig. 6.15 (a)]. When the skin effect increases, the function f_0 in Eq. (6.93) increases monotonically from $f_0(0) = 2d/D$ to $f_0(\infty) = 4$. When the shape of the metal grains is fixed and they are very oblate, that is, $D/d \gg 1$, the moments $M^E_{n,m}$ increase significantly by increasing the film thickness d.

Now, let us consider the far-IR, microwave, and radio frequency ranges, where the metal conductivity σ_m acquires its static value, that is, it is positive and does not depend on frequency. Then, it follows from Eqs. (6.32), (6.33), and (6.57) that Eq. (6.89) for the field moments acquires the following form

$$M^E_{n,m} \sim \left(\frac{8\pi\sigma_m}{\omega} \right)^{(n+m-1)/2} \tag{6.95}$$

Since typical metal conductivity is much lager than frequency ω in the microwave and radio bands, the moments remain large at these frequencies.

We proceed now to fluctuations of the local magnetic field $\mathbf{H}_1(\mathbf{r})$ in the reference plane $z = -d/2 - l_0$. The fluctuations of the field $\mathbf{H}(\mathbf{r})$ can still be neglected. Then, it follows from the second Eq. (6.38) that moments $M^H_{n,m} = \langle\, |\mathbf{H}_1(\mathbf{r})|^n \mathbf{H}_1^m(\mathbf{r})\rangle / (|\langle \mathbf{H}_1\rangle|^n \ \langle \mathbf{H}_1\rangle^m)$ of the local magnetic field is estimated as

$$M^H_{n,0} \equiv M^H_n \simeq (2\pi/c)^n \langle\, |\mathbf{j}_E(\mathbf{r})|^n\rangle / |\langle \mathbf{E}_1\rangle|^n \tag{6.96}$$

where the conditions $|\langle \mathbf{E}_1\rangle| = |\langle \mathbf{H}_1\rangle|$, discussed in connection with Eq. (6.62), are used. Thus the external electric field induces electric currents in a semicontinuous metal film and these currents, in turn, generate the strongly fluctuating local magnetic field.

To estimate the moments $\langle\, |\mathbf{j}_E(\mathbf{r})|^n\rangle$ of the electric current density in semicontinuous metal films, the approach suggested by Dykhne [50] was generalized for the nonlinear case [80]. Since in the considered case the electric current $\mathbf{j}_E$ is connected to the local field $\mathbf{E}$ via the first equation (6.27), the following equation $\langle |\mathbf{j}_E|^n\rangle = \alpha(u_m, u_d)\langle |\mathbf{E}(\mathbf{r})|^n\rangle$ can be written where the coefficient $\alpha(u_m, u_d)$ is a function of variables u_m and u_d.

Now, let us consider the concentration corresponding to the percolation threshold $p = p_c$ and set the percolation value as $p_c = \frac{1}{2}$. It is also supposed that statistical properties of the system do not change when interreplacing metal and dielectric. If all conductivities are increased by a factor k, than the average nonlinear current $\langle |\mathbf{j}_E|^n\rangle$also increases by a factor $|k|^n$; therefore the coefficient $\alpha(u_m, u_d)$ increases by $|k|^n$ times as well. Therefore, the coefficient $\alpha(u_m, u_d)$ has an important scaling property, namely, $\alpha(ku_m, ku_d) = |k|^n \alpha(u_m, u_d)$. By taking $k = 1/u_m$ the following equation is obtained

$$\alpha(u_m, u_d) = |u_m|^n \alpha_1(u_m/u_d) \tag{6.97}$$

Now, we perform the Dykhne transformation

$$\mathbf{j}^* = [\mathbf{n} \times \mathbf{E}] \tag{6.98}$$
$$\mathbf{E}^* = [\mathbf{n} \times \mathbf{j}_E] \tag{6.99}$$

It is easy to verify that the introduced field $\mathbf{E}^*$ is still a potential (i.e., $\nabla \times \mathbf{E}^* = 0$), whereas the current $\mathbf{j}^*$ is conserved (i.e., $\nabla \cdot \mathbf{j}^* = 0$). The current $\mathbf{j}^*$ is coupled to the field $\mathbf{E}^*$ by Ohm's law $\mathbf{j}^* = u^*\mathbf{E}^*$, where the "conductivity" u^* takes values $1/u_m$ and $1/u_d$. Therefore, the following equation$\langle |\mathbf{j}^*|^n\rangle = \alpha(1/u_m, 1/u_d)\langle |\mathbf{E}^*|^n\rangle$ holds, from which it follows that $\alpha(1/u_m, 1/u_d)\alpha(u_m, u_d) = 1$. Since we suppose that at the percolation threshold statistical properties of the system do not change when interreplacing metal and dielectric, the arguments in the first function can be changed to obtain $\alpha(1/u_d, 1/u_m)\alpha(u_m, u_d) = 1$. This equation, in turn, can be rewritten using Eq. (6.97) as $|u_m/u_d|^n \alpha_1^2(u_m/u_d) = 1$. Thus we find that $\alpha_1(u_m/u_d) = |u_d/u_m|^{n/2}$, and the final result is given by $\alpha(u_m, u_d) = |u_m\, u_d|^{n/2}$, that is, the following generalization of the Dykhne's formula is valid

$$\langle |\mathbf{j}_E|^n\rangle = |u_d u_m|^{n/2} \langle |\mathbf{E}|^n\rangle \tag{6.100}$$

This expression for $\langle|\mathbf{j}_E|^n\rangle$ is substituted into Eq. (6.96), that takes the following form

$$M^H_{n,m} = \left[\left(\frac{2\pi}{c}\right)^2 |u_d u_m|\right]^{n/2} M^E_{n,m} \tag{6.101}$$

In optical and IR spectral ranges, it is possible to simplify this equation as it has been done for Eq. (6.96). By again using the Drude formula (6.85) for the metal dielectric constant and by assuming that $\omega_\tau \ll \omega \ll \omega_p$, the following estimate is obtained

$$M^H_{n,m} = \left[\varepsilon'_d \frac{x \tanh}{(2d/D) + x \tanh x}\right]^{n/2} M^E_{n,m} \tag{6.102}$$

where the moment $M^E_{n,m}$ is given by Eq. (6.96) and $x = d/2\delta \simeq d\omega_p/2c$ has the same meaning as in Eq. (6.96). It follows from Eq. (6.102) that spatial moments of the local magnetic field $M^H_{n,m}$ are the same as moments of the local electric field $M^E_{n,m}$ in the limit of the strong skin effect when $x \gg 1$.

We can estimate the moments of the local electric and magnetic fields from Eqs. (6.96) and (6.102) for silver-on-glass semicontinuous film with $\omega_p = 9.1\,\text{eV}$, and $\omega_\tau = 0.021\,\text{eV}$. Thus the moments of the local electric field are equal to $M^E_{n,m} \sim (4 \cdot 10^2)^{n-1}$, so that the field fluctuations are huge and they are in agreement with the numerical results shown in Figs. 6.4–6.7 and Fig. 6.15. For a sufficiently strong skin effect ($x \gg 1$), the moments of the local magnetic field $M^H_{n,m} \sim M^E_{n,m}$, which is also in agreement with computer results.

At frequencies much smaller than the relaxation rate $\omega_\tau = 3.2 \times 10^{13}\,\text{s}^{-1}$, the silver conductivity acquires its static value $\omega_p^2/4\pi\omega_\tau \simeq 10^{18}\,\text{s}^{-1}$. In this case, the moments are given by Eq. (6.95). Thus for wavelength $\lambda = 3\,\text{cm}$ ($\omega/2\pi = \nu = 10\,\text{GHz}$) the moments are $M^H_{n,m} \sim M^E_{n,m} \sim (10^4)^n$. We conclude that the local electric and magnetic field strongly fluctuate in a very large frequency range from the optical down to the microwave and radio spectral ranges. The fluctuations become even stronger for the microwave and radio bands. The reason is that for the strong skin effect (when the penetration depth is much smaller than the size of a metal grain), losses are small in comparison with the EM field energy accumulated around the film. This opens the fascinating possibility of observing the Anderson localization of the surface plasmons in microwave experiments [20] with localization length in the centimeter scale.

6.8. ENHANCED LIGHT TRANSMISSION THROUGH METAL FILMS WITH NANOHOLES

Light transmission through a small (less than the wavelength) hole in an optically thick metal film is known to be small, $T \sim (d/\lambda)^4$ [81]. However, provided that there is an array of such subwavelength holes in the metal film, the optical transmission can be enhanced by several orders of magnitude. For instance, in a silver film of thickness $d = 200\,\text{nm}$ with an array of holes of diameter $D = 150\,\text{nm}$ and lattice constant $a_0 = 600\,\text{nm}$, large transmission peaks have been observed at wavelengths $\lambda \sim 300, 700$, and $950\,\text{nm}$. The maximum transmission exceeds unity when it is normalized to the area of the hole. This corresponds to the enhancement of nearly three orders of magnitude when compared to what one could expect for the same number of single holes [82–84].

Since the skin depth in silver $\delta \sim 20\,\text{nm}$ is much less than the film thickness, these experiments cannot be explained using the quasistatic approximation for the surface conductivity. We use the GOL approximation to calculate the transmittance of an array of subwavelength holes for the case of a strong skin effect. It is assumed, for simplicity, that the surface concentration of the holes $q = 1 - p$ is much smaller than 1, $q \ll 1$, which is typical for the experiments mentioned above. Then, the perturbation theory, developed originally for the effective dielectric constant [48], allows us to find the dimensionless effective Ohmic parameters $u'_e = (2\pi/c)u_e$ and $w'_e = (2\pi/c)w_e$ for the array of holes, namely,

$$u'_e = u'_m\left(1 + 2q\frac{u'_d - u'_m}{u'_d + u'_m}\right) \tag{6.103}$$

and

$$w'_e = w'_m\left(1 + 2q\frac{w'_d - w'_m}{w'_d + w'_m}\right) \tag{6.104}$$

where parameters $u'_m = (2\pi/c)u_m$, $w'_m = (2\pi/c)w_m$, and $u'_d = (2\pi/c)u_d$, $w'_d = (2\pi/c)w_d$ are given by Eqs. (6.32) and (6.33), respectively. For small holes with $D \ll \lambda$, the metal Ohmic parameter, which can be rewritten as $u'_m = 4i/Dk = 2i/(\pi\lambda D)$, is much larger in modulus than 1, $|u'_m| \gg 1$. The dielectric Ohmic parameter $u'_d = -i\pi\varepsilon' D/2\lambda$ in contrast is very small $|u'_d| \ll 1$. Then, Eq. (6.103) simplifies to

$$u'_e = u'_m(1 - 2q) \gg 1 \tag{6.105}$$

Equation (6.104) for the effective parameter w'_e also simplifies when the expressions (6.32) and (6.33) are substituted for the Ohmic parameters w'_m and w'_d. We also take into account that $w'_m = -1/u'_m$ and assume, for simplicity, that the reduced dielectric constant $\varepsilon'_d \approx 1$. Thus the following equation is obtained

$$w'_e = -\frac{1}{u'_m}\left(1 + q\frac{1}{1 + 4a}\right) \tag{6.106}$$

which results in the estimate $|w'_e| \ll 1$. Equation (6.44) for the film transmittance T simplifies to $T \simeq (1 + u'_e w'_e)/u'_e$ when the effective parameters $|u'_e| \gg 1$ and $|w'_e| \ll 1$. By substituting Eqs. (6.105) and (6.106) here, the following expression for T is obtained

$$T = \frac{\pi^2}{4}\left(\frac{D}{\lambda}\right)^2 \frac{d + 2D}{d + D}\frac{q}{1 - 2q} \tag{6.107}$$

Thus obtained transmittance is independent of the metal properties (e.g., it does not depend on the metal dielectric constant ε_m). It is an anticipated result for the strong skin effect when the penetration of the EM field in the metal can be neglected. The transmittance is proportional to the ratio of the hole size D and the wavelength λ squared. Therefore, the transmittance T is much larger than $T \sim (D/\lambda)^4$, which resulted from the Fraunhofer diffraction on single holes in the considered limit of $D/\lambda \ll 1$. By using the parameter values for the film studied in [83] we obtain $T \simeq 0.5\%$ for the wavelength $\lambda = 1\ \mu\text{m}$, which although smaller, is still close enough to the experimentally detected values.

For the wavelength $< \lambda = 1\ \mu\text{m}$, a set of maxima in the transmittance was observed in experiments. For these wavelengths the condition $|u'_m| \gg 1$ is not fulfilled and Eq. (6.107) cannot be used. Extrapolation of Eq. (6.103) to the short wavelengths λ reveals a resonance at $u'_m = -u'_d$, that is, when $\lambda/D = (\pi/2)\sqrt{\varepsilon'}$. To evaluate the effective parameter u'_e at the resonance condition, the Maxwell–Garnet (dipole) approximation [29,85] is used. This gives the following result

$$u'_e = u'_m \frac{u'_d + q\,u'_d + u'_m - q\,u'_m}{u'_d - q\,u'_d + u'_m + q\,u'_m} \tag{6.108}$$

instead of Eq. (6.103), which is obtained by the perturbation theory. According to Eq. (6.108), the effective parameter u'_e diverges when $u'_m = -u'_d\,(1 + q)/(1 - q)$. When the Ohmic parameter $u'_m = -u'_d\,(1 - q)/(1 + q)$ is used, the effective Ohmic parameter u'_e vanishes. Taking into account that the effective Ohmic parameter $w'_e \simeq i$ when $|u'_m| \sim |u'_d|$ [see Eqs. (6.32) and (6.33)] we obtain from Eq. (6.44) that the transmittance increases up to $T \simeq 1/(1 - w_e) \simeq 25\%$ for the wavelength corresponding to the condition $u'_e \approx 0$. It is interesting to note that the resonance transmittance remains the same for arbitrary surface concentration q of the holes. The condition for the described resonance, which can be dubbed as the "skin resonance", has a purely geometrical nature and does not depend on the metal dielectric function ε_m in contrast to the surface–plasmon resonance. It would be interesting to check experimentally a dependence of the resonance on the diameter D of the holes, when the surface hole concentration q is fixed. In this section, we restricted ourselves to emphasizing the role of the skin resonance and did not consider the surface plasmon waves (polaritons) in a periodic metal nanostructure, which can be important for reproducing the whole spectrum of the resonance transmittance.

6.9. SUMMARY AND CONCLUSIONS

In this chapter, a detailed theoretical consideration of the high-frequency response (optical, IR, and microwave) of thin-metal–dielectric films is presented. The GOL approximation based on direct solution of Maxwell's equations, without having to invoke the quasistatic approximation, is developed. In this approximation, the EM properties of semicontinuous metal films are described in terms of *two* parameters, u_e and w_e, in contrast to the usual description with a single complex conductivity. This approach allows us to calculate field distributions and optical properties of semicontinuous metal films in a wide frequency range: from optical to microwave and radio frequencies. The local electric and magnetic fields experience giant fluctuations near the percolation threshold. The local fields are associated with localized surface plasmons excited by an external field. The localization of the surface plasmons maps the Anderson localization in quantum mechanics. Starting from this point the theory begins to invoke ideas of the real-space renormalization group and gives a quantitative picture of the local field distribution.

It is shown that metal–dielectric films can exhibit especially interesting properties when there is a strong skin effect in metal grains. In particular, the theory predicts that the local magnetic fields as well as electric fields strongly fluctuate within a very large spectral range spread out from optical to radio frequencies. The obtained equations for the field distributions and the high-order moments for the electric and magnetic fields allow one to describe various optical phenomena in percolation films, both linear and nonlinear. For example, surface enhancement for Raman scattering is proportional to the fourth moment [39,40] and, therefore, it is strongly enhanced in a wide spectral range. The same is valid for the Kerr nonlinearity, which is also proportional to the fourth moment [41,45]. The giant electric field fluctuations near the percolation threshold in metal–dielectric films already have been observed in the microwave [20] and optical [21] ranges.

The giant electric and magnetic field fluctuations explain, in particular, a number of previously obtained phenomena that remained so far unclear. For example, the absorbance in a percolation film calculated using the quasistatic approximation predicts for the maximum $A \simeq 0.2$ [34], which is twice as small as the measured value [13–16,20] and the one following from the GOL equations [18]. This can occur because of neglecting the magnetic field in the quasistatic calculations. The energy of the electric field can be converted into the energy of the magnetic field, so that the magnetic component of the EM field is also responsible for the absorption. As seen in Figs. 6.5–6.7, the magnetic and electric fields can be comparable in magnitudes, even at a relatively moderate skin effect. This indicates that the magnetic field can carry out roughly the same amount of energy as the electric field. In accordance with this, Figs. 6.8–6.10 show that the absorbance in GOL beyond the quasistatic approximation reaches the value $A \simeq 0.45$ (in agreement with experiments), which is almost twice as large as that found in the quasistatic approximation.

We also considered the optical properties of a metal film with an array of subwavelength holes and showed that transmittance of the film is much larger than Fraunhofer diffraction predicts. This result is an agreement with recent experiments

[82–84]. For such films, a new effect, skin resonance, is predicted; at this resonance the transmittance can increase up to 25% regardless of the surface concentration of the holes.

ACKNOWLEDGMENTS

The studies described in this chapter were performed in close collaboration with the following individuals: Drs. Boccara, Brouers, Clerc, Gadenne, Lagarkov, Rivoal, and Mr. Shubin. The assistance of Mr. Podolskiy in preparation of some figures is also highly appreciated. This work was supported in part by NSF (DMR-9810183), PRF, NATO, and RFFI (98-02-17628).

REFERENCES

1. D. Stauffer and A. Aharony, *An Introduction to Percolation Theory*, 2nd ed., Taylor and Francis, London, 1994.
2. M. Sahimi, *Applications of Percolation Theory*, Taylor and Francis, London, 1994.
3. D. J. Bergman and D. Stroud, *Solid State Phys.* ***46***, 14 (1992).
4. J. P. Clerc, G. Giraud, and J. M. Luck, *Adv. Phys.* ***39***, 191 (1990)
5. P. Sheng, *Philos. Mag. B* ***65***, 357 (1992); T. Li, P. Sheng, *Phys. Rev. B* ***53***, 13268 (1996).
6. P. Sheng, *Introduction to Wave Scattering, Localization, and Mesoscopic Phenomena*, Academic, San Diego, CA, 1995.
7. A. K. Sarychev and F. Brouers, *Phys. Rev. Lett.* ***73***, 2895 (1994)
8. L. Tortet, J. R. Gavarri, J. Musso, G. Nihoul, J. P. Clerc, A. N. Lagarkov, and A. K. Sarychev, *Phys. Rev. B* ***58***, 5390 (1998).
9. A. B. Pakhomov, S. K. Wong, X. Yan, and X. X. Zhang, *Phys. Rev. B* ***58***, 13375 (1998).
10. R. W. Cohen, G. D. Cody, M. D. Coutts, and B. Abeles, *Phys. Rev. B* ***8***, 3689 (1973).
11. G. A. Niklasson and C. G. Granquist, *J. Appl. Phys.* ***55***, 3382 (1984).
12. L. C. Botten and R. C. McPhedran, *Opt. Acta* ***32***, 595 (1985); M. Gajdardziska-Josifovska, R. C. McPhedran, D. R. McKenzie, and R. E. Collins, *Appl. Opt.* ***28***, 2744 (1989); C. A. Davis, D. R. McKenzie, and R. C. McPhedran, *Optic Commun.* ***85***, 70 (1991).
13. P. Gadenne, A. Beghadi, and J. Lafait, *Opt. Commun.* ***65***, 17 (1988).
14. P. Gadenne, Y. Yagil, and G. Deutscher, *J. Appl. Phys.* ***66***, 3019 (1989).
15. Y. Yagil, M. Yosefin, D. J. Bergman, G. Deutscher, and P. Gadenne, *Phys. Rev. B* ***43***, 11342 (1991).
16. Y. Yagil, P. Gadenne, C. Julien, and G. Deutscher, *Phys. Rev.* ***46***, 2503 (1992).
17. T. W. Noh, P. H. Song, Sung-Il Lee, D. C. Harris, J. R. Gaines, and J. C. Garland, *Phys. Rev.* ***46***, 4212 (1992).
18. A. K. Sarychev, D. J. Bergman, and Y. Yagil, *Physica A* ***207***, 372 (1994); A. K. Sarychev, D. J. Bergman, and Y. Yagil, *Phys. Rev. B* ***51***, 5366 (1995).
19. Levy-Nathansohn and D. J. Bergman, *Physica A* ***241***, 166 (1997); *Phys. Rev. B* ***55***, 5425 (1997).

20. A. N. Lagarkov, K. N. Rozanov, A. K. Sarychev, and A. N. Simonov, *Physica A* ***241***, 199 (1997).
21. S. Grésillon, L. Aigouy, A. C. Boccara, J. C. Rivoal, X. Quelin, C. Desmarest, P. Gadenne, V. A. Shubin, A. K. Sarychev, and V. M. Shalaev, *Phys. Rev. Lett.* ***82***, 4520 (1999).
22. S. Gresillon, J. C. Rivoal, P. Gadenne, X. Quelin, V. M. Shalaev, and A. K. Sarychev, *Phys. Stat. Sol. A* ***175***, 337 (1999).
23. Lord Rayleigh, *The Theory of Sound*, 2nd ed., MacMillan, London, 1896.
24. J. A. Ogilvy, *Theory of Wave Scattering from Random Rough Surface*, Hilger, London 1991; A. S. Ilyinsky, G. Ya. Slepyan, and A. Ya. Slepyan, *Propagation, Scattering and Dissipation of Electromagnetic Waves*, IEE Electromagnetic Waves Series 36, Peregrinus, London, 1993.
25. M. V. Agranovich and D. L. Mills, Eds., *Surface Polaritons*, North-Holland, Amsterdam, The Netherlands, 1982; A. D. Boardman, Ed., *Electromagnetic Surface Modes*, Wiley, New York, 1992.
26. A. R. McGurn and A. A. Maradudin, *Phys. Rev. B* ***31***, 4866 (1985).
27. J. A. Sanches-Gil, A. A. Maradudin, J. Q. Lu, V. D. Freilikher, M. Pustilnik, and I. Yurkevich, *Phys. Rev. B* ***50***, 15353 (1994).
28. V. D. Freilikher, E. Kanzieper, and A. A. Maradudin, *Phys. Rep.* ***288***, 127 (1997).
29. J. C. Maxwell Garnett, *Philos. Trans. R. Soc. London* ***203***, 385 (1904).
30. D. A. G. Bruggeman, *Ann. Phys. (Leipzig)* ***24***, 636 (1935).
31. P. Sheng, *Phys. Rev. Lett.* ***45***, 60 (1980).
32. A. K. Sarychev and A. P. Vinogradov, *Phys. Stat. Sol. (b)* ***117***, K113 (1983).
33. F. Brouers, J. P. Clerc, and G. Giraud, *Phys. Rev. B* ***44***, 5299 (1991).
34. F. Brouers, J. P. Clerc, and G. Giraud, *Phys. Rev. B* ***47***, 666 (1993).
35. H. Ma, R. Xiao, and P. Sheng, *J. Opt. Soc. Am. B* ***15***, 1022 (1998).
36. A. P. Vinogradov, A. M. Karimov, and A. K. Sarychev, *Zh. Eksp. Teor. Fiz.* ***94***, 301 (1988) [*Sov. Phys. JETP* ***67***, 2129 (1988)].
37. G. Depardieu, P. Frioni, and S. Berthier, *Physica A* ***207***, 110 (1994).
38. F. Brouers, A. K. Sarychev, S. Blacher, and O. Lothaire, *Physica A* ***241***, 146 (1997).
39. F. Brouers, S. Blacher, A. N. Lagarkov, A. K. Sarychev, P. Gadenne, and V. M. Shalaev, *Phys. Rev. B* ***55***, 13234, (1997).
40. P. Gadenne, F. Brouers, V. M. Shalaev, and A. K. Sarychev, *J. Opt. Soc. Am. B* ***15***, 68 (1998).
41. V. M. Shalaev and A. K. Sarychev, *Phys. Rev. B* ***57***, 13265 (1998).
42. V. M. Shalaev, *Nonlinear Optics of Random Media: Fractal Composites and Metal-Dielectric Films*, Springer, Berlin, 2000.
43. F. Brouers, S. Blacher, and A. K. Sarychev, *Phys. Rev. B* **58**, 15897 (1998).
44. A. K. Sarychev, V. A. Shubin, and V. M. Shalaev, *Phys. Rev. E* ***59***, 7239 (1999)
45. A. K. Sarychev, V. A. Shubin, and V. M. Shalaev, *Phys. Rev. B* ***60***, 16389 (1999).
46. A. K. Sarychev, V. A. Shubin, and V. M. Shalaev, *Physica A* ***266***, 115 (1999).
47. D. S. Wiersma, P. Bartolini, A. Lagendijk, and R. Righini, *Nature (London)* ***390***, 671 (1997).
48. L. D. Landau, E. M. Lifshits, and L. P. Pitaevskii, *Electromagnetics of Continuous Media*, 2nd ed., Pergamon, Oxford, 1984.

49. J. D. Jackson, *Classical Electrodynamics*, 3rd ed., Wiley, New York, 1998.
50. A. M. Dykhne, *Zh. Eksp. Teor. Fiz.* ***59***, 110 (1970) [*Sov. Phys. JETP* ***32***, 348(1971)].
51. M. Golosovsky, M. Tsindlekht, and D. Davidov, *Phys. Rev. B* ***46***, 11439 (1992); M. Golosovsky, M. Tsindlekht, D. Davidov, and A. K. Sarychev, *Physica C* ***209***, 337 (1993).
52. A. A. Kalachev, S. M. Matitsin, K. N. Rosanov, and A. K. Sarychev, *Method for Measuring the Complex Dielectric Constant of Sheet Materials*, USSR Patent No. 1483394, 1987 (*USSR Bull. Izobr. Otkr.* 20, 1989).
53. A. A. Kalachev, I. V. Kukolev, S. M. Matitsin, L. N. Novogrudskiy, K. N. Rosanov, A. K. Sarychev, and A. V. Selesnev, in J. A. Emerson and J. M. Torkelson, Eds., *Optical and Electrical Properties of Polymers*, MRS 214, Pittsburg, 1991.
54. N. W. Ashcroft and N. D. Mermin, *Solid State Physics*, Holt, Rinehart, and Winston, New York, 1976.
55. D. J. Frank and C. J. Lobb, *Phys. Rev. B* ***37***, 302 (1988).
56. A. K. Sarychev, D. J. Bergman, and Y. Strelniker *Phys. Rev. B* ***48***, 3145 (1993).
57. P. J. Reynolds, W. Klein, and H. E. Stanley, *J. Phys. C* ***10***, L167 (1977).
58. A. K. Sarychev, *Sov. Phys. JETP* ***45***, 524 (1977) [*Zh. Eksp. Teor. Fiz.* ***72***, 1001 (1977)].
59. J. Bernasconi, *Phys. Rev. B* ***18***, 2185 (1978).
60. A. Aharony, *Physica A* ***205***, 330 (1994).
61. S. Blacher, F. Brouers, and A. K. Sarychev, in *Fractals in the Natural and Applied Sciences*, Chapman and Hall, London, 1995, Chapter. 24.
62. D. J. Bergman, E. Duering, and M. Murat, *J. Stat. Phys.* ***58***, 1 (1990).
63. E. D. Palik, Ed., *Handbook of Optical Constants of Solids*, Academic, New York, 1985; P. B. Johnson and R. W. Christy, *Phys. Rev. B* ***6***, 4370 (1972).
64. See, for example D. J. Semin, A. Lo, S. E. Roak, R. T. Skodje, and K. L. Rowlen, *J. Chem. Phys.* ***105***, 5542 (1996).
65. B. Kramer and A. MacKinnon, *Rep. Prog. Phys.* ***56***, 1469 (1993).
66. D. Belitz and T. R. Kirkpatrick, *Rev. Mod. Phys.* ***66***, 261 (1994).
67. M. V. Sadovskii, *Phys. Rep.* ***282***, 225 (1997).
68. K. B. Efetov, *Supersymmetry in Disorder and Chaos*, Cambridge University Press, UK, 1997.
69. J. A. Verges, *Phys. Rev. B* ***57***, 870 (1998).
70. A. Elimes, R. A. Romer, and M. Schreiber, *Eur. Phys. J. B* ***1***, 29 (1998).
71. T. Kawarabayashi, B. Kramer, and T. Ohtsuki, *Phys. Rev. B* ***57***, 11842 (1998).
72. V. I. Fal'ko and K. B. Efetov, *Phys. Rev. B* ***52***, 17413 (1995).
73. K. Muller et al., *Phys. Rev. Lett.* ***78***, 215 (1997).
74. M. V. Berry, *J. Phys. A* ***10***, 2083 (1977).
75. A. V. Andreev, O. Agam, B. D. Simons, and B. L. Altshuler, *Phys. Rev. Lett.* ***76***, 3947 (1996).
76. M. I. Stockman, L. N. Pandey, L. S. Muratov, and T. F. George, *Phys. Rev. Lett.* ***72***, 2486 (1994); M. I. Stockman, L. N. Pandey, L. S. Muratov, and T. F. George, *Phys. Rev. B* ***51***, 185 (1995); M. I. Stockman, L. N. Pandey, and T. F. George, *Phys. Rev. B* ***53***, 2183 (1996).
77. M. Kaveh and N. F. Mott, *J. Phys. A* ***14***, 259 (1981).

78. V. M. Shalaev, E. Y. Poliakov, and V. A. Markel, *Phys. Rev. B* ***53***, 2437 (1996); V. A. Markel, V. M. Shalaev, E. B. Stechel, W. Kim, and R. L. Armstrong, *Phys. Rev. B* ***53***, 2425 (1996).

79. M. I. Stockman, *Phys. Rev. E* ***56***, 6494 (1997), *Phys. Rev. Lett.* ***79***, 4562 (1997).

80. E. M. Baskin, M. V. Entin, A. K. Sarychev, and A. A. Snarskii, *Physica A* ***242***, 49 (1997).

81. H. A. Bethe, *Phys. Rev.* ***66***, 163 (1944).

82. T. W. Ebbesen, H. J. Lezec, H. F. Ghaemi, T. Thio, and P. A. Wolff, *Nature (London)* ***391***, 667 (1998).

83. H. F. Ghaemi, T. Thio, D. E. Grupp, T. W. Ebbesen, and H. J. Lezec, *Phys. Rev. B* ***58***, 6779 (1998).

84. T. J. Kim, T. Thio, T. W. Ebbesen, D. E. Grupp, and H. J. Lezec, *Opt. Lett.* ***24***, 256 (1999).

85. N. A. Nicorovici, R. C. McPhedran, and L. C. Botten, *Phys. Rev. Lett.* ***75***, 1507 (1995); *Phys. Rev. E* ***52***, 1135 (1995).

CHAPTER SEVEN

Optical Nonlinearities in Metal Colloidal Solutions

VLADIMIR P. SAFONOV and YULIA E. DANILOVA

Institute of Automation and Electrometry, Siberian Branch of the Russian Academy of Sciences, 630090 Novosibirsk, Russia

V. P. DRACHEV and S. V. PERMINOV

Institute of Semiconductor Physics, Siberian Branch of the Russian Academy of Sciences, 630090 Novosibirsk, Russia

7.1. INTRODUCTION

Enhanced optical responses in metal nanostructured materials have been intensively studied during the past two decades. Surface-enhanced Raman scattering (SERS), surface-enhanced coherent anti-Stokes Raman scattering, enhanced degenerate four-wave mixing (DFWM), enhanced harmonic generation, and other fascinating optical effects have been predicted and observed in rough metal films and colloidal solutions [1–6]. It was found that the giant enhancement is associated with a collective interaction of metal nanoparticles in a metal–dielectric composite, and therefore the enhancement factors depend on the geometrical structure of the material [7–9].

Convenient media for experimental studies of the enhanced optical responses and collective effects in nanocomposites are colloidal silver and gold solutions, where one can easily prepare colloids consisting of separated nanoparticles (monomers). These particles can aggregate and form clusters consisting of monomers from a few up to thousands. The distances between the centers of the nearest monomers in the cluster can be as small as a diameter of the nanoparticles. This means that particles with high polarizability (such as silver and gold particles) strongly interact via dipolar forces, or more generally, multipolar coupling. The interaction changes the linear [2,10] and nonlinear [3] optical properties of the medium substantialy. For example, the absorption spectrum of a single gold or silver particle in the visible range consists of one peak associated with the excitation of the surface plasmon resonance. In the simplest aggregate, a pair, plasmon oscillations in different

Optics of Nanostructured Materials, Edited by Vadim A. Markel and Thomas F. George
ISBN 0-471-34968-2

particles interact via restoring forces and the plasmon resonance splits onto two peaks. The long-wavelength peak corresponds to an electric vector of a light field parallel to the axis of a pair, and the short-wavelength corresponds to the perpendicular electric vector; hence, the pair is an anisotropic unit. A sample composed of an ensemble of the pairs with the different distances has a wide absorption spectrum. The electric field of a resonance light wave, acting inside of the particles in the given pair is stronger than the local field for a single particle. Corresponding enhancement factors E_i/E_0 for a local field E_i, in comparison with an incident field E_0, are $G = \varepsilon_1^2/3\varepsilon_0\varepsilon_2$ for a plasmon resonance in a pair [4] and $f_1 = 3\varepsilon_0/i\varepsilon_2$ for a surface plasmon resonance in a single particle [11]. Here $\varepsilon = \varepsilon_1 + i\varepsilon_2$ and ε_0 are the dielectric constants of a metal particle and a host medium. In metals, $|\varepsilon_1|$ grows in a long-wavelength range and the value of G increases considerably with the wavelength. A value of $G = 18$ was estimated for a pair of silver particles at $\lambda = 532\,\text{nm}$, and in the near-infrared (IR) it may be as high as 10^2. For a single silver particle, the maximum value of f_1 is at the surface plasmon resonance, $\lambda = 400\,\text{nm}$, and can be roughly estimated as $|f_1| = 2$. (The ε of a nanoparticle depends on its size [10].)

Large colloidal aggregates are typically characterized by their fractal geometry [12]. The number of monomers N in a fractal cluster with radius R_c is $N = (R_c/R_0)^D$, where R_0 is a parameter having the meaning of the characteristic distance between the nearest monomers, and D is the fractal dimension. The value of D depends on a kind of aggregation process. The process of cluster–cluster aggregation, one of the most typical of colloid solutions, results in $D = 1.78$ [13].

As a first approximation, a fractal aggregate may be considered as an ensemble of pairs [2]. The interaction of a monomer with its nearest neighbor is the most essential coupling in this approximation. The pair approximation was found to be reasonable to describe optical properties of metal fractal cluster at the long-wavelength wing of absorption [7]. Therefore, we will use simple estimates of the binary theory for a comparison with our experimental data. The binary theory also leads to an important conclusion: An optical response for a given frequency is due only to a certain fraction of pairs, constituting a fractal, or, in other words, optical excitation is localized on a part of the monomers.

A more advanced approach takes into account the light-induced dipole interactions between all monomers constituting a fractal [7]. It was found that these dipole couplings are a result of the localized plasmon eigenmodes. Computer simulations [8,14,15] show that the areas of large local field fluctuations may be localized in small parts of a fractal aggregate. Positions and sizes of these hot spots change with the wavelength. The simulations are in qualitative agreement with the experimental results obtained with the help of near-field optics [16] and the photomodification technique [17]. The localization of plasmon excitation is inhomogeneous in the sense that, at any wavelength, modes of different coherent radii are possible and the electric field of the eigenmode is distributed spikewise in a fractal [9]. The local enhancement of the electric field in a hot spot of a silver fractal cluster may be much more than for a pair. The spatial distribution of the local field intensity for a given frequency of external light wave depends strongly on the polarization of the exciting radiation. This means that the resonant configurations of the monomers are highly anisotropic [9]. The anisotropy is characteristic of the binary model, as mentioned above.

In an intense electromagnetic (EM) field, a dipole moment induced on a particle may be written as a power series:

$$\mathbf{d} = \alpha^{(1)}\mathbf{E}_0(\mathbf{r}) + \alpha^{(2)}[\mathbf{E}_0(\mathbf{r})]^2 + \alpha^{(3)}[\mathbf{E}_0(\mathbf{r})]^3 + \cdots \tag{7.1}$$

where $\alpha^{(1)}$ is the linear polarizability of a particle, $\alpha^{(2)}$ and $\alpha^{(3)}$ are the nonlinear polarizabilities, and $\mathbf{E}_0(\mathbf{r})$ is the local field at site $\mathbf{r}$. The dipole moment per unit volume, that is, a polarization, which is the source of a field in a medium, may be represented in analogous form with the coefficients, called susceptibilities. When the local field exceeds the applied field, $\mathbf{E}_0$, considerably huge enhancements of nonlinear optical responses occur. The enhancement factor $G^{(n)}$ for a nonlinear optical process of the nth order in an aggregated colloid solution is defined as the intensity ratio of the radiation generated by nonlinear process on a monomer incorporated into a metal aggregated on an isolated monomer. The value of $G^{(n)}$ averaged over various fractal cluster realizations can be estimated in the binary approximation [3] as $G^{(n)} \approx G^{2(n-1)}$. When the generated frequency occurs in the absorption band of the aggregates, its amplitude can be enhanced as well, compared to the "initial" field. Then, the enhancement factor reaches its maximum $G^{(n)} \approx G^{2n}$. By taking into account the estimate for G given above, one can conclude that for the DFWM ($n = 3$) in silver aggregates, it is possible to obtain an average enhancement $G^{(3)} \approx 10^6$. It is noteworthy that the local enhancement of a nonlinear scattering of light in a hot spot may be much stronger. The advantage of the near-field optics observation of the enhancements was demonstrated for the SERS [18] and for the photomodification process [19].

Nonlinear-optical experiments with aggregated metal nanocomposites were done up to now with macroscopic samples. The enhancement due to aggregation was found for DFWM [4], for second harmonic generation [20], and for nonlinear absorption and refraction [21,22]. A 30-fold increase in the nonlinear susceptibility of a (polymer)–(molecular J-aggregate) film under doping of gold colloidal aggregates was observed [23]. An increase of the quadratic electrooptic coefficient of polymethylmethacrylate due to doping by silver aggregates was also reported [24]. The observed enhancement factors are in qualitative agreement with numerical calculations of average enhancement for a corresponding nonlinear process [25]. Besides, the experiments have also shown that the photomodification of colloidal aggregates [26], thermal effects, and nonlocal effects have been very essential in the nonlinear interactions of light with a colloid solution. The role of nonlocal effects in nanocomposites [27] is as follows: The ratio of the size of a particle (10 nm) to the wavelength (500 nm) is much greater than for a ordinary molecular medium. The size of localization of the plasmon eigenmode in a fractal aggregate may be comparable with the aggregate size [9]. This circumstance is favorable for observing effects caused by the spatial dispersion in nonlinear optical processes in metal nanocomposites. For nonlocal effects, an important role is played not only by the enhancement of the local field but also by the increase in its gradient in hot spots.

Our chapter is concerned with the experimental study of DFWM, nonlinear absorption and refraction, polarization effects, and selective photomodification of colloidal aggregates. Particular attention is paid to new, polarization, nonlinear effects (local and nonlocal) in colloid aggregates that were recently observed [27,28]. New optical schemes were developed to obtain the results that we present.

7.2. PREPARATION OF COLLOIDAL SOLUTIONS AND THEIR PROPERTIES

It is known that fractal clusters may be formed by metallic particles aggregated in a colloid solution [12,29]. The most developed are the experimental techniques for noble metal colloids, which is why we chose to work with silver hydrosols.

We use several types of colloids:

Ag($NaBH_4$). The method proposed by Creighton et al. [30] is as follows. 1–3 mg of $NaBH_4$ is dissolved in 20 mL of cooled bidistilled H_2O, and 5 mL of such water is used to dissolve 2 mg of $AgNO_3$. Then, the silver nitrate solution is quickly added to the test tube with the $NaBH_4$ solution, and the mixture is intensively shaken. The resultant colloid is yellow colored. The spectrum of a fresh hydrosol has a peak near $\lambda = 400$ nm with (full width at half-maximum) (fwhm) ranging from 60 to 80 nm. The long-wavelength range of the spectrum does not manifest itself. The electron microscope pattern shows that most monomers are 10–20 nm in size. After several days, the hydrosol aggregates and changes from yellow to green or gray in color. During aggregation, extinction in the 450–1000 nm range grows, which agrees with [1]. Electron microscopy demonstrates cluster formation, with each cluster containing 30–1000-monomers. From the micrographs available, the fractal dimensionality of large aggregates is determined to be $D \approx 1.8$. Nonaggregated and aggregated gold colloids for our experiments are prepared by an analogous technique.

Ag(c). In the second method of sample preparation we used collargol, which is a mixture of silver with proteins that stabilize hydrosol. Use 1 mL of collargol dissolved in 10 mL of bidistilled H_2O to yielded a brown solution, whose extinction spectrum showed a peak at $\lambda = 405$ nm, which broadened toward the long-wavelength wing. The broadening is believed to be caused by the silver monomers combining with protein molecules to form complexes. As seen from the electron micrographs, in these complexes the monomers are spaced at distances comparable or somewhat larger than their diameters of $\sim$ 10 nm. Stabilized isolated monomers, showing no changes in the absorption spectra within a 1-month period, were prepared by fragmentation of protein molecules resulting from heating the collargol solution with a small addition of $NaNO_3$. Aggregation of the hydrosol obtained was initiated by adding a 1:10 10% NaOH solution. In 1–2 weeks, the hydrosol became dark gray, and its spectrum displayed a high long-wavelength wing.

Ag(PVP). According to [31], the silver is reduced by (a) monomeric residue of polyvinylpyrrolidone (PVP) in ethanol. The average diameter of microparticles is $d_{av} = 12$–15 nm in various realizations. Coagulation aggregates produced simultaneously with monomers have sizes up to 0.5 μm. The distances between neighbor particles vary from 2 to 4 nm. The absorption spectrum has two well pronounced peaks: one at 415 nm and a second at 560–580 nm. The silver concentration is 7×10^{-4} M.

Ag(Carey Lea). We found the recipe of this hydrosol in [32], where $AgNO_3$ was reduced by a mixture of $FeSO_4$ and Na_3(cit) solutions. We used 120 times less concentration of all solutions than in [32], which corresponds to 1.7×10^{-3} M of silver. The coagulating aggregates range up to several microns in size. The monomer's diameter is $d_{av} = 12 \pm 3$ nm, the gap between boundaries being ~ 1 nm. The absorption spectrum has an implicit two-peaked form with the second peak at ~ 700 nm.

Ag(EDTA). This hydrosol results from adding $AgNO_3$ solution to a boiling mixture of ethylenediaminetetraacetic acid (EDTA) and NaOH solutions [33]. The silver concentration is 10^{-3} M. The absorption spectrum of this colloid is characterized by a high central peak at 410 nm (due to a large fraction of isolated monomers and mini-aggregates of pairs), a smearing second peak at ≈ 580 nm and a moderate long-wavelength band. Electron microscopy shows both the coagulation and coalescence types of boundaries in the aggregates, whose sizes do not exceed 100 particles. The average diameter of the monomers is $d_{av} = 24 \pm 2$ nm.

We also have studied polymer (gelatin or PVP) films containing silver colloid aggregates. The gelatin films were done by the following technique. After the colloid

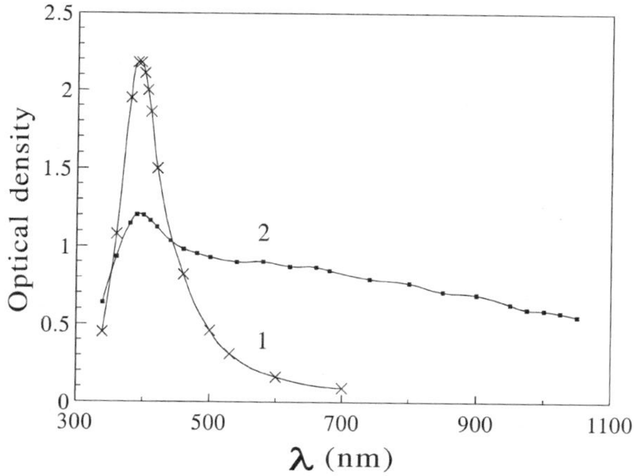

Figure 7.1. Typical absorption spectra of the weakly (curve 1) and strongly aggregated (curve 2) silver colloid solution, [Ag($NaBH_4$)].

had aggregated, we added 1% (by weight) of gelatin to the solution after which it was poured onto a glass substrate. Finally, the sample was dried for 1 week at room temperature. It formed a gelatin film with a thickness $l = 1$–25-μm doped with Ag fractal clusters.

Thus, the aggregation of all our colloids results in a broadening of absorption spectrum. Typical spectra are shown in Fig. 7.1.

The results of a calculation of the absorption spectrum due to the interaction between particles constituting the aggregates satisfactory describe experimental data for strongly aggregated colloids with sticked monomers, such as Ag($NaBH_4$) [8,34]. The calculations were done by the coupled-dipoles method [7].

7.3. SELECTIVE PHOTOMODIFICATION OF SILVER COLLOID AGGREGATES

We noted above that large fluctuations of the local field were found in computer simulations. The spatial distributions of the high-field regions are very sensitive to the frequency and polarization of the applied field, which is a reason for the effect of the selective photomodification of fractal aggregates [17,26]. Namely, the irradiation of the aggregates by a sufficiently powerful nano- or picosecond laser pulse results in local restructuring of resonance domains (hot spots, ranges of the high local field). This modification can be observed in a pulsed laser regime when the pulse energy per unit area, W, is higher that a certain threshold, W_{th}. Electron micrographs of colloidal silver aggregates before and after irradiation by a sequence of laser pulses at two different wavelengths are shown in Fig. 7.2.

Comparison of the micrographs of the cluster before and after irradiation at the laser wavelengths $\lambda_L = 1079$ and 540 nm shows that the structure of the cluster as a whole remains the same after irradiation, but monomers within small monometer-sized domains change their crystal structure, size, shape, and local arrangement. In [17], corresponding electron micrographs for the Ag(EDTA) colloid were presented. The minimum number of monomers in the region of modification was found to be 2–3 at $\lambda_L = 1079$ nm. Thus, the resonance domain at $\lambda_L = 1079$ nm can be as small as $\lambda_L/25$. Here, we present results for the Ag(Carey Lea) colloid. Although there are fluctuations in both shape and size of the modified domains, Fig. 7.2(b) reveals that "hot" zones associated with resonant excitation are highly localized, in accordance with theoretical predictions [2,7]. When the laser wavelength is close to the isolated monomer absorption peak, $\lambda_L = 540$ nm [Fig. 7.2(c)], localization of the optical excitations is much weaker and the sizes of the photomodified domains grow. We estimate that $\sim 30\%$ of all monomers are photomodified at $\lambda_L = 540$ nm [see Fig. 7.2 (c)], while only $\sim 14\%$ of the monomers are modified at $\lambda_L = 1079$ nm. The increase of localization of the optical excitations in fractals toward longer wavelengths (relative to the monomer absorption peak) was predicted theoretically in [7].

The photomodification leads, in turn, to a dichroic dip in the aggregate absorption spectrum in the vicinity of the laser wavelength [26,35]. The threshold dependence of the depth of the hole burnt near λ_L was observed in [26,35]. The experiments with

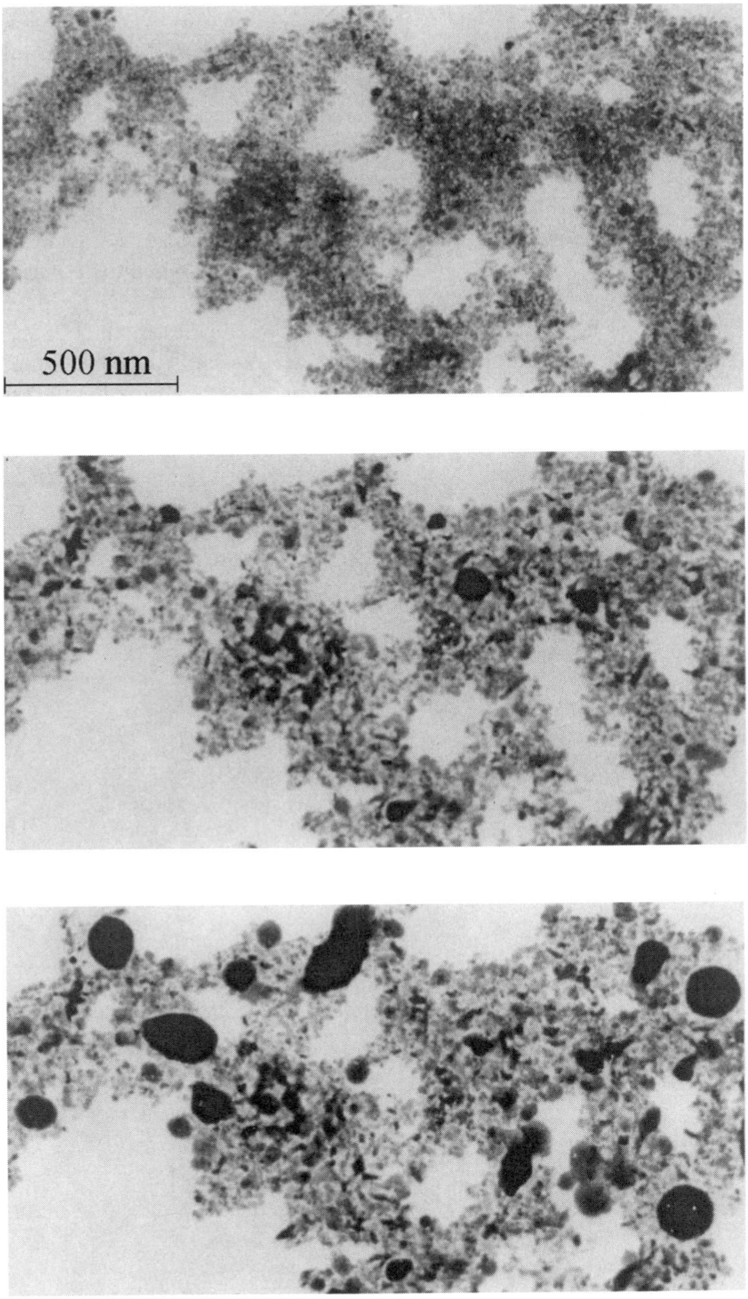

(a)

(b)

(c)

Figure 7.2. The electron micrographs of Ag(Carey Lea) aggregates. (a) Before irradiation; (b) after irradiation, $W = 21\ \text{mJ/cm}^2$, $\lambda_L = 1079\ \text{nm}$; and (c) irradiated at $W = 22.4\ \text{mJ/cm}^2$, $\lambda_L = 540\ \text{nm}$.

nano- and picosecond pulses showed that the threshold energy was approximately the same for both these cases. Two spectral holes and two thresholds are observed in this chapter. The existence of two dichroic spectral holes are in agreement with the binary theory discussed above and with the computer simulations [7].

An example of spectral hole burning is shown in Fig. 7.3. The dependence of the hole depth on laser intensity demonstrates a threshold character for each hole (see Fig. 7.4).

The first threshold (low energy) is connected with the appearance of the ultraviolet (UV) spectral holes. This spectral hole is observed when the polarization of the probe wave is perpendicular to that for a burning laser pulse. Increased absorption in the long-wavelength wing appears simultaneously with the UV spectral dip. Changes of the polycrystalline structure of some monomers were observed with the help of an electron microscope when the laser energy exceeds the first threshold. The observed changes might be connected with local heating of the sample by a laser

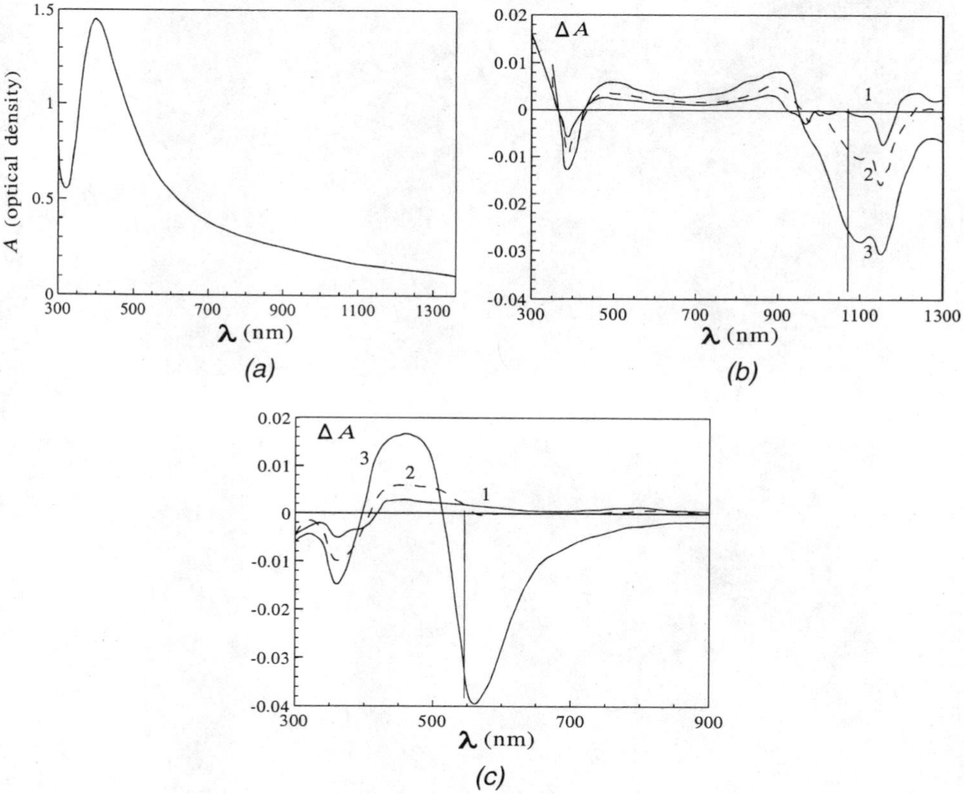

Figure 7.3. (a) Absorption spectrum of Ag($NaBH_4$) colloid. (b) Difference spectrum $\Delta A = A_{\text{after}} - A_{\text{before}}$ after irradiation of the sample at $\lambda_L = 1064$ nm and: 1, 2.9 mJ/cm^2; 2, 8.2 mJ/cm^2; 3, 16 mJ/cm^2. (c) Same as (b) for $\lambda_L = 532$ nm and: 1, 3.7 mJ/cm^2; 2, 9.2 mJ/cm^2; 3, 13.5 mJ/cm^2.

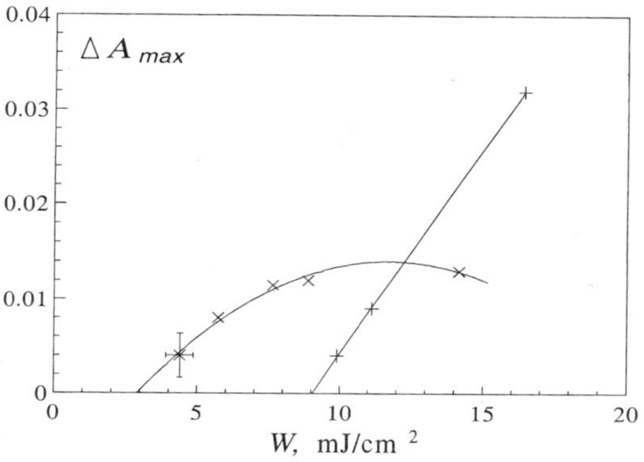

Figure 7.4. Dependence of the maximal spectral hole burnt ΔA_{max} on the energy W of the incident laser radiation at $\lambda_L = 532$ nm; $\times$, $\lambda = 360$ nm; $+$, $\lambda = 560$ nm.

pulse. We carried out the following estimation and experiment to support this interpretation. The increase in the nanoparticle Ag(EDTA) temperature in a gelatine film over the initial room temperature was estimated to be 150 K for the first threshold laser pulse energy. Following this estimate, we heated our sample in an oven. It was found that after heating up to 120°C and cooling of the sample to room temperature, absorbance in the visible range increased. An electron microscope study of silver aggregates after this heating revealed changing of the crystal structure of monomers in the aggregates without changes of the aggregates morphology. The thermal heating below 120°C did not result in the described features.

The second threshold is connected with the spectral hole near λ_L in the long-wavelength wing of absorption. This hole appears for the same linear polarizations of the probe and burning waves. The changes in fractal cluster morphology are due to irradiation with the energy above the second threshold. These changes in aggregate morphology were explained by sintering of monomers in the resonant domains because of laser heating of nanoparticles in the hot spots [17]. The estimation of laser heating of silver nanospheres in gelatine gave an increase in the temperature of the resonance monomers of 300 K for the second threshold energy. According to [36], sintering of metal nanoparticles starts when the temperature exceeds one-half of the melting point T_m (for Ag, $T_m = 1234$ K). One can see that the second threshold temperature is sufficient to initiate sintering in a silver colloid.

The phenomenon of a frequency-, polarization-, and space-selective photomodification is of interest, not only because of the information it provides concerning the localization of the optical excitations and the restructuring of nanoparticles, but also because of its possible application to dense optical data recording. The possibility of recording several spectral holes on the same wavelength-size fragment of a fractal film was demonstrated in [17]. Five spectral holes were recorded with light pulses of the same linear polarization. (The sequence of changing laser pulse wavelengths in that recording was from the IR to the green range.) So, the complex spectral information can be optically recorded inside λ-size regions of a fractal sample at room temperature.

In this section, we have shown that the resonant domains can be burnt by a laser pulse. Therefore, one can find the contribution of the resonant domains (hot spots) to the enhanced nonlinearities by measuring the nonlinear responses before and after photomodification. The decrease of local optical response in the hot spots after photomodification was observed with the help of near-field optics [19]. Note that the photoburning of the resonant domains is accompanied by the appearance of new hot spots at a given frequency, which is a result of the redistribution of a local field because of the change of an aggregate morphology.

7.4. DEGENERATE FOUR-WAVE MIXING

Measurements of the conversion efficiency of the probe wave into the signal versus the degree of hydrosol aggregation and radiation intensity were performed with the help of a Q-switched nanosecond laser in order to find the enhancement factor due to aggregation of silver particles into fractal clusters [4,37]. Two experimental geometries were tried: counterpropagating signal waves with a probe at an angle of 10^{-1}–10^{-2} rad to one of the signal beams and noncomplanar interaction used in the optical phase conjugation experiments [38]. The intensity of the probe beam I_p in counterpropagating geometry was 10% that of the pump I_0, while in the noncomplanar scheme it was $I_p = 0.3I_0$. All the beams were identically linearly polarized.

The probe and signal pulses were detected simultaneously by coaxial elements or photodiodes and supplied to an oscilloscope. The conversion efficiency of the probe beam into signal $\eta = I_s/I_p$ (I_s is the intensity of the signal generated by DFWM) was measured. Quartz cells containing hydrosol were $l = 1$, 2 and 3 mm thick. The experiments revealed the essential influence of silver particle absorption on the cell windows and cluster photomodification. To avoid this influence, the hydrosol was stirred after each pulse and the windows were wiped.

For the noncoplanar geometry, one can calculate the absolute value of $\chi^{(3)}(\omega)$ from the following equation:

$$|\chi^{(3)}(\omega)| = \frac{n_0^2 c\lambda}{48\pi^3} \cdot \frac{\alpha}{(1-T)\sqrt{T}} \cdot \frac{\sqrt{\eta}}{I_0} \tag{7.2}$$

where α is the absorption coefficient, $T = e^{-\alpha l}$ is the transmittance, λ is the wavelength of the signal, n_0 is the refractive index, and c is the speed of light. (Here, $\chi^{(3)}$ is defined in accordance with [39], so the obtained values of $\chi^{(3)}$ are three times less than in [21,22,37].)

7.4.1. Wavelength Dependence of the Nonlinear Susceptibility and Enhancement Factor

The large nonlinear optical susceptibility, $\hat{\chi}^{(3)}$, was reported in the nonaggregated nanocomposites with gold [11,40,41], silver [11,42], and cooper [43] particles at the surface plasmon resonance. High values of the susceptibility are caused by the enhancement of the local fields in metal particles at the surface plasmon resonance and by the high hyperpolarizabilities of metal particles [11]. Third-order polarizabilities of small metal particles can reach values as high as those of resonant atoms and, moreover, metal particles possess a wider spectral resonance than atoms. The mechanisms contributing to the third-order polarizability of the nanoparticles were considered in previous reports, including the nonequilibrium electron heating [41], the saturation of the absorption of the interband transition [41] and the sturation of the electron gas confined inside the particle [40,41,44].

Ways to increase the susceptibility are (a) increase the concentration of metal particles and (b) aggregate nanoparticles for achieving a large enhancement factor in a fractal structure. Let us compare these ways.

For the DFWM of nanosecond pulses at $\lambda = 532\,\text{nm}$, the susceptibility $|\chi^{(3)}| = 1.3 \times 10^{-7}$ esu was measured in a high-concentrated Au/silica ion implanted sample [45]. This value is several orders of magnitude larger than that measured for DFWM in the low-concentrated gold colloid solution [11] and gold-doped glass [41], and it is also larger than that previously reported for ion-implanted samples measured with picosecond pulses [46]. The corresponding susceptibility of gold particles $|\chi_m^{(3)}|$ which, according to [11,40,41], describes the nonlinear response to the internal field, is $|\chi_m^{(3)}| = |\chi^{(3)}|/(pf_1^2|f_1|^2) = 2.2 \times 10^{-7}$ esu, where p is the metal volume fraction. For 6-nm particles, $|f_1|$ is 1.7. The measured value of $|\chi_m^{(3)}|$ is close to the value of 1.1×10^{-7} esu obtained in [41] for the hot electron contribution.

The susceptibility of the non-aggregated samples drops sharply with detuning of the laser frequency from the surface plasmon resonance [11,40]. We observed quite different spectral dependence for the aggregated gold colloid. The measured values are $|\chi^{(3)}| = 1.2 \times 10^{-11}$ esu at 532 nm and 5.3×10^{-11} esu at 1064 nm (the metal volume fraction is $p = 5 \times 10^{-6}$). We measured the conversion efficiency for the DFWM in nonaggregated and aggregated gold colloids at $\lambda = 1064\,\text{nm}$. The laser beams were unfocused, and the pump intensity and duration were $5\,\text{MW/cm}^2$ and 14 ns, correspondingly. The enhancement factor for the susceptibility due to aggregation was found to be 36. These data are supposed to show the increase in the local field enhancement for the long-wavelength eigenmodes of gold aggregates in comparison with the monomer surface plasmon resonance.

Next, we describe the experimental results for a silver colloid that confirm this conclusion [4,37]. The results were obtained with the Ag(c) colloid at $\lambda = 532\,\text{nm}$. This wavelength is off the surface plasmon resonance for silver particles. Figure 7.5 shows the conversion efficiency measured versus the intensity of the strong pump radiation I_0 and the degree of hydrosol aggregation. The main result is that in colloids containing clusters, the same conversion efficiency as in nonaggregated monomers is attained at 10^3 times lower input intensities. This conclusion follows from comparison of the groups of points 1 and 3 in Fig. 7.5. The intensity dependence of the signal I_s is close to cubic, as is typical of four-photon scattering. By taking this cubic dependence into account, the enhancement factor for DFWM may be estimated as $\sim 10^6$, which agrees with the estimate of $G^{(3)}$ given above. Note that the DFWM in a nonaggregated colloid was measured in the focused laser

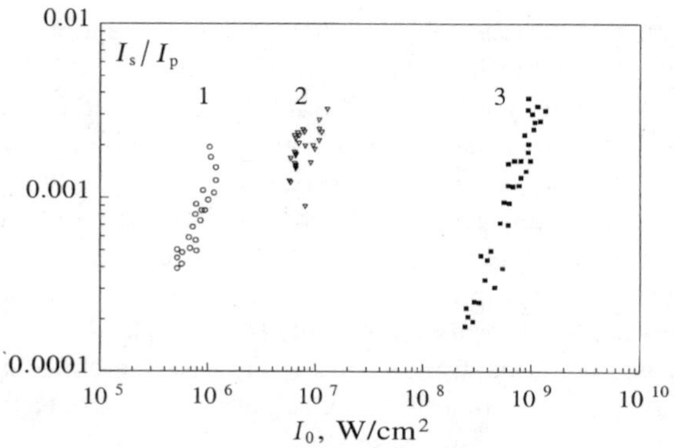

Figure 7.5. The DFWM conversion efficiency in the Ag(c) colloids with different aggregation degree. The parameter I_0 is the pump radiation intensity at wavelength 532 nm. (1) Nearly nonaggregated colloid; (2) weakly aggregated, and (3) strongly aggregated colloid. The cell is 3 mm thick.

beams. In unfocused beams with diameters of 2 mm, we observed the DFWM signal in a weakly aggregated colloid at a pump intensity of 10 MW/cm^2 (curve 2 in Fig. 7.5). Possibly this signal arose from a small number of aggregates in the irradiated volume.

At an increasing degree of hydrosol aggregation, the DFWM signal grows as follows. At $I_0 = 1$ MW/cm^2 in a freshly prepared Ag($NaBH_4$) hydrosol, the conversion efficiency η was below the detection level ($\eta \approx 2 \times 10^{-5}$), in 1 h it increased up to 4×10^{-5}, and in 7 h it reached $\eta \approx 10^{-3}$. In the most aggregated hydrosols doped by dye rhodamine 6G, $\eta \approx 0.04$.

We obtained $|\chi^{(3)}_{1111}| = 2 \times 10^{-10}$ esu for the most aggregated Ag($NaBH_4$) colloid with a volume fraction of metal, of $p = 5 \times 10^{-6}$, at $\lambda = 1064$ nm for a laser intensity of 1 MW/cm^2. This means that the susceptibility per unit volume of the metal is more than 10^{-4} esu. The measured $|\chi^{(3)}_{1111}|$ at $\lambda = 532$ nm is two times less. The figure of merit (FOM), $|\chi^{(3)}_{1111}|/\alpha$, is $\sim 10^{-9}$ esu $\cdot$ cm. The values of $|\chi^{(3)}_{1111}|/p$ and FOM for the aggregated colloid substantially exceeds (more than one to two orders of magnitude for nanosecond pulses) the corresponding values for nonaggregated composites at the plasmon resonance.

Measurements with picosecond pulses show the susceptibility to be two orders of magnitude less than with nanosecond pulses. This suggests that thermal effects play a large role, namely, heating of the crystal lattice [47] and host medium around the absorbing particle for a nanosecond pulse (see below).

7.4.2. Temporal Behavior of the Nonlinear Response

The response time of the cubic nonlinearity of nonaggregated metal nanocomposites was measured by the pulse delay technique in DFWM experiments [11,40,42,43,46]. It was found that there are fast and slow relaxation processes, and the fast decay time is shorter than the pulse duration, which was typically several picoseconds. The slow relaxation was attributed to the heat transfer from the gold particles to the host medium. Recent investigations of electron dynamics in metal films and nanoparticles by the pump–probe technique with femtosecond pulses revealed femto- and picosecond relaxation processes, connected with electron–electron and electron–phonon interactions, whose characteristic times were found to be dependent on the electron, T_e, and lattice, T_i, temperatures [48,49].

An existence of a fast nonlinear response may also be established with the help of experiments done in the frequency domain [50]. The nondegenerate four-wave mixing with frequency detunings of 10–100 cm^{-1} between the pump and probe waves was observed in nonaggregated gold nanocomposites [45,51]. The characteristic response times of 5.3 and 0.66 ps had been estimated based on the detuning curve [45]. The first of these was attributed to thermal diffusion from 6-nm particles to the host medium (silica), and the second to electron–phonon relaxation.

Fast nonlinear response of aggregated Ag(c) colloid was tested by the pulse delay technique in the DFWM scheme [37]. At low-pulse energies, the DFWM signal was doubled with the time delay increasing up to 30 ps, which coincides with the pulse duration.

A long tail (up to 200 ps) in the temporal dependence of the DFWM signal follows after the fast response of aggregated Ag(c). This tail can be explained by the influence of thermal effects [47]. The time of thermal diffusivity responsible for cooling of a single particle and heating of a host medium can be estimated as $t = C\rho r^2/3k_0 = 140$ ps (the values of the silver lattice specific heat $C = 0.25$ J/g·K, density $\rho = 10.8$ g/cm^3, and water thermal conductivity $k_0 = 0.006$ W/cm·K were used, where the radius of particle is $r = 10$ nm).

Additional evidence for the role of inertial effects was given by the observation of two-wave interactions in a silver colloid. It is known [52] that amplification of a probe wave by two-wave coupling is possible in the cases of inertial or nonlocal nonlinearity. We observed the gain of the 10-ns probe pulse at $\lambda = 532$ nm in an aggregated Ag(PVP) colloid. Both the pump and probe pulses had the same linear polarization. There are two maxima in the probe pulse after passing through the colloid. The first of them coincides with the maximum of the input pulse, and the second is delayed by 2–3 ns. The most amplification was found at the back part of the pulse envelope. This proves that the inertial component of the cubic nonlinearity of colloid aggregates is essential as well as the fast one discussed above.

7.4.3. Influence of Photomodification

It was found that near the photomodification threshold the DFWM signal with nanosecond excitation was diverted from the dependence of $I_S \propto I_0^2 \cdot I_p$. Slightly beyond the threshold, a jumping increase in the DFWM signal under a series of repeated nanosecond pulses was observed. This result may be interpreted as follows. When the replacement of the colloid aggregates for the time interval between subsequent pulses is less than the period of the lattice generated by the intersecting light beams, the photomodification generates a quasistatic holographic grating. The depth of modulation in such a grating grows under a series of light pulses, and the respective increase in the scattering signal is observed. In the experiments with picosecond pulses [37], it was found that when laser pulse energy exceeded the photomodification threshold, along with the noninertial component of the DFWM scattering, the inertial component arose at a time delay of 100–200 ps and more. This contribution prevailed over the usual thermal effect, which was observed at low intensity. A major contribution to the inertial component is associated with the photomodification of clusters. The calculated time of the sintering process for metal nanoparticles (100 ps) [36] coincides well with this measurement.

Well beyond the threshold, the photoburning leads to a substantial decrease in the nonlinear susceptibility. The efficiency of DFWM at $\lambda = 532$ nm was measured in Ag clusters in a 3-mm cell before the photoburning, and with a delay of 60 μs after the photoburning pulse with energy 10 times more than the threshold one. The result was that the signal decreased by a factor of 30 in comparison with the nonirradiated sample [22].

7.5. NONLINEAR ABSORPTION AND REFRACTION

To get further insight into the processes of nonlinear light interaction with aggregated colloid solution, we should separate out the amplitude and phase parts of the susceptibility. To measure the nonlinear refractive index and the nonlinear absorption coefficient, we used the Z-scan technique [53]. This method allowed us to determine the nonlinear correction to the refractive index, n_2, and to the absorption coefficient, α_2, from experimental data and to calculate, $\mathrm{Re}\,\chi^{(3)}$ and $\mathrm{Im}\,\chi^{(3)}$. It was found that for $\lambda = 540$ nm the aggregation of silver colloidal particles into fractal clusters was accompanied by an increase in nonlinear absorption from $\alpha_2 = -9 \times 10^{-10}$ cm/W to $\alpha_2 = -5 \times 10^{-7}$ cm/W, that is, the enhancement factor was 560 [22]. A value of >400 was obtained for the enhancement factor for the nonlinear refractive index. Estimated enhancement factor for the Kerr nonlinearity at $\lambda = 540$ nm are $\sim 10^3$, the square root of those for the DFWM. For aqueous colloids, the values obtained for $\chi^{(3)}$ were $\mathrm{Re}\,\chi^{(3)} = 10 \times 10^{-11}$ esu, $\mathrm{Im}\,\chi^{(3)} = -8.3 \times 10^{-11}$ esu for $\lambda = 540$ nm, and $\mathrm{Re}\,\chi^{(3)} = -3.5 \times 10^{-11}$ esu, $\mathrm{Im}\,\chi^{(3)} = -2.7 \times 10^{-11}$ esu for $\lambda = 1079$ nm. This finding suggests that the nonlinear absorption and nonlinear refraction provide nearly equal contributions to the nonlinearity in the green and near-IR ranges. This may be regarded as an additional advantage of the aggregated nanocomposite as a nonlinear medium with respect to nonaggregated. The latter is characterized by the predominantly imaginary susceptibility at the surface plasmon resonance [41]. The imaginary part of the cubic susceptibility is negative in the blue, green, and near-IR ranges, and positive in the red region [54].

We measured changes of the absorption spectra of aggregated $Ag(NaBH_4)$, Ag(c) colloids and polymer film doped by Ag(EDTA) aggregates under heating up to 90°C. Thermal heating results in a reversible decrease of the absorption coefficient in the studied spectral range (300–800 nm). The value of the decrease at $\lambda = 532$ nm corresponds to the dynamic photobleaching measured at the intensity 300 kW/cm^2 for the 10-ns pulse. An estimate shows that a silver nanoparticle with a diameter of 20 nm in water is heated up to 100°C at such fluence. This coincidence may be regarded as an indication of the essential role of a thermal mechanism of optical nonlinearity in colloid aggregates under nanosecond excitation.

Thus, we may conclude that both the absorptive and refractive nonlinearities of an aggregated silver colloid solution change their signs with wavelength. Moreover, we observed that in the green range the sign of the nonlinear coefficient changed during the 10-ns pulse. The sign changes with the wavelength may be related, on the one hand, to the variations of the signs of the enhancement factors, which were found in computer simulations [25]. On the other hand, the spectral and temporal sign changes may be due to the band structure of silver and a competition between two-photon absorption on interband transition, laser excitation of free electrons, and heating of crystal lattice. These factors were not taken into account in a considera-tion of nonlinear optical processes in fractal aggregates, but their impotance was revealed in studies of nanoparticles and metal films. Double–photon excitation was found to be essential in silver films [55]. The broadening of the surface plasmon resonance of a nonaggregated metal nanocomposite under laser excitation was observed in [49]. This broadening is governed by the time dependence of the electronic scattering rate and the temperature-induced smearing of the Fermi edge [56]. Plasmon excitation in the aggregates is very inhomogeneous. The electron and phonon temperatures in the hot zones on average grow higher than in an aggregate. As a result, the dielectric function of particles in the hot zones changes, and, under strong laser excitation, locations of the hot zones may change. This means that other monomers become resonant with the laser field during the laser pulse. Apparently, this is the reason for a shift and broadening of a spectral hole under the photo-modification by a nanosecond pulse versus a picosecond pulse, observed in [35]. The shift and broadening of a spectral hole were found to be dependent on the laser wavelength.

When the intensity exceeds 2–3 MW/cm^2, the results of the Z-scan fitting according to the procedure from [53] become quite unsatisfactory. This implies that the nonlinearity declines from the simple cubic law. Figure 7.6 presents the intensity dependence of the normalized nonlinear addition to the absorption coefficient $(-\Delta\alpha/\alpha)$ for an aggregated Ag($NaBH_4$) colloid (curve 2). Curve 1 shows the dependence of the persistent hole depth on the laser intensity. Curve 3 shows the calculated values of the saturated absorption $-\Delta\alpha/\alpha = (I/I_s)(1 + I/I_s)^{-1}$. Numerical fittings give us the best match of curve 3 with the experimental curve 2 for $I < I_{\text{th}}$ at the parameter value $I_s = 2.8 I_{\text{th}} = 1.4$ MW/cm^2. As may be inferred from com-parison of curves 2 and 3 in Fig. 7.6, the absolute value of nonlinear bleaching of the sample decreases for intensities $I > I_{\text{th}}$ [22]. The investigation of the interaction of the silver colloid solution with high-intensity radiation, significantly exceeding the photomodification threshold, showed changes of signs of the nonlinear absorption depending on the intensity [54].

As we noted before, it is difficult to apply the Z-scan technique in this case. We developed some new interferometric techniques [21,57] to measure the nonlinear

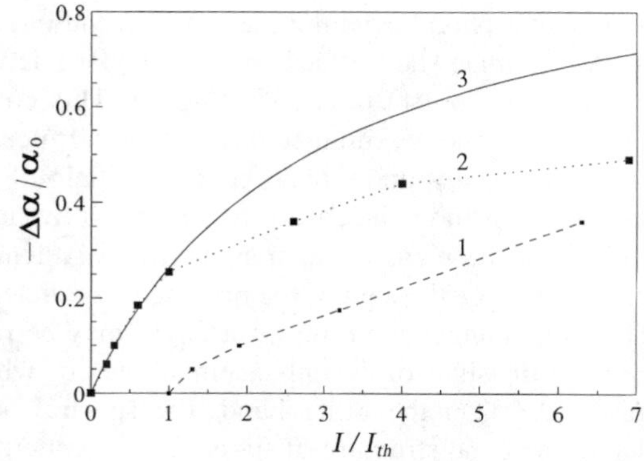

Figure 7.6. Intensity profiles: (1) depth of long-term photoburning hole at $\lambda_L = 540\,\text{nm}$ and $I_{\text{th}} = 0.5\,\text{MW/cm}^2$; (2) nonlinear contribution to the absorption coefficient at the same λ_L and a hydrosol cell thickness of 3 mm, ($\alpha_0 = 1.7\,\text{cm}^{-1}$); (3) calculated absorption saturation with $I_s = 2.8 I_{\text{th}}$.

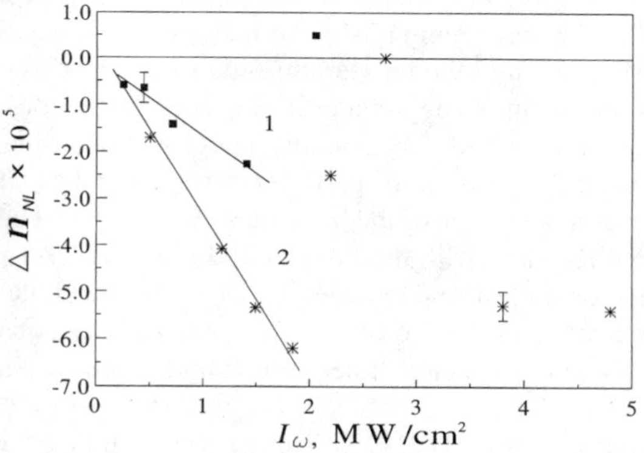

Figure 7.7. Dispersion of the nonlinear refractive index of (1), aqueous colloid Ag($NaBH_4$); and (2), ethanol colloid Ag(PVP). The exciting radiation I_ω was pulsed with duration $\approx 15\,\text{ns}$ and wavelength 1064 nm.

refraction in colloids under high intensity. The measurements that we have performed using the method of the nonlinear dispersion interferometer [21] showed a considerable decrease of the nonlinear addition to the refractive index when the photoburning occurred. In Fig. 7.7, we plot the dispersion of the nonlinear refractive index, $\Delta n = \Delta n(\omega) - \Delta n(2\omega)$, versus the incident intensity I_ω for both Ag(PVP) and Ag($NaBH_4$) colloids. The method of shearing interferometery allows one to separately obtain the value of $\Delta n(\omega)$ [58], where its dependence on the exciting intensity was found to have the same feature at $I_\omega \gtrsim 2\,\text{MW/cm}^2$. At lower intensity, there was a linear dependence, corresponding [for aqueous colloid, Ag($NaBH_4$)] to $\text{Re}\,\chi^{(3)} \approx -4.5 \times 10^{-10}$ esu, at $\lambda = 1064\,\text{nm}$.

7.6. NONLINEAR POLARIZATION PHENOMENA IN SILVER COLLOIDAL SOLUTIONS

In this section, we examine the polarization nonlinearities originating from both local and nonlocal responses. The third-order nonlinear polarization can be described by the following relation:

$$P_i^{(3)} = \chi_{ijkl}^{(3)} E_j E_k E_l + \Gamma_{ijklm}^{(3)} E_j E_k \nabla_m E_l \tag{7.3}$$

Here the local nonlinear response is represented by the tensor $\hat{\chi}^{(3)}$, and the tensor $\hat{\Gamma}^{(3)}$ is connected with the nonlocal response that is due to spatial dispersion effects. As it is known, in an isotropic medium the tensor $\hat{\Gamma}^{(3)}$ has one nonzero component g_1. In accordance with the condition of intrinsic permutation symmetry, the tensor $\chi^{(3)}$ for the case of isotropic materials can be represented as [39]

$$\begin{aligned}\chi_{ijkl}^{(3)}(\omega = \omega + \omega - \omega) = \chi_{1122}(\omega = \omega + \omega - \omega)(\delta_{ij}\delta_{kl} + \delta_{ik}\delta_{jl})\\ + \chi_{1221}(\omega = \omega + \omega - \omega)\delta_{il}\delta_{jk}\end{aligned} \tag{7.4}$$

If the optical field is of the form

$$\mathbf{E}(t, \mathbf{r}) = \mathbf{A}\exp[-i(\omega t - \mathbf{kr})] + \text{c.c.} \tag{7.5}$$

where c.c. is a complete conjugate, we can obtain the formula for the local part of $\mathbf{P}^{(3)}$ as

$$\begin{aligned}\mathbf{P}_{\text{loc}}^{(3)}(\omega) = \chi_1 \mathbf{A}(\mathbf{A}\mathbf{A}^*) + \chi_2 \mathbf{A}^*(\mathbf{A}\mathbf{A}) + \text{c.c.}\\ \chi_1 = 6\chi_{iijj}^{(3)} = 6\chi_{ijij}^{(3)}; \chi_2 = 3\chi_{ijji}^{(3)}; i,j = 1,2; i \neq j\end{aligned} \tag{7.6}$$

and that for the nonlocal one as[1]

$$\mathbf{P}_{\text{nonloc}}^{(3)}(\omega, \mathbf{k}) = -ig_1(\mathbf{A}\mathbf{A}^*)[\mathbf{kA}] + \mathbf{A}(\mathbf{A}^*[\mathbf{kA}]) + \text{c.c.} \tag{7.7}$$

Our goals now are to study four polarization effects for measurements of all nonzero components in (7.3). The first nonlinear gyrotropy [resulting from g_1 in (7.7)], which is the intensity-dependent rotation of the plane of polarization of a linearly polarized light beam. The next is the self-induced rotation of the polarization ellipse of the light (Re χ_{1221} component). By using a pump–probe configuration, we also studied the optical Kerr effect (OKE), that is, the nonlinear birefringence of a probe wave induced by a pump wave [corresponding to Re$(\chi_1 + 2\chi_2)$], and the

[1]The electric field in [27] was defined with the exponential factor $\exp[i(\omega t - \mathbf{kr})]$, so that the equation for $\mathbf{P}_{\text{nonloc}}^{(3)}$ had the opposite sign.

inverse Faraday effect (IFE), that is, the circular birefringence of a probe wave induced by a circularly polarized pump wave [corresponding to $\mathrm{Re}(\chi_1 - 2\chi_2)$].

7.6.1. Basic Relations

Gyrotropy and Rotation of Ellipse. Let us first proceed to the self-action of a strong wave (elliptically polarized, in the general case) propagating through the medium with susceptibilities determined by Eqs. (7.6), and (7.7). The interaction of a light wave with such a medium results in the rotation of the polarization ellipse of the wave, so that the azimuth of the ellipse is given by the following equation [59] (the z-axis is directed along the wavevector):

$$\begin{aligned} \alpha &\equiv \tfrac{1}{2}\arg(A_+A_-^*) \\ \frac{d\alpha}{dz} &= \rho_0' + \rho_1'(|A_+|^2 + |A_-|^2) + \sigma_2'(|A_-|^2 - |A_+|^2) \end{aligned} \tag{7.8}$$

Here, A_+ and A_- are the amplitudes of the right- and left-hand circular components of the wave, $A_\pm = (A_x \pm iA_y)/\sqrt{2}$; $\rho_0 = \rho_0' + i\rho_0''$ is the linear gyration constant, $\rho_1 = 2\pi g_1\omega^2/c^2 = \rho_1' + i\rho_1''$; $\sigma_{1,2} = 2\pi\omega^2\chi_{1,2}/kc^2 = \sigma_{1,2}' + i\sigma_{1,2}''$; and $k = (\omega/c)(\mathrm{Re}\,\varepsilon_0)^{1/2}$.

The first term on the right-hand side (rhs) of (7.8) corresponds to the natural gyrotropy of the medium. Note that this effect has not been found in our experiments. The other terms describe the nonlinear rotation of the polarization plane. The one containing ρ_1' is due solely to the nonlocal nature of the nonlinear response and takes place even if the wave is linearly polarized ($|A_+|^2 = |A_-|^2$). This effect is usually referred to as nonlinear gyrotropy, or nonlinear optical activity. This is unlike the term with σ_2', connected with the local nonlinearity, which vanishes in case of purely linear polarization, although it gives an additional rotation for elliptical case [60].

Now, we analyze the possibilities for studying polarization nonlinearities with the two-beam interaction technique. We assume a strong wave (pump),

$$E_i^{\mathrm{str}} = F_i \exp[-i(\omega t - \mathbf{Kr})] + \mathrm{c.c.} \tag{7.9}$$

and a weak one (probe),

$$E_i^{\mathrm{weak}} = S_i \exp[-i(\omega t - \mathbf{kr})] + \mathrm{c.c.} \tag{7.10}$$

We then assume the angle between $\mathbf{K}$ and $\mathbf{k}$ to be small enough so the components F_z and S_z can be neglected. If we substitute the total field ($\mathbf{E}^{\mathrm{str}} + \mathbf{E}^{\mathrm{weak}}$) into (7.3) and keep only the terms with the same frequency and wavevector direction as probe $\mathbf{E}^{\mathrm{weak}}$, we obtain

$$\begin{pmatrix} P_x \\ P_y \end{pmatrix}^{(3)} = 6\begin{pmatrix} \chi_{1111}|F_x|^2 + \chi_{1122}|F_y|^2 & \chi_{1122}F_xF_y^* + \chi_{1221}F_yF_x^* \\ \chi_{1122}F_yF_x^* + \chi_{1221}F_xF_y^* & \chi_{1111}|F_y|^2 + \chi_{1122}|F_x|^2 \end{pmatrix}\begin{pmatrix} S_x \\ S_y \end{pmatrix} \tag{7.11}$$

Here, we drop the terms containing $S_i S_j$ due to their smallness and use the known relation $\chi_{1111} = 2\chi_{1122} + \chi_{1221}$. For simplicity, we do not take into account the nonlocal contribution to nonlinear polarization [given by (7.7)].

Let us consider two classical cases of pump and probe wave polarizations: one corresponding to the optical Kerr effect and the other to the inverse Faraday effect.

Optical Kerr Effect. The strong wave is linearly polarized along the x axis: $F_y = 0$; the probe wave is also linearly polarized but rotated by 45° relative to strong wave: $S_x = S_y$. Equation (7.11) in this case becomes:

$$\begin{pmatrix} P_x \\ P_y \end{pmatrix}^{(3)} = 6|F_x|^2 \begin{pmatrix} \chi_{1111} & 0 \\ 0 & \chi_{1122} \end{pmatrix} \begin{pmatrix} S_x \\ S_y \end{pmatrix} \tag{7.12}$$

It is known (see, e.g., [39]) that this equation for the nonlinear polarization corresponds to induced birefringence on the probe wave, for the case of OKE we have,

$$\Delta n_x - \Delta n_y = \frac{12\pi}{n_0} |F_x|^2 \mathrm{Re}(\chi_{1122} + \chi_{1221}) \tag{7.13}$$

where n_0 is the linear refractive index.

Inverse Faraday Effect. In case of IFE, the pump wave has a circular polarization (right hand for definiteness) $F_x = iF_y$, and the probe one is linearly polarized, $S_y = 0$. In terms of the circular components of the complex amplitudes, the nonlinear polarization is given by

$$\begin{pmatrix} P_+ \\ P_- \end{pmatrix}^{(3)} = 6|F_+|^2 \begin{pmatrix} 2\chi_{1122} & 0 \\ 0 & \chi_{1122} + \chi_{1221} \end{pmatrix} \begin{pmatrix} S_+ \\ S_- \end{pmatrix} \tag{7.14}$$

that leads to a circular birefringence

$$\Delta n_+ - \Delta n_- = \frac{12\pi}{n_0} |F_+|^2 \mathrm{Re}(\chi_{1122} - \chi_{1221}) \tag{7.15}$$

This circular birefringence generally results in a rotation of the polarization plane. The rotation angle of the probe wave polarization plane, while the wave is propagating through the medium, is given by

$$\frac{d\alpha}{dz} = \frac{6\pi\omega}{cn_0} |F_+|^2 \mathrm{Re}(\chi_{1122} - \chi_{1221}) \tag{7.16}$$

and measured counterclockwise in the xy plane.

Thus, the given treatment shows that investigation of the polarization nonlinear phenomena allows one to determine all the nonlinear constants that characterize an isotropic medium. We have performed such an experimental study of the silver colloidal solution nonlinearity with special attention given to the changes of its nonlinear properties due to aggregation of the solution.

7.6.2. Experiments

For our experiments, we used an ethanol colloidal solution stabilized with PVP, prepared as described in [31]. Adding a small quantity of the alkali NaOH (5×10^{-5} parts by weight), we were able to change the degree of aggregation significantly. In Fig. 7.8, the linear absorption spectra are shown for two basic samples: strongly and weakly aggregated solutions.

Let us start with the experimental observation of *nonlinear gyrotropy* in the silver colloidal solution [27]. As can be seen from (7.8), for an elliptical polarization, which always takes place in real experimental conditions, both local and nonlocal nonlinear responses contribute to the nonlinear rotation of the polarization plane. The method proposed and implemented [27] to separate these effects consist in the following: We note that $|A_+|^2 - |A_-|^2 = 2|A_x||A_y|\sin(\phi_x - \phi_y)$, where $|A_{x,\,y}|$ and $\phi_{x,\,y}$ are the amplitudes and phases of the linearly polarized components of the field. By varying the phase difference $\phi_x - \phi_y$, it is possible to alter the contribution of the third term in (7.8), while that of the nonlocal response will remain constant.

The scheme of the polarization measurement is displayed in Fig. 7.9(a). We used the second harmonics of the YAG:Nd pulsed laser. The pulse shape and the transverse distribution in the beam are shown in Fig. 7.9(b).

The radiation passing through the polarizer had an ellipticity of $|A_y|^2/|A_x|^2 \approx 5 \times 10^{-5}$. The phase element (consisting of two identical wedges made of crystalline quartz cut so that its optical axis was directed along the y axis) introduces an

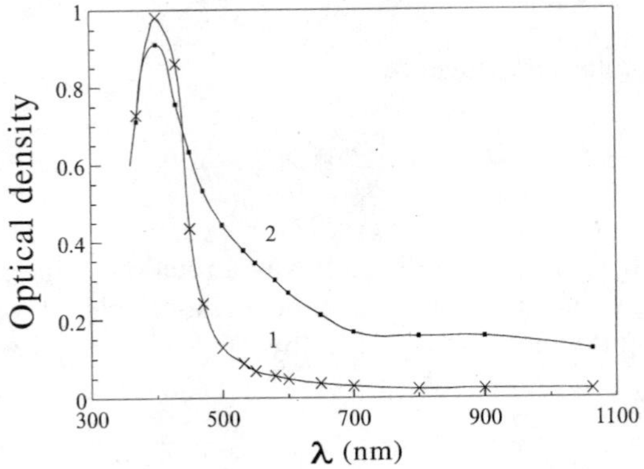

Figure 7.8. Linear absorption spectra of (1) weakly and (2) strongly aggregated colloid samples.

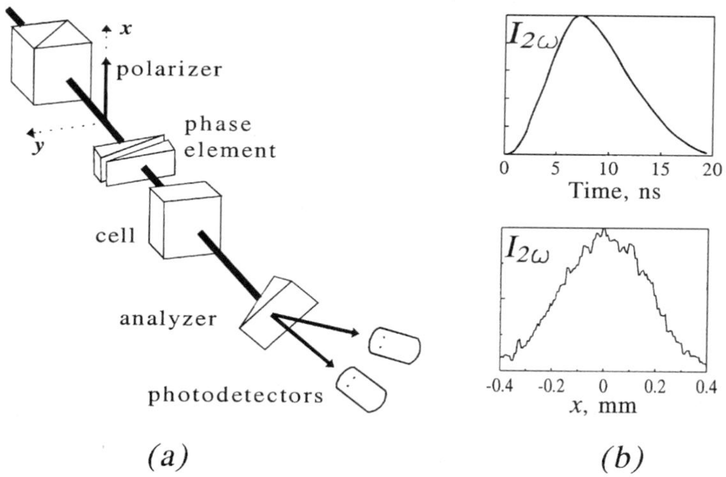

Figure 7.9. Scheme of the polarization measurement. The polarizer is a Glan prism, and the analyzer is a calcite wedge. The cell thickness is 3 mm. In part (b), the temporal and spatial properties of the incident radiation are also given.

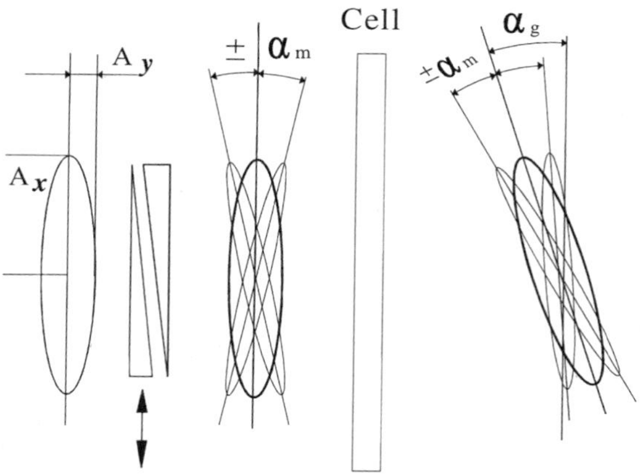

Figure 7.10. Explanation to the method of the nonlinear gyrotropy measurement. Shown is the evolution of the poloarization ellipse as the beam passes the phase element and the cell with the studied medium. Here, $\alpha_m = |A_y(0)|/|A_x(0)|$ is the magnitude of the azimuth variation due to the phase element, and α_g is the nonlinear rotation angle.

additional phase shift between the complex amplitudes,

$$\Delta\phi_0 = \phi_x(0) - \phi_y(0) \tag{7.17}$$

The effect of the phase element is thus to change the ratio of the semiaxis of the polarization ellipse [its azimuth $\alpha(0)$ is also varying], as schematically shown in Fig. 7.10.

With allowance for $|A_y|/|A_x| \ll 1$, the azimuth is given by

$$\alpha(0) = \frac{|A_y(0)|}{|A_x(0)|} \cos(\Delta\phi_0) \tag{7.18}$$

At the exit from the nonlinear medium, the azimuth of the ellipse $\alpha(l)$ also periodically changes with the variation of $\Delta\phi_0$, relatively to the mean point α_g, as

$$\alpha(l) = \alpha_g + b\cos(\Delta\phi_0 + c)$$
$$a_g = \rho_1' \int_0^l |A_x|^2 dz, \quad b = \frac{|A_y(0)|}{|A_x(0)|} \qquad c = 2\sigma_2' \frac{|A_x(0)|}{|A_y(0)|} \int_0^l |A_x||A_y| dz \tag{7.19}$$

As one can see, the local (σ_2') and nonlocal (ρ_1') parts can be separated since they result in different changes in $\alpha(l, \Delta\phi_0)$ as the intensity at the entrance varies.

In Fig. 7.11 the dependencies of the nonlinear rotation angle on the intensity at the entrance are plotted both for weakly and strongly aggregated colloids. The nonlinear optical activity is shown to be strongly dependent on the aggregation degree of the colloid. For an input intensity $I \lesssim 2\,\mathrm{MW/cm^2}$, the measurement data, being scaled to the same silver density in both samples, give the values of the nonlinear gyrotropy tensor as

$$\mathrm{Re}\,\Gamma_s^{(3)} \approx 0.9 \times 10^{-16}\ \mathrm{esu} \qquad \mathrm{Re}\,\Gamma_w^{(3)} \approx 1.1 \times 10^{-18}\ \mathrm{esu} \tag{7.20}$$

for strongly and weakly aggregated colloids, correspondingly. In both cases, the medium is levorotatory, that is, causes counterclockwise rotation, looking into the beam.

The strong amplification of the nonlinear gyration constant due to aggregation, as observed in our experiments, is in good agreement with the concept of large

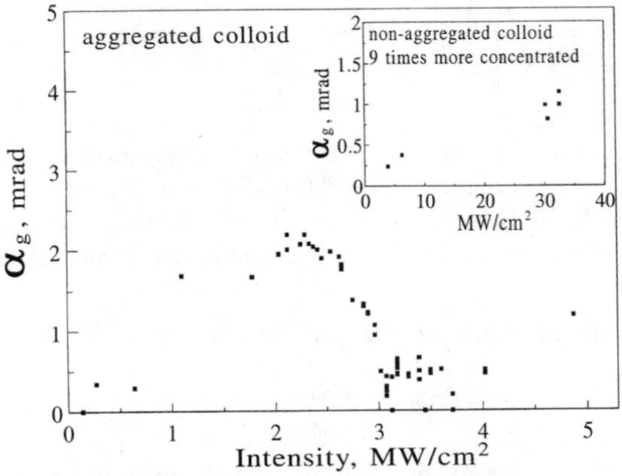

Figure 7.11. Dependencies of the nonlinear rotation angle on the incident intensity for the strongly aggregated Ag(PVP) colloid (main chart) and the weakly aggregated one (inset).

enhancement of the local field. The estimation gives a factor [27]

$$\Gamma_s^{(3)}/\Gamma_w^{(3)} \approx 100 \tag{7.21}$$

which is in accordance with the measured ratio.

We now describe the experiments devoted to the determination of the components of $\chi_{ijkl}^{(3)}$, denoted earlier [see (7.6)] as χ_1 and χ_2.

Rotation of Ellipse. It follows from (7.8) and (7.19) that the technique we just described is quite appropriate as well for measuring χ_2, provided that the incoming radiation would have a sufficient ellipticity. For this purpose, we used the same phase element (see Fig. 7.9), but rotated it by 45°. The phase element introduced the phase shift $\approx \pm\pi/4$, which made the polarization ellipse have a semiaxis ratio $|A_y|^2/|A_x|^2$ of $\sim$ 1:6, either clockwise or counterclockwise.

According to our experiments, the angles α_{SR} of self-induced rotation of the polarization ellipse were close in value and opposite in sign for left- and right-hand circular polarization of the wave. (We use the convention when the left-hand circular polarization corresponds to counterclockwise rotation looking into the beam.)

The value of the rotation in the strongly aggregated Ag solution, as the incident intensity increases from $\approx 0.7\,\text{MW/cm}^2$ to $\approx 5\,\text{MW/cm}^2$, reaches

$$\alpha_{SR} \approx (-1.8 \pm 0.2)\text{mrad} \quad \text{for right-hand polarized pump}$$
$$\alpha_{SR} \approx (2.1 \pm 0.2)\text{mrad} \quad \text{for left-hand polarized pump}$$

Taking the average angle for both polarizations of the pump, one can eliminate the influence of nonlinear gyrotropy, so we obtain Re $\chi_2 \approx 1.3 \times 10^{-11}$ esu. Here, we take into account also the concentration of the silver particles in this sample, namely, the concentration was 0.7 of that of the aggregated colloid whose spectrum is given in Fig. 7.8, curve 2.

One of the main goals of our experiments was an investigation of the dependence of the local nonlinear response versus incoming intensity, both for weakly and strongly aggregated colloid. For this purpose, we used a technique of two-beam coupling. The optical Kerr effect and inverse Faraday effect were observed.

Optical Kerr Effect. The configuration for OKE is schematically shown in Fig. 7.12. The probe wave polarization plane has an angle of 45° with respect to the x axis. The phase element introduces a phase shift $\Delta\phi_0$ between the components S_x and S_y. The analyzer splits the radiation into two channels polarized along $\pm$ 45° with respect to x axis, so the intensity in each channel is the result of the interference:

$$I_{1,2} = \tfrac{1}{2}[|S_x(l)|^2 + |S_y(l)|^2 \pm 2|S_x(l)||S_y(l)|\cos(\Delta\phi_0 + \Delta\phi_{\text{NL}})] \tag{7.22}$$

where $S_x(l)$ and $S_y(l)$ are the Cartesian components at the exit of the studied medium (of length l). Since the interference dependencies in these channels are shifted by π,

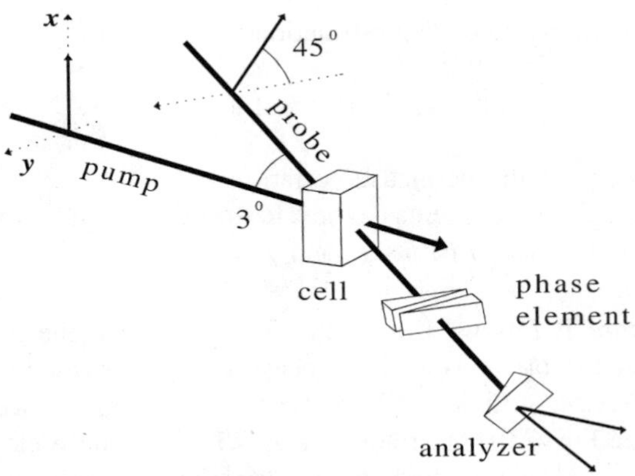

Figure 7.12. Experimental setup for OKE observation. A complanar pump–probe configuration is used, where both beams are linearly polarized, as shown by the arrows. The phase element (two crystalline quartz wedges) is aligned along the x and y axes. The analyzer (calcite wedge) is turned through 45° relative to the x axis.

it was natural to register a differential signal from the two channels at the point $\Delta\phi_0 = \pi/2$:

$$\Delta I \equiv I_1 - I_2 \approx 2|S_x(l)||S_y(l)|\Delta\phi_{NL} \tag{7.23}$$

The nonlinear phase shift $\Delta\phi_{\rm NL}$ is acquired within the medium due to the birefringence induced by the pump wave:

$$\Delta\phi_{\rm NL} = \frac{2\pi l(\Delta n_x - \Delta n_y)}{\lambda} \tag{7.24}$$

The calibration of the scheme was performed by means of the phase element.

The results of the experiments are given in Fig. 7.13.

Inverse Faraday Effect. The experimental setup for IFE is close to that for OKE, but the phase element was placed in this case into the pump wave and adjusted in such a way that the latter had right-hand circular polarization. The probe wave was linearly polarized along the x direction. The measured value was the angle $\alpha_{\rm IFE}$ of rotation of the probe wave polarization plane in the presence of the pump wave. This angle can be expressed in terms of induced circular birefringence as follows:

$$\alpha_{\rm IFE} = \frac{\pi l(\Delta n_+ - \Delta n_-)}{\lambda} \tag{7.25}$$

In Fig. 7.13(a) and (b), the values of the Cartesian and circular nonlinear birefringence are plotted versus the incident intensity, both for weakly and strongly

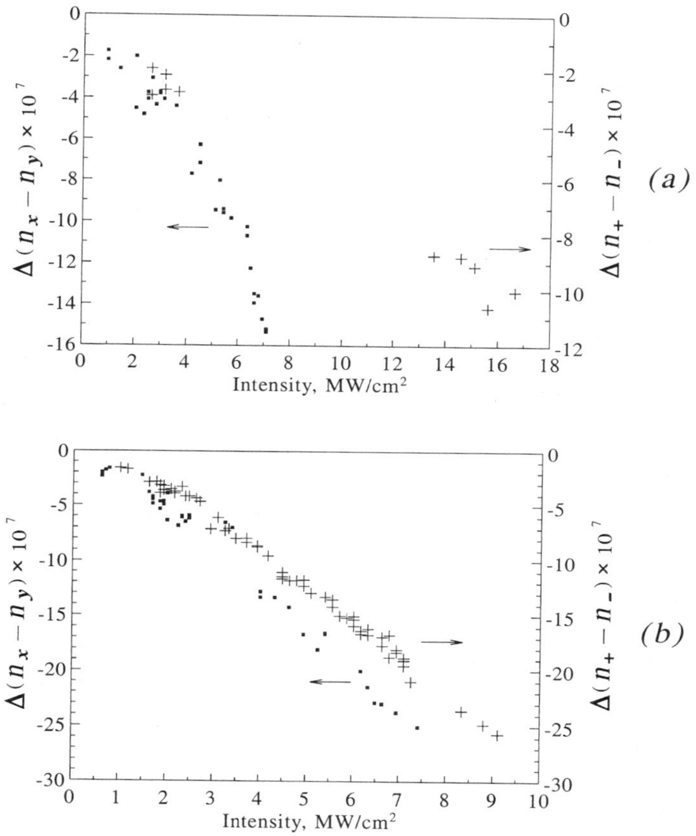

Figure 7.13. Results of IFE and OKE experiments. Each chart contains the dependence of induced birefringence, both Cartesian (solid points, referred to left axis) and circular (+ sign, right axis) for weakly (a) and strongly (b) aggregated Ag(PVP) colloid. The incident radiation had the same parameters: $\lambda = 532\,\text{nm}$, $\tau \approx 10\,\text{ns}$.

aggregated colloids. The linear dependencies exhibited are in a good agreement with those predicted by (7.15) and (7.13).

If we combine the data of IFE and OKE measurements, we obtain the final expressions (in terms of χ_1 and χ_2 introduced earlier) for the $\hat{\chi}^{(3)}$ components:

$$\text{Re}\,\chi_1(\text{esu}) = 2\sqrt{2}\,\frac{(\Delta n_x - \Delta n_y) + (\Delta n_+ - \Delta n_-)}{8\pi^2 10^{13} I_{\text{eff}}(\text{MW/cm}^2)}\,n_0^2 c(\text{esu}) \tag{7.26}$$

and analogously for 2Re χ_2 with a "$-$" sign in the numerator. Here $I_{\text{eff}} = [\int_0^l I(z)dz]/l$ is the effective value of the intensity inside the cell with the colloid, taking into account both linear and nonlinear absorption. The factor $2\sqrt{2}$ originates from the temporal and spatial averaging that took place in the experiments because we registered the entire beam by slow photodetectors. Here, we assume a Gaussian distribution of the radiation both in the time and in space domains [see Fig. 7.9(b)]. For strongly aggregated colloid, one can obtain from our measurements Re $\chi_1 \approx -1.8 \times 10^{-10}$ esu, taking $I_{\rm eff} \approx 0.72I(0)$.

The value of χ_2 resulting from the IFE and OKE experiments is of the same order of magnitude as that determined via self-rotation of the polarization ellipse. However, the values of $(\Delta n_x - \Delta n_y)$ and $(\Delta n_+ - \Delta n_-)$ happened to be very close, so that the accuracy of the χ_2 determination here is much lower than in case of ellipse self-rotation.

7.7. SUMMARY

Our experimental studies have confirmed the predicted giant enhancement of optical nonlinearities of metal fractal aggregates compared to isolated monomers. The enhancement is due to the existence of high local fields in localized fractal plasmon eigenmodes. Huge enhancement has been observed for the DFWM process, nonlinear absorption, refraction, and nonlinear gyrotropy in aggregated silver colloids. The enhancement of eigen and impurity nonlinearities has been found experimentally. The most promising results were obtained recently with fractal aggregate/dye/microcavity components. Low-threshold CW lasing and multiplicative enhancement of Raman scattering as high as $\sim 10^{10}$ were demonstrated [61]. Fast and slow components of nonlinear response were separated in aggregated colloid solutions. Giant local and nonlocal polarization nonlinearities of silver colloids were found. Nonlinear optical activity due to spatial dispersion was observed. The nonlinear coefficients for the optical Kerr effect, inverse Faraday effect, and self-rotation of the polarization ellipse were measured. The treated effects exhibit different dependencies on the degree of aggregation, so that the different mechanisms are supposed to be responsible for DFWM, nonlinear absorption and refraction, and polarization effects. The mechanisms should be a subject of further experimental and theoretical studies.

Thus, aggregated metal nanocomposites are effective, fast, and spectral wideband nonlinear media. But limiting processes are very essential in these media. The photoburning of resonance domains in fractal aggregates leads to a significant decrease in the nonlinear responses of samples. Investigations of nonlinear optical properties of colloidal aggregates are in progress now. We suppose the list of the observed effects and the values of nonlinear responses reported in this chapter show that metal aggregates are interesting objects for optics and for applications in photonics, nonlinear optical devices, and microanalysis.

ACKNOWLEDGMENTS

The authors are grateful to S.G. Rautian for very useful discussions of the results and E. Khaliullin for help in this work. This work was supported by RFBR, Grants No. 96-02-19331, No. 99-02-16670, No. 96-15-96642.

REFERENCES

1. R. K. Chang and T. E. Furtak, Eds., "Surface-enhanced Raman scattering," Plenum, New York, 1982.

2. V. M. Shalaev and M. I. Stockman, "Optical properties of fractal clusters (susceptibility, giant combination scattering by impurities)," *Sov. Phys. JETP* ***65***, 287 (1987); [*Zh. Eksp. Teor. Fiz.* ***92***, 509 (1987)].

3. A. V. Butenko, V. M. Shalaev, and M. I. Stockman, "Giant impurity nonlinearities in optics of fractal clusters," *Sov. Phys. JETP* ***67***, 60 (1988); [*Zh. Eksp. Teor. Fiz.* ***94***, 107 (1988)].

4. S. G. Rautian, V. P. Safonov, P. A. Chubakov, V. M. Shalaev, and M. I. Stockman, "Surface-enhanced parametric scattering of light by silver clusters," *JETP Lett.* ***47***, 243 (1988); [*Pis'ma Zh. Eksp. Teor. Fiz.* ***47***, 200 (1988)].

5. V. M. Shalaev, "Electromagnetic properties of small-particle composites," *Phys. Rep.* ***272***, 61 (1996).

6. V. M. Shalaev and M. Moskovits, Eds., "Nanostructured materials," ACS Symp. Ser. 679, American Chemical Society, Washington, DC, 1997.

7. V. A. Markel, L. A. Muratov, M. A. Stockman, and T. F. George, "Theory and numerical simulation of optical properties of fractal clusters," *Phys. Rev. B* ***43***, 8183 (1991).

8. V. A. Markel, V. M. Shalaev, E. B. Stechel, W. Kim, and R. L. Armstrong, "Small-particle composites. I. Linear optical properties," *Phys. Rev. B* ***53***, 2425 (1996).

9. M. I. Stockman, "Inhomogeneous eigenmode localization, chaos, and correlations in large disordered clusters," *Phys. Rev. E* ***56***, 6494 (1997).

10. U. Kreibig and M. Vollmer, "Optical properties of metal clusters," Springer-Verlag, Berlin-Heidelberg-New York, 1995.

11. D. Richard, P. Roussignol, and C. Flytzanis, "Surface-mediated enhancement of optical phase conjugation in metal colloids," *Opt. Lett.* ***10***, 511 (1985).

12. D. Weitz and M. Oliveria, "Fractal structures formed by kinetic aggregation of aqueous gold colloids," *Phys. Rev. Lett.* ***52***, 1433 (1984).

13. R. Jullien and R. Botet, "Aggregation and fractal aggregates," World Scientific, Singapore, 1987.

14. V. M. Shalaev, R. Botet, and A. V. Butenko, "Localization of collective diploe excitations on fractals," *Phys. Rev B* ***48***, 6662 (1993).

15. M. I. Stockman, L. N. Pandey, L. S. Muratov, and T. F. George, "Optical absorption and localization of eigenmodes in disordered clusters," *Phys. Rev. B* ***51***, 185 (1995).

16. V. M. Shalaev, R. Botet, D. P. Tsai, J. Kovacs, and M. Moskovits, "Fractals—localization of dipole excitations and giant optical polarizabilities," *Physica A* ***207***, 197 (1994).

17. V. P. Safonov, V. M. Shalaev, V. A. Markel, Yu. E. Danilova, N. N. Lepeshkin, W. Kim, S. G. Rautian, and R. L. Armstrong, "Spectral dependence of selective photomodification in fractal aggregates of colloidal particles," *Phys. Rev. Lett.* ***80***, 1102 (1998).

18. S. M. Nie and S. R. Emory, "Probing single molecules and single nanoparticles by surface-enhanced Raman scattering," *Science* ***275***, 1102 (1997).

19. W. D. Bragg, V. P. Safonov, V. M. Shalaev, W. Kim, Z. C. Ying, K. Banerijee, R. L. Armstrong, and V. A. Markel, "Near-field observation of selective photomodification in fractal aggregates," Conference on Lasers and Electro-Optics, Baltimore, May 23–28, 1999. Technical Digest, CWO7, Optical Society of America, Washington, DC, 1999.

20. I. A. Akimov, A. V. Baranov, V. M. Dubkov, V. I. Petrov, and E. A. Sulabe, "The influence of the shape and aggregation of silver particles on amplification of Raman scattering and second harmonics spectra," *Op. Spektrosk.* ***63***, 1276 (1987).
21. Yu. E. Danilova, V. P. Drachev, S. V. Perminov, and V. P. Safonov, "The nonlinearity of refractive index and absorption coefficient of fractal clusters in colloidal solutions," *Bull. Russ. Acad. Sci., Phys.* ***60***, 342 (1996); [*Izv. RAN Ser. Fiz.* ***60***, 18 (1996)].
22. Yu. E. Danilova, N. N. Lepeshkin, S. G. Rautian, and V. P. Safonov, "Excitation localization and nonlinear optical processes in colloidal silver aggregates," *Physica A*, ***241***, 231 (1997).
23. F. M. Zhuravlev, N. A. Orlova, V. V. Shelkovnikov, A. I. Plekhanov, S. G. Rautian, and V. P. Safonov, "Giant nonlinear susceptibility of thin films with (molecular J-aggregate)–(metal cluster) complexes," *Sov. JETP Lett.* ***56***, 260 (1992); [*Pis'ma Zh. Eksp. Teor. Fiz.* ***56***, 264 (1992)].
24. M. P. Andrews, M. G. Kuzyk, and F. Ghbremichael, "Local field enhancement of the cubic optical nonlinearity in fractal silver nanosphere/poly(methylmethacrylate) composites," *Nonlinear Opt.* ***6***, 103 (1993).
25. V. M. Shalaev, E. Y. Poliakov, and V. A. Markel, "Small-particle composites. II. Nonlinear optical properties," *Phys. Rev. B* **53**, 2437 (1996).
26. A. V. Karpov, A. K. Popov, S. G. Rautian, V. P. Safonov, V. V. Slabko, V. M. Shalaev, and M. I. Stockman, "Revealing of wavelength- and polarization-selective photomodification of silver clusters," *Sov. JETP Lett.* ***48***, 571 (1988); [*Pis'ma Zh. Eksp. Teor. Fiz.* ***48***, 528 (1988)].
27. V. P. Drachev, S. V. Perminov, S. G. Rautian and V. P. Safonov, "Giant nonlinear optical activity in an aggregated silver nanocomposite," *JETP Lett.* ***68***, 651 (1998); [*Pis'ma ZhETF* ***68***, 618 (1998)].
28. V. P. Drachev, S. V. Perminov, S. G. Rautian, and V. P. Safonov, "Enhanced polarization nonlinearity of silver fractal clusters," Quantum electronics and Laser Science Conf., Baltimore, May 23–28, 1999. Technical Digest QME6, Optical Society of America, Washington, DC, 1999, p. 26.
29. B. M. Smirnov, "Fractal clusters," *Sov. Adv. Phys.* ***149***, 177 (1986).
30. J. A. Creighton, C. G. Blatchford, and M. G. Albrecht, "Plasma-resonance enhancement of Raman scattering by pyridine adsorbed on silver or gold sol particles of size comparable to the excitation wavelength," *J. Chem. Soc. Faraday Trans. 2* ***75***, 790 (1979).
31. H. Hirai, "Formation and catalytic functionality of synthetic polymer-noble metal colloid," *J. Macrom. Sci-Chem.* ***A13***, 633 (1979).
32. G. Frens, and Th. G. Overbeek, "Carey Lea's colloidal silver," *Kolloid Z.Z. Polym.* ***233***, 922 (1969).
33. S. M. Heard, F. Griezer, C. G. Barraclough, and J. V. Sanders, "The characterization of Ag sols by electron microscopy, optical absorption and electrophoresis," *J. Coll. Interf. Sci.* ***93***, 545 (1983).
34. Yu. E Danilova, V. A. Markel, and V. P. Safonov, "The absorption of light by the random silver clusters," *Opt. Atm. Okeana* ***6***, 1436 (1993).
35. Yu. E. Danilova and V. P. Safonov, "Absorption spectra and photomodification of silver fractal clusters," in M. M. Novak, Ed., *Fractal Reviews in the Natural and Applied Sciences*, Chapman and Hall, London, 1995, p. 101.

36. H. Zhu and R. S. Averback, "Sintering process of two nanoparticles: a study by molecular dynamics simulations," *Philos. Mag. Lett.* ***73***, 27 (1996).

37. A. V. Butenko, Yu. E. Danilova, S. V. Karpov et al., "Nonlinear optics of metal fractal clusters," *Z. Phys. D* ***17***, 283 (1990).

38. B. Ya Zeldovich, N. F. Pilipetsky, and V. V. Shkunov, *Optical Phase-Conjugation*, Moscow, Nauka, 1985.

39. R. W. Boyd, *Nonlinear Optics*, Academic, San Diego, CA, 1992.

40. F. Hache, D. Ricard, and C. Flytzanis, "Optical nonlineariries of small metal particles: surface-mediated resonance and quantum size effects," *J. Opt. Soc. Am. B* ***3***, 1647 (1986).

41. F. Hache, D. Ricard, C. Flytzanis, and U. Kreifig, "The optical Kerr effect in small metal particles and metal colloids—the case of gold," *Appl. Phys. A* ***47***, 347 (1988).

42. K. Uchida, S. Kaneko, S. Omi, C. Hate, H. Tanji, Y. Asahara, A. J. Ikushima, T. Tokizaki, and A. Nakamura, "Optical nonlinearities of a high-concentration of small metal particles dispersed in glass-copper and silver particles," *J. Opt. Soc. Am. B* ***11***, 1236 (1994).

43. L. Yang, K. Becker, F. M. Smith, R. H. Magruder III, R. F. Haglund, Jr., L. Yang, R. Dorsinville, R. R. Alfano, and R. A. Zuhr, "Size dependence of the 3rd-order susceptibility of copper nanoclusters investigated by 4-wave-mixing," *J. Opt. Soc. Am. B* ***11***, 457 (1994).

44. S. G. Rautian, "Nonlinear saturation spectroscopy of the degenerate electron gas in spherical metallic particles," *Sov. JETP* ***85***, 451 (1997).

45. V. P. Safonov, J. G. Zhu, N. N. Lepeshkin, R. L. Armstrong, V. M. Shalaev, C. W. White, R. A. Zuhr, and Yu. E. Danilova, "Spectral dependence of cubic nonlinearity in gold nanoparticles and their fractal aggregates," Quant. Electronics and Laser Science Conference, Baltimore, May 23–28, 1999. Technical Digest QME5, Optical Society of America, Washington, DC, 1999.

46. R. H. Magruder, III, Li Yang, R. F. Haglung Jr., C. W. White, L. Yang, R. Dorsinville, and R. R. Altano, "Optical-properties of gold nanocluster composites formed by deep ion-implantation in silica," *Appl. Phys. Lett.* ***62***, 1730 (1993).

47. M. J. Bloemer, J. W. Haus, and P. R. Ashley, "Degenerate four-wave mixing in colloidal gold as a function of particle size," *J. Opt. Soc. Am. B* ***7***, 790 (1990).

48. R. H. M. Groeneveld, R. Sprik, and A. Lagendijk, "Femtosecond spectroscopy of electron-electron and electron-phonon energy relaxation in Ag and Au," *Phys. Rev. B* ***51***, 11433 (1995).

49. M. Perner, P. Post, U. Lemmer, G. von Plessen, J. Feldmann, U. Becker, M. Mennig, M. Schmitt, and H. Schmidt, "Optically induced damping of the surface plasmon resonance in gold colloids," *Phys. Rev. Lett.* ***78***, 2192 (1997).

50. T. Yajima, "Nonlinear optical spectroscopy of an inhomogeneously broadened resonant transition by means of 3-wave mixing," *Opt. Commun.* ***14***, 378 (1975).

51. E. J. Heilweil and R. M. Hochstrasser, "Nonlinear spectroscopy and picosecond transient grating study of colloidal gold," *J. Chem. Phys.* ***82***, 4762 (1985).

52. V. L. Vinetskii, N. V. Kukhtarev, S. G. Odulov, and M. S. Soskin, "Dynamic self-diffraction of coherent-light beams," *Sov. Phys. Usp.* ***22***, 742 (1979); [*Usp. Fiz. Nauk* ***129***, 113 (1979)].

53. M. Sheik-Bahae, A. A. Said, T. H. Wei, D. J. Hagan, E. W. Van Stryland, "Sensitive measurement of optical nonlinearities using a single beam," *IEEE J. Quantum Electron.* ***26***, 760 (1990).

54. R. L. Armstrong, V. P. Safonov, N. N. Lepeshkin, W. Kim, and V. M. Shalaev, "Giant optical nonlinearities of fractal colloid aggregates," in *Nonlinear Optical Liquids and Power Limiters*, *SPIE Proc.* ***3146***, 107 (1997).
55. C. Douketis, V. M. Shalaev, T. L. Haslet, Z. Wang, and M. Moskovits, "The role of localized surface plasmons in photoemission from silver films: direct and indirect transition channels," *J. Electron Spectrosc. Related Phenomena* ***64/65***, 167 (1993).
56. M. Perner, G. von Plessen, and J. Feldmann, "Reply on comment on "Optically induced damping of the surface plasmon resonance in gold colloids," *Phys. Rev. Lett.* ***82***, 3188 (1999).
57. V. P. Drachev, S. V. Ertsenkin, S. V. Perminov, V. P. Safonov, and P. A. Chubakov, "Shearing interferometer based on second-harmonic generation and a novel technique of n_2 direct measurement," *Appl. Opt.* ***36***, 8622 (1997).
58. V. P. Drachev, S. V. Perminov, and V. P. Safonov, "New methods of n_2 measurement based on dispersion interferometry with applications to KTP and silver colloidal solutions," 11th Int. Vavilov Conference on Nonlinear Optics, S. G. Rautian, Ed., *SPIE Proc.* ***3485***, 35 (1998).
59. S. A. Akhmanov, G. A. Lyakhov, V. A. Makarov, and V. I. Zharikov, "Theory of nonlinear optical activity in isotropic media and liquid crystals," *Opt. Acta* ***29***, 1359 (1982).
60. P. Maker, R. Terhune, and C. Savage, "Intensity-dependent changes in the refraction index of liquids," *Phys. Rev. Lett.* ***12***, 507 (1964).
61. W. Kim, V. P. Safonov, V. M. Shalaev, and R. L. Armstrong, "Fractals in microcavities: giant coupled, multiplicative enhancement of optical responses," *Phys. Rev. Lett.* ***82***, 4811 (1999).

CHAPTER EIGHT

Local Fields' Localization and Chaos and Nonlinear-Optical Enhancement in Clusters and Composites

MARK I. STOCKMAN

Department of Physics and Astronomy, Georgia State University, Atlanta, GA 30303

This chapter is devoted to linear and nonlinear optical properties of disordered clusters and nanocomposites. Linear and nonlinear optical polarizabilities of large disordered clusters, fractal clusters in particular, and susceptibilities of nanocomposites are found analytically and calculated numerically. A spectral theory with dipole interaction is used to obtain quantitative numerical results. Major properties of systems under consideration are giant fluctuations, inhomogeneous localization, and chaos of local fields that cause strong enhancement (by many orders of magnitude) of nonlinear optical responses. The enhancement and fluctuations properties of the local fields are intimately interrelated to the inhomogeneous localization of the systems' eigenmodes (surface plasmons). Due to these fluctuations, mean-field theory completely fails to describe nonlinear optical responses.

8.1. INTRODUCTION

Clusters and nanocomposites belong to so-called nanostructured materials. Typically, such materials are nanoparticles either bound to each other by covalent or van der Waals bonds, or dispersed in a host medium. Description of electromagnetic (EM) properties of such a system is a long-standing problem going back to such names as Maxwell Garnett [1], Lorentz [2], and Bruggeman [3].

Properties of such materials may be dramatically different from those of bulk materials with identical chemical composition. A characteristic property of such systems is confinement of electrons, phonons, electric fields, and so on, in small spatial regions. Such a confinement, in particular, modifies spectral properties (shifts quantum levels and changes transition probabilities), and also changes the interaction

Optics of Nanostructured Materials, Edited by Vadim A. Markel and Thomas F. George
ISBN 0-471-34968-2

between the constituent particles. As we will discuss in this chapter, local (near-zone) EM fields are strongly fluctuating in space. Their magnitude is greatly (by orders of magnitude) enhanced with respect to external (exciting) fields.

A phenomenon closely related to the enhancement and fluctuations of local fields is localization of elementary excitations (eigenmodes) in the composites [4–8]. The relevant excitations are polar waves that are traditionally called plasmons (this term originates from theory of metallic nanoparticles containing electron plasma, but is now often used in application to other nanocomposites). Plasmon–resonant properties leading to enhancement of local fields are especially pronounced in some metallic (especially silver, gold, or platinum) colloidal clusters, metal nanocomposites, and rough surfaces. A typical example of such responses is surface-enhanced Raman scattering (see, e.g., a review of [9] and reference cited therein).

The most pronounced effect of the fluctuating local fields is on nonlinear optical susceptibilities. The reason for that can be understood qualitatively. Imagine two fields with the same average intensity $I_1 \propto \langle \mathbf{E}^2 \rangle$. For the sake of argument, let us say, the first field has the same constant intensity I_1 in $N \gg 1$ points, and the second is strongly localized at one point where its intensity then should be $I_2 = \langle NI_1 \rangle$. Consider an nth-order nonlinearity where the nonlinear response is proportional to $\langle \mathbf{E}^{2n} \rangle \propto \langle I^n \rangle$. The magnitude of the nonlinear response for the first (constant) field is proportional to $\frac{1}{N}(NI_1^n) = I_1^n$. In contrast, the response to the second (strongly localized) field is $\frac{1}{N}(NI_1)^n = N^{n-1}I_1^n$. In such a way, the enhancement coefficient (the ratio of the nonlinear response in the second case to that in the first case) is N^{n-1}. Hence, the localization has a potential to bring about strongly enhanced nonlinear responses where the enhancement increases with the order of nonlinearity and the degree of localization (spatial fluctuations).

To maximize this effect, our goal is to find systems with the maximum spatial fluctuations of the local densities. We certainly expect that the density fluctuations will cause correspondingly large fluctuations of the local fields.

There exists a class of systems that stands out in this respect. These are self-similar (fractal) systems, which (on average) repeat themselves at different scales. In other words, looking at such a system and not seeing its boundaries (neither at the maximum or a minimum scales), one cannot say what fraction of the system is observed, and what is the actual size of the objects seen. For such systems, the number N of constituent particles (monomers) contained within a radius R scales as

$$N \cong \left(\frac{R}{R_0}\right)^D \tag{8.1}$$

where R_0 is a typical distance between monomers, and D is the fractal (Hausdorff) dimension of the cluster. The density of monomers as given by Eq. (8.1) is asymptotically zero for large clusters

$$\rho \cong R_0^{-3}\left(\frac{R}{R_0}\right)^{D-3} \to 0 \tag{8.2}$$

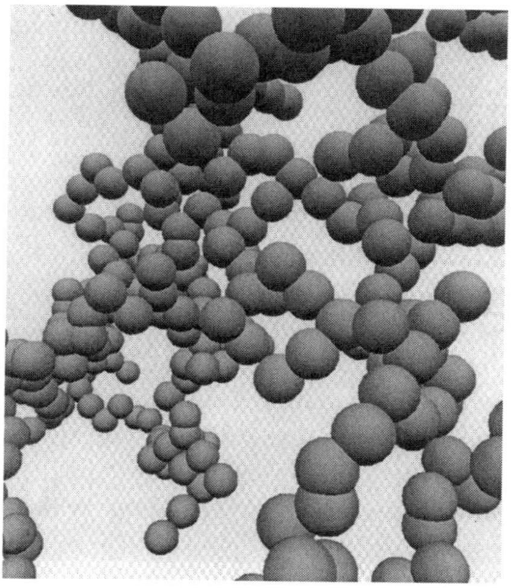

Figure 8.1. Cluster–cluster aggregate of $N = 1000$ monomers.

However, this does not mean that the interaction between monomers can be neglected. The underlying reason is that there is a strong correlation between monomers in a cluster with the pair–pair correlation function scaling similar to Eq. (8.2). Thus we have a unique system whose macroscopic density is asymptotically zero, but the interaction inside the system is strong. This idea was proposed by us in earlier papers. [4,10].

An example of a fractal cluster, obtained by cluster–cluster aggregation [11] (CCA) is shown in Fig. 8.1. In this figure, one can trace the rarefied nature of fractal systems (represented by voids of density) and strong fluctuations of the density of constituent particles (monomers).

To avoid possible misunderstanding, we point out that other, nonfractal systems also possess significantly enhanced optical responses, especially those that are tailored to have optimally–chosen dielectric properties changing in space [12]. The physical origin of the enhancement in this case is the same as above. As an example of such systems we will consider a random Maxwell Garnett (MG) composite, where dielectric or metallic spheres are embedded in a host medium at random positions. Such a composite was considered earlier in a mean-field approximation [13]. Below, we will consider a model of such composites where the inclusion spheres are positioned on a cubic lattice in a host medium and call it a random lattice gas (RLG). We will use RLG as a model of random but not fractal composites.

There has been an increase of interest in the optical properties (both linear and, especially, nonlinear) of composites during the last decade [12–21]. This revival of interest is due to improved theoretical understanding of the origin of the optical enhancement in composites and the demonstrated possibility of engineering composites whose desired nonlinear properties are better than those of their constituents. Theoretical advances in this field are based to a significant degree on spectral methods. A general spectral method [22–25] and the dipolar spectral method [5,6] have proved to be very useful in both analytical theory and numerical computations. In particular, the spectral method has been used efficiently in the theory of electrorheological fluids, that is, liquid composites whose hydrodynamic properties depend on the applied electric field [26,27].

In Section 8.2, we present the coupled dipole equations governing optical responses and summarize the dipolar spectral theory. In that section, we also summarize some predictions of scaling. In Section 8.3, on the basis of the spectral theory, we consider linear optical polarizabilities of clusters and susceptibilities of composites. Section 8.4 is devoted to the problem of localization–delocalization of the elementary excitations (eigenmodes) of disordered clusters and composites. We summarize computations that have led us to introduce a concept of inhomogeneous localization of eigenmodes. Section 8.5 deals with a phenomenon of giant fluctuations of local electric fields in clusters and composites, which contribute to the strong enhancement of their nonlinear optical responses. Chaos of eigenmodes considered is Section 8.6 is a phenomenon similar to quantum chaos. Individual eigenmodes (surface plasmons) and even their averaged correlation functions exhibit very strong fluctuations on all scales. In Section 8.7, we consider enhancement of Raman scattering and nonlinear parametric mixing in clusters. Nonlinear polarizabilities of MG composites in spectral theory compared with a mean-field theory are discussed in Section 8.8. Concluding remarks are presented in Section 8.9.

8.2. EQUATIONS GOVERNING OPTICAL (DIPOLAR) RESPONSES, SPECTRAL REPRESENTATION, AND SCALING

We concentrate on the dipole–dipole interaction, which is a universal interaction between polarizable particles at large distances. We consider a cluster (or a composite) whose particles (called below monomers) are positioned at points $\mathbf{r}_i$. Let us assume that the system (a cluster or composite) is subjected to the electric field $\mathbf{E}$ of the incident optical wave. This field induces the dipole $d_{i\alpha}$ at an ith monomer (here $\alpha = x, y, z$ denotes the Cartesian components of the vector, and similar notations will be used for other vectors). The dipole moments satisfy a well-known system of equations

$$\alpha_0^{-1} d_{i\alpha} = E_{i\alpha}^{(0)} - \sum_{j=1}^{N} \left(\delta_{\alpha\beta} - 3 \frac{(r_{ij})_\alpha (r_{ij})_\beta}{r_{ij}^2} \right) \frac{d_{j\beta}}{r_{ij}^3} \tag{8.3}$$

Here, $E_{i\alpha}^{(0)}$ is the wave-field amplitude at the ith monomer, $\mathbf{r}_{ij} = \mathbf{r}_i - \mathbf{r}_j$ is the relative vector between the ith and jth monomers, and α_0 is the dipole polarizability of the monomer. We assume that the size of the system is much less than the wavelength of the exciting wave, and therefore the exciting field $E_\alpha^{(0)}$ is the same for all the monomers of a cluster. We note that the dipole interaction is not valid in the close vicinity of a monomer. Our choice of interaction is justified if intermediate-to-large scales predominantly contribute to the properties under consideration.

The dipolar spectral theory of the optical response of fractal clusters has been developed in [5] and [6]. We note that a similar spectral approach has been independently introduced by Fuchs and co-workers [28,29]. The material properties of the system enter Eq. (8.3) only via the combination Zr_{ij}^3 where we have introduced the notation $Z \equiv \alpha_0^{-1}$. This, along with the (approximate) self-similarity of the system, is a prerequisite for scaling in terms of the spectral variable Z. A principal requirement for the scaling of a certain physical quantity F is that the system eigenmodes contributing to F should have their localization radii L intermediate between the maximum scale (size of the cluster R_c) and the minimum scale, R_0. Then, this quantity will not depend on any external length, leading to scaling.

Because the quantity F is not sensitive to the maximum scale R_c, it should have the functional dependence $F = F(ZR_0^3)$. On the other hand, the eigenmodes contributing to F are insensitive to a much smaller minimum scale. Therefore, the dependence on R_0 can be only power (scaling), and consequently

$$F(Z) \propto \left(ZR_0^3\right)^{\gamma} \tag{8.4}$$

where γ is some scaling index.

In accord with the above arguments, it is convenient to express all results not in terms of frequency, but in terms of Z, separating the imaginary and real parts, $Z = -X - i\delta$. The choice of signs in this expression makes the dissipation parameter δ positive, while the spectral parameter X is positive when the frequency is blue shifted from the plasmon resonance, and negative otherwise. For the sake of reference, here we give the expressions of X and δ for a metallic nanosphere in the Drude model,

$$X = \frac{1}{R_m^3}\frac{(\varepsilon_0 + 2\varepsilon_h)^{3/2}}{3\varepsilon_h\omega_p}(\omega - \omega_s) \qquad \delta = \frac{1}{R_m^3}\frac{(\varepsilon_0 + 2\varepsilon_h)^{3/2}}{3\varepsilon_h\omega_p}\frac{\gamma}{2} \tag{8.5}$$

where ε_0 is the interband dielectric constant of the metal, ε_h is the dielectric constant of the ambient medium (host), ω_p is the metal plasma frequency, $\omega_s = \omega_p/\sqrt{\varepsilon_0 + 2\varepsilon_h}$ is the surface plasmon frequency, and R_m is the nanosphere's radius.

The spectral dependence of X and δ for silver is illustrated in Fig. 8.2. The most important feature in the figure is that in the yellow–red region of visible light, the real part of the polarizability greatly exceeds its imaginary part. Their ratio has the meaning of the quality factor of the surface–plasmon oscillations Q defined as

$$Q \equiv \frac{|X|}{\delta} \tag{8.6}$$

This factor shows how many times the order of magnitude of the amplitude of the local field in the vicinity of a resonant monomer exceeds that of the exciting field. This important fact is considered in detail below [see Section 8.5, the discussion of Eqs. (8.19)–(8.21)]. The fact that Q may be large (as large as $\cong 10^2$) for

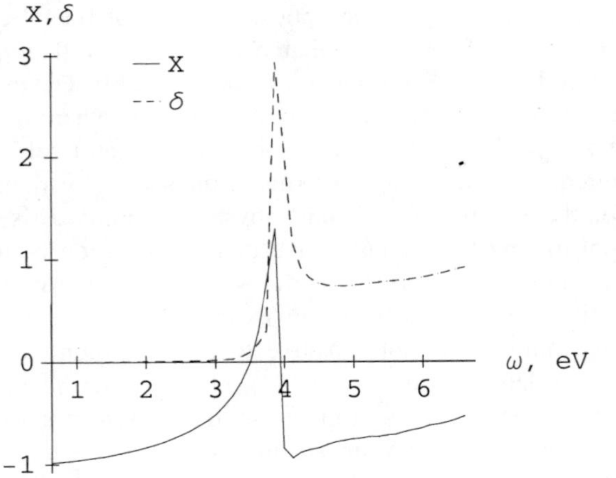

Figure 8.2. Dependence of the spectral parameter X and the dissipation parameter δ (in the units of R_0^3) on light frequency expressed as photon energy.

many metals plays an important role in this theory since the enhancement of optical responses is a resonant phenomenon, and strong dissipation would completely suppress it.

We note that in some earlier work the quality factor has been defined differently, $Q' \equiv R_0^{-3}/\delta$. This definition is convenient because in the Drude model [see Eq. (8.5)] this factor, $Q' = (R_m^3)/R_0^3(6\varepsilon_h\omega_p)/[\gamma(\varepsilon_0 + 2\varepsilon_h)^{3/2}]$, is a constant. In reality, Q' is not a constant, because the dissipation parameter δ depends on frequency (cf. Fig. 8.2 and its discussion). In this chapter we use the definition of Eq. (8.6) because it directly determines the enhancement factor of the local fields M_n [cf. Eq. (8.21) and its discussion], which is the most important characteristic for the present theory.

To introduce the spectral representation, [5,6], we will rewrite Eq. (8.3) as one equation in the $3N$-dimensional space. To do so, we introduce $3N$-dimensional vectors $|d)$, $|E^{(0)})$, ..., whose projections give the physical vectors

$$(i\alpha|d) = d_{i\alpha}, \qquad (i\alpha|E^{(0)}) = E_{i\alpha}^{(0)}, \ldots \tag{8.7}$$

Equation (8.3) then acquires the form

$$(Z + W)|d) = |E^{(0)}) \tag{8.8}$$

where W is the dipole–dipole interaction operator with the matrix elements

$$(i\alpha|W|j\beta) = \begin{cases} \left(\delta_{\alpha\beta} - 3\dfrac{(r_{ij})_\alpha (r_{ij})_\beta}{r_{ij}^2}\right)\dfrac{1}{r_{ij}^3} & i \neq j \\ 0 & i = j \end{cases} \tag{8.9}$$

We introduce the eigenmodes (plasmons) $|n)$ $(n = 1, \ldots, 3N)$ as the eigenvectors of the W operator,

$$W|n) = w_n|n) \tag{8.10}$$

where w_n are the corresponding eigenvalues. Practically, the eigenvalue problem (8.10) can be solved numerically for any given cluster. Having done so, one can calculate Green's function

$$\mathbf{G}_{i\alpha,j\beta} = \sum_{n=1}^{3N} \frac{(i\alpha|n)(j\beta|n)}{Z + w_n} \tag{8.11}$$

which carries the maximum information on the spectrum and linear response of the system. We note that due to the time-reversal symmetry of the system (absence of a magnetic field), all eigenvectors can be chosen and will be assumed real. Therefore, all amplitudes are symmetric, in particular, $(i\alpha|n) = (n|i\alpha)$.

8.3. LINEAR OPTICAL RESPONSES

The polarizability of a cluster or finite volume of a composite α and its density of eigenmodes ρ are expressed in terms of $\mathbf{G}$ as

$$\alpha_{\alpha\beta} = \sum_i \alpha^{(i)}_{\hat{a}\hat{a}} \qquad \alpha^{(i)}_{\hat{a}\hat{a}} = \sum_j \mathbf{G}_{i\alpha,j\beta} \qquad \rho = \sum_{i,j} \mathbf{G}_{i\alpha,i\alpha} \tag{8.12}$$

where $\alpha^{(i)}_{\hat{a}\hat{a}}$ is a polarizability of an ith monomer in the cluster (or a composite), and summation over repeated vector indexes is implied. The dielectric constant of a cluster (composite) is given by

$$\varepsilon_c = \varepsilon_h \left(1 + 4\pi\alpha \frac{N}{V}\right) \tag{8.13}$$

where ε_h is the dielectric constant of a host, V is the volume occupied by the cluster (composite), and $\alpha = \langle (1/3N) \sum_{i,j} \mathbf{G}_{i\beta,j\beta} \rangle$ is the polarizability of a monomer in the cluster (composite). Next, we will consider results of numerical computations using Eqs. (8.12) and (8.13) and will compare them to some analytical predictions.

First, let us consider scaling predictions. For this purpose, we have to invoke a large magnitude of the quality factor of the optical resonance (8.6), $Q \gg 1$. In this case, a dependence of type (8.4) becomes $F(X) \propto \left(R_0^3|X|\right)^{\gamma}$. We have introduced [5,6] such a dependence for $\alpha_{\alpha\beta}(X)$ and $\rho(X)$ and argued that the two quantities have the same scaling,

$$\mathrm{Im}\,\alpha(X) \cong \rho(X) \cong R_0^3 \left|R_0^3 X\right|^{d_o - 1} \tag{8.14}$$

where d_o is an index that we called the optical spectral dimension. We have also argued that the physical range of d_o is $1 \geq d_o \geq 0$.

The strong localization has been essential for the derivation of Eq. (8.14). It implies that all eigenmodes (at least all contributing eigenmodes) of a cluster are strongly localized The strong localization, as discussed by Alexander [30], means that for any given frequency parameter X there exists only one characteristic length L_X of these eigenmodes playing the role of simultaneously their wavelength and their localization length. By the use of scale invariance arguments, we have shown [5,6] that L_X should scale as

$$L_X \cong R_0 \left| R_0^3 X \right|^{d_o - 1/(3-D)} \tag{8.15}$$

We have subjected the scaling predictions of Eqs. (8.14) and (8.15) to an extensive comparison with the results of large-scale computations [7]. Similar results have been obtained [7] for other types of clusters. These results are quite unexpected. One of those, a polarizability and eigenmode density for CCA, is shown in Fig. 8.3. The conclusion that one can draw from this figure is that neither the polarizability nor density of eigenmodes scale. Interestingly enough, they still appear to be quite close to each other, supporting the conclusion of [5] and [6] that all eigenmodes of a fractal cluster contribute (almost) equally to its optical absorption. This conclusion can be understood physically from an idea that a fractal is disordered and does not possess any geometric (point-group) symmetry on all intermediate scales (between R_0 and R_c). Consequently, such strong disorder does not impose any selection rules that would otherwise govern the contribution of a specific eigenmode to optical absorption.

Another relation to check is that of Eq. (8.15). First, one has to formulate how to calculate the localization radius. We use the definition of [7],

$$L_X = \frac{\sum_n \rho_n L_n}{\sum \rho_n} \qquad \text{where} \qquad L_n = \sum_{i\beta} \mathbf{r}_i^2 (n|i\alpha)^2 - \left(\sum_{i\beta} \mathbf{r}_i (n|i\alpha)^2 \right)^2 \tag{8.16}$$

and where $\rho_n = [(X - w_n)^2 + \delta^2]^{-1}$, L_n is the localization radius of a given eigenmode, and L_X is the localization radius at a given frequency. The computed dependence of L_X is shown in Fig. 8.4. As one can clearly see, there is no scaling in these data either. This finding is in contradiction with the conclusion of [31] (precision of our calculations is much higher than that of [31]).

Evident failure of the scaling implies that at least one of the assumptions lying at its foundation is incorrect. Because we used the same model (dipole–dipole) for both the scaling theory and the numerical computations, the nonapplicability of the model to system is out of the question. In our consideration, we consistently used high values of Q, so that the condition $Q \gg 1$ is also satisfied. The only cause for the failure of scaling appears to be the strong localization assumption.

Now let us discuss the linear polarizability of a MG composite, as calculated in [32]. In Fig. 8.5, we show the results of a computation of the linear dielectric

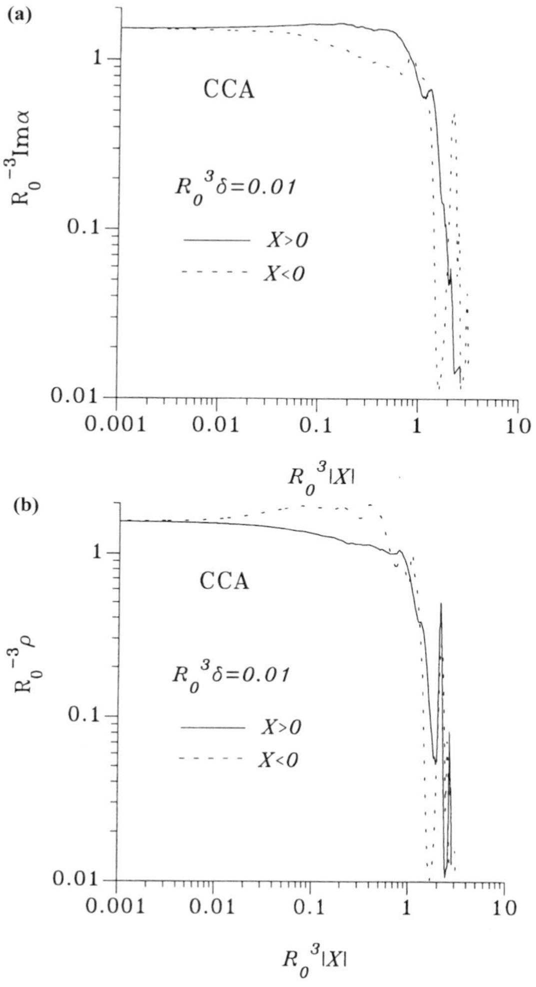

Figure 8.3. Numerically obtained absorption (Im α) (a) and density of eigenmodes (b) for CCAs.

constant for a composite consisting of silver nanospheres in a dielectric host with a dielectric constant of $\varepsilon_h = 2.0$. This figure displays both the results of the spectral theory computations accordingly to Eq. (8.13) and those of a mean-field approximation known as the MG formula (or an equivalent Lorentz–Lorenz formula), (see, e.g., [13]). As we see, the mean-field theory gives a satisfactory description in the wings of the spectral contour, but fails in the region of the resonant absorption of the inclusions, where it considerably overestimates ε_h.

Undoubtedly, there should be reasons for the failure of both the scaling and mean-field theory. As we now understand, two interrelated phenomena can be blamed for

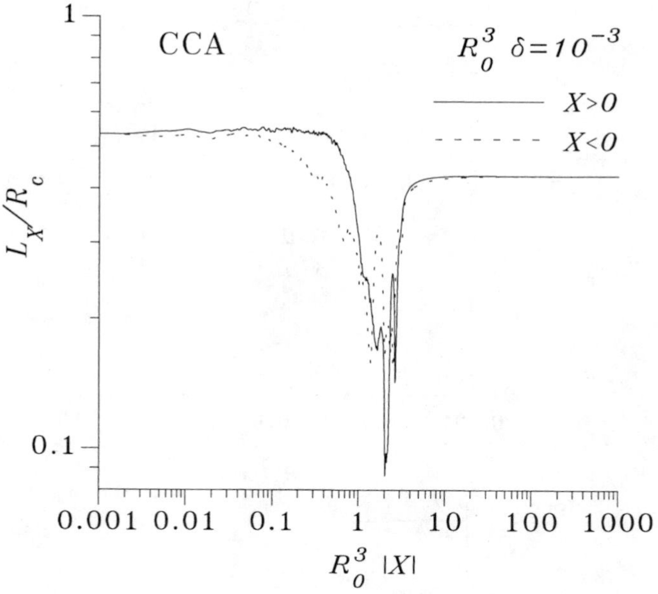

Figure 8.4. Averaged localization (coherence) radius L_X as a function of the spectral parameter X.

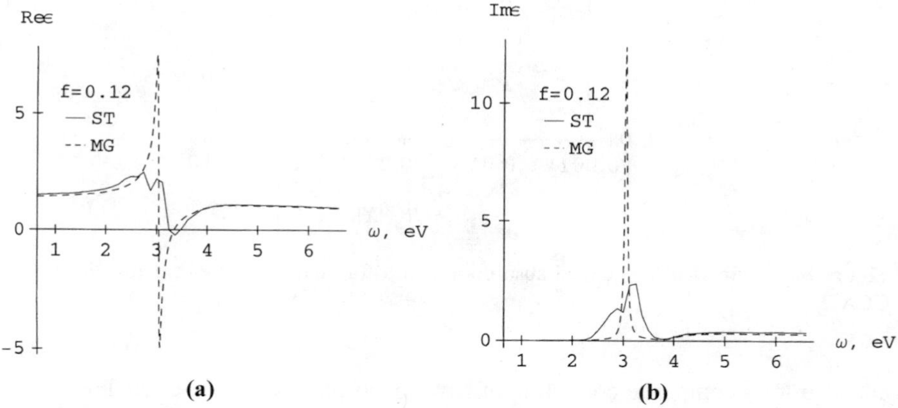

Figure 8.5. Comparison of the dielectric susceptibility of a composite in present spectral theory (ST) (a) with the result of the MG formula (b). The relative dielectric constant of a composite ε is shown as a function of the photon frequency in energy units (eV).

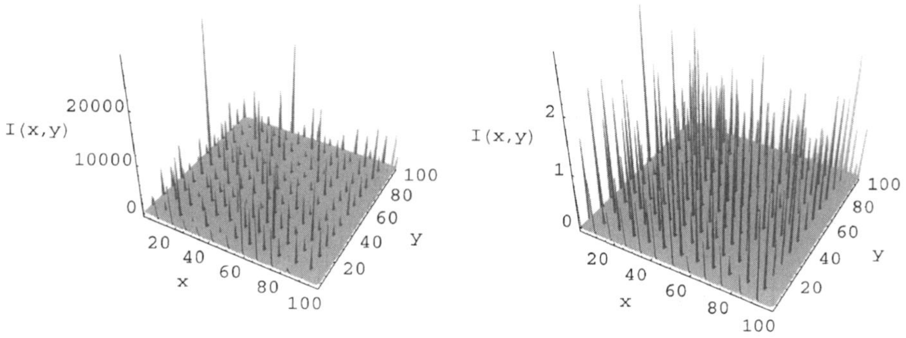

(a) **(b)**

Figure 8.6. Spatial distribution of the intensity of induced dipoles at different inclusions for a MG composite. This distribution is plotted in the following way: A composite is generated and local dipole polarizabilities $\alpha_{\beta\gamma}^{(i)}$ are found for each inclusion from Eq. (8.12). Then, the inclusions are projected onto the *xy* plane. If there are several inclusions projected onto the same site, one of them is randomly left. The square of the local polarizability $|\alpha_{\beta z}^{(i)}|^2$ for each of the inclusions left is plotted as the vertical coordinate. (a) Shows the distribution for the resonant region (in terms of the spectral parameter, $R_0^3 X = -0.01$, while (b) is for the off-resonant region (spectral wing, $R_0^3 X = -1.0$).

this failure. These are inhomogeneous localization of eigenmodes and giant fluctuations of the local fields in space, see Sections 8.4 and 8.5. In the case of the inhomogeneous localization, there are eigenmodes of all localization radii from the minimum distance between monomers (inclusions) R_0 to the total size of the system R_c. Both of these two extremes render the scaling theory inapplicable. It is important to emphasize that the eigenmodes with such vastly different localization radii coexist at the *same frequency* (Section 8.4). Obviously, the strong fluctuations contradict the basic assumption of the mean-field theory.

While we consider the inhomogeneous localization and giant fluctuations for clusters in Sections 8.4 and 8.5, here, in Fig. 8.6, we demonstrate fluctuations of the local fields for the MG composite. As one can clearly see, the spatial fluctuations (i.e., a change from an inclusion to an inclusion particle) of the local fields are significant in the resonant region [see Fig. 8.6(b)], where the intensity of local fields changes in space by orders of magnitude. These fluctuations cause the mean-field theory to fail. As Fig. 8.6(a) shows, in the spectral wings (in the off-resonant region), these fluctuations are much smaller. Consequently, the mean-field approximation describes the polarizability reasonably well.

8.4. INHOMOGENEOUS LOCALIZATION OF EIGENMODES

To introduce the inhomogeneous localization [8,33], (see also [34] and [35]), we consider all eigenmodes of a single cluster. In Fig. 8.7, we show a special plot where each eigenmode is represented by a point in the coordinates: its localization length

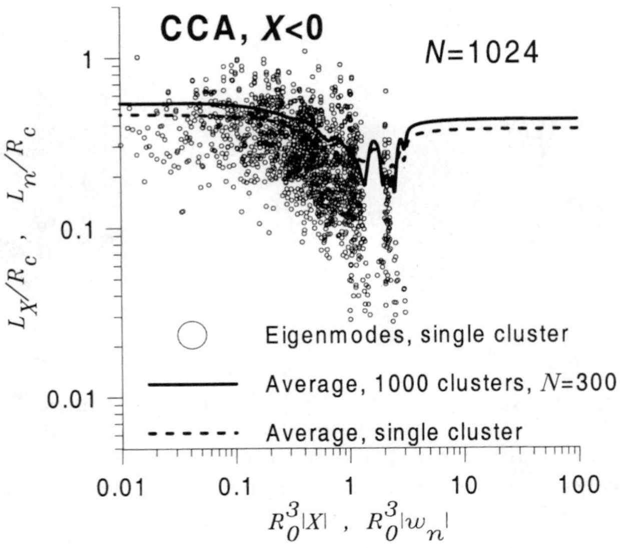

Figure 8.7. All eigenmodes with negative eigenvalues for a CCA cluster of $N = 1024$ monomers.

L_n versus the spectral parameter X. As one can see, at any frequency (value of X) within a wide range of X, the eigenmodes have a very broad spectrum of their localization radii L_X, from the minimum scale $\cong R_0$ of the distance between the monomers to the maximum scale $\cong R_c$ of the cluster total radius. As already mentioned in Section 8.3, either of these extremes ($L_n \cong R_c$ and $L_n \cong R_0$) violates a necessary scaling condition $R_0 \gg L_n \gg R_c$, causing failure of scaling for at least some quantities, such as linear responses (see Section 8.3). However some other quantities, as we will see below in Sections 8.5–8.7, still do scale, because they are insensitive to those extremes).

It is also useful to take a closer look at some members of the ensemble of eigenmodes of a cluster. We take two pairs of eigenmodes that have almost equal frequencies (within a pair). The spatial distribution of the intensity of these eigenmodes is presented in Fig. 8.8. As we can see, at comparatively large eigenvalues ($R_0^3 w_n = 1.029$), the eigenmode is indeed very well localized, demonstrating a very sharp peak over just a few monomers. The localization radius for this eigenmode is indeed very small, $L_n/R_c = 0.034$. Counterintuitively, an eigenmode with a very close frequency ($R_0^3 w_n = 1.033$) is almost completely delocalized (for this eigenmode, $L_n/R_c = 0.75$). However, the examination of its spatial profile shows that in reality this eigenmode consists of two sharp peaks separated by a distance on the order of the total size of the cluster. As we go closer to the plasmon resonance (smaller $|w_n|$), the minimum width of the peaks increases, as one would expect in view of Eq. (8.15). At the same time, the internal structure of the eigenmode remains highly irregular with large spatial fluctuations. Again, the eigenmodes at the same frequency possess very different localization radii.

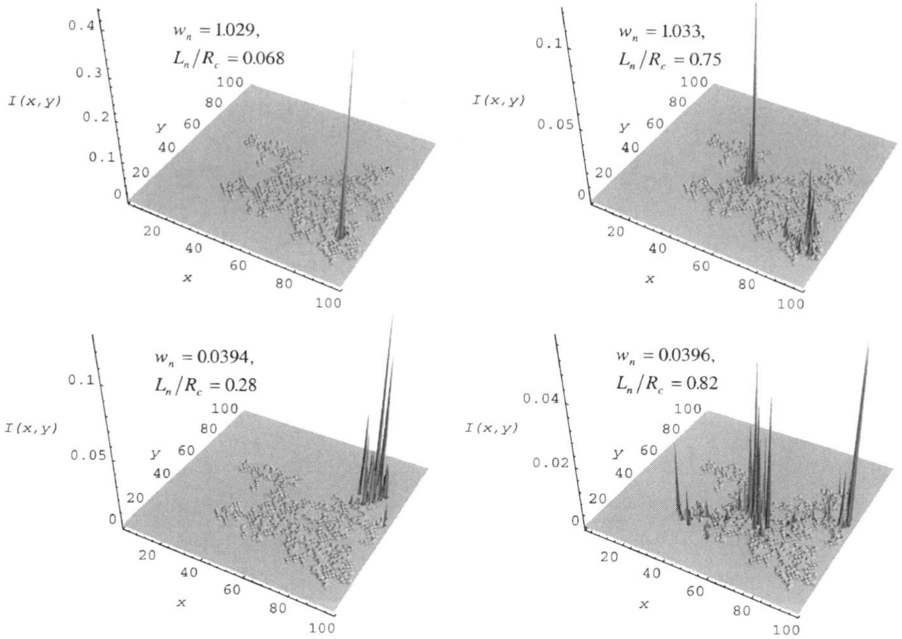

Figure 8.8. Spatial distribution of the local field intensities for an individual CCA cluster ($N = 1500$) shown over the two-dimensional (2D) projection of the cluster for the eigenvalues w_n (in the units of R_0^{-3}) as indicated. The coordinates are shown in units of the lattice spacing R_0. The value of the gyration radius of the individual eigenmodes is given relative to the cluster radius R_c.

The typical minimum size of an eigenmode (its "core") is of interest by itself. It is understandable that a small core implies large spatial fluctuations of the eigenmode intensities. These bring about strong enhancement of nonlinear photoprocesses, as we have already noted in Section 8.1 and discuss in Sections 8.5 and 8.7. Below in this section, in conjunction with Fig. 8.9, we argue that the characteristic size of this core l_X scales as a function of the spectral variable X.

The behavior described above is very unusual. For the known localization patterns, eigenmodes are strongly localized for short wavelength (high frequencies) and delocalized in the long-wavelength wing. The physical reason for this is that a long wave (wavelength $\lambda \gg$ the typical size of the scattering inhomogeneities) sees almost homogeneous medium and propagates almost freely. In contrast, a short wave is strongly scattered from inhomogeneities with sizes on the order of λ (strongly here means that the scattering length itself is on the order of λ). Thus, there exists the mobility edge, that is, a frequency above which waves are localized and below which they propagate.

This logic, which leads to the existence of the mobility edge, is obviously inapplicable to fractals. They are self-similar systems and, therefore, do not possess

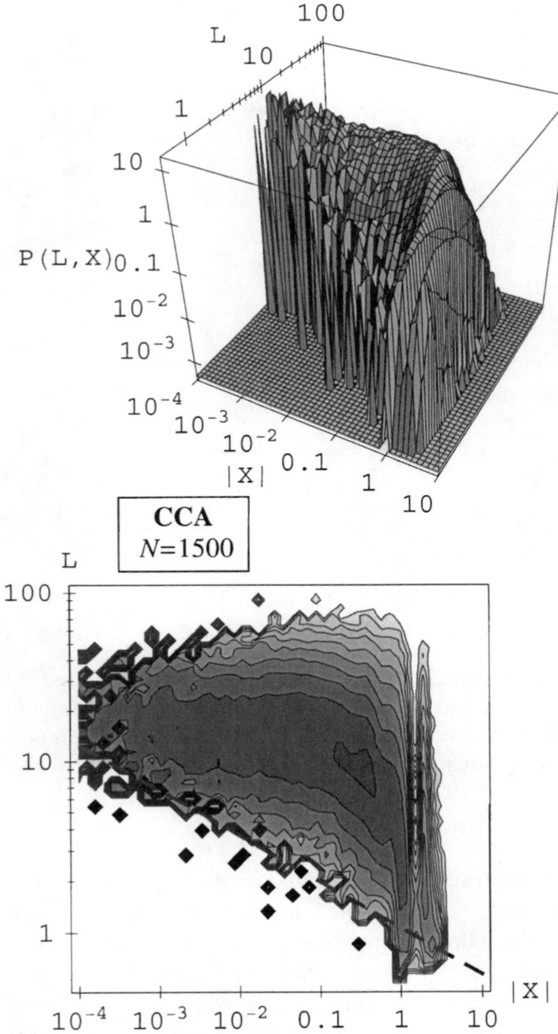

Figure 8.9. Localization-length distribution $P(L, X)$ of eigenmodes for CCA clusters ($N = 1500$). The position of the lower X cutoff is qualitatively illustrated by the dashed bold line.

any characteristic scattering length. For any eigenmode wavelength λ, there always exist inhomogeneities of the sizes comparable to λ. This may suggest that all of the eigenmodes are strongly localized, as assumed in [5] and [6]. However, the result presented above and all numerical modeling of [7,8], and [33–35] have shown that the strong localization is not the case. In actuality, inhomogeneous localization [8,33] takes place, where the eigenmodes in a wide range of the localization radii coexist at any given frequency.

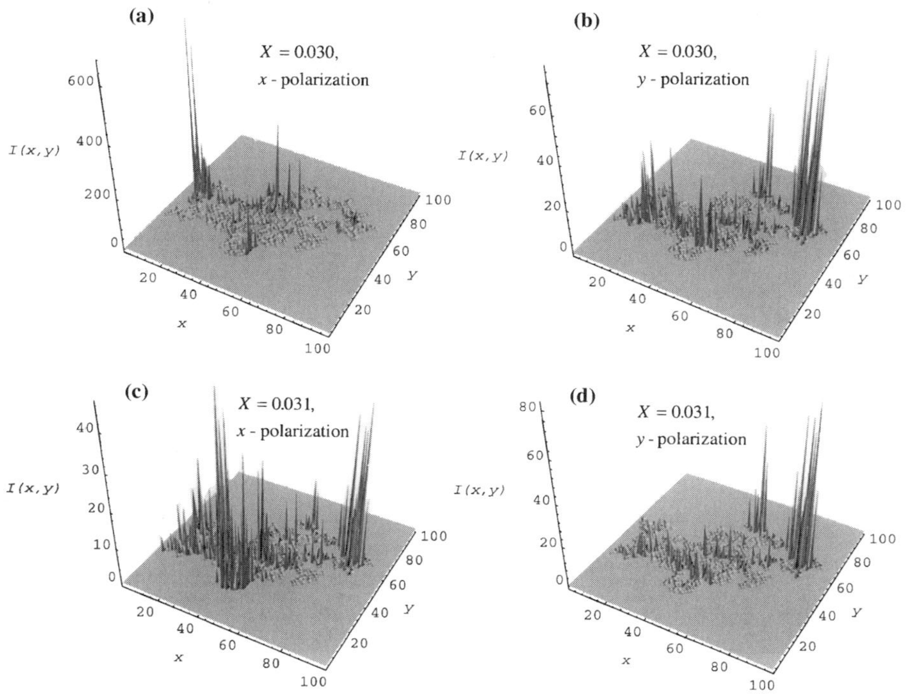

Figure 8.10. Spatial distribution of the local field intensities for external excitation of an individual CCA cluster ($N = 1500$) for the values of the spectral parameter X and the polarization of the exciting radiation shown in the figure. The value of the dissipation parameter is $R_0^3\delta = 10^{-3}$.

Apart from the individual eigenmodes shown in Figure 8.8, it is of interest to study a distribution of local intensities induced by an *external* exciting field, similar to what we have shown for a nonfractal MG composite in Figure 8.6. For fractal CCA clusters, an example of such a distribution is shown in Figure 8.10 for two frequencies (parameters X) that are very close to each other, and two perpendicular linear polarizations of the exciting light. These distributions are indeed extremely singular and fluctuating in space, even between the nearest-neighbor monomers. This property of the local fields is the reason underlying the giant fluctuations of the local fields (see Section 8.5). The overall width of the distribution is of the order of the total cluster size. This is explained by the fact that the external radiation at a given frequency excites a group of individual eigenmodes, within which there always are delocalized modes. Because the interaction is very long-ranged *and* the clusters are self-similar, there is no intrinsic length scale characteristic of the problem. Consequently, the spatial extent of the intensity distribution is limited only by the clusters' size.

Change of polarization of the exciting radiation at a given frequency brings about a dramatic redistribution of local intensities and change in the maximum intensities [cf. Fig. 8.10(a), (c) and (b), (d)]. The physical reason for this is that the resonant configurations of the monomers in most cases are highly anisotropic. This explains the high selectivity of the cluster photomodification in the radiation polarization observed experimentally [36–38]. The change of frequency of the exciting radiation by $< 1\%$ also brings about pronounced changes in the intensity spatial distributions [cf. Fig. 8.10(a), (b) and (c), (d)]. Generally, the observed intensity distributions are in good qualitative agreement with the direct experimental observation by Moskovits and co-workers [39] of the near-field optical fields in large silver clusters. Recently, a similar behavior that is qualitatively consistent with the inhomogeneous localization has been observed in more detail in silver colloids having fractal (supposedly, self-affine) geometry in near-field scanning optical microscope experiments [40–42].

(A quantitative comparison is not possible because the distributions for individual clusters are inherently chaotic, strongly fluctuating from one cluster to another.) However, the conclusion of [39] that the observed phenomena support the strong localization hypothesis contradicts our conclusions. We have commented [33] that the observations of [39] do not support its strong-localization conclusions. In fact, these observations do support the inhomogeneous localization picture described in this chapter (see Section 8.4), which is incompatible with the strong localization.

The patterns of the local fields discussed above show that the inhomogeneous localization scenario of polar excitations (plasmons) in large self-similar clusters is principally distinct from both the strong and weak localization scenarios of non-polar excitations. The above-discussed individual eigenvectors (eigenstates) are chaotic. Consequently, they are difficult to compare quantitatively with each other. Therefore, we consider next statistical characteristics (measures) of the eigenvectors.

To examine the statistical properties of an eigenmode distribution, we introduce the distribution function $P(L, X)$, which is the probability density that an eigenmode at a given X has the localization radius L

$$P(L,X) = \left\langle \sum_n \delta(L - L_n)\delta(X - w_n) \right\rangle \tag{8.17}$$

We show this distribution calculated for CCA clusters in Fig. 8.9 and for MG composites modeled as a RLG in Fig. 8.11.

The most conspicuous feature of the distribution of Fig. 8.9 is its very large width. This width extends from almost the total size of the system R_c to some minimum cut-off size l_X, which is a function of frequency $\omega(X)$. The cutoff is clearly seen in Fig. 8.9 where it is also indicated in the lower panel by a thick dotted line. This cutoff is seen in the intensity spatial distributions of individual eigenmodes as a core, that is, the characteristic minimum width of an eigenmode [cf. Figs. 8.8 and 8.10 and their discussion in Section 8.4].

The width of the distribution $P(L, X)$ is so large that its characterization by a single dispersion relation L_X [see Eq. (8.16)] is absolutely insufficient. For most of the spectral region, the cutoff length l_X by magnitude is intermediate between the maximum and minimum scales, R_c and R_0. This, along with the self-similarity of the clusters, suggests that l_X scales with X, that is, $l_X \propto |X|^\lambda$. Indeed, Fig. 8.9 supports

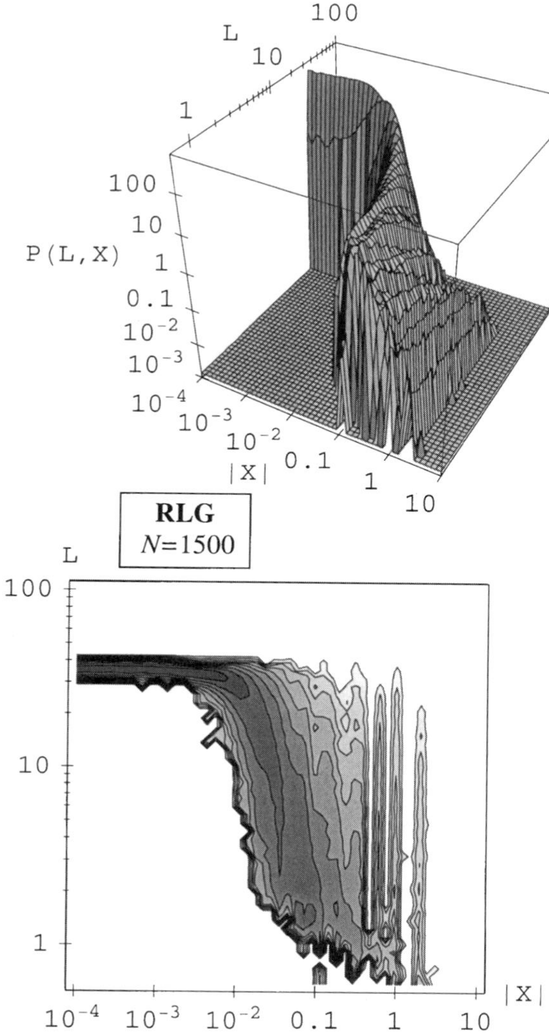

Figure 8.11. Same as in Fig. 8.9 but for a MG composite, simulated as a RLG.

the possibility of such a scaling with the corresponding index $\lambda \approx -0.25$. This illustrates the general property of the inhomogeneous localization of eigenmodes for fractal (self-consistent) clusters and composites.

A different situation exists for nonfractal composites, as one can see with an example of the MG composite (simulated as RLG) shown in Fig. 8.11. The distribution for $|X|0.1$ is similar to that of CCA (Fig. 8.9), which is characteristic of inhomogeneous localization. The major distinction from Fig. 8.9 is that the distribution in Fig. 8.11 shows the complete delocalization of the eigenmodes for $|X| \leq 0.01$ that appears in a narrow range. Such a delocalization is expected for the low $|X|$ part of the spectrum, that is, at frequencies close to the plasmon resonance of the individual inclusions (monomers). In contrast, there is no such delocalization for fractal (CCA) clusters, as seen in Fig. 8.9.

8.5. GIANT FLUCTUATIONS OF LOCAL FIELDS AND ENHANCEMENT OF NONRADIATIVE PHOTOPROCESSES

The picture of the intensities in any individual eigenmode (see Fig. 8.8) shows very large random changes of the intensity from one monomer to another, that is, fluctuations in space. When a cluster is subjected to an external exciting radiation, its response is due to the excitations of eigenmodes. Therefore, we may expect that the eigenmode fluctuations will cause strong fluctuations of the local fields at individual monomers. We have already discussed spatial distribution of the local fields in Sections 8.3 and 8.4. In this Section, following [43], we discuss distributions of the externally induced local fields over their intensity. This distribution determines enhancement of optically nonlinear incoherent processes (e.g., nonlinear photochemical reactions, optical modification, and local melting).

The local field at an ith monomer is expressed in terms of Green's function (8.12) as

$$E_{i\alpha} = Z\sum_{j\beta} \mathbf{G}_{i\alpha,j\beta} E_{\beta}^{(0)} \tag{8.18}$$

In terms of this field, the local field-intensity enhancement coefficient G_i for an ith monomer and the corresponding distribution function $P(G)$ are defined as

$$G_i = \frac{|\mathbf{E}_i|^2}{|\mathbf{E}^{(0)}|^2}, \qquad P(G) = \left\langle \frac{1}{N}\sum_i \delta(G - G_i) \right\rangle \tag{8.19}$$

In addition, we introduce an nth moment of this distribution:

$$M_n = \langle G^n \rangle = \int P(G) G^n dG \tag{8.20}$$

By its physical meaning, M_n is the enhancement coefficient of an nth-order nonradiative (i.e., without emissions of photons) nonlinear photoprocess. If, for instance, a molecule is attached to a monomer of a cluster, then M_n shows how many times the rate of its n-photon optical excitation exceeds such a quantity for an isolated molecule. A similar estimate is valid for the enhancement of a composite consisting of the nonlinear matrix and resonant inclusion clusters.

The spectral dependencies of M_n for different combination of the degree of nonlinearity n and the dissipation parameter δ are shown in Fig. 8.12. The data in this figure are scaled by the factor $R_0^3 \delta Q^{-2(n-1)}$, with the resonant quality factor given by Eq. (8.6). The most remarkable feature of Fig. 8.12 is an almost perfect collapse of the data into a universal curve in an intermediate region of X for the case

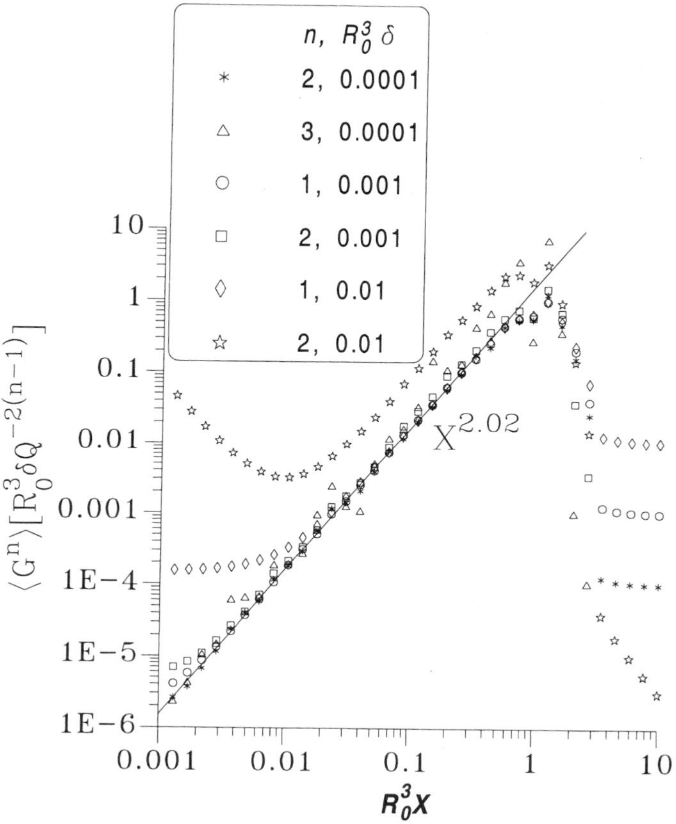

Figure 8.12. Normalized enhancement factors $\langle G^n\rangle\delta Q^{-2(n-1)}$ as functions of X for CCA clusters for the values of δ and n shown.

of very low dissipation ($R_0^3\delta < 0.01$). Moreover, this curve is actually close to a straight line in the intermediate region indicating a scaling behavior of $M_n(X)$. We conjecture this scaling as the dependence $M_n \cong Q^{2(n-1)}M_1\lambda$.

For the first moment M_1, we have previously obtained [5,6] the *exact* relation $M_1 = (X^2 + \delta^2)\mathrm{Im}\,\alpha/\delta$. Because the absorption $\mathrm{Im}\alpha$ does not scale in X (see Section 8.3), the enhancement coefficient M_n should not scale either. However, the dependence $\mathrm{Im}\alpha(X)$ in the intermediate region of X is flat (see Fig. 8.3). Therefore, for $Q \gg 1$ the apparent scaling in X takes place with a trivial index of $2n$,

$$M_n \cong Q^{2n-1}|X|\mathrm{Im}\,\alpha = \frac{X^{2n}}{\delta^{2n-1}}\mathrm{Im}\,\alpha \propto X^{2n} \tag{8.21}$$

in agreement with Fig. 8.12.

A major result of [43], given by Eq. (8.21), is that the excitation rate of a nonradiative nth-order nonlinear photoprocess in the vicinity of a disordered cluster is resonantly enhanced by a factor of $M_n \cong Q^{2n-1}$. This quantity can be understood qualitatively in the following way. For each of the n photons absorbed by a resonant monomer, the excitation probability (rate) is increased by a factor of $\cong Q^2$ (proportional to the local field *intensity*), therefore the total rate is increased by a factor of $\cong Q^{2n}$. However, the fraction of monomers that are resonant is small, $\cong Q^{-1}$. Consequently, the resulting enhancement factor is $M_n \cong Q^{2n-1}$, which is in agreement with Eq. (8.21).

For instance, for silver in the red spectral region $Q \cong 30$ (the optical constants for silver are adopted from [44], each succeeding order of the nonlinearity gives enhancement by a factor of $Q^2 \cong 1000$. We emphasize that the origin of this enhancement is the high-quality optical resonance in the monomers modified (shifted significantly to the red) by the structure of a cluster. Among the interesting effects related to the enhanced nonradiative excitation, we mention one, the selective photomodification of silver clusters [36–38].

We have considered above the moments (averaged powers) of the local fields. Now, we consider another characteristic of the fluctuations, the distribution function $P(G)$ (8.19) of the local-field intensity. Because the change of the minimum scale R_0 also implies the change of the local fields, it is possible that the distribution function scales in some intermediate range,

$$P(G) \cong G^{-\varepsilon} \tag{8.22}$$

Under the scale-invariance assumptions, the index ε does not depend on the minimum scale R_0. Consequently, ε does not depend on frequency (the spectral parameter X) either.

A simple model that allows one to calculate the scaling index ε is the binary approximation [5,6]. In this approximation, the eigenmode is localized at only a pair of the monomers. In this case, we have found [43] the enhancement factor G_i for a pair separated by a distance r, located at an angle Θ to the direction of the exciting field,

$$G_i(r, \Theta) = \frac{\delta^2 \sin^2\Theta}{(X - r^{-3})^2 + \delta^2} + \frac{\delta^2 \cos^2\Theta}{(X + 2r^{-3})^2 + \delta^2} \tag{8.23}$$

By using this equation, we can calculate the distribution function as

$$P(G) = \int \delta(G - G_i(r, \Theta)) C(r) d^3 r \tag{8.24}$$

Here, $C(r) \propto r^{D-1}$ is the density correlation function of the cluster. If we take into account that large values of G are of interest, we obtain from Eq. (8.24)

$$P(G) \propto G^{-3/2} \tag{8.25}$$

Thus, in the binary approximation we obtain a *universal* scaling index of $-3/2$. Surprisingly, this index does not depend on the cluster's dimension D. Its value is determined merely by the vector nature of the fields.

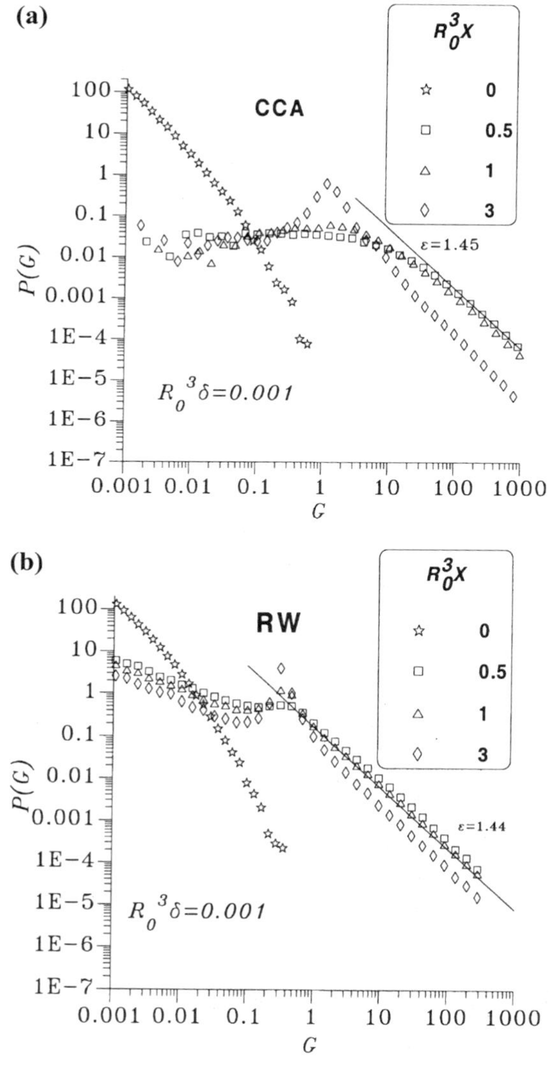

Figure 8.13. Distribution function of the local field intensity $P(G)$ calculated for $R_0^3\delta = 0.001$, for the values of X shown. The data in (a) are for CCA clusters and in (b) for random walk (RW) clusters.

It is interesting to compare both the scaling prediction and the calculated value of the index $\varepsilon = 3/2$ with the numerical results. These are shown in Fig. 8.13 for CCA and random walk (RW) clusters. The main feature is an unusually wide distribution. The local intensities are on the order of the exciting intensity ($G = 1$), as well as three orders of magnitude smaller or greater. This feature is referred to as giant fluctuations of the local field. The regions of high intensity are responsible for enhanced nonlinear responses. We note that the local regions of high intensity (hot spots) have also been observed directly with the scanning photon-tunneling (near-field optical) microscope [39] (see also a comment [33] in [39]). The value of the index ε is indeed almost independent from the frequency (parameter X), as expected from the scaling theory. Interestingly enough, these values ($\varepsilon = 1.45$ for CCA and $\varepsilon = 1.44$ for RW) are quite close to those predicted by the binary theory ($\varepsilon = 3/2$). This agreement is unexpected because there is no reason to believe that the binary theory is applicable in a wide range of frequencies.

8.6. CHAOS OF EIGENMODES

The eigenmode equation (8.10) has the same form as the quantum mechanical Schrödinger equation. In quantum mechanics, it is not uncommon that highly excited states or states of complex systems possess chaotic behavior (see, e.g., [45,46]). One may expect a similar situation for eigenmodes of large disordered clusters and composites. The extreme sensitivity of the individual eigenmodes to a very small change of their frequency, which is discussed in Section 8.4 and illustrated by Fig. 8.8, is a direct indication of such chaos. Even more than individual eigenmodes, statistical properties of chaotic eigenstates are of great interest. The giant fluctuations of the intensities of local fields discussed in Section 8.5. provide one of the statistical descriptions. In this section, we will consider spatial correlations of the chaotic eigenmodes [34,35].

A principal property that distinguishes this problem from quantum mechanical chaos is the long-range nature of the dipole–dipole interaction. A similar tight-binding problem of quantum mechanics (Anderson model) is usually formulated with only next-neighbor hopping. In studied quantum mechanical problems, chaotic quantum states do not possess long-range spatial correlations [45,47].

The long-ranged interaction on one hand tends to induce the long-range spatial correlations. On the other hand, it may tend to establish a mean field, suppress fluctuations, and eliminate the chaos. As we demonstrate below, either of those trends may dominate, depending on the system geometry and spectral region.

We expect that chaos is most pronounced in clusters and composites with fractal geometry. The rationale for it is the following. A mean field is established and spatial chaos is eliminated when the correlation range of eigenmodes exceeds a characteristic size of the density variations in the system. However, fractal (self-similar) geometry implies that the system repeat itself on all spatial scales and, consequently, there exists no such characteristic spatial scale. This is a prerequisite for the coexistence of chaos and long-range correlations.

To characterize the spatial correlations of eigenmode amplitudes, we introduce the amplitude correlation function (also called dynamic form factor),

$$S_{\alpha\beta}(\mathbf{r}, X) = \left\langle \sum_{n,i,j} (i\alpha|n)(j\beta|n)\delta(\mathbf{r} - \mathbf{r}_{ij})\delta(X - w_n) \right\rangle \tag{8.26}$$

where $\delta(\cdots)$ is the Dirac's δ function (not to be confused with the dissipation parameter δ). It is useful to note that in the limit $\delta \ll |X|$,the correlator $S_{\alpha\beta}(\mathbf{r}, X)$ is related to Green's function (8.11),

$$S_{\alpha\beta}(\mathbf{r}, X) = \frac{1}{\pi} \mathrm{Im} \sum_{i,j} G_{i\alpha, j\beta} \delta(\mathbf{r} - \mathbf{r}_{ij}) \tag{8.27}$$

This function expresses the correlation factor of amplitudes $(i\alpha|n)$ and $(j\beta|n)$ of two eigenmodes with polarizations α and β at two points separated by a spatial interval of $\mathbf{r}$ at an eigenvalue (frequency) of X, averaged over an ensemble of the systems. Similarly, we introduce a correlation function of the intensities of eigenmodes

$$C(\mathbf{r}, X) = \left\langle \sum_{n,i,j} (i\alpha|n)^2 (j\beta|n)^2 \delta(\mathbf{r} - \mathbf{r}_{ij}) \delta(X - w_n) \right\rangle \tag{8.28}$$

where the summation in the repeated indexes α and β is implied as is done throughout this chapter. In a similar way, this function yields a correlation of two eigenmode intensities, $(i\alpha|n)^2$ and $(j\beta|n)^2$.

We have calculated [34,35] the above-defined correlation functions for an ensemble of CCA clusters (composites) with $N = 1500$ monomers (inclusions). The corresponding result for $S(\mathbf{r}, X) = \frac{1}{3} S_{\beta\beta}(\mathbf{r}, X)$ is shown in Fig. 8.14. We see a developed pattern of irregular, chaotic correlations in almost the whole spatial-spectral region. The landscape observed in Fig. 8.14 well deserves the name of a "devil's hill", where narrow regions of positive and negative correlations are interwoven, resulting in a turbulence-like pattern. This chaotic behavior is indeed a reflection of the chaos of individual eigenmodes. However, this chaos is in some sense stronger, because it is in a quantity averaged over an ensemble of statistically independent composites (consequently, it is fully reproducible). The deterministically chaotic pattern of Fig. 8.14 is likely to depend on a specific topology of the composite (CCA clusters). The straight lines shown in Fig. 8.14 are given by the binary–ternary approximation, see the discussion of Fig. 8.16 below.

To distinguish between fluctuations of phase and amplitude in the formation of the devil's hill in Fig. 8.14, we will compare it to the second-order correlation function $C(\mathbf{r}, X)$ shown in Fig. 8.15. We see that that this function differs dramatically from the dynamic form factor. The relief in Fig. 8.15 is very smooth, in contrast to Fig. 8.14. This implies that the devil's hill is formed due to spatial fluctuations of the phases of individual eigenmodes, while their amplitudes are smooth functions in space-frequency domain. Another important feature present in Fig. 8.15 is the long range of the correlation. The correlation decay in r is indeed weaker than any power.

As we discussed above, self-similarity (fractality) of the CCA composite is of principal importance for the coexistence of chaos and long-range correlations of the eigenmodes. To emphasize this point, we show in Fig. 8.16 the results for the

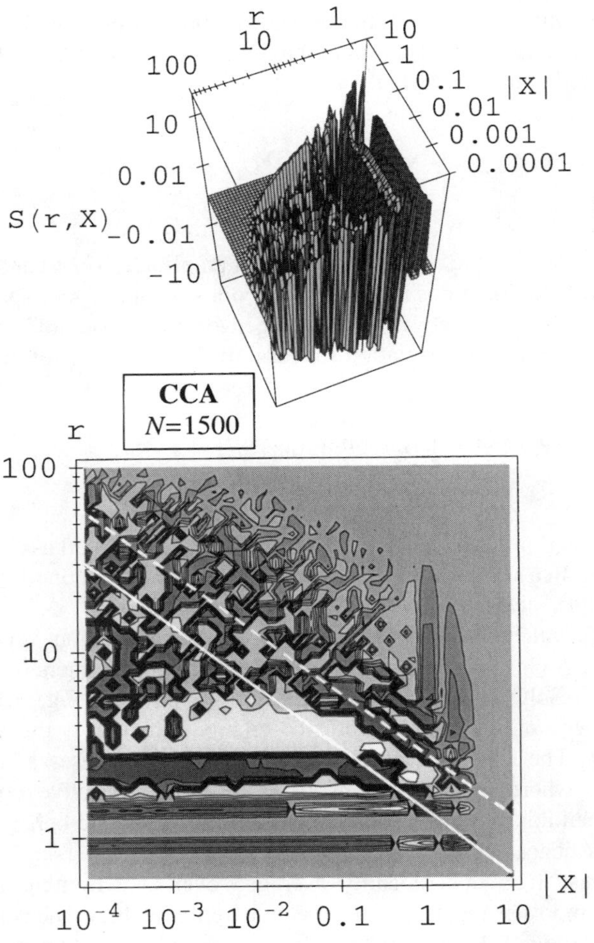

Figure 8.14. Dynamic form factor $S(r, X)$ for CCA composite (the number of inclusions $N = 1500$, averaged over an ensemble of 300 composites). (a) Shows a three-dimensional (3D) representation of the function, while (b) is the corresponding contour map. The vertical scale is pseudologarithmic to show positive and negative values of $S(r, X)$ simultaneously. To obtain it, a small region of the plot for $|S(r, X)| \leq 10^{-4}$ is removed. The function plotted is $\log[10^4|S(r, X)|]\text{sgn}[S(r, X)]$. The horizontal scales are logarithmic.

dynamic form factor *S(r,X)* for RLG composites. We see that for extremely large eigenvalues, $|X| \geq 1$, the form factor *S(r,X)* has a quasirandom structure that is reproducible under the ensemble averaging. This structure is similar to what is observed for CCA composites (cf. Fig. 8.14). In the middle of the spectral region, $1 \geq |X| \geq 0.003$, the form factor is dominated by a few branches of excitation. The strongest one is marked on Fig. 8.16(b) by a solid white line $X = -2r^{-3}$ that is the corresponding branch of binary approximation [5,6]. In this approximation,

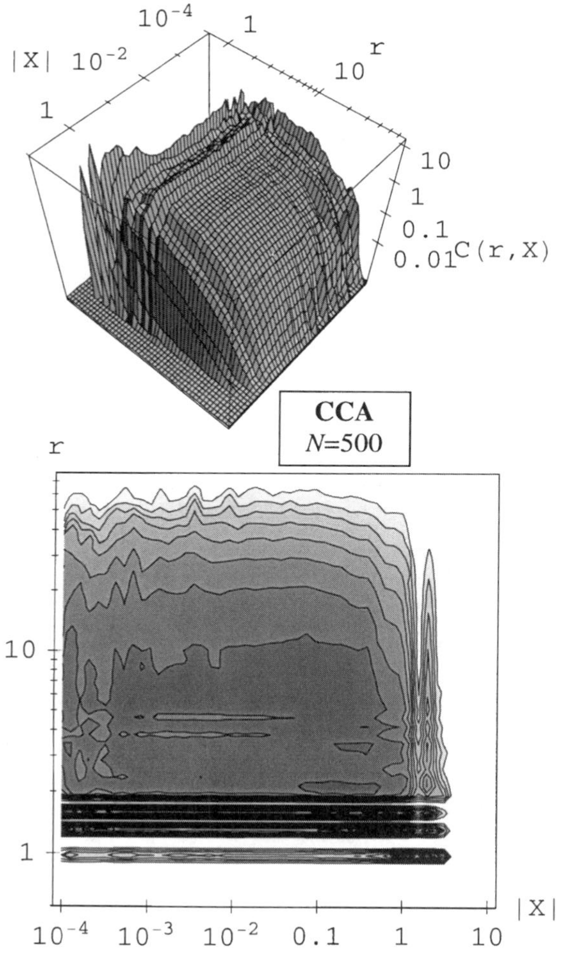

Figure 8.15. Intensity correlation function $C(r, X)$ calculated for CCA composite ($N = 1500$) averaged over ensemble of 300 systems. (a) Shows the function plotted in the triple-logarithmic scale and (b) is the corresponding contour map.

each eigenmode consists of two excited regions (hot spots) and the form factor is given by

$$S(r,X) = \delta(r)\rho(X) + f(r)[\delta(X + 2r^{-3}) - \delta(X - 2r^{-3}) + 2\delta(X - r^{-3}) - 2\delta(X + r^{-3})] \tag{8.29}$$

where $f(r)$ is a smooth distribution of the intermonomer distances, $f(r) = N\langle\delta(r - r_{ij})\rangle$. The weaker branch, marked by the dotted line, $X = -(1 + \sqrt{57})r^{-3}$, is due to the ternary excitations, that is, eigenvectors consisting of three hot spots.

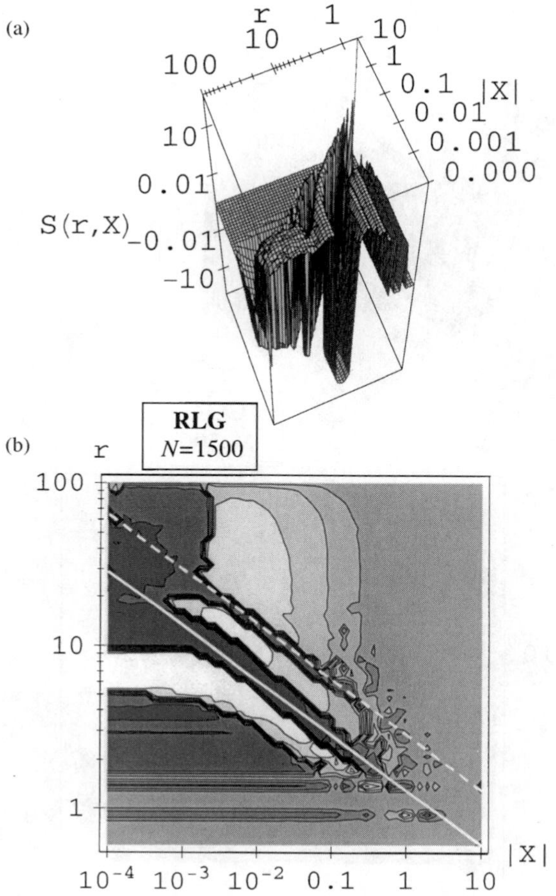

Figure 8.16. The same as in Fig. 8.14, but for RLG composite.

At $X \approx 0.003$, we see a sharp transition to delocalization, where the positive-correlation region becomes uniform, spreading over most of the system. This transition occurs when the correlation length becomes comparable to the size of the system R_c, that is, at $X \cong R_c^{-3} \propto N$. Consequently, this transition is mesoscopic, that is, it phases out as the size of the system becomes very large. This is in contrast to an Anderson transition where the local density of scatters is the determining parameter. This difference from Anderson localization is due to the long-range interaction in our case. As a result, most of the eigenvectors are not propagating waves, but change from binary–ternary excitations to delocalized surface plasmons as $|X|$ decreases.

To briefly conclude this section, chaos of eigenmodes is present for both conventional and fractal geometries of the composites. However, fractal composites demonstrate this chaos in the whole spectral region, while conventional composites demonstrate only in the extremes of the spectrum. For conventional geometries

(RLG composites in particular), the eigenmodes are dominated by binary–ternary excitations with a mesoscopic transition to uniform surface plasmons near plasmon resonance of the inclusion particles (monomers), that is, at $|X| \to 0$. In Section 8.7, we will study the role of the fluctuations and chaos of eigenmodes on the nonlinear radiative photoprocesses.

8.7. ENHANCEMENT OF RADIATIVE PHOTOPROCESSES AND NONLINEAR POLARIZABILITIES IN CLUSTERS

Among enhanced radiative photoprocesses, one of the best studied experimentally is surface-enhanced Raman scattering (SERS) [9]. The intensity of Raman scattering for molecules adsorbed at rough surfaces or colloidal-metal particles is known to be greatly enhanced, by a factor of up to 10^6. There are two limiting cases of SERS. In the first case, the Raman shift is very large, much greater than the homogeneous width of the monomer's absorption spectrum. In this case, as we have shown [48] the exact expression of the SERS enhancement factor G^{RS} can be obtained as

$$G^{\mathrm{RS}} = \frac{X^2 + \delta^2}{\delta} \operatorname{Im} \alpha \approx Q|X| \operatorname{Im} \alpha \tag{8.30}$$

This result predicts that on an the order of magnitude the enhancement is the same (by a factor of $\cong Q$) as for nonradiative photoprocesses of the second order. Qualitatively, we may interpret this result in the following way. For a large frequency shift, the outgoing photon is out of resonance and does not considerably interact with the system. Correspondingly, the entire enhancement is due to the local field of the absorbed photons, the same as for nonradiative processes [cf. the discussion after Eq. (8.21)].

A much more dramatic enhancement is found [48] for the second case, namely, that of Raman frequency shifts smaller than monomer's line width. This situation is characteristic of the most typical and interesting cases of SERS. We have shown [48] that in this case there exists an approximate expression for the enhancement coefficient G^{RS} of SERS,

$$G^{\mathrm{RS}} \approx \frac{1}{2} Q^3 R_0^3 |X| \operatorname{Im} \alpha = \frac{1}{2} \left(\frac{X}{\delta}\right)^3 R_0^3 |X| \operatorname{Im} \alpha \tag{8.31}$$

In this case, we have $G^{\mathrm{RS}} \cong Q^3$, where the physical interpretation is the following. Enhancement by $\cong Q^2$ is due to the first power of the local-field intensity. Another power $\cong Q^2$ appears because the outgoing photon is not emitted freely, but in the resonant environment. Finally, one power of Q vanishes because only a small fraction $\cong 1/Q$ of the monomers is resonant to the exciting radiation.

We compare the functional dependence predicted by Eq. (8.31) with the numerical simulation in Fig. 8.17. As we see, this equation describes the numerically found dependencies quite well, especially the dependence on X.

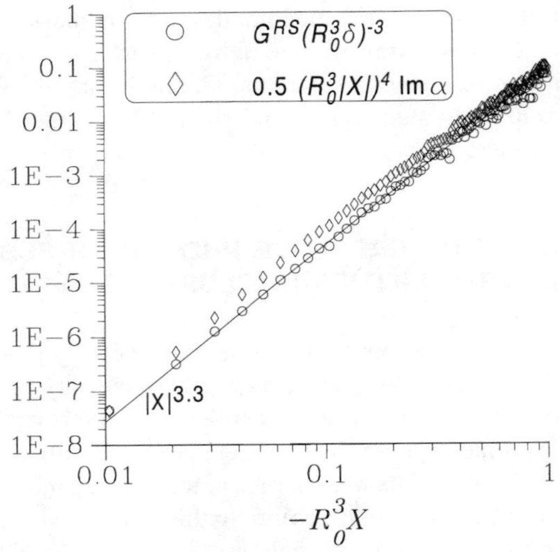

Figure 8.17. Scaled enhancement coefficient of SERS from silver colloid clusters in comparison with the prediction of Eq. (8.31). The theoretical dependence is calculated for CCA clusters.

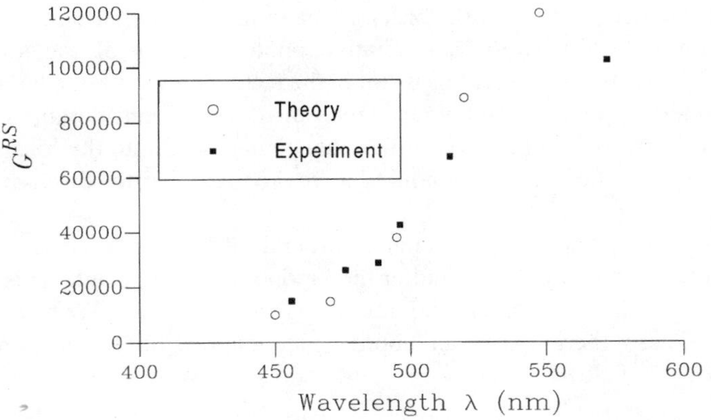

Figure 8.18. Theoretical and experimental spectral (in terms of wavelength) dependencies of the enhancement coefficient G^{RS} for silver colloid clusters.

Equation (8.31) shows that the Raman scattering is enhanced for an adsorbed molecule by a factor of $G^{RS} \cong Q^3$. For noble metals in the red region of visible light, we have $Q \cong 30-100$, so the predicted enhancement is very large, $G^{RS} \cong 10^4-10^6$. This explains the range of enhancements found experimentally [9]. The comparison [48] of theoretical spectral dependence with experiment is shown in Fig. 8.18. As we can see, the theory explains the experimental observations qualitatively. Note, however, that recently an extremely high enhancement of the Raman scattering has been observed [49] with $G^{RS} \cong 10^{10}$. This enhancement allowed observation of Raman scattering from a single molecule. Such an enhancement is not understood in the framework of the EM theory of [48]. It is possible that the so-called chemical enhancement mechanism, in addition to the EM one, contributes to yield enhancements so high.

Now, we will very briefly discuss coherent, or parametric, nonlinear photoprocesses. These processes are due to nonlinear wave mixing. One of the most interesting is frequency-degenerate four-wave mixing (a third-order process), which is responsible for such an interesting effect as phase conjugation. We have found the corresponding enhancement coefficient in the binary approximation in [10] and numerically in [50]. Further extensive computations of linear and nonlinear responses of clusters were done later in [51] and [52].

Physically, the enhancement of an nth-order parametric photoprocess can be estimated as we have suggested in [42]. The enhancement of the amplitude of each of the $n+1$ participating photons (including the emitted photon) is $\cong Q$. This yields the enhancement of the amplitude of the process as $\cong Q^{n+1}$. For any coherent process, the amplitude, not probability, is the quantity to average. The averaging is done by multiplying by the fraction of resonant monomers, which is found to be $\delta \mathrm{Im}\,\alpha \equiv (X/Q)\mathrm{Im}\,\alpha$ (see derivation of Eq. (8.5) in [42]). This leads to a mean amplitude $\cong Q^n X\,\mathrm{Im}\,\alpha$. Finally, the amplitude enhancement should be squared, yielding the intensity-enhancement coefficient

$$G^{(n)} \approx Q^{2n}(X\,\mathrm{Im}\,\alpha)^2 \tag{8.32}$$

For $n = 3$, this reduces to

$$G^{(3)} \approx (X\,\delta)^6 (X\,\mathrm{Im}\,\alpha)^2 \tag{8.33}$$

in agreement with the corresponding result of [50]. The comparison of Eq. (8.33) to the numerical calculations shown in Fig. 8.19 supports this result.

A remarkable feature of Eq. (8.33) is that the predicted enhancement is quite large. For the realistic dielectric parameters of silver in the visible range, $Q \cong 10$–30, and correspondingly $G^{(3)} \cong 10^4$–10^7. The experimental investigation [53] has indeed found a very strong enhancement for the phase conjugation, with $G^{(3)} \cong 10^6$, confirming the theoretical predictions.

8.8. ENHANCED NONLINEAR SUSCEPTIBILITIES OF COMPOSITES

8.8.1. Macroscopic and Mesoscopic Fields and Integral Formulas for Optical Responses of Composites

The enhancement discussed in Section 8.7 is calculated for nonlinear-optical *polarizabilities* of molecules adsorbed at the surface of monomers of a cluster. In this section, based on [32], we follow [25], [26], and [54] and consider *susceptibilities* of

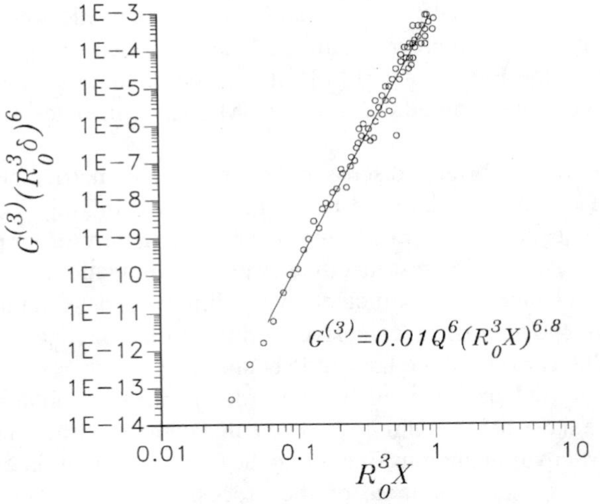

Figure 8.19. Scaled enhancement coefficient for third-order degenerate parametric process, calculated for CCA.

composites that possess the conventional nonfractal geometry. For such composites, one can introduce *mesoscopic* electric field $\mathbf{e}(\mathbf{r})$ and induction $\mathbf{d}(\mathbf{r})$ varying in space due to the inhomogeneity of the composite's material. The conventional averaging of these fields yields the corresponding macroscopic quantities

$$\mathbf{E} = \frac{1}{V}\int_V \mathbf{e}(\mathbf{r})d^3r \qquad \mathbf{D} = \frac{1}{V}\int_V \mathbf{d}(\mathbf{r})d^3r \tag{8.34}$$

where V is a sample volume of the composite.

We will concentrate on MG composites [1]. Such a composite is a suspension of nanospheres in a uniform host. These nanospheres (monomers) in the theory of composites are conventionally called inclusions. Here we will not consider another type, Bruggeman composites [3], which are a mixture of two continuous materials whose geometry may be similar or identical. We consider the third-order nonlinear susceptibilities of the MG composites. Note that we have already discussed linear dielectric constants of the MG composites above in Section 8.3.

We assume that the constituent materials of the composite are isotropic. Then, the third-order optical nonlinearity is defined in a general case by the following material relation between the induction and field:

$$\begin{aligned}\mathbf{d}(\mathbf{r}) &= \varepsilon(\mathbf{r})\mathbf{e}(\mathbf{r}) + 4\pi[A(\mathbf{r})|\mathbf{e}(\mathbf{r})|^2\mathbf{e}(\mathbf{r}) + \tfrac{1}{2}B(\mathbf{r})\mathbf{e}(\mathbf{r})^2\mathbf{e}^*(\mathbf{r})]\\ \mathbf{D} &= \varepsilon_c\mathbf{E} + 4\pi[A_c|\mathbf{E}|^2\mathbf{E} + \tfrac{1}{2}B_c\mathbf{E}^2\mathbf{E}^*]\end{aligned} \tag{8.35}$$

where $\varepsilon(\mathbf{r}), A(\mathbf{r})$, and $B(\mathbf{r})$ are functions of the coordinates that acquire the values ε_i, A_i, and B_i in the inclusion material and the values ε_h, A_h, and B_h in the host. These parameters for the host are assumed for simplicity to be coordinate independent, and for the inclusions they may depend on the coordinate, while similar macroscopic (averaged) quantities for the composite ε_c, A_c, and B_c are always coordinate independent. As before, we consider optical (oscillating in time) electric fields. By $\mathbf{e}(\mathbf{r}), \mathbf{d}(\mathbf{r}), \mathbf{E}$, and $\mathbf{D}$, we understand the time-independent *amplitudes* of those fields.

In what follows, we assume the linear polarization of the exciting radiation. Then, the third-order nonlinear susceptibility (hypersusceptibility) of an isotropic composite is completely characterized by one constant,

$$\chi_c^{(3)} = A_c + \tfrac{1}{2} B_c \tag{8.36}$$

We impose a standard boundary condition on the mesoscopic electrostatic potential at the surface of the composite, where the mesoscopic field becomes the macroscopic field,

$$\varphi(\mathbf{r}) = -\mathbf{E}\mathbf{r} \qquad \text{for } \mathbf{r} \in S \tag{8.37}$$

where $\varphi(\mathbf{r})$ is the mesoscopic potential. From this, transforming from a bulk-to-surface integral and back in Eq. (8.34), one derives an exact equation for the induction,

$$D = \frac{1}{VE} \int_V \mathbf{d}(\mathbf{r})\mathbf{e}(\mathbf{r}) d^3 r \tag{8.38}$$

By using Eq. (8.38) and applying a generalized Gauss theorem along with the boundary conditions (8.37) in first order in the field $\mathbf{E}$, one obtains an expression for the linear dielectric constant of the composite

$$\varepsilon_c = \frac{1}{V\mathbf{E}^2} \int_V \varepsilon(\mathbf{r}) (\mathbf{e}^{(1)}(\mathbf{r}))^2 d^3 r \tag{8.39}$$

Here and below, the order of optical nonlinearity (in the external field) is denoted in parentheses as a superscript. Alternatively, this dielectric constant ε_c can be found directly from the macroscopic first-order field (induction) as

$$\varepsilon_c = \frac{1}{VE} \int_V \varepsilon(\mathbf{r}) \mathbf{e}_z^{(1)}(\mathbf{r}) d^3 r \tag{8.40}$$

To interpret this expression, we define the dipole moment induced on an ath inclusion as

$$\mathbf{d}_a = -\frac{1}{4\pi} \int_{V_a} \frac{1}{s(\mathbf{r})} \mathbf{e}^{(1)}(\mathbf{r}) d^3 r \qquad s(\mathbf{r}) \equiv \frac{\varepsilon_h}{\varepsilon_h - \varepsilon_i} \tag{8.41}$$

where V_a is the volume of this ath inclusion. In this section and below, we will number the inclusions with the indexes a, b, ... to prevent possible confusion with the notation i for the inclusions. We introduce by definition the polarizability tensor of an ath inclusion in the composite $a_{\beta\lambda}^{(a)}$ by the relation

$$d_{a\beta} = \alpha_{\beta\lambda}^{(a)} E_\gamma \tag{8.42}$$

The inclusion's number (a) in the superscript should not be confused with the order in the field. This order may, in principle, be 1, 2 or 3. It is easy to see that Eq. (8.40) is equivalent to Eq. (8.13), which was used to compute ε_c in Section 8.3.

To compute the nonlinear susceptibility (8.36) of the composite, we collect terms of the third order in optical nonlinearity. Then, we use the generalized Gauss theorem and the boundary conditions (8.37) to eliminate the unknown third-order field $\mathbf{e}^{(3)}(\mathbf{r})$. As a result, one obtains an *exact* expression [25,54] for the third-order susceptibility of the composite

$$\chi_c^{(3)} = \frac{1}{V|\mathbf{E}|^2\mathbf{E}^2} \int_V \chi^{(3)}(\mathbf{r}) |\mathbf{e}^{(1)}(\mathbf{r})|^2 (\mathbf{e}^{(1)}(\mathbf{r}))^2 d^3 r \tag{8.43}$$

where the hypersusceptibility of the (isotropic) constituent materials of the composite is defined as

$$\chi^{(3)}(\mathbf{r}) = A(\mathbf{r}) + \tfrac{1}{2} B(\mathbf{r}) \tag{8.44}$$

The most intriguing property of Eq. (8.43) is that $\mathbf{e}^{(3)}(\mathbf{r})$ does not enter at all, thus greatly simplifying the task of computing the hypersusceptibility of the composite $\chi_c^{(3)}$. The price to pay for this advantage is that one cannot compute each of the constants A_c and B_c characterizing the most general response, but only their combination (8.44).

From Eq. (8.43), it follows that the hypersusceptibility of the composite is a sum of two contributions. The first contribution is due to nonlinearity in the inclusions, and the second is due to the nonlinearity in the host. These contributions can be considered independently.

8.8.2. Hypersusceptibility of a Composite for the Case of Nonlinearity in Inclusions

For a MG composite, the inclusions are spheres. We assume the characteristic distance between the inclusions to greatly exceed the radius R_0 of a sphere. This ensures applicability of the dipole interaction approximation. Also, it allows one to find the linear (first-order) field in an ath inclusion particle,

$$e_{a\beta}^{(1)} = q\alpha_{\beta\lambda}^{(a)} E_\gamma \qquad q \equiv \frac{1}{R_0^3} \frac{3\varepsilon_h}{\varepsilon_i - \varepsilon_h} \tag{8.45}$$

This field, of course, is uniform in this approximation. If we substitute Eq. (8.45) into Eq. (8.43), we obtain a closed expression for the hypersusceptibility of the composite

$$\chi_c^{(3)} = \chi_i^{(3)} f |q|^2 q^2 \langle |\alpha_{\beta z}^{(a)}|^2 (\alpha_{\beta z}^{(a)})^2 \rangle \tag{8.46}$$

where f is a fill factor (a fraction of the total volume occupied by the inclusions) and

$$\alpha_{\beta z}^{(a)} = \sum_b \mathbf{G}_{a\beta,bz} \tag{8.47}$$

is the polarizability of an ath inclusion in the composite. Formulas of Eqs. (8.46) and (8.47) give in a closed form a general expression for the third-order (frequency-degenerate) hypersusceptibility in the case of nonlinear inclusions.

Apart from the general expression obtained above, we would like to obtain a compact formula for $\chi_c^{(3)}$ in a mean-field approximation, similar to what has been done in [14]. To do so, we note that a mean-field approximation implies that variations of the local fields are neglected (cf. the discussion of Fig. 8.6). By assuming that and by taking into account the spherical symmetry of an inclusion, we can model the composite in the following way. A dielectric sphere of radius R_0 of an inclusion particle with a dielectric constant of ε_i is surrounded by a spherical shell of the host material of radius R with a dielectric constant ε_h. The space outside of this shell is the homogeneous composite with the dielectric constant ε_c. At all interfaces, the conventional boundary conditions of continuity of potential and the normal component of the induction are imposed.

The solution of this model for the field inside an inclusion (for $r \leq R_0$) is

$$\mathbf{e}^{(1)} = \frac{3\varepsilon_h}{\varepsilon_i + 2\varepsilon_h} \mathbf{E}_l \qquad \mathbf{E}_l \equiv \frac{\varepsilon_c + 2\varepsilon_h}{3\varepsilon_h} \mathbf{E} \tag{8.48}$$

where we have introduced the effective local (Lorentz) field $\mathbf{E}$. The solution outside of the inclusion $r \leq R_0$ is

$$\mathbf{e}^{(1)}(\mathbf{r}) = \mathbf{E}_l + \alpha_0 \frac{3(\mathbf{r}\mathbf{E}_l)\mathbf{r} - r^2 \mathbf{E}_l}{r^5} \qquad \alpha_0 \equiv R_0^3 \frac{\varepsilon_i - \varepsilon_h}{\varepsilon_i + 2\varepsilon_h} \tag{8.49}$$

where we have introduced the polarizability of an isolated inclusion α_0. By substituting Eqs. (8.48) and (8.49) into Eq. (8.40), we obtain the familiar MG formula that we write in the explicit form

$$\varepsilon_c = \frac{1 + 2f\beta}{1 - f\beta} \qquad \beta \equiv \frac{\alpha_0}{R_0^3} = \frac{\varepsilon_i - \varepsilon_h}{\varepsilon_i + 2\varepsilon_h} \qquad f \equiv \frac{R_0^3}{R^3} \tag{8.50}$$

where we introduce the fill factor f (i.e., a fraction of the volume of a composite occupied by inclusions).

If we substitute the internal field given by Eq. (8.48) into Eq. (8.43), we obtain the required expression for the hypersusceptibility in the case of nonlinearity in the inclusions,

$$\chi_c^{(3)} = \chi_i^{(3)} f \left| \frac{\varepsilon_c + 2\varepsilon_h}{\varepsilon_i + 2\varepsilon_h} \right|^2 \left(\frac{\varepsilon_c + 2\varepsilon_h}{\varepsilon_i + 2\varepsilon_h} \right)^2 \tag{8.51}$$

This expression precisely coincides with the corresponding result of [13].

We model the MG composite as a RLG of inclusions embedded in a uniform host. For the sake of numerical computations, we realistically assume that the inclusions are silver nanospheres whose dielectric constant ε_i is close to the bulk values [44], and the embedding medium (the host) has the dielectric constant of $\varepsilon_h = 2.0$.

We have numerically calculated the enhancement coefficient $g^{(3)} = \chi_c^{(3)}/\chi_i^{(3)}$ for the third-order susceptibility of a MG composite using both the ST [Eq. (8.46)] and the MF theory [Eq. (8.51)]. The results are shown in Fig. 8.20. As one can immediately conclude, the MF theory completely fails to reproduce the results of the spectral theory that is exact within the framework of the dipole approximation. The underlying cause of this failure is in strong fluctuation and chaos of the eigenmodes and local fields discussed throughout this chapter (see Sections 8.4–8.6). The degree of failure of the MF theory for nonlinear processes is much more dramatic than for linear susceptibility (cf. Fig. 8.5 with Fig. 8.20). It is understandable, because nonlinear processes are much more sensitive to fluctuations of the local fields. This is due to the fact that a nonlinear response is proportional to a high power of the acting field.

The difference between the results of the present theory and the MF theory is so large that it is impossible to show both simultaneously on the same linear scale. Therefore, we show in Fig. 8.21 these results on the logarithmic scale. As we can

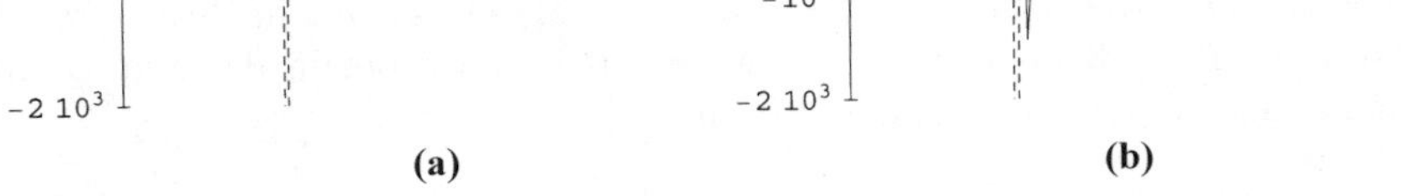

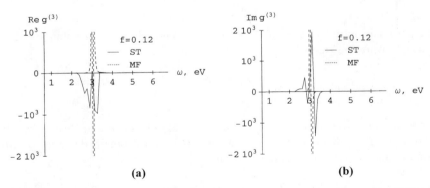

Figure 8.20. Real (a) and imaginary (b) parts of the enhancement coefficient $g^{(3)} = \chi_c^{(3)}/\chi_i^{(3)}$ for the case of nonlinearity in inclusions. The curves are shown for a MG composite with a fill factor of $f = 0.12$ as functions of the light frequency (expressed as the photon energy). The results of the present spectral theory are indicated by ST and those of the mean-field theory by MF.

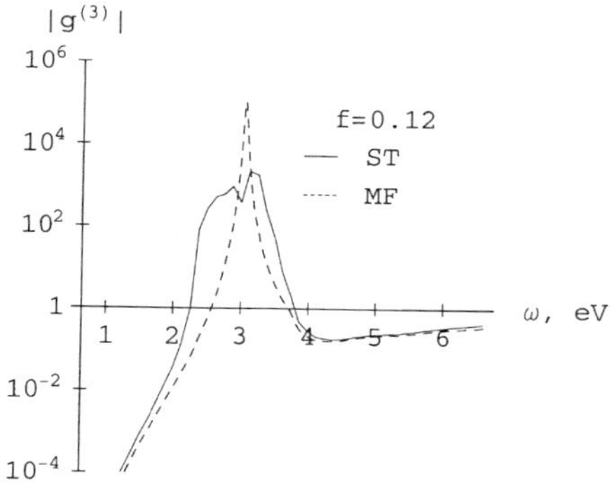

Figure 8.21. Magnitude $|g^{(3)}| = |\chi_c^{(3)}/\chi_i^{(3)}|$ of the enhancement coefficient for a MG composite ($f = 0.12$) computed in the present ST and MF approximation for the case of nonlinearity in inclusions. Note the logarithmic scale.

see, the spectral profiles of $\chi_c^{(3)}$ predicted by these two theories are very different. The MF theory overestimates the enhancement in its peak (close to the surface plasmon resonance) by more then two orders of magnitude and underestimates the enhancement off the plasmon resonance frequency by about the same order of magnitude. The spectral profile in the MF theory is much narrower than what follows from the ST. The cause of these differences is in the neglect of the spatial fluctuations of the local field in the MF theory.

An important feature seen in Figs. 8.20 and 8.21 is a strong enhancement of the third-order hypersusceptibility by over three orders of magnitude. Still, this enhancement is much smaller than the one predicted and observed for fractal clusters (see Section 8.7). The reason is that the fluctuations of the local fields are much stronger in fractals, leading to the much stronger enhancement of the nonlinear optical responses.

8.8.3. Hypersusceptibility of a Composite for the Case of Nonlinearity in the Host

In this case, to compute the hypersusceptibility from Eq. (8.43), one needs to know the mesoscopic field in the host (outside of the inclusion particles). This is obtained from Eqs. (8.3) and (8.8) as

$$e_\beta^{(1)}(\mathbf{r}) = E_\beta - \sum_b W_{\beta\gamma}(\mathbf{r} - \mathbf{r}_b) d_{b\gamma}$$
$$W_{\beta\gamma}(\mathbf{r}) \equiv \left[\delta_{\beta\gamma} - 3 r_\beta r_\gamma r^{-2}\right] r^{-3} \qquad (8.52)$$

The hypersusceptibility of the composite in this case has been computed by substituting (8.52) into Eq. (8.43) and performing the numerical integration by the Monte Carlo method.

Similarly, the hypersusceptibility in the MF approximation has been obtained by integrating an expression that follows from Eq. (8.43) in our coated-sphere model of the composite unit cell,

$$\chi_c^{(3)} = \chi_h^{(3)} \frac{3}{4\pi R^3} \int_{R_0}^{R} \left| \frac{\mathbf{e}^{(1)}(\mathbf{r})}{E} \right|^2 \left(\frac{\mathbf{e}^{(1)}(\mathbf{r})}{E} \right)^2 d^3 r \tag{8.53}$$

which yields

$$\begin{aligned} g^{(3)} \equiv \frac{\chi_c^{(3)}}{\chi_h^{(3)}} &= \frac{1}{5} |p|^2 p^2 (1-f)[8f(1+f+f^2)|\beta|^2 \beta^2 \\ &\quad + 6f(1+f)|\beta|^2\beta + 2f(1+f)\beta^3 + 18f(|\beta|^2 + \beta^2) + 5] \\ p &\equiv \frac{\varepsilon_c + 2\varepsilon_h}{3\varepsilon_h} \end{aligned} \tag{8.54}$$

This result reduces to the corresponding formula of [13], if one retains only the lowest powers of f. This difference is due to the nature of approximations made in [13].

In Fig. 8.22, we show the real and imaginary parts of the third-order enhancement coefficient $g^{(3)} = \chi_c^{(3)}/\chi_h^{(3)}$ for the case of the nonlinear host under consideration. As we can see, there is a disagreement between the present ST and the MF approximation by orders of magnitude. The enhancement in this case is significantly larger than for nonlinear inclusions, and the spectral dependence is different (cf. Section 8.8.2 and Fig. 8.20).

Because the difference between the ST and the MF theory is too large to compare them on the same plot, we show the magnitude of the enhancement coefficient $|g^{(3)}| = |\chi_c^{(3)}/\chi_i^{(3)}|$ for both the theories in Fig. 8.23. We can see that the magnitude of the enhancement is about an order of magnitude higher than for the case of nonlinearity in the inclusions (cf. Fig. 8.21). There is enhancement also in the wings

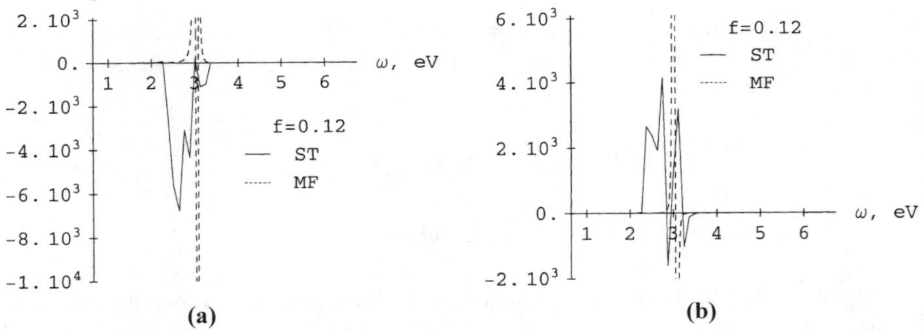

(a) (b)

Figure 8.22. Same as in Fig. 8.20, but for the case of nonlinearity in the host.

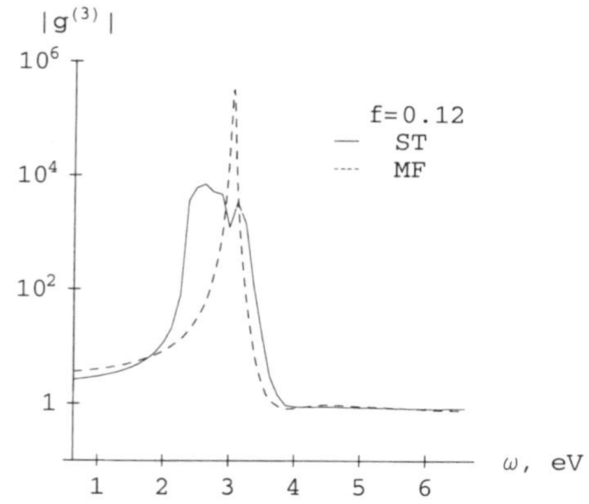

Figure 8.23. Same as in Fig. 8.21, but for the case of nonlinearity in the host.

of the spectral profile, while for the case of internal nonlinearity the composite has lower $\chi_c^{(3)}$ in the spectral wing than that of its nonlinear constituent. The difference is due to a different spectral dependence of the fields inside and outside of an inclusion. Specifically, a large value of the dielectric constant close to the plasmon resonance leads to small values of the fields inside the inclusions (dielectric screening), but not outside.

Similar to Section 8.8.2, the MF enhancement has a much narrower and higher spectral profile peaking about the plasmon resonance frequency of the composite. In contrast, the actual enhancement, as given by the present ST, has a much wider and lower spectral profile. The reason is that it is formed by many chaotic eigenmodes widely distributed in frequency. This feature is interdependent with the giant spatial fluctuations of the local fields (see Sections 8.4–8.6).

Finally, we point out a principal distinction between the optical enhancement in fractal systems (clusters) and nonfractal systems (MG composites). As already mentioned, for the MG composites, the enhancement is peaked about the plasmon resonance frequency, though the enhancement profile is very wide (see, e.g., Figs. 8.21 and 8.23). In contrast, the enhancement for fractal clusters increases from the plasmon resonance frequency toward the extreme spectral wings (cf. Figs. 8.12, 8.18, and 8.19). The root of this difference is in much larger (giant) fluctuations of the local fields in fractals, which increase in a scaling manner toward spectral wings.

8.9. CONCLUDING REMARKS

We have considered a variety of the photoprocesses mediated by disordered clusters and composites. We employed models with both fractal (self-similar) and conventional (nonfractal) geometries to simulate different existing systems.

A remarkable feature of the disordered clusters and composites is giant fluctuations of the local fields that bring about their enhanced nonlinear-optical responses.

Namely, the enhancement is due not only to the high averaged value of the local fields, but principally due to their fluctuations in space, from one monomer (inclusion particle) to another. Nonlinearity, causing a higher power of the local fields to be an acting parameter, enhances the effects of these fluctuations. Therefore, the enhancements are found to increase dramatically with the order of nonlinearity.

The chaos of the local fields bears many similarities to quantum chaos, but differs due to the long range of the dipole interaction. Not only individual eigenmodes are chaotic, but also their spatial correlation factors. These chaos and fluctuations are responsible for the dramatic failure of the MF theory to describe the nonlinear susceptibility of composites.

Both fractal clusters and nonfractal composites exhibit strong enhancement of the radiative and nonradiative nonlinear optical responses. However, there are significant qualitative differences and strong quantitative differences between these two classes of systems. Specifically, the enhancement in fractals is much greater. It also has a different spectral dependence: For fractals the enhancement increases monotonically from the plasmon resonance frequency toward the red (infrared) edge of the spectral contour. In contrast, for nonfractal composites, the enhancement is a broad peak around the plasmon resonance. The underlying reason for these differences is giant (spatial) fluctuations of local fields dominating the optical properties of fractals. At the same time, similar fluctuations for conventional (nonfractal) geometries are much smaller.

Finally, many of the effects predicted have been verified experimentally, and we have mentioned some of them. However, many very interesting effects are not discussed due to space limitations. Among them, we recognize the enhanced laser production of plasma by colloidal-metal clusters [55].

REFERENCES

1. J. C. Maxwell Garnett, *Philos. Trans. R. Soc. London* **203**, 385 (1906).
2. H. A. Lorentz, *Theory of Electrons*, Dover, New York, 1952.
3. D. A. G. Bruggeman, *Ann. Phys. (Leipzig)* **24**, 636 (1935).
4. V. M. Shalaev and M. I. Stockman, "Optical properties of fractal clusters (susceptibility, surface enhanced raman scattering by impurities)," *Zh. Eksp. Teor. Fiz.* ***92***, 509 (1987) [*Sov. Phys. JETP* ***65***, 287–294 (1987)].
5. V. A. Markel, L. S. Muratov, and M. I. Stockman, "Theory and numerical simulation of the optical properties of fractal clusters," *Zh. Eksp. Teor. Fiz.* ***98***, 819 (1990) [*Sov. Phys. JETP* ***71***, 455 (1990)].
6. V. A. Markel, L. S. Muratov, M. I. Stockman, and T. F. George, "Theory and numerical simulation of optical properties of fractal clusters," *Phys. Rev. B* ***43***, 8183 (1991).
7. M. I. Stockman, L. N. Pandey, L. S. Muratov, and T. F. George, "Optical absorption and localization of eigenmodes in disordered clusters," *Phys. Rev. B* ***51***, 185 (1995).
8. M. I. Stockman, L. N. Pandey, and T. F. George, "Inhomogeneous localization of polar eigenmodes in fractals," *Phys. Rev. B* ***53***, 2183 (1996).
9. M. Moskovits, "Surface enhanced spectroscopy," *Rev. Mod. Phys.* ***57***, 785 (1985).

10. A. V. Butenko, V. M. Shalaev, and M. I. Stockman, "Giant impurity nonlinearities in optics of fractal clusters," *Zh. Eksp. Teor. Fiz.* ***94***, 107 (1988) [*Sov. Phys. JETP* ***67***, 60 (1988)].
11. T. A. Witten and L. M. Sander, "Diffusion-limited aggregation, a kinetic critical phenomenon," *Phys. Rev. Lett.* ***47***, 1400 (1981).
12. G. L. Fischer, R. W. Boyd, R. J. Gehr, S. A. Jenekhe, J. A. Osaheni, J. E. Sipe, and L. A. Wellerbrophy, "Enhanced nonlinear-optical response of composite materials," *Phys. Rev. Lett.* ***74***, 1871 (1995).
13. J. E. Sipe and R. W. Boyd, "Nonlinear susceptibility of composite optical materials in the Maxwell Garnett model," *Phys. Rev. B* ***46***, 1614 (1992).
14. R. W. Boyd and J. E. Sipe, "Nonlinear susceptibility of layered composite materials," *J. Opt. Soc. Am. B* ***11***, 297 (1994).
15. R. W. Boyd, "Influence of local field effects on the nonlinear optical properties of composite materials," *Proc. SPIE* ***2524***, 136 (1996).
16. R. J. Gehr, G. L. Fisher, and R. W. Boyd, "Nonlinear optical response of layered composite materials," *Phys. Rev. A* ***53***, 2792 (1996).
17. R. J. Gehr and R. W. Boyd, "Optical properties of nanostructured optical materials," *Chem. Materials* ***8***, 1807 (1996).
18. R. W. Boyd, R. J. Gehr, G. L. Fisher, and J. E. Sipe, "Nonlinear optical properties of nanocomposite materials," *Pure Appl. Opt.* ***5***, 505 (1996).
19. R. J. Gehr, G. L. Fisher, and R. W. Boyd, "Nonlinear optical response of porous-glass-based materials," *J. Opt. Soc. Am. B* ***14***, 2310 (1997).
20. D. B. Smith, G. L. Fisher, R. W. Boyd, and D. A. Gregory, "Cancellation of photoinduced absorption in metal nanoparticle composites through a counterintuitive consequence of local field effects," *J. Opt. Soc. Am. B* ***14***, 1625 (1997).
21. G. L. Fisher and R. W. Boyd, "Third-order nonlinear-optical properties of selected composites," *ACS Symp. Ser.* ***679***, 108 (1997).
22. G. W. Milton, "Bounds on the complex permittivity of a two-component composite material," *J. Appl. Phys.* ***52***, 5286 (1981).
23. G. W. Milton, "Bounds on the transport and optical properties of a two-component composite material," *J. Appl. Phys.* ***52***, 5294 (1981).
24. G. W. Milton, "Bounds on the electromagnetic, elastic, and other properties of two-component composites," *Phys. Rev. Lett., 46542* (1981).
25. D. J. Bergman and D. Stroud, in H. Ehrenreich and D. Turnbull, Ed., *Solid State Physics*, Vol. 46, Academic, Boston, 1992, p. 147.
26. H. Ma, W. Wen, W. Y. Tam, and P. Sheng, "Frequency dependent electrorheological properties: Origins and bounds," *Phys. Rev. Lett.* ***77***, 2499 (1996).
27. W. Y. Tam, G. H. Yi, W. Wen, H. Ma, M. M. T. Loy, and P. Sheng, "New electro-rheological fluid: Theory and experiment," *Phys. Rev. Lett.* ***78***, 2987 (1997).
28. K. Ghosh and R. Fuchs, "Spectral theory for two-component porous media," *Phys. Rev. B* ***38***, 5222 (1988).
29. R. Fuchs and F. Claro, "Spectral representation for the polarizability of a collection of dielectric spheres," *Phys. Rev. B* ***39***, 3875 (1989).
30. S. Alexander, "The vibration of fractals and scattering from aerogels," *Phys. Rev. B* ***40***, 7953 (1989).

31. V. M. Shalaev, R. Botet, and A. V. Butenko, "Localization of collective dipole excitation on fractals," *Phys. Rev. B* ***48***, 6662 (1993).
32. M. I. Stockman, K. B. Kurlayev, and T. F. George, "Linear and nonlinear optical susceptibilities of Maxwell Garnett composites: Dipolar spectral theory," *Phys. Rev. B* ***60***, 17071 (1999).
33. M. I. Stockman, L. N. Pandey, L. S. Muratov, and T. F. George, "Comment on photon scanning tunneling microscopy images of optical excitations of fractal metal colloid clusters," *Phys. Rev. Lett.* ***75***, 2450 (1995).
34. M. I. Stockman, "Chaos and spatial correlations for dipolar eigenmodes," *Phys. Rev. Lett.* ***79***, 4562 (1997).
35. M. I. Stockman, "Inhomogeneous eigenmode localization, chaos and correlations for large disordered clusters," *Phys. Rev. E* ***56***, 6494 (1997).
36. A. V. Karpov, A. K. Popov, S. G. Rautian, V. P. Safonov, V. V. Slabko, V. M. Shalaev, and M. I. Stockman. "Observation of a wavelength- and polarization-selective photomodification of silver clusters," *Pis'ma Zh. Eksp. Teor. Fiz.* ***48***, 528 (1988) [*JETP Lett.* ***48***, 571 (1988)].
37. Yu. E. Danilova, A. I. Plekhanov, and V. P. Safonov, "Experimental study of polarization-selective holes burned in absorption spectra of metal fractal clusters," *Physica A* ***185***, 61 (1992).
38. V. P. Safonov, V. M. Shalaev, V. A. Markel, Yu. E. Danilova, N. N. Lepeshkin, W. Kim, S. G. Rautian, and R. L. Armstrong, "Spectral dependence of selective photomodification in fractal aggregates of colloidal particles," *Phys. Rev. Lett.* ***80***, 1102 (1998).
39. D. P. Tsai, J. Kovacs, Z. Wang, M. Moskovits, V. M. Shalaev, J. S. Suh, and R. Botet, "Photon scanning tunneling microscopy images of optical excitations of fractal metal colloid clusters," *Phys. Rev. Lett.* ***72***, 4149 (1994).
40. P. Zhang, T. L. Haslett, C. Douketis, and M. Moskovits, "Mode localization in self-affine fractal interfaces observed by near-field microscopy," *Phys. Rev.* ***57***, 15513 (1998).
41. V. A. Markel, V. M. Shalaev, P. Zhang, W. Huynh, L. Tay, T. L. Haslett, and M. Moskovits, "Near-field optical spectroscopy of individual surface-plasmon modes in colloid clusters," *Phys. Rev. B* ***59***, 10903 (1999).
42. S. I. Bozhevolnyi, V. A. Markel, V. Coello, W. Kim, and V. M. Shalaev, "Direct observation of localized dipolar excitations on rough nanostructured surfaces," *Phys. Rev. B* ***58***, 11441 (1998).
43. M. I. Stockman, L. N. Pandey, L. S. Muratov, and T. F. George, "Giant fluctuations of local optical fields in fractal clusters," *Phys. Rev. Lett.* ***72***, 2486 (1994).
44. P. B. Johnson and R. W. Christy, "Optical constants of noble metals," *Phys. Rev. B* ***6***, 4370 (1972).
45. M. V. Berry, "Regular and irregular semiclassical wavefunctions," *J. Phys. A* ***10***, 2083 (1977).
46. M. C. Gutzwiller, *Chaos in Classical and Quantum Mechanics*, Springer-Verlag, New York, 1991.
47. V. I. Fal'co and K. B. Efetov, "Long range correlations in the wave functions of chaotic systems," *Phys. Rev. Lett.* ***77***, 912 (1996).
48. M. I. Stockman, V. M. Shalaev, M. Moskovits, R. Botet, and T. F. George, "Enhanced Raman scattering by fractal clusters: Scale invariant theory," *Phys. Rev. B* ***46***, 2821 (1992).

49. K. Kneipp, Y. Wang, H. Kneipp, L. T. Perelman, I. Itzkan, R. D. Dasari, and M. Feld, "Single molecule detection using surface-enhanced Raman scattering (SERS)," *Phys. Rev. Lett.* ***78***, 1667 (1997).

50. V. M. Shalaev and M. I. Stockman, "Resonant excitation and nonlinear optics of fractals," *Physica A* ***185***, 181 (1992).

51. V. A. Markel, V. M. Shalaev, E. B. Stechel, W. Kim, and R. L. Armstrong, "Small-particle composites. I. Linear optical properties," *Phys. Rev. B* ***53***, 2425 (1996).

52. V. M. Shalaev, E. Y. Polyakov, and V. A. Markel, "Small-particle composites. II. Nonlinear optical properties," *Phys. Rev. B* ***53***, 2437 (1996).

53. A. V. Butenko, P. A. Chubakov, Yu. E. Danilova, S. V. Karpov, A. K. Popov, S. G. Rautian, V. P. Safonov, V. V. Slabko, V. M. Shalaev, and M. I. Stockman, "Nonlinear optics of metal fractal clusters," *Z. Phys. D* ***17***, 283 (1990).

54. D. Stroud and P. M. Hui, "Nonlinear susceptibilities of granular matter," *Phys. Rev. B* ***37***, 8719 (1988).

55. M. M. Murnane, H. C. Kapteyn, S. P. Gordon, J. Bokor, E. N. Glytsis, and R. Falcone, "Efficient coupling of high-intensity subpicosecond laser pulses into solids," *Appl. Phys. Lett.* ***62***, 1068 (1993).

CHAPTER NINE

Some Theoretical and Numerical Approaches to the Optics of Fractal Smoke

VADIM A. MARKEL

Department of Electrical Engineering, Washington University, St. Louis, MO 63130

VLADIMIR M. SHALAEV

Department of Physics, New Mexico State University, Las Cruces, NM 88003

THOMAS F. GEORGE

University of Wisconsin-Stevens Point, Stevens Point, WI 54481

9.1. INTRODUCTION

Investigation of the optical properties of carbonaceous smoke produced by incomplete combustion of different types of fuels or wild fires has practical importance for many application areas, such as climate research and remote sensing of fires, to name just a few. Very active and vigorous research into the optical properties of smoke has been conducted over the past 30 years. Many of the experimental and theoretical questions are now resolved. The geometrical structure and chemical composition of soot aggregates has been studied in detail (see Chapter 10), and many analytical and numerical methods for calculating the optical characteristics and for obtaining physical properties of smoke from optical measurements have been developed. However, there are several factors that preclude this topic from being closed.

Most of the quantitatively accurate results in the visible and near-infrared (IR) spectral regions appeared only recently, as increasingly powerful computers became available. However, when the wavelength is further increased, the electromagnetic (EM) interaction of small carbon nanospheres that comprise soot particles becomes stronger and more important, and so becomes important the geometrical structure of soot. This importance was demonstrated by Bruce et al. [1] who measured optical

Optics of Nanostructured Materials, Edited by Vadim A. Markel and Thomas F. George
ISBN 0-471-34968-2

characteristics of diesel soot from the visible to the centimeter wavelength range. Unfortunately, most analytical and numerical methods become less effective, or even not applicable, when the interaction of primary nanospheres becomes strong. This fact is, of course, evident for perturbative methods that treat such interaction as a perturbation. But this is also true for nonperturbative numerical methods based on the multipole expansion of fields scattered by each nanosphere and satisfying boundary conditions at each surface of discontinuity. Generally, such approaches lead to an infinite system of linear equations with respect to the unknown expansion coefficients, which has to be truncated. After the truncation, its dimensionality is $\sim N(L+1)^2$, where N is the number of monomers in a soot cluster and L is the maximum order of the spherical harmonics involved in the expansion. As will be illustrated below, the value of L required for satisfactory convergence of the method grows with the wavelength, and eventually makes obtaining a numerical solution not feasible. In this chapter, we will discuss a nonperturbative method based on the geometrical renormalization of clusters. This method allows one to stay within the dipole approximation ($L = 0$) and, therefore, lacks the numerical complexity of the full multipole expansion.

Another reason why research into the optics of carbonaceous soot is far from being completed is the complexity of the object. Indeed, most of the results obtained in the literature assume the simplest geometrical structure and composition of soot clusters. In practice, soot is much more complicated, both geometrically and chemically. In the atmosphere, soot can interact and form agglomerates with moisture and other chemical elements, which can lead to restructuring and a significant change in optical properties. Chapter 10 is largely devoted to this circle of problems, as well as to experimental and theoretical aspects of studying the soot structure variability and its implications for optical properties.

In this chapter, we focus on analytical and numerical approaches to calculating optical properties of fractal smoke with a simple fractal structure and optical constants that are assumed to be known. Thus, we will consider a purely EM problem, leaving a lot of complexity that is characteristic to the physical properties of soot out of our discussion. However, two introductory sections give a brief review of the geometrical properties and optical constants of fractal soot, since they are used in numerical examples throughout the chapter. In Sections 9.5 and 9.6 we will consider two topics that are, in a sense, nontraditional: fluctuations of light intensity scattered by random smoke aggregates and the absorption of light by the smoke clusters placed inside water microdroplets, both in the first Born approximation.

9.2. GEOMETRICAL PROPERTIES

It has been long recognized that smoke usually consists of agglomerates of hundreds or thousand of small, nearly spherical particles (monomers) with typical radii varying from 10 to 50 nm, depending on the origin of the smoke [1–6]. The distribution of monomer sizes for a specific type of smoke is, however, significantly more narrow,

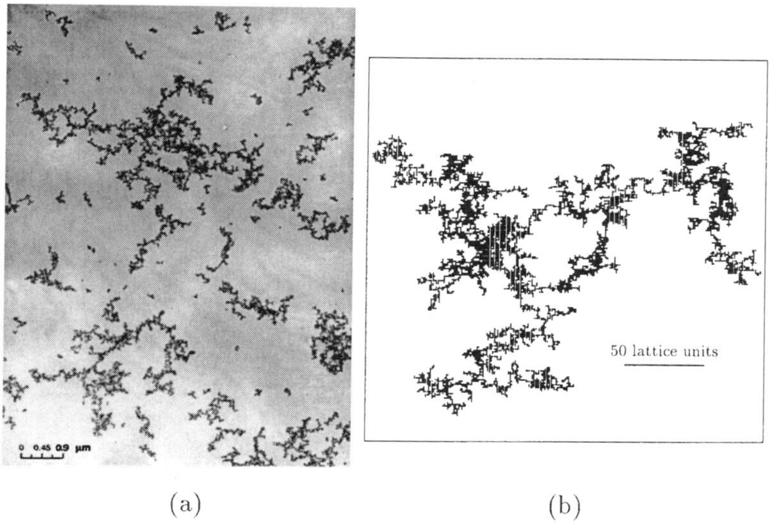

Figure 9.1. Experimental (a) (Courtesy of E. F. Mikhailov, S. S. Vlasenko, and A. A. Kiselev) and computer-generated (b) smoke clusters.

and it is customary to assume that the diameters of the spherical monomers are the same. A sample micrograph of several smoke agglomerates is shown in Fig. 9.1(a). The reader will recognize that smoke clusters have fractal geometry. This fact was verified experimentally by digitization of electron micrograph images similar to the one shown in Fig. 9.1(a) [3,5,7–9] and by scattering experiments [7,8,10,11]. The value of fractal dimension D was shown to be close to 1.8.

The geometrical structure of smoke aggregates is most often simulated using the cluster–cluster aggregation model introduced by Meakin [12] and Jullien et al. [13]. In this model, monomers are sparsely and randomly distributed in space at the initial moment of time and, then, allowed to move via Brownian trajectories, sticking on contact. The subclusters formed in this process continue to move, colliding and sticking with other subclusters and isolated monomers, until large agglomerates are formed. A sample computer-generated cluster–cluster aggregate is shown in Fig. 9.1(b). The Meakin model accurately describes the statistical properties of soot because it captures the most important features of the real aggregation process: subclusters of various sizes and individual particles move in space simultaneously and independently, there is no fixed center of aggregation, and the dependence of mobility of individual subclusters on their mass can be easily taken into account in simulations.

Although the visual resemblance of the experimental and computer-generated samples in Fig. 9.1 is more or less apparent, a comparison is complicated due to the random nature of the smoke agglomerates. Therefore, it is essential to study the statistical characteristics of the clusters. One of the most important of such characteristics, from the point of view of optical properties, is the pair density–density correlation function $p(r)$, which can be defined as the probability density to find a pair of distinct monomers separated by the distance r. The Fourier transform (FT) of $p(r)$ gives the optical structure factor [14]. For fractals, this function obeys the power-law dependence on r in the so-called intermediate asymptote region $l \ll r \ll R_g$, where l is the distance between two neighboring monomers (referred to as the "lattice unit" below), and R_g is the cluster radius of gyration, defined as $R_g = \sqrt{\langle(\mathbf{r}_i - \mathbf{R}_{\rm cm})^2\rangle}$, with $\mathbf{r}_i$ and $\mathbf{R}_{\rm cm}$ being the radius vectors of the ith monomer in a cluster and of the cluster's center of mass. In fact, the above inequalities do not need to be especially strong (usually, the factor of 2 is sufficient), and since most of the physically important integrals involving $p(r)$ converge at the lower limit, the region of applicability of the scaling formula can be extended to $r = 0$. Then, for a monodisperse ensemble of random soot clusters, $p(r)$ can be written in the most general form as

$$p(r) = \frac{ar^{D-1}}{Nl^D} f\left[\frac{r}{R_g(N)}\right] \qquad \int_0^\infty p(r)dr = 1 \tag{9.1}$$

where N is the number of monomers in a cluster, a is a numerical constant of the order of unity, D is the fractal dimension, and $f(x)$ is the cutoff function, such that $f(0) = 1$, $|df(0)/dx| < \infty$. The two-point correlation function found numerically in [15] is shown in Fig. 9.2(a).

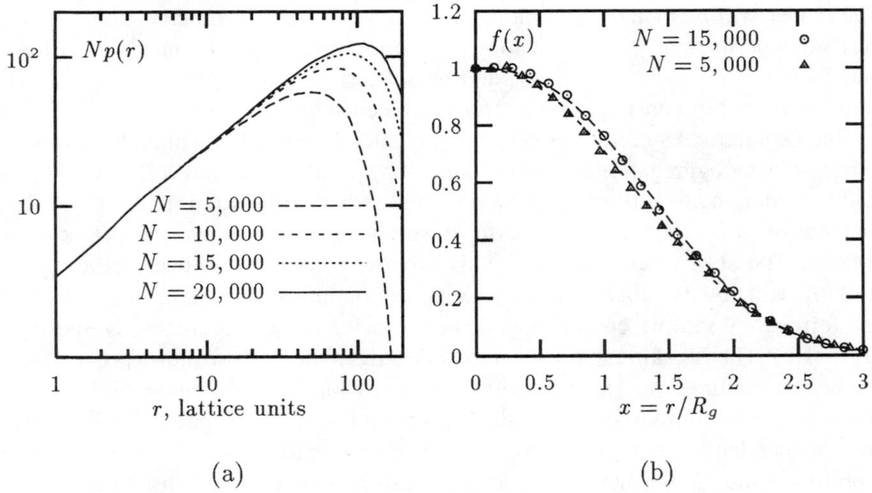

Figure 9.2. Two-point correlation functions $p(r)$ for cluster–cluster aggregates with different numbers of particles N (a) and the corresponding cutoff function $f(x)$ (b). In (b), dashed lines correspond to the generalized exponential cutoff of the form $f(x) = \exp(-\alpha x^\beta)$, and the centered symbols (circles and triangles) correspond to the numerical calculations. The values of the constants are $\alpha = 0.344$; $\beta = 2.238$ for $N = 5000$; and $\alpha = 0.273$, $\beta = 2.489$ for $N = 15,000$.

The dependence of the gyration radius on N is also governed by the fractal dimension,

$$R_g(N) = blN^{1/D} \tag{9.2}$$

where b is another dimensionless constant.

In a recent numerical study [15], we found that the constants D determined from Eqs. (9.1) and (9.2) can be slightly different.[1] We have also evaluated the constants a and b numerically for computer-generated cluster–cluster aggregates and found that $a \approx 4$ and $b \approx 0.6$, which is in qualitative agreement with other studies. We estimate from the data of Mountain and Mulholland [17], Cai et al. [8,11], and Oh and Sorensen [18], that $b \approx 0.4$. The values of a constant related to b (and of the fractal dimension) obtained from numerical simulations by different authors were reviewed by Wu and Friedlander [19], with b varying in the range from 0.25 to 0.5. It should be noted that while the formulas (9.1) and (9.2) are quite universal, the specific values of a and b can vary depending on the regime of aggregation. They are also sensitive to the number of primary particles employed in numerical calculations. In our calculations, the maximum value of N employed for the numerical fitting of (9.2) was 20,000, which is significantly larger than the maximum N in the set of data presented by Wu and Friedlander [19] ($N_{\max} = 500$). The dependence on N can be explained by the phenomenon of multiscaling, when the characteristic constants can slowly depend on N.

Significantly less information can be found on the constant a. However, when the form of the cutoff function $f(x)$ is specified, a is not an independent constant because of the normalization condition. The combination ab^D is fixed and can be calculated from the form of the cutoff function. We have estimated that $ab^D \approx 1.6$ [15].

When the physically important integrals involving $p(r)$ converge at the upper limit while r is still smaller than R_g, and the value of the cutoff function in (9.1) does not deviate significantly from unity, the knowledge of the exact form of the cutoff function $f(x)$ is not necessary. This was illustrated by Berry and Percival [20] in the frame of the mean-field approximation, and we will see such examples below. But in general, the form of $f(x)$ influences the optical properties. The well-known example is the first Born approximation for the differential scattering cross-section for the "intermediate" values of the transmitted wavevector $q = |\mathbf{k} - \mathbf{k}'| \sim 1/R_g$ [14]. The most frequently discussed forms of $f(x)$ are the generalized exponential $f(x) = \exp(-\alpha x^\beta)$ (the Gaussian cutoff is the particular case $\beta = 2$) and the so-called overlapping spheres cutoff $f(x) = (x - x_0)^2(x + 2x_0)/2x_0^3$ if $x < x_0$, which is the exact analytic cutoff for a random nonfractal gas of particles enclosed in a spherical volume.

Mountain and Mulholland found numerically that $f(x)$ is of a generalized exponential form with $\alpha = 0.2$ and $\beta = 2.5$ [17]. Sorensen and co-workers studied the cutoff functions by analyzing electron micrographs of soot clusters [11] and also indirectly by analyzing light scattering data [21,22] (with the interpretation of the

[1]This phenomenon is related to multiscaling [16].

scattering data based on the first Born approximation) and found that $f(x)$ is decaying with x much faster than exponentially, with the Gaussian cutoff being a fairly good approximation. Our numerical results, which are in qualitative agreement with the above references, also confirm the generalized exponential form of $f(x)$ with coefficients α and β exhibiting a slow systematic dependence on the number of monomers N. An example of generalized exponential fit to the numerically calculated $f(x)$ is shown in Fig. 9.2(b) (see the figure caption for details).

To conclude this section, we briefly discuss higher order correlation functions. They naturally appear in higher orders of the perturbation theory, or when fluctuations of the optical characteristics (rather than the ensemble-average quantities) are considered [15,23]. In a Gaussian medium, all higher order correlation functions can be expressed analytically through the second-order correlators [24]. This fact significantly simplifies the diagrammatic technique in the perturbation expansion of the mean field for wave propagation in a random Gaussian medium [25]. However, as we have verified numerically, fractal cluster–cluster aggregates are not Gaussian [15]. Because of the many similarities between the computer-generated cluster–cluster aggregates and real smoke clusters, it is reasonable to believe that this conclusion is also true for real smoke. As an example, we have studied in detail the reduced four-point correlation function $p_4(r)$, which is important for describing the deviations of scattered intensity from the average due to the random nature of clusters (see Section 9.5). The correlator $p_4(r)$ is defined as the probability density to find the distance $|\mathbf{r}_i - \mathbf{r}_j + \mathbf{r}_k - \mathbf{r}_l|$ $[i \neq j,\ k \neq l$, and any of the pair of indexes $(i,\ j)$ can coincide with any of the pair $(k,\ l)]$. It was found that $p_4(r)$ is not described by a scaling formula with a cutoff similar to (9.1), but is given by a

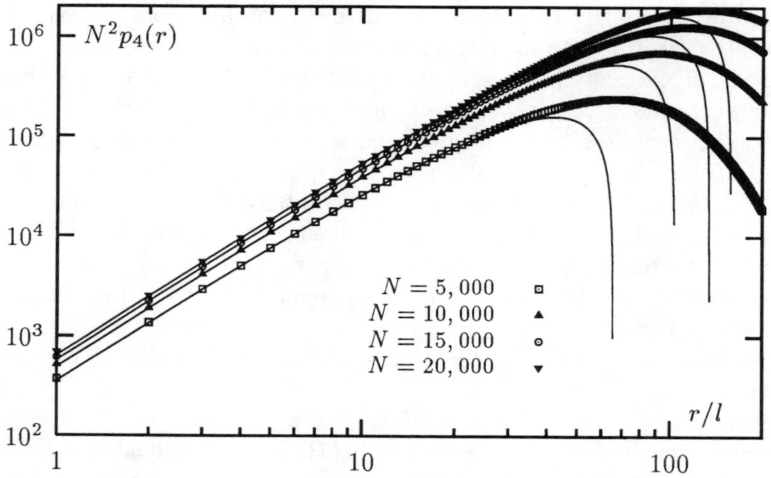

Figure 9.3. Four-point correlation function $p_4(r)$ calculated numerically (centered symbols) and from the analytic expression (1.3) with $n = 2$, $c_0 \approx c_1 \approx 23$ (solid line) for different N. The gyration radius R_g changes from $\approx 50l$ for $N = 5000$ to $\approx 100l$ for $N = 20{,}000$.

series of the form

$$p_4(r) = \frac{r^2}{N^{3/D} l^3} \sum_{k=0}^{n} (-1)^k c_k \left(\frac{r}{R_g} \right)^{k(2D-3)} \tag{9.3}$$

where c_k are dimensionless positive coefficients that have to be determined numerically. Approximation of $p_4(r)$ by the expression (9.3) is illustrated in Fig. 9.3. By increasing the number of terms n in (9.3), it is possible to fit $p_4(r)$ for increasingly higher values of r. The important feature of (9.3) is that the coefficients c_k do not depend on N (apart from a very weak multiscaling dependence) and are in that sense universal, which distinguishes (9.3) from an arbitrary power expansion. Another important feature of Eq. (9.3) is that $p_4(r)$ cannot be described by a universal scaling behavior with a cutoff function of the type $p_4(r) = r^2 f(r/ R_g)/N^{3/D} l^3$. Although a function $f(x)$ can be found from (9.3), its first derivative diverges at $x = 0$ for $D < 2$ (one of the requirements for the cutoff function is a finite derivative at $x = 0$).

9.3. OPTICAL CONSTANTS OF CARBON SMOKE

Dalzell and Sarofim [26] studied the optical constants of several soots using reflectance measurements. They suggested a dispersion formula that describes experimental measurements quite accurately and is based on the well-known quantum expression for the complex dielectric function

$$\varepsilon(\omega) = 1 - \sum_n \frac{f_n^2}{\omega^2 - \omega_n^2 + i\gamma_n \omega} \tag{9.4}$$

Earlier, Taft and Philipp [27] identified experimentally three optical resonances in graphite, two of which correspond to bound electrons and one to a conduction electron. The resonance frequencies are $\omega_c = 0$ (conduction electron), $\omega_1 = 1.25 \times 10^{15}\ \mathrm{s}^{-1}$ and $\omega_2 = 7.25 \times 10^{15}\ \mathrm{s}^{-1}$ (or corresponding wavelengths: $\lambda_c = \infty$, $\lambda_1 = 1.51\ \mu\mathrm{m}$, $\lambda_2 = 0.26\ \mu\mathrm{m}$). The values of the relaxation constants were found to be $\gamma_c = \gamma_1 = 6.00 \times 10^{15}\ \mathrm{s}^{-1}$, $\gamma_2 = 7.25 \times 10^{15}\ \mathrm{s}^{-1}$. Dalzell and Sarofim [26] assumed that the same electronic transitions contribute to the dielectric constant of carbon soot and used the above values of ω_n, γ_n to fit the formula (9.4) to their experimental data treating f_n, which depend on the concentration of optically active electrons, as free parameters. A very accurate fit to the experimental data for propane soot was achieved for the following values of f_n: $f_c = 4.04 \times 10^{15}\ \mathrm{s}^{-1}$, $f_1 = 2.93 \times 10^{15}\ \mathrm{s}^{-1}$, $f_2 = 9.54 \times 10^{15}\ \mathrm{s}^{-1}$ in the spectral range $0.4\ \mu\mathrm{m} < \lambda < 10\ \mu\mathrm{m}$.

The real and imaginary parts of the complex refraction index $m = \sqrt{\varepsilon} = n + ik$ calculated from formula (9.4) with the constants specified above are illustrated in Fig. 9.4(a). The wavelength range in Fig. 9.4(a) is somewhat expanded compared to the range of experimental measurements by Dalzel and Sarofim (from 0.4 to 10 μm) toward shorter wavelengths, so that the high-frequency resonance at $\lambda = 0.26\ \mu\mathrm{m}$ is clearly visible.

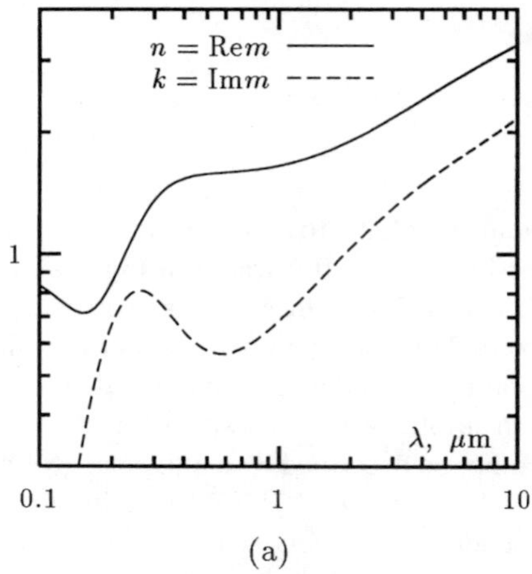

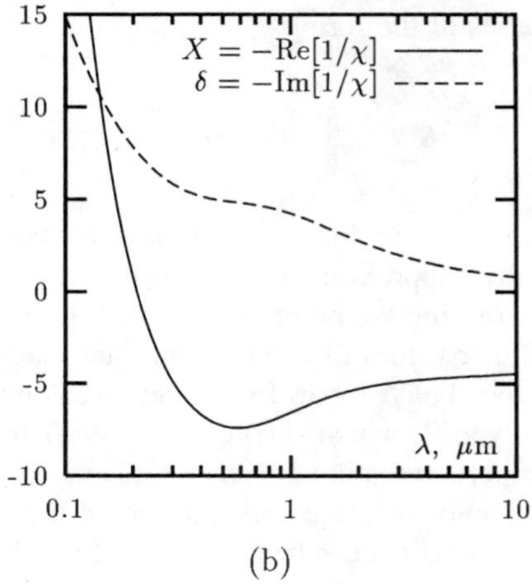

Figure 9.4. (a): Real and imaginary parts of the complex refraction index $m = n + ik$ of carbon calculated from the data of Dalzell and Sarofim [26]. (b): Spectral dependence of the parameters X and δ defined by $1/\chi = -(X + i\delta), \chi = (3/4\pi)(\varepsilon - 1)/(\varepsilon + 2)$.

Analogous three-electron dispersion formulas were used by Habib and Vervisch [28] to describe optical constants of smoke at the flame temperatures. The temperature dependence is mainly governed by the temperature dependence of the conduction electron relaxation constant: $\gamma_c \propto T^{1/2}$ (Lee and Tien [29]). Another issue discussed in the literature is the dependence of the free-electron concentration on the H/C ratio of the fuel [26,28,29].

We use formula (9.4) with the values of constants specified above in all our numerical examples.

In Fig. 9.4(b) we also show the spectral dependence of two important optical parameters, X and δ, which were originally introduced in [30,31].[2] They are defined as $X = -\text{Re}[1/\chi]$, $\delta = -\text{Im}[1/\chi]$, where $\chi = (3/4\pi)(\varepsilon - 1)/(\varepsilon + 2)$. The physical meaning of these parameters is that X is the generalized detuning from the resonance and δ which is the generalized absorption strength. As we will see below, analysis of the spectral dependence of these variables can be useful even when the number of optical resonances is > 1 (at least three in the case of black carbon) and the spectral dependence of ε is more complicated than in the simple Drude model with one optically active electron. Note that X and δ, as defined above, are dimensionless and independent of the sample geometry.

9.4. OPTICAL PROPERTIES OF SMOKE

In this section, we review some of the theoretical and numerical approaches to the calculation of optical characteristics of smoke clusters. In particular, we will

[2]The parameters X and δ used in [30,31] differ from the dimensionless parameters defined below by a multiplicative factor $\pi l^3/6$, which has the dimensionality of length cubed.

discuss optical cross-sections of linear scattering, absorption, and extinction. A special emphasis will be made on the extinction, and numerical examples in this section will be restricted to illustration of the extinction cross-section. A review of the structure factor and scattering in the Born approximation can be found in Chapter 10.

9.4.1. Basic Equations

A very convenient starting point for solving the problem of the linear interaction of EM waves with smoke clusters is the Lippman–Schwinger formulation or, more specifically, the Maxwell equations in the integral form, written for the polarization function $\mathbf{P}(\mathbf{r})$ inside the soot material:

$$\mathbf{P}(\mathbf{r}) = \chi\left[\mathbf{E}_{\text{inc}}(\mathbf{r}) + \sum_{i=1}^{N}\int_{V_i} \boldsymbol{G}_R(\mathbf{r}-\mathbf{r}')\mathbf{P}(\mathbf{r}')d^3r'\right] \qquad \mathbf{r} \in V_k \tag{9.5}$$

$$\chi = (3/4\pi)[(\varepsilon - 1)/(\varepsilon + 2)] \tag{9.6}$$

In the next few paragraphs, we explain the notations used in (9.5) and their physical meaning.

First, since we are considering only the linear interaction between the EM fields and matter, Eq. (9.5) is written in the frequency domain for just one (but yet unspecified) value of the frequency ω. The time-dependence factor, $e^{-i\omega t}$, is common to all time-varying quantities and will be omitted everywhere below.

In principle, the incident field $\mathbf{E}_{\text{inc}}$ can be arbitrary as long as it satisfies the free-space Maxwell equations. In most practical cases, the curvature of the wave front of the incident radiation is much larger than the characteristic system size, and it is sufficient to consider incident plane waves of the form

$$\mathbf{E}_{\text{inc}}(\mathbf{r}) = \mathbf{E}_0 \exp(i\mathbf{k}\cdot\mathbf{r}) \tag{9.7}$$

where $k = \omega/c$ is the free-space wavenumber. In this section, we will consider only plane incident waves. However, in Section 9.6 the smoke aggregates will be placed inside water microdroplets and the incident field will be replaced by vector spherical harmonics.

Next, $\boldsymbol{G}_R(\mathbf{r})$ is the *regular* part of the free-space Green's function for the vector wave equation. If there is a point dipole $\mathbf{d}$ at the origin, the electric field at a point $\mathbf{r} \neq 0$ is given by $\mathbf{E}(\mathbf{r}) = \boldsymbol{G}_R(\mathbf{r})\mathbf{d}$. Green's function is a tensor (dyadic) because it transforms one vector into another, which is, in general, not collinear with the first one. The complete Green's function $\boldsymbol{G}(\mathbf{r})$ contains both *regular* and *singular* parts:

$$\boldsymbol{G}(\mathbf{r}) = \boldsymbol{G}_R(\mathbf{r}) - \frac{4\pi}{3}\,\delta(\mathbf{r})\boldsymbol{I} \tag{9.8}$$

where $\boldsymbol{I}$ is the unity tensor and $\delta(\mathbf{r})$ the delta function. The coordinate representation of $\boldsymbol{G}_R(\mathbf{r})$ is given by

$$(G_R(\mathbf{r}))_{\alpha\beta} = k^3[A(kr)\delta_{\alpha\beta} + B(kr)r_\alpha r_\beta/r^2] \tag{9.9}$$

$$A(x) = [x^{-1} + ix^{-2} - x^{-3}]\exp(ix) \tag{9.10}$$

$$B(x) = [-x^{-1} - 3ix^{-2} + 3x^{-3}]\exp(ix) \tag{9.11}$$

where the Greek indexes denote the Cartesian components and $\delta_{\alpha\beta}$ is the Kronecker delta symbol. However, calculation of spatial integrals that arise in the perturbation expansion considered in Section 9.4.2 is much easier with the use of the following representation of the *complete* Green's function:

$$\boldsymbol{G}(\mathbf{r}) = \left(\boldsymbol{I} + \frac{1}{k^2}\nabla\nabla\right)g(\mathbf{r}) \tag{9.12}$$

$$g(\mathbf{r}) = \frac{k^2 e^{ikr}}{r} \tag{9.13}$$

where $g(\mathbf{r})$ is the scalar Green's function. The notation $\nabla\nabla$ can be understood as $\nabla\nabla\mathbf{F} = \nabla(\nabla \cdot \mathbf{F})$.

The integral in (9.5) is taken over the region occupied by the soot material, which is composed from many spherical regions denoted by V_i $(i = 1, \ldots, N)$. We will also denote by V_{tot} the space region occupied by all monomers, that is, $V_{\text{tot}} = V_1 \cup V_2 \cup \ldots \cup V_N$. The center of each spherical region is located at the point $\mathbf{r}_i$, and its radius is R_m (same for all monomers). We denote the monomer volume by v $[v = (4\pi/3)R_m^3]$ and the total volume of all monomers by v_{tot} $(v_{\text{tot}} = Nv)$. Thus, the capital letter V will be used to denote the spatial regions, while small v is their volumes. Below, we will sometimes write $\int_{V_{\text{tot}}}$ as a shortcut for $\sum_{i=1}^{N} \int_{V_i}$.

We also assume that $R_m \ll \lambda$ throughout this chapter. This is a fundamental assumption used in all derivations and numerical examples below. Although it is usually quite accurate, the ratio R_m/λ can become large in the visible spectral range for some types of smoke produced by huge fires (in which case the size of primary spheres tends to be larger), and, of course, R_m/λ cannot be considered as small for shorter wavelengths. The influence of the finite monomer size was studied by Mulholland and co-workers [32,33].

Finally, the coupling constant χ (9.6) is, in fact, the dielectric susceptibility of a sphere in the quasistatic limit. However, no quasistatic approximations were made in Eq. (9.5).

The polarization function $\mathbf{P}(\mathbf{r})$ can be used to calculate all optical properties of the smoke clusters. The scattering amplitude $\mathbf{f}(\mathbf{k}')$ is given by

$$\mathbf{f}(\mathbf{k}') = k^2 \int_{V_{\text{tot}}} \left\{\mathbf{P}(\mathbf{r}) - \frac{1}{k^2}\,[\mathbf{P}(\mathbf{r}) \cdot \mathbf{k}']\mathbf{k}'\right\}\exp(-i\mathbf{k}' \cdot \mathbf{r})d^3r \tag{9.14}$$

with $\mathbf{k}'$ being the scattered wavevector and $\mathbf{k}$ the incident wavevector, where $|\mathbf{k}| = |\mathbf{k}'| = k = \omega/c$.

The differential scattering cross-section is given by

$$\frac{d\sigma_s}{d\Omega} = \frac{|\mathbf{f}(\mathbf{k}')|^2}{|\mathbf{E}_0|^2} \tag{9.15}$$

and the integral extinction, scattering, and absorption cross-sections, σ_e, σ_s, and σ_a, respectively, can be found from the optical theorem:

$$\sigma_e = \frac{4\pi \mathrm{Im}[\mathbf{f}(\mathbf{k}) \cdot \mathbf{E}_0^*]}{k|\mathbf{E}_0|^2} \tag{9.16}$$

$$\sigma_s = \frac{1}{|\mathbf{E}_0|^2} \int |\mathbf{f}(\mathbf{k}')|^2 d\Omega \tag{9.17}$$

$$\sigma_a = \sigma_e - \sigma_s \tag{9.18}$$

where we have assumed excitation by a plane wave of the form (9.7).

The expression for the extinction cross-section follows readily from (9.16) and (9.14):

$$\sigma_e = \frac{4\pi k}{|\mathbf{E}_0|^2} \mathrm{Im} \int_{V_{\mathrm{tot}}} \mathbf{P}(\mathbf{r}) \cdot \mathbf{E}_{\mathrm{inc}}^*(\mathbf{r}) d^3 r \tag{9.19}$$

The expressions for the integral scattering and absorption cross-sections contain double volume integration, since σ_s is quadratic in $\mathbf{f}$. However, the angular integration in (9.17) and one of the volume integrals can be calculated in the most general form with the use of the main equation (9.5) [34,35], which leads to expressions for the integral cross-sections that contain only one volume integration. The result for the absorption has a more compact form:

$$\sigma_a = \frac{4\pi k\delta}{|\mathbf{E}_0|^2} \int_{V_{\mathrm{tot}}} \mathbf{P}^*(\mathbf{r}) \cdot \mathbf{P}(\mathbf{r}) d^3 r \tag{9.20}$$

where the parameter $\delta = -\mathrm{Im}[1/\chi]$ was introduced in Section 9.3. The integral scattering cross-section can be found as the difference between (9.19) and (9.20).

The above formulas take an elegant form if we introduce operator notations. First, we notice that the integral transformation on the right-hand side (rhs) of Eq. (9.5) has the form of a linear integral operator acting on what can be viewed as an element of an infinite-dimensional Hilbert space $L_2(V_{\mathrm{tot}})$ of vector functions that are square integrable in V_{tot}. Thus, we can introduce a linear operator W that acts on an arbitrary element $|f\rangle$ of $L_2(V_{\mathrm{tot}})$ according to the rule

$$W|f\rangle \rightarrow \int_{V_{\mathrm{tot}}} G_R(\mathbf{r} - \mathbf{r}')\mathbf{f}(\mathbf{r}')d^3 r' \qquad \mathbf{r} \in V_{\mathrm{tot}} \tag{9.21}$$

The operator W is an infinite-dimensional symmetrical operator. It is a "mixed" operator in the sense that it is both tensorial and integral. We can easily verify that if $|f\rangle$ is an element of $L_2(V_{\text{tot}})$, then $W|f\rangle$ is also an element of $L_2(V_{\text{tot}})$. Further, we can define a scalar product of two vectors $|f\rangle$ and $|g\rangle$ and a norm in $L_2(V_{\text{tot}})$ as

$$\langle g|f\rangle = \int_{V_{\text{tot}}} \mathbf{g}^*(\mathbf{r}) \cdot \mathbf{f}(\mathbf{r}) d^3r \tag{9.22}$$

$$\|f\| = \sqrt{\langle f|f\rangle} \tag{9.23}$$

respectively. Using the above notations, we can rewrite (9.5) as

$$|P\rangle = \chi[|E_{\text{inc}}\rangle + W|P\rangle] \tag{9.24}$$

where the vector $|P\rangle$ corresponds to the polarization function $\mathbf{P}(\mathbf{r})$ and $|E_{\text{inc}}\rangle$ to the incident field. The expressions for the optical cross-sections take the following forms:

$$\sigma_e = \frac{4\pi k}{|\mathbf{E}_0|^2} \operatorname{Im}\langle E_{\text{inc}}|P\rangle \qquad \sigma_a = \frac{4\pi k\delta}{|\mathbf{E}_0|^2} \langle P|P\rangle \tag{9.25}$$

To conclude this section, we raise the consideration to a slightly higher level of abstraction. The solution to the operator equation (9.24) can be written in symbolic form as

$$|P\rangle = [1/\chi - W]^{-1}|E_{\text{inc}}\rangle \tag{9.26}$$

The operator $R(E) = [E - W]^{-1}$, where E is an arbitrary (complex) scalar is called the *resolvent* of the operator W. The extinction cross-section is given by the diagonal matrix element of the resolvent, $\langle E_{\text{inc}}|R(1/\chi)|E_{\text{inc}}\rangle$.

For an arbitrary Hermitian operator H, the resolvent can be expanded in terms of the eigenvectors of H as

$$R(E) = \frac{1}{E - H} = \sum_n \frac{|n\rangle\langle n|}{E - E_n} \tag{9.27}$$

where $|n\rangle$ and E_n are the eigenvectors and eigenvalues of H. However, W is not Hermitian but *complex and symmetrical*. The symmetry of W should be understood as the following property of the kernel $\boldsymbol{G}_R$:

$$\boldsymbol{G}_R(\mathbf{r}) = \boldsymbol{G}_R(-\mathbf{r}), \ (G_R(\mathbf{r}))_{\alpha\beta} = (G_R(\mathbf{r}))_{\beta\alpha} \tag{9.28}$$

These two equalities provide that, for two arbitrary elements of $L_2(V_{\text{tot}})$, $|f\rangle$ and $|g\rangle$,

$$\langle f^*|W|g\rangle = \langle g^*|W|f\rangle \tag{9.29}$$

where the asterisk denotes the complex conjugation (c.c.) of the corresponding function.[3] Equation (9.29) can be viewed as a generalized symmetry condition for W. It can be used to prove that if $|f\rangle$ and $|g\rangle$ are two different (linear independent) eigenvectors of W, they obey $\langle f^*|g\rangle = 0$ [34], which is an analog of the orthogonality condition for eigenfunctions of Hermitian operators, $\langle f|g\rangle = 0$. This property, in turn, can be used to write an analog of expansion (9.27) for the non-Hermitian operator W:

$$R(E) = \frac{1}{E - W} = \sum_n \frac{|n\rangle\langle n^*|}{\langle n^*|n\rangle[E - w_n]} \tag{9.30}$$

Analogously to the notations of Eq. (9.27), we denote the eigenvectors and eigenvalues of W by $|n\rangle$ and w_n. However, now w_n is, generally, a complex number, as well as the factor $\langle n^*|n\rangle$ in the denominator of (9.30). We emphasize that $\langle n^*|n\rangle \neq \langle n|n\rangle$. The latter value is equal to unity for normalized eigenvectors, while the former is a complex number.

Now, we recall that the generic variable E has to be substituted by $1/\chi$ in (9.30). This fact and the structure of Eq. (9.30) emphasize the importance of the parameters X and δ introduced at the end of Section 9.3. According to the definition, $1/\chi = -(X + i\delta)$. If the interaction between monomers is turned off, for example, by disaggregating a smoke cluster and moving the monomers far from each other, all the eigenvalues w_n turn to zero. Then, X plays the role of the generalized detuning from the resonance of an isolated (noninteracting) monomer, while δ is the energy loss parameter (again, in the absence of interaction). In the presence of interaction, the eigenvalues w_n become nonzero. The real parts of w_n's describe frequency shifts of collective resonances, while the imaginary parts can change the collective radiative losses due to constructive or destructive interference.

In the quasistatic approximation, when the system size is much smaller than the wavelength, the operator W becomes Hermitian, and all w_n's are real and independent of the optical frequency [31]. The only source of the spectral dependence of the solution is in this case the spectral dependence of X and δ on λ.

9.4.2. Perturbative Methods

It is convenient to build the perturbation expansions of the optical cross-sections starting from the operator form (9.24) of the integral Maxwell equations. The Born expansion for the polarization function $|P\rangle$ is obtained by iterating (9.24):

$$|P\rangle = \chi \sum_{k=0}^{\infty} (\chi W)^k |E_{\text{inc}}\rangle \tag{9.31}$$

[3]In our notations, the symbol $\langle f|$ stands for the c.c. of the function **f**, and hence, $\langle f^*|$ is, in fact, the function **f** itself. This system of notations may seem to be artificial, but we have decided to follow the standard Dirac notations, though they are more appropriate for Hermitian operators, while the operator W is not Hermitian.

The corresponding expansion for the extinction cross-section follows from (9.25) and (9.31):

$$\sigma_e = 4\pi k v_{\text{tot}} \ \text{Im}\left[\chi \sum_{k=0}^{\infty} B_k \chi^k\right] \qquad B_k \equiv \frac{\langle E_{\text{inc}}|W^k|E_{\text{inc}}\rangle}{\langle E_{\text{inc}}|E_{\text{inc}}\rangle} \tag{9.32}$$

where we have taken into account that $\langle E_{\text{inc}}|E_{\text{inc}}\rangle = v_{\text{tot}}|\mathbf{E}_0|^2$. Because W is an integral operator [see (9.21)], a calculation of the coefficients B_k requires a calculation of $k+1$ volume integrals over V_{tot}. The approximation in which only the first nonzero term in (9.32) is left is often called the first Born approximation (the corresponding coefficient is B_0 in our notations). Multiple scattering is completely neglected in this approximation. In the next order (second Born approximation), the coefficient B_1 is retained. Physically, this is equivalent to taking into account double scattering. The convergence condition for the Born expansion is $\max[|\chi w_n|] < 1$.

One can improve the convergence of the perturbation expansion for σ_e by adopting a more sophisticated approach. According to (9.25), (9.26), $\sigma_e \propto \text{Im}\langle E_{\text{inc}}|R(1/\chi)|E_{\text{inc}}\rangle$, where $R(E) = [E - W]^{-1}$. To obtain an expansion of this matrix element of the resolvent, we build an infinite sequence of vectors $|P_k\rangle$ and complex numbers Q_k such that († stands for Hermitian conjugation)

$$W^\dagger|P_k\rangle = Q_{k+1}^*|E_{\text{inc}}\rangle + |P_{k+1}\rangle, \quad k = 0,\ 1,\ 2,\ \ldots \tag{9.33}$$

$$\langle P_k|E_{\text{inc}}\rangle = 0, \forall k > 0 \tag{9.34}$$

$$|P_0\rangle = |E_{\text{inc}}\rangle \tag{9.35}$$

It is straightforward to show that recursion (9.33) defines a unique set of $|P_k\rangle$ and Q_k and

$$|P_k\rangle = (TW^\dagger)^k|E_{\text{inc}}\rangle \qquad k = 0,\ 1,\ 2,\ \ldots \tag{9.36}$$

$$Q_k = \frac{\langle E_{\text{inc}}|W(TW)^{k-1}|E_{\text{inc}}\rangle}{\langle E_{\text{inc}}|E_{\text{inc}}\rangle} \qquad k = 1,\ 2,\ 3,\ \ldots \tag{9.37}$$

$$T \equiv 1 - \frac{|E_{\text{inc}}\rangle\langle E_{\text{inc}}|}{\langle E_{\text{inc}}|E_{\text{inc}}\rangle} \tag{9.38}$$

Here, T is the projection operator with respect to the vector $|E_{\text{inc}}\rangle$. Note that the vectors $|P_k\rangle$ are, in general, not normalized or mutually orthogonal.

Next, we define a vector $|\psi\rangle \equiv [1/\chi - W]^{-1}|E_{\text{inc}}\rangle$, so that $\langle E_{\text{inc}}|R(1/\chi)|E_{\text{inc}}\rangle = \langle E_{\text{inc}}|\psi\rangle$, and notice that $\langle E_{\text{inc}}|\psi\rangle$ must satisfy the equation

$$\langle E_{\text{inc}}|\psi\rangle = \chi[\langle E_{\text{inc}}|E_{\text{inc}}\rangle + \langle E_{\text{inc}}|W|\psi\rangle] \tag{9.39}$$

It also follows from the definition of $|\psi\rangle$ that

$$\langle P_k|W|\psi\rangle = Q_{k+1}\langle E_{\text{inc}}|\psi\rangle + \chi\langle P_{k+1}|W|\psi\rangle, \ k \geq 0 \tag{9.40}$$

If we act repeatedly n times by the operator W to the left in Eq. (9.39) and use the recursion (9.40), we obtain

$$\begin{aligned}\langle E_{\text{inc}}|\psi\rangle = \chi[\langle E_{\text{inc}}|E_{\text{inc}}\rangle + (Q_1 + Q_2\chi + Q_3\chi^2 + \ldots \\ + Q_n\chi^{n-1})\langle E_{\text{inc}}|\psi\rangle + \chi^n\langle P_n|W|\psi\rangle]\end{aligned} \tag{9.41}$$

Equation (9.41) can be rewritten as

$$\langle E_{\text{inc}}|R(1/\chi)|E_{\text{inc}}\rangle = \langle E_{\text{inc}}|\psi\rangle = \frac{\chi\langle E_{\text{inc}}|E_{\text{inc}}\rangle}{1 - \sum_{k=1}^{n} Q_k\chi^k} + \xi_n \tag{9.42}$$

The residual term, ξ_n, still depends on the unknown vector $|\psi\rangle$ and is given by

$$\xi_n = \chi^n\langle P_n|\psi\rangle\left[1 - \sum_{k=1}^{n} Q_k\chi^k\right]^{-1} \tag{9.43}$$

If we neglect ξ_n, we can write the expansion for the extinction cross-section:

$$\sigma_e = 4\pi k v_{\text{tot}}\text{Im}\left[\frac{\langle E_{\text{inc}}|R(E)|E_{\text{inc}}\rangle}{\langle E_{\text{inc}}|E_{\text{inc}}\rangle}\right] = 4\pi k v_{\text{tot}}\text{Im}\left[\frac{\chi}{1 - \sum_{k=1}^{\infty} Q_k\chi^k}\right] \tag{9.44}$$

Equivalently, this can be rewritten as

$$\sigma_e = 4\pi k v_{\text{tot}}\text{Im}\left[\frac{1}{1/\chi - Q_1 - \Sigma(\chi)}\right] \tag{9.45}$$

where the self-energy $\Sigma(\chi)$ is given by $\Sigma(\chi) = \sum_{k=1}^{\infty} Q_{k+1}\chi^k$. The expansion (9.45) has the form of the Dyson equation. In the first order, by neglecting the self-energy, we recover the result of the mean-field approximation, which was introduced by Berry and Percival [20] for the problem of scattering of light by fractal smoke clusters (the constant Q_1 in (9.45) is analogous to the constant P in [20]). In fact, the mean-field result serves as the first-order approximation for the above expansion.

The expansion coefficients Q_k can be easily expressed in terms of the corresponding coefficients for the Born expansion, B_k. For the first few terms, $Q_1 = B_1$, $Q_2 = B_2 - B_1^2$, $Q_3 = B_3 - 2B_1B_2 + B_1^3$. Therefore, in order to build the expansion (9.45), it is sufficient to calculate the volume integrals involved in the calculation of the B_k's.

It is instructive to compare the expansion (9.44) with the Born expansion. One can show (the details of the derivation are omitted) that the Taylor expansion of (9.44) with respect to powers of χ, where only the n first Q_k's are left in the denominator, exactly coincides with the Born expansion (9.32) up to the order $k = n$ in χ. For $k > n$, the absolute value of the difference between the coefficients in these two expansions is not greater than the absolute values of the corresponding coefficients B_k. For example, in the $n = 2$ case, the difference in the third-order coefficients is $B_3 - 2Q_1Q_2 - Q_3^3 = \langle E_{\text{inc}}|W|E_{\text{inc}}\rangle^3$, and $|\langle E_{\text{inc}}|W|E_{\text{inc}}\rangle|^3 \leq |B_3| = |\langle E_{\text{inc}}|W^3|E_{\text{inc}}\rangle|$ (generalized Hölder inequality). Practically, this means that (1) if the expansion (9.32) converges, the expansion (9.44) also converges and (2) the convergence of (9.44) is at least as fast as that of (9.32). Even in the first order, the expansion (9.44) contains infinite orders of multiple scattering, which is known to be true for the mean-field approximation [20].

Now, we turn our attention to the calculation of the coefficients B_k. The first coefficient is trivial, $B_0 = 1$. Next, we recall that W is an integral operator defined by (9.21) and write B_1 as

$$B_1 = \frac{\langle E_{\text{inc}}|W|E_{\text{inc}}\rangle}{v_{\text{tot}}|\mathbf{E}_0|^2} \tag{9.46}$$

$$\langle E_{\text{inc}}|W|E_{\text{inc}}\rangle = \sum_{i,\,j=1}^{N} \int_{V_i} d^3r \int_{V_j} d^3r' [\mathbf{E}_{\text{inc}}^*(\mathbf{r}) \cdot \boldsymbol{G}_R(\mathbf{r} - \mathbf{r}')\mathbf{E}_{\text{inc}}(\mathbf{r}')] \tag{9.47}$$

By using the fundamental assumption of small primary particles $kR_m \ll 1$ (R_m is the radius of the spherical volumes V_i, V_j) and by using Eq. (9.7), we replace $\mathbf{E}_{\text{inc}}(\mathbf{r}')$ by $\mathbf{E}_0 \exp(i\mathbf{k} \cdot \mathbf{r}_j)$ and $\mathbf{E}_{\text{inc}}^*(\mathbf{r})$ — by $\mathbf{E}_0 \exp(-i\mathbf{k} \cdot \mathbf{r}_i)$ in the above integral, and group the terms with $i = j$ together to obtain

$$\begin{aligned}\langle E_{\text{inc}}|W|E_{\text{inc}}\rangle &= \sum_{i} \int_{V_i} d^3r \int_{V_i} d^3r' [\mathbf{E}_0^* \cdot \boldsymbol{G}_R(\mathbf{r} - \mathbf{r}')\mathbf{E}_0] \\ &\quad + \sum_{i \neq j} e^{i\mathbf{k}\cdot(\mathbf{r}_j - \mathbf{r}_i)} \int_{V_i} d^3r \int_{V_j} d^3r' [\mathbf{E}_0^* \cdot \boldsymbol{G}_R(\mathbf{r} - \mathbf{r}')\mathbf{E}_0]\end{aligned} \tag{9.48}$$

The double integral over the same volume V_i in the first term of (9.48) turns to zero in the limit $R_m \ll \lambda$. This can be easily illustrated by considering Eq. (9.5) for a single sphere case ($N = 1$) and observing that in this limit $\mathbf{P} = \chi\mathbf{E}_0 = \text{const}$ inside the sphere; hence, the integral on the rhs of (9.5) must go to zero. The integral over the different volumes V_i and V_j can be easily evaluated using (9.12), (9.13), and the spherical harmonic expansion of $g(\mathbf{r} - \mathbf{r}')$ [36]:

$$g(\mathbf{r} - \mathbf{r}') = 4\pi i k^3 \sum_{l=0}^{\infty} \sum_{m=-l}^{l} j_l(kr_<) h_l^{(1)}(kr_>) Y_{lm}^*(\Omega_{\mathbf{r}'}) Y_{lm}(\Omega_{\mathbf{r}}) \tag{9.49}$$

where $j_l(x)$, $h_l^{(1)}(x)$ are the spherical Bessel and Hankel functions, $r_< = \min(r,\ r')$, $r_> = \max(r,\ r')$, and $Y_{lm}(\Omega_{\mathbf{r}})$ are spherical harmonics. By choosing the origin at the center of the volume V_j, we find that $r_> = r$, $r_< = r'$, and

$$\int_{V_j} g(\mathbf{r} - \mathbf{r}') d^3r' = 4\pi i k^3 \sum_{l=0}^{\infty} \sum_{m=-l}^{l} h_l^{(1)}(kr) Y_{lm}(\Omega_{\mathbf{r}}) \int_0^{R_m} j_l(kr')(r')^2 dr' \int Y_{lm}^*(\Omega_{\mathbf{r}'}) d\Omega_{\mathbf{r}'} \tag{9.50}$$

If we take account of the fact that $\int Y^*_{lm}(\Omega_{\mathbf{r}'})d\Omega_{\mathbf{r}'} = \sqrt{4\pi}\delta_{l0}\delta_{m0}$, this is simplified to

$$\int_{V_j} g(\mathbf{r}-\mathbf{r}')d^3r' = 4\pi i h_0^{(1)}(kr)\int_0^{kR_m} j_0(x)x^2dx = v\beta(kR_m)g(\mathbf{r}) \tag{9.51}$$

$$\beta(x) = 3j_1(x)/x = 3(\sin x - x\cos x)/x^3 \tag{9.52}$$

The Taylor expansion of $\beta(x)$ near $x = 0$ is $\beta(x) = 1 - x^2/10 + \ldots$. Since we already neglected the phase dependence of the incident field over the volume of integration, keeping the terms of the order of x^2 amounts to excessive precision. Therefore, we set $\beta = 1$. The integration was performed in a reference frame, where $\mathbf{r}_j = 0$. In a general reference frame, one has

$$\int_{V_j} g(\mathbf{r}-\mathbf{r}')d^3r' = vg(\mathbf{r}-\mathbf{r}_j) \qquad \mathbf{r} \notin V_j \tag{9.53}$$

Consequently, for the integral of the tensor Green's function $\boldsymbol{G}_R$ one has

$$\int_{V_j} \boldsymbol{G}_R(\mathbf{r}-\mathbf{r}')d^3r' = v\left(\boldsymbol{I} + \frac{1}{k^2}\hat{\nabla}_{\mathbf{r}}\hat{\nabla}_{\mathbf{r}}\right)\int_{V_j} g(\mathbf{r}-\mathbf{r}')d^3r' = v\boldsymbol{G}_R(\mathbf{r}-\mathbf{r}_j), \qquad \mathbf{r} \notin V_j \tag{9.54}$$

Repeating analogous integration over the variable $\mathbf{r} \in V_i$, we obtain

$$\int_{V_i} d^3r \int_{V_j} d^3r' \boldsymbol{G}_R(\mathbf{r}-\mathbf{r}') = v^2\boldsymbol{G}_R(\mathbf{r}_i - \mathbf{r}_j) \qquad i \neq j \tag{9.55}$$

The above result seems to be obvious for two spheres that are separated by a distance much larger than their radii, but much less so for two touching spheres. However, it is exact in the limit $kR_m \to 0$. The physical interpretation of Eq. (9.55) is that in this particular order of the perturbation expansion each sphere can be adequately represented by a point dipole moment located in its center. In other words, the integral equations can be replaced by a set of discrete equations with respect to the dipole moments of the monomers, which constitutes the essence of the dipole approximation. It might seem that the dipole approximation must work simply because kR_m is small. However, this is known not to be the case for interacting spheres in close vicinity of each other (we will return to this in Section 9.4.3). In fact, the dipole approximation breaks down in the next order of the perturbation expansion.

Returning to calculation of B_1, we find

$$B_1 = \frac{v}{N|\mathbf{E}_0|^2}\sum_{i\neq j} e^{-i\mathbf{k}\cdot\mathbf{r}_{ij}}\mathbf{E}_0^* \cdot \boldsymbol{G}_R(\mathbf{r}_{ij})\mathbf{E}_0 \qquad \mathbf{r}_{ij} = \mathbf{r}_i - \mathbf{r}_j \tag{9.56}$$

Now, we proceed with statistical averaging of (9.56). This averaging can be introduced in two different ways. First, if a soot cluster is large enough, the probability distribution for the absolute values r_{ij} must be given by the pair correlation function $p(r_{ij})$ that was discussed in Section 9.2, while all the spatial orientations of $\mathbf{r}_{ij}$ are equiprobable (clusters are spherically symmetrical on average). This can be called self-averaging. The other approach is ensemble averaging over a distribution of different realizations of random clusters. The orientational averaging can be easily carried out in a spherical system of coordinates where the direction of vector $\mathbf{k}$ coincides with the z-axis and the direction of $\mathbf{E}_0$ (assuming linear polarization) with the x-axis. Then, using the tensor structure of $\boldsymbol{G}_R$ (9.9–9.11), we obtain

$$B_1 = k^3 v\,(N-1)\langle e^{-ikr_{ij}\cos\theta}[A(kr_{ij}) + B(kr_{ij})\sin^2\theta\cos^2\phi]\rangle \tag{9.57}$$

where we have taken into account that the total number of terms in the sum (9.56) is $N(N-1)$ and $\langle\cdots\rangle$ stands for statistical averaging. Orientational averaging is easily achieved by integrating (9.57) over $\sin\theta\, d\theta\, d\phi/(4\pi)$. The radial averaging is done with the use of the correlation function $p(r)$ (9.1) which is, by definition, the probability density to find a distinct pair of monomers in a cluster separated by the distance r. If we use the functional form (9.1) for $p(r)$, we arrive at

$$B_1 = \frac{\pi a}{6}(kl)^{3-D}\int_0^\infty x^{D-1} f\left(\frac{x}{kR_g}\right)F(x)dx \tag{9.58}$$

where $a \approx 4$ is the numerical constant, $f(x)$ is the cutoff function discussed in Section 9.2, we have used $v = (\pi/6)l^3$, and $F(x)$ is the result of angular integration of (9.57):

$$F(x) = \frac{\sin x}{x}A(x) - \left(\frac{\cos x}{x^2} - \frac{\sin x}{x^3}\right)B(x) \tag{9.59}$$

and $A(x)$, $B(x)$ are defined by (9.10) and (9.11). The power series expansion of $F(x)$ near $x = 0$ is $F(x) = (11/15)x^{-1} + 2i/3 - (46/105)x - (2i/9)x^2 + O[x^3]$. Therefore, the radial integral in (9.58) converges at the lower limit for $D > 1$, that is, for any physically reasonable fractal dimension. As was discussed in Section 9.2, this fact justifies the extension of the region of applicability of the scaling formula (9.1) to $r = 0$.

Convergence at $r = \infty$ is guaranteed by the cutoff function $f(x)$. However, for $D < 2$, (9.58) converges at the upper limit of integration even if we set $f(x/kR_g) = 1$. Therefore, for large clusters with $kR_g \gg 1$, the integral (9.58) converges while the cutoff function f is still close to unity. This means that for sufficiently large clusters with $D < 2$ (which is usually the case), the particular form of the cutoff function is not important, and we can calculate the integral (9.58) analytically [20]. The final result of integration is

$$B_1 = \frac{\pi a}{24}(2kl)^{3-D}\exp\left[\frac{i\pi(D+1)}{2}\right]K(D) \qquad 1 < D < 2,$$

$$kR_g \gg 1 \tag{9.60}$$

$$K(D) = \frac{\Gamma(D)}{(D-1)(2-D)(3-D)}\left[\frac{4(5D-18)}{(4-D)(6-D)} + D + 1\right] \tag{9.61}$$

where $\Gamma(x)$ is the gamma function.

Note that the function $K(D)$ in (9.61) diverges as D approaches 2. This result indicates that the perturbation expansion for large clusters becomes less accurate as D approaches 2. However, there is no real divergence even for $D > 2$ because of the cutoff function $f(x/kR_g)$, which was set to unity for the derivation of (9.60). Therefore, the result (9.60) should be used with caution. In particular, progressively larger values of kR_g are required for convergence of the integral when D approaches 2. In general, when kR_g is not sufficiently large, or $D > 2$, the value of B_1 depends on the exact form of $f(x)$ and on the gyration radius of the cluster, R_g. The dependence of the integral of the type (9.58) on D for the simple exponential cutoff was considered by Berry and Percival [20] and by Shalaev et al. [37].

The result (9.60) was obtained in the "intermediate" wavelength limit $R_m \ll \lambda \ll R_g$. It is also possible to calculate B_1 in the long-wavelength limit $R_m \ll R_g \ll \lambda$ (the quasistatic approximation). This can be easily done by observing that, when $kR_g \gg 1$, the integral (9.58) converges for small values of x. Therefore, we can keep only the first two terms in the power series expansion of $F(x)$ near $x = 0$, which is necessary to calculate both the real and imaginary parts of B_1: $F(x) \approx (11/15)x^{-1} + 2i/3$. By substituting this expression into (9.58), and by using (9.2), we obtain

$$B_1 = ab^D k^3 v_{\text{tot}}\left[\frac{11}{15kR_g}\int_0^\infty x^{D-2}f(x)dx + \frac{2i}{3}\int_0^\infty x^{D-1}f(x)dx\right] \tag{9.62}$$

In expression (9.62), a, b are the numerical coefficients [see formulas (9.1) and (9.2) for the definitions]. As was mentioned in Section 9.2, the dimensionless combination ab^D is $\simeq 1.6$ for cluster–cluster aggregates. The integrals on the rhs of (9.62) are simple numbers, and can be evaluated numerically given a specific form of the cutoff function $f(x)$. If $f(x)$ is given by the generalized exponential formula with constants described in the caption of Fig. 9.2, the integrals are $\simeq 1.6$ and 1.2, respectively. As can be seen from (9.62), the coefficient B_1 becomes purely real in the limit $\lambda \to \infty$.

As was mentioned above, the calculation of higher coefficients B_k cannot be performed in the "dipole approximation". In practice, this means that the chain integrals of the kind

$$\int_{V_{i_1}} d^3r_1 \int_{V_{i_2}} d^3r_2 \cdots \int_{V_{i_n}} d^3r_n \boldsymbol{G}_R(\mathbf{r}_1 - \mathbf{r}_2)\boldsymbol{G}_R(\mathbf{r}_2 - \mathbf{r}_3)\cdots \boldsymbol{G}_R(\mathbf{r}_{n-1} - \mathbf{r}_n)$$

cannot be represented as

$$v^n G_R(\mathbf{r}_{i_1} - \mathbf{r}_{i_2}) G_R(\mathbf{r}_{i_2} - \mathbf{r}_{i_3}) \cdots G_R(\mathbf{r}_{i_{n-1}} - \mathbf{r}_{i_n})$$

Generally, this simple integration rule can only be applied to the "end of chain" integration variables ($\mathbf{r}_1$ and $\mathbf{r}_n$ in the above example). In the case of the double scattering coefficient B_1, both integration variables are, effectively, "end of chain," and the dipole approximation works. The integration variable $\mathbf{r}_2$ in the above example is "middle of the chain," in other words, it appears in two Green's functions instead of one. Therefore, the integration over $\mathbf{r}_2$ cannot be so easily performed.

9.4.3. Nonperturbative Methods

As we saw in Section 9.4.2, the dipole approximation is accurate up to the second order of the Born expansion. This indicates that when the interaction is weak, both the perturbation expansion and dipole approximation become accurate. Of course, it is also possible to formulate the dipole approximation nonperturbatively, which is done below.

In the dipole approximation, each monomer in a cluster is considered to be a point dipole with polarizability α, located at the point $\mathbf{r}_i$ (at the center of the respective spherical monomer). The dipole moment of the ith monomer, $\mathbf{d}_i = \int_{V_i} \mathbf{P}(\mathbf{r}) d^3 r$, is proportional to the external electric field at the point $\mathbf{r}_i$, which is a superposition of the incident field and all the secondary fields scattered by other dipoles. Therefore, the dipole moments of the monomers are coupled to the incident field and to each other as described by the coupled dipole equation (CDE):

$$\mathbf{d}_i = \alpha \left[\mathbf{E}_{\text{inc}}(\mathbf{r}_i) + \sum_{j \neq i}^{N} G_R(\mathbf{r}_i - \mathbf{r}_j) \mathbf{d}_j \right] \tag{9.63}$$

which is simply a discrete version of the integral equation (9.5). It was introduced in the context of the discrete dipole approximation by Purcell and Pennypacker [38], and for fractal clusters by Markel et al. [30,31].[4] The CDE is a system of $3N$ linear equations that can be solved to find the dipole moments $\mathbf{d}_i$. All the optical cross sections can be found in complete analogy with (9.14)–(9.20):

$$\mathbf{f}(\mathbf{k}') = k^2 \sum_{i=1}^{N} \left[\mathbf{d}_i - (\mathbf{d}_i \cdot \mathbf{k}')\mathbf{k}'/k^2 \right] \exp(-i\mathbf{k}' \cdot \mathbf{r}_i) \tag{9.64}$$

$$\sigma_e = \frac{4\pi k}{|\mathbf{E}_0|^2} \operatorname{Im} \sum_{i=1}^{N} \mathbf{d}_i \cdot \mathbf{E}^*_{\text{inc}}(\mathbf{r}_i) \tag{9.65}$$

[4]The discrete dipole approximation, although leading to similar equations, is used to solve a problem essentially different from the one described in this chapter. For more references, see [39,40].

$$\sigma_a = \frac{4\pi k y_a}{|\mathbf{E}_0|^2} \sum_{i=1}^{N} |\mathbf{d}_i|^2 \tag{9.66}$$

$$y_a \equiv -\mathrm{Im}\left(\frac{1}{\alpha}\right) - \frac{2k^3}{3} = \delta/v \geq 0 \tag{9.67}$$

The last formula (9.67) needs an explanation. The definition of the absorption parameter y_a in (9.66) follows from rigorous integration of the scattering amplitude (9.64) [34]. It must be nonnegatively defined for any physically reasonable polarizability α [41]. However, if we use the usual relation between the polarizability α and susceptibility χ, $\alpha = v\chi$, this condition [as well as the second equality in (9.67)] can be violated for purely real values of ε when χ is also real. As was shown by Draine [39], the above relation between α and χ should be modified to take into account radiative reaction. The corrected formula is $1/\alpha = 1/v\chi - i2k^3/3$, which, taking into account $\delta = -\mathrm{Im}[1/\chi]$, immediately leads to the second equality in (9.67). For strongly absorbing carbon, the radiative corrections are negligibly small.

The advantage of the dipole approximation is simplicity: An integral equation is replaced by a finite system of linear equations. However, we already saw that the dipole approximation is not accurate in the third, and all the higher, orders of the perturbation expansion. In fact, the general nonapplicability of the dipole approximation was recognized and verified both theoretically [42,43] and experimentally [44]. A simple physical explanation of why the dipole approximation fails was provided, for example, by Sansonetti and Furdyna [44]. First, in the dipole approximation, the local field acting on a certain dipole is evaluated at the center of the corresponding monomer. However, the field produced by neighboring monomers is highly nonuniform over the volume of the first monomer, and cannot be replaced by a single value. And second, the dipole approximation neglects higher multipole moments of the monomers, which is a good approximation for those monomers which are far away from each other, but not for nearest neighbors. Effectively, by replacing two touching spheres by two point dipoles located at their centers, we underestimate the strength of their interaction.

To overcome the limitations of the dipole approximation, a rigorous numerical approach has been developed by Gerardy and Ausloos [42] (in the long-wavelength limit), Claro [43,45,46] and Claro and co-workers [47,48], Mackowski [49], Mackowski and Mischenko [50], Fuller [51,52], and Xu [53]. The essence of this method is to expand the EM field inside each sphere and the field scattered by each sphere in vector spherical harmonics, and to match the boundary condition on all surfaces of discontinuity. Generally, this method leads to an infinite - dimensional system of linear equations with respect to the expansion coefficients. In order to solve this system, one needs to truncate it by assuming that all the expansion coefficients for spherical harmonics of the order larger than L are zero. Then, the total number of equation scales (for large values of L) as NL^2.

Although a detailed description of the above methods is beyond the scope of this chapter, in Section 9.4.4, we will illustrate with a numerical example an important trend: When the interaction of monomers in a cluster becomes stronger and the perturbation expansion, correspondingly, less accurate (or even diverges), the maximum number L required for attaining accurate results tends to increase. This makes the "coupled multipoles" method computationally applicable only for situations with either a small number of monomers or weak interaction.

To overcome the inadequacy of the dipole approximation and the overwhelming computational complexity of the "coupled multipole" method, we have suggested a phenomenological procedure that can be referred to as the cluster renormalization approach [54,55]. This approach allows one to stay in the frame of the dipole approximation, and is described in detail below. The following two factors are important for understanding the renormalization approach.

First, we note that most calculations employ computer - generated samples. The geometry of these samples does not coincide with that of experimental soot exactly, which is obviously impossible, but rather reproduces certain statistical geometrical properties of real soot. Among such properties are density correlation functions, total volume of the material, and average radius of gyration, R_g. However, such characteristics as the number of monomers in a cluster, N, and monomer radius, R_m, might be considered as *not essential*. It is known, for example, that the real carbon monomers are not actually spherical, and nearest neighbors touch each other at more than just one geometrical point, so that the model of touching spheres is only an idealization. Second, as mentioned above, the dipole approximation in its pure form underestimates the interaction strength. In particular, it predicts the shift of the resonance frequency in small clusters of spheres to be significantly less than is experimentally measured [44]. In order to correct the interaction strength of the dipole approximation, it is tempting to move the monomers closer to each other (of course, this relates to computer-generated samples) by allowing them to intersect geometrically. However, doing this will evidently reduce the overall system size (R_g), which is an essential parameter of the problem. The other possible way to introduce the intersections is to increase the radii of the spheres (R_m) while keeping distance between nearest neighbors (l) unchanged. This will, however, lead to an increase of the total volume of the material. Luckily, for fractal clusters, it is possible to introduce a simultaneous renormalization of the sphere radii (R_m), the total number of monomers (N) and the distance between nearest neighbors (l) in such a way that the overall volume (v_{tot}) and the gyration radius (R_g) are unchanged, and to introduce an arbitrary geometrical intersection of neighboring spheres. The transformation is

$$R'_m = R_m \left(\frac{\xi}{2}\right)^{D/(3-D)} \tag{9.68}$$

$$N' = N \left(\frac{2}{\xi}\right)^{3D/(3-D)} \tag{9.69}$$

$$l' = \xi R'_m \tag{9.70}$$

where ξ is a phenomenological intersection parameter ($1 < \xi \leq 2$, $\xi = 2$ for touching spheres and $\xi < 2$ for geometrically intersecting spheres).

Thus, the main idea of the renormalization approach is to model an ensemble of real clusters with experimental values of R_m and N (and $l = 2R_m$) by a computer-generated "renormalized" ensemble with corresponding parameters R'_m, N', and with the geometrical intersection of neighboring spheres: $l' = \xi R'_m < 2R'_m$. The intersection parameter ξ is phenomenological and must be adjusted. The initial value for ξ can be obtained from the following simple considerations, which can be also used to justify the physical plausibility of the renormalization method.

It can be shown [56] that a linear chain of intersecting spheres has the same depolarization coefficients as an infinite cylinder for $\xi = [4\sum_{k=1}^{\infty} k^{-3}]^{1/3} \approx 1.688$. It is important to note that two independent depolarization coefficients can simultaneously be "tuned" to correct values by adjusting only one free parameter ξ. As is well known, the depolarization coefficients in ellipsoids (an infinite cylinder being a particular case) determine the spectral positions of the resonances. Thus, the renormalization procedure gives the correct spectral locations of the optical resonances for a one-dimensional (1D) chain. The line shape of each resonance can still be described incorrectly. However, in the situation of a large fractal cluster, typical absorption and extinction spectra are superpositions of many collective resonances, and the line shapes of individual resonance are of little importance.

Another approach to estimating the parameter ξ is by analogy with the discrete dipole approximation (see [38–40]) in which bulk nonspherical particles are modeled by arrays of point dipoles located on a cubic lattice. In the first approximation, the polarizability of the dipoles is taken to be equal to that of an equivalent sphere with the radius R_m such that its volume is equal to the volume of the lattice cell, that is, $(4\pi/3)R_m^3 = l^3$. From this equality we find $\xi = l/R_m = (4\pi/3)^{1/3} \approx 1.612$.

Given a computer-generated renormalized ensemble of clusters, we can build the CDE (9.63) and solve it numerically to obtain all desirable optical constants. The CDE (9.63) can be written in the operator form [30,31], analogously to (9.24), except that the Hilbert space now has a finite dimensionality $3N$. We denote by $|d\rangle$ the $3N$-dimensional vector of dipole moments, and write

$$|d\rangle = \alpha(|E_{\text{inc}}\rangle + W_m|d\rangle) \tag{9.71}$$

The operator W_m in (9.71) is a square $3N \times 3N$ matrix rather than an integral operator W in (9.24), which is being emphasized here by using the subscript "m". In a basis of vectors $|i\alpha\rangle$, such that $d_{i\alpha} = \langle i\alpha|d\rangle$ (the Greek indexes denote the Cartesian components of the vectors), the matrix elements of W_m are $\langle i\alpha|W_m|j\beta\rangle = k^3[A(kr_{ij})\delta_{\alpha\beta} + B(kr_{ij})r_{ij,\alpha}r_{ij,\beta}/r_{ij}^2]$.

The eigenvector expansion of the solution to (9.71) was proposed in [30,31] for the quasistatic case ($R_g \ll \lambda$) and in [34] in the general case, and is similar to (9.30):

$$|d\rangle = \sum_n \frac{|n\rangle\langle n^*|E_{\text{inc}}\rangle}{\langle n^*|n\rangle[1/\alpha - w_n]} \tag{9.72}$$

where $|n\rangle$ are now the eigenvectors of W_m.

If we use (9.65) with $\sum_{i=1}^{N} \mathbf{d}_i \cdot \mathbf{E}^*_{\text{inc}}(\mathbf{r}_i) = \langle E_{\text{inc}}|d\rangle$, we can write for the extinction cross-section:

$$\sigma_e = \frac{4\pi k}{|\mathbf{E}_0|^2} \text{Im} \sum_n \frac{\langle E_{\text{inc}}|n\rangle\langle n^*|E_{\text{inc}}\rangle}{\langle n^*|n\rangle[1/\alpha - w_n]} \tag{9.73}$$

Now, we recall that $1/\alpha = 1/v\chi - i2k^3/3$ and $1/\chi = -(X + i\delta)$, where the spectral dependence of X and δ are illustrated in Fig. 9.4 (b), and write

$$\sigma_e = -\frac{4\pi k v}{|\mathbf{E}_0|^2} \text{Im} \sum_n \frac{\langle E_{\text{inc}}|n\rangle\langle n^*|E_{\text{inc}}\rangle}{\langle n^*|n\rangle[X + i(\delta + 2k^3 v/3) + v w_n]} \tag{9.74}$$

Equation (9.74) illustrates the importance of the parameters X and δ. The interaction of the monomers is weak and can be neglected when the "normalized" dimensionless eigenvalues $v w_n$ are small compared to $X + i(\delta + 2k^3 v/3)$. By neglecting the eigenvalues in the denominator of (9.74), we immediately recover the first Born approximation. There are two distinct cases when such an approximation is valid. The first case is a nonresonant interaction, when $|X| \gg v \text{Re}\, w_n$. But even if $X + \text{Re}\, w_n$ can turn exactly to zero (resonant interaction), the absorption parameter δ can be still sufficiently large to make the first Born approximation accurate.

Further, we can introduce the "weighted" density of states $\Gamma(w', w'')$ according to

$$\Gamma(w', w'') = \sum_n \frac{\langle E_{\text{inc}}|n\rangle\langle n^*|E_{\text{inc}}\rangle}{\langle n^*|n\rangle} \delta(w' - \text{Re}\, w_n)\delta(w'' - \text{Im}\, w_n) \tag{9.75}$$

and rewrite (9.74) as

$$\sigma_e = -\frac{4\pi k v}{|\mathbf{E}_0|^2} \text{Im} \int \frac{\Gamma(w', w'')dw'dw''}{X + i(\delta + 2k^3 v/3) + v(w' + iw'')} \tag{9.76}$$

This formula is a convenient starting point for a family of analytical approximations. The familiar mean-field approximation can be obtained by assuming that the eigenvalues of W_m all lie in a small bound region in the complex plane, while the complex variable $X + i(\delta + 2k^3 v/3)$ is far from this region. Then, the density of states can be approximated as $\Gamma(w', w'') = N|\mathbf{E}_0|^2 \delta(w' - \text{Re}\, Q_1)\delta(w'' - \text{Im}\, Q_1)$, and the integration in (9.76) results in a formula similar to (9.45) without the self-energy term. The requirement for applicability of the mean-field approximation is, obviously, different than that for the first Born approximation. Namely, the eigenvalues do not need to be small, but rather quasidegenerate. Higher order approximations can be built by making the form of $\Gamma(w', w'')$ more complicated. As the first step, $\Gamma(w', w'')$ can be assumed to be constant in a certain bound rectangular area in the complex plane and zero outside, with the dispersion (first moments) determined from numerical diagonalization of a typical matrix W_m. This functional form still allows one to integrate (9.76) analytically. At higher levels of approximation, other moments of $\Gamma(w', w'')$ can be specified as well.

Note that for highly absorbing carbon material, the parameter δ can be much larger than the imaginary parts of the eigenvalues $v \,\mathrm{Im}\, w_n$ (as well as the factor $2k^3 v/3$). In this case, it is sufficient to consider a 1D function $\Gamma(w')$. This is a good approximation in the quasistatic limit $\lambda \gg R_g$, when the imaginary parts of eigenvalues are proportional to the small factor $2k^3/3$ [34] and can be neglected. The complex eigenvalues of W_m are discussed in much more detail in Chapter 5, for the case of nonfractal, random and spherically symmetrical distribution of dipoles with the average interparticle distance $\sim \lambda$. The results of Chapter 5 are indicative of the fact that the imaginary parts of the eigenvalues in this case are still of the order of $2k^3/3$, and can be neglected for strongly absorbing soot. However, a calculation of the imaginary parts of the eigenvalues for a self-supporting fractal cluster, which is not small compared to λ, has not been performed to the best of our knowledge.

An important remark should be made regarding the influence of the renormalization procedure (9.68)–(9.70) on expressions of the form (9.74) and (9.76). The parameters X and δ do not depend on the geometry of the problem and, therefore, are not affected by the renormalization. However, the eigenvalues vw_n are changed as the result of renormalization. In particular, it is easy to see from (9.68)–(9.70) that the renormalized volume is $v' = v(\xi/2)^{3D/(3-D)}$. In general, the eigenvalues of the interaction matrix W_m do not scale with the parameter l, and it is impossible to write a similar relation between w_n and w'_n. However, this becomes possible in the quasistatic limit $\lambda \gg R_g$, when the intermediate- and far-zone terms in (9.10) and (9.11) can be neglected and the exponential factor $\exp(ix)$ can be set to unity. Then, from the form of the interaction matrix, it follows that $w'_n = w_n(l/l')^3 = w_n(2/\xi)^{9/(3-D)}$. By combining these two expressions, we obtain in the quasistatic limit: $v'w'_n = vw_n(2/\xi)^3$. We recall that ξ is the intersection parameter, and $1 < \xi \leq 2$. Thus, the intersection procedure effectively increases the normalized eigenvalues and, consequently, the interaction strength. The same tendency holds beyond the quasistatic limit, although the ratio $v'w'_n/vw_n$ becomes different for different n in this case.

Due to volume limitations, we have skipped the important question of orientational averaging in the dipole approximation. This averaging is trivial in the quasistatic case, and can be performed by averaging results for three orthogonal polarizations of the incident wave. However, the averaging becomes much more complicated in the case of finite k; see [57] for a further reference.

9.4.4. Numerical Examples

In this section, we illustrate the methods described above with a few numerical examples. We start with the perturbation expansion for the extinction cross-section.

In Fig. 9.5, we plot the results of perturbative calculations of the extinction efficiency Q_e defined as

$$Q_e = \frac{\sigma_e}{kv_{\mathrm{tot}}} \tag{9.77}$$

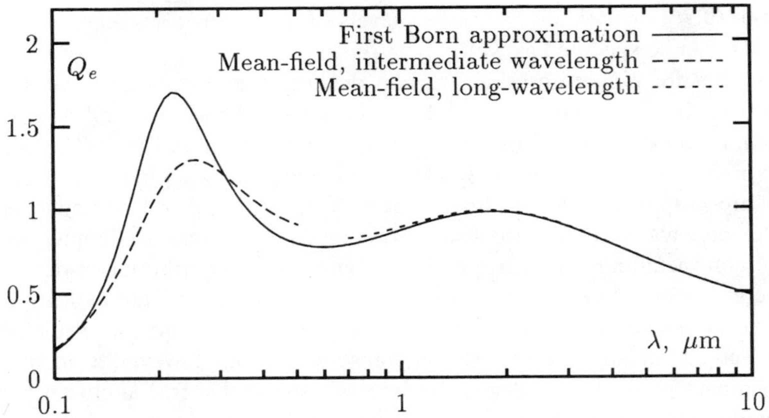

Figure 9.5. Perturbative calculations of the extinction efficiency Q_e for a cluster with $N = 2500$, $l = 2R_m = 0.02\,\mu m$, and $R_g \approx 33l = 0.66\,\mu m$.

The extinction efficiency is calculated for an ensemble of clusters with $N = 2500$, $l = 2R_m = 0.02\,\mu m$, and $R_g \approx 33l = 0.66\,\mu m$. In this figure, we compare the first Born approximation for Q_e obtained from (9.32) by retaining only the $k = 0$ term in the summation, compared to the first-order mean-field approximation (9.45) obtained by setting $\Sigma = 0$ in (9.45). We used the optical constants for carbon described in Section 9.3. The constant $Q_1 = B_1$ that is used in the mean-field approximation was calculated from the analytical formulas for the "intermediate" wavelength regime ($R_m \ll \lambda \ll R_g$) (9.60) and (9.61), and for the long-wavelength regime ($R_m \ll R_g \ll \lambda$) (9.62). The corresponding curves are plotted in the spectral regions where these regimes are valid. In the intermediate region $\lambda = R_g$, both analytical expressions for B_1 become inaccurate and numerical integration according to (9.58) should be performed. However, it is plausible to assume from the figure that the two curves will smoothly connect to each other near $\lambda = R_g$. Note that in the limit $kR_m \ll 1$, the first Born approximation also gives the "noninteracting" value of Q_e, that is, calculated for isolated spherical monomers.

While the mean-field approximation gives significantly different results from the first Born approximation for $\lambda \ll 1\,\mu m$, the difference becomes small for larger wavelengths. This might seem to be an indication of fast convergence of the perturbation series for large λ's. However, it is not the case. In fact, the coefficient B_1 becomes small in the long-wavelength limit [see Eq.(9.3)] due to the special symmetry of the dipole–dipole interaction. However, as we saw above, the dipole approximation is not accurate in the higher orders of the expansion, and the higher coefficients B_k can be large.

The deficiency of the mean-field approximation in the long-wavelength limit is most easily demonstrated with quasistatic calculations, in the limit $\lambda \gg R_g$. In Fig. 9.6, we compare the results of the first Born and the mean-field approximations [with the long-wavelength version of B_1 calculated according to (9.62)] to the numerical nonperturbative solution based on the expansion of all scattered fields into spherical

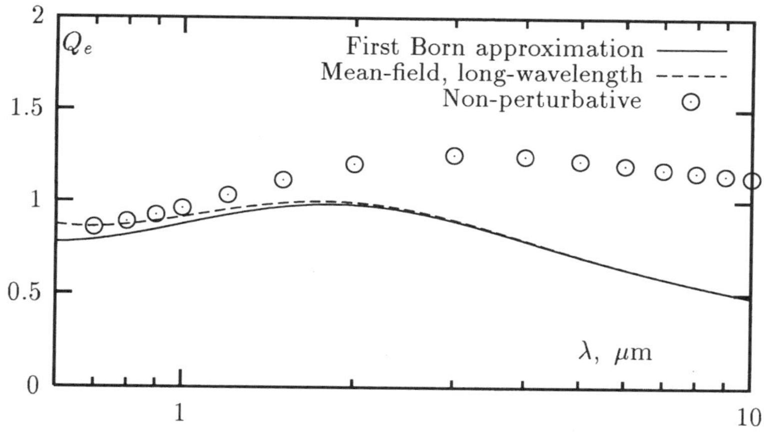

Figure 9.6. Perturbative and nonperturbative calculations of the extinction efficiency Q_e for a cluster with $N = 100$, $l = 2R_m = 0.02\,\mu\text{m}$, and $R_g \approx 10.6l = 0.2\,\mu\text{m}$.

harmonics and considering boundary conditions at each spherical surface.[5] The calculations are done for a small cluster with the following parameters: $N = 100$, $l = 2R_m = 0.02\,\mu\text{m}$, and $R_g \approx 10.6l = 0.2\,\mu\text{m}$, so that the condition $\lambda \gg R_g$ is fulfilled everywhere in the spectral region shown in Fig. 9.6.

As can be seen from this figure, the perturbation expansion gives a decent agreement with the nonperturbative results for $\lambda \leq 1\,\mu\text{m}$. However, at $\lambda = 1\,\mu\text{m}$, the non-perturbative solution is approximately two times larger. In general, the nonperturbative solution decreases much more slowly with λ. Note that the spectral dependence of the extinction cross section σ_e differs from that for the extinction efficiency Q_e by the factor $k \propto 1/\lambda$.

As the discrepancy between the perturbative and nonperturbative solutions increases in the long-wavelength spectral range, the number of spherical harmonics required for obtaining an accurate nonperturbative solution also grows. This tendency is illustrated in Fig. 9.7, where we plot the extinction efficiency as a function of the maximum order of the spherical harmonics, L. We see that near $\lambda = 10\,\mu\text{m}$, accurate results are obtained for $L \sim 10$. This value grows for larger lambda, as the optical properties of carbon become more metallic. However, even a calculation with only $N = 100$ and $L = 10$ requires ~ 600 Mb of memory in the quasistatic case and twice as much for finite k's. Since the memory requirement grows as $\sim N^2L^4$, calculations with significantly larger L's or N's seem to be problematic. This is especially true for clusters of metallic particles. Our estimates show that for silver colloidal clusters in the visible and near-IR spectral ranges, the required L is on the order of 100 (data not shown).

As an alternative method, we consider the dipole approximation coupled to the geometrical renormalization of clusters described in Section 9.4.3. In Fig. 9.8, we plot the results of calculations of Q_e in the dipole approximation for different values

[5]Fortran codes courtesy of D. Mackowski.

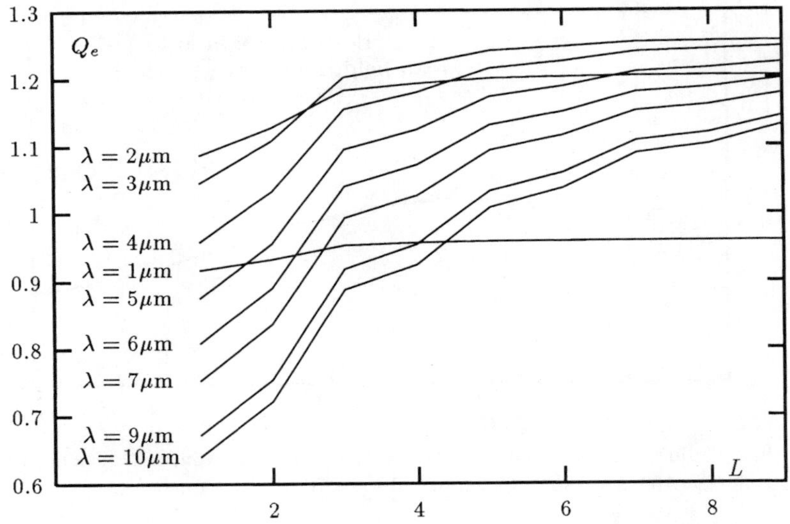

Figure 9.7. Extinction efficiency Q_e as a function of the maximum order L of the spherical harmonics involved in the nonperturbative calculation based on the expansion of all scattered fields into spherical harmonics and considering boundary conditions at each spherical surface (Fortran codes courtesy of D. Mackowski for more details see [49]). The calculations were performed in the quasistatic limit for a cluster with $N = 100$, $l = 2R_m = 0.02\,\mu\text{m}$ and $R_g \approx 10.6l = 0.2\,\mu\text{m}$, and for different values of λ, from 1 to 10 μm.

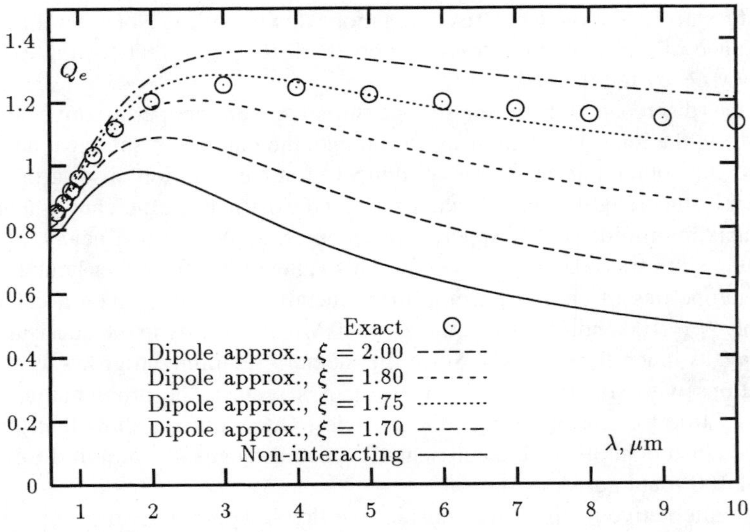

Figure 9.8. Extinction efficiency Q_e as a function of λ in the quasistatic limit. Solid curve — fist Born approximation (noninteracting limit). The $\xi = 2$ curve is the dipole approximation without geometrical renormalization of clusters. The centered symbols are the "exact" solution based on the multipole expansion.

of the intersection parameter ξ compared to the calculations based on the multipole expansion (referred to as "exact" in the figure caption) and to the first Born approximation. Since the latter can be obtained by considering isolated monomers, it is also referred to as the "non-interacting" approximation. The case $\xi = 2$ corresponds to the usual dipole approximation without renormalization.

Even without renormalization, the dipole approximation gives more accurate results than the first Born (noninteracting) approximation. But the introduction of geometrical intersections allows one to achieve much better accuracy. A good fit is obtained for $\xi = 1.75$. Note that, due to computational limitations, these results were obtained for single random realizations of computer-generated (renormalized) fractal clusters. However, the renormalization approach is statistical in nature, and we believe that ensemble averaging will increase the quality of the fit for a properly adjusted intersection parameter ξ. Nevertheless, the maximum deviation of the $\xi = 1.75$ curve from the "exact" result is only 2%.

Now, we turn our attention to the "weighted" density of states, $\Gamma(w', w'')$ defined by (9.75), and the analytical approximations that can be derived from simplification of the form of $\Gamma(w', w'')$. We restrict our consideration to the quasistatic limit, when the imaginary parts of all eigenvalues are small, and it is sufficient to consider a 1D function $\Gamma(w')$. The quasistatic analog of (9.76) is

$$\sigma_e = -\frac{4\pi k v}{|\mathbf{E}_0|^2} \operatorname{Im} \int \frac{\Gamma(w')dw'}{X + i\delta + vw'} \tag{9.78}$$

where we have also neglected the small term $2k^3 v/3$. This can be rewritten for the efficiency Q_e as

$$Q_e = \frac{4\pi\delta}{N|\mathbf{E}_0|^2} \int \frac{\Gamma(w')dw'}{(X + vw')^2 + \delta^2} \tag{9.79}$$

A typical quasistatic density of states, calculated for an ensemble of clusters with $N = 1000$, is illustrated in Fig. 9.9. It is normalized by the condition $\int \Gamma(w')dw' = N|\mathbf{E}_0|^2$. The step-like function shown in Fig. 9.9 has the same normalization, first and second moments as the numerical $\Gamma(w')$.

By comparing Figs. 9.4 and 9.9, we can conclude that the complex variable $-1/\chi = X + i\delta$ always lies far in the complex plane from the region on the real axis occupied by the normalized eigenvalues vw_n [the region where $\Gamma(w')$ is not zero]. This, in turn, leads to the idea that the fine structure of $\Gamma(w')$ is not important. As we mentioned in Section 9.4.3, replacing $\Gamma(w')$ by a delta function with the same normalization and first moment results in the mean-field approximation. However, it can be seen from Fig. 9.9 that the first moment of $\Gamma(w')$ is equal to zero. This, indeed, follows from the expression (9.62) for $B_1 = Q_1$ in the limit $k \to 0$, and is a consequence of the spherical symmetry of the clusters and the tensor properties of the dipole–dipole interaction. Therefore, the mean-field approximation in the quasistatic limit is, essentially, equivalent to the first Born (noninteracting) approximation. This fact is also illustrated in Fig. 9.6.

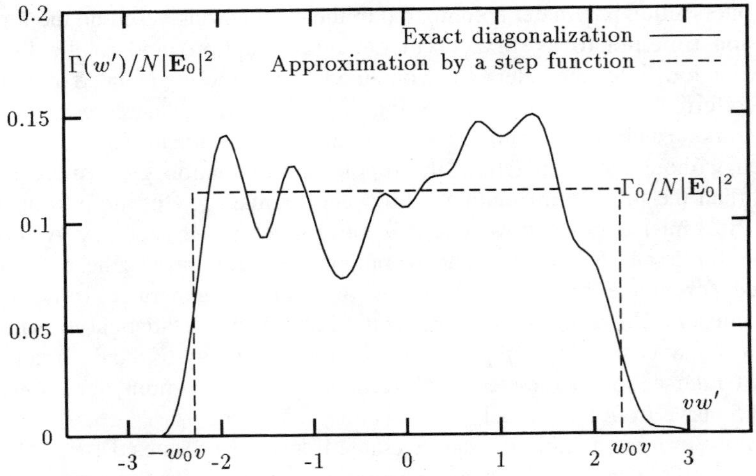

Figure 9.9. "Weighted" density of states $\Gamma(w')$ in the quasistatic limit, and its approximation by a step function with the equivalent normalization, first and second moments. The numerical diagonalization is performed for an ensemble of 10 clusters with $N = 1000$. The values of the constants are $vw_0 = 2.29$ and $\Gamma_0 = N|\mathbf{E}_0|^2/2w_0$.

To go beyond the mean-field approximation, we replace $\Gamma(w')$ by a step-like function, which preserves the second moment of $\Gamma(w')$ in addition to the first moment and normalization, as shown in Fig. 9.9 (see the figure caption for numerical values of the constants). By using the step-like function in (9.79), we obtain

$$Q_e = \frac{\pi\delta}{w_0} \int_{-w_0}^{w_0} \frac{dw'}{(X + vw')^2 + \delta^2} \tag{9.80}$$

The integral can be easily evaluated, and results in

$$Q_e = \frac{2\pi}{vw_0} \left[\arctan\left(\frac{X + vw_0}{\delta} \right) - \arctan\left(\frac{X - vw_0}{\delta} \right) \right] \tag{9.81}$$

For clusters of touching spheres (without geometrical renormalization), vw_0 is a constant. From our calculations for computer-generated cluster–cluster aggregates, $vw_0 \approx 2.29$. Now, we recall that the renormalization results in $v'w'_0 = (2/\xi)^3 vw_0$. Therefore, for a renormalized cluster, the extinction efficiency can be approximately written as

$$Q_e = \frac{2\pi}{2.29(2/\xi)^3} \left[\arctan\left(\frac{X + 2.29(2/\xi)^3}{\delta} \right) - \arctan\left(\frac{X - 2.29(2/\xi)^3}{\delta} \right) \right] \tag{9.82}$$

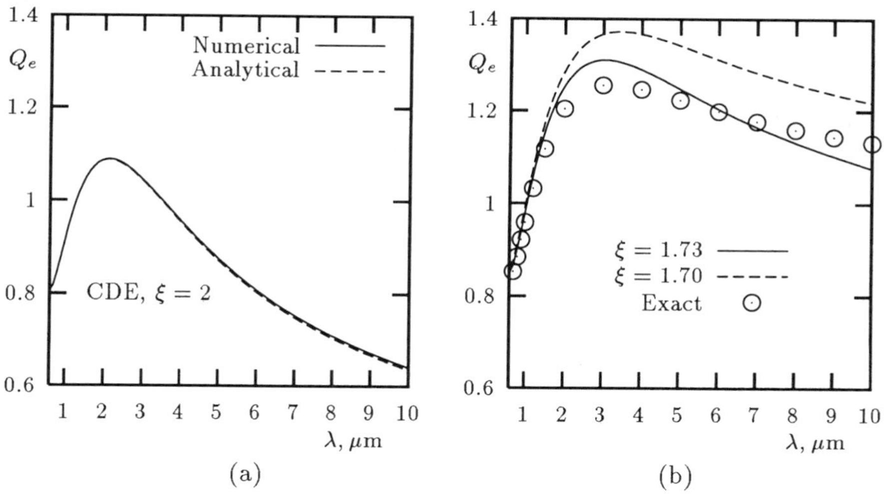

Figure 9.10. (a) Analytical expression (9.82) compared to numerical calculations in the dipole approximation without cluster renormalization ($\xi = 2$). (b) Analytical expression (9.82) for two different values of $\xi < 2$ compared to the exact result.

The results of calculations according to formula (9.82) are shown in Fig 9.10. First, in Fig 9.10 (a), we compare the analytical expression (9.82) with the results of numerical calculations in the dipole approximation without the geometrical renormalization ($\xi = 2$). The analytical and numerical data fit very accurately. In Fig 9.10(b), we plot Q_e given by (9.82) for two different values of $\xi < 2$ compared to the "exact" result (the one shown in Fig. 9.6 by circles). We see that the closest fit is achieved for $\xi \sim 1.73$. The curve with $\xi = 1.7$ gives a less accurate approximation of Q_e, but better reproduces the λ dependence in the long-wavelength region (up to a multiplicative constant). When comparing the results of the dipole approximation and the analytical formula (9.82) with the exact calculations, which are shown in Fig. 9.10 (b) by circles, keep in mind that the latter were obtained by truncation of the maximum order of spherical harmonics, L. As seen in Fig. 9.7, the calculated value of Q_e still continues to grow for $L \sim 10$ and $\lambda > 2\ \mu$m. Our calculations were truncated at $L = 9$ due to computational limitations. But it can be stated with a reasonable amount of confidence that the "true" results for Q_e are somewhat larger than those shown, for example, in Fig. 9.10(b). This follows from the monotonic growth of Q_e as a function of L illustrated in Fig. 9.7. Therefore, the curve with $\xi = 1.7$ might be actually more accurate than the one with $\xi = 1.73$.

In conclusion of this section, we note that the value vw_0 [or the second moment of $\Gamma(w')$] can depend on the fractal dimension of the clusters. The fine features of $\Gamma(w')$ can also depend on less essential properties of the clusters, such as the type of lattice used in numerical calculations. For nonresonant carbon, these fine details of the density of states are largely insignificant. However, they become important in the resonance situation, when the denominator in the formulas (9.76) and (9.78) can become purely imaginary and small. The resonance EM interaction in fractal clusters is considered theoretically in Chapter 8, and from the experimental point of view given in Chapter 7. Danilova discussed the phenomenological intersection parameter ξ for the case of resonance interaction and compared analytical and experimental results in [58].

9.5. FLUCTUATIONS OF LIGHT SCATTERED BY RANDOM SMOKE CLUSTERS

The well-known result (e.g., [14]) for the differential scattering cross-section of fractal clusters in the first Born approximation is

$$\frac{d\sigma_s}{d\Omega} \propto q^{-D} \qquad \text{if} \quad q \gg 1/R_g \tag{9.83}$$

where $\mathbf{q} = \mathbf{k} - \mathbf{k}'$ is the transmitted wavevector. This result is obtained by statistical averaging using the density–density correlation function (9.1), and is statistical in nature. The intensity of light scattered from a single random cluster can be different from (9.83). The statistical averaging implied in the derivation of (9.83) can be understood in two different ways. The first is ensemble averaging, over an ensemble of random realizations of clusters. One can also hope that if a single cluster is large enough, it's differential scattering cross-section approaches the ensemble average value, and the deviations decrease with the cluster size as $1/\sqrt{N}$ (self-averaging).

In this section, we show that the self-averaging can occur for random nonfractal clusters, but not for fractal aggregates with long-range correlations. In fact, we will show numerically that the relative dispersion of the scattered intensity in an ensemble of fractal cluster–cluster aggregates is always close to unity. A more detailed account is published in [23].

9.5.1. General Relations

The differential scattering cross-section can be calculated using the general definition (9.15) and the expression for the scattering amplitude $f(\mathbf{k}')$ either in the integral form (9.14) or the discretized version (9.64). In the first Born approximation, which will be used throughout this section, both formulas for $\mathbf{f}$ lead to the same result. For simplicity, we start from the discretized expression (9.64), and, by substituting $\mathbf{d}_i = \alpha \mathbf{E}_{\text{inc}}(\mathbf{r}_i) = \alpha \mathbf{E}_0 \exp(i\mathbf{k} \cdot \mathbf{r}_i)$, we obtain

$$\mathbf{f}(\mathbf{k}') = k^2 \alpha \left[\mathbf{E}_0 - \frac{(\mathbf{E}_0 \cdot \mathbf{k}')\mathbf{k}'}{k^2} \right] \sum_{i=1}^{N} \exp(i\mathbf{q} \cdot \mathbf{r}_i) \tag{9.84}$$

where $\mathbf{q} = \mathbf{k} - \mathbf{k}'$. The differential scattering cross-section can be easily obtained from the above expression

$$\frac{d\sigma_s}{d\Omega} = k^4|\alpha|^2|\mathbf{E}_0|^2\sin^2[\psi(\mathbf{E}_0, \mathbf{k}')]\left|\sum_{i=1}^{N} \exp(i\mathbf{q} \cdot \mathbf{r}_i)\right|^2 \tag{9.85}$$

where $\psi(\mathbf{E}_0, \mathbf{k}')$ denotes the angle between $\mathbf{E}_0$ and $\mathbf{k}'$. The prefactor $k^4|\alpha|^2 \times |\mathbf{E}_0|^2\sin^2[\psi(\mathbf{E}_0, \mathbf{k}')]$ does not depend on a random realization of the set $\{\mathbf{r}_i\}$ and, therefore, is the same for all clusters. In the case of scattering of a depolarized wave, we must replace $\sin^2[\psi(\mathbf{E}_0, \mathbf{k}')]$ by $\langle\sin^2\psi\rangle = 0.5$, and the above factor becomes simply a constant. In contrast, the factor $\left|\sum_{i=1}^{N} \exp(i\mathbf{q} \cdot \mathbf{r}_i)\right|^2$ in (9.85) is random and can vary from cluster to cluster.

When considering fluctuations of scattered light by different random clusters, we do not need to keep a factor that is common to all of them. Therefore, it is convenient to define the intensity of light scattered by some individual cluster as

$$I(\theta, \phi) = I(\mathbf{q}) = \left|\sum_{i=1}^{N} \exp(i\mathbf{q} \cdot \mathbf{r}_i)\right|^2 \tag{9.86}$$

where θ is the angle between the direction of the incident wavevector $\mathbf{k}$ and the direction of scattering, and ϕ is the azimuthal angle. The absolute value of $\mathbf{q}$ depends on the scattering angle θ as

$$q = k\sqrt{2(1 - \cos\theta)} \tag{9.87}$$

The intensity (9.86) coincides with the "real" intensity of the scattered light up to some constant in the case of a depolarized incident wave, and up to the ψ-dependent factor, $\sin^2[\psi(\mathbf{E}_0, \mathbf{k}')]$, for a polarized wave. The definition of scattered intensity (9.86) is suitable for the calculation of *relative* fluctuations, that is, for the dispersion of scattered intensity divided by the average scattered intensity. If we want to calculate *absolute* fluctuations, we need, of course, to keep all the prefactors. In this section, we will focus on relative fluctuations. The absolute value of fluctuations can always be reconstructed, provided the average scattered intensity is known.

Let us consider the intensity of light scattered by some number of fractal clusters randomly distributed in a certain volume. The distance between clusters is supposed to be large compared to the wavelength of the incident radiation, λ, and the distribution of clusters in space to be random and uncorrelated. Then, we can add the intensities of light scattered by each cluster, rather than the amplitudes.

The average scattered intensity $\langle I\rangle$ is defined as

$$\langle I\rangle = \langle I(\theta, \phi)\rangle = \lim_{M\to\infty} \frac{1}{M}\sum_{k=1}^{M} I_k(\theta, \phi) \tag{9.88}$$

where $I_k(\theta, \phi)$ is the intensity scattered by the kth cluster and M is the total number of clusters that scatter the light. With the use of (9.86), we can rewrite (9.88) as

$$\langle I \rangle = \lim_{M\to\infty} \frac{1}{M} \sum_{k=1}^{M} \sum_{i,j=1}^{N_k} \exp[i\mathbf{q} \cdot (\mathbf{r}_i^{(k)} - \mathbf{r}_j^{(k)})] \tag{9.89}$$

where N_k is the number of monomers in the kth cluster, and $\mathbf{r}_i^{(k)}$ is the coordinate of the ith monomer in kth cluster.

For an ensemble of spherically symmetrical (on average) clusters, the dependence of $\langle I \rangle$ on ϕ is weak (it vanishes for an infinite ensemble); therefore, we will use the notation $\langle I \rangle = \langle I(\theta) \rangle = \langle I(q) \rangle$, where q is defined by (9.87).

If we detect the scattered light from just one cluster, it can be much different from $\langle I \rangle$. A convenient measure of these variations is the standard deviation (dispersion), σ_I:

$$\sigma_I^2 = \langle I^2 \rangle - \langle I \rangle^2 \tag{9.90}$$

The value of σ_I characterizes possible deviations of I_k from $\langle I \rangle$ calculated for an infinite ensemble of clusters and has a simple mathematical meaning: The probability that an individual I_k lies within the interval $\langle I \rangle \pm \sigma_I$ is approximately two-thirds.

In the case of a finite M, one can be interested in a measure of fluctuations of the average value (9.88) itself [the "lim" sign in this case should, of course, be omitted in (9.88) and (9.89)]. If we register the scattered light from different ensembles of clusters consisting of some finite number of clusters M, we will come up with different results. We can define the standard deviation $\sigma_I^{(M)}$ of these random values in the usual way. The relation between $\sigma_I^{(M)}$ and $\sigma_I \equiv \sigma_I^{(1)}$ is well-known from mathematical statistics:

$$\sigma_I^{(M)} = \frac{\sigma_I}{\sqrt{M}} \tag{9.91}$$

The actual value of M depends on the scheme of the experiment. In one possible setting, the scattering volume is small enough (e.g., due to focusing a laser beam) and contains only one cluster at a time. Because of the random motion of clusters, it contains different clusters in different moments of time. In this case, one can register scattered radiation for some large period of time (excluding the periods when the volume contains no clusters at all and the signal is zero) and calculate the time-averaged intensity and its standard deviation, which coincides with σ_I. If the volume contains an average of M clusters at a given time, the measured standard deviation would be $\sigma_I^{(M)}$. As will be shown numerically in Section 9.5.2, σ_I is universal for cluster–cluster aggregates over a wide range of scattering angles. The relation (9.91) can be used to find the average number of clusters in the scattering volume (and, hence, the number density of clusters).

9.5.2. Monodisperse Clusters

First, we consider monodisperse ensembles of clusters consisting of N monomers each. The task of calculating σ_I includes finding two average values: $\langle I \rangle$ and $\langle I^2 \rangle$. Apart from calculating the dispersion (Eq. 9.90), $\langle I \rangle$ is interesting by itself and is experimentally measurable. It is well known that the pair correlation function $p(r)$ (9.1) can be used to calculate $\langle I \rangle$. Indeed, for a monodisperse ensemble, (9.89) can be simplified to

$$\langle I \rangle = N + N(N-1)\langle \exp(i\mathbf{q} \cdot \mathbf{r}_{ij}) \rangle \tag{9.92}$$

where $\mathbf{r}_{ij} = \mathbf{r}_i - \mathbf{r}_j$, and only distinct monomers belonging to the same cluster are considered. Now, we can use the function $p(r)$ to calculate $\langle \exp(i\mathbf{q} \cdot \mathbf{r}_{ij}) \rangle$:

$$\langle \exp(i\mathbf{q} \cdot \mathbf{r}_{ij}) \rangle = \int_0^\infty p(r) \exp(i\mathbf{q} \cdot \mathbf{r}_{ij}) \frac{dr \sin\theta \, d\theta \, d\phi}{4\pi} \tag{9.93}$$

After performing the angular integration, (9.93) simplifies to

$$\langle \exp(i\mathbf{q} \cdot \mathbf{r}_{ij}) \rangle = \int_0^\infty p(r) \frac{\sin qr}{qr} dr \tag{9.94}$$

In the case of $D < 2$ and $q \gg R_g^{-1}$, we can calculate the integral (9.94) without specifying the form of the cutoff function for $p(r)$, similarly to the calculation of the coefficient B_1 in Section 9.4.2. Indeed, in this case the cutoff function can be set to unity, and the result is

$$\langle \exp(i\mathbf{q} \cdot \mathbf{r}_{ij}) \rangle = \frac{a\Gamma(D-1)\sin[\pi(D-1)/2]}{N(ql)^D} \tag{9.95}$$

In the other limiting case, $q \ll R_g^{-1}$, sin (qr) in (9.94) can be expanded in a power series, and the result of integration up to the lowest nonzero power of q is

$$\langle \exp(i\mathbf{q} \cdot \mathbf{r}_{ij}) \rangle = 1 - (qR_g)^2/3 \tag{9.96}$$

As follows from (9.96) and (9.92), $\langle I(\theta = 0) \rangle = N^2$, which means that the forward scattering is always coherent in the first Born approximation.

From (9.95) and (9.92), it can be concluded that the minimum possible value of $\langle I \rangle$ is N, which can be reached for large values of q. For the backscattering, when the value of q is maximum, the expression for $\langle I \rangle$ becomes $\langle I \rangle \approx N[1 + 5 \times 10^{-2}(\lambda/l)^D]$, where we used the numerical values for all the coefficients (assuming $D = 1.8$). The characteristic value of λ is $\lambda_c \approx 5.4l$, so that $\langle I \rangle$ approaches its lower bound for $\lambda \ll \lambda_c$. However, the above inequality contradicts the fundamental assumption of this chapter that λ is much larger than monomer size R_m (or lattice unit, l). Therefore, in the spectral region where the monomers are optically small, $kR_m \ll 1$, the first term in (9.92) can be neglected.

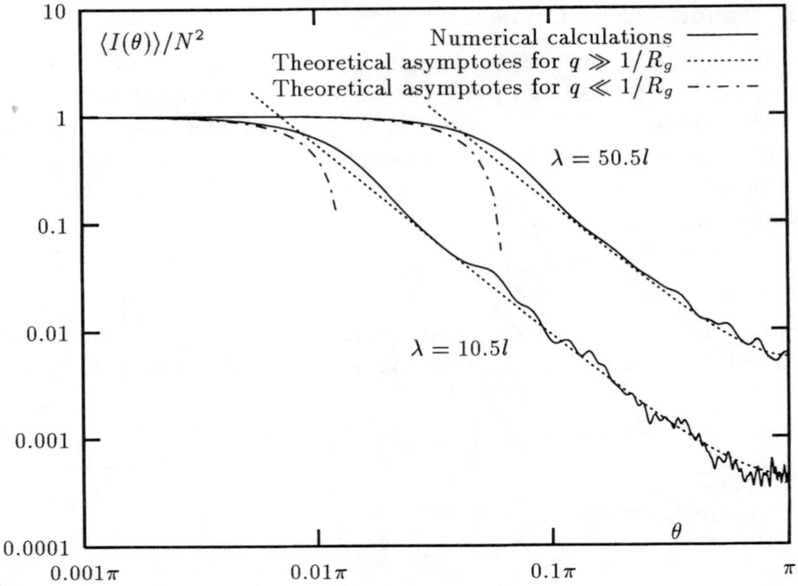

Figure 9.11. Average intensity of the scattered light as a function of the scattering angle θ for $\lambda = 10.5l$ and $\lambda = 50.5l$. Calculations were performed for a computer-generated monodisperse ensemble of 40 random cluster–cluster aggregates consisting of $N = 10{,}000$ monomers each, similar to the one shown in Fig. 9.1. Noninteger values of λ/l are chosen to avoid lattice effects. The definition of I is given in Eq. (9.86).

The theoretical asymptotes (9.95) and (9.96), along with the results of numerical calculations for $\langle I \rangle$ for different values of λ, are illustrated in Fig. 9.11 (see the figure caption for details).

Whereas $\langle I \rangle$ is defined by $p(r)$, one needs a higher order correlation function for the calculation of $\langle I^2 \rangle$. Indeed, the definition of $\langle I^2 \rangle$, analogous to (9.89), contains a fourfold summation, which, after grouping together the terms with different indexes matching each other, turns to

$$\begin{aligned}\langle I^2 \rangle &= N(2N-1) + 4N(N-1)^2 \langle \exp(i\mathbf{q} \cdot \mathbf{r}_{ij}) \rangle \\ &\quad + N(N-1)(N^2 - 3N + 3) \langle \exp(i\mathbf{q} \cdot \mathbf{r}_{ijkl}) \rangle \end{aligned} \tag{9.97}$$

where $\mathbf{r}_{ijkl} = \mathbf{r}_{ij} - \mathbf{r}_{kl}, i \neq j, k \neq l$ and any of the pair of indexes (i,j) can coincide with any of the pair (k,l). It is easy to show that $\langle \exp(i\mathbf{q} \cdot \mathbf{r}_{ijkl}) \rangle$ is expressed through the four-point correlation function, $p_4(r)$, which was introduced in Section 9.2 (see also Fig. 9.3), exactly in the same form as in Eq. (9.94) with $\mathbf{r}_{ij}$ being replaced by $\mathbf{r}_{ijkl}$ and p by p_4.

We now turn to the numerical results for fluctuations that are presented in Fig. 9.12. In this calculation, we allowed θ to change from 0 to 2π, so that the "observer"

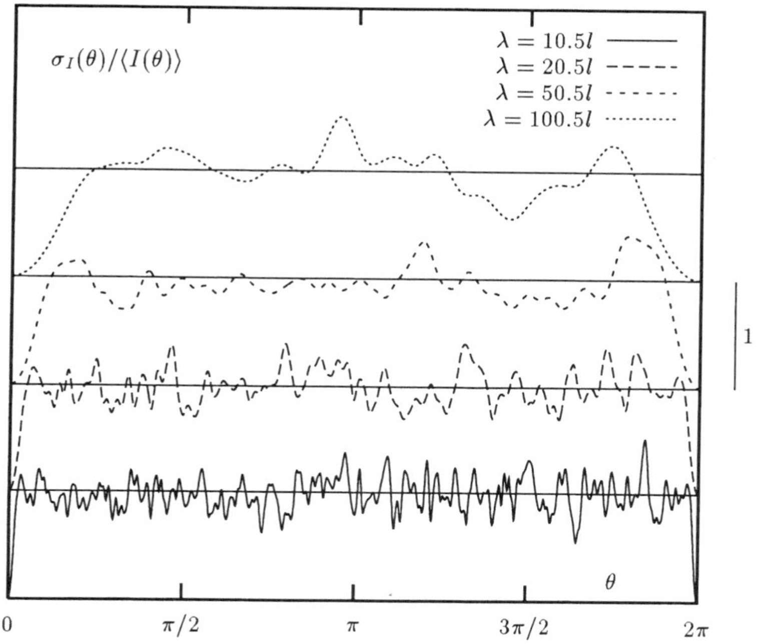

Figure 9.12. Relative fluctuations $\sigma_I(\theta)/\langle I(\theta)\rangle$ for different wavelengths as functions of the scattering angle θ. For each curve, the horizontal line corresponds to the level $\sigma_I/\langle I\rangle = 1$; the distance between the nearest horizontal lines is 1; and $\sigma_I(0)/\langle I(0)\rangle = \sigma_I(2\pi)/\langle I(2\pi)\rangle = 0$. Calculations are performed for the same ensemble of computer-generated clusters as in Fig. 9.11. Note that θ is allowed to change from 0 to 2π (unlike the usual spherical system of coordinates where $0 \le \theta \le \pi$), so that the "observer" makes a complete revolution from the "forward' direction of scattering to the "backward" direction and back to "forward".

makes a whole revolution from the "forward" direction of scattering to the "backward" direction and back to "forward". In the usual spherical system of coordinates, this corresponds to θ varying from 0 to π, then changing ϕ to $\phi + \pi$, and varying θ back from π to 0. Note that, for a finite ensemble of random clusters, the result is not necessarily symmetrical with respect to the point $\theta = \pi$. However, it must be symmetrical for an infinite ensemble of spherically symmetrical (on average) clusters; this follows from the fact that neither σ_I nor $\langle I\rangle$ can depend on ϕ in this case.

First, we consider the domains of θ where the asymptote (9.95) is valid. The characteristic values of θ_c [defined from the condition $q(\theta_c) = R_g^{-1}$] are $5.4 \times 10^{-3}\pi$ for $\lambda = 10.5l$, $1.0 \cdot 10^{-2}\pi$ for $\lambda = 20.5l$, $2.6 \times 10^{-2}\pi$ for $\lambda = 50.5l$, and 0.16π for $\lambda = 100.5l$. One can easily see that the value of $\sigma_I/\langle I\rangle$ fluctuates near unity if $\theta_c \ll \theta \ll 2\pi - \theta_c$.[6] It should be noted that for a finite ensemble, $\sigma_I/\langle I\rangle$ is a random quantity itself. Since there is no noticeable systematic dependence on θ in the domain

[6]As can be seen from Figs. 9.11 and 9.12, there is no need for *strong* inequalities here.

defined above, we can perform additional averaging of $\sigma_I/\langle I\rangle$ over θ, the results for this averaging are (up to the third significant figure): 0.98 for $\lambda = 10.5l$, 1.00 for $\lambda = 20.5l$, 0.96 for $\lambda = 50.5l$, and 1.01 for $\lambda = 100.5l$.

The numerical data suggest that the value of the relative fluctuations of the intensity of light scattered by cluster–cluster aggregates is very close to unity and statistically independent of the scattering angle θ, as long as θ lies in the domain defined above. This is true for a wide range of wavelengths λ. However, for very large λ, the domain of θ shrinks and becomes essentially empty when $\lambda = 4\pi R_g$.

9.5.3. Polydisperse Clusters

Now, we consider a polydisperse ensemble of clusters, that is, an ensemble containing clusters with different N's. We first look at the case of large q, when the condition $q \gg R_g^{-1}$ is fulfilled for almost every cluster in the ensemble.

We can calculate $\langle I\rangle$ by performing an additional averaging over N in Eq. (9.92). In the case of large q, this averaging leads to

$$\langle I\rangle = \langle N\rangle\{1 + a\Gamma(D-1)\sin[\pi(D-1)/2]/(ql)^D\} \tag{9.98}$$

It is natural to assume that the intensity scattered by some individual cluster I_k can be represented as

$$I_k = N_k J_k \tag{9.99}$$

where N_k and J_k are statistically independent random variables, and

$$\langle J\rangle = 1 + a\Gamma(D-1)\sin[\pi(D-1)/2]/(ql)^D \tag{9.100}$$

Then, ensemble averaging of (9.99) results in (9.98).

For a monodisperse ensemble, J_k coincide with I_k up to some constant, common for each cluster. Therefore, the relative dispersion of J, $\sigma_J/\langle J\rangle$, coincides with the relative dispersion of I in a monodisperse ensemble.

Further, we can use (9.99) to calculate the relative dispersion of scattered intensity in a polydisperse ensemble in terms of that in a monodisperse ensemble and the dispersion of the random variable N. Straightforward algebra yields

$$\frac{\sigma_I}{\langle I\rangle} = \frac{\sigma_J}{\langle J\rangle}\sqrt{\frac{\sigma_N^2}{\langle N\rangle^2}\left(1 + \frac{\langle J\rangle^2}{\sigma_J^2}\right) + 1} \tag{9.101}$$

From the numerical results of Section 9.5.2, we know that $\sigma_J/\langle J\rangle$ is very close to unity. By substituting this value into (9.101), we obtain

$$\frac{\sigma_I}{\langle I\rangle} = \sqrt{2\frac{\sigma_N^2}{\langle N\rangle^2} + 1} \tag{9.102}$$

It follows from formula (9.102) that $\sigma_I/\langle I\rangle$ is always close to unity, even for very polydisperse ensembles. The value of $\sigma_N/\langle N\rangle$ cannot be much > 1 for any physically reasonable distribution of N. For example, if N is uniformly distributed from 0 to $N_{\max}$, this value is equal to $1/\sqrt{3}$. If the distribution has two sharp peaks of equal height near N_1 and N_2, it is equal to $|N_1 - N_2|/(N_1 + N_2)$.

In order to verify Eq. (9.101), we calculated $\sigma_I/\langle I\rangle$ for a polydisperse ensemble of 100 clusters. The number of particles in a particular cluster was found from the Gaussian probability distribution with the average $\langle N\rangle = 5000$ and the dispersion[7] $\sigma_N = \sqrt{\langle N^2\rangle - \langle N\rangle^2} = 2000$. The ratio $\sigma_N/\langle N\rangle$ for this ensemble is $\simeq 0.37$. The calculations were done for two different values of λ. After additional averaging over angles (as described in Section 9.5.2), the results obtained are as follows: $\sigma_I/\langle I\rangle = 1.109$ for $\lambda = 10.5$, and $\sigma_I/\langle I\rangle = 1.087$ for $\lambda = 20.5$. The results following from the theoretical formula (9.101) and the corresponding results for a monodisperse ensemble ($\sigma_J/\langle J\rangle$) are 1.109 and 1.120, respectively. As we see, the results match closely. For the case of $\lambda = 10.5$, the difference is only in the fifth figure.

9.5.4. Fluctuations of Light Scattered by Trivial (Nonfractal) Clusters

It is interesting to compare the fluctuations of light scattered by fractal and by trivial ($D = 3$) clusters. In this Section, we discuss fluctuations in light intensity scattered from such systems, restricting consideration to only monodisperse ensembles.

To model random nonfractal clusters, we use the algorithm of randomly close-packed hard spheres. In this algorithm, one first chooses a volume to be occupied by a cluster. In our simulations it is a sphere of radius R_s ("s" standing for "sphere"), since we intended to build clusters that are spherically symmetrical. Then, monomers are randomly placed inside the volume. At each step, the intersection condition is checked: If the newly placed monomer approaches any of the previously placed ones closer than the distance l, this step is rejected and the next random position is tried. In this way, each monomer can be thought of as a hard sphere of radius $l/2$. The procedure stops when a large number of tries is consequently rejected. In our simulations, this number was chosen to be 2×10^7. This algorithm allows one to achieve a fairly dense packaging. Consequently, we have packed 40 different clusters with an average of 9200 monomers per cluster into a spherical volume with radius $R_s = 14.2l$. The volume fraction occupied by the particles was ≈ 0.40. For comparison, it is ≈ 0.52 in the case of a simple cubic lattice and can be even lower for some other types of lattices. The minimum distance from a given monomer to its nearest neighbor was very close to l; the maximum distance varied from $1.2l$ to $1.3l$. Although this ensemble of clusters was not completely monodisperse, the variation of N was very small: The ratio of the standard deviation of N and the mean was equal to 2.4×10^{-3}.

[7] These parameters characterize the porbability distribution according to which the values of N were picked for each cluster. The actual parameters of the ensemble were slightly different: $\langle N\rangle = 5343$, $\sigma_N = 1953$.

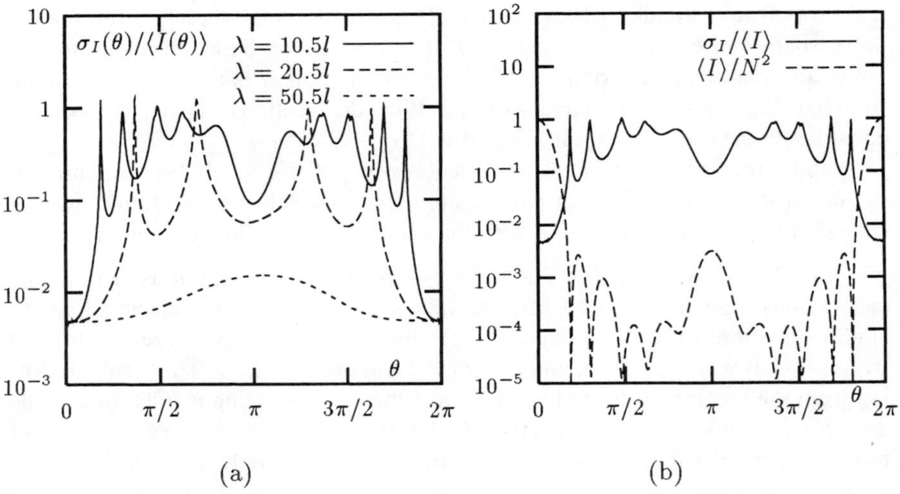

Figure 9.13. (a) Relative fluctuations $\sigma_I(\theta)/\langle I(\theta)\rangle$ for random close-packed nonfractal clusters ($D = 3$) packed in a sphere of the radius $R_s = 14.2l$, for different wavelengths. (b) Relative fluctuations, $\sigma_I(\theta)/\langle I(\theta)\rangle$, compared to the average scattered intensity $\langle I\rangle$ for the same ensemble of clusters and $\lambda = 10.5l$.

The results of numerical simulations of $\sigma_I/\langle I\rangle$ for the ensemble of 40 clusters described above are shown in Fig. 9.13(a). As in Section 9.5.3, the scattering angle θ varies from 0 to 2π. First, we notice the strong and systematic dependence of $\sigma_I/\langle I\rangle$ on θ. [For fractal clusters, this dependence looks like statistical noise (cf Fig. 9.12)]. Second, for most angles the value of $\sigma_I/\langle I\rangle$ is significantly < 1 and decreases when λ grows. This dependence on λ is anticipated, because if there are many monomers in the volume λ^3, a cluster becomes optically similar to a dielectric sphere, and its random structure is of no importance. But this is not the case for fractal clusters; they are geometrically different and random on all scales up to the maximum scale R_g. As seen from Fig. 9.12, $\sigma_I/\langle I\rangle$ for fractal clusters is on the order of 1, even for $\lambda = 100.5l$. But for nonfractal clusters, $\sigma_I/\langle I\rangle$ is much smaller, on the order of 10^{-2} for $\lambda = 50.5l$.

The second feature of Fig. 9.13(a) is the presence of sharp maxima in $\sigma_I/\langle I\rangle$, where it becomes on the order of 1. These maxima occur for the angles θ at which $\langle I(\theta)\rangle$ has a minima [see Fig. 9.13(b)].

The problem of fluctuations can be solved exactly for spherically symmetrical random clusters, provided the positions of monomers in clusters are absolutely uncorrelated. This is not the case for the close-packed clusters discussed above, because in this model monomers cannot approach each other closer than l, which brings about short-range correlations. It is clear that the model of totally uncorrelated clusters (random gas) is not exact since the monomers act like hard spheres during aggregation. However, theoretical results for uncorrelated clusters help explain the main features shown in Fig. 9.13.

Consider a "random gas" of uncorrelated particles inside a spherical volume of radius R_s. The ensemble-average quantities $\langle \exp(i\mathbf{q}\cdot\mathbf{r}_i)\rangle$, $\langle \exp(i\mathbf{q}\cdot\mathbf{r}_{ij})\rangle$, and $\langle \exp(i\mathbf{q}\cdot\mathbf{r}_{ijkl})\rangle$ can be obtained from straightforward integration and are as follows:

$$\langle \exp(i\mathbf{q}\cdot\mathbf{r}_i)\rangle = \varphi(qR_s) \equiv \frac{3}{(qR_s)^3}[\sin(qR_s) - qR_s\cos(qR_s)] \tag{9.103}$$

$$\langle \exp(i\mathbf{q}\cdot\mathbf{r}_{ij})\rangle = \varphi^2(qR_s) \tag{9.104}$$

$$\langle \exp(i\mathbf{q}\cdot\mathbf{r}_{ijkl})\rangle = \varphi^4(qR_s) \tag{9.105}$$

The values of $\langle I\rangle$ and $\langle I^2\rangle$ can be found according to (9.92) and (9.97), with the use of (9.103)–(9.105). The expression for $\langle I\rangle$ is

$$\langle I\rangle = N + N(N-1)\varphi^2(qR_s) \tag{9.106}$$

and the expression for $\sigma_I/\langle I\rangle$ is (in the limit of large N)

$$\frac{\sigma_I}{\langle I\rangle} = \frac{\sqrt{1 - 4\varphi^2 + 3\varphi^4 + 2\varphi^2(1-\varphi^2)N}}{1 + \varphi^2 N} \tag{9.107}$$

If $\varphi(qR_s)$ turns to zero for some value of q, this means that $\sigma_I/\langle I\rangle$ has a maximum and is on the order of 1 for this q. At the same time, the average scattered intensity (9.106) has a minimum.

The function $\varphi(x)$ becomes exactly zero if x is a solution to $\tan(x) = x$. The first root of this equation is $x \approx 1.43\pi$. The corresponding scattering angle is defined by $\cos\theta = 1 - 0.26(\lambda/R_s)^2$. This equation has a solution only if $\lambda < 2.8R_s$. In Fig. 9.13, we have sharp maxima in $\sigma_I/\langle I\rangle$ for $\lambda = 10.5l$ and $\lambda = 20.5l$, but there are no sharp maxima for $\lambda = 50.5l$. For the clusters considered, $R_s = 14.2l$, and the critical value of λ is $39.7l$. We see that $\lambda = 50l$ exceeds the critical value and, therefore, the corresponding curve in Fig. 9.13 has no sharp maxima.

Now, we analyze the expression (9.107) in more detail. First, when $N \to \infty$, this expression assumes the form

$$\frac{\sigma_I}{\langle I\rangle} = \sqrt{\frac{2(1/\varphi^2 - 1)}{N}} \tag{9.108}$$

As one would expect, the relative fluctuations are proportional to $1/\sqrt{N}$. To obtain (9.108), we took the limit $N\varphi^2 \gg 1$. This condition can be expressed in terms of the density ν of monomers in clusters, where $N = 4\pi R_s^3\nu/3$. By using (9.103), we find that, in order to obtain (9.108), the following inequalities must hold

$$\nu \gg \frac{1}{12\pi}q^3 \qquad \text{if} \quad qR_s \sim 1 \tag{9.109}$$

$$\nu \gg \frac{qR_s}{12\pi}q^3 \qquad \text{if} \quad qR_s \gg 1 \tag{9.110}$$

The condition is always fulfilled if $qR_s \ll 1$, since $\varphi(0) = 1$. Note that in order to derive (9.109) and (9.110), we assumed that $\tan(qR_s) \neq qR_s$ and $\sin(qR_s)$, $\cos(qR_s) \sim 1$. As discussed above, if $\tan(qR_s) = qR_s$, $\varphi(qR_s)$ turns exactly to zero, the condition $N\varphi^2 \gg 1$ cannot be fulfilled.

The above inequalities show that in order to observe the $1/\sqrt{N}$ dependence for the fluctuations, one needs to have many monomers in the volume q^{-3}. This condition depends on the value of qR_s and is stronger when $qR_s \gg 1$. We emphasize that for fractal clusters we can never obtain the $1/\sqrt{N}$ dependence for relative fluctuations (see, e.g., the curve in Fig. 9.12 for $\lambda = 100.5l$). The reason is that the fractal clusters are disordered on all scales up to the maximum scale R_g, whereas trivial random clusters become homogeneous on scales larger than $1/\sqrt[3]{\nu}$.

Now, we turn our attention to the nature of the sharp maxima in $\sigma_I/\langle I \rangle$, which are seen in Fig. 9.13(a). As mentioned above, these maxima coincide with the diffraction minima of the average scattered intensity. The diffraction minima occur because within the first Born approximation, and for certain scattering angles, the EM fields produced by monomers in a cluster almost exactly cancel each other due to destructive interference. As a result, the scattered field for these scattering angles is produced, in fact, by a very few monomers, rather than by the whole cluster. This results in strong relative fluctuations.

9.6. ABSORPTION OF LIGHT BY SMOKE CLUSTERS PLACED IN A WATER DROPLET

Soot clusters often form agglomerates with water microdroplets, especially in the clouds [59–63]. Naturally, this might be expected to lead to dramatic changes in the optical characteristics. In this section, we obtain qualitative results concerning the *absorption* cross-section of such composite microdroplets in the first Born approximation, with an account of the fractal morphology of carbon soot clusters. Our consideration includes the nonfractal homogeneous distribution of carbon inclusions as a limiting case. A more detailed account can be found in [64].

The theoretical treatment in this section is somewhat different from the rest of the chapter. The small parameter of the perturbation expansion will not be χ, as above, but the volume fraction of carbon soot inside a water droplet. The microdroplet radius will be denoted by R_d ("d" standing for "droplet"), and its volume by $v_d = (4\pi/3)R_d^3$, so that the small parameter of the expansion is v_{tot}/v_d, with v_{tot} still being the total volume of carbon inclusions. It is not assumed, however, that the carbon inclusions in a microdroplet form one self-supporting cluster, or are built from spherical monomers. The only important quantity entering the calculations will be the average density of inclusions, $\langle \rho(\mathbf{r}) \rangle$, where the averaging is performed over an ensemble of droplets of the same radius with random carbon inclusions inside. Correspondingly, the results obtained in Section 9.6.1 are of a statistical nature.

9.6.1. Introductory Remarks and Review

When soot particles are placed inside a water droplet, they are no longer exited by a plane wave, but rather by internal modes of a high-quality optical resonator. To complicate things further, the resonator modes can effectively couple to the modes of clusters themselves.

There have been a considerable number of experimental [65–70] and theoretical [51,71–78] studies of scattering and absorbing properties of inhomogeneous spheres. The simplest model for a water droplet with an inclusion inside is a spherical dielectric particle with an eccentric spherical inclusion. An exact formal solution to the problem of light scattering and absorption by such composite spheres was obtained by Borghese et al. [72] and generalized for the case of multiple arbitrarily positioned spherical inclusions by Borghese et al. [75] and Fuller [51,52,76,79]. The solutions were obtained by the vector spherical harmonic (VSH) expansion of electrical fields inside the homogeneous spherical regions and satisfying the boundary conditions at all the discontinuity surfaces. Even in the case of one spherical inclusion, the solution must be obtained from an infinite-order system of linear equations. As discussed in Section 9.4.3, the VSH expansion is truncated at some maximum order L, and the system contains $\sim L^2$ equations [72]. When multiple inclusions are considered, the number of equations is further increased, which makes the problem very complicated numerically. Also, the approach based on the consideration of the exact boundary conditions requires a knowledge of the exact geometry of the problem before the time-extensive calculations. This fact complicates the averaging of solutions over a random distribution of inclusions inside water droplets.

An alternative approach based on perturbation theory was developed by Kerker et al. [71] and Hill et al. [77]. According to their method, the dielectric function of an inhomogeneous sphere is represented as a sum of a constant (unperturbed) value and a small coordinate-dependent perturbation. In the zeroth-order approximation, the field inside the droplet is calculated within the assumption that the perturbation of the dielectric function is equal to zero. This field is given by the Mie expansion in terms of the VSH. In the next iteration, the field in the zeroth-order approximation induces some additional polarization (or, equivalently, current) in the volume, proportional to the perturbation of the dielectric function. This additional polarization can be used to calculate changes of scattering and absorbing characteristics of the inhomogeneous sphere as compared to the homogeneous (unperturbed) one. A big advantage of this method is that it allows one to perform averaging over random perturbations. However, it has a drawback. As was pointed out by Hill et al. [77], the internal field must be computed iteratively. That is, the additional polarization calculated in the first iteration described above should produce some additional internal electrical field, which, in turn, gives rise to additional polarization (now proportional to the unperturbed dielectric function), and so on. Physically, this means that the modes of a spherical resonator are coupled to the modes of the perturbation of the dielectric function. In order for a finite-order approximation to be accurate, it is necessary that the perturbation expansion of any physical quantity under consideration converges. In Section 9.6.2, we show that, in general, this is not the case. More specifically, this expansion always diverges for physical quantities related to scattering (such as the differential scattering cross-section). However, the perturbation expansion converges for the absorption cross-section when the imaginary part of the unperturbed dielectric function is zero (or sufficiently small).

We will use the above fact to calculate absorption cross-sections of carbon smoke particles inside spherical water droplets in the first order of the perturbation theory. The perturbation expansion is mathematically similar to that of Kerker et al. [71] and Hill et al. [77]. The water itself is assumed to be nonabsorbing. We perform calculations for a fractal distribution of carbon inclusions with a power-law dependence of the density on the distance from the center of a water droplet; the case of trivial (nonfractal) geometry is considered as a limiting case when $D = 3$.

The approach developed below applies to any spherical highly transparent microcavities doped with strongly absorbing inclusions with the fractal dimension from 1 to 3, not just to carbon soot inside water droplets. However, the numerical results are strongly dependent on the refractive index of the host. The difference between microdroplets with the refractive index of water (~ 1.33) and of sulfate (~ 1.52) was demonstrated by Fuller [52,79].

9.6.2. Formulation of the Model

Consider a plane monochromatic wave of the form (9.7) incident on a spherical water droplet of a radius R_d containing a carbon soot cluster inside. The physical system under consideration can be characterized by a dielectric function of the form

$$\varepsilon(\mathbf{r}) = \begin{cases} \varepsilon_1 + (\varepsilon_2 - \varepsilon_1)\rho(\mathbf{r}) & r \leq R_d \\ 1, & r > R_d \end{cases} \tag{9.111}$$

Here ε_1 and ε_2 are the dielectric constants of water and carbon, respectively, and $\rho(\mathbf{r})$ is the density of carbon inclusions inside the droplet, normalized by the condition

$$\int_{V_d} \rho(\mathbf{r}) d^3\mathbf{r} = v_{\text{tot}} \tag{9.112}$$

where v_{tot} is the total volume occupied by carbon and $\int_{V_d}$ denotes integration over the spatial area defined by $r \leq R_d$ ($\int_{V_d} d^3\mathbf{r} = v_d$). We assume that the volume fraction of carbon is small, so that the small parameter of the problem is v_{tot}/v_d. We also assume that $\rho(\mathbf{r}) = 0$ for $r > R_d$, that is, the soot cluster is completely covered by water.

In our notations, $\rho(\mathbf{r})$ denotes the exact density of carbon inclusions for some given random realization of a soot cluster. As such, $\rho(\mathbf{r}) = 1$ if the radius vector $\mathbf{r}$ lies in the area occupied by carbon, and $\rho(\mathbf{r}) = 0$ otherwise. We will see that for a calculation of some average physical characteristics, such as absorption, one needs to average $\rho(\mathbf{r})$ over random realizations of carbon soot clusters. We denote the average density by $\langle\rho(\mathbf{r})\rangle$; it can be interpreted as the probability to find some given point $\mathbf{r}$ inside a droplet occupied by carbon. If $\langle\rho(\mathbf{r})\rangle$ is bound everywhere inside the sphere, the condition $v_{\text{tot}}/v_d \ll 1$ implies that $\langle\rho(\mathbf{r})\rangle \ll 1\forall\mathbf{r}$.

The integral equation for the electric field $\mathbf{E}(\mathbf{r})$ analogous to (9.5) has the form

$$\mathbf{E}(\mathbf{r}) = \mathbf{E}_{\text{inc}}(\mathbf{r}) + \int_{V_d} \boldsymbol{G}(\mathbf{r}-\mathbf{r}')\frac{\varepsilon(\mathbf{r}')-1}{4\pi}\mathbf{E}(\mathbf{r}')d^3\mathbf{r}' \tag{9.113}$$

This equation differs from (9.5) is that the coupling constant is now coordinate-dependent and cannot be moved out of the integral. Therefore, it is more convenient to write it explicitly as a function of $\varepsilon(\mathbf{r}')$. Note also that the equation is written for the electric field rather than for polarization $\mathbf{P}(\mathbf{r}) = [(\varepsilon(\mathbf{r})-1)/4\pi]\mathbf{E}(\mathbf{r})$.

At the next step, we represent the electrical field inside the sphere as a sum of two contributions:

$$\mathbf{E}(\mathbf{r}) = \mathbf{E}_s(\mathbf{r}) + \mathbf{E}_c(\mathbf{r}) \tag{9.114}$$

where $\mathbf{E}_s(\mathbf{r})$ is the solution to Eq. (9.113) with $\varepsilon_2 = \varepsilon_1$, that is,

$$\mathbf{E}_s(\mathbf{r}) = \mathbf{E}_{\text{inc}}(\mathbf{r}) + \frac{\varepsilon_1 - 1}{4\pi}\int_{V_d} \boldsymbol{G}(\mathbf{r}-\mathbf{r}')\mathbf{E}_s(\mathbf{r}')d^3\mathbf{r}' \tag{9.115}$$

and $\mathbf{E}_c(\mathbf{r})$ is the additional term, which appears because of the presence of a carbon cluster. The parameter $\mathbf{E}_s(\mathbf{r})$ is given by the Mie solution for a dielectric sphere and we assume that it is known. Substituting $\mathbf{E}(\mathbf{r})$ in the form (9.114) into (9.113), we find the equation for $\mathbf{E}_c(\mathbf{r})$:

$$\begin{aligned}\mathbf{E}_c(\mathbf{r}) = {} & \frac{\varepsilon_2 - \varepsilon_1}{4\pi}\int_{V_d} \rho(\mathbf{r}')\boldsymbol{G}(\mathbf{r}-\mathbf{r}')\mathbf{E}_s(\mathbf{r}')d^3\mathbf{r}' \\ & + \int_{V_d} \frac{\varepsilon_1 - 1 + (\varepsilon_2-\varepsilon_1)\rho(\mathbf{r}')}{4\pi}\boldsymbol{G}(\mathbf{r}-\mathbf{r}')\mathbf{E}_c(\mathbf{r}')d^3\mathbf{r}'\end{aligned} \tag{9.116}$$

The first term in (9.116) with the known function $\mathbf{E}_s(\mathbf{r})$ serves as a free term for the integral equation (9.116).

For many practical problems, a knowledge of the ensemble-averaged internal field is sufficient. (Evidently, this class of problems does not include the problems of nonlinear optics that require consideration of fluctuations of the local field.) We cannot perform direct averaging of Eq. (9.116) over random realizations of inclusions, because such averaging would add an additional unknown term $\langle\rho(\mathbf{r})\mathbf{E}_c(\mathbf{r})\rangle$. In the general case, we cannot factorize this correlator as $\langle\rho(\mathbf{r})\mathbf{E}_c(\mathbf{r})\rangle = \langle\rho(\mathbf{r})\rangle\langle\mathbf{E}_c(\mathbf{r})\rangle$. However, in the linear (in v_{tot}/v_d) approximation we can neglect the above term as a higher-order correction. Then, it becomes possible to write an equation for the ensemble-average value $\langle \mathbf{E}_c(\mathbf{r}) \rangle$:

$$\begin{aligned}\langle \mathbf{E}_c(\mathbf{r}) \rangle = & \frac{\varepsilon_2 - \varepsilon_1}{4\pi} \int_{V_d} \langle \rho(\mathbf{r}') \rangle \boldsymbol{G}(\mathbf{r} - \mathbf{r}') \mathbf{E}_s(\mathbf{r}') d^3\mathbf{r}' \\ & + \frac{\varepsilon_1 - 1}{4\pi} \int_{V_d} \boldsymbol{G}(\mathbf{r} - \mathbf{r}') \langle \mathbf{E}_c(\mathbf{r}') \rangle d^3\mathbf{r}' \end{aligned} \tag{9.117}$$

We can draw two important conclusions from the general form of (9.117). First, the ratio of $|\langle \mathbf{E}_c \rangle| / |\mathbf{E}_s|$ is of the same order of magnitude as v_{tot}/v_d. This can be seen by multiplying $\langle \rho(\mathbf{r}') \rangle$ in (9.117) by some arbitrary constant α. The average field $\langle \mathbf{E}_c(\mathbf{r}) \rangle$ is also multiplied by the same factor α. This means that $|\langle \mathbf{E}_c(\mathbf{r}) \rangle| / |\mathbf{E}_s(\mathbf{r})| \sim \langle \rho(\mathbf{r}) \rangle \sim v_{\mathrm{tot}}/v_d$. A similar result is readily obtained for the exact field $\mathbf{E}_c(\mathbf{r})$ (before the averaging).

Second, it is generally impossible to apply the Born expansion or similar perturbation expansion to a calculation of $\langle \mathbf{E}_c(\mathbf{r}) \rangle$. Indeed, both terms on the rhs of (9.117) are of the same order of magnitude (proportional to v_{tot}/v_d). Suppose we start from the zeroth-order approximation $\langle \mathbf{E}_c^{(0)}(\mathbf{r}) \rangle = 0$, and substitute it into (9.117) to obtain the first-order approximation, and so on. It is easy to see that all the terms in the generated expansion will be of the same order of magnitude, and thus convergence cannot be reached.

The above fact makes the general scattering problem for a water droplet containing a cluster inside very complicated. Indeed, the only small parameter of the problem, v_{tot}/v_d, cannot be used to generate a converging expansion for $\mathbf{E}_c(\mathbf{r})$. However, as we show below, we can use the fact that $|\mathbf{E}_c(\mathbf{r})| / |\mathbf{E}_s(\mathbf{r})| \sim v_{\mathrm{tot}}/v_d$ to calculate the absorption cross-section when water itself is weakly absorbing.

The formula for the absorption cross-section in terms of the polarization function can be obtained from the optical theorem and direct integration of the scattering amplitude [34,35], and is analogous to (9.20), except the dielectric function $\varepsilon(\mathbf{r})$ now is not constant inside the integration volume:

$$\sigma_a = \frac{16\pi^2 k}{|\mathbf{E}_0|^2} \int_{V_d} \frac{\operatorname{Im} \varepsilon(\mathbf{r})}{|\varepsilon(\mathbf{r}) - 1|^2} |\mathbf{P}(\mathbf{r})|^2 d^3\mathbf{r} = \frac{k}{|\mathbf{E}_0|^2} \int_{V_d} \operatorname{Im}[\varepsilon(\mathbf{r})] |\mathbf{E}(\mathbf{r})|^2 d^3\mathbf{r} \tag{9.118}$$

By using formulas (9.111) for $\varepsilon(\mathbf{r})$ and (9.114) for $\mathbf{E}(\mathbf{r})$, we can rewrite Eq. (9.118) for the absorption cross-section as

$$\begin{aligned}\sigma_a = & \frac{k \operatorname{Im} \varepsilon_1}{|\mathbf{E}_0|^2} \int_{V_d} |\mathbf{E}_s(\mathbf{r})|^2 d^3\mathbf{r} + \frac{k \operatorname{Im} (\varepsilon_2 - \varepsilon_1)}{|\mathbf{E}_0|^2} \int_{V_d} \rho(\mathbf{r}) |\mathbf{E}_s(\mathbf{r})|^2 d^3\mathbf{r} \\ & + \frac{k \operatorname{Im} \varepsilon_1}{|\mathbf{E}_0|^2} \int_{V_d} \{2\operatorname{Re}[\mathbf{E}_s(\mathbf{r}) \cdot \mathbf{E}_c^*(\mathbf{r})] + |\mathbf{E}_c(\mathbf{r})|^2\} d^3\mathbf{r} \\ & + \frac{k \operatorname{Im} (\varepsilon_2 - \varepsilon_1)}{|\mathbf{E}_0|^2} \int_{V_d} \rho(\mathbf{r}) \{2\operatorname{Re}[\mathbf{E}_s(\mathbf{r}) \cdot \mathbf{E}_c^*(\mathbf{r})] + |\mathbf{E}_c(\mathbf{r})|^2\} d^3\mathbf{r} \end{aligned} \tag{9.119}$$

Now, we analyze the terms on the rhs of (9.119). The first term gives the absorption cross-section by a water droplet without inclusions. It is given by the well-known Mie solution and, consequently, is of no interest to us. Taking into account that $\langle\rho(\mathbf{r})\rangle \sim |\mathbf{E}_c|/|\mathbf{E}_s| \sim v_{\text{tot}}/v_d$, we find that the second and the third terms are of the same order of magnitude and give the first-order correction to the absorption cross-sections. Finally, the fourth term is of the order of $(v_{\text{tot}}/v_d)^2$, and can be neglected in the first approximation.

Even in the first approximation, the expression for the absorption cross-section contains the unknown field $\mathbf{E}_c(\mathbf{r})$ in the third term of (9.119). However, for the particular case of carbon and water, $\text{Im}\,\varepsilon_2 \gg \text{Im}\,\varepsilon_1$. This additional factor allows one to neglect the third term in the expansion (9.119). In principle, the first term can be still large or comparable to the second one due to the large factor v_d/v_{tot}, but this fact does not complicate further derivations.

Finally, we can represent the absorption cross-section as $\sigma_a = \sigma_{a,\text{water}} + \sigma_{a,\text{carbon}}$, where $\sigma_{a,\text{water}}$ is given by the first term in (9.119), and

$$\langle\sigma_{a,\text{carbon}}\rangle = \frac{k\,\text{Im}\,\varepsilon_2}{|\mathbf{E}_0|^2}\int_{V_d}\rho(\mathbf{r})|\mathbf{E}_s(\mathbf{r})|^2 d^3\mathbf{r} \tag{9.120}$$

Formula (9.20) gives the absorption cross-section associated with carbon inclusions in first order in v_{tot}/v_d; the higher corrections are of the order of $(v_{\text{tot}}/v_d)^2$. In the ideal case of $\text{Im}\,\varepsilon_1 = 0$, this formula gives the total absorption of a composite droplet. Below, we will assume for simplicity that ε_1 is a real number.

Since ρ and $\mathbf{E}_s$ are statistically independent, we can perform direct averaging of (9.120) over random realizations of carbon soot inclusions:

$$\langle\sigma_{a,\text{carbon}}\rangle = \frac{k\,\text{Im}\,\varepsilon_2}{|\mathbf{E}_0|^2}\int_{V_d}\langle\rho(\mathbf{r})\rangle|\mathbf{E}_s(\mathbf{r})|^2 d^3\mathbf{r} \tag{9.121}$$

Note that in the above averaging the radius of a water droplet is fixed.

9.6.3. Enhancement Factor

We define the enhancement factor G as the ratio of the absorption cross-section of a carbon soot cluster in a water microdroplet , defined by (9.121) to that in vacuum:

$$G = \frac{\langle\sigma_{a,\text{carbon}}\rangle}{\langle\sigma^{(0)}_{a,\text{carbon}}\rangle} \tag{9.122}$$

where $\langle\sigma^{(0)}_{a,\text{carbon}}\rangle$ is the average absorption cross-section of carbon soot in vacuum. The latter can be easily calculated using Eq. (9.118) and replacing $\mathbf{E}(\mathbf{r})$ by $\mathbf{E}_{\text{inc}}(\mathbf{r})$. By taking into account that $|\mathbf{E}_{\text{inc}}(\mathbf{r})|^2 = |\mathbf{E}_0|^2$ and $\varepsilon(\mathbf{r}) = 1 + (\varepsilon_2 - 1)\rho(\mathbf{r})$ for carbon soot in vacuum, we find that $\langle \sigma^{(0)}_{a,\text{carbon}} \rangle = kv \operatorname{Im} \varepsilon_2$ and

$$G = \frac{1}{v_{\text{tot}} |\mathbf{E}_0|^2} \int_{V_d} \langle \rho(\mathbf{r}) \rangle |\mathbf{E}_s(\mathbf{r})|^2 d^3\mathbf{r} \tag{9.123}$$

The average density of carbon inclusions $\langle \rho(\mathbf{r}) \rangle$ must be spherically symmetrical: $\langle \rho(\mathbf{r}) \rangle = \langle \rho(r) \rangle$. Therefore, the angular integration in (9.123) can be done in the most general form, without specifying $\langle \rho \rangle$:

$$G = \frac{1}{v_{tot} |\mathbf{E}_0|^2} \int_0^{R_d} r^2 \langle \rho(r) \rangle dr \int |\mathbf{E}_s(\mathbf{r})|^2 d\Omega \tag{9.124}$$

The internal field $\mathbf{E}_s$ is given by the expansion in terms of the VSHs, $\mathbf{M}_{omn}$, $\mathbf{M}_{emn}$, $\mathbf{N}_{omn}$, and $\mathbf{N}_{emn}$ (for a detailed description of the VSH expansion, see [80]). For a plane incident wave, only the VSHs with $m = 1$ are left in this expansion. Further, if the incident wave is polarized along the x axis, $\mathbf{M}_{e1n}$ and $\mathbf{N}_{o1n}$ are not excited.

For linear absorption, it is sufficient to consider a linear polarization of the incident wave. An elliptical polarization can be described as a superposition of two linearly polarized waves; the absorbed power due to these two waves is added arithmetically because of the linear nature of the interaction. Below, we will adopt the linear polarization of the incident wave along the x axis ($\mathbf{E}_0 = \mathbf{e}_x E_0$), and will use the following simplified notations for the VSHs that can be excited in this particular case: $\mathbf{M}_n \equiv \mathbf{M}_{o1n}$ and $\mathbf{N}_n \equiv \mathbf{N}_{e1n}$. Then, the expansion for the $\mathbf{E}_s$ field takes the form

$$\mathbf{E}_s = \sum_{n=1}^{\infty} i^n \frac{E_0(2n+1)}{n(n+1)} (c_n \mathbf{M}_n - i d_n \mathbf{N}_n) \tag{9.125}$$

Here c_n and d_n are the internal field coefficients [80] defined by

$$c_n = \frac{j_n(x)[x h_n^{(1)}(x)]' - h_n^{(1)}(x)[x j_n(x)]'}{j_n(x_1)[x h_n^{(1)}(x)]' - h_n^{(1)}(x)[x_1 j_n(x_1)]'} \tag{9.126}$$

$$d_n = \frac{j_n(x)[x h_n^{(1)}(x)]' - h_n^{(1)}(x)[x j_n(x)]'}{(x_1/x) j_n(x_1)[x h_n^{(1)}(x)]' - (x/x_1) h_n^{(1)}(x)[x_1 j_n(x_1)]'} \tag{9.127}$$

$$x = kR_d \qquad x_1 = k_1 R_d \qquad k_1 = \sqrt{\varepsilon_1} k \tag{9.128}$$

where $j_n(x)$ and $h^{(1)}(x)$ are the spherical Bessel and Hankel functions of the first kind, respectively, and the prime denotes differentiation with respect to the argument in parentheses.

If we take into account the mutual orthogonality of the VSHs, the angular integral in (9.124) can be written as

$$\int |\mathbf{E}_s(\mathbf{r})|^2 d\Omega = \sum_{n=1}^{\infty} \frac{|E_0|^2 (2n+1)^2}{n^2 (n+1)^2} \left[|c_n|^2 \int \mathbf{M}_n^2 d\Omega + |d_n|^2 \int \mathbf{N}_n^2 d\Omega \right] \tag{9.129}$$

Note that for a purely real dielectric constant ε_1 the VSHs are also real; this is why $|\mathbf{M}_n|^2$ and $|\mathbf{N}_n|^2$ were replaced by $\mathbf{M}_n^2$ and $\mathbf{N}_n^2$ in (9.129).

Integration according to (9.129) can be performed directly using the normalization formulas for the VSHs (the details omitted), and the result is

$$\int |\mathbf{E}_s(\mathbf{r})|^2 d\Omega = 2\pi |E_0|^2 \sum_{n=1}^{\infty} (2n+1) \times \left\{ |c_n|^2 j_n^2(k_1 r) + |d_n|^2 \left[n(n+1) \left(\frac{j_n(k_1 r)}{k_1 r} \right)^2 + \left(\frac{j_n(k_1 r)}{k_1 r} + j_n'(k_1 r) \right)^2 \right] \right\} \tag{9.130}$$

Further calculations require specifying the form of $\langle \rho(r) \rangle$. Below, we consider two cases: fractal distribution of the inclusion density and homogeneous distribution.

An important question is how the carbon inclusions are located inside the microdroplets. This can be influenced by many factors such as the chemical composition of soot particles, surface-tension forces, temperature, and so on. The formation of agglomerates of soot clusters and water can change the geometrical properties of the clusters due to the action of surface-tension forces [81,82]. The average density of inclusions must be spherically symmetrical if there is no distinguished direction in space. We also assume that, in accordance with the fractal density distribution, it obeys a power law with the scaling parameter D according to

$$\langle \rho(r) \rangle = \frac{v_{\text{tot}} D}{4\pi R_d^D} r^{D-3} \quad \text{if} \quad r < R_d \tag{9.131}$$

Here the radius of the microdroplet, R_d, serves as the cutoff, and the density function (9.131) satisfies the normalization (9.112).

Note that, according to its physical meaning as the probability of finding a spot at the distance r from the droplet center occupied by carbon, $\langle \rho(r) \rangle < 1$. In fact, the perturbation expansion used above relies on the assumption that $\langle \rho(r) \rangle \ll 1$. The formula (9.131) may seem to contradict this assumption when $r \to 0$. However, the divergence of $\langle \rho(r) \rangle$ at small r is not significant since all the physically important radial integrals converge fast enough in this limit (see below); thus the actual value of $\langle \rho(0) \rangle$ is not important. The small parameter of the perturbation expansion, v_{tot}, is obviously present in the definition (9.131).

The case $D = 1$ corresponds to inclusions in the form of long linear sticks, while $D = 3$ corresponds to a homogeneous distribution of inclusions. If $D = 3$ (trivial geometry), the problem becomes mathematically equivalent to the Mie problem for a homogeneous dielectric sphere with some effective dielectric constant ε_{eff}. A nonperturbative analytic solution can be obtained in this case. However , this method has certain difficulties. First, the form of ε_{eff} is not obvious. For carbon inclusions of spherical shape and small concentration, one can use $\varepsilon_{\text{eff}} = \varepsilon_1 + (3v_{\text{tot}}/v_d)$

$(\varepsilon_2 - \varepsilon_1)/(\varepsilon_2 + 2\varepsilon_1)$ [83]. But this formula is not applicable when the inclusions are not of spherical shape or form clusters of touching particles. An approach based on determining the effective dielectric function was used by Chowdhury et al. [74,78] who suggested averaging of the ε with the weight that includes the local intensity of the *unperturbed* electric field inside the sphere. Chowdhury et al. define two different averaged dielectric constants, one of which is used for computation of the internal (or external) field coefficients and the other for the effective absorption (or gain). This method is somewhat similar to the perturbative approach used here in that it uses the unperturbed electric field to compute the effective ε. Different effective medium approximations were also used by Videen and Chylek [84]. The second difficulty is that the extinction and scattering cross-sections in the analytical Mie solution are expressed as infinite series involving the scattering coefficient a_n, c_n; the absorption cross-section must be calculated as the difference between these two values. When the absorption is small, such a calculation involves a numerical procedure of finding a small difference between two large numbers, and the round-off errors become very significant. Instead of finding an analytical solution based on some definition of ε_{eff}, we will use the perturbative approach developed above for the $D = 3$ case. This approach is also valid for $D < 3$ (fractal geometry), when an analytical solution cannot be obtained, thus allowing us to maintain self-consistency of the results.

9.6.4. Numerical Results

By using the average density function (9.131), and the result of the angular integration of $|\mathbf{E}_s(\mathbf{r})|^2$ (9.130), one can express the absorption enhancement factor (9.124) in terms of simple radial integrals involving spherical Bessel functions. By inserting the expressions (9.130) and (9.131) into (9.124) and taking the integrals containing derivatives of the spherical Bessel functions by parts, we arrive, after some rearrangement of terms, at the following result:

$$G = \frac{D}{2(k_1 R_d)^D} \sum_{n=1}^{\infty} (2n+1) \Bigg\{ |c_n|^2 I_n(1) + |d_n|^2 \Bigg[\frac{5-D}{2} x_1^{D-2} j_n^2(x_1) + x_1^{D-1} j_n'(x_1) j_n(x_1) + I_n(1) + \frac{(4-D)(3-D)}{2} I_n(3) \Bigg] \Bigg\} \tag{9.132}$$

$$I_n(\alpha) = \int_0^{x_1} x^{D-\alpha} j_n^2(x)\,dx \tag{9.133}$$

The integrals $I_n(\alpha)$ converge for all physically interesting values of the parameters and must be evaluated numerically, except in the trivial case $D = 3$.

As was pointed out by Bohren and Huffmen [80], the diffraction parameter $x = kR_d$ (or $x_1 = \sqrt{\varepsilon}x$) cannot be, in general, viewed as the only independent variable of the problem, although it may seem so from the mathematical form of Eqs. (9.123) and (9.132). Indeed, when ε_1 depends on λ, $x_1/x \neq$ const. Instead, there are two physically independent parameters that completely define the solution to the scattering problem: R_d and λ. However, when ε_1 does not depend on λ, $x_1/x = \varepsilon_1 =$ const, and the diffraction parameter x becomes the only independent variable. In this case, we do not need to know whether x changes due to a change in R_d or in λ.

For the particular case of water, the assumption $x_1/x =$ const is a good approximation in the spectral range from 0.3 to 2.0μm [85]. We have set $\sqrt{\varepsilon_1} = 1.33 =$ const, which allowed us to perform numerical calculation of the enhancement factor G as a function of one independent variable, $x = kR_d$. We allowed x to change from 0 to 1000. This range of x includes most of the practical values of R_d and λ. Thus, for $\lambda = 0.4\,\mu$m, R_d can vary from 0 to $\approx 60\,\mu$m.

Now, we turn to the calculation of the internal field coefficients, c_n and d_n. We calculated the Bessel functions and their first derivatives that are used in the definitions (9.126) and (9.127) of the internal field coefficients, using the three-point recursion relation [86]. The maximum order n that gives significant contribution to the optical cross-sections can be roughly estimated [80] as $n_{\max} \approx x = kR_d$. The internal field coefficients $|c_n|^2$ and $|d_n|^2$ decrease dramatically for $n > n_{\max}$, as illustrated in Fig. 9.14. In Fig. 9.14(a), we plot the internal field coefficients for $\sqrt{\varepsilon_1} = 1.33$ and $x = 259.664$. The specific value of x was chosen from the condition that the absorption cross-section has a resonance. In terms of the VSHs, the resonance occurs for the order $n = 131$, when $|c_n|^2$ reaches the value of $\approx 3.23 \times 10^7$; there is also a large number of weaker resonances of $|c_n|^2$. (Note that $|d_n|^2$ has no strong resonances.) Since the total number of VSHs that contribute to the absorption is of the order of 300, and $|c_{131}|^2$ is more than five orders of magnitude larger than the average background, we can conclude that the resonant VSH gives the prevailing input to the optical cross-sections. For comparison, we plot in Fig. 9.14(b) the internal field coefficients for the same refraction index, but for an off-resonant value of $x = 260.400$. Both pictures look very similar, apart from the resonance order $n = 131$ in Fig. 9.14(a).

The numerical results for the absorption enhancement factor $G(x)$ are shown in Fig. 9.15 for $D = 1.1$ [Fig. 9.15(a)] and $D = 3.0$ [Fig. 9.15(b)]. As can be seen in Fig. 9.15, $G(x)$ has a large number of quasirandom morphology-dependent resonances (due to the presence of resonances in the internal field coefficients illustrated in Fig. 9.14), but only a very slight systematic dependence of $G(x)$ on x can be seen in the interval $10 < x < 1000$. The slight systematic increase of $G(x)$ can be attributed to an increase in the average resonance quality with the size parameter x. It can be also seen from a comparison of Figs. 9.15(a) and (b) that the enhancement factor is larger, on average, for $D = 1.1$ than for $D = 3.0$.

In a polydisperse ensemble of microdroplets with size parameters in the wide range $10 < x < 1000$ the individual resonances are smoothed out and the average absorption enhancement factor, $\langle G \rangle$, which is practically important, is given by the averaging of $G(x)$ over x. We performed such averaging in the interval of x specified above for different D ($1 \leq D \leq 3$) and the results are shown in Fig. 9.16. The averaging was performed with the step size in x equal to 0.1. This step size was small enough so that most resonances were visually resolved. Averaging with a larger resolution resulted in a smaller value of $\langle G \rangle$ because the resonances in $G(x)$ are very

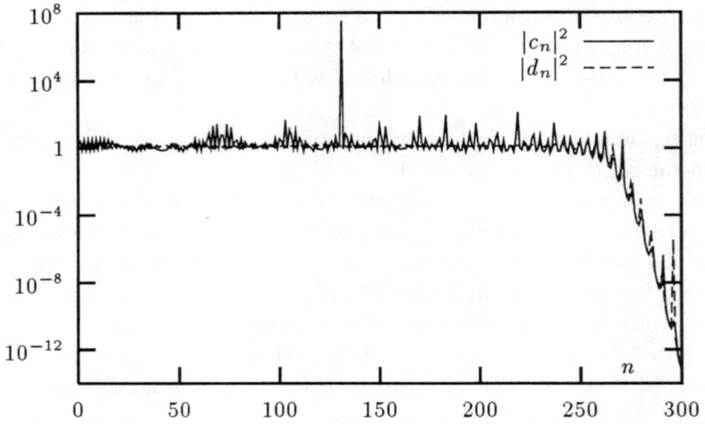

(a)

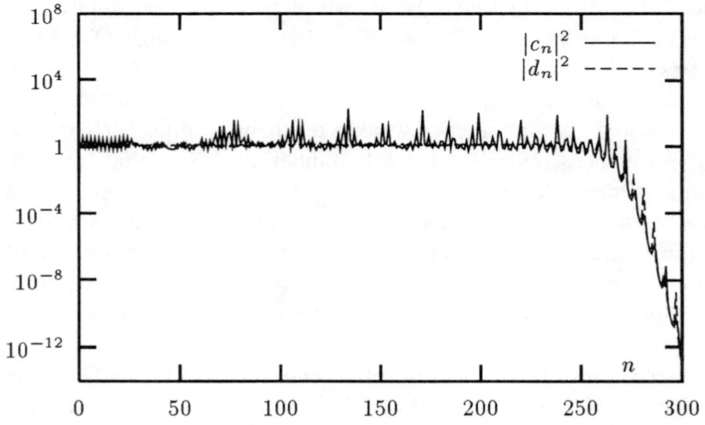

(b)

Figure 9.14. Internal field coefficients, $|c_n|^2$ and $|d_n|^2$, as functions of the VSH order, n. (a): $x = 259.664$ (resonance order $n = 131$); (b): $x = 260.400$ (no pronounced resonances).

narrow. It is important to emphasize the significance of the averaging process. For a randomly chosen x, $G(x)$ is, with a large probability, less than $\langle G \rangle$ by a factor of 4 or 5. Thus the resonances of $G(x)$ play an important role and should not be ignored. It should be noted that the averaging was performed in the region of $10 < x < 1000$, where there is no pronounced systematic dependence of $G(x)$ on x. For $x < 10$, the averaged G is considerably smaller.

As can be seen in Fig. 9.16, $\langle G \rangle$ is maximum for $D = 1$ and decreases toward $D = 3$. For the practically important value $D = 1.8$, $\langle G \rangle \approx 16$, and the maximum variation of $\langle G \rangle$ with D, does not exceed ± 6. The dependence of $\langle G \rangle$ on D can be

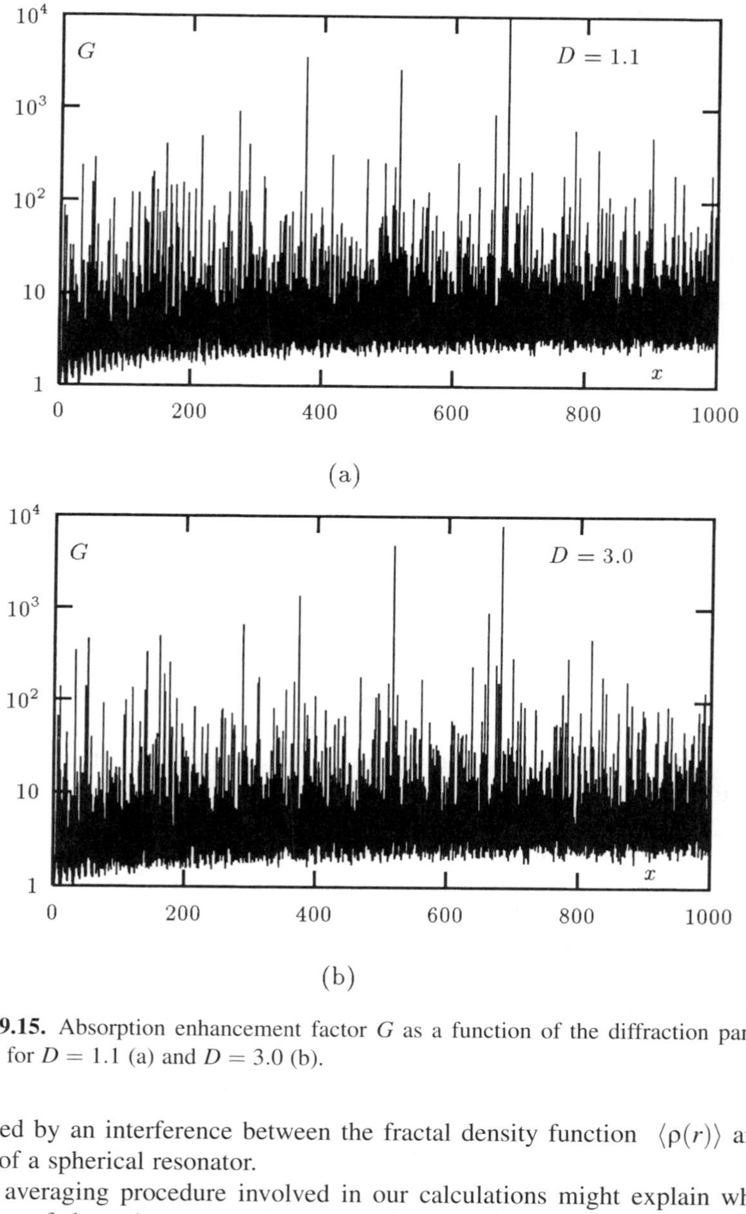

Figure 9.15. Absorption enhancement factor G as a function of the diffraction parameter $x = kR_d$ for $D = 1.1$ (a) and $D = 3.0$ (b).

explained by an interference between the fractal density function $\langle \rho(r) \rangle$ and the modes of a spherical resonator.

The averaging procedure involved in our calculations might explain why our estimates of the enhancement factor are significantly larger than those reported earlier [52,79,84,87]. Fuller calculated the specific absorption cross-section for a single spherical carbon grain located near the surface of a water droplet [52] and inside the water droplet [79] as a function of the grain's position. Although Fuller's data are not averaged over the whole volume of the microdroplet, they indicate that the volume-averaged absorption enhancement factor is < 14. Chylek et al. [87]

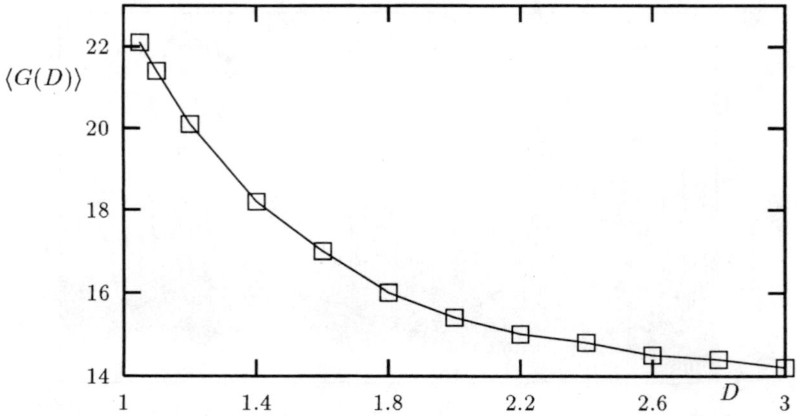

Figure 9.16. Average absorption enhancement factor $\langle G\rangle$ as a function of the fractal dimension, *D*.

averaged the same quantity for the carbon inclusion location distributed evenly within a spherical cone with the axis collinear to the incident wave propagation direction and over the whole volume [84]. In the first case, the authors estimate the enhancement factor to be ≈ 4, and in the second ≈ 2. However, all of the above calculations were performed for a fixed value of the diffraction parameter x. Because the resonances are very narrow, it is unlikely that a randomly selected value of x will lie within a resonance. Our calculations indicate that if x is chosen exactly in resonance, the volume-averaged enhancement factor can be as large as 10^4. Our calculations indicate that the averaged absorption factor is larger by a factor of $\sim 4-5$ than that calculated for a randomly selected value of x. For the trivial distribution of carbon inclusions ($D = 3$), we obtain the averaged enhancement factor of 14, while for a randomly selected value of x the typical (most probable) enhancement factor is from 2 to 4. This suggests that, although the resonances in x are very narrow, they are not small in the integral sense, and should be taken into account.

In conclusion, we note that within the framework of the first Born approximation that was used throughout this section, the absorption cross-section of a free carbon soot cluster excited by a plane wave is proportional to the total volume of carbon and does not depend on the cluster's geometrical configuration. However, this is not the case when the cluster is excited by the inhomogeneous modes of a spherical resonator instead of plane waves. In this case, the absorption is stronger, on average, if the inclusions tend to concentrate in the spatial regions where the intensity of local fields is higher.

ACKNOWLEDGMENTS

This research was supported by Battelle under Contract DAAH04-96-C-0086. It was partially supported by National Computational Science Alliance under Grant PHY980006N and utilized the NCSA HP/Convex Exemplar SPP-2000. The authors are grateful to D. Mackowski for making his Fortran codes available and to E. Mikhailov, S. Vlasenko, and A. Kiselev for providing digital images of fractal soot aggregates.

REFERENCES

1. C. W. Bruce, T. F. Stromberg, K. P. Gurton, and J. B. Mozer, *Appl. Opt.* ***30***, 1537 (1991).
2. W. H. Dalzel, G. C. Williams, and H. C. Hottel, *Combust. Flame* ***14***, 161 (1970).
3. S. R. Forrest and T. A. Witten, *J. Phys. A* ***12***, L109 (1979).
4. R. D. Mountain and G. W. Mulholland, in F. Family and D. P. Landau, Eds., *Kinetics, of Aggregation and Gelation*, North-Holland, Amsterdam, 1984, pp. 83–86.
5. U. O. Koylu and G. M. Faeth, *Combust. and Flame* ***89***, 140 (1992).
6. E. F. Mikhailov and S. S. Vlasenko, *Phys. Usp.* ***165***, 253 (1995).
7. H. X. Zhang et al., *Langmuir* ***4***, 867 (1988).
8. J. Cai, N. Lu, and C. M. Sorensen, *Langmuir* ***9***, 2861 (1993).
9. U. O. Koylu, G. M. Faeth, T. L. Farias, and M. G. Carvalho, *Combust. Flame* ***100***, 621 (1995).
10. T. T. Charalampopoulos and H. Chang, *Combust. Flame* ***87***, 89 (1991).
11. J. Cai, N. Lu, and C. M. Sorensen, *J. Colloid Interface Sci.* ***171***, 470 (1995).
12. P. Meakin, *Phys. Rev. Lett.* ***51***, 1119 (1983).
13. R. Jullien, M. Kolb, and R. Botet, *J. Phys. Lett. (Paris)* ***45***, L211 (1984).
14. J. E. Martin and A. J. Hurd, *J. Appl. Crystallogr.* ***20***, 61 (1987).
15. V. A. Markel, V. M. Shalaev, E. Y. Poliakov, and T. F. George, *Phys. Rev. E* ***55***, 7313 (1997).
16. C. Amitrano, A. Coniglio, P. Meakin, and M. Zannetti, *Phys. Rev. B* ***44***, 4974 (1991).
17. R. D. Mountain and G. W. Mulholland, *Langmuir* ***4***, 1321 (1988).
18. C. Oh and C. M. Sorensen, *J. Colloid Interface Sci.* ***193***, 17 (1997).
19. M. K. Wu and S. K. Friedlander, *J. Colloid Interface Sci.* ***159***, 246 (1993).
20. M. V. Berry and I. C. Percival, *Opt. Acta* ***33***, 577 (1986).
21. C. M. Sorensen, N. Lu, and J. Cai, *J. Colloid Interface Sci.* ***174***, 456 (1995).
22. C. M. Sorensen, C. Oh, P. W. Schmidt, and T. P. Rieker, *Phys. Rev. E* ***58***, 4666 (1998).
23. V. A. Markel, V. M. Shalaev, E. Y. Poliakov, and T. F. George, *J. Opt. Soc. Am. A* ***14***, 60 (1997).
24. S. M. Rytov, Y. A. Kravtsov, and V. I. Tatarskii, *Elements of Random Fields*, Vol. 3 of *Principles of Statistical Radiophysics*, Springer-Verlag, Berlin, 1989.
25. S. M. Rytov, Y. A. Kravtsov, and V. I. Tatarskii, *Wave Propagation through Random Media*, Vol. 4 of *Principles of Statistical Radiophysics*, Springer-Verlag, Berlin, 1989.
26. W. H. Dalzell and A. F. Sarofim, *J. Heat Transfer* ***91***, 100 (1969).
27. E. A. Taft and E. A. Philipp, *Phys. Rep.* ***138***, A197 (1965).
28. Z. G. Habib and P. Vervisch, *Combust. Sci. Technol.* ***59***, 261 (1988).
29. S. C. Lee and C. L. Tien, in *Eighteenth Symposium (International) on Combustion*, The Combustion Institute, 1981, pp. 1159–1166.

30. V. A. Markel, L. S. Muratov, and M. I. Stockman, *Sov. Phys. JETP* ***71***, 455 (1990).
31. V. A. Markel, L. S. Muratov, M. I. Stockman, and T. F. George, *Phys. Rev. B* ***43***, 8183 (1991).
32. G. W. Mulholland, C. F. Bohren, and K. A. Fuller, *Langmuir* ***10***, 2533 (1994).
33. G. W. Mulholland and R. D. Mountain, *Combust. Flame* ***119***, 56 (1999).
34. V. A. Markel, *J. Opt. Soc. Am. B* ***12***, 1783 (1995).
35. V. A. Markel and E. Y. Poliakov, *Philos. Mag. B* ***76***, 895 (1997).
36. J. D. Jackson, *Classical Electrodynamics*, Wiley, New York, 1975.
37. V. M. Shalaev, R. Botet, and R. Jullien, *Phys. Rev. B* ***44***, 12216 (1991).
38. E. M. Purcell and C. R. Pennypacker, *Astrophys. J.* ***186***, 705 (1973).
39. B. T. Draine, *Astrophys. J.* ***333***, 848 (1988).
40. B. Draine and P. Flatau, *J. Opt. Soc. Am. A* ***11***, 1491 (1994).
41. V. A. Markel, *J. Mod. Opt.* ***39***, 853 (1992).
42. J. M. Gerardy and M. Ausloos, *Phys. Rev. B* ***22***, 4950 (1980).
43. F. Claro, *Phys. Rev. B* ***25***, 7875 (1982).
44. J. E. Sansonetti and J. K. Furdyna, *Phys. Rev. B* ***22***, 2866 (1980).
45. F. Claro, *Phys. Rev. B* ***30***, 4989 (1984).
46. F. Claro, *Solid State Commun.* ***49***, 229 (1984).
47. R. Rojas and F. Claro, *Phys. Rev. B* ***34***, 3730 (1986).
48. R. Fuchs and F. Claro, *Phys. Rev. B* ***35***, 3722 (1987).
49. D. W. Mackowski, *Appl. Opt.* ***34***, 3535 (1995).
50. D. W. Mackowski and M. Mischenko, *J. Opt. Soc. Am. A* ***13***, 2266 (1996).
51. K. A. Fuller, *J. Opt. Soc. Am. A* ***11***, 3251 (1994).
52. K. A. Fuller, *J. Opt. Soc. Am. A* ***12***, 881 (1995).
53. Y.-l. Xu, *Appl. Opt.* ***34***, 4573 (1995).
54. V. A. Markel et al., *Phys. Rev. B* ***53***, 2425 (1996).
55. V. A. Markel and V. M. Shalaev, in D. A. Jelski and T. F. George, Eds., *Computational Studies of New Materials*, World Scientific, Singapore, 1999, pp. 210–243.
56. V. A. Markel, *J. Mod. Opt.* ***40***, 2281 (1993).
57. M. K. Singham, S. B. Singham, and G. C. Salzman, *J. Chem. Phys.* ***85***, 3807 (1986).
58. Y. E. Danilova, Ph.D. Thesis, Institute of Automation and Electrometry, SB RAS, Novosibirsk, 1999.
59. P. Chylek, V. Ramaswamy, and R. J. Cheng, *J. Atm. Sci.* ***41***, 3076 (1984).
60. R. E. Danielson, D. R. Moore, and H. C. Van Hulst, *J. Atm. Sci.* ***26***, 1078 (1969).
61. H. Grassl, *Contrib. Atm. Phys.* ***48***, 199 (1975).
62. S. Twomey, *J. Atm. Sci.* ***33***, 1087 (1976).
63. K. Y. Kondratyev, V. I. Binenko, and O. P. Petrenchuk, *Izv. Akad. Nauk (USSR) Fiz. Atm. Okeana* ***17***, 122 (1981).
64. V. A. Markel and V. M. Shalaev, *J. Quant. Spectrosc. Rad. Transfer* ***63***, 321 (1999).
65. H.-B. Lin et al., *Opt. Lett.* ***17***, 970 (1992).
66. P. Chylek, D. Ngo, and R. G. Pinnik, *J. Opt. Soc. Am. A* ***9***, 775 (1992).
67. R. L. Armstrong et al., *Opt. Lett.* ***18***, 119 (1993).

68. J.-G. Xie, T. E. Ruekgauer, R. L. Armstrong, and R. G. Pinnik, *Opt. Lett.* ***18***, 340 (1993).
69. J. Gu, T. E. Ruekgauer, J.-G. Xie, and R. L. Armstrong, *Opt. Lett.* ***18***, 1293 (1993).
70. D. Ngo and R. G. Pinnik, *J. Opt. Soc. Am. A* ***11***, 1352 (1994).
71. M. Kerker, D. D. Cooke, H. Chew, and P. J. McNulty, *J. Opt. Soc. Am.* ***68***, 592 (1978).
72. F. Borghese, P. Denti, R. Saija, and O. I. Sindoni, *J. Opt. Soc. Am. A* ***9***, 1327 (1992).
73. M. M. Mazumder, S. C. Hill, and P. W. Barber, *J. Opt. Soc. Am. A* ***9***, 1844 (1992).
74. D. Q. Chowdhury, S. C. Hill, and M. M. Mazumder, *IEEE J. Quantum Electron.* ***29***, 2553 (1993).
75. F. Borghese, P. Denti, and R. Saija, *Appl. Opt.* ***33***, 484 (1994).
76. K. A. Fuller, *Opt. Lett.* ***19***, 1272 (1994).
77. S. C. Hill, H. I. Saleheen, and K. A. Fuller, *J. Opt. Soc. Am. A* ***12***, 905 (1995).
78. D. Q. Chowdhury, S. C. Hill, and M. M. Mazumder, *Opt. Commun.* ***131***, 343 (1996).
79. K. A. Fuller, *J. Opt. Soc. Am. A* ***12***, 893 (1995).
80. C. F. Bohren and D. R. Huffman, *Absorption and Scattering of Light by Small Particles*, Wiley, New York, 1983.
81. E. F. Mikhailov, S. S. Vlasenko, T. I. Ryshkevitch, and A. A. Kiselev, *J. Aerosol Sci.* ***27***, S709 (1996), suppl. 1.
82. E. F. Mikhailov, S. S. Vlasenko, A. A. Kiselev, and T. I. Ryshkevich, in M. M. Novak and T. G. Dewey, Eds., *Fractal Frontiers*, World Scientific, Singapore, 1997, pp. 393–402.
83. L. D. Landau and L. P. Lifshitz, *Electrodynamics of Continuous Media*, Pergamon Press, Oxford, 1984.
84. G. Videen and P. Chylek, *Opt. Commun.* ***158***, 1 (1998).
85. I. Thormahlen, J. Straub, and U. Grigull, *J. Phys. Chem. Ref. Data* ***14***, 933 (1985).
86. H. Bateman, *Higher Transcendental Functions*, Vol. 2, McGraw-Hill, New York, 1953.
87. P. Chylek et al., *J. Geophys. Res.* ***101***, 23365 (1996).

CHAPTER TEN

Optics and Structure of Carbonaceous Soot Aggregates

EUGENE F. MIKHAILOV, SEGEY S. VLASENKO, and ALEXEI A. KISELEV

Physics Research Center, Department of Atmospheric Physics,
St. Petersburg State University, Petrodvoretz,
Ulyanovskaya 1, 198904 St. Petersburg, Russia

10.1. INTRODUCTION

The optical properties of carbonaceous aerosol particles have been the subject of intent concern for the last decade. This interest was stimulated by the high optical activity of the carbonaceous smoke in the processes of radiative transfer in the atmosphere, and by the necessity to understand the radiative properties of soot containing flames. A wide range of practical applications requires reliable data on optical properties and on the amount and structure of soot aerosol in atmospheric air, including the remote sensing of atmospheric pollution, visibility determination, and so on. The data are also needed to evaluate the possible climatic impact of smoke clouds in the case of large-scale fires, and to calculate various scenarios of a nuclear winter [1–5]. In this case, the prediction accuracy will directly depend on the precision of the data.

Study of the optical properties of soot is a challenging problem, due to the complexity of the structure of soot. In general, soot consists of small spherical primary particles combined into branched aggregates. The primary particles, or monomers, have a mean size of tens of nanometers and are composed of amorphous elemental carbon mixed with some amount of high molecular hydrocarbons. The mass ratio of these components fluctuates for the different fuel types and burning conditions [6].

The most pronounced feature of a soot aerosol is its strong variability, that is, the liability to change its composition and morphology with coagulation, sedimentation, and restructuring under the action of external force fields. This is expected to have a striking effect on the prediction of its optical characteristics. This liability to

Optics of Nanostructured Materials, Edited by Vadim A. Markel and Thomas F. George
ISBN 0-471-34968-2

structural changes is determined by the weakness and flexibility of interparticle bonds and by the low mechanical strength of brunched structure.

The effect of restructuring on the optical properties of soot will be most significant in a humid atmosphere, because water condensation on soot particles could produce very unpredictable results. Thus, capturing of soot particles by water droplets can increase the total absorption of the atmosphere. On the other hand, the presence of soot particles as a nucleation center could trigger the water nucleation and cloud formation, and therefore can increase the total albedo [7–9]. Together with the specified effects, nucleation of water on the surface of soot aggregates will transform the structure of a cluster by means of capillary forces, which in turn will change its own optical properties [10].

If we consider the outlined problems, there is no doubt that a correct account of soot structure transformation is absolutely necessary to provide accurate optical models of atmospheric aerosol. For that reason, numerous laboratory studies of structural variability of carbonaceous aerosol particles, as well as the development of improved theoretical methods, have been reported in recent years.

The objective of this chapter is to briefly discuss the basic problems that arise when one studies the optics of soot aggregates, both theoretically and experimentally. There is a brief review of the experimental results, obtained in our laboratory and elsewhere, which concerns the structure of soot aggregates in relation to their optical properties. These results are aimed at showing that these objects are very far from the ideal model structures that are the subject of numerical analysis so widespread in modern days. We also tried to demonstrate the complexity of those carbonaceous structures from different points of view: from the structure of the inner core of a primary particle to the dispersion composition of an aerosol particle ensemble. Thus this chapter is divided into an Introduction, three section and a conclusion: Section 10.2 contains data on soot formation in combustion processes and methods of its structural diagnostics; Section 10.3 describes various theoretical approaches to characterization of soot optical properties; and Section 10.4 concentrates on the investigation of soot variability as revealed in practical experiments, as well as to the *in situ* optical measurements of soot aerosol subjected to the morphological transformation.

10.2. THE STRUCTURE OF SOOT AGGREGATES

10.2.1. Formation of Soot Aggregates in Hydrocarbon Fuel Burning Processes

Soot is a complicated object in a structural sense. The process of soot aggregates growth starts in the flame zone during the high-temperature pyrolysis of the fuel, which gives birth to the polyaromatic hydrocarbons C_nH_m (so-called PAH precursors). According to [11], in a highly heated environment with insufficient oxygen content, the processes of radical polymerization develop rather rapidly, resulting in the precursors mass growth and simultaneous raise of the C/H ratio. Mass spectra of the products of acetylene–oxygen and benzene–oxygen flames provide some knowledge about the size and mass of these precursors [12], which appears to be $\sim 2 \times 10^3$ amu for the acetylene flame, and 1.5×10^3 amu for the benzene flame, the sample being taken during the last stage of the pyrolysis. It corresponds to 120–160 carbon atoms per particle.

If the precursors leave the high-temperature zone, they become involved in the process of diffusional coagulation. At this stage of evolution, the growth of soot particles is determined by the settling of precursors on the surface of a germ. The size of the resulting particles, which we will refer to as "primary", varies broadly in the range of 10–80 nm and is highly dependent on the fuel type and total amount of oxidant in the burning zone [13,14].

For a long time primary soot particles were thought to be spherical globules, consisting of crystallite—small packs of graphitic plane fragments with the mutual normal [15,16]. The degree of order in such a globule is supposed to be growing toward the surface. The crystallite consists of 3–6 layers for a great majority of soot types, whose size is 1.5–3.0 nm along the graphitic plane and 1.0–2.0 nm across the pack. A schematic view of this model representation of soot primary particle microstructure is shown in Fig. 10.1(a). Recently, Ishiguro [17] showed that in fact it has a more complicated organization. By means of phase-contrast high-resolution microscopy and the hollow-cone beam method, two significantly different spatial regions were found inside the soot globule: the inner core and the outer shell. The core comprises several smaller fine particles, from 3 to 4 nm in diameter, each in turn consisting of a 1-nm nucleus covered by several carbon layers with a distorted structure. The outer shell is composed of microcrystallites with a periodic orientation of carbon sheets, representing the so-called graphitic structure.

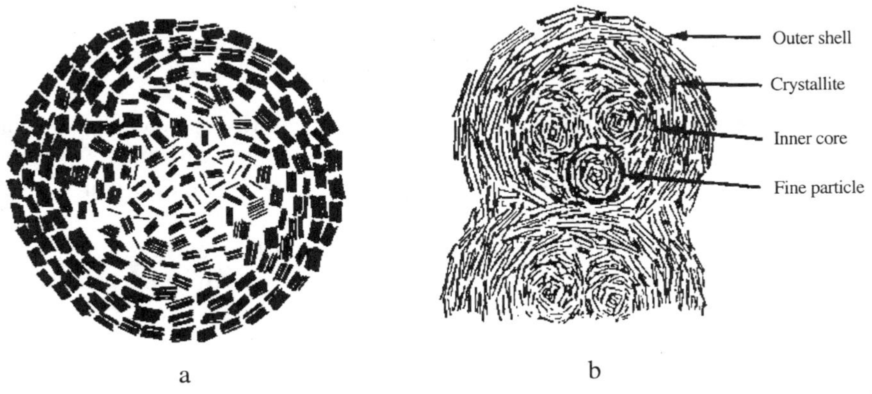

Figure 10.1. Inner structure of a primary particle. (a) Obsolete model: disordered core covered with parallel layers of crystallites; and (b) modern concept: sevetal ultrafine particles within the inner core, with the outer shell composed of parallel layers of crystallites. The bridge between the two particles is overgrown by condensing matter, thus strengthening the interparticle bond.

From the point of view of the stability of the resulting aggregate, it is essential that the growth of primary particles goes simultaneously with their coagulation into multiconnected chains. Connection of spherules due to electrostatic forces before the completion of the growth cycle calls forth the conversion of the interparticle contact type from the weak coagulational one to the mechanically strong phase-type contact [Fig 10.1(b)].

Inside the flame zone, the preliminary coagulation runs chiefly in a kinetic regime, when the particles are moving along the linear trajectories [18]. Small clusters grow rapidly due to the capture of single particles. This stage of growth is characterized by the cluster–particle type of aggregation with the hybrid trajectory of particle motion: either linear or Brownian, correspondingly, reaction-limited (RLA) or diffusion-limited (DLA) models of aggregation [19]. Since the early stage of coagulation is conducted in a weakly ionized media at the approximate temperature of 1000 K, some primary particles carry a charge with them. Due to the presence of uncompensated charges, the clusters become oriented under the action of electrostatic forces, which results in the preferential growth of elongated thread-like aggregates in coagulation [Fig. 10.2(a)]. With the gradual depletion of single particles, the cluster–cluster type of aggregation (CCA model) prevails [19]. The resulting clusters are highly rarefied, chain-like structures with multiple connections. An example of such a cluster is shown in Fig. 10.2(b).

If we consider the mechanical stability of resulting aggregates, it is worth noting that the CCA is going chiefly outside the flame zone. Here, the temperature is low and the concentration of loose carbon compounds is poor. Thus the coagulational contacts remain almost pointed, and therefore weak compared to the contacts that arise during the early stage of aggregation. As it will further be show, this condition determines the strong changeability of CCA aggregates under the action of external force fields.

10.2.2. Fractal Structure of Soot Aggregates: Methods of Determination

At present, aerodisperse aggregates resulting in the stochastic coagulation of primary particles are known to be objects with complicated geometrical shapes. They are characterized by self-similarity or, equivalently, scaling invariance, which is the same thing. To provide a quantitative description of an object like that, one often uses the fractal approach, which means that we regard the aggregate as a fractal set characterized by fractal dimensions (plus some other structural parameters). Correspondent aggregates are called fractal clusters. According to Mandelbrot [20], who is often regarded as the father of the fractal approach to natural phenomena, fractality is yet another extensive parameter of a cluster (e.g., perimeter, projectional area, and occupied volume), which can be applied when the set has the fractional value of the Hausdorf–Bezikovich (HB) dimension. The dimension of HB for a set enclosed in a k-dimensional space is introduced as follows:

$$N = L^{-D_H} \tag{10.1}$$

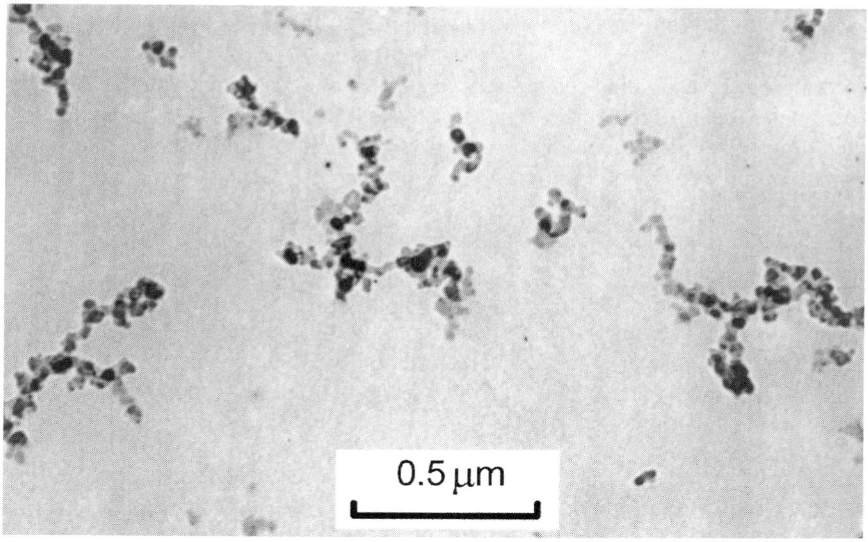

(a)

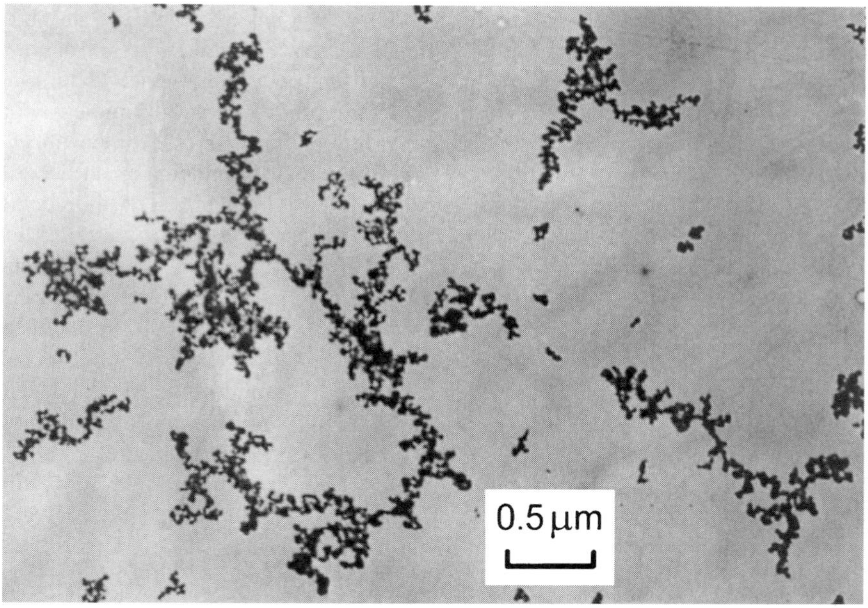

(b)

Figure 10.2. Representative view of soot aggregates, formed in the various combustion processes. (a) In the early stage of aggregation, small clusters grow by cluster–particle sticking. (b) In the next part of aggregation, the aggregates are formed of the several clusters.

where N is a number of k-dimensional spheres of diameter L necessary to cover the set.

Numerous investigations have shown the effectiveness of the fractal dimension as a parameter, bearing information on the physical properties of aerodisperse aggregates. First of all, the value of the fractal dimension is highly sensitive to cluster formation and growth conditions, which gives us the opportunity to recover the origin of the cluster and its history from the fractal analysis data. Another advantage of the fractal dimension is that it can be unambiguously associated with the values of other physical characteristics of the cluster: mobility, mechanical strength, optical cross-sections, and so on. In other words, to get the reliable description of an aerodisperse system of aggregates one should combine the knowledge of traditional mean size of the cluster with the characteristic fractal dimension.

In the last few years, which were marked by an intensive outgrowth of fractal theory, several varieties of fractal dimensions were introduced, and many new methods of its determination were developed.

Note here that various types of fractal indexes have a different physical sense, and hence have different values for the same aggregate. Thus, the fractal dimension of the cluster perimeter is not equal to the fractal dimension revealed from the projectional area analysis or from the relative distribution of elements inside the cluster. As pointed out by Stanley et al. [21], a more detailed classification of fractal dimensions is needed, where each physically relevant surface of an object would have a correspondent fractal dimension. For example, it is possible to distinguish an "accessible", or unshielded cluster perimeter, which determines the rate of the coagulational growth of the cluster, and the total perimeter, which governs the adsorptional processes. An individual value of fractal dimension will be found for each perimeter. The choice of structural parameters for an accurate description of the aggregate should be based on the knowledge of the specific physical properties in the current environment. Moreover, values of fractal dimension for the same geometrical set of points inside the cluster were obtained experimentally by different methods and could diverge significantly, while fractal theory gives a uniform value. This phenomena is a consequence of the fact that real clusters are *not exactly fractal* in the mathematical sense. The scaling relations of this type (10.1) hold only approximately, and only in a limited range of scales. That is why a more accurate term, "fractal-like objects", became more abundant in recent periodicals. Accordingly, the type of fractal dimension and the method of its determination should always be kept in mind when using this term. The possible distinction is extremely essential for a comparison of results, which were obtained by different authors or when different methods were used to validate experimental data.

The following scaling relationship between the number of particles in a cluster and its gyration radii is used most often to calculate the fractal dimension:

$$N = k_f \left(\frac{R_g}{d_p} \right)^D \tag{10.2}$$

where N is the number of primary particles in a cluster, R_g is the gyration radius, D is the fractal dimension, d_p is the diameter of primary particles, and k_f is the proportionality coefficient—the so-called fractal prefactor (lacunarity), which characterize the packing density of spherules in an aggregate. The gyration radius obeys the following relationship:

$$R_g^2 = \sum_{i=1}^{N} \nu_i \mathbf{r}_i^2 \Big/ \sum_{i=1}^{N} \nu_i \tag{10.3}$$

where $\mathbf{r}_i$ is the radius vector of the ith spherule in relation to the cluster center of mass, and ν_i is the volume of the same spherule. By defining the gyration radius in this way we can account for the polydispersity of spherules.

The theory of fractal objects specifies that dimension D in relation (10.2) coincides with the HB dimension, although it should be noted that to determine D one should consider a set of aggregates of various sizes, supposing that all of them have the same fractal dimension. The last assumption appears to be a rather rough approximation, as different stages of coagulation growth correspond to different patterns of aggregation, therefore producing clusters of different dimensions. It means that practical usage of relation (10.2) provides some kind of net fractal dimension for an ensemble of polydisperse clusters.

When the internal structure of a single cluster or distribution of clusters over the fractal dimension is under investigation, the scaling relationship of the following form is generally used

$$N \sim R^D \tag{10.4}$$

where N is the number of spherules whose distance from the center of mass is not more than R. According to this relationship, the number density of primary particles inside the cluster decreases toward the boundary as

$$\rho(R) \sim R^{D_E - D} \tag{10.5}$$

where D_E is the dimension of enclosing space, for example, equal to 3 for the three-dimensional (3D) aggregates. A more accurate description of scaling invariance for disposition of spherules inside the cluster could be provided by the pair-correlation function, which is defined as a weighted average density of spherules at a given distance inside the cluster. For a finite-size cluster comprising N primary particles, the pair correlation function is given as follows:

$$C(\mathbf{R}) = \frac{1}{2\pi RN} \sum_{j=1}^{N} \langle \rho(\mathbf{R}_j + \mathbf{R}) \rho(\mathbf{R}_j) \rangle \tag{10.6}$$

Here, $\mathbf{R}_j$ is the radius vector of the particle j inside the cluster and $\langle \cdots \rangle$ denotes averaging over all possible orientations of the reference point vector $\mathbf{R}$. The pair correlation function of fractal structure obeys the law

$$C(\mathbf{R}) \sim R^{D_E - D} \tag{10.7}$$

where D_E is the dimension of the enclosing space, and D is the HB dimension.

Utilization of the fractal approach in application to existing aerodisperse objects allowed a notable extension to the traditional morphological and dispersion analysis, which provided additional information on the internal structure of particles. On the other hand, it provoked the problems of accurate determination of the fractal dimension and its further interpretation. If we consider the accumulated 10-years of experience in this field, one can note that there are two basic methods of fractal analysis of aerosol aggregates when speaking about carbonaceous nanometer-size particles: electron microscopy and the scattering of electromagnetic (EM) radiation by fractals (this includes X-ray and neutron ray small-angle scattering). *In situ* methods of indirect determination of the fractal nature of carbonaceous aerosol, based on the fractal-dependent behavior of physical properties, are also rapidly evolving.

10.2.2.1. The Method of Electron Microscopy. The method of electron microscopy is based on the visualization of the cluster's structure by producing two-dimensional (2D) [with transmission electron microscopes (TEM)] or 3D [with scanning electron microscopes (SEM)] images of clusters under investigation, with the following procession based on the known scaling relations. The TEM method was the first used for fractal analysis of aerosol particles and, so far, the majority of our information was obtained with it. Certain features of this method are to be kept in mind when applying it to practice.

First, TEM requires sampling of aerosol particles on a special support, which means the possibility of misrepresentative sampling, or distortion of morphological and dispersion composition of sampled aerosols in comparison with the original system. There is also the possibility of violation of the aggregates structure due to collision with the support [22]. Second, this method implies discretization and digitalization of microscope images, when the whole field is divided into a finite number of pixels of different brightness. The original continuous image on the plane is projected onto a discrete array of elements, in most cases arranged on a orthogonal grid. The minimal size of the cell l_0 is limited by the resolving power of the analyzing unit, and, in turn determines the quality of the fractal analysis. Both the equality of l_0 and the size of monomer r provide the best conditions for further procession. Besides, the total number of cells should be large ($\sim 10^4$), to confirm the scaling invariance of a cluster over a wide range of scales. The brightness of a cell carries information as to whether it belongs to the object or not. In practice, however, it is never possible to determine brightness exactly due to noisiness, luminosity, unevenness, and so on. So, the procedure of object distinction (binarization of image) involves some preliminary frame processing, (e.g., filtration, background smoothing) and therefore should employ certain heuristic algorithms. The resulting set of pixels is a mathematical representation of the original cluster, and naturally, as ideally as possible, do not completely reflect the true structure of the object. Third, the routine calculation of the fractal dimension for digital images implicates usage of discrete quantities instead of continuous terms of equations (10.1)–(10.5), which also can bring some distortion to the final result. Fourth, the TEM method deal with the 2D images of 3D objects—projections of clusters onto the plane of an image. It is obvious that structural parameters of an image are related to parameters of the original cluster, but direct transition of their values onto the characteristics of a 3D cluster could be ambiguous. This question was studied thoroughly in the past years. In particular, Meakin et al. [23] found that for so-called "optically transparent" fractal clusters ($D < 2$) the fractal dimension of the image almost coincide with the true fractal dimension of the original cluster. For a more accurate account of the projectional distortion, a special correction coefficient is introduced that could be calculated by a numerical simulation of fractal clusters [24–26].

For evaluation of a fractal dimension of clusters by their 2D images, several practical routines are used, which are based on the sequential algorithms of planar fractal analysis.

The Successive Squares Method This method employs Eq. (10.4) for the analysis of a digitized image. In the case of a constant HB dimension for the whole image, the number of image elements within the square (circle) with a side L, which belong to the cluster, is given by the relationship

$$N \sim L^D \tag{10.8}$$

The routine begins with the constructing of successively growing squares with the center located in the center of mass of the cluster, and is followed by evaluation of the number of elements belonging to the image of the cluster located inside each square. The parameter D is calculated from the slope of the $N(L)$ plot in logarithmic scale, since it should be linear for the fractal aggregate. For an actual fractal-like aggregate, the linearity of $\log N(\log L)$ dependence will hold only within a certain range of $L : d_p < L < d_a$, where d_p and d_a are diameters of the monomer and the aggregate, respectively. This linear part of the plot is used for determination of a fractal dimension $D = \Delta \ln N(L)]/\Delta(\ln L)$. Thus a defined fractal dimension is often called a *mass fractal dimension*.

The Covering Set Method This method is based on the direct application of the HB dimension to the digital image of a fractal cluster. At each step of the algorithm, the whole image is divided into an array of equal squares with the side ε. Then, the number of squares that contain at least one pixel of the cluster image are counted. According to (10.1) the number N_b of such squares for fractal objects obeys the law

$$N_b(\varepsilon) \sim \varepsilon^{-D_b} \tag{10.9}$$

If we change the size of a square at each step and count the number of occupied squares N_b, we can find D_b in the same fashion as the successive squares method. This kind of fractal dimension is called a *cell fractal dimension*. Comparative calculations of these two fractal dimensions for the same set of clusters showed [27] that the cell fractal dimension is featured by a lower dispersion of values, which makes it preferable for the characterization of the fractal geometry of soot aggregates.

The Pair-Correlation Method The pair-correlation function (10.6) can be applied to the digital image of a cluster. In this case, N is the total number of occupied pixels, for example, pixels belonging to the image, $\rho(\mathbf{R}_j)$ and $\rho(\mathbf{R}_j + \mathbf{R})$ are the densities at a reference point vector $\mathbf{R}_j$ and at vector $\mathbf{R}$, respectively, equal to 1 if the pixel belongs to the image and 0 otherwise. The summation goes over all occupied pixels and the dimension of the enclosing space D_E is equal to 2.

The Particle Counting Method If clusters are not too large for the employment of the above methods, it is possible to determine the fractal dimension making use of the scaling relationship (10.2) for the whole ensemble of aggregates within view of a microscope or located on the photograph. As noted earlier, N is the total number of pixels belonging to the cluster image and R_g is determined by Eq. (10.3), where $\nu_i = 1$ and the summation is going over the cluster's elements. When the set of N and R_g is calculated for the whole ensemble, it is easy to obtain D through the linear fit of $\log N(\log R_g)$ plot. To simplify the numerical routine, instead of using the R_g in formula (10.2), other characteristic sizes are often used, for example, the Ferret diameter $D_F = \sqrt{LW}$, where L and W are, respectively, the length and width of the rectangle circumscribed around the cluster.

10.2.3. *In situ* Methods for Determination of Fractal Dimension of Soot Aggregates

In situ methods are highly desirable for two reasons. First, they are able to provide remote sensing in a hostile environments or hard-to-reach places. Second, they represent unintrusive techniques allowing us to avoid possible damage to the object under investigation. Among other reviews of the *in situ* methods the optical measurements have a very long history, which dates back to the pioneering work of Guinier as early as 1937 [29].

10.2.3.1. Angle Light Scattering. Together with the TEM methods, analysis of angle light scattering (ALS) is the basic method for recovering a fractal dimension from the scaling experiments. This method comprises visible or near-infrared (IR) light scattering (ALS), small angle X-ray scattering (SAXS), and small angle neutron scattering (SANS). The last two techniques are not *in-situ*, as far the soot clusters are concerned, because bulk materials are required to obtain a reasonable value of scattered intensity. Though it is not a problem for laboratory generated soot, the sampling of a necessary quantity of aerosol from ambient air could be an unfeasible task. Besides, close packing of clusters in a bulk sample will corrupt the branched outer part of aggregates, which could possibly lead to the altering of the average structural parameters. Nevertheless, collection of experimental methods of determination of fractal dimension would not be complete without SAXS and SANS techniques, because of their ability to penetrate into the hidden part of the cluster and inside the core of a primary particle.

In this section, we shall trace the basics of the ALS method in detail, because it has the most important bearing on the understanding of coupling between the optical and structural properties of carbonaceous aggregates.

The main idea of angle light scattering [28] involves the relation of the scattered radiation $I(q)$ to the structural factor $S_N(q)$,

$$I(q) = I_1(q)N^2 S_N(q) \tag{10.10}$$

where $I_1(q)$ is the intensity scattered by a single particle and $q = (4\pi/\lambda)\sin(\theta/2)$ is a scattered wave vector. For isotropic aggregates consisting only of identical particles, $S_N(q)$ is related to the pair correlation function $g_N(r)$ as follows:

$$S_N(q) = \frac{1}{N}\left[1 + \int_0^\infty g_N(r)4\pi r^2 \frac{\sin(qr)}{(qr)} dr\right] \tag{10.11}$$

$g_N(r)$ is normalized by

$$\int_0^\infty g_N(r)4\pi r^2 dr = N - 1 \tag{10.12}$$

from which it follows that $g_N(r)$ is the mean number density of monomers in a spherical layer of thickness dr at a distance r from the chosen monomer, that is,

$$dN(r) = 4\pi r^2 g_N(r)dr \tag{10.13}$$

On the other side, from the scaling relationship (10.5) it follows that

$$dN(r) = \frac{D_f}{d_p^{D_f}} r^{D_f - 1} dr \qquad d_p \ll r \ll R_g \tag{10.14}$$

where d_p and R_g are, respectively, the primary particle diameter and the gyration radius. By combining (10.13) and (10.14) we have

$$g_N(r) = \frac{D_f}{4\pi r_p^{D_f}} r^{D_f - 3} f\left(\frac{r}{R}\right) \qquad r \gg d_p \tag{10.15}$$

Here $f(r/R_g)$ is a cutoff function introduced in (10.15) to account for the finite size of a cluster. It has a sigmoidal form and is close to unity when $r \ll R_g$ and must tend to zero more quickly than any power law for $r \gg R_g$. One has to subdue temptation to regard f as a density cutoff in a real-space, though it is characteristic of the external surface of a fractal. For this reason, some simple choices are forbidden, for example, the Yeaviside function, which leads to possible negative values for $I(q)$. The parameter, $f(r/R_g)$ must exhibit some "width" even for an object with a sharp boundary.

The small q, Eq. (10.11) can be expanded up to second order in q

$$S_N(q) = 1 - \frac{q^2}{6N}\int_0^\infty 4\pi r^4 g_N(r)dr + \cdots \tag{10.16}$$

By introducing R_g^2 as the mean of the squared distances from the center of mass and expanding $(\mathbf{r}_i - \mathbf{r}_j)^2$, we easily have

$$R_g^2 = \frac{1}{2N^2}\sum_{i=1}^{N}\sum_{j=1}^{N}(\mathbf{r}_i - \mathbf{r}_j)^2 = \frac{1}{2N}\int_0^\infty 4\pi r^4 g_N(r)dr \tag{10.17}$$

By making use of this form of R_g^2 in (10.16) and by substituting the result into (10.10), we have the q-dependent relation for $I(q)$ in the so-called Guinier regime

$$I(q) = I_1(q)N^2\left(1 - \frac{q^2R_g^2}{3} + \cdots\right) \qquad qR_g \ll 1 \tag{10.18}$$

Experimental measurement in a low q regime allows us to calculate the mean R_g for the system of clusters independent of their fractal nature, keeping in mind that the actual aerosol system is always a polydisperse one.

For the intermediate range of q-values, $(d_p^{-1} \gg q \gg R^{-1})$, the structure factor $S_N(q)$ could be determined by combining Eqs. (10.15) and (10.11)

$$S_N(q) = \frac{1}{N}\left[1 + \frac{D_f}{r_p^{D_f}}\int_0^\infty r^{D_f-1}e^{-(r/R)^\gamma}\frac{\sin(qr)}{qr}dr\right] \qquad qd_p \ll 1 \tag{10.19}$$

where $f(r/R_g)$ is taken in the exponential form $f(r/R) = e^{-(r/R)^\gamma}$ with $\gamma > 0$ [29, 30]. For $qR \gg 1$, (10.19) transforms to

$$S_N(q) = \frac{b}{N}q^{-D_f} \qquad b = \left(\frac{d_p}{2}\right)^{D_f} D_f\Gamma(D_f - 1)\sin\left[\frac{\pi}{2}(D_f - 1)\right] \tag{10.20}$$

where $\Gamma(x)$ denotes the gamma function. In this case, the scattered intensity will have the following form:

$$I(q) = NI_1(q)bq^{-D_f} \qquad qd_p \ll 1 \qquad qR_g \gg 1 \tag{10.21}$$

This formula defines the fractal, or power-law regime of the scattering of light on fractal-like clusters. The parameter D_f could be recovered experimentally from the slope of the reduced intensity versus q plot in double logarithmic coordinates. Note that the choice of a q range for recovering the fractal dimension must be done with a great care: as it is shown by Jullien [29] that the precise form of a cutoff function has an influence on the shape of a scattering intensity curve that is especially noticeable near the inflexion point located between the Guinier regime and the fractal regime. Here, we have taken a step onto the ground of optical modeling, for any cutoff function is just a mathematical abstraction, together with the assumption of monodispersity of primary particles. The only way to make a reasonable estimation of the fractal dimension from a log–log plot of $I(q)$ is to stay well within the large q fractal-like regime of scattering, avoiding the inflexion region (as well as the large q Bragg-like region), and thus escaping the need to make any theoretical reduction concerning the results obtained on the basis of the approximate model.

10.2.3.2. Other In-Situ Techniques. Another known technique of the *in situ* measuring of structural parameters for an aerodisperse system is based on the study of the aerodynamic mobility of aerosol particles. In this case, formula (10.2) reads as the scaling relationship of the aggregate's mass M to its gyration radius R_g, which in turn is linearly related to the aerodynamic redius R_a [31]. The aerodynamic radius of a nonspherical particle is defined as the radius of a sphere with the same aerodynamic mobility. Hence, the relation $\mathbf{M} \sim \mathbf{R}_a^D$ is valid in a wide range of sizes for fractal-like aggregates. If we make use of this relation, the fractal dimension of the aggregate could be determined, provided that independent measurements of M and R_a are possible. Several practical realizations of this method are known at present, although most of them considered aggregates of more noble materials than soot (silver agglomerates, magnesium oxide). Mobility is often measured with the help of a differential mobility analyzer (DMA), which requires the charging of aerosol particles in order to obtain the linear relation of their velocity and the electric field strength E. Another method is to determine the mobility through measurements of the settling velocity of particles, taking into account their coagulation [32]. This technique is rather complex and utilizes a charge-coupled device camera together with a video recording of a coagulational process.

The mass of floating in gas aggregates could be obtained with a low-pressure impactor, implying inertial settling of particles onto the support [33]. By varying the pressure inside the impactor, it is possible to alter the terminal mass of settling particles, thus evaluating the mean mass of the particles within a certain size range. In the work of Weber [34] and Weber et al. [34a], the usage of inductively coupled plasma optical emission spectroscopy was reported for the determination of an aggregate mass. The method relies on the measurements of the photoemission spectra of particles injected into the plenum with a stable high-temperature plasma. This method was also used by Nyeki and Colbeck [35,49], in combination with their method of balancing the single aggregate with an electric field in a modified

TABLE 10.1. Fractal Dimension D_f and Prefactor k_f Determined by Different Methods

Method	Flame Type	D_f	k_f	Reference
ALS	Various turbulent	1.86	2.25	[24]
ALS	Laminar acetylene	1.75	2.78	[24]
2D TEM	Laminar Acetylene	1.66	2.35	[24]
2D TEM	Various turbulent	1.82	8.5	[24]
2D TEM	Laminar ethylene	1.74	1.80	[70]
2D TEM	Laminar ethylene	1.62	2.18	[70]
In situ	C_4H_{10} flame	1.87–2.19		[35]
Aerodynamic	Acetylene	1.97		[32]
3D TEM	Laminar acetylene	1.40	9.4	[25]

Millikan cell. Note that all of these techniques require extremely accurate calibration with the test objects, which in itself is not an easy task.

As already mentioned, the different techniques often yield different values of fractal dimension and prefactor. Moreover, the magnitudes can vary notably even within one method. To illustrate this phenomena, several data sets were collected in Table 10.1.

Although the values are close, especially within one method of determination, they never coincide when obtained by different methods. This is the principal difference, specified on one hand by the different nature of the yard stick employed in the method of determination, and on the other hand by the different nature of the set of surface elements involved in the scaling process. Another is that the soot aggregates system is hardly a representative object to investigate. It is extremely difficult, from an experimental point of view, to produce soot particles under precisely uniform conditions. The small variation of input parameters (carbon/oxygen ratio of fuel, laminarity, environment conditions, etc.) can alter the result significantly. In this context, the measurements of fractal dimension are valuable not in the absolute sense, but as an instrument to reveal the relative tendency of structure modification.

10.3. THEORETICAL CONSIDERATION OF CARBONACEOUS SOOT OPTICAL PROPERTIES WITH ACCOUNT FOR ITS STRUCTURAL FEATURES

Until now we regarded the internal structure of a cluster as a set of independent parameters, required to describe completely its physical and chemical properties. Since the subject of this work is limited by the optical and structural properties of soot aerosol particles, it is now necessary to decide, which parameters to use and to what extent they influence the optical characteristics of objects under investigation. The problem is in fact that the choice of structural parameters needed in each particular case is dependent, on one hand, on the specific features of the theoretical model intended to be used, and on the other hand, on the structure itself, for it is subject to change together with current environmental conditions. For example, if the optical model deals with the large-scale aerosol system, it utilizes the set of structural parameters related to the statistical characteristics of that system, such as distribution of interparticle distances, mean size, dispersion, *N*th moments of size distribution, and average values of optical constants. If the optical properties of a single particle are to be described, another set of optical parameters is being utilized: the diffraction factor $\rho = \pi a/\lambda$ (characteristic size in relation to the wavelength), shape characteristics, optical constants of the particulate matter $n + ik$ for the given wavelength, or relative refraction index (in relation to the surrounding media). In certain cases, some additional parameters are needed for a correct account of the particles structure: anisotropy of the refraction index relative to the orientation of the incident EM field and surface characteristics [degree of roughness, characteristic size of roughness, surface fractal dimension (or multifractal spectrum), volume distribution of material density inside the particle, porosity, or cavitation (essential for the X-ray optics)]. One common attribute can be traced here, that is, the practical value of all these parameters is relative to the characteristics of the surrounding media (including the incident EM radiation). Thus, shape characteristics are important only when the size of the particle can be compared with the wavelength, while the surface pattern or volume distribution of a substance inside the particle body should be considered only if the wavelength is very small (e.g., X-rays).

In the case of soot aggregates, the term "structure" becomes even more dependable on the environmental conditions (temperature, humidity, residence time in the gaseous media), and exhibits more hierarchy in itself. The known tendency of the soot primary particles to form large clusters makes it necessary to add an additional stage into the sequence of structural description of the soot aerosol:

1. *A single primary particle (monomer).* The whole set of structural characteristics could be engaged to characterize its optical properties.
2. *An aggregate of primary particles (cluster).* Aggregates are featured by the presence of interparticle contacts. The primary particles are fixed in their places and can exhibit some collective effects. The distances between the particles are small, and hence they could not be considered as independent in the processes of scattering and absorption of EM radiation (at least within the correlation length). The role of multiple scattering should be considered in each particular case. The spatial disposition of monomers inside the cluster is described by a pair-correlation function, which for the case of a large aggregate exhibit a scaling fractal-like behavior. Usage of this phenomena allows one to abandon presetting of the interparticle disposition in the explicit form.
3. *A system of aggregates.* Clusters are floating in the gaseous phase far from each other (at a distance of more than 10 gyration radii) and are independent in optical processes. A complete system description requires presetting of the size distribution function or particle number distribution, which are the same if clusters in the system comply with the fractal behavior. Since not every

cluster in the system is genuinely fractal (e.g., too small), only a statistical approach to the whole system allows one to use the numerous advantages of the fractal approach.

We have already considered *in situ* optical methods for measuring the fractal dimension of aerodisperse aggregates. More complex optical measurements must be performed to recover complete information about the internal structure of clusters. In these cases, the inverse problem is to be solved in order to determine structural parameters from the measurements of optical characteristics (optical cross-sections, angle dependencies of light scattering, etc.) This could be done only on the base of a model, which reliably associates optical and structural characteristics of the aggregate. To create a model like this, one has to obtain *a priori* knowledge about the inner structure by independent methods. Rapidly developing electron microscopy techniques and computational possibilities provide the necessary information, and therefore stimulates the need for fuller and more comprehensive models of soot.

Given the above hierarchy scale of the soot aerosol structure, theoretical models of its optical properties can be classified in a to account for structural features. Two distinct classes can be observed: In the methods for the first class, the cluster organization of soot is neglected, while the second class comprises theoretical models that take the aggregate nature of soot particles into consideration. Computational models of the first type usually employ the concept of the equivalent particle (the particle that has a simple spherical shape and some equivalent physical parameter, e.g., aerodynamic diameter or volume) plus some additional assumption about its properties to make the result agreeable. It could be a specially chosen value of the net refraction index or model of concentric layers with a radially dependent value of density and/or refraction index [37].

Methods for the second type include consideration of the cooperative effects arising from the close location of primary particles, though various approximations, concerning monomer sizes or their location inside the cluster, are also present. The exact methods of computing the optical properties of an aggregate should be included into this class, also. These methods utilize the exact location of each monomer inside the cluster and interative solution of the wave equation. Though properly characterizing single aggregate optical cross-sections, these methods often suffer from computational difficulties when there is a need to calculate some optical characteristic for the system of polydisperse aggregates.

10.3.1. Semiempirical Models

Most of the optical models from the first class of proposed classification are based on the well-known theory of absorption and scattering of EM radiation by a spherical particle of arbitrary size, which was elaborated on at the beginning of the century by Mie [37]. The exact solution of the wave equation in this theory involves expansion of the internal and the scattered EM field into the spherical vector wave functions [38]. The possible application of this theory to the case of soot clusters has the following features.

Mie theory is applicable to spherical particles. Although some reductions were made, which allowed us to consider particles with lower symmetry (ellipsoids, spheroids), the basic requirement of the theory is the simplicity of the particle form. Hence, a correct account of cluster shape is impossible. We could substitute the volume equivalent compact spherical particle for the cluster, which looks like an attractive idea, considering the low overall density of the porous structure. Although a well-developed soot aggregate could reach the size of 2–5 μm, the density of 0.2–0.3 g/cm^3 provides a small (comparing to the original) diameter of a compact equivalent sphere having the density of amorphous carbon. When speaking of the visible and IR spectral region, this also means applying the Raleigh approximation—the small size cutoff of the Mie theory, when the particle is regarded as a single dipole and the above mentioned expansion is being cut after the second term of summation. Unfortunately, as was confirmed in numerous practical measurements ([10,39], and references cited therein), the obtained optical characteristics do not have the slightest resemblance to those of real soot particles. Utilization of a dense particle with a diameter equal to the size of a cluster does not improve the situation, not to mention the fact that a dense particle of this size will not stay in the atmosphere for long.

The models approximating a soot cluster by a sphere with the radial distribution of structural quantities were more successful, especially when describing composite soot particles, covered, for example, by a water film or enclosed inside a drop [40, 41]. Note that the model implying concentric layers with a consequently changing refractive index is lacking physical content, although the exact form of a radial distribution of refractive indexes and density could be chosen to match the measured integral characteristics of a single aggregate. Still, the inverse problem, that is, recovery of structural parameters from optic measurements, could not be solved, considering the polydispersity of structural parameters in the real systems.

In fact, the boundaries between the theories of the first and second type are quite fuzzy. The well-known Raleigh–Debye–Ganz (RDG) approximation allows us to calculate the scattering and absorption of light on the particles with an arbitrary size and shape without employment of the sophisticated Mie theory or some exact methods [36,42,43]. The basic requirements of this approximation are formulated as follows [38]:

$$|m - 1| \ll 1 \qquad qR_g|m - 1| \ll 1 \tag{10.22}$$

where $m = n + ik$ is the relative complex refraction index of a particle. Within the integral formulation of the scattering problem, it could be strictly shown that one obtains the RDG approximation when substituting the electric field inside the particle with the field of the incident wave. The first condition is then read as a demand for low reflectance on the particle–media border, and the second for a demand for small phase differences of scattered radiation from various primary particles. The succeeding scheme for solving the scattering problem is utterly simple. By dividing the elements of the total amplitude scattering matrix by volume and then approaching zero volume, one can regard the resulting quantities as scattering matrix elements for the elementary volume. Within the accepted assumptions, the elements of the total scattering matrix for a particle of arbitrary size and shape could be recovered by integrating over the volume of the particle. The shape account will be concentrated then in the integral

$$f(E, \varphi) = \frac{1}{v}\int_v e^{i\sigma} dv \tag{10.23}$$

which has the meaning of the structure factor and where $\sigma = \kappa \mathbf{R} \cdot (\mathbf{e}_z - \mathbf{e}_r), k = 2\pi/\lambda$ is the wave number, $\mathbf{e}_z$ and $\mathbf{e}_r$ are the unit vectors in the direction of the incident wave propagation, and the scattering direction, respectively. Here, $\mathbf{R}$ is the radius vector of the given elementary volume and v is the total particle volume. One can easily see that this factor is equal to 1 for the forward scattering ($\Theta = 0, \mathbf{e}_z = \mathbf{e}_r$) for a particle of any shape. This fact could be useful for the determination of integral structure characteristics, such as mass, and so on, from SAXS.

If the aggregates involve small primary particles with negligible multiple scattering, and if conditions (10.22) are satisfied, the aggregate cross-sections become [38]

$$C^a_{\text{abs}} = NC^p_{\text{abs}} \qquad C^a_{\text{sca}} = N^2 C^p_{\text{sca}} \qquad C^a_{\text{pp}}(\theta) = N^2 C^p_{\text{pp}}(\theta) \tag{10.24}$$

In this equations, $C_{\text{abs}}, C_{\text{sca}} C_{\text{pp}}$, denote, respectively, absorption, scattering, and differential scattering cross-sections, with the subscript pp standing for the vertical vv or horizontal hh polarization. The superscript a means that the quantity pertains to the aggregate, and p refers to the single isolated primary particle. Specifically, C^p_{sca} is the scattering cross-section of the primary particle in the Raleigh limit. For this approximation, absorption is the same as for independently scattering particles; however, the total scattering from an aggregate is N times higher than scattering from an aggregate in a pure Raleigh limit (when the whole aggregate is considered small in comparison with the wavelength).

Yet, we have to verify the conformity of optical constants and typical sizes of soot aggregates with the demands of the RDG approximation. There is still no absolute data about what values should be used in optical applications, which is just another confirmation of soot's strong variability, not only in the structural, but also in the chemical sense. Since a simple evaluation is our goal here, we shall use some average values typical for soot (in the visible spectrum): $m = 1.5 \pm 0.6i$ for $\lambda = 435$ nm [44,45], and $R_g = 0.5\ \mu$m. It could easily be seen that condition $|m - 1| \ll 1$ does not hold in all cases, and $qR_g|m - 1| \ll 1$ is valid only within a very narrow region, up to 5°.

Nevertheless, the RDG approximation, adapted for use with the fractal approach, is most effectively exploited in the calculation of soot optical properties. Next, we shall briefly discuss a few works, where this approximation was appropriately used.

One specially interesting method was introduced in the work of Dobbins and Megaridis [36]. Having taken into account the cluster porosity and statistical sphericity, they postulated that a randomly oriented aggregate with a material refractive index m can be replaced by a porous sphere of diameter D_{ps} and the effective refractive index m_{ps}. Here, m_{ps} could be calculated within the terms of Lorenz–Lorentz relation $G(m_{\text{ps}}) = \eta_v G(m)$, where $G(m_{\text{ps}}) = (m^2 - 1)/(m^2 + 2)$, and η_v is the volume fraction or loading efficiency of the spherical envelope occupied by the porous material. The loading efficiency of an array of N primary spheres of uniform diameter d_p, enclosed within a sphere of diameter D_{ps} is given by $\eta_v = nd_{\text{ps}}^3/D_{\text{ps}}^3$. In conjunction with Eq. (10.2), which relates the number of primary particles and gyration radius of a fractal cluster, one obtains $\eta_v \sim R_g^{D_f - 3}$, where D_f, as seen earlier, is the mass fractal dimension. The fact that D_f could not exceed 3 indicates that the loading efficiency will decrease with the growth of the cluster's diameter, which in turn will result in a decrease of m_{ps}. The hypothesis of the porous sphere implies that the demands of the RDG approximation would be fulfilled if D_{ps} and m_{ps} replace the cluster diameter and material refractive index. Indeed, with the introduced above typical values of R_g and m, the values of $qD_{\text{ps}}|m_{\text{ps}} - 1|$ and $|m_{\text{ps}} - 1|$ are almost zero. With D_{ps} and m_{ps} as input parameters instead of R_g and m, the porous sphere model can be applied to the optical cross-sections given by the RDG theory. Computational simulations of light scattering have shown that the porous sphere model adequately describes the total scattering cross-section in the low qR_g regime, and underestimates these quantities when $qR_g > 1$. It also gives reasonable values of single scattering albedo and volumetric extinction, in comparison with the data obtained with the help of more advanced methods. Note, however, that the value of D_{ps} cannot be found independently of m_{ps} from simple geometrical considerations, and therefore requires a determination from the comparison of optical characteristics calculated with the porous sphere model by calculations from some exact computational procedure. The model also lacks some specifically fractal effects of scattering, for example, saturation of the scattering/absorption ratio at large N, as predicted by the exact methods.

In our opinion, the principal benefit of this approach lies in the demonstration of physical concepts, which allow an efficient use of the RDG approximation without a formal compliance with the imposed requirements of small absorption and refraction by particles. It also challenges the adequacy of applying the material refraction index of the bulk substance to the porous flake-life structures as soot.

In fact, most "semiempirical" optical models of carbonaceous smokes utilize the RDG approach (see, e.g., [24,39,45–48]). While the formal requirement $qR_g|m-1| \ll 1$ can be overpassed within the concept of a porous sphere, there are some additional demands, namely, negligible multiple scattering and self-interaction. Berry and Percival [50] evaluated the adequacy of the RDG theory for fractal aggregates, using mean-field theory, and argue that effects of multiple scattering are negligible for $D_f < 2$, when $N \ll x^{-D_f}$, $x = \pi d_p/\lambda$ being the size parameter for the individual primary particle. If we take the typical value of $d_p = 20\,\text{nm}$ and the middle of the visible spectrum, we readily obtain that N should be $\ll 10$ monomers per cluster, which is hardly so. In this case, theory provides another criteria:

$$x \ll \left[\frac{|\varepsilon - 1| \cdot 2^{1-D_f} (D_f(D_f + 1))^{D_f/2}}{3(D_f - 1)(2 - D_f)} \right]^{1/(D_f - 3)} \tag{10.25}$$

where ε is dielectric constant, and which requires $x \ll 0.15$, or, for our typical values, $d_p \ll 10\,\text{nm}$. Thus, no conditions that allow the application of RDG theory to the calculation of carbonaceous soot optical properties are formally fulfilled.

A further improvement on the approximate theory of RDG can be obtained by applying correction factors for the phase and multiple scattering effects for real aggregates to the expressions of cross-sections of Eq. (10.24):

$$C^a_{\text{abs}} = NC^p_{\text{abs}} \qquad C^a_{\text{sca}} = N^2 C^p_{\text{sca}} g(k, R_g) \qquad C^a_{\text{pp}}(\theta) = N^2 C^p_{\text{pp}}(\theta) S_N(q, R_g) \tag{10.26}$$

where $S_N(q, R_g)$ is a form factor, and $g(k, R_g)$ is the aggregate scattering factor.

The idea of utilization of a form factor is much the same as in the method of the angle light scattering investigation of structure parameters, which was described in Section 10.2. In the RDG approximation for fractal aggregates, (RDG–FA) it is used to account for the power-law behavior of the pair correlation function, which arises in summation of scattering amplitudes from the elementary volumes comprising the whole particle. In the case of soot, these elementary volumes could be naturally associated with the primary particles, as they do satisfy the requirement of being small. The expression for the form factor has the same form as in the ALS method:

$$S_N(q) = \begin{cases} 1 - \dfrac{q^2 R_g^2}{3} & \text{if} \quad q^2 R_g^2 \ll 1 \\ \dfrac{b}{N} q^{-D_f} & \text{if} \quad q^2 R_g^2 \gg 1 \end{cases} \tag{10.27}$$

Equation (10.27) properly provides $C^a_{\text{pp}}(\theta) = N^2 C^p_{\text{pp}}(\theta)$, when (θ) approaches zero, and the power-law scattering in the regime of large qR_g. The inflexion point can be found by equalizing the Guinier and power-law form factors and their derivatives.

As was pointed out by Köylü and Faeth [48], this approximation provides a most agreeable result, when it is compared with the exact computational results and experimental measurements. It is worth noting here that direct calculation of optical properties when the exact location of primary particles inside the cluster is prescribed, is often limited to the clusters of a few tens of monomers due to computational difficulties. It means that comparison of the results obtained by the approximate methods with the exact data, concerns only this narrow range of sizes, while the most fascinating fractal effects take place when the aggregate is really large. On the other hand, when speaking about verification of computational results by practical measurements, one has to keep in mind the diversity of soot properties. Thus, complexity of the soot formation process, involving high-temperature pyrolysis, heterogeneous nucleation, and coagulational growth, gives rise to the development of polydispersity. Polydispersity in the system of soot particles manifests itself as a particle number distribution of clusters and as the size distribution of primary particles. This circumstance motivates the need for the distribution function to be introduced into their corresponding equations. The evidence for a polydispersity account is determined also by the strong size dependency of optical cross-sections: as the Raleigh approximation is used for the primary particle cross-section, the scattering cross-sections are related to the sixth power of size variation, and absorption cross-section, to the third.

Farias et al. [51] numerically analyzed the influence of a polydispersity of aggregates onto the cross-sections of particles, introducing the log normal size distribution function $p(N)$ into the expression for calculation of the mean optical cross-section:

$$C_j = \int_{N=1}^{\infty} C_j(N)p(N)\,dN \qquad j = \mathrm{vv(hh), sca, abs} \tag{10.28}$$

The clusters with 16–256 particles and a standard deviation of $\sigma_N = 2$ were used in calculations. The results were compared to those of a monodispersed aggregate $(\sigma_N = 1)$. The most notable deviation from the monodispersed case was found for the Guinier scattering regime, for $C_{vv}(\theta)$ in this range is dependent on the second moment of the particle number distribution $\overline{N^2}$. For the forward scattering, the cross-section exceed the monodispersed case by a factor of 2.4, while the absorption cross-section increased only by a factor of 1.5. The data of numerical modeling also showed that with an increase in the size of the aggregate the effect of a polydispersity gradually vanishes. This finding is clearly due to the transition of scattering in a power-law regime.

The influence of a polydispersity of primary particles to the optical properties of soot clusters was evaluated by imposing the normal law of size distribution. The relative arrangement of particles of various size was not taken into account as this effect is compensated by statistical averaging of a scattered field over the various orientations of aggregates. Note the possible relative deviation of monomer size from the average value does not exceed 20–30%, while the same quantity for the cluster polydispersity can reach hundreds percents. This fact determines the relative insignificance of monomer polydispersity in optical modeling. Thus, displacement of the optical characteristics into an area of higher values is explained exclusively by the increasing mean volume of particles $\overline{v_p}(\sigma_{d_p}, \overline{d_p}) > v_p\,(\sigma_{d_p} = 0; d_p)$. The data obtained for forward scattering, total scattering, and absorption have shown that this effect is determined by the magnitude of the standard deviation σ_{d_p}.

Summarizing, Farias et al. [51] come to the conclusion that the polydisperse version of the theory (RDG–PFA) quite adequately described the optical characteristics of soot aggregates by taking into account both their structural features and effects of polydispersity.

10.3.2. Exact Methods

Until now we considered approximate or semiapproximate methods of calculation of the optical characteristics of ensembles of soot aggregates. It is necessary to remark that only these types of computational techniques are suitable for the solution of inverse problems. Indeed, the processes of interaction of radiation with aerosol systems essentially displays a statistical pattern, and there is no necessity to recover precise coordinates of each monomer in a cluster. For atmospheric optical problems the integral structural parameters, such as a mean gyration radius, the mean size of a monomer, the fractal dimension, the mean optical constants, and statistical characteristics (e.g., cluster size distribution moments) are of vital importance. However, exact methods are necessary as a tool for evaluation of the applicability of approximate techniques and for creation of a reference model, which should include in it all the possible effects of structure features without any limitations.

Among these methods are the Iscander–Chen–Penner (ICP) solution [52,53] and the method of T-matrices [54,55]. Since a detailed description of these methods is completely beyond the scope of this chapter, we shall only mention the basic ideas of the methods.

The ICP method provides an exact calculation of the optical cross-sections of soot aggregates, taking into account the contribution of multiple scattering from the primary particles of a cluster down to the third order, as well as the contribution of an own field of particles (self-interaction term). The ICP method considers aggregates as an array of small cells, where each is a dipole in the Raleigh approximation. The problem of scattering of EM waves by an aggregate of N spherical particles of diameter d_p can be solved by the calculation of the internal field for each primary particle, from a system of $3N \times 3N$ linear equations.

The T-matrix method is based on a construction of a matrix of linear transformations, relating expansion coefficients of incident and scattered fields into vector spherical harmonics. The basic virtue of the method is that the elements of a T-matrix for a given aggregate depend exclusively on its structural properties. Thus, once having calculated the T-matrix for a cluster, it can be used for calculation of a scattered field for any orientation of a cluster and the configuration of an incident wave.

When speaking about exact methods, one should keep in mind this problem, which arises when some comparison of numerical simulation with experimental measurements is needed. To make the theoretical predictions worthy of interpretation, one should provide the averaging over all dispersion parameters matching the system of actual aggregates. Obviously, the best set of model dispersion parameters are those obtained from independent experiment or measured *in situ*. Here again we arrive at the necessity of complete and reliable knowledge of the actual structure of the aggregates under investigation.

For example, Farias et al. [56] analyzed the number of realizations required in relation to the number of primary particles in a cluster, complex refraction index, scattering angles, and diameter of monomers. By conducting an averaging over 64 various aggregates with an identical number of particles, N, and 16 various orientations per cluster, an uncertainty in numerical calculations $<10\%$ could be achieved. The necessity for averaging is chiefly governed by a strong sensitivity of differential scattering cross-sections to the orientation of aggregates on the large angles. A smaller number of realizations is necessary to evaluate the integral magnitude of scattering and absorption.

10.4. EXPERIMENTAL DATA OF SOOT STRUCTURE VARIABILITY

In spite of the fact that the use of the fractal approach and the advances of the exact numerical methods have allowed us to take one more step into the understanding of the optical properties of complex aerodisperse systems, we are still very far from the exact description of an actual structure of carbonaceous soot, and, therefore, its exact optical characterization. The reason is that there is a strong variability of soot aerosol in interaction with the natural environment.

The first researches conducted in this area showed that the problem of simulation of optical properties requires a detailed study of the particle transformation processes, that is, the structural changes that aggregates undergo during their stay in the atmosphere. The cause of this variability lies in the morphological features of soot particles. The soot aggregates represent a metastable structure with a rather weak energy of coagulational contacts, a developed surface, and an extremely low density. It is clear from *a priori* considerations that objects of this type will tend to occupy a more stable energy state under the influence of environmental factors (e.g., temperature and humidity).

Within the fractal approach, structure changing will be exhibited by alteration of the structural parameters R_g, k_f, and D_f. In a number of cases, these changes can be fundamentally significant, limiting the area of the application of a certain theory. In particular, formation of more dense clusters with $D_f > 2$ due to restructuring requires utilization of more elaborate models to take into account the effects of multiple scattering.

For these reasons, there is a need for a detailed study of the particle restructuring effect.

10.4.1. Restructuring of Soot Aggregates

Now, we present some of the results of our experimental study, which were devoted to the solution of this problem, together with data on the scattering and extinction of light obtained for the particles with a various set of structural parameters. In all experiments, the effect of a structure transformation was evaluated by electron microscopic analysis of sampled particles with consequent computer processing of digitized image (see Section 10.2). For all soot types, the particle number distribution $\Delta N/N = f(R)$, the total projection area of a cluster S, the shape anisotropy factor, the radius of gyration R_g, the fractal dimension D_f, and the mean diameter of primary particles d_p were determined. The averaging of parameters was conducted over an ensemble containing hundreds of clusters.

Since the main focus of this chapter is to demonstrate the correlation between optical and structural features of carbonaceous soot aggregates rather than verification of existing optical models of soot we had no need to use an aerosol with strictly determined properties.

On the contrary, the soot aerosol with a maximum resemblance to that arising in the natural and anthropogenic sources seemed to us to be the most interesting subject for research. The technique of soot aerosol production for further modification and study was determined by necessity to have a long-time constant source with a large mass output of soot and a wide adjustment range. Therefore, we have chosen the laminar acetylene burner with an aerodynamic focusing of a flame, whose design was explicitly described by Samson et al. [25].

In Section 10.4.1, first we shall discuss the pure structural effects that arise due to the heating of an aerosol, and in the processes of water vapor condensation on the soot clusters. In Section 10.4.2, we shall outline some optical phenomena related to these structural modifications, which were observed in the experiment.

10.4.1.1. Thermal Effect. For the study of soot clusters restructuring due to thermal heating, we utilized a simple technique, where the aerosol flow was heated in the in-line tube furnace followed by the thermoprecipitator sampling for TEM examination. Figure 10.3 below shows the changing of soot clusters structural parameters with temperature, as revealed by the TEM analysis [57].

First of all, the soot heating leads to a decreasing of the mean aggregates gyration radius from 1.4 μm at 200°C to 0.2 μm at the highest temperatures, as illustrated by Figure 10.3(a). This decreasing can be caused both by the breaking of the large aggregates into smaller fragments and by the compression of the soot aggregates structure.

Figure 10.3(c) and (d) represents the temperature behavior of the soot aerosol fractal dimension D_f and the prefactor k_f. It is evident that the fractal dimension remains practically constant for temperatures below 700°C and then slightly decreases while the temperature rises. The fractal prefactor demonstrates a similar behavior but its decrease at high temperature is more pronounced. These data allow us to conclude that soot aggregates do not exhibit noticeable internal structure modification with heating up to 700°C (though the mean cluster gyration radius is reduced by 70%). The decrease of D_f can be understood in the following way: with fragmentation of the original clusters the percentage of true fractal aggregates decreases, and the small prefractal aggregates dominate in the averaging result. While formally they still satisfy the scaling relationship, the decrease of the fractal dimension indicate that the system is out of the fractal range. These facts, together with the decrease of mean R_g, indicate that the fragmentation process, rather than the compression of clusters, prevails in the thermal restructuring of soot.

With more intensive heating, soot primary particles in aggregates became packed closely together and flaky clusters transform into compact mainly anisotropic agglomerates. Note that the intense heating of soot (> 900°C) leads to the diminishing of primary soot particles caused by the oxidation of carbon from the surface of particles. Figure 10.3(b) illustrates this fact.

The fragmentation of clusters becomes possible due to the variation of force strength, which binds the primary particles together. According to high-resolution TEM data, there are four typical interparticle contacts present in the soot of natural origin (see Fig. 10.4). These types are the coagulational type (A), the coagulational type with an interlayer of an organic substance (B), the heterocoagulational contact between the amorphous carbon particle and the layer of high molecular substance (C), and the phase-type contact (D).

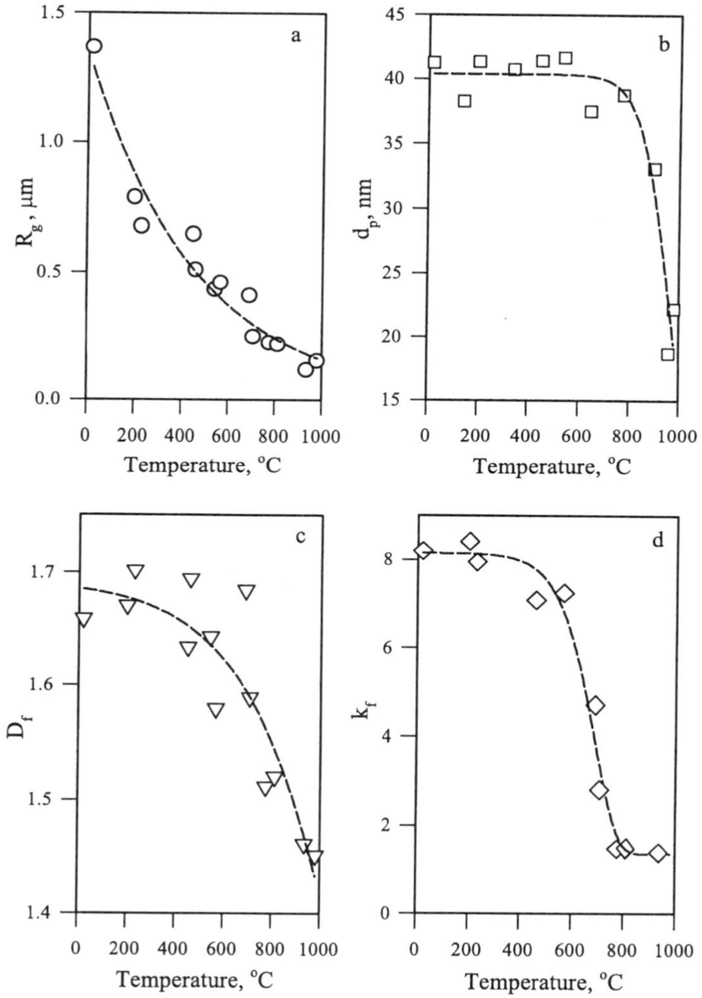

Figure 10.3. Changing of the structural parameters of soot clusters with heating. (a) Gyration radius; (b) primary particle diameter; (c) average weighted fractal dimension, and (d) lacunarity (fractal prefactor).

Coagulational contact arises in the touching of carbon primary particles and appears to be of the most general type [Fig. 10.5(a)]. In this type of contact, the binding force depends on the particle shape and temperature. In the highly heated environment, the surface diffusion of the substance leads to the sintering of particles, with the transformation of the point contact into a surface one. In the ultimate case, the strong phase contact arises when there is no phase discontinuity between the particles. At an early stage of growth, the mutual diffusion of the substance in the contact area is accompanied by the contact zone overgrowing with condensed

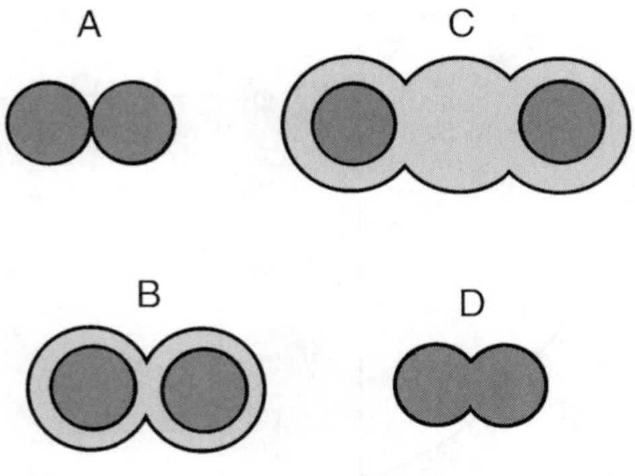

Figure 10.4. Types of contacts between the particles comprising the soot aggregate. In order of increasing strength of contact: A is the coagulational type, B is the coagulational type with the interlayer of an organic substance, C is the heterocoagulational contact between the amorphous carbon particle and the layer of high molecular substance, and D is the phase contact.

carbon, thus making the phase contact even stronger (type D). At the same time, the coagulational contact appears to be considerably weakened by the organic compounds, which are adsorbed on the surface of primary particles (type B). The high molecular resinous interlayer with a comparatively weak force of surface tension reduces the interparticle binding area approximately by an order of magnitude, providing additional flexibility to the cluster branches.

The effect of bond loosening is even more pronounced when the bond contain a heterogeneous contact (type C). Clusters with this type of contact arise at long residence times by the cluster–particle mechanism of aggregation, and are very unstable to the influence of ambient factors. Fragmentation takes place at arbitrary temperatures and leads to formation of funiculus, followed by a complete break-off (Figs. 10.5(b) and 10.6). These are the contacts responsible for the fragmentation of clusters at arbitrary temperatures and hence, the decrease of R_g.

Thus, depending on the nature of the interparticle contacts, the mechanical strength of the aggregates of soot, and consequently, their stability to environmental effects will strongly differ. Obviously, soot aerosol comprising clusters with the prevailing B and C types of interparticle contacts will be more liable to the effect of external forces.

10.4.1.2. Restructuring under the Action of Capillary Forces. For a better understanding of the influence of carbon particles on atmospheric processes the effect of transformation of soot aggregates in the environment of a condensing water vapor is of special interest. The influence of humidity on optical properties of soot particles have been reported earlier (see [58–60]). In particular, Colbeck et al. [10] showed that in the environment of saturated and supersaturated vapor, the fractal dimension of soot clusters D_f increases from 1.78 to 2.0–2.5 as a result of compaction of aggregates due to capillary condensation.

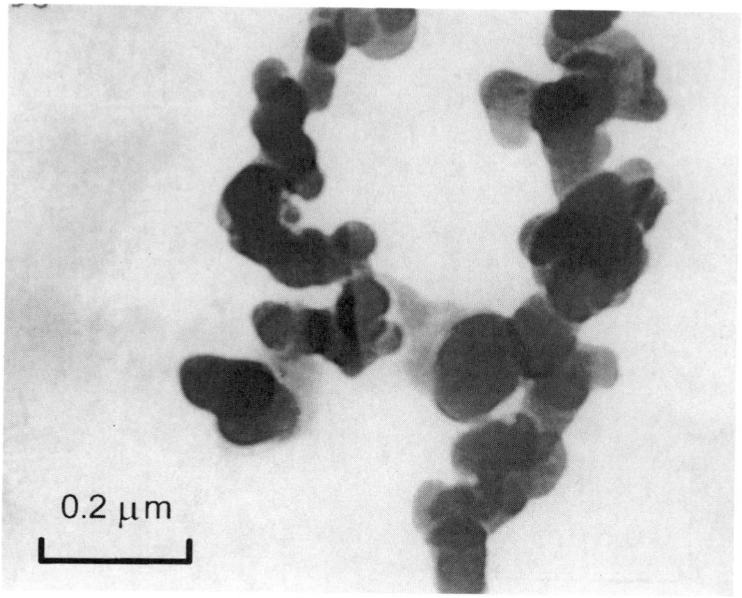

(a)

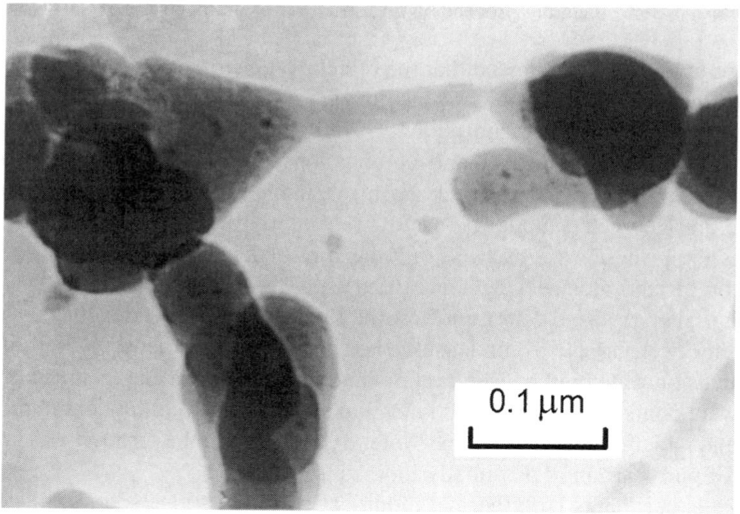

(b)

Figure 10.5. TEM images of interparticle contacts, strong magnification. (a) Contact of the coagulational type and (b) heterocoagulational contact, image taken in the instance of breaking.

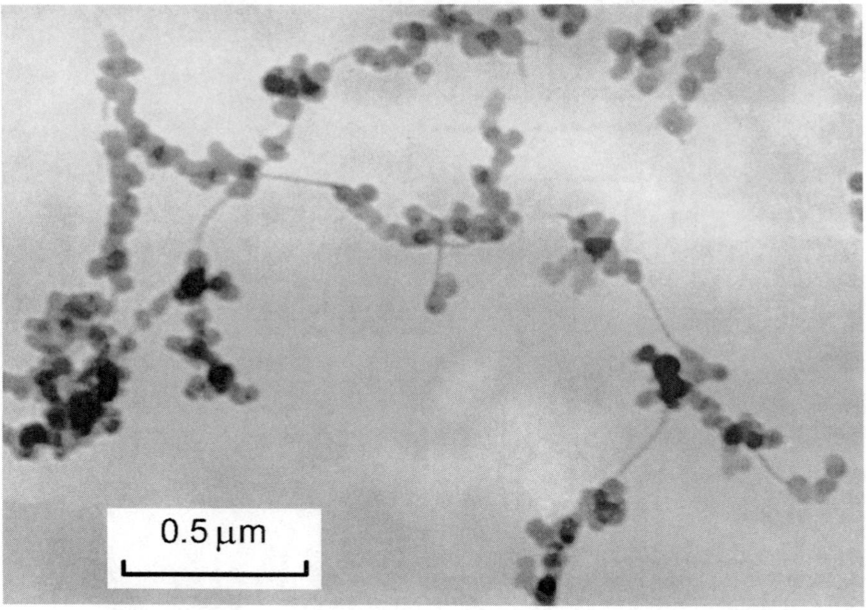

Figure 10.6. Fragmentation of soot aggregates, going with the breaking of heterocoagulational contacts. Bond breaking proceeds with formation of filaments.

Recent investigations showed that the effect of the structure modification of soot particles under the action of condensing water vapor is immediately connected to their hygroscopicity and morphology. Both factors are directly related to the composition of the initial fuel and combustion conditions. As natural soot is a complex system, not only structurally but also chemically, objects with a controllable structure and geometry should be used for the study of restructuring mechanisms. This was the principle we used (within the limits of practical applicability) in the realization of laboratory experiments [61,61a].

Our initial goal was to determine how the restructuring effect depends on a degree of particles wetting by a condensate. For this purpose, the set of comparative experiments on water and benzene vapor condensation on the fractal clusters of pure carbon were conducted. Benzene is known to be a lyophilic component in relation to carbon and moistens it well, whereas water exhibits lyophobic properties, that is, the surface of pure carbon is practically inert to a water.

Unlike experiments with thermal modification of acetylene soot, in this case we used the arc generator of soot particles, where a material of pure graphite electrodes vaporized into the flow of an inert gas (He) with the following condensation and coagulation, this generating a desirable quantity of disperse carbon. The resulting clusters are much the same as those produced in the acetylene burner or open oil fire. The advantage of this technique is that the resultant clusters consist exclusively of pure carbon, without additives of high molecular organic and mineral substances. Clusters produced in the generator were directed immediately into the thermally

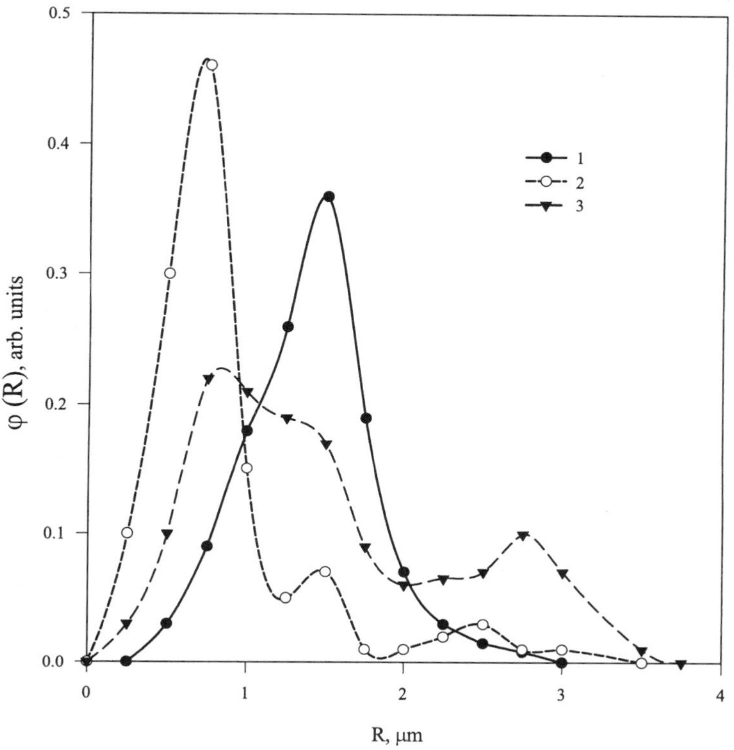

Figure 10.7. Evolution of size spectrum of pure carbon clusters subjected to the action of benzene and water vapor. Curve 1 is the initial size distribution; Curve 2 is the size distribution after benzene action (saturated benzene vapor); and Curve 3 is the size distribution after water action (supersaturated water vapor).

stabilized reaction chamber, where they were mixed up with the vapors of water or benzene, which was poured on the bottom. The reaction chamber was a glass flask connected to the in-line cooler downstream. By changing the liquid temperature, the vapors partial pressure in plenum could be adjusted to the desired value, thus changing the saturation magnitude in the cooler. Vapor condensation on the particles and the corresponding structure modification took place in the cooler zone at a temperature of $\sim$ 293 K.

Figure 10.7 displays the size distribution functions of clusters, reflecting the effect of water and benzene vapor condensation on an aerodisperse system of pure carbon clusters. A spectrum of the sizes of aggregates after condensation of the vapor undergoes noticeable change, which is clear from the reduction of the characteristic size of the clusters.

One can see from comparison of the curves that the transformation of size spectrum is expressed more clearly in the benzene vapor than in water. In particular, the modal size of clusters decreases approximately by a factor of 2 after condensation

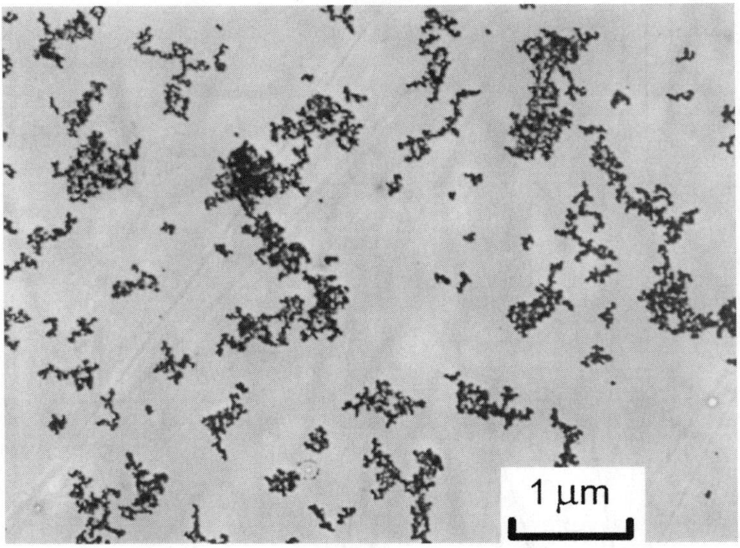

(a)

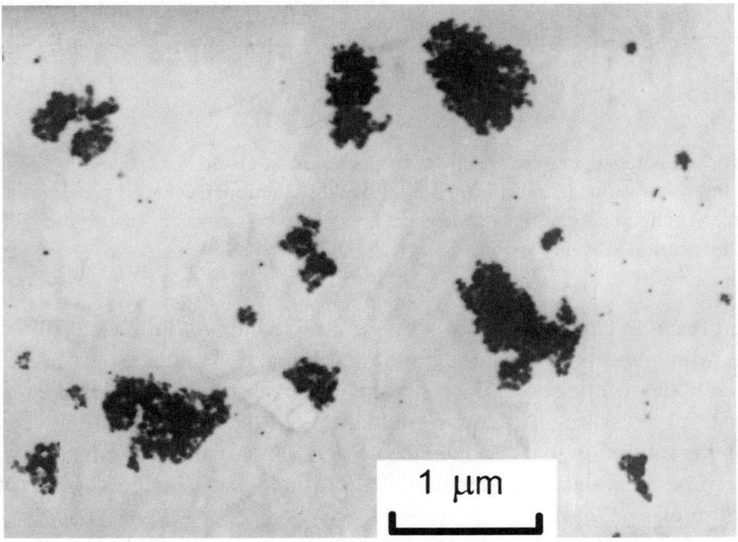

(b)

Figure 10.8. Restruction of pure carbon aggregates in the environment of benzene vapor. (a) The restructuring effect after the action of a low saturated benzene vapor with a thermostat temperature of 60°C; (b) the restructuring effect after the action of a saturated benzene vapor with a thermostat temperature of 80°C.

of benzene and only by a factor of 1.3 after condensation of water. The structure changes of soot aggregates can be traced visually on microphotographs of particles shown in Fig. 10.8(a) and (b).

Note that the changes observed in a spectrum happen mainly due to compression of large aggregates. The compression starts with formation of dense almost spherical globules, consisting of densely packed primary particles and situated inside the branched cluster structure. If further condensation proceeds, the whole cluster will be involved in the process of collapse, evolving into the single dense globule. The action of surface tension arising with condensation of the liquid phase in the cluster's inner part is supposed to be the cause.

A remarkable fact is that despite the practically absolute passivity of pure carbon to water, and, therefore, the extremely low efficiency of such aerosols as condensation nuclei, the clusters also exhibit dense globule formation in the water vapor environment. This phenomena indicates the existence of specific active centers reducing the work needed to form a germ of a drop, and thereby increasing the probability of vapor condensation [62]. By taking into account that the initial stage of soot generation takes place in the ionized media, it would be reasonable to suppose that such centers are the electrical charges localized inside carbon clusters [18]. Formation and distribution of the liquid phase in such a system, and consequently, its resultant effect, depend both on the structure of the clusters, and the charge distribution pattern in them. Microphotographs of large clusters with a pronounced effect of charge presence are shown in Fig. 10.9.

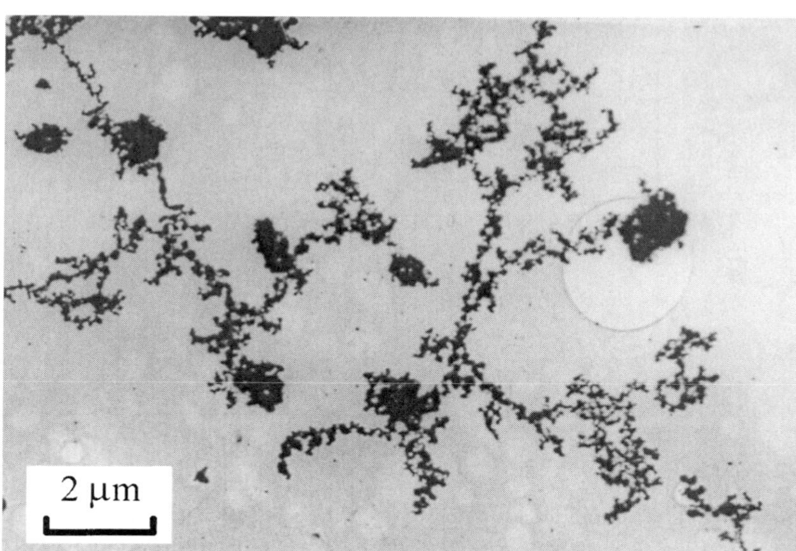

Figure 10.9. Restructuring of pure carbon aggregates in the environment of a supersaturated water vapor. Note the globules forming in the sites of charge localization.

Results concerning carbon clusters behavior in the benzene environment deserve a more detailed consideration. The data of Fig. 10.8 show that the compressive effect of benzene vapor condensation is significantly higher than that of water vapor. This finding is well understood in terms of the surface phenomena theory: The process of formation of a new benzene droplet on a wet surface does not require passage through a potential barrier. Also, because there is no need to condense nuclei, the process starts in several points of the cluster simultaneously. While growing, droplets coalesce into larger ones, covering the entire volume of the cluster, compressing its structure into one dense globule of a more or less spherical form. When considering condensation on the lyophilic surfaces one has to remember that a soot cluster exhibits a highly developed system of micropores, and their filling with the liquid phase begins when the partial vapor pressure is still under the critical value of saturation. It is possible that the restructuring effect of the capillarity will manifest itself in this range of pressures. To verify this hypothesis, a thin layer of carbon clusters deposited on an organic film was placed into sorbtograph (ADS-1B, Japan), where the preset vapor pressure was maintained for 1 h, beginning with the moment adsorption equilibrium was established. The result of benzene vapor action on the cluster structure is visualized by a shift of the fractal dimension distribution measured at two different values of the partial pressure of the benzene vapor, as shown in Fig. 10.10. The shift of the spectrum to the higher values of D or, what is the same, the densification of the internal structure of clusters, occurs when the relative partial pressure of benzene vapor approaches the value of 0.8 and proceeds with pressure gaining.

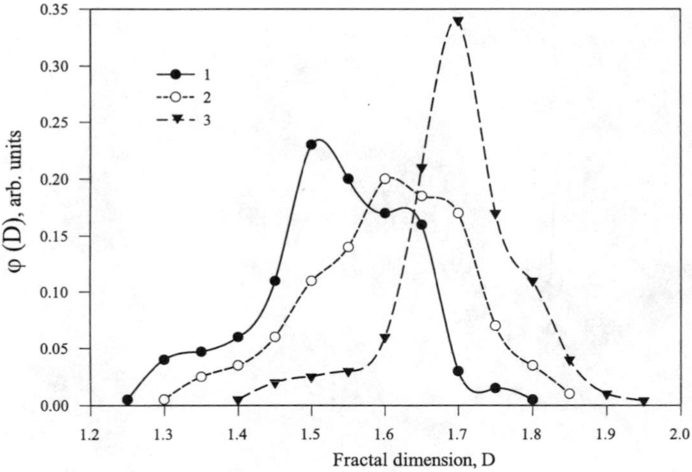

Figure 10.10. Evolution of the spectrum of a fractal dimension for the soot aggregates exposed to the action of benzene vapor. Curve 1 is the initial spectrum; Curve 2 is the spectrum after the action of a low saturated benzene vapor, with a partial pressure of 0.8; and Curve 3 is the spectrum after the action of a saturated benzene vapor with a partial pressure of 0.99.

According to Kelvin's formula, the partial pressure interval 0.8–0.99 is responsible for the hysteresis of the capillary condensation in the pores of the effective size range of 20–400 nm. Apparently, structural changes in aggregate are to take place on the same size scale. Besides, soot clusters are very sparse and thus provide an easy access to the entire surface, which make possible the simultaneous compression of the structure through the whole volume of the cluster. Flexible linear bonds with a packing index 2 are expected to break apart. This will result in integration of the neighboring branches into multiple-connected chains. The transformation leads to the increment of packing index and gradual strengthening of structure. When the balance between the elastic and capillary forces is achieved, the restructuring process comes to a halt. Finally, the branches transform from the loose chains of monomers into more compact and more stable formations of closely packed particles (see Fig. 10.11). With the growth of the partial vapor pressure, the larger parts of the cluster become involved in the deformation process, making the whole aggregate more and more compact. This very effect is demonstrated by the shift of the fractal spectrum to the region of higher values of D with the pressure growth (see Fig. 10.10). In the ultimate case of supersaturation, the capillary condensation induces formation of compact globes with the projective fractal dimension equal to 2 (Fig. 10.8(b)).

At the same time, no restructuring effect was found in similar experiments with unsaturated water vapor, which confirms the significance of the surface lyophility factor in condensation processes on fractal structures.

A soot particles produced in hydrocarbon fuel combustion are mainly a bicomponent system consisting of elementary and organic carbon, it is natural to expect, that the chemical nature of an organic phase will influence the wetting effect, and consequently, the processes of cluster restructuring. The ratio of organic (OC) and elemental (EC) carbon in a flame depends on the kind of fuel and the conditions of its combustion. In particular, the data of Hildemann et al. [63] yields the average value of the OC/EC ratio for diesel soot equal to 0.80, and for smoke formed at the combustion of foliaceous and coniferous wood accordingly, 9.3 and 16.9.

In practice, the chemical structure of soot aggregates becomes inhomogeneous from the moment of their formation, when the products of the incomplete combustion of fuel (aliphatic and aromatic hydrocarbons with a molecular mass 300–500 g/mol) are being deposited on a surface of primary carbon particles [6]. To evaluate the impact of the organic fraction of soot on the restructuring effects, we have subjected the condensing water vapor to the soot samples obtained by burning off different hydrocarbon fuel: methane, petroleum, and coniferous wood. For each sample, the simultaneous measurements of adsorption capacity were executed in order to evaluate their hydrophilic properties.

Shown in Fig. 10.12 are the particle size distribution spectra, measured after the condensation of water vapor on the in-line operated setup with a carrier flow of 2 L/min. This corresponds to $\sim$2s passage time inside the condensation zone. Apparently, most significant changes of size spectrum occur for the soot of natural sources. At the same time, a very slight displacement of the spectrum in the case of pure carbon aggregates is observed, though the data of Figure 10.9 shows that some special restructuring in the inner structure of the aggregate does occur.

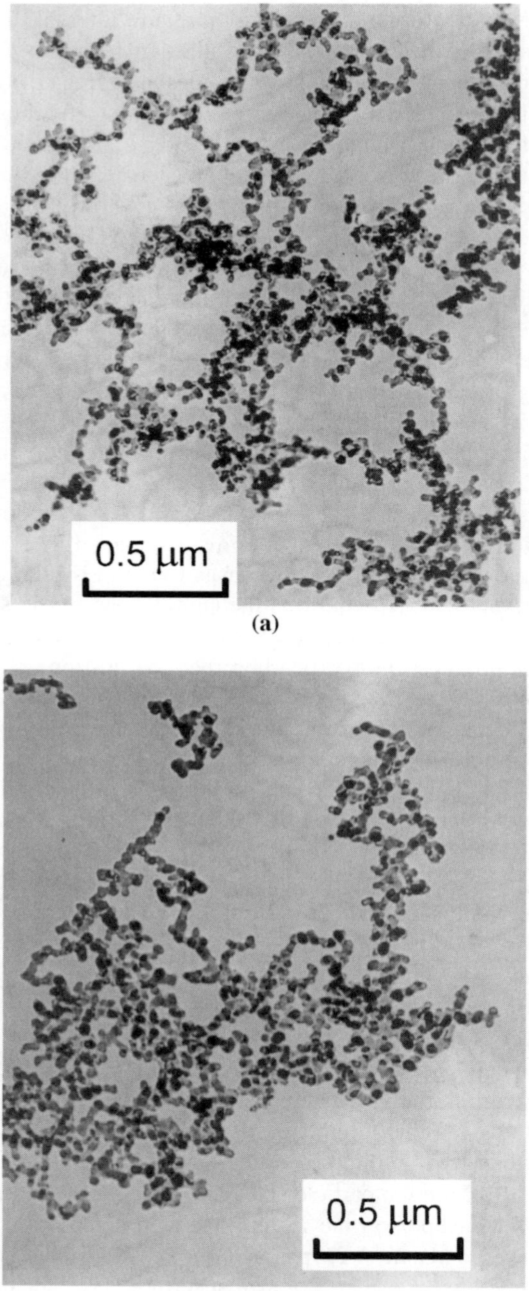

(a)

(b)

Figure 10.11. Visualization of the carbon cluster structure after the long-term action of a saturated benzene vapor, whose partial pressure is 0.99. Note that both images depict the same cluster. (a) Before action and (b) after action.

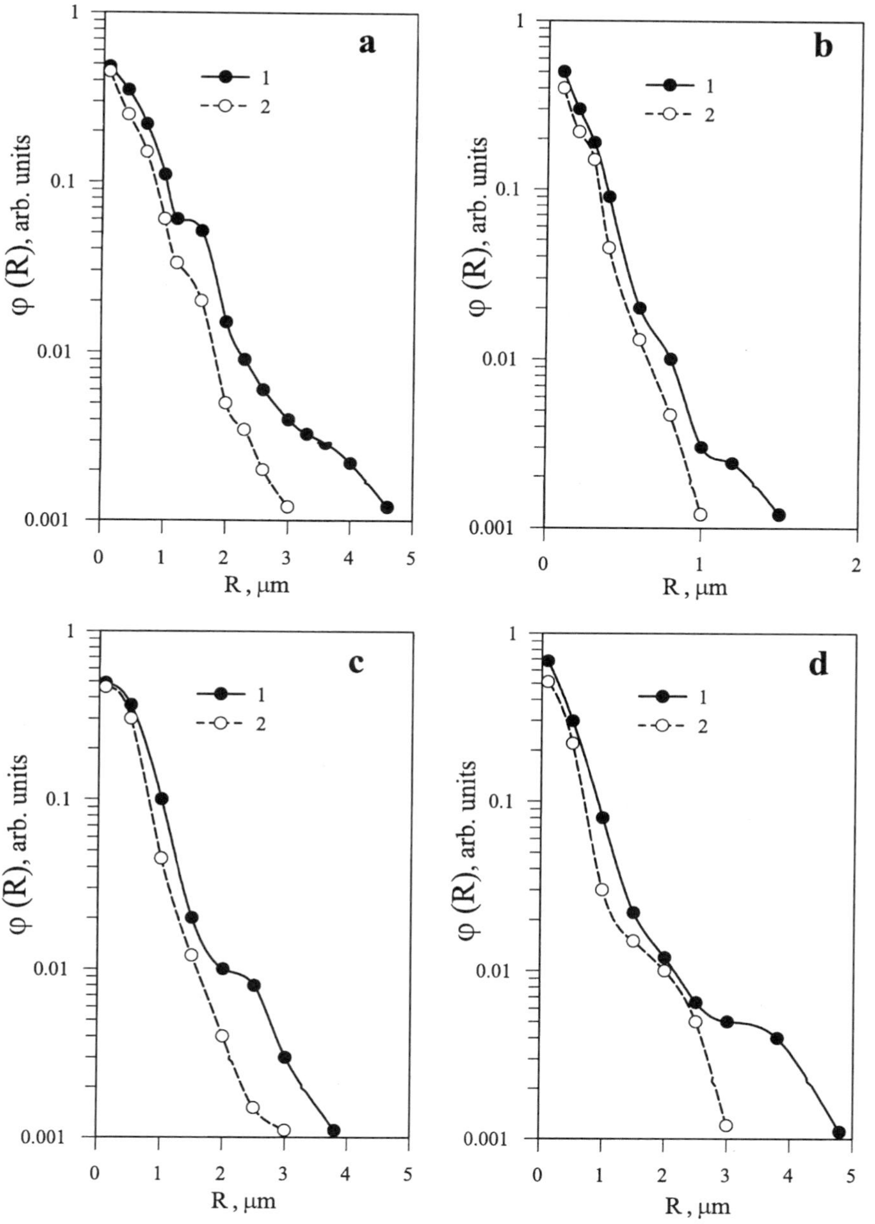

Figure 10.12. Evolution of the size spectra of aggregates of different soots exposed to the short-term action of water vapor. In all cases, curve 1 corresponds to the initial size distribution and curve 2 corresponds to the resulting spectrum. Soot obtained by burning: (a) pine wood; (b) natural gas; (c) oil, open fire; and (d) pure carbon evaporated in a He environment.

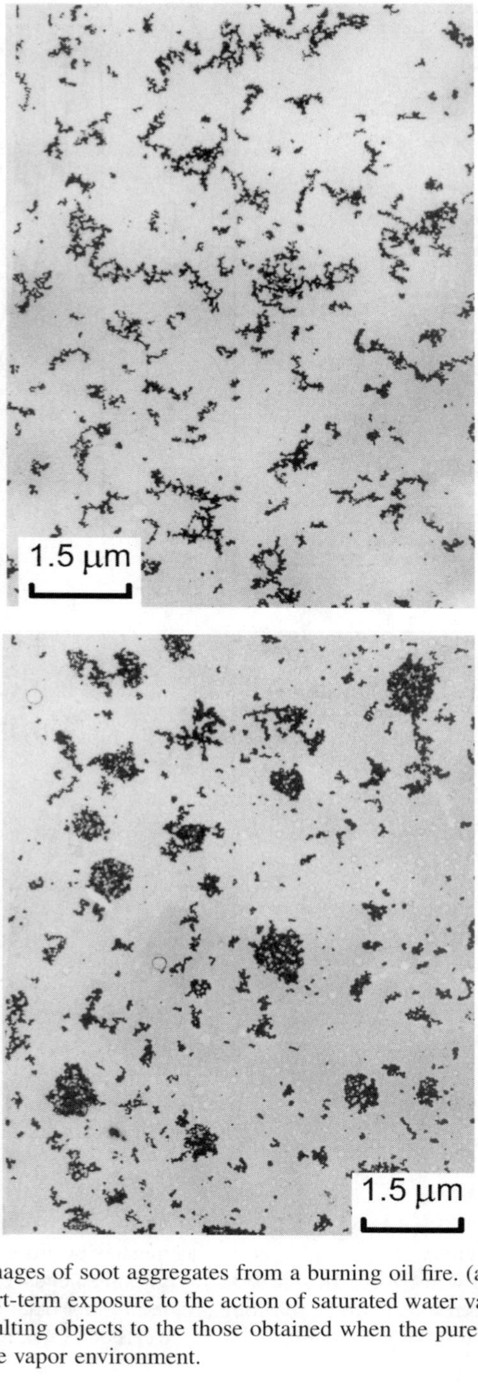

Figure 10.13. TEM images of soot aggregates from a burning oil fire. (a) Initial aggregates and (b) soot after a short-term exposure to the action of saturated water vapor. Note the close resemblance of the resulting objects to the those obtained when the pure carbon clusters are modified in the benzene vapor environment.

TABLE 10.2. Structure Parameters of Soot Aggregates Before and After Water Vapor Action

	Fractal Dimension D_β		Mean Size R(μm)	
Type of Fuel	Before Action	After Action	Before Action	After Action
Wood	1.76	1.88	0.59	0.47
Oil	1.72	1.85	0.53	0.40
Pure carbon	1.74	1.76	0.45	0.45
Natural gas[a]			0.19	0.19

[a] The fractal dimensions were not recovered due to insufficient data.

One can see that substantial displacement of spectra into the region of smaller sizes occurs even for short reaction times. Table 10.2 presents the corresponding microphysical characteristics of soot particles. The average effect is demonstrated by growth of fractal dimension, which, in conjunction with reduction of the mean size of aggregates, denotes the densification of clusters. For soot of natural origin, that is, generated in open fires of oil, wood, and gas, the effect of size reduction reaches 25%. The modification of soot aggregates obtained from the open oil fire can be visually traced in the microphotographs of Fig. 10.13. Note the close resemblance of resulting objects to those obtained when the pure carbon clusters modify in the benzene vapor environment. This fact obviously proves that lyophility properties of the surface of a soot particle play a dominant role in restructuring processes.

Special attention was drawn to the fact that the compression of aggregates is followed by the reduction of the clusters' anisotropy. The data of Fig. 10.14 reveal the fact that this effect is most prominent for the soot clusters of 1-μm size and forms larger clusters in natural sources, which obviously supports the conclusion concerning its high liability to structure transformation.

Capillary induced restructuring can be realized only for wettable particles, that is, for particles able to hold the liquid phase on the surface. A known technique to reveal the pattern of surface–vapor interaction is the measurement of the adsorption capacity, which is characterized by the formation of an adsorption isotherm. Correspondent adsorption–desorption isotherms for the different types of soot are shown in Fig. 10.15.

A comparative study of the curve shows that the quantity of vapor adsorbed by samples of natural soot is almost an order of magnitude higher than that adsorbed by pure carbon samples. The presence of carboxylic and hydroxylic groups containing oxygen in the resinous additives of natural soots, enhance the interaction energy of vapor molecules with the particle surface, and therefore enhanced the wettability of a surface [64]. Thus, the adsorption capacity is the most characteristic factor of the soot aggregate restructuring liability.

By comparing the thermal and condensational effects of structure change, one can see that they are of a completely opposite nature. While the aggregate exposed to the thermal heating breaks into fragments with a strong anisotropy, the particles experiencing the action of the condensing vapor become partially or completely

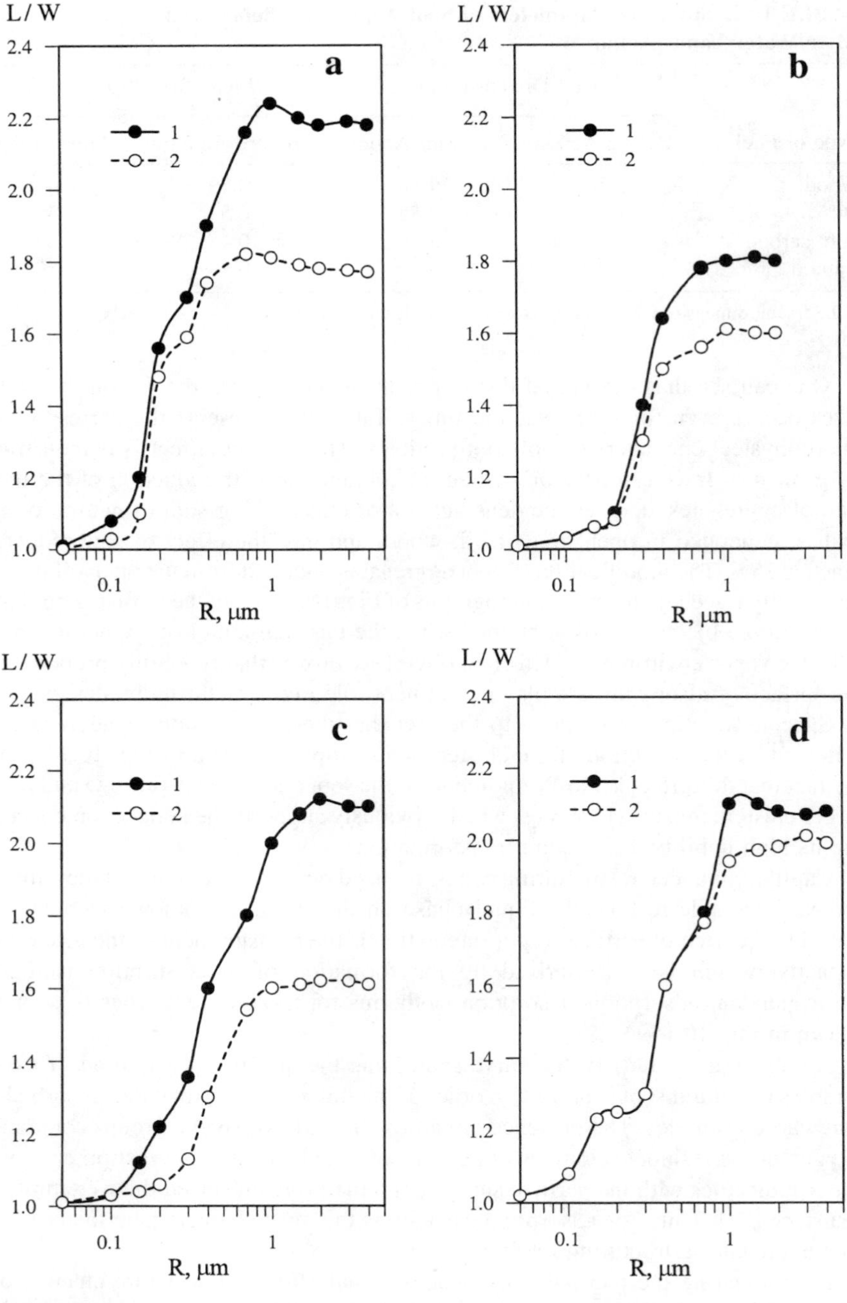

Figure 10.14. Dependence of the cluster anisotropy ratio on cluster size. Soot obtained by burning (a) pine wood, stove; (b) natural gas; (c) oil, open fire; and (d) pure carbon evaporated in a He environment.

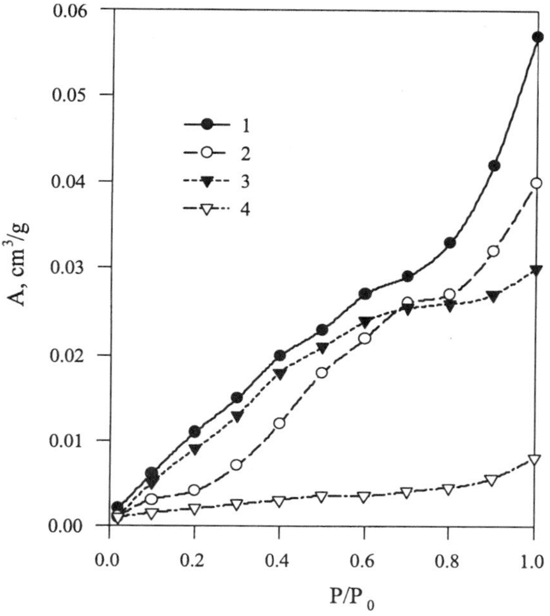

Figure 10.15. Isotherms of adsorption of water on the bulk sample of soot of a different origin. Soot obtained by burning: curve 1 is pine wood, stove; curve 2 is oil, open fire; curve 3 is natural gas; and curve 4 is pure carbon evaporated in a He environment.

collapsed structures with considerably low anisotropy. The resulting clusters will have essentially different physical and optical properties. We arrive at the conclusion that in order to predict the optical properties of the soot aerosol system that arises in some natural or anthropogenic eruption (forest fire, oil plant fire, etc.), we have to take into account the dynamic change of environmental conditions, that is, temperature, humidity, presence of clouds, and so on. Structural properties will never stay constant in real atmospheric conditions. This finding will have a crucial effect on the optical properties of soot aggregates and will be demonstrated in Section 10.4.2.

10.4.2. Optical Measurements of Modified Soot Aggregates

In general, laboratory measurements of the optical characteristics of a soot aerosol have two main objectives: validation of theoretical and empirical optical models and determination of the structure parameters by optical methods. Another objective can be classified as an accumulation of experimental data in order to establish certain relations between a structure of aggregates and their optical properties. Next, we shall consider some features of laboratory optical measurements concerning the aerodisperse systems.

The validation of theoretical models is possible only if there is a stable source of a soot aerosol delivering clusters with well known, reproducible, and preset structural parameters. It is also desirable that the system of aggregates not be too polydisperse, or at least that all structral parameters not be scattered over a wide range of values. This condition is associated with the following circumstances: a majority of models successfully describe the variation of optic characteristics when only one structure parameter (e.g., the number of monomers N) is being adjusted, while simultaneous alteration of two and more parameters will essentially obstruct the calculation and will make the data interpretation ambiguous. A series of works by Köylü and Faeth [39,47,48] and Dobbins et al. [65], presents a thorough investigation of the subject and should be recommended for a better insight into the problem.

Determination of structure parameters by optical methods represents a class of inverse problems of aerosol optics. Historically, these problems were devoted to the determination of the mean geometric characteristics of aerosol particles, and included, apart from the measurements of optical cross-sections, determination of a backscattering coefficient, a degree of light polarization, an angle distribution of scattered light, and also spectral measurements. Recovery of structure characteristics and optical constants were made either with the help of numerous empirical regularities detected as a result of large data sets processing, or by means of Mie theory and its computing approximations.

Since the "discovery" of the aggregate and fractal-like nature of the soot particles, the circle of problems in this field was narrowed down to a problem of recovery of the fractal dimension from the plot of intensity of scattered light versus the wavevector q in double logarithmic coordinates. The use of the relationship $I(q) \sim q^{-D}$, as was already mentioned in connection with ALS techniques, implicates the supposition about a negligibility of multiple scattering, which is not justified for a general type of soot aerosol, especially in the case of dense clusters with a mean fractal dimension $D_f > 2$. For an almost ideal system of fractal aggregates, the measurement of an angle light distribution allows us to find the mean R_g from the point on the plot where the Guinier scattering regime $qR_g < 1$ crosses over into the fractal law regime $qR_g > 1$ [29,66]. In practice, the nonideality of the fractal system (appreciable presence of nonfractal objects) leads to a strong uncertainty of this point location. The transition from one scattering regime to another becomes smooth and indistinct, or cannot be identified at all.

The scope of Section 10.4.2.1 covers the demonstration of the effect of nonideal fractality and variability of physical and chemical properties of real carbonaceous soot aggregates. As was pointed out, two ultimate types of objects evolve as a result of soot cluster modification under the action of ambient factors. These are the small chain structures, resulting in the thermal fragmentation of original clusters, and the dense spherical globules resulting in processes of vapor condensation. The experimental evidence for transition of optical properties from those of the original developed clusters into optics of marginal structures is presented in Section 10.4.2.1.

10.4.2.1. Scattering and Extinction of Light by Soot Clusters after Thermally Fragmented Clusters. The experimental technique for thermal modification was described above, with the circumstantiation that the measurements were conducted with soot from an acetylene burner as the only reliable steady source of long operation. The whole setup is comprised of an acetylene burner, a buffer chamber, an in-line furnace, a polar nephelometer, and a sampling device. The polar nephelometer was able to operate in the in-line regime, thus providing nonstop measurements of angle scattering along with the structural transformation. A simple procedure allowed us to rebuild the free scattering volume into the cell for measurements of extinction.

To eliminate the effect of the incident beam on the detector, the diagrams were measured from 10°. The measurements were performed at the wavelength 435 nm and utilized vertically polarized light to simplify the data procession. Reduced differential scattering cross-sections corresponding to the different temperatures of the furnace are presented in Fig. 10.16. The scale of the axes is chosen so that the difference between the scattering on the soot aggregates with different internal structure parameters becomes more evident.

A linear section can be clearly detected on each curve, the section where angular distribution of scattered light obeys the $I(q) \sim (qR_g)^{d_f}$ law. We deliberately change the notation here to express that d_f is not the *mass fractal dimension*. The power of the exponent retrieved from this curve appeared to be practically constant and equal to $d_f = 2.1 \pm 0.1$ for soot aggregates subjected to temperatures $< 780°C$, and to $d_f = 2.6 \pm 0.1$, when heating temperature $> 900°C$. Like the mechanisms of soot restructuring during thermal attack discussed above, it means that fragments of clusters preserve the fractal nature up to a certain limit, although the size of the

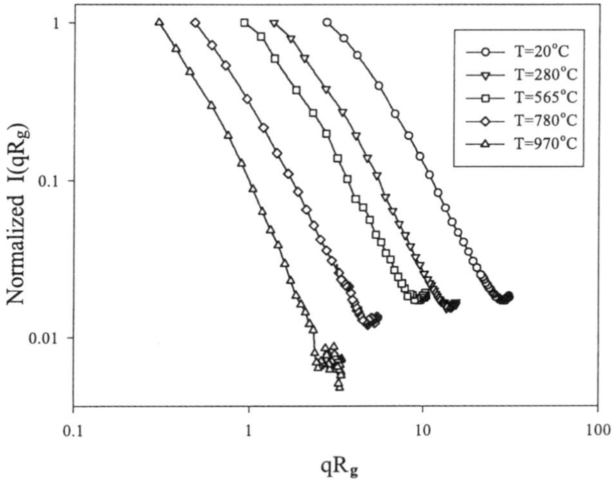

Figure 10.16. Reduced differential scattering cross-sections for the soot clusters subjected to heating at different temperatures.

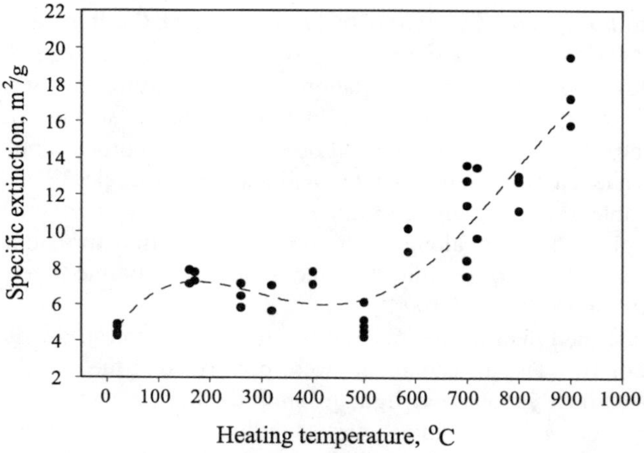

Figure 10.17. Changing of the specific extinction cross-section of the heated soot aerosol.

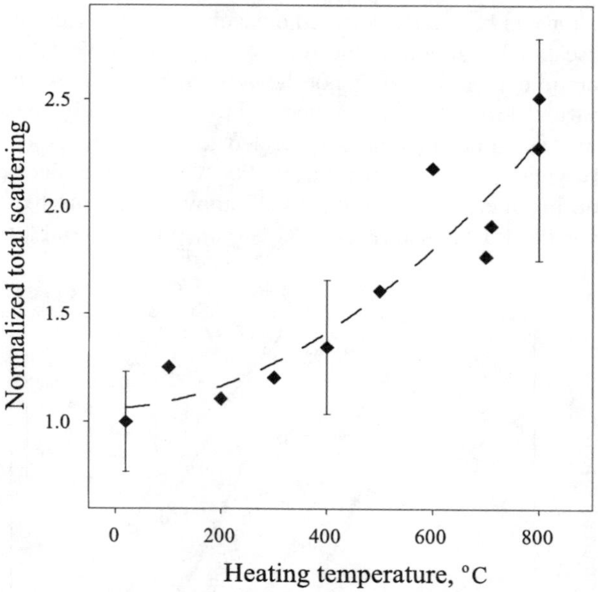

Figure 10.18. Changing of the total scattering cross-section of the heated soot aerosol.

clusters and the number of primary particles in them vary over a wide range (see Fig. 10.3).

Synchronous measurement of light extinction in the laminar aerosol flow demonstrated the steep rise of a specific extinction cross-section from 7 to 15 m^2/g. The anomalous behavior of the extinction is supported by the measurements of the total scattering into the 10°–150° range of the scattering angle, which is normalized by the mass of the scattering soot (see Fig. 10.17 and 10.18). Apparently, soot cluster disintegration clears the way for an incident wave to irradiate the previously inaccessible internal parts of the cluster, so that more of the primary particles become involved in the absorption–scattering processes.

10.4.2.2. Scattering of Light by Soot Aggregates Subjected to the Condensing Water Vapor. The data of soot clusters restructuring in the environment of saturated water vapor, as discussed above, provide a mean for controlling the alteration of soot aerosol particles that can be used for the optical measurements of modified soot.

In order to investigate the processes of a condensation restructuring of soot particles, it is obviously necessary to have the possibility of controling a degree of the supersaturation or undersaturation of water vapor and the resident time of particles in the zone with a high water vapor content. For the optical measurements of a modified aerosol, it is also desirable to have a flow operating system that, unlike the close-cycled system, is not subject to the effects of a natural relaxation.

To conduct water condensation on the soot particles, the flow diffusion nucleation chamber (FDNC) was designed, which operates on the principle described in detail by Vohra and Heist [67]. Its design was slightly changed to make it suitable for the large soot aggregates. Soot aerosol, produced in the acetylene burner, mixed with water vapor in a certain proportion, first pass through the preheater and then through the tubular nucleation chamber, whose walls are cooled down with the vapor of the liquid nitrogen. Water vapor was produced by the bubbling of clean air through water at a certain temperature. By varying the water temperature, one is able to regulate the initial vapor concentration. The cooling of the vapor–aerosol mixed stream results in supersaturation of the vapor in a region downstream of the chamber inlet, and therefore abrupt condensation of water on the centers of nucleation provided by soot aggregates. In the stationary regime, the constant humidity distribution is being set inside the cooler, with effective humidity (i.e., calculated on the assumption of the absence of particles) reaching as much as 200% in the coaxial zone. By adjusting the temperatures of the evaporator, preheater, and cooler, the desired mean value of supersaturation can be achieved. The exact profile of the initial saturation can be calculated analytically by cooperatively solving the equations of heat transfer and diffusion of vapor.

Dried aerosol was dispatched into the cell for attenuation measurements or into the scattering volume for angle scattering measurements. To eliminate the possible condensation of residual water vapor on the cell's windows, the latter were slightly heated. The angle scattering measurements were carried out by registering the intensity of light scattered from the vertical jet of the aerosol issued into the open air, and then collected on the fiberglass filter for mass analysis. The attenuation and scattering were measured for a wavelength of 435 nm of a vertically polarized collimated light beam. All optical measurements were accompanied with the sampling of aerosol particles by a thermoprecipitator for the TEM analysis.

The result of the interaction between the supersaturated water vapor and the soot aggregates is manifested in the modification of the aggregates' structure, that is

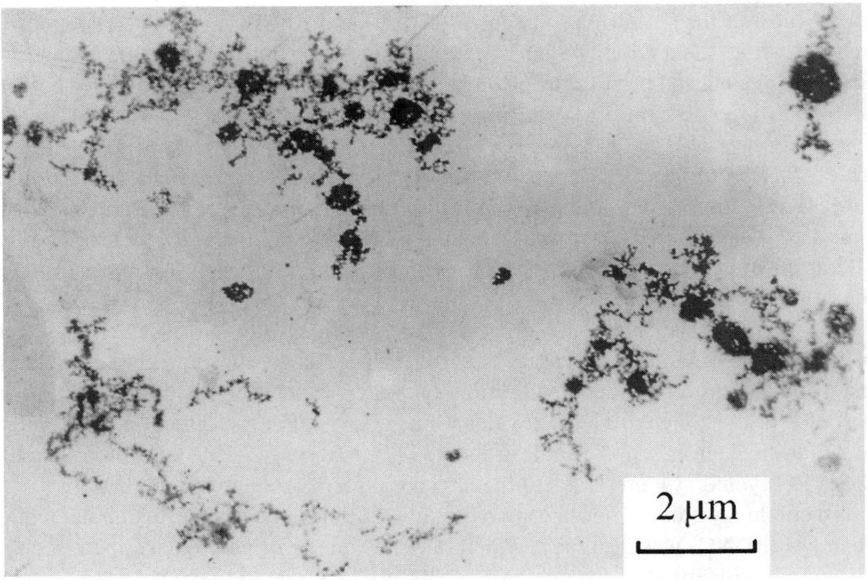

(a)

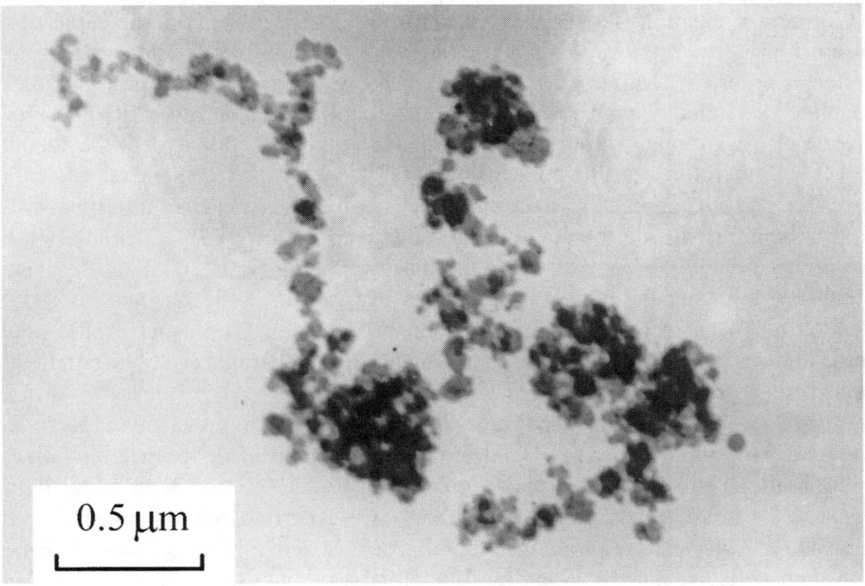

(b)

Figure 10.19. Acetylene soot aggregates after the action of water vapor in the FDNC. Compact globules correspond to the sites of droplet condensation. (a) A general survey of the aggregates after evaporation of condensed water and (b) a close view of the globule. It is evident that each globule consists of the primary particles, collapsed as a result of the action of capillary forces.

clearly seen from the photographs in the Figure 10.19. The dense spherical objects (globules), which correspond to the sites of droplet nucleation, are well resolved on the image. The collapse of the initially rarefied cluster's structure under the action of capillary forces during nucleation and subsequent evaporation, is seen as a mechanism leading to the globules formation. Still, the size of the globules is generally smaller than the size of an aggregate, though there are exclusions, for examples, when the aggregate serving as a nucleation center is small (consisting of a few primary particles). This means that under these conditions the droplets are not growing to the sizes, where they incorporate the soot aggregates. These conditions could be realized when the soot particles are serving as nucleation centers, but are not externally mixed with droplets containing atmosphere.

The visual survey of the restructuring pattern reveals that the globules grow in almost the same way as for the pure carbon aggregates (Fig. 10.9). This is so because the acetylene soot is reported to contain a very small amount of high molecular hydrocarbons, and this is supported by the general view of the restructuring pattern.

Still, note that in the case of acetylene soot the condensation starts at the earlier stages of humidity growth, while for pure carbon soot, full-scale supersaturation was needed to give yield a globule formation. The possible explanation of this disagreement is that acetylene soot contains some, though very small, amount of adsorbed oxygen or oxygen containing lyophilic OC [68]. Progressive densification can be explained by condensation inside the small-scale pores and the sites of the chain junctions. Therefore, though restructuring starts in the same way as in the case of the natural soot with a high OC content, its efficiency is rather low, and does not provide a complete collapse of an aggregate. At the same time, several repetitions of the condensation–evaporation cycle will probably accomplish the process of densification. Apparently, the more wettable the particles are, the lesser number of cycles will be needed.

Figure 10.20 depicts the size distribution of globules and the comparative change of the soot clusters size distribution after the condensation–evaporation cycle took place. One could argue, however, that the conditions imitated in the FDNC are not representative in the atmosphere. Indeed, this high supersaturation for the short resident time is never achieved in natural conditions. On the other hand, the low density and flake-like structure of soot aggregates provide for the long resident time of the particles in the airborne state—for ~ 1 week. For that period, the slow condensation under low supersaturation (of $\sim 1\%$) could be reproduced several times, with the result that it is thought to be very close to that obtained in the FDNC. From this point of view, the simulated process could be regarded as a step in the process of soot ageing in the atmosphere due to its interaction with the water vapor environment. Finally, this process will result in complete densification of airborne soot particles, which are often observed in the samples taken from urban air.

Apart from the question of whether they constitute a significant fraction of soot aerosol, these objects will obviously differ in their optical characteristics from the initial fractal clusters due to the presence of compact globules. The corresponding

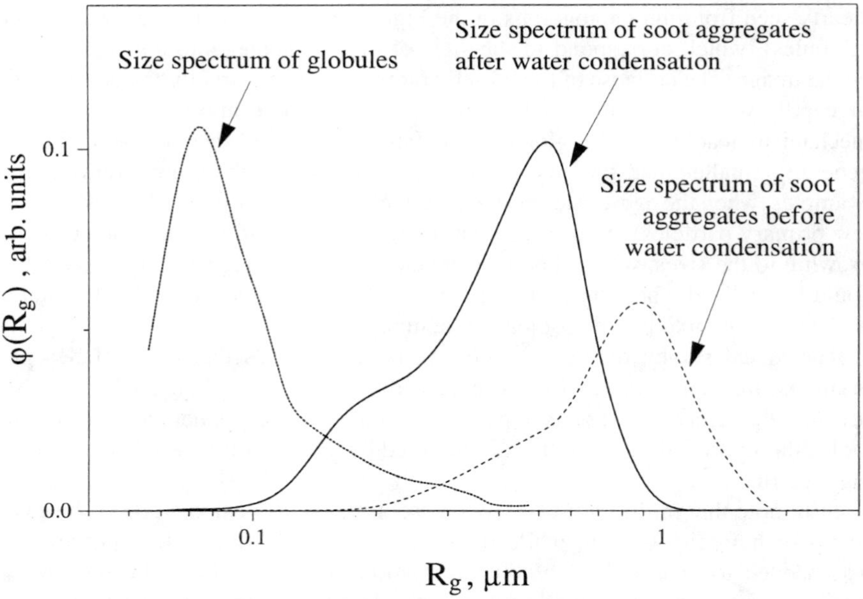

Figure 10.20. Size spectra of acetylene soot aggregates before and after passage through FDNC, and the size spectrum of globules formed in the process of condensation–evaporation. Obtained from the analysis of TEM images.

change of differential scattering cross-sections for the modified aerosol is shown in Figure 10.21. It could be clearly seen that the measured scattering curve for the aggregates with globules (open squares) essentially diverges from the linear pattern (in the log–log coordinates) for initial clusters (black dots), which is characteristic for scattering on the fractal objects. Figure 10.21 also is comprised of results of $I(q)$ computation, carried out on the base of Raleigh–Debye–Gans approximation for fractal aggregates (RDG-FA) [49] (curve 1). To make this result comparable to the measured values, we used distributions of particles over the gyration radius R_g, the number of primary particles per aggregate N, and the projectional fractal dimension D_f, measured from a series of TEM images as the input parameters for model calculations.

We used the primary particle diameter equal to 40 nm and the complex refractive index of the amorphous carbon black from the well-known work of Dalzell and Sarofim [44] $m = 1.54 + 0.48i$.

Although initial aggregates reveal the fractal structure, which is demonstrated by the linear part of the scattering curve, the incline of this part does not coincide with the $-D_f$, taken from the TEM measurements (line 1). As was mentioned earlier, this discrepancy is caused by the difference in the nature of the fractal dimension, retrived from the measurements of a different physical nature. This means that even for dimensions < 2, the fractal dimension retrieved from scattering

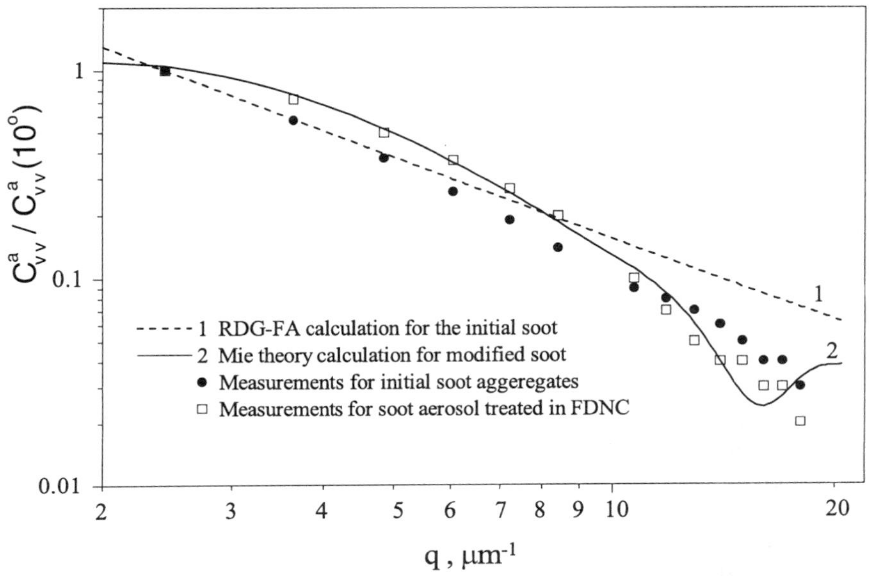

Figure 10.21. The q dependence of differential scattering cross-sections for the initial system of aggregates and the aggregates passed through the FDNC.

measurements is not equal to the one obtained through the box-counting processing of the digitized 2D image. They should be equal only for an absolutely ideal fractal object (computer simulated), which is clearly not the case for the disperse soot aggregates.

The appearance of soot aggregates after water vapor condensation in the FDNC inicates that their structure does not preserve its fractal nature. We now have a two-component system, with dense spherical globules embedded into the rarefied framework of the fractal cluster. It looks natural to calculate the optical properties of such a system by using two different approaches—scattering on the fractal for the residuals of the initial structure and Mie scattering for the globules. The result of this joint approach is represented by curve 2 in Figure 10.21. As in the previous case, we used a measured size distribution of globules (see Fig 10.20) as input parameters for the Mie calculation, and the effective value of the refractive index, calculated with the Maxwell–Garnett approach (see, e.g., [45]) taking into account the fact that the globules are formed from some number of spherical primary particles of uniform size. By considering the volume fraction of compactly packed primary particles in a globule to be $\pi/6$, we have the effective refractive index as follows:

$$\frac{m_e^2 - 1}{m_e^2 + 2} = \frac{\pi}{6}\frac{m^2 - 1}{m^2 + 2}$$

which gives $m_e = 1.33 + 0.25i$.

TABLE 10.3. Specific Extinction Cross-Sections for the Acetylene Soot Aggregates

Method of Determination of Specific Extinction Cross-Sections	Initial Particles $\sigma_{ext}(m^2/g)$	Particles after FDNC Treatment at Relative Humidity 180% $\sigma_{ext}(m^2/g)$
1 RDG–FA calculation	5.3 ± 0.2	
2 Mie calculation with account for globules' size spectrum (Fig 10.20)		8.5 ± 0.2
3 Weighted average value for the double-component system		7.5 ± 0.5
4 Measured attenuation	4.0 ± 0.4	5.5 ± 0.5

As seen from Figure 10.20, curve 2 makes a good fit of experimental data for the case of modified soot. Note, though, that the curves are normalized to unity, and therefore do not provide any data on the absolute values of the differential scattering cross-sections. Because we were not able to measure the total scattering efficiencies, we may only talk about the general resemblance of the calculated scattering pattern to the measured one. On this ground, the suggestion could be made that the scattering is governed mostly by globular component of the aerosol system. In fact, scattering from the fractal part of the system brings in a very small variation of observed dependence. This can be understood by noting that the mass fraction of the globular component constitutes in our case (according to the TEM analysis) $\sim 80\%$ of the total mass of the aerosol matter.

Together with the variation of differential optical parameters, some changes, though not as pronounced, were observed in the extinction efficiency. Measured and calculated specific extinction cross-sections of the soot aerosol in question, before and after the treatment in the FDNC, are listed in the Table 10.3.

The weighted average value of specific extinction, calculated with the joint use of the RDG–FA and Mie approaches (line 3 of the Table 10.3) in 40% higher than the value of the initial soot, calculated solely by RDG–FA. This is almost the same gain of magnitude that is observed in experiment (line 4). The discrepancy of the absolute values of the specific extinction cross-section and their distinction (especially for the measured values) from the commonly accepted value of $\sim 9\text{–}10\,m^2/g$ are not crucial, since it is the relative behavior of this magnitude that is of practical interest. Besides, some data indicate that the possible limits of specific extinction variation could be very wide in the case of carbonaceous aerosols (see, e.g., Liousse et al., [69], who reported the variation of σ_{ext} from 4 to $20\,m^2/g$).

In this way, the interaction of water vapor with the soot aerosol particles (in the case when condensation starts *inside* the aggregates) could lead not only to formation of large droplets with soot inclusions, but also to formation of small compact globules embedded inside the framework of the initial fractal cluster. In this case, the optical properties, namely, the differential scattering cross-section and the specific extinction, exhibit essential changes. In particular, angle scattering dependence deviates from the power-law regime of scattering, and the specific extinction grows by $\sim 40\%$. For a system such as this, it appears impossible to use the fractal approach to calculate differential characteristics, and some arbitrary approach should be used to incorporate the presence of nonfractal objects. The joint use of the RDG–FA and Mie scattering theories, although they fail to give a precise quantitative description of this phenomena, reproduce a similar behavior of optical characteristics, which is observed in experiment.

10.5. CONCLUSIONS

In this chapter, the structural aspects of the optical properties of carbonaceous soot particles are considered. It is shown that the coagulation of primary nanoparticles in larger aerosol structures with a complex geometry considerably changes their optical properties, which can no longer be described by the classical theory of scattering. For optical characterization of these objects the fractal approach should be used, which requires profound knowledge of internal structure of soot aggregates. Thus, by using the methods of TEM and angle scattering of light, it is possible, on the base of developed algorithms, to recover a complete set of structural parameters, that are important for the creation of updated optical models. The methods of electron microscopy are preferable, because they can provide an independent determination of the structural parameters. On the other hand, light scattering is a more representative method in a way because EM wave theory senses the whole ensemble of aggregates simultaneously, in comparison with TEM techniques, where the set of aggregates under investigation is limited by the characteristics of the sampling device and the area of the sample grid. The combination of the TEM and ALS methods is a possible way to yield the most complete set of structural parameters needed as input quantities for any optical model.

It is rather difficult to choose the best approach to calculation of optical properties of soot aggregates. Each model has its own advantages and disadvantages, and should be used in the proper situation. Sometimes the exact methods are the only possible approach to interpret observed phenomena, and sometimes the combination of the classical and the fractal approach should be chosen, even if some of basic assumptions of these methods do not comply with the real structural characteristic of soot aggregates. As was demonstrated, the decision always relys on the knowledge of the real structure of the investigated object.

By considering the use of this or that optical model for prediction of optical properties of soot aerosol in the atmospheric environment, one must take into account the significant discrepancy between the structural parameters used in the model simulations and the real structure of carbonaceous soot particles. This discrepancy is caused mainly by processes of "ageing" of soot under the influence of ambient atmospheric factors, which comprise humidity and temperature, background aerosol and gases, cloud formations, and so on.

The laboratory experiments on structural modification have shown that the restructuring degree of aggregates first of all depends on the nature of the interparticle associations. Formation of soot particles at different stages of coagulational growth is featured by the different energy of contact between the particles. Furthermore, the presence of high-molecular substances adsorbed on a surface of soot particles sharply increases the mobility of connections, thus lowering the activation barrier of a restructuring [71].

The specific features of the aggregate structure determines their different behavior under the action of different forces. Thus, thermal impact results in a breaking of the weak interparticle bonds, which leads to a fragmentation of the aggregate into smaller anisotropic structures. At the same time, the condensation of water vapor on the soot particles results in formation of dense globular structures, apparently due to the central symmetry of capillary forces.

The data of optical measurements exhibit strong correlation with structural features. From the point of view of possible climatic impact, it is noteworthy that, given the constant chemical composition, the changes of structural parameters R_g, k_f, D_f result in a significant increase (by a factor of 2.5) of the total scattering and extinction cross-sections, while the pattern of scattering preserves its fractal-like nature practically up to the limit, where complete disintegration and oxidation of clusters take place. At the same time, scattering measurements of aggregates exposed to condensational effect reveal the essential altering of the scattering pattern, featured by gradual vanishing of the power-law regime with the densification of clusters.

The impact of structural and chemical variability on the optical properties of soot aggregates and, therefore, on the optical characteristics of the soot aerosol system can, in our opinion, be exploited to explain the wide dispersion of observed optical parameters known from the periodicals. For example, Liousse et al. [69] reported the known values of the extinction cross-section from 4 to 20 m^2/g.

Undoubtedly, the further development of optics of nanoobjects, especially of optically active aggregates of carbonaceous particles, will not be possible without elaboration on the quantitative methods, providing a description of their complex "multiscale" structure. The fractal approach is thought to be the most promising in this case, because it allows us to deal with very large statistical ensembles of similar objects. At the same time, the possible diversity of real objects from the mathematical abstraction, utilized by the fractal approach, should always be kept in mind.

REFERENCES

1. W. F. Cooke and J. N. Wilson, "A global black carbon aerosol model," *J. Geophys. Res.* ***101***, 395 (1996).
2. A. B. Pittock, T. P. Ackerman, P. J. Crutzen, M. C. MacCracen, C. S. Shapiro, and R. P. Turco, *Environmental Consequence of Nuclear War: Physical and Atmospheric Effects*, Vol. 1, Wiley, Chichester, UK, 1986.
3. H. Grassl, "What are the radiative and climatic consequences of the changing concentration of atmospheric aerosol particles?" in F. S. Rowland and I. S. A. Isaksen, Eds., *The Changing Atmosphere*, Vol. 1, Wiley, Chichester, 1988.
4. G. S. Golitsyn, A. H. Shukurov, and A. S. Ginzburg. "Complex investigation of microphysical and optical properties of smoke aerosol," *Izv. AN USSR, Phyz. Atm. Oceana* ***24***, 227 (1988).

5. J. M. Haywood, D. L. Roberts, A. Slingo, J. M. Edwards, and K. P. Shine, "General circulation model calculations of the direct radiative forcing by antropogenic sulfate and fossil-fuel soot aerosol," *J. Climate* ***10***, 1562 (1997).
6. S. R. McDow, M. Jang, Y. Hong, and R. M. Kamens, "An approach to studying the effect of organic composition on atmospheric aerosol photochemistry," *J. Geophy. Res.* ***101***, 593 (1996).
7. P. Chylek and J. Hallett, "Enhanced absorption of solar radiation by cloud droplets containing soot particles in their surface," *Q. Jl. R. Met. Soc.* ***118***, 167 (1992).
8. S. Twomey, M. Piepgrass, and T. L. Wolfe, "An assessment of the impact of pollution on the global cloud albedo," *Tellus* ***36B***, 356 (1984).
9. E. J. Jensen and O. B. Toon, "The potential impact of soot particles from aircraft exhaust on cirrus clouds," *Geophys. Res. Lett.* ***24***, 249 (1997).
10. I. Colbeck, L. Appleby, E. J. Hardman, and R. M. Harrison, "The optical properties and morphology of cloud-processed carbonaceous smoke," *J. Aerosol Sci.* ***21***, 527 (1990).
11. J. Lahaye, "Particulate carbon from the gas phase," *Carbon* ***30***, 309 (1992).
12. K. H. Homann, in H. Janger and H. G. Wagner Eds., *Soot Formation in Combustion*, Vandenhoeck and Ruprecht, Göttingen, Vol. 3, 1990, pp. 101–124.
13. P. H. Chambrion, H. Jander, N. Petereit, and H. Gg. Wagner, "Soot growth in atmospheric C_2H_4/Air/O_2 flames. Influence of the fuel carbon density," *Z. Phys. Chem.* ***194***, 1 (1996).
14. R. J. Santoro and J. H. Miller, "Soot formation in laminar diffusion flames," *Langmuir* ***3***, 244 (1987).
15. S. I. Vojutsky and S. S. Rubina, "Modern knowlege about size, shape and structure of soot particles," *Usp. Khim.* ***21(1)***, 84 (1952).
16. F. A. Heckman and D. E. Harling, "Carbon structure," *Rubber Chem. Technol.* ***39***, 1 (1966).
17. T. Ishiguro, Y. Takatory, and K. Akihama, "Microstructure of diesel soot probed by electron microscopy: First observation of inner core and outer shell," *Combust. Flame* ***108***, 231 (1997).
18. E. F. Mikhailov and S. S. Vlasenko, "The generation of fractal structures in gaseous phase," *Phys.-Usp.* ***38***, 252 (1995).
19. P. Meakin, "Fractal aggregates," *Adv. Colloid Interface Sci.* ***102***, 249 (1988).
20. B. B. Mandelbrot, *The Fractal Geometry of Nature*, Freeman, New York, 1983.
21. H. E. Stanley, in L. Pietronero and E. Tosatti, Eds., *Fractals in Physics*, North-Holland, Amsterdam, The Netherlands, 1986.
22. R. A. Dobbins and C. M. Megaridis, "Morphology of flame-generated soot as determined by thermophoretic sampling," *Langmuir* ***3***, 254 (1987).
23. P. Meakin, in C. Domb and J. L. Lebowitz, Eds., *Phase Transitions and Critical Phenomena*, Vol. 12, Academic, London, 1988, pp. 336, 442.
24. Ü. Ö. Köylü, Y. Xing, and D. Rosner, "Fractal morphology analysis of combustio-generated aggergates using angular light scattering and electron microscope images," *Langmuir* ***11***, 4848 (1995).
25. R. J. Samson, G. M. Mulholland, and J. W. Gentry, "Structural analysis of soot agglomerates," *Langmuir* ***3***, 272 (1987).

26. J. Cai, N. Lu, and C. M. Sorensen, "Comparision of size and morphology of soot aggregates as determined by light scattering and electron microscope analysis," *Langmuir* ***9***, 2861 (1993).

27. E. F. Mikhailov and S. S. Vlasenko, "Effect of particle anisotropy on the growth of fractal clusters PbI_2 in gas phase," *Sov. J. Chem. Phys.* ***10(7)***, 1574 (1993).

28. T. Nicolai, D. Durand, and J-C. Gimel, in W. Brown, Ed., *Light Scattering. Principles and Development*, Oxford University Press, Oxford, UK, 1996, pp. 201, 231.

29. R. Jullien, "From Guinier to fractals," *J. Phys. France* **2**, 759 (1992).

30. A. Pearson and R. W. Anderson, "Long-range pair correlation and its role in small-angle scattering from fractal clusters," *Phys. Rev. B* ***48***, 5865 (1993).

31. A. Schmidt-Ott, "New approaches to *in situ* characterization of ultrafine agglomerates," *J. Aeros. Sci.* ***19***, 553 (1988).

32. Wu, Zhangfa, I. Colbeck, and S. Simons, "Kinematic coagulation, aerosol agglomerates and the fractal dimension. Aerosols: Their generation, behavior and applicaton," *Aerosol Soc.* 24 (1994).

33. B. Schleicher, S. Kunzel, and H. Burtscher, "*In situ* measurement of size and density of submicron aerosol particles," *J. Appl. Phys.* ***78***, 4416 (1995).

34. A. P. Weber, Ph.D. Thesis, Paul Scherrer institute, Villigex, Switzerland.

34a. A. P. Weber, U. Baltensperger, H. W. Gaggeler, and A. Schimidt-Ott, "*In situ* characterization and structure modification of agglomerated aerosol particles," *J. Aerosol Sci.* ***27***, 915 (1996).

35. S. Nyeki and I. Colbeck, "Fractal dimension analysis of single, *in situ*, restructured carbonaceous aggregates," *Aerosol Sci. Tech.* ***23***, 109 (1995).

36. R. A. Dobbins and C. M. Megaridis, "Absorption and scattering of light by polydisperse aggregates," *Appl. Opt.* ***30***, 4747 (1991).

37. V. N. Kuzmin, "The effect of athmospheric soot aggregation on its optical properties in the fractal cluster model," *Izv. USSR, Phyz. Atm. Oceana* ***28***, 309 (1992).

37a. G. Mie, "Beiträge zur Optik Früber Medsen Speziell Kolloidalen Metallösungen," *Ann. Phys.* ***25***, 377 (1908).

38. C. F. Bohren and D. R. Huffman, *Absorption and Scattering of Light by Small Particles*, Wiley, New York, 1983.

39. Ü. Ö. Köylü and G. M. Faeth, "Radiative properties of flame-generated soot," *J. Heat Transfer* ***115***, 409 (1993).

40. A. L. Aden and M. Kerker, "Scattering of electromagnetic waves from two concentric spheres," *J. Appl. Phys.* ***22***, 1242 (1951).

41. A. Fassi-Fihri, K. Suhre, and R. Rosset, "Internal and external mixing in atmospheric aerosols by coagulation: Impact on the optical and hydroscopic properties of the sulphate-soot system," *Atm. Environ.* ***31***, 1393 (1997).

42. R. Jullien and R. Botet, *Aggregation and Fractal Aggregates*, World Scientific Publishing, Singapore, 1987, pp. 46–50.

43. J. E. Martin and A. J. Hard, "Scattering from fractals," *J. Appl. Crystallogr.* ***20***, 61 (1987).

44. W. H. Dalzell and A. F. Sarofim, "Optical constants of soot and their application to heat-flux calculations," *J. Heat Transfer* ***91***, 100 (1969).

45. Ü. Ö. Köylü and G. M. Faeth, "Spectral extinction coefficients of soot aggregates from turbulent diffusion flames," *J. Heat transfer* ***118***, 415 (1996).

46. A. R. Jones, "Scattering efficiency factors for agglomerates of small spheres," *J. Phys. D: Appl. Phys.* ***12***, 1661 (1979).

47. Ü. Ö. Köylü and G. M. Faeth, "Optical properties of overfire soot in buoyant turbulent diffusion flames at long residence times," *J. Heat. Transfer* ***116***, 152 (1994a).

48. Ü. Ö. Köylü and G. M. Faeth, "Optical properties of soot in buoyant laminar diffusion flames," *J. Heat Transfer* ***116***, 971 (1994).

49. S. Neyki and I. Colbeck, "Optical and dynamical investigations of fractal aggregates," *Sci. Prog. Oxford* ***76***, 149 (1992).

50. M. V. Berry and I. Percival, "Optics of fractal clusters such as smoke," *Opt. Acta* ***33***, 571 (1986).

51. T. L. Farias, Ü. Ö. Köylü, and M. G. Carvalho, "Effects of polydispersity of aggregates and primary particles on radiative properties of simulated soot," *J. Quant. Spectrosc. Radiat. Transfer* ***55***, 357 (1996).

52. M. F. Iskander, H. Y. Chen, and J. E. Penner, "Optical scattering and absorption by branched chaines of aerosols," *Appl. Opt.* ***28***, 3083 (1989).

53. H. Y. Chen, M. F. Iscander, and J. E. Penner, "Light scattering and absorption of by fractal agglomerates and coagulations of smoke aerosols," *J. Mod. Opt.* ***37***, 171 (1990).

54. P. C. Waterman, "Matric methods in potential theory and electromagnetics scattering," *J. Appl. Phys.* ***50***, 4550 (1979).

55. M. I. Mishchenko, L. D. Travis, and D. W. Mackowski, "T-Matrix computations of light scattering by nonspherical particles: A review," *J. Quant. Spectrosc. Radiat. Transfer* ***55***, 535 (1996).

56. T. L. Farias, M. G. Carvalho, Ü. Ö. Köylü, and G. M. Faeth, "Computational evaluation of approximate Rayleigh–Debye–Gans/fractal aggregate theory for the absorption and scattering properties of soot," *J. Heat Transfer* ***117***, 152 (1995).

57. E. F. Mikhailov, S. S. Vlasenko, and A. A. Kiselev, in M. M. Novak, Ed., *Fractals and Beyond*, World Scientific Publishing, Singapore, 1998, pp. 317–329.

58. J. Hallett, J. G. Hudson, and C. F. Rogers, "Characterization of combustion aerosols for haze and cloud formation," *Aerosol Sci. Technol.* ***10***, 73 (1989).

59. P-F. Huang, B. J. Turpin, M. J. Pipho, D. B. Kittelson, and P. H. McMurry, "Effects of water condensation and evaporation on diesel chain-agglomerate Morphology," *J. Aerosol Sci.* ***25***, 447 (1994).

60. E. Wiengarther, H. Burtscher, and U. Baltensperger, "Hygroscopic properties of carbon and diesel soot particles," *Atm. Environ.* ***31***, 2311 (1997).

61. E. F. Mikhailov, S. S. Vlasenko, A. A Kiselev, and T. I. Ryshkevich, "The structural changes in fractal particles of carbon black under effect of capillary forces: Experimental results," *Colloid J.* ***59***, 176 (1997).

61a. E. F. Mikhailov, S. S. Vlasenko, A. A. Kiselev, and T. I. Ryshkevich, in M. M. Novak and T. G. Dewey Eds., *Fractal Frontiers*, World Scientific, Singapore, 1997, pp. 393–402.

62. A. W. Castleman, P. M. Holland, and R. G. Keesee, "The properties of Ion clusters and their relationship to heteromolecular nucleation," *J. Chem. Phys.* ***68***, 1760 (1978).

63. F. A. Hildeman, G. R. Merkovski, and G. R. Cass, "Chemical composition of emissions from urban sources of fine organic aerosols," *Environ. Sci. Technol.* ***25***, 744 (1992).

64. H. P. Boehm, "Some aspects of the surface chemistry of carbon blacks and other carbons," *Carbon* ***32***, 759 (1994).
65. R. A. Dobbins, G. W. Mulholland, and N. P. Bruner, "Comparison of a fractal smoke optics model with light extinction measurements," *Atm. Environ.* ***28***, 889 (1994).
66. C. M. Sorensen, J. Cai, and N. Lu, "Light-scattering measurements of monomer size, monomers per aggregate, and fractal dimension for soot aggregates in flames," *Appl. Opt.* ***31***, 6547 (1992).
67. V. Vohra and R. H. Heist, "The flow diffusion nucleation chamber: A quantitive tool for nucleation research," *J. Chem. Phys.* ***104***, 382 (1996).
68. A. R. Chugtai, M. E. Brooks, and D. M. Smith, "Hydration of black carbon," *J. Geophys. Res.* ***101***, 19505 (1996).
69. C. Liousse, H. Cachier, and S. G. Jennings, "Optical and thermal measurements of black carbon aerosol content in different environments: Variation of the specific cross-section," *Atm. Environ.* ***27A***, 1203 (1993).
70. C. M. Megaridis and R. A. Dobbins, "Morphological description of flame-generated materials," *Combust. Sci. Technol.* ***71***, 95 (1990).
71. A. P. Weber and S. K. Friedlander, "*In situ* determination of the activation energy for restructuring of nanometer aerosol agglomerates," *J. Aerosol Sci.* ***28***, 179 (1997).

CHAPTER ELEVEN

Optoelectronic Properties of Quantum Wires

ALEXANDER A. BALANDIN

Department of Electrical Engineering,
University of California-Riverside, Riverside, CA 92521

FRANK L. MADARASZ

Center for Applied Optics, University of Alabama in Huntsville,
Huntsville, AL 35899

FRANK SZMULOWICZ

Air Force Research Laboratory, Materials and Manufacturing Directorate (AFRL/MLPO),
Wright Patterson AFB, OH 45433

SUPRIYO BANDYOPADHYAY

Department of Electrical Engineering, University of Nebraska—Lincoln,
Lincoln, NE 68588

Low-dimensional semiconductor quantum nanostructures, particularly quantum wires, have been predicted to exhibit many properties that can be used to realize novel optoelectronic devices. The potential advantage of these structures arises from spatial confinement of photoexcited electrons and holes that modifies their optical absorption, emission, and other related characteristics. This chapter presents a review of optical and electronic properties of semiconductor quantum wires crucial for optoelectronic applications. As a prototype, we consider quasi-one-dimensional (1D) structures with lateral dimensions comparable to the exciton Bohr radius (100–400 Å) in Group IIIA (14)–VA (15) and IIB (12)–VIA (16) compound semiconductors. Most of the results presented here are relevant to $Al_xGa_{1-x}As/GaAs$ heterostructure quantum wires but can be easily extended to other similar material systems. Throughout this chapter, we concentrate only on optoelectronic properties of excitonic/biexcitonic origin and how they are modified in a quantum wire. Spatial confinement of carriers in quasi-1D systems is expected to increase binding energies of all excitonic complexes and, thus, make their contribution to the optical response

Optics of Nanostructured Materials, Edited by Vadim A. Markel and Thomas F. George
ISBN 0-471-34968-2

more pronounced. Additionally, the theoretical formalism in this chapter is general enough to include the effect of dielectric confinement of excitons, final barrier height in embedded quantum wires as opposed to free-standing wires, and barrier-wire interdiffusion. In the second half of this chapter, we give a detailed account of the effects of a magnetic field on the excitonic optical properties of quantum wires. A magnetic field causes additional spatial confinement of carriers, and hence enhances the effect of quasi-1D confinement. Moreover, it provides an external "handle" to vary the degree of confinement and thus "tune" the optoelectronic properties. As a pathological example, we show in the later parts of the chapter how the third-order dielectric susceptibility (responsible for optical nonlinearity, nonlinear differential refractive index, and differential absorption) can be modulated with a magnetic field.

11.1. INTRODUCTION

The development of sophisticated growth and self-assembly techniques for synthesizing quasi-1D semiconductor structures (quantum wires) [1–3] has stimulated a large body of work pertinent to these systems. The InGaAs/InP quantum wires with widths down to 100 Å and size fluctuations of 20 Å or less have been fabricated by regular electron beam lithography and wet etching [4]. Clever techniques employing molecular beam epitoxy (MBE) growth of a narrow gap semiconductor on V-groove surfaces of a wider gap semiconductor have yielded quantum wires of unprecedented quality [5]. Additionally, free-standing quantum wires of 6-nm diameter have been realized by impregnating naturally occurring opal with molten semiconductors [6]. Carbon nanotubes are another example of quantum wires. The routine availability of such systems has resulted in a commensurately increased interest in their optoelectronic properties.

Much of the interest in quantum wires has been stimulated by the possibility of novel low-dimensional physics related to spatial confinement of carriers [7] and phonons [8–10], as well as applications in optoelectronic devices [11]. Dramatic sharpening of absorption peaks in quantum wires, the possibility to shift and modulate these peaks in amplitude and frequency with an applied external field, as well as strong optical nonlinearity of excitonic origin [predicted $\chi^{(3)}$ up to 1.2 esu] make quasi-1D structures very attractive for use in optoelectronic devices such as modulators, limiters, switches, and wavelength converters [12–13].

Owing to the van Hove singularities in the electron density of states (DOS) in 1D structures, quantum wire-based lasers clearly show enhanced differential gain [14]. Recently, it was found that the gain spectrum of GaInAsP/InP quantum-wire lasers is narrower than that of the quantum-well lasers [11]. These results were obtained for compressively strained quantum-wire lasers with a wire width of 200–250 Å, fabricated by electron beam lithography and two-step organometallic vapor phase epitaxial growth. Recently, GaAs/AlGaAs quantum wire laser arrays were fabricated by metallorganic chemical vapor deposition on nonplanar substrates [15]. Despite the fact that the effective width of quantum wires was small ($\sim$ 6 nm), the linear light output power of the laser arrays reached over 100 mW under pulsed conditions (1 μs pulses at 1 kHz) at room temperature.

In laser applications, the benefits of sharper DOS spectrum in quantum wires are usually offset by slower carrier energy relaxation (phonon bottleneck) and rapid k-space filling by photoexcited carriers and excitons. However, there are a number of applications where these features are a plus. Rapid k-space filling results in rapid onset of optical nonlinearity or, equivalently, a lower threshold pumping power to induce optical nonlinearity in a given material system. This is a significant advantage in high-density optoelectronic chips where heat dissipation is a serious problem.

In quantum wires, the giant third-order nonlinear susceptibility $\chi^{(3)}$ arises primarily due to quasi-1D excitons and biexcitons (excitonic molecules) [16–18]. It was shown in [17] that the biexciton binding energy contributes directly to $\chi^{(3)}$. An external magnetic field can increase this binding energy leading to stronger optical nonlinearities. Additionally, the field can act as an agent to modulate the nonlinear absorption/gain in quantum wires, which opens up the possibility of realizing externally tunable devices. For this reason, a significant portion of this chapter will be dedicated to the analysis of excitonic optical nonlinearities in quantum wires and the effects produced by a magnetic field.

11.2. EXCITONIC OPTICAL PROPERTIES

There have been relatively few theoretical studies of excitons in quantum wires and even fewer dealing with biexcitons [19–21]. To our knowledge, the influence of a magnetic field on these entities has not been explored in sufficient detail. Experimental data in this area are sparse and only recently has the effect of a magnetic field on exciton absorption and magnetophotoluminescence in quantum wires been reported [22,23]. Unambiguous signatures of biexcitons in quantum wires were also observed very recently [24,25] and it is an important development since the mere existence of these entities in 1D structures has been quite moot [26].

11.2.1. Theoretical Formulation

For quantum wells [quasi two-dimensional (2D) system], it is possible to construct an analytical model that incorporates accurately the physics of excitonic complexes and the geometry of a structure [27–28], finite barrier potentials and band off-sets, barrier penetration, parabolic shaped wells, and so on, provided only the lowest electron and hole subbands are occupied. Quantum wires of rectangular cross-section, on the other hand, present a theoretical impasse. In general, because the carriers are confined in two directions, the Hamiltonian cannot be broken up into separable parts unless the confining potentials are infinite [29,30]. Because of this problem, most theoretical models assume strictly 1D wires or cylindrical wires with an infinite potential at the circumference of the wire. When computing the biexciton binding energy, which is needed for the calculation of $\chi^{(3)}$, an "effective" 1D

Coulomb potential in the radial direction is often employed [31]; still, $\chi^{(3)}$ was not explicitly calculated. On the other hand, Glutsch and Bechstedt [29] approximately treated finite confining potentials in a square wire under the condition that the potential in one direction is infinite but in the other direction is finite. Their interest though, was only in calculating the spatially nonlocal absorption, $\chi^{(1)}$, by single excitons; no attempt was made to determine $\chi^{(3)}$.

The biexciton problem in the context of quantum wires is even more complex than the exciton problem and beset with ambiguities. Ivanov and Haug [26] claimed that biexcitons do not exist in 1D wires because of the strong influence of the polariton effect. Physically, this result was attributed to the "topological" feature of a 1D structure; the exciton inside the biexciton cannot make a complete oscillation. However, this argument breaks down for realistic quantum wires with nonzero lateral dimensions. Recent experimental measurements of the biexciton binding energies in quantum wires gave solid proof for the existence of biexcitons in quantum wires and showed excellent agreement with our theoretical predictions for the biexciton binding energy [24].

Another problem with biexcitons is the lack of a standard model. While there are standard models [27,32–35] for excitons, the paucity of experimental data for comparison has inhibited the emergence of a single, commonly accepted theoretical model and numerical approach for the calculation of biexcitonic states. In most cases, the biexciton problem is treated variationally. The archetypal examples of such calculations are works by Brinkman et al. [36] (bulk), Kleinman [37] (quantum well), and Banyai et al. [31] (quantum wire). In our study, we calculated exciton and biexciton binding energies variationally for rectangular GaAs quantum wires treating the Coulomb interaction terms exactly in their full three-dimensional (3D) form [18,38–40]. The trial wave function was chosen to be of the Heitler–London type. However, Ishida et al. [41] presented a study of biexcitons within the framework of a 1D tight-binding model and showed that in their model calculation, the effect of particle correlation over the lattice-constant length scale is important. In our particular case, the correlation effect is not critical for several reasons. First, unlike Ishida et al. [41], we are primarily interested in the regime of moderate quantum confinement (in our case wire width $L = 100$–700 Å and wire thickness $W = 100$–300 Å, which is characteristic of standard samples). Ishida et al. [41] considered strongly confined 1D wires with wire width smaller than the size (effective Bohr radius) of an exciton (~ 100 Å for GaAs). For this type of narrow 1D systems, the exciton wave function becomes very compact and the behavior of electrons and holes on the atomic length scale assumes critical importance. Obviously, this is not the regime that we address. Another objection raised by Ishida pertained to the treatment of the Coulomb interaction with its characteristic $1/r$ singularity. Usually, the singularity is dealt with by an arbitrary numerical truncation procedure as in [31]. It was this truncation that was criticized. However, we treat the Coulomb interaction in its full 3D form, which allows us to treat the $1/r$ singularity without the application of any ad hoc truncation procedure. Finally, the discrepancy between the Ishida result and the conventional variational result based on a Heitler–London scheme becomes significant ($\sim 50\%$) when the electron and hole effective masses are similar ($\sigma = m_e/m_h \approx 1$). In our case, we are dealing with effective mass ratios $\sigma \approx 0.1$, which makes the discrepancy much smaller.

In the next sections, we present our theoretical formalism, which is based on a combination of the variational approach and exact numerical solution of the Schrödinger equation. The binding energies of excitons and biexcitons that we compute are used for determining nonlinear optical response of quantum wires. Comparison of our theoretical results with available experimental data is also given.

11.2.2. Excitons in Quantum Wires

Throughout this chapter, we consider a generic GaAs quantum wire surrounded by an AlGaAs cladding. The unconfined (free) direction is along the x axis and the finite lateral dimensions are along the z and y axes. In this section, we will outline the basic theory, which includes finite lateral dimensions of the wire, finite barrier height, and Al interdiffusion into the wire from the surrounding cladding material. The interdiffusion is in the lateral direction, here taken as the z direction (see Fig. 11.1). Since the potential in the orthogonal direction is of square shape and we are interested only in the lowest lying states, we may approximate it, relative to the lateral direction, as infinite. The electron and hole potential energies, V_e and V_h, respectively, are defined via the finite conduction, ΔE_c^0, and valence, ΔE_v^0, band offsets that are assumed to bear a ratio 60:40. All quantities are graded in accordance with the interdiffused Al composition distribution $x(z)$. Following the work of Madarasz and Szmulowicz [42] on Cd interdiffusion across HgCdTe heterojunctions, the Al grading distribution will be taken as

$$x(z) = x_0 + x_c + \left[\frac{x_c - x_0}{2}\right]\left\{\mathrm{erf}\left[\frac{z - z_0^+}{C}\right] - \mathrm{erf}\left[\frac{z - z_0^-}{C}\right]\right\} \tag{11.1}$$

where x_0 (here 0) and x_c are the Al compositions in the wire and cladding material, respectively ($z_0^\pm = \pm W/2$ and $2C$ is the Al grading width in which the Al composition changes by 85%).

The Schrödinger equation for the single particle states $F_i^{(n)}$, in the lateral direction is

$$-\frac{\hbar^2}{2}\frac{d}{dz}\left[\frac{1}{m_i^*[x(z)]}\frac{dF_i^{(n)}(z)}{dz}\right] + V_i[x(z)]F_i^{(n)}(z) = E_i^{(n)}F_i^{(n)}(z) \tag{11.2}$$

where $i = e, h; n = 1, 2, ..., N; E_i^{(n)}$ is the eigenenergy of the nth subband. Equation (11.2) must be solved numerically. We can do this with a "modified" fourth-order Runge–Kutta routine, which accounts for the values of the variable coefficients, $m_i^*[x(z)]$ and $V_i[x(z)]$, between the node points of the integration mesh. Here m_i^* and V_i are written as functions of $x(z)$ to emphasize their implicit dependence on the Al compositional grading; $x(z)$ should not be confused with the coordinate x.

Finite Potential Well and Aluminum Interdiffusion

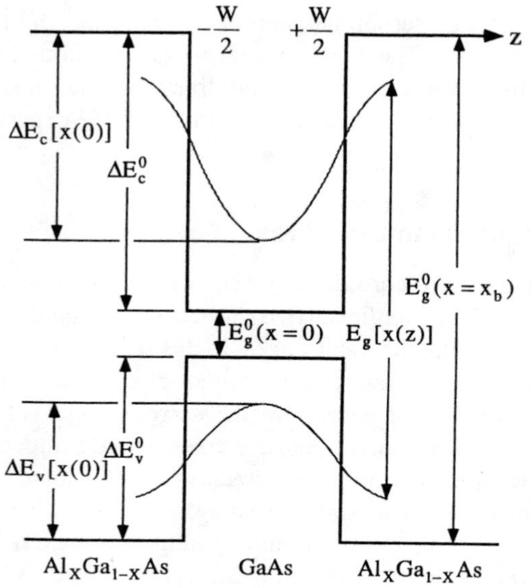

Graded Potential Energies

$$V_e(z) = 0.6\left[E_g[x(z)] - E_g[x(0)]\right] + E_g^0(x = 0)$$

$$V_h(z) = -0.4\left[E_g[x(z)] - E_g[x(0)]\right]$$

Figure 11.1. Definition of quantities used in the calculation of the finite Al-graded band-offset potential energies along the z axis. Potential along the y axis is assumed to be infinite.

In the envelope function approximation, the Hamiltonian of a Wannier exciton is given by [20]

$$H^X = \frac{p_{y_e}^2}{2m_c^*} + \frac{p_{y_h}^2}{2m_\perp^*} + \frac{p_x^2}{2\mu} - \frac{e^2}{\varepsilon\sqrt{x^2 + \rho^2}} + E_g^0 + V_{e,h}(y_{e,h}, z_{e,h}) \tag{11.3}$$

where m_c^* is the electron conduction-band effective mass, $m_\perp^*$ is the heavy-hole valence band effective mass perpendicular to the axis of the wire, μ is the heavy-hole exciton reduced mass, e is the charge of an electron, ε is the static dielectric constant, E_g^0 is the fundamental band gap of the wire material, $V_{e,h}(y_{e,h}, z_{e,h})$ is the confinement potential for electrons and holes along lateral dimensions (L and W), and $\rho^2 = (y_e - y_h)^2 + (z_e - z_h)^2$ is the square of the exciton cylindrical radius in the confined directions. In terms of the Luttinger parameters, the expressions for the hole and reduced masses are

$$\frac{1}{m_\perp^*} = \frac{1}{m_0}(\overline{\gamma_1} - 2\overline{\gamma_0}) \tag{11.4}$$

and

$$\frac{1}{\mu} = \frac{1}{m_c^*} + \frac{1}{m_{||}^*} \tag{11.5}$$

where m_0 is the free-electron mass, $\overline{\gamma_0} \equiv \overline{\gamma_2} = \overline{\gamma_3}$. The quantity $m_{||}^*$ is the heavy-hole effective mass parallel to the axis of the wire.

No exact solution is possible even for this relatively simple Hamiltonian . The envelope function is then chosen to be a trial wave function in a variational procedure [18]

$$\Psi = \frac{2}{L} g_t(x; \eta_x) \cos(k_y y_e) \cos(k_y y_h) u_e^{(1)}(z_e) u_h^{(1)}(z_h) \tag{11.6}$$

where $g_t(x; \eta_x)$ is a Gaussian-type "orbital" function,

$$g_t(x; \eta_x) = \frac{1}{\sqrt{\eta_x}} \left(\frac{2}{\pi}\right)^{1/4} e^{-(x/\eta_x)^2} \tag{11.7}$$

in which η_x is the e–h variational parameter and $k_y = \pi/L$.

The variational parameter η_x is determined by minimizing the expectation value of the exciton Hamiltonian with respect to η_x; its value is the function of the dimensions of the wire and specific values of the material parameters used. The only components of the Hamiltonian that depend on η_x are the kinetic energy of relative motion and the e–h interaction term. Together, they define the exciton binding energy. The kinetic and potential energy terms can be written as [19]

$$E_K^X = \left\langle \Psi \left| \frac{p_x^2}{2\mu} \right| \Psi \right\rangle = \frac{\hbar^2}{2\mu} \frac{1}{\eta_x^2} \tag{11.8}$$

and

$$V_{\text{eh}} = V(|\boldsymbol{x}i + \boldsymbol{\rho}_e - \boldsymbol{\rho}_h|) = \left\langle \Psi \left| \frac{(-)e^2}{\varepsilon\sqrt{x^2 + \rho^2}} \right| \Psi \right\rangle \tag{11.9}$$

Clearly, the potential term is the more difficult of the two to deal with. In evaluating the interaction integral, we first employ a 2D Fourier transform (FT) of the Coulomb potential

$$\frac{1}{\sqrt{x^2 + \rho^2}} = \frac{1}{2\pi} \int_{-\infty}^{+\infty} dq_y \int_{-\infty}^{+\infty} dq_z \, e^{-i\mathbf{Q}\boldsymbol{\rho}} \frac{e^{-i|\mathbf{Q}||x|}}{|\mathbf{Q}|} \tag{11.10}$$

where $\mathbf{Q} = q_y\boldsymbol{j} + q_z\boldsymbol{k}$. By using this transformation, we can integrate analytically over all real-space variables, which gives us

$$V_{\text{eh}} = -\frac{e^2}{2\pi\varepsilon\eta_x}\int_0^{2\pi} d\phi \int_0^{+\infty} dQ_{\eta_x} I(Q_{\eta_x}) H_{LW}(Q_{\eta_x}, \phi) \tag{11.11}$$

where we have changed **q**-space coordinates to polar coordinates and have defined a dimensionless quantity $Q_{\eta_x} = \eta_x Q$. The integrand I is given by

$$I(Q_{\eta_x}) = \exp(Q_{\eta_x}/2\sqrt{2})^2[1 - \text{erf}(Q_{\eta_x}/2\sqrt{2})] \tag{11.12}$$

where erf is an error function. The factor $H_{LW} = H_L H_W$, is defined as

$$H_L(Q_{\eta_x}, \phi) = \frac{\pi^4}{\{\pi^2 - [(L/2\eta_x)Q_{\eta_x}\sin\phi]^2\}^2} \times \frac{\sin^2[(L/2\eta_x)Q_{\eta_x}\sin\phi]}{[(L/2\eta_x)Q_{\eta_x}\sin\phi]^2} \tag{11.13}$$

and

$$\begin{aligned} H_W(Q_{\eta_x}, \phi) = & \int_{-\infty}^{+\infty} dz_e \exp\left(-i\frac{Q_{\eta_x}}{\eta_x}\cos\phi z_e\right)|u_e^{(1)}(z_e)|^2 \\ & \times \int_{-\infty}^{+\infty} dz_h \exp\left(-i\frac{Q_{\eta_x}}{\eta_x}\cos\phi z_h\right)|u_h^{(1)}(z_h)|^2 \end{aligned} \tag{11.14}$$

One should note here that in the case of infinite potential barriers, the factor H_W can be further transformed to the above analytical expression as shown in [19]. The original five-dimensional integration in the interaction has been reduced to the merely 2D integration, which can be evaluated numerically. Unlike the real-space integral, which exhibits numerical singularities at $r \to 0$, our expression clearly has no singularities since the potentially troublesome factors are well behaved, that is,

$$\lim_{\theta \to 0} \frac{\sin\theta}{\theta} \to 1 \tag{11.15}$$

and

$$\lim_{\theta \to \pi} \frac{\sin\theta}{\pi - \theta} \to 1 \tag{11.16}$$

Ground-state exciton binding energies E_B^X can now be found using the relation

$$E_B^X = E_{\text{conf}}^{e1} + E_{\text{conf}}^{hh1} - \min_{\eta_x}\langle\psi|\mathscr{H}^X|\psi\rangle \tag{11.17}$$

where $E_{\text{conf}}^{e1}, E_{\text{conf}}^{hh1}$ are the lowest electron and the highest heavy hole magneto-electric subband bottom energies in a quantum wire measured from the bottom of the bulk conduction band and the top of the bulk valence band. The exciton length is also simply $\eta_{opt}/2^{1/4}$, where η_{opt} is the value of η_x that minimizes the expectation value in Eq. (11.17). In addition to the binding energies, we calculated the radius (or more correctly the "length") of the exciton along the x axis given analytically as $\sqrt{\langle x^2 \rangle} = \eta_x/2^{1/4}$ for different values of wire dimensions. These quantities will be further used for the evaluation of the exciton absorption in quantum wires.

11.2.3. Exciton Absorption

Optical absorption of excitons, as two-particle systems, is conceptually different from optical absorption in the one-electron picture. Its treatment in general form requires definition of coupled exciton–photon degenerate eigenstates (polaritons) [43,44] and introduction of some type of energy dissipation process that is usually chosen to be inelastic scattering by phonons [45,46]. Exciton–polariton formalism is particularly important for extremely narrow quantum wires approaching the true 1D limit, for excitation pulses with temporal width greater than the inverse polariton gap frequency, and systems with exciton–photon interaction stronger than the exciton damping [26].

In our analysis, we consider realistic quantum wires with finite lateral dimensions and strong exciton damping. In this approximation, we replace polaritons by "bare" photons and excitons. Within this approximation we can still use Fermi's golden rule to calculate exciton optical absorption. The general expression for the linear absorption coefficient in the dipole approximation is given by [47]

$$\alpha(\omega) = \frac{1}{V} \frac{4\pi^2 e^2}{ncm_0^2\omega} \sum_{if} |\langle i|\boldsymbol{\varepsilon} \cdot \mathbf{P}|f\rangle|^2 \delta(\varepsilon_f - \varepsilon_i - \hbar\omega) \tag{11.18}$$

where ω is the angular frequency of the incident monochromatic radiation, V is the volume of the sample, n is the index of refraction of the material, c is the speed of light, is the unit polarization vector of the incident radiation, $\mathbf{P}$ is a many-electron momentum operator, $|i\rangle = |0\rangle$ is the initial (ground) state of the system made up of Slater determinants for N electrons in the valence subband, and $|f\rangle = |g\rangle$ is the final (exciton ground) state in the form of a wave packet summed over Slater determinants for $N - 1$ electrons in the valence band and one electron in the conduction subband.

For the quantum wire excitons, we can further write [19,48]

$$|\langle 0|\boldsymbol{\varepsilon} \cdot \mathbf{P}|g\rangle|^2 = |\langle u_v| \cdot \mathbf{p}|u_c\rangle|^2 |\langle \chi(y_h)|\chi(y_e)\rangle|^2 |\langle \chi(z_h)|\chi(z_e)\rangle|^2 \delta_{K_x,0} T\, |g_t(x=0)|^2 \tag{11.19}$$

where $|u_v\rangle$ and $|u_c\rangle$ are the zone-center, periodic Bloch valence- and conduction-band wave functions, respectively, the χ's are the normalized functions given by $\chi(y_{e,h}) = \sqrt{2/L}\cos[\pi(y_{e,h}/L)]$, T is the length of the wire, and $K_x = k_{x_c} - k_{x_v}$ is the difference between the conduction- and valence-band electron wavevectors. Moreover, the smallness of the photon momentum in the direction parallel to the wire axis dictates that, in the exciton absorption process, the electron momentum in the x direction be conserved. This fact is reflected by the Kronecker δ function. The factor $g_t(x=0)$ shows that nonvanishing absorption is possible only for excitons with nonzero amplitudes at $x=0$.

Following the choice of (J, M_J)-representation basis vectors for the $J=\frac{3}{2}$ multiplet [48], the allowed optical transitions between the $HH1$ and $E1$ subbands are given by

$$|\langle u_v|p_x|u_c\rangle|^2 = \frac{m_0}{4}E_P \tag{11.20}$$

where E_P is the Kane matrix element, which is related to the familiar Kane momentum matrix element P by $E_P = (2/m_0)P^2$. The selection rules allow nonzero transition probabilities for the following light propagation directions K and polarization vectors ε,

$$\mathbf{K}||\boldsymbol{x}, \varepsilon = \varepsilon_y \qquad \mathbf{K}||\boldsymbol{y} \qquad \varepsilon = \varepsilon_x \qquad \mathbf{K}||z, \varepsilon = \varepsilon_{x,y} \tag{11.21}$$

By making the necessary substitutions in the expression for the absorption coefficient, Eq. (11.18), one gets the final expression for the $HH11 - E1$ exciton peak absorption

$$\alpha_{\text{ex}}(\omega_{g0}) = \frac{2\sqrt{2\pi}}{LW\eta_x}\frac{e^2}{ncm_0\omega_{g0}}\frac{E_P}{\Gamma_0\sqrt{2\pi}} \qquad \text{(esu)} \tag{11.22}$$

where Γ_0 is a broadening parameter obtained by replacing the δ function in Eq. (11.19) by a Gaussian-shaped function [49], and $\hbar\omega_{g0}$ is the energy required to excite the system from its ground state $|0\rangle$, in which the valence band is full and the conduction band is empty, to the final state $|g\rangle$, which contains a $HH11 - E1$ exciton in its ground state. In Fig. 11.2, we present the peak exciton absorption coefficient as a function of wire dimensions. In order to determine enhancement of exciton optical nonlinerity in quantum wires, we need to calculate the strength of exciton–exciton interaction, which manifests itself via biexciton binding energy.

11.2.4. Biexcitons in Quantum Wires

The biexciton Hamiltonian is considerably more complicated since it involves the interaction of four particles. The kinetic energy operator is

$$\begin{aligned} E_K^{xx} = &-\frac{\hbar^2}{2\mu}\left[\frac{\partial^2}{\partial x_{1a}^2}+\frac{\partial^2}{\partial x_{2b}^2}\right] - \frac{\hbar^2}{\mu_{||}^*}\left[\frac{\partial}{\partial x_{ba}}+\frac{\partial}{\partial x_{1a}}-\frac{\partial}{\partial x_{2b}}\right]\frac{\partial}{\partial x_{ba}} - \frac{\hbar^2}{2m_c^*}\left[\frac{\partial^2}{\partial y_1^2}+\frac{\partial^2}{\partial y_2^2}\right] \\ &-\frac{\hbar^2}{2m_\perp^*}\left[\frac{\partial^2}{\partial y_a^2}+\frac{\partial^2}{\partial y_b^2}\right] - \frac{\hbar^2}{2}\frac{\partial}{\partial z}\left[\frac{1}{m_c^*(z)}\frac{\partial}{\partial z}\right] - \frac{\hbar^2}{2}\frac{\partial}{\partial z}\left[\frac{1}{m_\perp^*(z)}\frac{\partial}{\partial z}\right] \end{aligned} \tag{11.23}$$

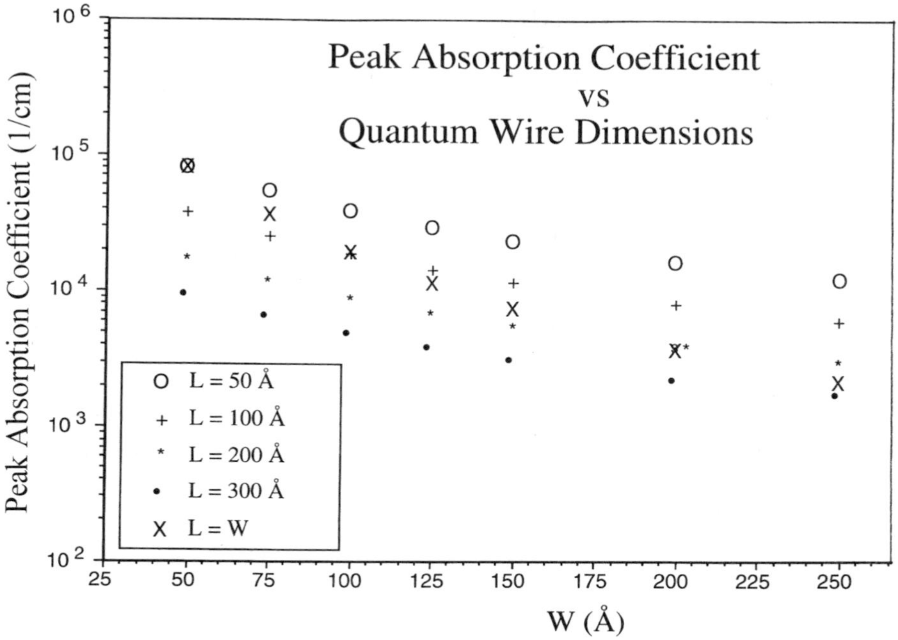

Figure 11.2. Peak absorption coefficient for resonant excitation as a function of the wire dimensions. The results are shown for infinite potential barriers.

where $m_\parallel^*$ is the effective mass parallel to the wire axis. The potential energy (Coulomb interaction) is given by

$$V_C = \frac{e^2}{4\pi\varepsilon}\left\{\frac{(-1)}{\sqrt{x_{1a}^2+\rho_{1a}^2}}+\frac{(-1)}{\sqrt{x_{1b}^2+\rho_{1b}^2}}+\frac{(-1)}{\sqrt{x_{2b}^2+\rho_{2b}^2}}\right.$$
$$\left.+\frac{(-1)}{\sqrt{x_{2a}^2+\rho_{2a}^2}}+\frac{1}{\sqrt{x_{ab}^2+\rho_{ab}^2}}+\frac{1}{\sqrt{x_{12}^2+\rho_{12}^2}}\right\} \tag{11.24}$$

where

$$\rho_{\alpha\gamma}^2 = (y_\alpha - y_\gamma)^2 + (z_\alpha - z_\gamma)^2$$
$$x_{\alpha,\gamma} = x_\alpha - x_\gamma$$

with $\alpha, \gamma = 1, 2, a, b$, and $\alpha \neq \gamma$. Since it is not possible to solve a four-body problem exactly, again a variational approach is employed. In choosing the variational wave function, we are motivated by many considerations. In the regime of weak quantum confinement, where the lateral dimensions L, W of the wire are much larger than the effective Bohr radius a_B^* of a free exciton, the Coulomb interaction energy is dominant over the confinement energy. In that case, one may apply the "free boson" model to determine the different electron–hole pair states. The "free boson" model is based on the Heitler–London approximation in which the biexciton wave function is written as a linear combination of atomic orbitals in analogy with the hydrogen molecule. The linear combination is chosen symmetric in space so that it corresponds to the singlet state, which is known to be lower in energy than the triplet state [26].

In the regime of strong quantum confinement, L and W are smaller than the effective Bohr radius a_B^*. In this case, the confinement energy is the dominant contribution. Electrons and holes are confined separately, and the Coulomb potential may be neglected in comparison with the large kinetic energy caused by the strong confinement. In our case, we are primarily interested in the regime of moderate quantum confinement. Therefore, we have to adopt a physically realistic trial wave function that would allow us to use the Heitler–London approach while taking into account the difference in the motion along confined and "free" directions. We achieve this by considering the electrons and holes to be independently quantized along y and z directions, while applying the Heitler–London approximation along the x direction. Following the approach in [18,19] we choose a trial wave function as a singlet state with the electron–hole pair contributions given by the Gaussian-type "orbital" functions:

$$\Psi = \frac{1}{S(x_{ba})}\{\psi_{1a}\psi_{2b} + \psi_{2a}\psi_{1b}\}g_{ba}(x_{ba}) \equiv \Theta(x_{ba})g_{ba} \tag{11.25}$$

where

$$\psi_{\alpha\gamma} = g_{\alpha\gamma}\phi_\alpha(y_\alpha)\chi_\alpha(z_\alpha)\phi_\gamma(y_\gamma)\chi_\gamma(z_\gamma) \tag{11.26}$$

with

$$g_{ba}(x_{ba}) = \frac{1}{\sqrt{\xi}}\left(\frac{2}{\pi}\right)^{1/4} e^{-(x_{ba}/\xi)^2} \tag{11.27}$$

and

$$g_{1a} = e^{-x_{1a}^2/\eta^2} \tag{11.28}$$

$$g_{2b} = e^{-x_{2b}^2/\eta^2} \tag{11.29}$$

$$g_{2a} = e^{-(x_{2b}-x_{ab})^2/\eta^2} \tag{11.30}$$

$$g_{1b} = e^{-(x_{1a}+x_{ab})^2/\eta^2} \tag{11.31}$$

The wave functions $g_{\alpha\gamma}$ are Gaussian orbitals whose "spread" η and ξ are variational parameters. It is clear that these parameters physically correspond to the electron–hole and hole–hole separations along the length of the quantum wire. The quantity S is a normalization constant that can be evaluated as described in [19]. The quantities

$\phi_{e,h}(y_{e,h})$, and $\chi_{e,h}(z_{e,h})$ are the y and z components of the wave functions of the independently confined electrons and holes in the quantum wire. The y components are represented by regular particle-in-a-box states

$$\phi_{e,h}(y_{e,h}) = \sqrt{\frac{2}{L}} \cos\left(\pi \frac{y_{e,h}}{L}\right) \tag{11.32}$$

The z component of the electron and hole wave functions, $\chi_{e,h}(z_{e,h})$, are affected by the finite barrier height and Al grading and have to be found numerically by solving the one-particle Schrödinger equation given in Section 11.2.2 [here, $\chi_{e,h}(z_{e,h})$ correspond to $F^{(1)}_{e,h}(z_{e,h})$].

In Eq. (11.27), ξ is the hole–hole variational parameter and the normalization is such that $(g_{ba}, g_{ba})_{x_{ba}} = 1$. Also, we define the normalization factor $S(x_{ba})$ by $(\Theta, \Theta)_{\mathbf{x}} = 1$, where the integration is over $\mathbf{x} = \{x_{1a}, x_{2b}, y_\alpha, y_\beta\}$. From these definitions, we find the normalization factor to be given by

$$S^2(x_{ba}) = c^2_{LW} \pi \eta^2 (1 + e^{-x^2_{ba}/\eta^2}) \tag{11.33}$$

where $c_{LW} = \sqrt{N} = LW/4$.

The final expressions for the expectation values of both the kinetic and potential energies are long and complicated. They may be found in their entirety in the papers by Madarasz et al. [19,38]. What is important for this chapter is just their functional forms, which are represented by K and P. The kinetic energy and potential energies are

$$E^{XX}_K = 2E^X_K + K(\eta; \xi) \tag{11.34}$$

where E^X_K is the total kinetic energy of a single exciton, and

$$\begin{aligned} V_C = {} & \frac{e^2}{2\pi\varepsilon} \frac{1}{\eta} \int_0^{+\infty} dQ_\eta \\ & \times \int_0^{2\pi} d\phi \{P[I_1^{(ba)}; I_1^{(1a)}; I_2^{(1a)}; I_3^{(1a)}; I_0^{(21)}; I_1^{(21)}; I_2^{(21)}] H_{LW}(Q_{\eta_{xx}}, \phi)\} \end{aligned} \tag{11.35}$$

where the I's are functions of Q_η, η, and ξ, and each is made up of two terms similar in form to Eq. (11.12). The parameter $H_{LW}(Q_\eta, \phi)$ is identical to Eqs. (11.13)–(11.14), but with $\eta_x \rightarrow \eta$. As in the case of the single exciton, the biexciton Hamiltonian must be minimized with respect to the variational parameters in order to obtain the biexciton binding energy. Since the biexciton wave function of Eq. (11.25) is made up of single particle states, the Al-graded potential will also affect the biexciton binding energy.

By using the formalism developed in Sections 11.2.2 and 11.2.3, we numerically calculate exciton and biexciton binding energies for different wire dimensions. In Fig. 11.3, we compare the exciton and biexciton binding energies when calculated

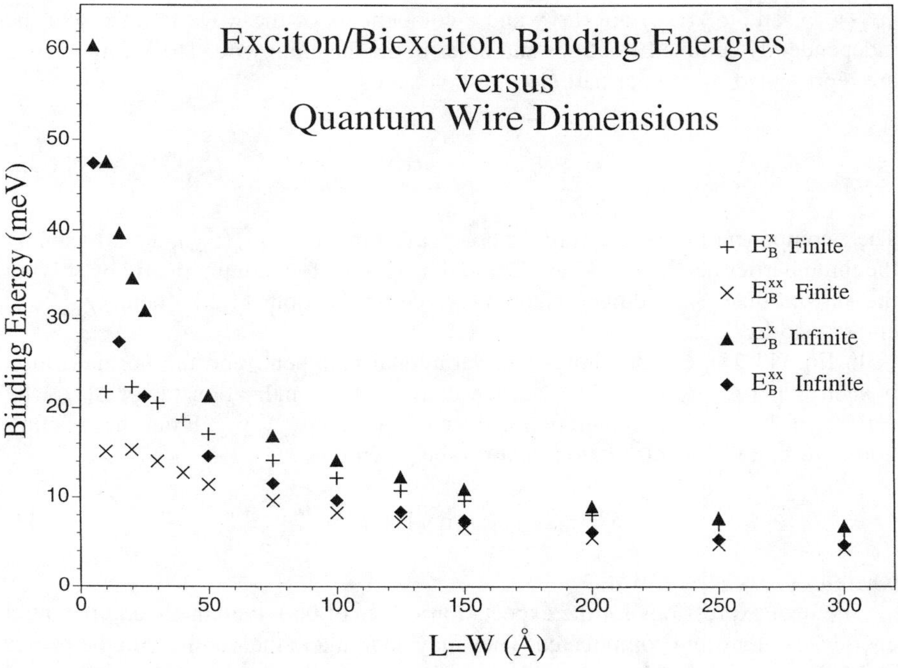

Figure 11.3. Exciton–biexciton (E_B^X/E_B^{XX}) binding energies in square GaAs/Al$_{0.3}$Ga$_{0.7}$As quantum wires, with and without finite band offset potentials included.

with finite band offset potentials and with infinite potentials in a square, $L = W$, wire. The exciton energies are plotted with blackened triangles and plus signs representing infinite potential barriers and finite band offset potential barriers, respectively. The biexciton energies are plotted with blackened squares and crosses representing infinite potential barriers and finite band offset potential barriers, respectively. In the case of the infinite potential barriers, as $L = W$ approaches zero, both exciton and biexciton binding energies tend to infinity as would be expected in analogy to the 1D hydrogen atom [18,20], However, when finite band offsets and wave function barrier penetration are included, even for just 1D, dramatic effects are seen, especially at the smaller dimensions. For example, at $L = W = 25$ Å the percent differences before and after the finite band offsets have been included are 44.8 and 45.4% for the exciton and biexciton, respectively.

The roll-over of both curves at the smaller dimensions is reminiscent of a quasi-bidimensional quantum well system with finite band offsets. The binding energy of the exciton first increases with decreasing dimension. However, when either particle possesses a confinement energy that is close to the height of its respective band offset potential barrier, the bidimensional character is lost and the binding energy begins to decrease. In the present case, it is the unidimensional character that is lost. For the 10×10 Å wire, the single-particle electron and hole ground-state energies in the direction of the finite band offsets, are 186.9 and 91.3 meV, respectively. These are comparable to the respective conduction and valence band offsets of 232 meV and 155 meV.

Numerical results presented in this section for the biexciton binding energy have recently found experimental confirmation [24]. Baars et al. [24] reported spectrally resolved four-wave mixing experiments on $In_xGa_{1-x}As/GaAs$ quantum wires for a wide range of lateral sizes ($L = 200 - 1000$ Å). Due to the polarization dependence of the four-wave mixing signal, beats in the decay of the signal and an additional emission line in the spectrum was attributed to biexcitons. It was found that the biexciton binding energy depends on both lateral dimensions L and W of the wire. The biexciton binding energy in a quantum wire (50×300 Å) is ~ 2.6 meV, which is close to the value predicted by our model. Some discrepancy (theory usually gives higher binding energies) is probably due to the underestimated barrier penetration of the wave function. Moreover, the variational procedures always overestimate true eigenvalues.

11.2.5. Nonlinear Optical Properties

Pump–probe spectroscopy is used experimentally to distinguish excitonic optical nonlinearities from other nonlinearities in semiconducting materials [3]. We restrict our calculation to the optical nonlinearity arising from the population saturation of the exciton state. This type of nonlinearity is expected to be enhanced by spatial confinement in quantum wires. For resonant excitation, the expression for $\chi^{(3)}$ in the rotating wave approximation is given by [19,20]. Here, we wish to investigate the frequency dependence of $\chi^{(3)}$ in a narrow range about the resonant exciton absorption line. The general derivation of $\chi^{(3)}$ for low-density exciton–biexciton systems is given in [19]: it is based on the summation over 16 double Feynman diagrams. The spurious size dependence that generally accompanies the derivation of $\chi^{(3)}$ for a local two-level system has been eliminated by taking into account Pauli's exclusion principle, that is one physical site is not allowed to be doubly excited [17,50].

In the frequency range of interest, the lowest lying states are the major contributors to $\chi^{(3)}$. Accordingly, the general expression reduces to

$$\begin{aligned}\chi^{(3)} &= \frac{(-)2}{\pi\sqrt{2\pi}}\frac{n_0}{\eta_x}\frac{\xi}{\eta_x}\frac{e^4}{m_0^2\omega_{g0}^4}E_P^2 \\ &\times\left[\frac{1}{(\omega_1-\omega_{g0}+i\Gamma_{g0})}-\frac{1}{(\omega_1-\omega_{g0}+\omega_{bx}(\eta,\xi)+i\Gamma_{bg})}\right] \\ &\times\left[\sum_{r=1}^{2}\left\{\frac{1}{\hbar^3(\omega_r-\omega_2+i\gamma)}\left[\frac{1}{(\omega_{g0}-\omega_2+i\Gamma_{g0})}+\frac{1}{(\omega_r-\omega_{g0}+i\Gamma_{g0})}\right]\right\}\right. \\ &+\frac{1}{(\omega_1+\omega_2-2\omega_{g0}+\omega_{bx}(\eta,\xi)+i\Gamma_{b0})} \\ &\left.\times\left\{\frac{1}{(\omega_1-\omega_{g0}+i\Gamma_{g0})}+\frac{1}{(\omega_2-\omega_{g0}+i\Gamma_{g0})}\right\}\right]\end{aligned} \tag{11.36}$$

TABLE 11.1. Physical Parameters Used for Simulation

x^a	0.0	0.3
E_G^0(eV)	1.5212	1.899
$m_\perp^*(m_0)$	0.450	0.576
$m_\parallel^*(m_0)$	0.027	0.027
$m_c^*(m_0)$	0.067	0.092
ε	12.5	12.5
$n_0(1/\mathrm{cm}^2)$	7.89×10^{14}	7.89×10^{14}
Ep(eV)	23.0	23.0

[a] $E_G^0(x) = 1.512 + 1.155x + 0.37x^2$

where ω_2 and ω_1 are the pump–probe frequencies, respectively, n_o is the average area density of unit cells, m_o is the rest mass of an electron, e is the charge on an electron, E_P is the Kane matrix element, and $\hbar\omega_{bx}$ is the biexciton binding energy. The parameters γ and Γ_{ij} are longitudinal and transverse relaxation/broadening parameters, respectively. The ij indexes refer to the 0-system ground state; g-exciton ground state; and b-biexciton ground state. The parameter $1/\gamma = 1/\Gamma_{ii}$ is the population decay time for state i, and $1/\Gamma_{ij}$ is the dephasing or lifetime of the coherent superpositions between states i and j. Note that if all the transverse relaxation parameters are assumed to be equal, the model reduces to the independent boson model. Then, when the biexciton binding energy approaches zero so does the third-order susceptibility, because of the first bracketed factor in Eq. (11.36), as it should. In order to calculate the exciton and biexciton binding- and ground-state energies and optical susceptibilities, we employed the physical parameters summarized in Table 11.1. All parameters needed but not listed in Table 11.1 are shown on the figures.

The Al interdiffusion in the lateral direction requires a numerical solution of Eq. (11.2) for the single-particle electron and hole states. As a part of this numerical approach one must search for the integration range, that is, numerical infinity, in which the wave function appropriately tends to zero and yields the correct corresponding eigenvalue: This must be done for every set of cross-sectional dimensions and for each characteristic Al grading width $2C$. In turn, the values of the three variational parameters are changed; one must further search for an acceptable starting value of the exciton variational parameter in order to start the minimization process. Needless to say, the whole process is time consuming. Thus, we have limited our dimensional parametric study to the smallest grading, $C = W/16$, corresponding to an almost abrupt junction. And, for illustrative purposes, we have limited ourselves to one set of wire dimensions $(L \times W = 125 \times 75\ \text{Å})$ and four characteristic diffusion lengths ($C = W/16, W/8, W/4, W/2$). This particular set of wire dimensions was chosen because it is on the order of the bulk exciton radius and corresponds to a peak third-order optical susceptibility calculated for the infinite barrier model. The results of the electronic part of the calculation are summarized in Table 11.2.

TABLE 11.2. Infinite Potential Results versus Finite Graded Potential Results[a,b]

Parameter (unit)	Infinite	$W/16^c$	$W/8^c$	$W/4^c$	$W/2^c$
ε_1^e(meV)	99.75	38.67	41.69	51.38	47.24
ε_1^h(meV)	14.85	9.66	11.11	16.73	16.76
$\eta_x(a_0)$	362.20	395.45	393.58	389.95	395.12
E_B^X(meV)	−14.23	−12.27	−12.35	−12.54	−12.29
V_{eh}(meV)	−19.69	−16.85	−16.97	−17.25	−16.88
E_{g0}^x(eV)	1.662	1.597	1.602	1.617	1.668
$\xi(a_0)$	1024.43	1116.79	1111.62	1101.55	1115.88
$\eta(a_0)$	293.38	320.36	318.86	313.96	318.09
E_B^{XX}(meV)	−9.71	−8.36	−8.42	−8.55	−8.38
V^{XX}(meV)	−53.61	−45.88	−46.22	−46.97	−45.97
E_{g0}^{XX}(eV)	3.31	3.186	3.195	3.225	3.328
$z(0.01u_{e\max}^{(1)})$(Å)		97.00	99.00	100.00	115.00
$z(0.01u_{h\max}^{(1)})$(Å)		61.00	62.00	63.00	69.00

[a](GaAs Wire—$Al_{0.3}Ga_{0.7}As$ Cladding). $L \times W = 125 \times 75$ Å.
[b]Band offsets—0.6 conduction, 0.4 valence.
[c]One-half A1 grading width.

With reference to Table 11.2, we first note that the inclusion of finite band offsets lowers both electron and hole single-particle subband energies for the narrowest grading (almost abrupt), $W/16$, giving $\sim 61\%$ difference for the electron and 35% difference for the hole. And, of course, this is precisely what is expected to happen. Fixing our attention on just the electron states for the moment, we see that, by increasing the diffusion length from $W/16$ to $W/8$ and then from $W/8$ to $W/4$, the subband energies rise. The reason for this trend is clear especially when we consider the graded well structure: As the diffusion width is increased the bottom of the well narrows pushing the states up. However, when the diffusion length is increased from $W/4$ to $W/2$, the trend appears to be reversed. The cause of this reversal is also clear: The well is now "over graded". That is, under this condition, there is enough Al concentration located at the center of the well to change the fundamental gap from that of pure GaAs to some percentage AlGaAs. As a result, the gap is increased and the well, as it rises, begins to flatten out. Relative to the bottom of the graded well, the subband energy has indeed become smaller, but relative to the bottom of the GaAs well, it has continued to rise. In fact, as the fraction of Al in the center of the graded well approaches $x = 0.3$, the subband(s) coalesce(s) at the bottom of the well and form the conduction band in bulk AlGaAs. Thereafter, the lowering of subband energies is to be expected.

A similar argument can be made for the hole subband energies. However, it is apparent from Table 11.2 that when the diffusion length reaches $W/4$, the subband energy exceeds that of the infinite barrier model. The cause of such a result is not totally unexpected. The barrier height for the holes is two-thirds of that for the electrons. This means that the corresponding Al grading of the valence band produces a well that is narrower for $z < W/2$ and wider for $z > W/2$ than that of the conduction band. One then expects a larger percentage increase in energy for the lower lying hole subbands bands. For example, changing the diffusion length from $W/8$ to $W/4$ produces an $\sim 19\%$ increase in the electron subband energy and an $\sim 34\%$ increase in the hole subband energy.

The exciton and biexciton binding energies are lowered with the inclusion of the finite band offsets by $\sim 14\%$ for the smallest diffusion length of $W/16$. The evolution of their respective values with increasing Al-grading width can easily be explained in terms of the arguments given above for the electron and hole subbands. Similarly, the exciton and biexciton ground-state energies are lowered with the inclusion of the finite band offsets. Since each is dominated by the fundamental band-gap energy, their change is rather small, both being $\sim 4\%$ for the smallest diffusion length of $W/16$. Because of such a small change in the exciton ground-state energy, the effect on the magnitude of optical susceptibility is negligible. However, at higher densities of excitons and biexcitons than considered here, the lower binding energies do affect the stability of the system and will thus limit the operational conditions under which peak $\chi^{(3)}$ values may be obtained and maintained.

The structure in $\chi^{(3)}$ is a strong function of the values of the longitudinal and transverse relaxation parameters. These parameters are quite difficult to obtain accurately by experiment and there are no first principle theoretical models of which we are aware. In addition, they are probably strong functions of the confinement

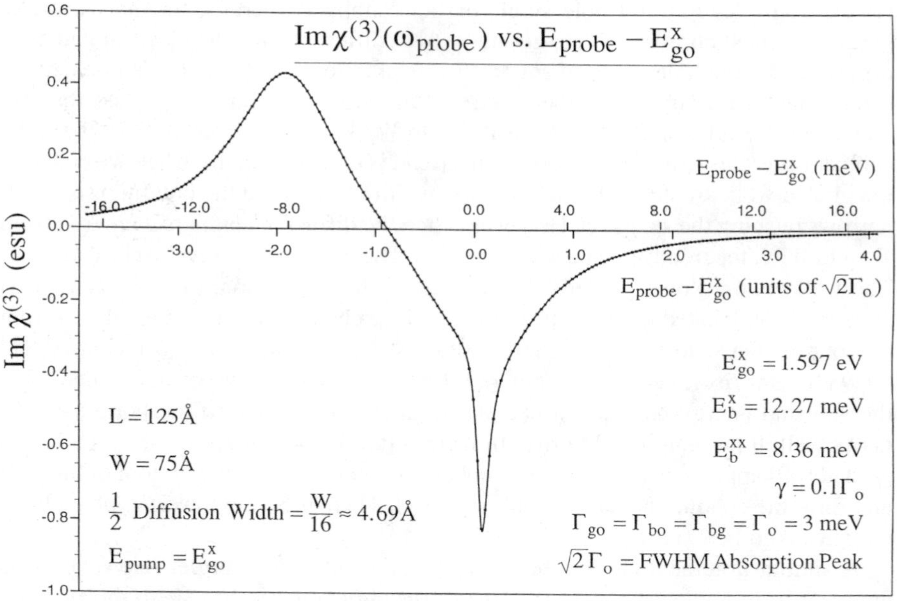

Figure 11.4. The imaginary part of the third-order optical susceptibility as a function of the probe energy for a dual beam pump–probe experiment. The pump is set at the exciton resonance and the longitudinal broadening parameter is one-tenth the value of the transverse broadening parameters.

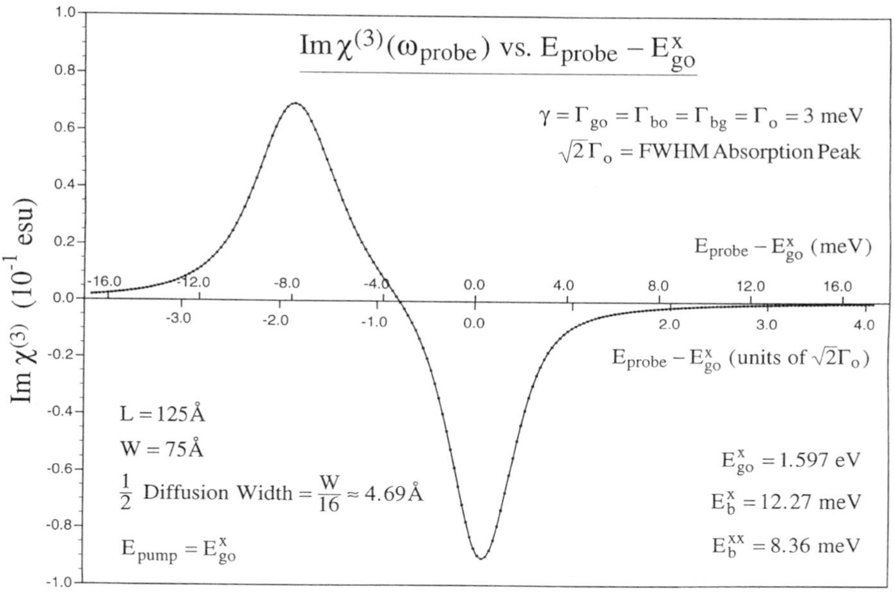

Figure 11.5. Same as Fig. 11.4, but with the longitudinal broadening parameter is equal to the transverse broadening parameters.

dimensions as well as the population density of the excitons and temperature [51,52]. Consequently, some researchers have chosen them to be equal (e.g., [53,54]), while some have used them as fitting parameters (e.g., [55]). There is, however, good reason to believe that the longitudinal parameter is perhaps as much as an order of magnitude smaller than the transverse parameters [56]. Accordingly, for comparison, we have calculated $\chi^{(3)}$ for values $\gamma = 0.1\Gamma_0$ and all $\Gamma_{ij} = \Gamma_0$; and for $\gamma = \Gamma_{ij} = \Gamma_0$, where $\sqrt{2}\Gamma_0$ corresponds to the full width at half maximum (fwhm) of the Gaussian representing the exciton linear absorption.

In Figs. 11.4–11.9, we have calculated $\chi^{(3)}$ for a two-beam experiment in which one beam, the pump, is fixed and the other, the probe, is allowed to vary over a frequency range in which the pump is fixed. Specifically, in Figs. 11.4 and 11.5 the pump is fixed right on the exciton resonance, and in Figs. 11.6 and 11.7, and 11.8 and 11.9, the pump is detuned from the exciton resonance by $+\sqrt{2}\Gamma_0$ and $-\sqrt{2}\Gamma_0$, respectively. The abscissas on all plots are in dual energy units: On top are the more conventional millielectronvolts units, and on the bottom are the fwhm units of $\sqrt{2}\Gamma_0$. We have chosen the fwhm units in order to give a measure of the relative strengths of the optical susceptibility and exciton absorption. The peak value of exciton absorption for all plots given here is $1.94 \times 10^4\ \text{cm}^{-1}$. Figs. 11.3, 11.5 and 11.7 have values of $\gamma = 0.1\Gamma_0$, while Figs. 11.4, 11.6 and 11.8 have values of $\gamma = \Gamma_{ij} = \Gamma_0$. Note that $-\text{Im}\chi^{(3)}$ is proportional to $-\Delta\alpha$, the differential change in optical transmission.

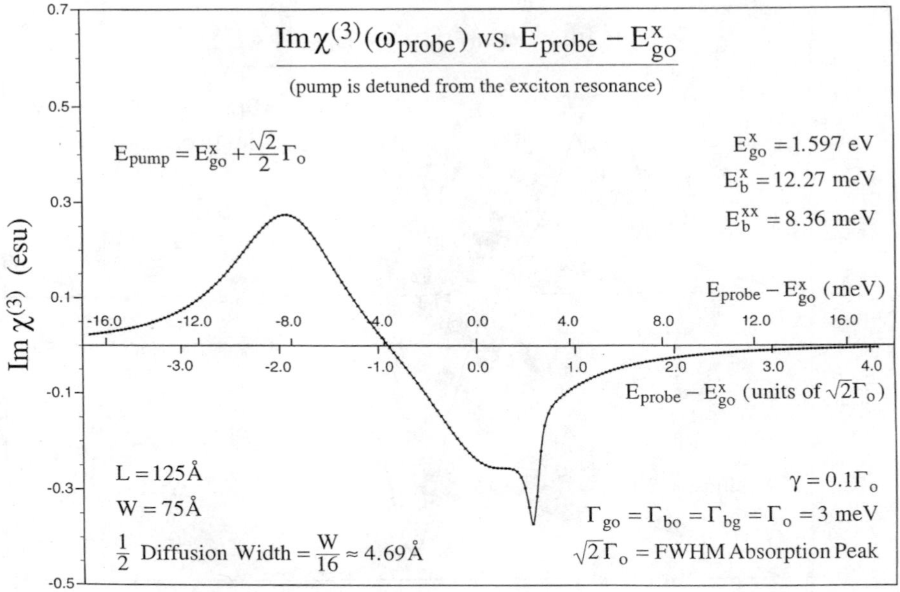

Figure 11.6. Same as Fig. 11.4, but with the pump detuned slightly above the exciton resonance.

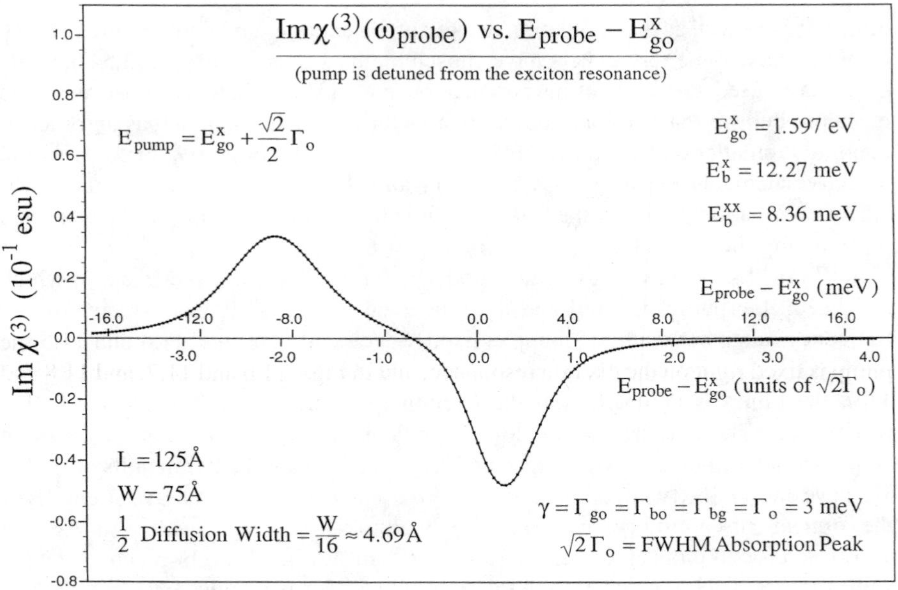

Figure 11.7. Same as Fig. 11.6, but with the the longitudinal broadening parameter equal to the transverse broadening parameters.

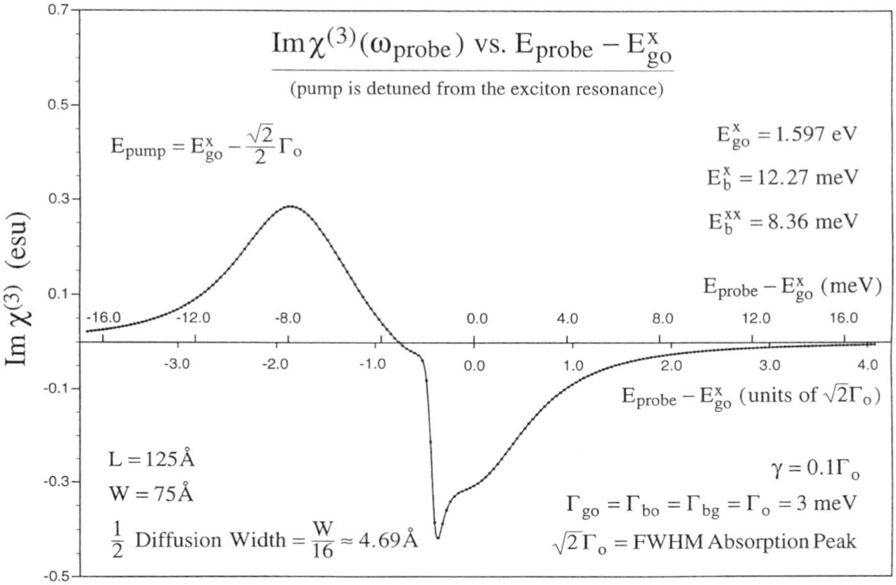

Figure 11.8. Same as Fig. 11.6, but with the pump detuned slightly below the exciton resonance.

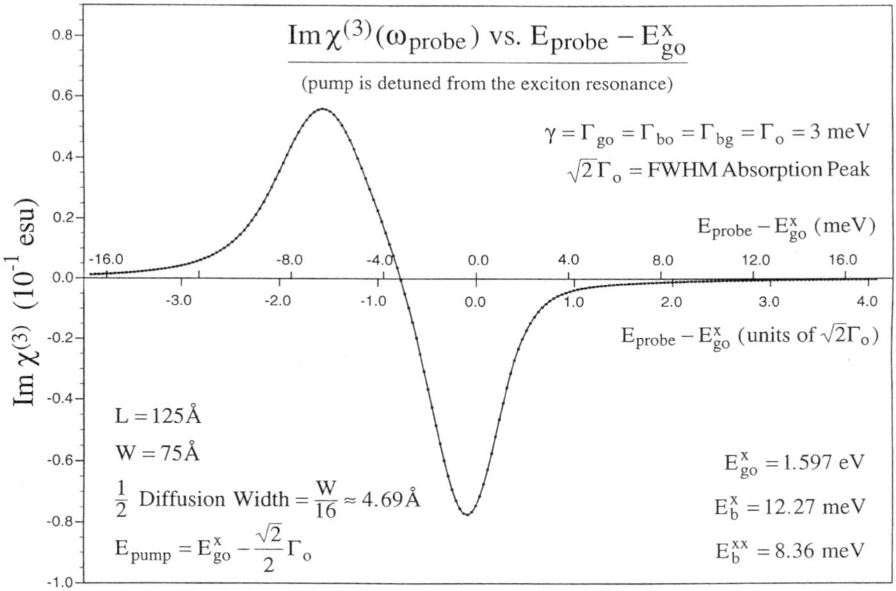

Figure 11.9. Same as Fig. 11.8, but with the the longitudinal broadening parameter equal to the transverse broadening parameters.

The first important thing to notice about Figs. 11.4–11.7 is the giant values of $\chi^{(3)}$ with respect to the bulk material. Comparing Figs. 11.4 with 11.5, 11.6 with 11.7, and 11.8 with 11.9 it is immediately apparent that Figs. 11.4, 11.6, and 11.8 all posses a rather abrupt, narrow negative peak for probe energies near or equal to the pump energies. The genesis of these peaks may be directly traced back to the $1/[\omega_r - \omega_2 + i\gamma]$ factor in Eq. (11.36) for the susceptibility. When $\chi^{(3)}$ is separated into real and imaginary parts, this factor leads to a resonance factor in both parts that goes as $1/[(\omega_r - \omega_2)^2 + \gamma^2]$. The frequency ω_2 corresponds to the pump and the index r on ω_r is summed over values of 1 (probe) and 2. Since ω_1 is the probe, it is varied, and, when in resonance with ω_2, the resonance factor becomes dominant. Its strength, however, is extremely sensitive to the magnitude of γ: when $\gamma = 0.1\Gamma_0$ its strength is 10^2, or two orders of magnitude larger then when $\gamma = \Gamma_0$, as is the case in Figs. 11.5, 11.7, and 11.9. Physically, γ is related to the population decay rate of the exciton state. A smaller γ, then, means the lifetime of the exciton state is larger and that the exciton system is more stable. In turn, there is a higher probability of forming an excitonic molecule in a two-step photon absorption process. The resonance spiking is only significant if the detuning of the pump lies within one to two fwhm of the exciton absorption peak. Note that in Figs. 11.6 and 11.8 the curves tend to have a relative minimum at peak exciton absorption. Of course, when the pump is in resonance with the peak exciton absorption, the spiking is amplified even further resulting in the curve displayed in Fig. 11.4. On the other hand, the curves displayed in Figs. 11.5, 11.7, and 11.9, those for which $\gamma = \Gamma_0$, show no sign of the abrupt spiking: The resonance factor is now ~ 10 times smaller than before. The larger γ not only smears out the resonance spiking, but it reduces the overall magnitude of each curve (note the scales of these curves are a factor of 10 times smaller).

The negative peak in all of these spectra, indicating transmission, is due to a bleaching (saturation) of the one-pair exciton transition. Physically, the initial exciton population created by the pump beam tends to amplify the probe beam, by way of stimulated emission, when the probe energy is tuned at or near the exciton linear absorption peak. Another feature in all of the curves is the optical absorption—the region of positive $\text{Im}\chi^{(3)}$. The absorption may be attributed to the formation of the excitonic molecule [53–55]. The initial exciton population enables the probe to be more strongly absorbed when its energy matchs the exciton–biexciton transition energy, $\hbar\omega_{g0} - \hbar\omega_{bx}$. In the present case, the biexciton binding energy is 8.36 meV. The maximum of each curve occurs at energies slightly greater than $-2.0\sqrt{2}\Gamma_0 = -9.3$ meV, in other words, very near the biexciton binding energy. Calculations for a wire of dimensions $L \times W = 225 \times 175$ Å substantiate this interpretation. In that calculation, the biexciton binding energy was -5.55 meV, and the maxima of the $\text{Im}\chi^{(3)}$ curves occurred at energies slightly less than $-1.3\sqrt{2}\Gamma_0 = -6.61$ meV.

In the last two plots, Figs. 11.10 and 11.11, we show typical curves for a single-beam pump–probe experiment, where $\gamma = 0.1\Gamma_0$ and $\gamma = \Gamma_0$, respectively. Since the pump and probe beams are always in resonance, the factor $1/[(\omega_r - \omega_2)^2 + \gamma^2]$ becomes just $1/\gamma^2$, which is constant, and it now acts to modulate the magnitude of $\text{Im}\chi^{(3)}$. Both curves exhibit similar structure but, as expected, the curve for which $\gamma = 0.1\Gamma_0$ is approximately an order of magnitude larger.

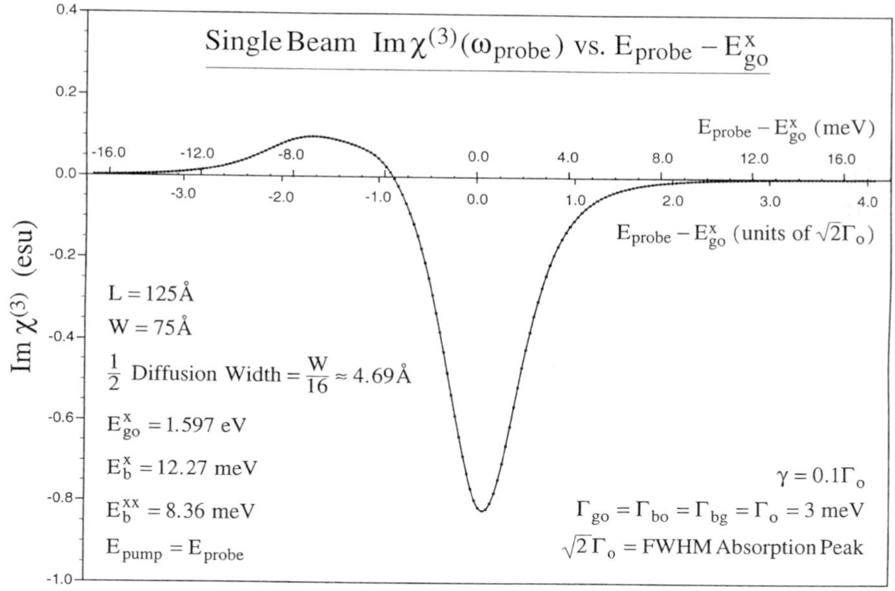

Figure 11.10. The imaginary part of the third-order optical susceptibility as a function of the energy for a single-beam pump–probe experiment. The longitudinal broadening parameter is one-tenth the value of the transverse broadening.

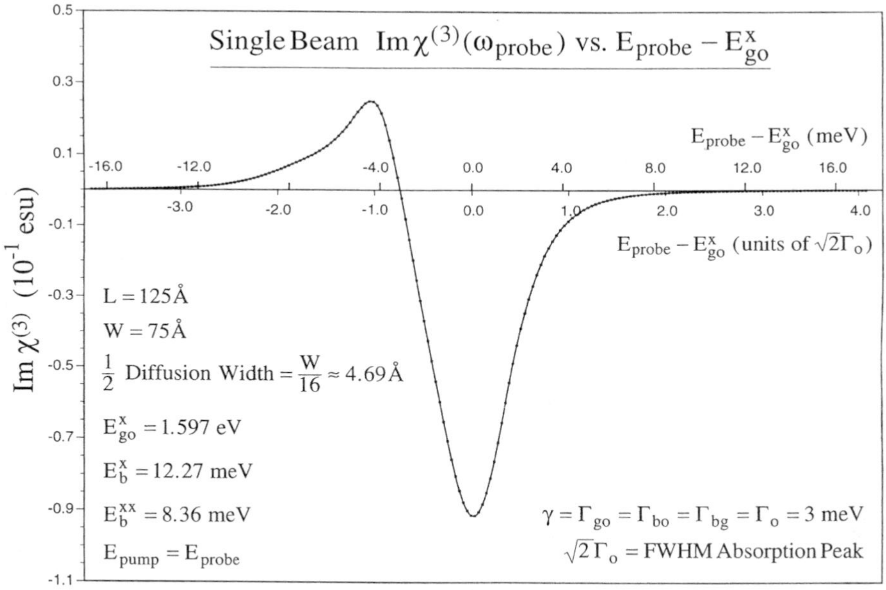

Figure 11.11. Same as Fig. 11.10, but with the longitudinal broadening parameter equal to the transverse broadening parameters.

11.3. EFFECT OF A MAGNETIC FIELD

In the past, the effects of a magnetic field on excitonic complexes in low-dimensional structures have been largely ignored since a magnetic field is technologically less significant than an electric field in device applications. However, a magnetic field has two important advantages. First, it increases the exciton–biexciton binding energy by squeezing the electrons and holes together, while an electric field decreases the energy by spatially separating oppositely charged particles. Therefore, a magnetic field can cause improved nonlinear optical properties. Second, it offers a far richer insight into the physics of excitonic complexes, especially for low-dimensional systems.

In our review of magnetooptical properties of quantum wires, we will concentrate our attention on the regime of intermediate fields characterized by the dimensionless parameter $\beta \equiv \hbar eB/\mu R^* \approx 1$. Here, B is the magnetic flux density, and R^* is the excitonic Rydberg energy. The wire lateral dimensions will be chosen in such a way that the electrostatic confinement of excitons is on the order of magnitude of the magnetostatic confinement, that is, magnetic length $l_B \equiv \sqrt{(\hbar/eB)} \approx L, W$.

We consider a generic quantum wire of rectangular cross-section as in Section 11.2. The only difference is that now we consider infinite potential barriers located at $y = \pm L/2$ and $z = \pm W/2$. A magnetic field of flux density B is applied along the z direction.

11.3.1. Magnetoexcitons in a Quantum Wire

For nondegenerate and isotropic bands, the Hamiltonian of a free Wannier exciton in this system is given within the envelope-function approximation by

$$\begin{aligned}\mathscr{H}^X = & \frac{1}{2m_e}(p_{x_e} - eBy_e)^2 + \frac{1}{2m_h}(p_{x_h} + eBy_h)^2 \\ & + \frac{p_{y_e}^2 + p_{z_e}^2}{2m_e} + \frac{p_{y_h}^2 + p_{z_h}^2}{2m_h} \\ & - \frac{e^2}{\varepsilon[(x_e - x_h)^2 + (y_e - y_h)^2 + (z_e - z_h)^2]^{1/2}} \\ & + V_C(y_{e,h}, z_{e,h}) \end{aligned} \tag{11.37}$$

where we have chosen the Landau gauge

$$\mathbf{A} = (-By, 0, 0)$$

The quantities $m_e, m_h, (x_{e,h}, y_{e,h}, z_{e,h})$ are the effective masses and coordinates of electrons and holes, respectively, ε is the dielectric constant, and $V_{e,h}(y_{e,h}, z_{e,h})$ are the confinement potentials for electrons and holes along y and z directions. For convenience, we replace $x_{e,h}$ coordinates by the center-of-mass (X) and relative coordinates (x). This is accomplished by using the quantum mechanical definition of momentum operators and taking into account that in a center-of-mass and relative coordinate system

$$p_{x_{e,h}} = -i\hbar \frac{m_{e,h}}{M}\frac{\partial}{\partial X} \mp i\hbar \frac{\partial}{\partial x} \tag{11.38}$$

$$p_{x_{e,h}}^2 = -\hbar^2 \left(\frac{m_{e,h}}{M}\right)^2 \frac{\partial^2}{\partial X^2} \mp 2\hbar^2 \frac{m_{e,h}}{M}\frac{\partial^2}{\partial X \partial x} - \hbar^2 \frac{\partial^2}{\partial x^2} \tag{11.39}$$

By defining

$$\begin{aligned} P_X &\equiv -i\hbar \frac{\partial}{\partial X} \\ p_x &\equiv -i\hbar \frac{\partial}{\partial x} \end{aligned} \tag{11.40}$$

we obtain

$$\begin{aligned} \mathscr{H}^X = {} & \frac{P_X^2}{2M} + \frac{p_x^2}{2\mu} + \frac{p_{y_e}^2 + p_{z_e}^2}{2m_e} + \frac{p_{y_h}^2 + p_{z_h}^2}{2m_h} \\ & + \frac{eB(y_e - y_h)}{M} P_X + eB(y_e/m_e + y_h/m_h)p_x \\ & + \frac{e^2B^2}{2}(y_e^2/m_e + y_h^2/m_h) + V_{e,h}(y_{e,h}, z_{e,h}) \\ & - \frac{e^2}{4\pi\varepsilon\left(x^2 + (y_e - y_h)^2 + (z_e - z_h)^2\right)^{1/2}} \end{aligned} \tag{11.41}$$

where

$$\begin{aligned} \frac{1}{\mu} &= \frac{1}{m_e} + \frac{1}{m_h} \\ M\mathbf{R} &= m_e\mathbf{r}_e + m_h\mathbf{r}_h \\ \mathbf{r}_{e,h} &= x_{e,h}\boldsymbol{x} + y_{e,h}\boldsymbol{y} + z_{e,h}\boldsymbol{z} \\ \mathbf{R} &= \boldsymbol{x}X + \boldsymbol{y}Y + \boldsymbol{z}Z \end{aligned}$$

We use the variational approach outlined in Section 11.2 with the modification that allows us to take a magnetic field into consideration [57,58]. Since the Hamiltonian does not depend on X, P_X is a good quantum number. By dropping the term associated with P_X, we take a trial wave function analogous to the one used in Section 11.2.1

$$\Psi = \frac{2}{L} g_t(x, \eta_x)\phi_e(y_e)\phi_h(y_h)\chi_e(z_e)\chi_h(z_h) \tag{11.42}$$

where $g_t(x, \eta_x)$ is the same the Gaussian-type "orbital" function as in Section 11.2.1. The variables $\chi_{e,h}(z_{e,h})$ are the z components of the wave functions that are not affected by the magnetic field. They are given by particle-in-a-box states $\chi_{e,h}(z_{e,h}) = \cos(\pi z_{e,h}/W)$. The electron and hole wave functions along the y direction, $\phi_{e,h}(y_{e,h})$, are to be calculated numerically when a magnetic field is present. This is done by solving the Schrödinger equation directly following the prescription given in [59]. This combination of the variational and exact numerical solution was initially developed in [58,38].

It is important to note that there are really two different cases of exciton quantization: (1) an electron–hole droplet whereby the exciton is considered to be a particle by itself, and (2) an entity consisting of an independently confined electron and hole. According to [60], the criterion for this separation is $L, W = 3a_{\mathrm{B}}^*$, where $a_{\mathrm{B}}^*$is the effective Bohr radius in the bulk. The trial wave function of Eq. (11.42) implicitly assumes the electron and the hole are independently confined along the y and z directions, which corresponds to the case

$$L, W < 3a_{\mathrm{B}}^* \tag{11.43}$$

The wave function in Eq. (11.42) involves the variational parameter η_x, which is evaluated by minimizing the expectation value of the Hamiltonian in Eq. (11.41) (with given trial wave functions) with respect to η_x. Once this is accomplished, one can find the exciton binding energies and the exciton length for different values of magnetic field and the wire width. The functional to be minimized can be written as follows [58]

$$\begin{aligned}
\langle\psi|\mathscr{H}^X|\psi\rangle = {} & \frac{\hbar^2}{2\mu\eta_x^2} + \frac{\hbar^2}{2\mu W^2} + \frac{\hbar^2}{2m_e}\int_{-L/2}^{L/2}(\phi_e')^2 dy_e \\
& + \frac{\hbar^2}{2m_h}\int_{-L/2}^{L/2}(\phi_h')^2 dy_h + \frac{e^2B^2}{2m_e}\int_{-L/2}^{L/2}(\phi_e y_e)^2 dy_e \\
& + \frac{e^2B^2}{2m_h}\int_{-L/2}^{L/2}(\phi_h y_h)^2 dy_h \\
& - \frac{e^2}{\varepsilon}\int_{\Omega}\frac{g_t^2(x,\eta)\phi_e^2\phi_h^2\chi_e^2\chi_h^2}{[x^2 + (y_e - y_h)^2 + (z_e - z_h)^2]^{1/2}}d\zeta
\end{aligned} \tag{11.44}$$

where $d\zeta = dx dy_e dy_h dz_e dz_h$. The integration of the last (Coulomb) term is carried out over a hyper-rectangle Ω that has an infinite interval along the x direction and is limited by $\pm L/2$ and $\pm W/2$ along the y and z directions, respectively. To obtain (11.6), we have made use of the boundary conditions $\phi_{e,h}(\pm L) = 0$, which allowed us to integrate some of the terms analytically using integration by parts. Note that the expectation value of the non-Hermitean operator $eB(y_e/m_e + y_h/m_h)p_x$, which arises in the presence of a magnetic field, is identically zero for the chosen trial wave function, which makes the expectation value in Eq. (11.44) strictly real and shows that the trial wave function space is admissible. Equation (11.44) allows us to treat

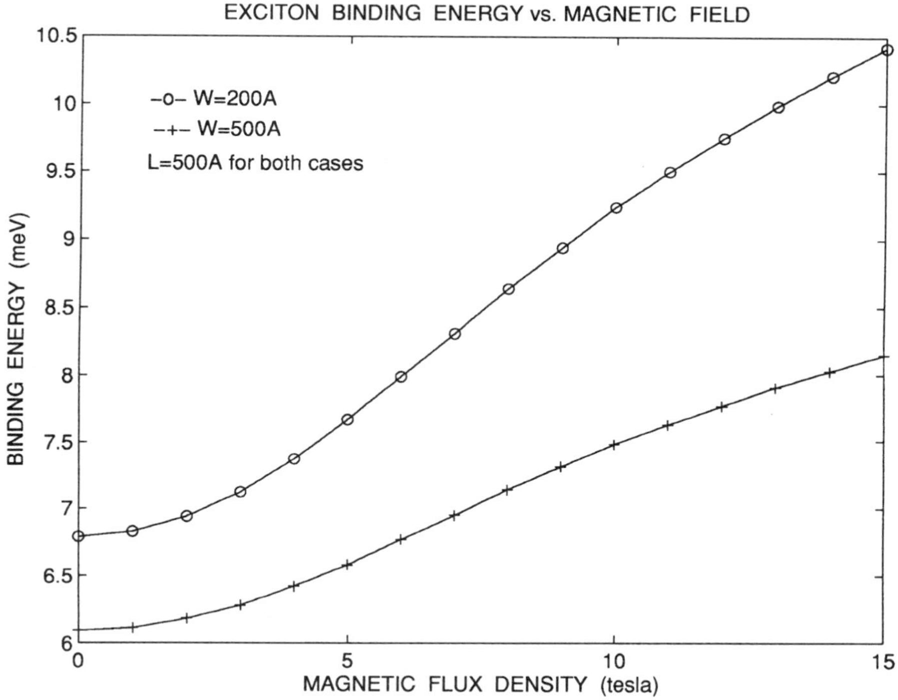

Figure 11.12. Magnetic field dependence of exciton binding energy in a GaAs quantum wire. The results are shown for different wire thickness W.

the Coulomb interaction term exactly in its full 3D form throughout the calculation, which is physically more realistic than the earlier approaches.

Ground state exciton binding energies E_B^X can now be found using the same definition given in Section 11.2.1. Figure 11.12 shows both the exciton binding energy and the exciton radius as a function of the magnetic field. Binding energy increases with the magnetic field for all wire widths, which is in qualitative agreement with the results obtained for 2D systems [34], except that while the increase is sublinear in 2D systems, it is superlinear in 1D systems. It is interesting to note that the effect of the magnetic field is much more pronounced for the wider wire $L = 500$ Å than for the narrower wire $L = 300$ Å. This finding can be explained in the following way. A magnetic field squeezes the electron and hole wave functions causing these states to condense into cyclotron (Landau) orbits whose radii shrink with increasing magnetic fields. As long as the wire width L is smaller than the magnetic length or the lowest cyclotron radius l_B $(= \sqrt{\hbar/eB})$, the effect of the magnetic field is not very pronounced and the geometric confinement predominates. It is only when $L > l_B$ that the effect of the magnetic field becomes predominant. Therefore, a wider wire will show a stronger magnetic field induced effect.

It is interesting to note that the effect of localization of the exciton wave function by a magnetic field can be observed not only along lateral dimensions but also along

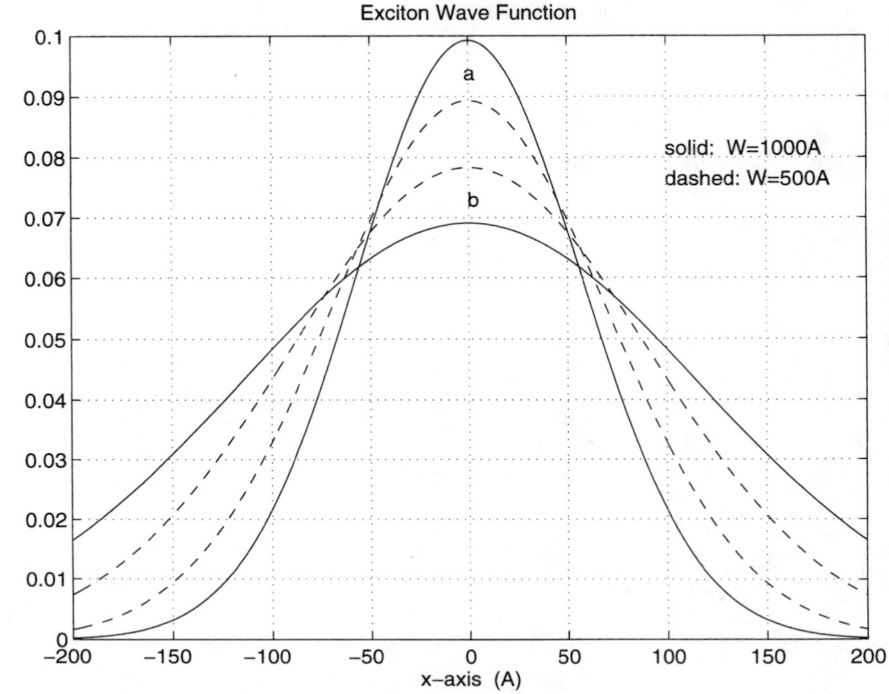

Figure 11.13. The x component of the exciton wave function $g_t[x, \eta_{\text{opt}}(B)]$. The two curves grouped as "a" correspond to a magnetic flux density of 10 T while those grouped as "b" correspond to 0 T. This figure illustrates that the exciton wave function is squeezed by a magnetic field not only along lateral dimensions but along "unconfined" x axis as well.

the "free" x direction. Figure 11.13 shows the x component of the exciton wave function $g_t[x, \eta_x(B)]$ for two different wire widths, with and without a magnetic field.

Recently, Someya et al. [23] reported the effect of an external magnetic field on the exciton binding energy and radius in a GaAs quantum wire by measuring the photoluminescence spectra and comparing them with those of quantum wells. They found that a magnetic field squeezes the exciton wave function to a size that is far below what can be achieved in quantum wells. This is in excellent agreement with our theoretical predictions [40,58], and is very promising for potential optoelectronic applications of quantum wires.

11.3.2. Magnetobiexcitons in Quantum Wires

For nondegenerate and isotropic bands, the Hamiltonian of a biexciton in a quantum wire subjected to a magnetic field is given in the envelope-function approximation by [21]

$$
\begin{aligned}
\mathscr{H}^{XX} = & \frac{1}{2m_e}(p_{x_1}^2 + p_{x_2}^2) + \frac{1}{2m_h}(p_{x_a}^2 + p_{x_b}^2) \\
& + \frac{1}{2m_e}(p_{y_1}^2 + p_{y_2}^2) + \frac{1}{2m_h}(p_{y_a}^2 + p_{y_b}^2) \\
& + \frac{1}{2m_e}(p_{z_1}^2 + p_{z_2}^2) + \frac{1}{2m_h}(p_{z_a}^2 + p_{z_b}^2) \\
& + \frac{1}{2m_e}[e^2B^2(y_1^2 + y_2^2) - 2eB(y_1 p_{x_1} + y_2 p_{x_2})] \\
& + \frac{1}{2m_h}[e^2B^2(y_a^2 + y_b^2) + 2eB(y_a p_{x_a} + y_b p_{x_b})] \\
& + V_C + V_{e,h}(y_{e,h}, z_{e,h})
\end{aligned}
\tag{11.45}
$$

where V_C is the Coulomb interaction between various charged entities. The electron coordinates are subscripted by the numerics 1 and 2 while hole coordinates are subscripted by the letters a and b. In Eq. (11.45) we neglected the Zeeman splitting since the Landé g factor is small for GaAs.

For weak Coulomb interaction (characteristic of relatively narrow gap materials with large dielectric constant), the wave function of the biexciton can be written as a product of particle-in-a-box states along the z direction, magnetoelectric states along the y direction, and a Heitler-London type symmetric linear combination of Gaussian orbitals (involving variational parameters) along the x direction [40]. Because of this particular choice of the wave function, it is convenient to use the following relative and center-of-mass coordinates to transform the Hamiltonian in Eq. (11.45):

$$
\begin{cases}
X = [m_e(x_1 + x_2) + m_h(x_a + x_b)]/2(m_e + m_h) \\
x_{1a} = x_1 - x_a \\
x_{2b} = x_2 - x_b \\
x_{ab} = x_a - x_b
\end{cases}
$$

To rewrite the biexciton Hamiltonian in the new coordinate system, we utilize the usual canonical transformations for the momentum operators. We also drop the kinetic energy operators associated with the center-of-mass motion since the Hamiltonian is invariant in X so that P_X is a good quantum number. We are only interested in obtaining biexcitonic states that can be accessed optically and under optical excitation the center-of-mass motion can be neglected since the photon momentum is too small to create states of significant center-of-mass kinetic energy. Consequently, the transformed Hamiltonian becomes

$$
\begin{aligned}
\mathscr{H}^{XX} = & -\frac{\hbar^2}{2\mu}\left(\frac{\partial^2}{\partial x_{1a}^2} + \frac{\partial^2}{\partial x_{2b}^2}\right) - \frac{\hbar^2}{m_h}\left(\frac{\partial^2}{\partial x_{ab}^2} - \frac{\partial^2}{\partial x_{1a}\partial x_{ab}} + \frac{\partial^2}{\partial x_{2b}\partial x_{ab}}\right) \\
& -\frac{\hbar^2}{2m_e}\left(\frac{\partial^2}{\partial y_1^2} + \frac{\partial^2}{\partial y_2^2} + \frac{\partial^2}{\partial z_1^2} + \frac{\partial^2}{\partial z_2^2}\right)
\end{aligned}
$$

$$
\begin{aligned}
&-\frac{\hbar^2}{2m_h}\left(\frac{\partial^2}{\partial y_a^2}+\frac{\partial^2}{\partial y_b^2}+\frac{\partial^2}{\partial z_a^2}+\frac{\partial^2}{\partial z_b^2}\right)\\
&+\frac{e^2B^2}{2}\left(\frac{y_1^2+y_2^2}{m_e}+\frac{y_a^2+y_b^2}{m_h}\right)+\frac{eBi\hbar}{m_e}\left(y_1\frac{\partial^2}{\partial x_{1a}}+y_2\frac{\partial^2}{\partial x_{2b}}\right)\\
&+\frac{eBi\hbar}{m_h}\left[y_a\frac{\partial^2}{\partial x_{1a}}+(y_b-y_a)\frac{\partial^2}{\partial x_{ab}}+y_b\frac{\partial^2}{\partial x_{2b}}\right]+V_C+V_{e,h}(y_{e,h},z_{e,h}) \qquad (11.46)
\end{aligned}
$$

The Coulomb interaction term in Eq. (11.46) is the same as in Eq. (11.24).

The trial wave functions that is used for the variational procedure is the same as the one given by Eq. (11.6) with one principal difference. The quantities $\phi_{e,h}(y_{e,h})$, are y components of the wave functions of independently confined electrons and holes in the quantum wire that are affected by a magnetic field. They are magneto-electric states and are to be calculated independently by solving the Schrödinger equation for an electron in a quantum wire subjected to a magnetic field. This is achieved by following the prescription given in [59]. The quantities $\chi_{e,h}(z_{e,h})$ are the z components of the wave function of independently confined electrons and holes. These quantities are not affected by the magnetic field (the magnetic field is oriented along the z direction) and are therefore again represented by particle-in-a-box states. The chosen trial wave function implies that the "true" wave function is more sensitive to the separation between the holes than between the electrons. Since holes have higher effective masses, this assumption is physically realistic.

The biexciton binding energy E_B^{XX} is given by

$$E_B^{XX}=\min\langle\Psi|\mathscr{H}^{XX}|\Psi\rangle-2\min\langle\psi|\mathscr{H}^{X}|\psi\rangle \qquad (11.47)$$

where min $\langle\Psi|\mathscr{H}^{XX}|\Psi\rangle$ is found by minimizing the expectation value of the Hamiltonian $\mathscr{H}^{XX}$ with respect to the hole–hole variational parameter, and min $\langle\psi|\mathscr{H}^{X}|\psi\rangle$ is found by minimizing the expectation value of the exciton Hamiltonian with respect to the electron–hole variational parameter η as described in Section 11.3.1. As pointed out by Klepfer et al. [20], introduction of a third variational parameter, which allows for relaxation of the electron-hole pair within the biexciton, would not seriously change the biexciton binding energy when there is no magnetic field present. We tacitly assume that the same is true when a magnetic field is present. Mathematical details of computing the expectation values of the exciton and biexciton Hamiltonian were provided by Balandin and Bandyopadhyay [21]. One should note that every imaginary term in the expectation value of the Hamiltonian vanishes exactly, which makes the expectation value strictly real and shows that the trial wave function space is admissible.

Before presenting results, we should make a few remarks about the computational details. The evaluation of the integrals in the Coulomb term of the expectation value is not straightforward owing to the $1/r$ singularity. The authors in [18,19] avoided dealing with this by applying a FT of the Coulomb term. This is not possible in our case since we have a magnetic field present that transforms particle-in-a-box states along the width of the quantum wire (y direction) into magnetoelectric states whose wave functions $\phi_{e,h}(y_{e,h})$ are calculated numerically. Fortunately, unlike the case of the strictly 1D Coulomb potential, the Coulomb term treated in its full 3D form is not pathological, and can be integrated in real space using the extended (multidimensional) definition of the improper integral that exists for this situation. This procedure involves evaluation of five- and seven-dimensional integrals in real-space.We checked for the convergence of the integrals by setting all but the free coordinates to zero and letting the relative coordinates along the x direction gradually vanish over a hyper-sphere in order to ensure that the volume element approaches zero faster than the singularity.

In Fig. 11.14, we present the biexciton binding energy as a function of wire width L for three different values of the magnetic field. The binding energy always decreases with increasing wire width as expected because of decreasing electrostatic confinement. For narrow widths, $L \leq 100$ Å, the magnetic field has little effect. Only at large widths does it cause the rate of decrease to decelerate. These features can be explained by invoking the complementary roles of electrostatic and magnetostatic confinement. At narrow widths, the electrostatic confinement is predominant and the

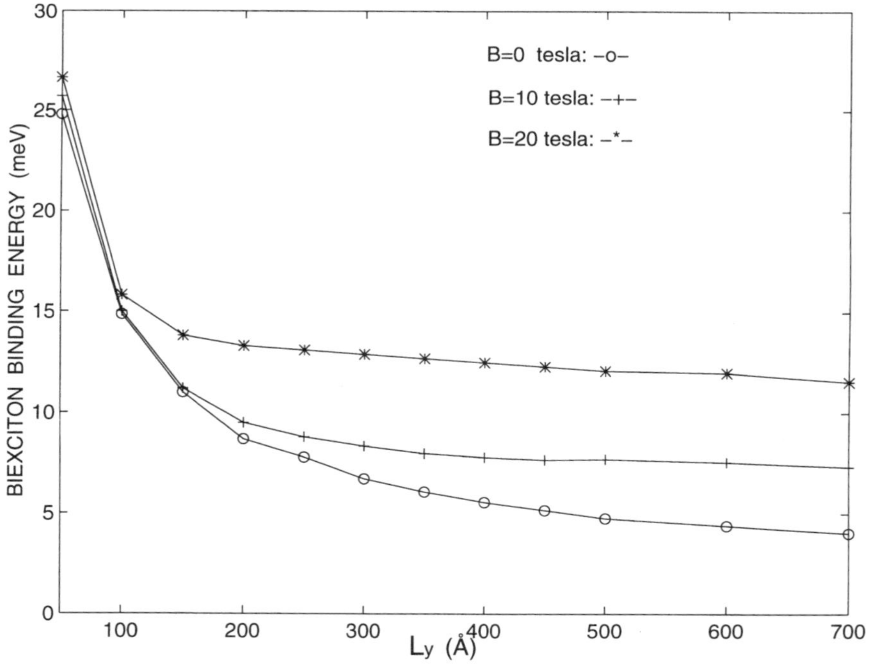

Figure 11.14. Biexciton binding energy in a GaAs quantum wire as a function of wire width L for three different values of the magnetic flux density B. When there is no field present, the binding energy decreases monotonically with increasing wire width. When the field is present, there the binding energy curves tend to saturate beyond a wire width of $\sim$ 300 Å. The thickness along the z direction is $W = 200$ Å.

magnetic field has little or no effect. Once the wire width significantly exceeds the magnetic length l_B, the electrostatic confinement becomes weaker and yields dominant sway to the magnetostatic confinement. Since the latter depends only on the magnetic field strength and is independent of the wire width, the dependence of the binding energy on the wire width becomes much weaker. In the limit of very wide wires (approaching the 2D limit), the electrostatic confinement will gradually vanish. The magnetoelectric states will condense into pure Landau orbits and the confinement will be entirely of magnetostatic origin. In that case, we expect the binding energy to become independent of the wire width so that the top two curves in Fig. 11.14 will saturate at a constant value. This is exactly what happens asymptotically as $L \rightarrow \infty$.

Figure 11.15 shows the biexciton binding energy as a function of the magnetic field for different values of wire width L. For relatively wide wires ($L = 700$ Å), the binding energy increases sublinearly with the magnetic field, whereas for narrower wires ($L = 500$ Å), the increase is superlinear. For wider wires, the magnetostatic confinement takes over the dominant role at relatively low magnetic field strengths, and hence increasing the magnetic field to very high values will not have a dramatic effect. This causes the sublinearity. For narrower wires, the increase is superlinear

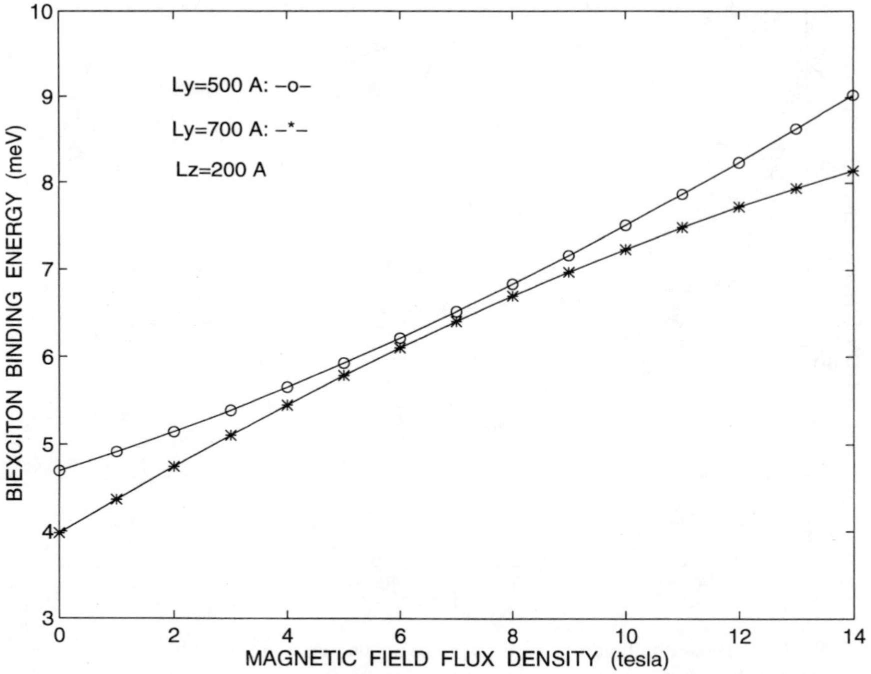

Figure 11.15. Biexciton binding energy in a GaAs quantum wire as a function of magnetic flux density for two values of the wire width: $L = 500$ Å and $L = 700$ Å. The thickness along the z direction is $W = 200$ Å.

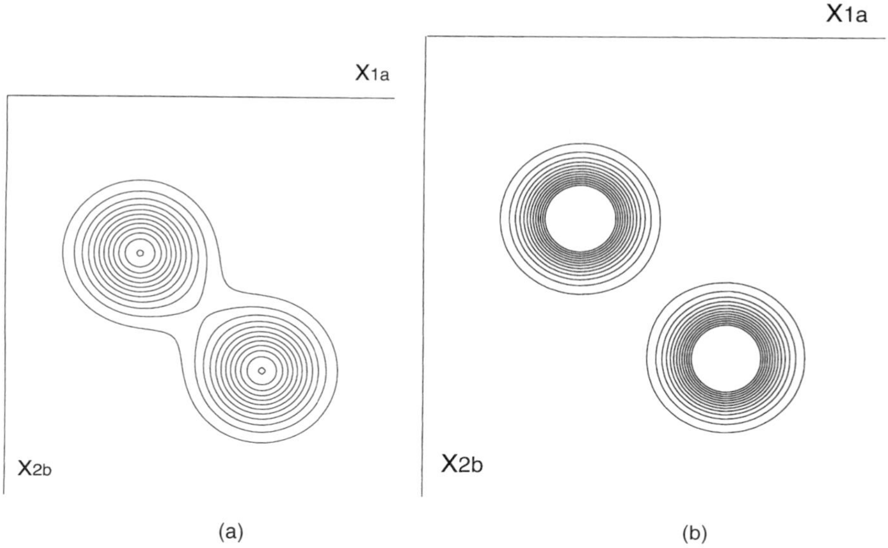

Figure 11.16. Contour plot of the biexciton probability density as a function of relative coordinates x_{1a}, x_{2b} for a quantum wire with (a) and without (b) magnetic field. The wire dimensions are $W = 200$ Å, $L = 700$ Å.

since at low values of the field, the electrostatic confinement is dominant and the field has no effect. Only at relatively high fields, the binding energy increases as the magnetostatic confinement becomes increasingly dominant.

Figure 11.16 shows the probability density distribution of the biexciton as a function of relative coordinates x_{1a} and x_{2b} with and without magnetic field. These two relative coordinates are the separations between one electron and one hole, and between the other electron and the other hole. The lateral dimensions of the wire are chosen to be $W = 200$ Å, $L = 700$ Å, that is, much larger than the magnetic length at 10 T so that the effect of the magnetic field is well observed. We can see that the magnetic field squeezes the biexciton wave function along the free (x axis) direction for both relative coordinates.

The dependence of the ratio of biexciton to exciton binding energy E_B^{XX}/E_B^X on the mass ratio $\sigma = m_e/m_h$ has traditionally been a subject of significant interest and has been investigated by various authors for quasi-1D systems [31,41]; for 2D systems [37]; and for bulk [36,61–65]. The primary findings till now have been that (1) the ratio decreases monotonically with increasing σ from $\sigma = 0$ (hydrogen molecule limit) to $\sigma = 1$ (positronium limit), (2) the ratio versus σ plot is symmetric about $\sigma = 1$, (3) the ratio at any value of σ decreases with increasing dimensionality (quantum wire → quantum well → bulk), and (4) the plot has zero slope at $\sigma = 1$. All of these features are unaffected by a magnetic field as shown in Fig. 11.17. The slight nonzero slope at $\sigma = 1$ is a consequence of numerics and also the Heitler–London type approximation for the variational wave function.

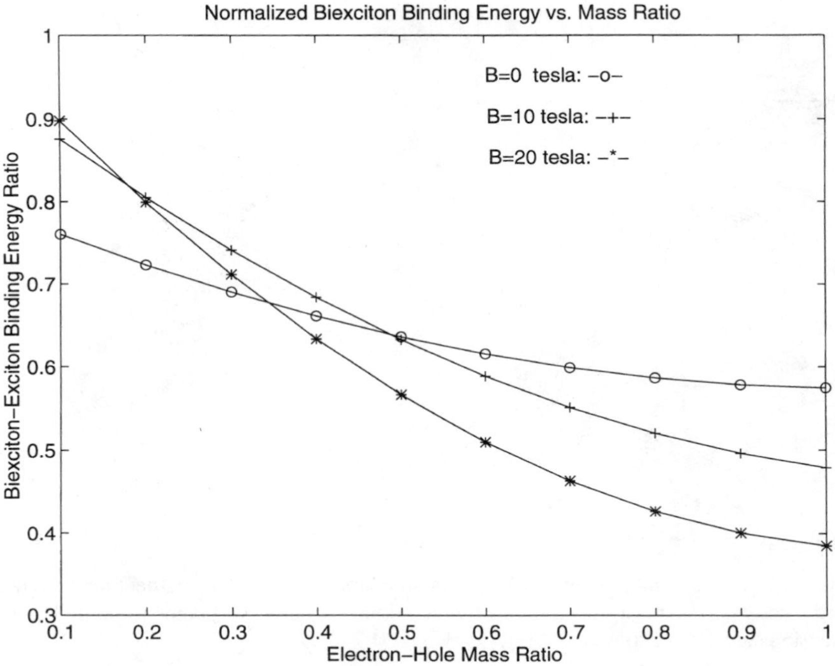

Figure 11.17. Biexciton–exciton binding energy ratio as a function of electron–hole mass ratio σ for different values of magnetic flux density. The wire dimensions are $W = 200\,\text{Å}$, $L = 700\,\text{Å}$.

The magnitude of the energy ratio is of course material dependent since it depends on the Coulomb interaction, and hence on the dielectric constant. The dependence of the ratio on σ is strongest (i.e., the slope of the plot is largest) for the highest magnetic field and the intercept is also largest for the highest field. Because of these two features, there are crossings between the three characteristics in Fig. 11.17. The ratio of the binding energies increases with the magnetic field for small values of σ, but decreases with the magnetic field as σ approaches unity (i.e., the electron and hole effective masses become equal).

11.3.3. Modulation of Refraction and Absorption Indexes

We wish to determine effects of the external magnetic field on the differential refractive index and absorption associated with the third-order nonlinear susceptibility $\chi^{(3)}$ in quantum wires. We assume nearly resonant pumping of the excitonic state in a nondegenerate pump–probe spectroscopy experiment as in Section 11.1 and calculate the changes in refractive index Δn and absorption $\Delta\alpha$ relevant to this situation. The actual measurable quantities in such an experiment are usually the transmission in the absence (Π_0) and in the presence (Π) of the pump. The differential transmission spectra can be found from these quantities as $D = (\Pi - \Pi_0)/\Pi_0$. For small values of the differential transmission (well below unity), D is proportional to the differential absorption $\Delta\alpha$. In fact, $D \approx -\Delta\alpha d$, where d is the wire thickness along the direction of the optical beam.

The nonlinear differential refractive index and absorption can be evaluated theoretically [66,67]. These quantities are given by

$$\Delta n = \frac{2\pi}{\sqrt{\varepsilon_r}} \mathrm{Re}\chi^{(3)} \tag{11.48}$$

and

$$\Delta\alpha = \frac{4\pi\omega}{c\sqrt{\varepsilon_r}} \mathrm{Im}\chi^{(3)} \tag{11.49}$$

where c is a speed of light, ε_r is a relative dielectric constant of the material, ω is a near resonant frequency of pump beam, and $\mathrm{Im}\chi^{(3)}$, $\mathrm{Re}\chi^{(3)}$ are the imaginary and real parts of the nonlinear third-order susceptibility $\chi^{(3)}$, which need to be calculated. The general expression for $\chi^{(3)}$ at relatively low density of excitonic complexes is given in Section 11.2.5.

Since here we are interested in the modulation of the differential refractive index and absorption of quantum wires with a magnetic field, the influence of the field on all parameters in Eq. (11.36) is especially important. The value of Γ in quantum wires is primarily determined by carrier–phonon interactions [68–70]. As shown in [68], the scattering rates associated with these interactions can be affected by a magnetic field at any given kinetic energy of an electron or hole. However, when the rates are averaged over energy, the magnetic field dependence turns out to be quite weak. As a first approximation, we can therefore consider the rates to be independent of the magnetic field. We also neglect thermal broadening of the damping parameters since it is less important in quantum confined systems than in bulk [52].

By using the formalism developed in this chapter for calculation of parameters of excitonic complexes in quantum wires, we can plot $\chi^{(3)}$ as a function of a magnetic field and pump–probe frequencies. In Fig. 11.18, we present a 3D plot of $\mathrm{Im}\chi^{(3)}$ for a two beam experiment in which the frequency of one beam, the pump, is fixed and that of the other, the probe, is allowed to vary over a frequency range of $\hbar\Delta\omega = 40$ meV centered around the pump frequency. The pump frequency is chosen to be slightly detuned from the exciton resonance by a frequency $-(\sqrt{2}/2)\Gamma/\hbar$. The quantum wire dimensions that have been used to plot this figure are $L = 500$ Å, $W = 200$ Å. The longitudinal broadening parameter γ is chosen to be one-tenth that of the transverse broadening parameter Γ, which is a physically reasonable ratio. As we already pointed out, a pronounced negative peak is due to a saturation (or bleaching) of the excitonic state. A magnetic field makes the peak deeper, without significant broadening, thus enhancing transmission further. The region of positive $\mathrm{Im}\chi^{(3)}$ is attributed to the formation of excitonic molecules [71,72].

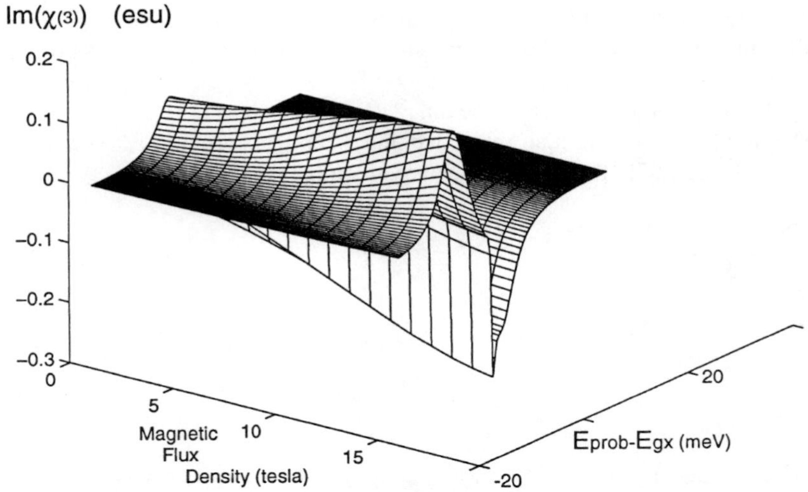

Figure 11.18. The imaginary part of the third-order nonlinear susceptibility as a function of pump and probe detuning energy and magnetic flux density. The pump is tuned slightly below the exciton resonance for each value of the magnetic field and the longitudinal broadening parameter is assumed to be one-tenth that of the transverse broadening parameter. The wire dimension is $W = 200\,\text{Å}$ and $L = 500\,\text{Å}$.

The same basic features are repeated in the absorption spectrum presented in Fig. 11.19. Here we plot the differential absorption $\Delta\alpha$ as a function of the pump—probe detuning frequencies when the longitudinal broadening parameter γ is one-tenth of the transverse broadening parameter Γ. As we can see, when the pump frequency is nearly resonant with the excitonic absorption, the swing in the differential absorption $\Delta\alpha$ is very large (0.5×10^5/cm – 10^5/cm). Another feature to note is that the frequency separation between the positive and negative peaks (associated with biexciton formation and exciton bleaching) is quite sensitive to the magnetic field. This separation is not sensitive to damping (values of γ and Γ) or slight detuning of the pump. Therefore, we can use a magnetic field to tune this separation, thus realizing magnetooptical devices. In Fig. 11.20, we show the differential refractive index Δn as a function of pump–probe detuning frequency and magnetic field. The parameter Δn exhibits more complicated behavior with a strong negative peak occurring at the energy in between the positive and negative resonances in the absorption change. The negative peak is related to the fact that $\Delta\alpha$ has a positive dispersive peak on its low-energy side.

Although not shown in this chapter, we also found that damping has a deleterious effect on the nonlinearity. As the damping parameter γ increases from 0.1Γ to Γ, the swing in Δn drops from 0.4 to 0.05 when no magnetic field is present, resulting in a 20-fold reduction in the nonlinearity. However, when a magnetic flux density of 10 T is present, Δn drops by only a factor of 6. Therefore, a magnetic field makes the nonlinearity less sensitive to damping.

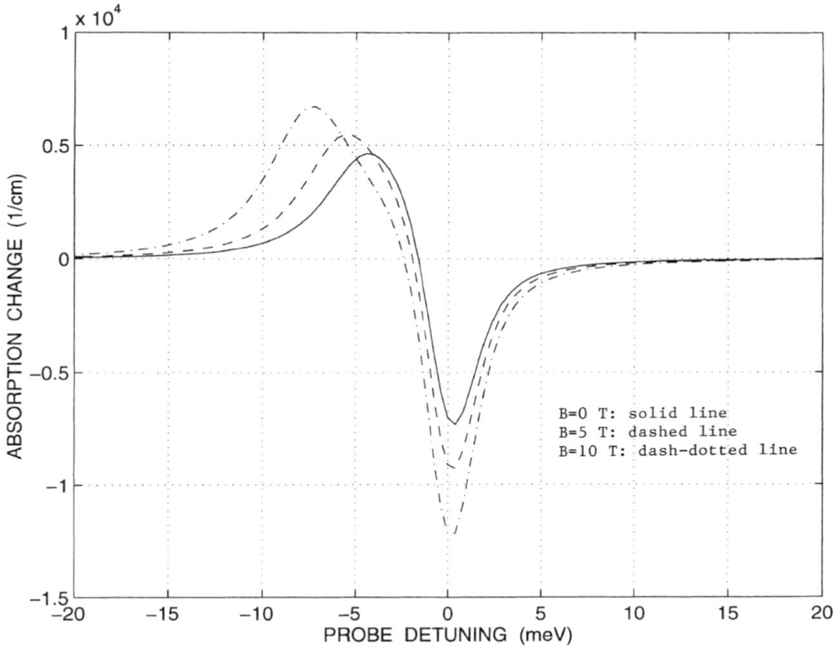

Figure 11.19. The differential absorption $\Delta\alpha$ as a function of pump–probe detuning energies for different values of a magnetic field. The pump is set at exciton resonance for each value of a magnetic field. The longitudinal broadening parameter is one-tenth that of the transverse broadening parameter.

The strong dependence of Δn and $\Delta\alpha$ on an external magnetic field has an important consequence for device applications [73]. One possible application of band gap resonant optical nonlinearities in quantum confined systems is optical bistability and switching devices associated with it. It was pointed out in [74] that in order to achieve optical bistability, one should provide a large refractive index swing at a relatively low absorption level. For bistable etalons using quantum wells, the relationship between minimum index change and absorption in the material for bistability to be observable can be written as $\Delta n/\alpha\lambda > \sqrt{3}/6\pi$, where λ is the wavelength of the pump beam. By using this criterion, it was concluded in [35] that bistability is not achievable in quantum well etalons from excitonic mechanisms alone since in the region of large Δn, excitonic absorption is also very high. However, in quantum wires, the criterion for bistability can be met, especially in the presence of a magnetic field. This advantage is significant.

11.4. DIELECTRIC CONFINEMENT EFFECTS

As we have already discussed, spatial confinement significantly enhances exciton and biexciton binding energies in quantum wires. The quantum confinement can be

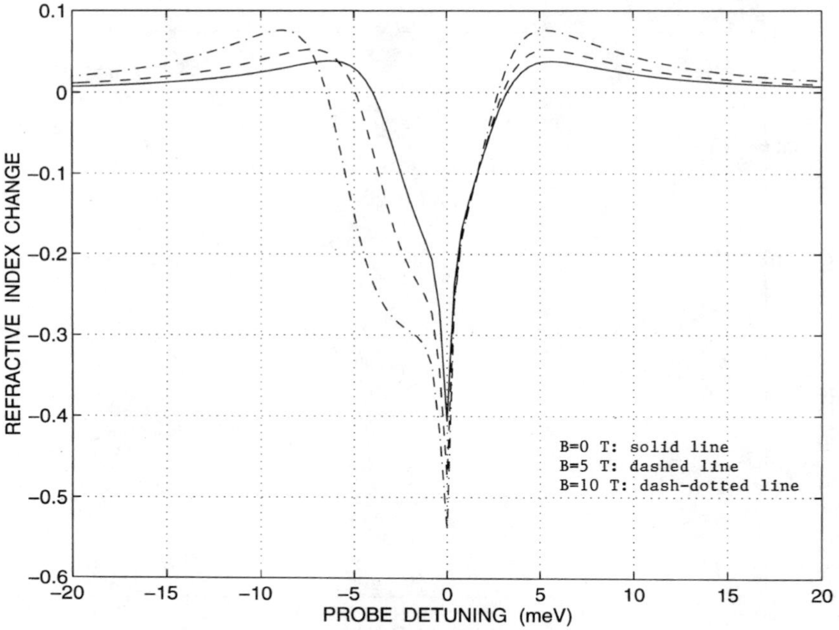

Figure 11.20. The differential refractive index Δn as a function of pump and probe detuning energies for different values of a magnetic field. All parameters and conditions are the same as in Fig. 11.20.

increased further by a magnetic field that "squeezes" the electron and hole wave functions into edge states or cyclotron orbits. Moreover, if the medium surrounding the wire has a smaller dielectric constant, then the effective dielectric constant of the entire system is reduced, which then reduces the screening of the attractive interaction between an electron and a hole. As a result, an electron and a hole become even more tightly bound and this increases the binding energy and oscillator strengths further [75–77]. The latter can have very important consequencies for optoelectronic applications of quantum wires. In this section, we examine this dielectric confinement effect in the presence of a magnetic field, and the interplay between magnetostatic and dielectric confinement.

While dielectric confinement effects in quantum wells and dots have been studied [77,78], scant attention has been given to quantum wires. So far, the very few theoretical treatments that have been reported for wires have concentrated narrowly on specific carrier density regimes and geometries, for example, wedge shaped wire [79].

The exciton binding energy correction due to the dielectric constant misfit on a wire-barrier boundary can be found using the same Hamiltonian as in Section 11.2.4 with additional dielectric confinement potential term $U_D \neq 0$. To calculate U_D, we adopt the image-charge method that is known to provide good qualitative agreement with experiments [80]. The image-charge method, which is well established in electrostatics, accounts for the electric field induced by charged particles in parallel geometries by means of imaginary charges placed in surrounding media. The values of the image charges are determined from the continuity conditions for the electrostatic potentials and the normal components of the displacement vector at the interfaces. Following [73], we can write

$$U_D \equiv U_{\rm self}^{(e)} + U_{\rm self}^{(h)} + U_{\rm attr}^{(e-h)} \tag{11.50}$$

where

$$U_{\rm self}^{(e)} = \frac{e^2}{2\varepsilon_w} \sum_{l=-\infty}^{\infty} \sum_{m=-\infty}^{\infty} \frac{\zeta^{|l|+|m|}}{[(y_e - y_{e,l,m})^2 + (z_e - z_{e,l,m})^2]^{1/2}} \tag{11.51}$$

$$U_{\rm self}^{(h)} = \frac{e^2}{2\varepsilon_w} \sum_{l=-\infty}^{\infty} \sum_{m=-\infty}^{\infty} \frac{\zeta^{|l|+|m|}}{[(y_h - y_{h,l,m})^2 + (z_h - z_{h,l,m})^2]^{1/2}} \tag{11.52}$$

$$U_{\rm attr}^{(e-h)} = -\frac{e^2}{\varepsilon_w} \sum_{l=-\infty}^{\infty} \sum_{m=-\infty}^{\infty} \frac{\zeta^{|l|+|m|}}{[(x_e - x_h)^2 + (y_e - y_{h,l,m})^2 + (z_e - z_{h,l,m})^2]^{1/2}} \tag{11.53}$$

and

$$(x_\alpha, y_{\alpha,l,m}, z_{\alpha,l,m}) = (x_\alpha, Ll + (-1)^l y_\alpha, Wm + (-1)^m z_\alpha)$$

with $\alpha = e$ or h; and $l, m = -\infty, \ldots, \infty$. The $l = m = 0$ term is excluded from the summation. The dielectric constant of the surrounding medium is assumed to be less than the dielectric constant of the wire material ($\varepsilon_b < \varepsilon_w$), which is usually the case in real systems. Finally, the quantity $\zeta = (\varepsilon_w - \varepsilon_b)/(\varepsilon_w + \varepsilon_b)$ is a measure of the *dielectric misfit* between the quantum wire material (dielectric constant ε_w) and the boundary material (dielectric constant ε_b).

One can see from Eq. (11.50) that the dielectric confinement term, U_D, is written as a sum of two positive self-energy terms resulting from repulsive interactions of electron–"electron image" (hole–"hole image"); and a negative term that is due to attractive interaction between electron–"hole image" (hole–"electron image"). The summations over the infinite series in Eqs. (11.51–11.53) come about because any charge image, in turn, creates another image in the opposite flat boundary of the wire. For materials with a small dielectric misfit one can retain only the first-order terms ($|l| + |m| = 1$); for material systems with higher dielectric misfit (such as GaAs surrounded by air for which $\zeta \approx 0.9$), higher order terms are required. In order to write the first terms in U_D explicitly, one has to carry out the following procedure. By setting $l = 0$ and $m = \pm 1$, one obtains the terms that correspond to interactions of the electron and hole with their images in the flat boundaries along the z direction. Similarly, setting $m = 0$ and $l = \pm 1$ one obtains the terms that correspond to interactions of the electron and hole with their images in the flat boundaries along the y direction.

The ground-state exciton binding energies E_B^X can now be found using the relation

$$E_B^X = E_{\text{conf}}^{e1} + E_{\text{conf}}^{hh1} + E_{\text{self}}^{(e)} + E_{\text{self}}^{(h)} - E^{\text{tot}} \tag{11.54}$$

The total energy E^{tot} is determined by minimizing the expectation value of the total Hamiltonian by varying the parameter η_x, that is,

$$E^{\text{tot}} = \min_{\eta_x} \langle \Psi | H^X | \Psi \rangle \tag{11.55}$$

and the electron self-energy $E_{\text{self}}^{(e)}$ is given as

$$E_{\text{self}}^{(e)} = \langle \psi_e(y_e, z_e) | U_{\text{self}}^{(e)} | \psi_e(y_e, z_e) \rangle \tag{11.56}$$

The hole self-energy is obtained from the last expression by substituting the hole index "h" in place of the electron index "e" field. Analogous formalism can be developed for biexcitons in a quantum wire embedded within material of different dielectric constant [73].

The dependence of the exciton binding energy on the dielectric misfit is shown in Figure 11.21. It is interesting to note that the dependence is almost exactly *linear*: for

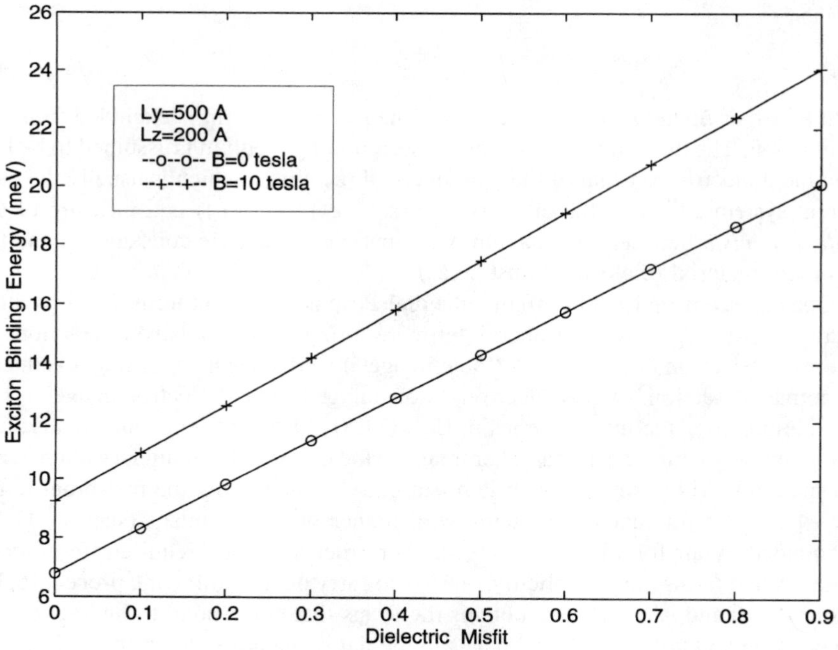

Figure 11.21. Exciton binding energy as a function of the dielectric misfit for two values of the magnetic field: (a) $B = 0\,\text{T}$ ("o"); (b) $B = 10\,\text{T}$ ("+"). The thickness along the z direction is 200 Å while the width is 500 Å.

the magnetic flux density of 10 T, it can be approximated by the interpolation formula $E_B^X \approx 9.240 + 16.376\zeta + 0.089\zeta^2$, where the prefactor of the quadratic term is more than two-orders of magnitude smaller than that of the linear term. The increase in the binding energy and the decrease in the radius as a result of dielectric confinement can be simply ascribed to a lowering of the effective dielectric constant (and hence the screening) in the entire system due to the image charges. A magnetic field reduces the image-charge effect because it squeezes the electron and hole together thereby tending to condense them into an uncharged single particle. Consequently, a magnetic field quenches the dielectric confinement.

Figure 11.22 shows the oscillator strength of the exciton transition as a function of a magnetic flux density [73]. The magnetic field induced quenching of the dielectric confinement effect is more pronounced in this figure than in the plot for the exciton binding energy. Since

$$f_{1s} \propto \int \alpha(\hbar\omega)d\omega \tag{11.57}$$

it appears that the integrated absorption, a quantity easily measured experimentally, is quite sensitive to the magnetic field [$\alpha(\hbar\omega)$ is the absorption coefficient].

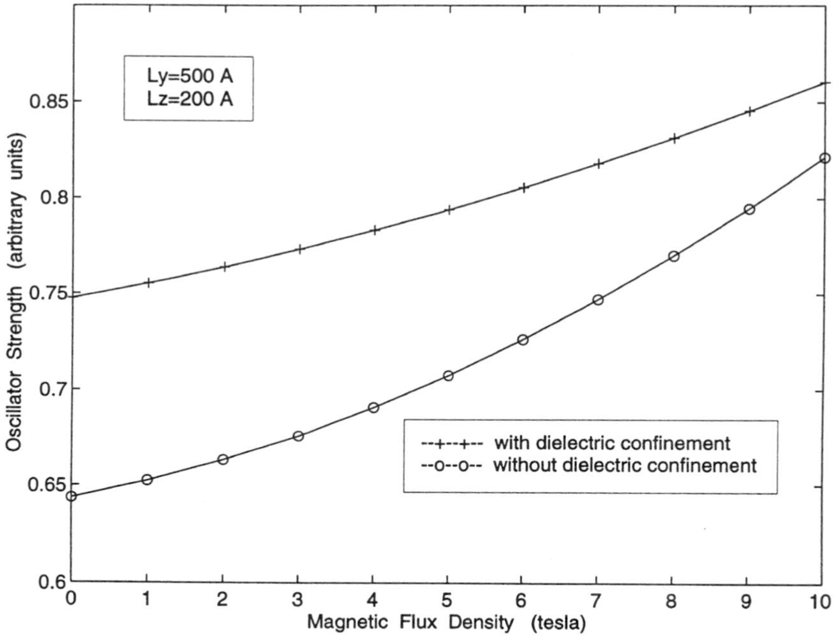

Figure 11.22. Exciton oscillator strength as a function of a magnetic flux density with (+) and without (o) dielectric confinement. The thickness along the z direction is 200 Å and the width is 500 Å.

11.5. BAND-GAP RENORMALIZATION IN QUANTUM WIRES

Another major source of optical nonlinearity in a semiconductor structure is band-gap renormalization (BGR), which shifts the optical band-gap (and hence optical spectra in frequency) with increasing excitation intensity. For many of the opto-electronic applications, one has to know the exact spectral position of the band-to-band absorption or emission, which depend on BGR value. The excitons created by the incident radiation ionize into free carriers, which reduce the band-gap due to many-body interactions. The BGR effects can be roughly divided into two types arising (1) from direct plasma screening mediated through electron–hole plasma and (2) from exciton screening (so-called "dielectric" screening), which is caused by the very existence of the excitons themselves. In this chapter, we restrict ourselves to the regime when optically excited excitons have not dissociated into free electron–hole pairs to create a plasma. Since plasma screening is less effective in systems of reduced dimensionality, this picture is valid over a wide range of conditions. Band-gap renormalization in an electron–hole liquid formed by excitons that have undergone Mott transition is beyond the scope of this section.

In order to evaluate BGR in a semiconductor quantum wire subjected to a magnetic field, we employ the Zimmermann's model [81], which is based on the random-phase approximation and utilizes the fact that in quantum wires, the sum of exchange and correlation energies is almost independent of band characteristics and depends on carrier density alone [82,83]. The BGR energy in this model can be written as [81]

$$\Delta E_G \equiv \Delta E_G(n, T_c, B, L_y, L_z) = -\frac{3.24 r_s^{-3/4}}{(1+0.0478 r_s^3 \theta^2)^{1/4}} E_B^X \tag{11.58}$$

where $\theta = kT_c/E_B^X$ is the reduced temperature, T_c is the average carrier temperature, and $E_B^X \equiv E_B^X(B, L_y, L_z)$ is the exciton ground-state binding energy, and $r_s = (3/4\pi n)^{1/3}(1/a_B^*)$ is the interparticle distance normalized to the exciton Bohr radius a_B^*. For our numerical simulation we used the average carrier temperatures (the same for both electrons and holes) in the range $T_c = 70-150\,\text{K}$, which corresponds to the value calculated in [84] for a 25 Å wide GaAs quantum well with a sheet carrier density of $n_{2D} = 2.5 \times 10^{10}\,\text{cm}^{-2}$. This value was obtained for a low ambient temperature ($T = 10$ K) and under the assumption that electron–hole pairs are created by a 0.25-ps laser pulse.

In order to find an absolute position of the exciton peak, we used the following approximation for the bulk band-gap energy [85]

$$E_G(T) = E_G(0) - \alpha T^2/(T+\beta) \tag{11.59}$$

where, for GaAs, $E_G(0) = 1.519$ meV, $\alpha = 5.405 \times 10^{-4}$, $\beta = 204$, and T is an ambient temperature in kelvin. The spectral position of the lowest exciton peak can be found then as $E_{\text{exc}} = E_G(T) + E_{\text{conf}}^{e1} + E_{\text{conf}}^{hh1} - E_B^X + \Delta E_G$.

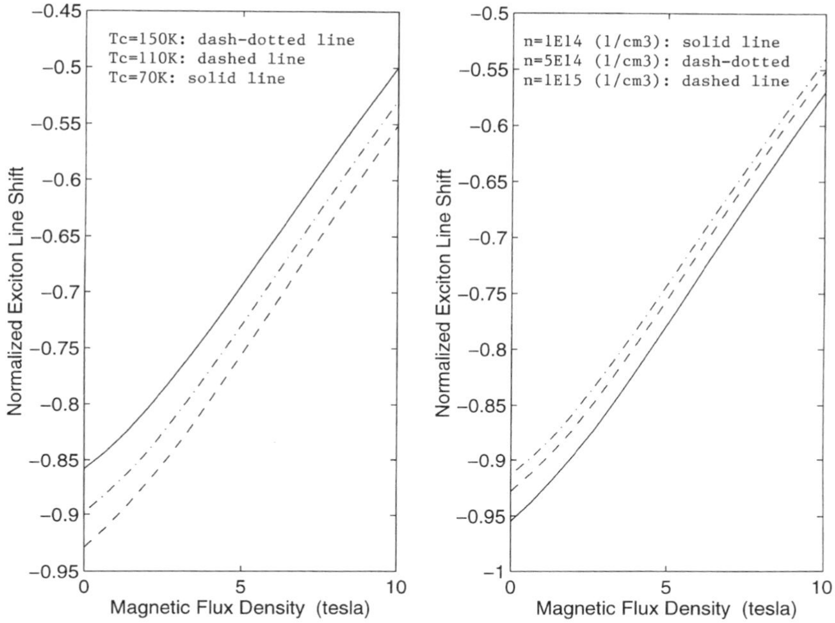

Figure 11.23. The exciton line shift versus magnetic flux density. The left part corresponds to the different carrier temperatures: $T_c = 70$ K (solid), $T_c = 110$ K (dashed), and $T_c = 150$ K (dashed–dotted). The right part corresponds to the different carrier densities $n = 1 \times 10^{14}\ \text{cm}^{-3}$ (solid), $n = 5 \times 10^{14}\ \text{cm}^{-3}$ (dashed–dotted), and $n = 1 \times 10^{15}\ \text{cm}^{-3}$ (dashed). The ambient temperature, $T = 10$ K, is the same for both parts.

Figure 11.23 shows the energy shift of the lowest exciton line with respect to the fundamental band gap of the bulk material as a function of magnetic flux density. The left part corresponds to a constant carrier density, $n = 1 \times 10^{15}\ \text{cm}^{-3}$, and different carrier temperatures; while the right part corresponds to constant temperature, $T_c = 110$K and different carrier densities. All results pertain to a 500 Å wide and 200 Å thick free-standing GaAs quantum wire. One can observe a large absolute band-gap red-shift due to many-body effects, which is somewhat offset (up to 3 meV) by the blue shift caused by the magnetic field. The blue shift of the exciton peak from its zero-field position indicates that the confinement energy due to the magnetic field is larger than the increased exciton binding energy due to quasi 1D Coulomb interaction. The absolute value of the red shift (with respect to the bulk band-gap energy) is higher for a system with higher carrier density and lower temperature.

11.6. CONCLUSIONS

In this chapter, we presented a review of optical and electronic properties of semiconductor quantum wires important for optoelectronic applications. Electron and hole confinement in quasi 1D structures modifies the density of states and enhances the Coulomb interaction of electrons and holes, thus bringing about a variety of effects that are very interesting from the viewpoint of both physics and device application. The potential advantage of these structures for optoelectronic applications arises from the spatial confinement of photoexcited electrons and holes, which modifies their optical absorption, emission, and increases the oscillator strengths. Spatial confinement of carriers in quasi-1D systems has been shown to increase binding energies of all excitonic complexes and, thus, make their contribution to the optical response more pronounced. The third-order susceptibility values as high as 1.0–1.2 esu were predicted for 12-nm thick GaAs quantum wires. The theoretical formalism presented in this chapter is general enough to include the effect of dielectric confinement of excitons, final barrier height in embedded quantum wires, barrier–wire interdiffusion, and the effects of a magnetic field. The validity of our predictions based on this formalism was recently confirmed by measurements of the biexciton binding energy in quantum wires. Rapid progress in nanometerscale fabrication technology may soon allow for practical utilization of many novel physical phenomena examined in this chapter.

ACKNOWLEDGMENTS

The authors are indebted to K. L. Wang (UCLA), V. Dneprovskii (Moscow State University), V. Fomin (University of Antverpen), and A. Miller (University of Notre Dame) for many illuminating discussions on quantum wires. The work of F.S. was supported by the Air Force Contract at the Wright Laboratory, Wright-Patterson Air Force Base, Ohio.

REFERENCES

1. T. Kojima, J. Xue-Ying, Y. Hayafune, and S. Tamura, *Jpn. J. Appl. Phys. 1* ***37***, 5961 (1998).
2. H. Mariette, D. Brinkman, G. Fishman, C. Gourgon, L. S. Dang, and A. Loffler, *J. Crystal Growth* ***159***, 418 (1996).
3. V. S. Dneprovskii, E. A. Zhukov, E. A. Muljarov, and S. G. Tikhodeev, *JETP* ***87***, 382 (1998).
4. A. Forchel, P. Ils, K. H. Wang, and O. Schilling, *Microelectron. Eng.* ***32***, 317 (1996).
5. See, for example, E. Kapon, in G. Abstreiter, A. Aydinli, and J.-P. Leburton, Eds., *Optical Spectroscopy of Low Dimensional Semiconductors*, NATO ASI Series E, Vol. 344, Kluwer Academic Press, Dordrecht, The Netherlands, 1997, pp. 99–126.
6. V. V. Poborchii, M. S. Ivanova, and I. A. Salamatina, *Superlattices Microstruct.* ***16***, 133 (1994).
7. A. Svizhenko, A. Balandin, and S. Bandyopadhyay, *J. Appl. Phys.* ***81***, 7927 (1997).
8. A. Svizhenko, A. Balandin, S. Bandyopadhyay, and M. A. Stroscio, *Phys. Rev. B* ***57***, 4687 (1998).

9. E. P. Pokatilov, V. M. Fomin, S. N. Balaban, S. N. Klimin, L. C. Fai, and J. T. Devreese, *Superlattices Microstruct.* ***23***, 331 (1998).
10. V. M. Fomin, V. N. Gladilin, J. T. Devreese, C. Van Haesendonck, and G. Neuttiens, *Solid State Commun.* ***106***, 293 (1998).
11. T. Kojima, S. Tanaka, H. Yasumoto, and H. Nakaya, *Jpn. J. Appl. Phys. 2* ***37***, L1386 (1998).
12. E. Kapon, *Proceedings of Integrated Photonics Research*, Vol. 6, Optical Society of America, Washington, DC, 1996, p. 284.
13. C. Weisbuch and B. Vinter, *Quantum Semiconductor Structures: Fundamentals and Applications*, Academic, San Diego, CA, 1991.
14. H. Schweizer, U. A. Griesinger, V. Harle, and F. Adler, *Optoelectronic Nanostructures: Physics and Technology*, Proceedings of the SPIE, Vol. 2399, 1995, p. 407.
15. Q. Yi, Z. Jinming, X. Zuntu, and C. Lianghui, *Chin. J. Semiconduct.* ***17***, 155 (1996).
16. A. A. Golovin and E. I. Rashba, *JETP Lett.* ***17***, 690 (1973).
17. T. Ishihara, *Phys. Status. Solidi.* ***159***, 371 (1990).
18. F. L. Madarasz, F. Szmulowicz, F. K. Hopkins, and D. L. Dorsey, *J. Appl. Phys.* ***75***, 639 (1994).
19. F. L. Madarasz, F. Szmulowicz, F. K. Hopkins, and D. L. Dorsey, *Phys. Rev. B* ***49***, 13528 (1994).
20. R. O. Klepfer, F. L. Madarasz, and F. Szmulowicz, *Phys. Rev. B* ***51***, 4633 (1995).
21. A. Balandin and S. Bandyopadhyay, *Phys. Rev. B* ***54***, 5712 (1996).
22. R. Rinaldi, R. Cingolani, M. Lepore, M. Ferrara, I. M. Catalano, F. Rossi, L. Rota, E. Molinari, P. Lugli, U. Marti, D. Martin, F. Morier-Gemoud, P. Ruterana, and F. K. Reinhart, *Phys. Rev. Lett.* ***73***, 2899 (1994).
23. T. Someya, H. Akiyama, and H. Sakaki, *Phys. Rev. Lett.* ***74***, 3664 (1995).
24. T. Baars, W. Braun, M. Bayer, and A. Forchel, *Phys. Rev. B* ***58***, R1750 (1998).
25. F. Vouilloz, D. Y. Oberli, S. Wiesendanger, and B. Dwir, *Phys. Status Solidi A* ***164***, 259 (1997).
26. A. L. Ivanov and H. Haug, *Phys. Status Solidi B* ***188***, 61 (1995).
27. D. A. B. Miller, D. S. Chemla, T. C. Damen, A. C. Gossard, E. Wiegmann, T. H. Wood, and C. A. Burrus, *Phys. Rev. B* ***32***, 1043 (1985).
28. D. S. Chemla, D. A. B. Miller, and S. Schmitt-Rink, in H. Haug, Ed., *Optical Nonlinearities and Instabilities in Semiconductors*, Academic, New York, 1988, pp. 38–120; and pp. 325–359 and references cited therein.
29. S. Glutsch and F. Bechstedt, *Phys. Rev. B* ***47***, 4315 (1993); 6385(1993).
30. U. Bockelmann and G. Bastard, *Europhys. Lett.* ***15***, 215 (1991).
31. L. Banyai, I. Galbraith, C. Ell, and H. Haug, *Phys. Rev. B* ***36***, 6099 (1987).
32. R. C. Miller, D. A. Kleinman, W. T. Tsang, and A. C. Gossard *Phys. Rev. B* ***24***, 1134 (1981).
33. G. Bastard, E. E. Mendez, L. L. Chang, and L. Esaki, *Phys. Rev. B* ***26***, 1974 (1982).
34. A. V. Kavokin, A. I. Nesvizhskii, and R. P. Seisyan, *Semiconductors* ***27***, 530 (1993).
35. D. A. B. Miller, D. S. Chemla, T. C. Damen, A. C. Gossard, W. Weigmann, T. H. Wood, and C. A. Burrus, *Phys. Rev. Lett* ***53***, 2173 (1984).
36. W. F. Brinkman, T. M. Rice, and B. Bell, *Phys. Rev. B* ***8***, 1570 (1973).

37. D. A. Kleinman, *Phys. Rev. B* ***28***, 871 (1983).
38. F. L. Madarasz, F. Szmulowicz, and F. K. Hopkins, *Phys. Rev. B* ***52***, 8964 (1995).
39. F. L. Madarasz, F. Szmulowicz, F. K. Hopkins, in M. Cahay et al., Eds., *Quantum Confinement III: Physics and Applications*, The Electrochemical Society, Inc., Pennington, NJ, 1996, pp. 70–90.
40. A. Balandin and S. Bandyopadhyay, in M. Cahay et al., *Quantum Confinement III: Physics and Applications*, The Electrochemical Society, Inc., Pennington, NJ, 1996, pp. 117–128.
41. K. Ishida, H. Aoki, and T. Ogawa, *Phys. Rev. B* ***52***, 8980 (1995).
42. F. L. Madarasz and F. Szmulowicz, *J. Appl. Phys. Rev.* ***62***, 3267 (1987).
43. J. J. Hopfield and D. S. Thomas, *Phys. Rev.* ***132***, 563 (1963); J. J. Hopfield and D. S. Thomas, *Phys. Rev.* ***182***, 945 (1969).
44. K. Oimatsu, T. Iida, H. Nishimura, K. Ogawa, and T. Katsuyama, *J. Lumin.* ***48, 49***, 713 (1991).
45. E. S. Koteles, in E. I. Rashba and M. D. Struge, Eds., *Excitons*, Modern Problems in Condensed Matter Sciences, North-Holland, Amsterdam, The Netherlands, 1982, p. 83.
46. L. V. Keldysh, *Phys. Status. Solidi B* ***188***, 11 (1995).
47. F. Bassani and G. P. Parravicini, *Electronic States and Optical Transitions in Solids*, Pergamon, New York, 1975, pp. 152–154.
48. G. Bastard, *Wave Mechanics Applied to Semiconductor Heterostructures*, Halsted, New York, 1988.
49. I. Suemune and L. A. Coldren, *IEEE J. Quantum Electron.* ***24***, 1778 (1988).
50. H. Ishihara and K. Cho, *Phys. Rev. B* ***42***, 1724 (1990).
51. T. Takagahara, *Solid State Commun.* ***28***, 279 (1991).
52. H. Qiang, F. H. Pollack, C. M. S. Torres, W. Leitch, A. H. Kean, M. A. Stroscio, G. Iafrate, and K. W. Kim, *Appl. Phys. Lett.* ***61***, 1411 (1992).
53. Y. Z. Hu, M. Lindberg, and S. W. Koch, *Phys. Rev. B* ***42***, 1713 (1990).
54. L. Belleguie and L. Bnyai, *Phys. Rev. B* ***47***, 4498 (1993).
55. T. Takagahara, *Phys. Rev. B* ***39***, ***10*** 206 (1989).
56. L. Banyai, I. Galbraith, and H. Haug, *Phys. Rev. B* ***38***, 3931 (1988).
57. A. Balandin and S. Bandyopadhyay, *Superlattices Microstruct.* ***19***, 97 (1996).
58. A. Balandin and S. Bandyopadhyay, *Phys. Rev. B* ***52***, 8312 (1995).
59. A. Balandin and S. Bandyopadhyay, *J. Appl. Phys.* ***77***, 5924 (1995).
60. A. D'Andrea and R. Del Sole, *Phys. Rev. B* ***46***, 2363 (1992).
61. R. K. Wehner, *Solid State Commun.* ***7***, 457 (1969).
62. M. A. Lee, P. Vashista, and R. K. Kalia, *Phys. Rev. Lett.* ***51***, 2422 (1983).
63. A. C. Cancio and Y.-C. Chang, *Phys. Rev. B* ***42***, 11317 (1990).
64. J. Adamowski, S. Bednarek, and M. Suffczynski, *Solid State Commun.* ***9***, 2037 (1971).
65. O. Akimoto and E. Hanamura, *Solid State Commun.* ***10***, 253 (1972).
66. A. Balandin and S. Bandyopadhyay, *Superlattices Microstruct.* ***23***, 1197 (1998).
67. A. Balandin and S. Bandyopadhyay, *Phys. Rev. B* ***54***, 5721 (1996).
68. N. Telang and S. Bandyopadhyay, *Phys. Rev. B* ***48***, 18002 (1993).

69. N. Telang and S. Bandyopadhyay, *Appl. Phys. Lett.* ***62*** 3161 (1993).

70. N. Telang and S. Bandyopadhyay, *Phys. Low-Dim Struc.* ***9/10*** 63 (1996).

71. A. Balandin, S. Bandyopadhyay, and A. Svizhenko, in Zh. Alferov et al., Eds., *Nanostructures: Physics and Technology*, Russian Academy of Sciences, St. Petersburg, Russia, 1996, pp. 294–297.

72. S. Bandyopadhyay, A. Balandin, and A. Svizhenko, in *Proceedings of SPIE*, Vol. 3404, ALT'97, International Conference on Laser Surface Processing, V. I. Pustovoy, Ed., SPIE, May 1998, pp. 302–311.

73. A. Balandin and S. Bandyopadhyay, in M. Cahay et al., Ed., *Quantum Confinement IV: Nanoscale Materials, Devices and Systems*, The Electrochemical Society, Pennington, NJ 1997, pp. 242–254.

74. A. Miller, P. K. Milsom, R. J. Manning, in S. Martellucci and A. N. Chester, Eds., *Nonlinear Optics and Optical Computing*, Plenum Press, New York, 1990.

75. L. V. Keldysh, *JETP Lett.* ***29***, 658 (1965).

76. N. S. Rytova, *Dokl. Akad. Nauk* ***163***, 118 (1965).

77. M. Kumagai and T. Takagahara, *Phys. Rev. B* ***40***, 1235 (1989).

78. T. Takagahara, *Phys. Rev. B* ***47***, 4569 (1993).

79. E. A. Andryushin and A. P. Silin, *Sov. Phys. Solid State* ***35***, 971 (1993).

80. A. L. Yablonskii, A. B. Dzyubenko, S. G. Tikhodeev, L. V. Kulik, and V. D. Kulakovskii, *JETP Lett.* ***64***, 51 (1996).

81. R. Zimmermann, *Phys. Status Solidi B* ***146***, 371 (1988).

82. P. Vashishta and R. K. Kalia, *Phys. Rev. B* ***25***, 6492 (1982).

83. K. Bohnert, H. Kalt, A. L. Smirl, and D. P. Norwood, *Phys. Rev. Lett.* ***60***, 37 (1988).

84. M. C. Marchetti and W. Pötz, *Phys. Rev. B* ***40***, 12391 (1989).

85. S. M. Sze, *Physics of Semiconductor Devices*, Wiley, New York, 1981, p. 15.

CHAPTER TWELVE

Quantum Dots: Physics and Applications

KANG L. WANG

Device Research Laboratory, Department of Electrical Engineering, University of California—Los Angeles, Los Angeles, CA 90095

ALEXANDER A. BALANDIN

Department of Electrical Engineering, University of California—Riverside, Riverside, CA 92521

Continuous progress in fabrication techniques for semiconductor quantum dots resulted in significant achievements in both understanding the physical processes in quasi-zero-dimensional (0D) structures and in their practical applications. In this chapter, we outline the basic physical properties of quantum dots, present a review of the state-of-the-art fabrication techniques such as different types of self-assembly and nanostamping, and discuss current and possible future applications of quantum dots. We particularly focus our attention on utilization of quantum dots for improving infrared (IR) photodetection, second-harmonic generation, and wavelength conversion. The second part of this chapter presents a rather detailed feasibility study of the possibility of use of quantum dots for future nanoelectronic and quantum computing architectures.

12.1. INTRODUCTION

Semiconductor quantum dots (QD) are three-dimensionally (3D) confined systems. They can be fabricated by a variety of techniques such as self-assembled molecular beam epitaxy (MBE) growth [1–5], direct e-beam lithography [6], nanostamping [7], inclusion of small spherical semiconductor nanocrystals in a dielectric matrix, and some other techniques [8]. The confining potential at the interface of a QD gives rise to a series of discrete, atomic-like energy levels for the conduction band electron and valence band hole states of the semiconductor. The typical dimensions of QDs are in the range from a few nanometers to a fraction of a micron, and their size, shape, and

Optics of Nanostructured Materials, Edited by Vadim A. Markel and Thomas F George
ISBN 0-471-34968-2

interactions can be controlled through the use of advanced nanofabrication technology. Electron and hole confinement in quasi-0D structures modifies the density of states and enhances the Coulomb interactions of electrons and holes, thus bringing about a variety of effects that are interesting from the viewpoint of physics as well as device applications.

12.1.1. Basic Physical Properties

In recent years, substantial progress has been achieved in both theoretical description and fabrication of QDs. To give a general insight into the effects of quantum confinement on the conduction band electron or valence band hole (single-particle) states, it is useful to consider first the idealized case of a cuboidal QD limited by infinite potential barriers. We assume that the dot's dimensions L_x, L_y, and L_z are comparable to the de-Broglie wavelength or the exciton Bohr radius in the material of choice. In this case, the electron (hole) wave function can be written in the envelope function approximation as

$$\Psi^{e,h} = \psi_x^{e,h}(x)\psi_y^{e,h}(y)\psi_z^{e,h}(z) \tag{12.1}$$

where $\Psi_x^{e,h} = (2/L_x)\frac{1}{2}\ \sin(\pi nx/L_x)$ is the confined envelope along the x direction. The envelopes along other directions are obtained by the appropriate change of the subscript. These wave functions form a complete orthogonal set with energy given by the expression

$$E_{n,k,l}^{e,h} = E_G + \frac{\pi^2\hbar^2}{2m_{e,h}^*}\left(\frac{n_x^2}{L_x^2} + \frac{n_y^2}{L_y^2} + \frac{n_z^2}{L_z^2}\right) \tag{12.2}$$

Here E_G is the band gap energy and $m_{e,h}^*$ is the effective mass of an electron (hole); n_x, n_y, and n_z are electron and hole quantum numbers; $\hbar = h/2\pi$ is the Plank's constant. The corresponding quasi-0D density of states (DOS), which gives QDs most of their exotic properties, can be written as a set of δ functions,

$$D(\omega) = 2\sum_{n_x,n_y,n_z} \delta(\hbar\omega - E_{n_x,n_y,n_z}^{e,h}) \tag{12.3}$$

Due to the orthogonality of the envelope wave functions, optical transitions obey certain selection rules, such as that the electron and hole quantum numbers should be equal for interband transitions [9].

The excitonic properties of QDs are also different from those of bulk materials (or systems of higher dimensionality) and should be treated by taking into account spatial confinement of electron (hole) wave functions and increased Coulomb interaction. The wave function of an exciton can be expressed as a linear combination of the electron (hole) states of Eq. (12.1). The ground-state exciton binding energy in a low-dimensional structure can then be found using the variational techniques (see

Chapter 11). In order to obtain the higher exciton energy states in a QD, one has to solve the Schrodinger equation by the matrix diagonalization method [9,10]. In addition to the discreteness of energy levels in a quantum dot, the oscillator strength, which is distributed over continuum states in bulk materials, becomes concentrated on the sharp exciton transitions in 0D systems. Consequently, the excitonic optical nonlinearity becomes enhanced and the saturation power reduces relative to that in the bulk semiconductor [10,11].

The exciton binding energy in the bulk semiconductor can be written as

$$E_B^X = \frac{e^2}{2\varepsilon a_{\mathrm{B}}} = \frac{\mu e^4}{2\varepsilon^2 \hbar^2} \tag{12.4}$$

where a_{B} is the exciton Bohr radius, μ is the exciton reduced mass, and ε is the dielectric constant. From Eq. (12.4), the exciton binding energy is expected to increase not only via increasing the reduced mass but also by reducing the dielectric constant ε. As a result, the exciton binding energy in QDs can be enhanced even further by embedding them in insulating material with small dielectric constant. This dielectric confinement effect is caused by the effective reduction of ε due to the penetration of the electric field into the barrier medium [12]. It is shown to be significant for QDs based on a variety of material systems (see Section 12.2.3 and Section 11.4).

Modification of the electron (hole) properties and carrier DOS is not the only major effect the spatial confinement brings about. Partial (or complete) confinement of the phonon wave function to the volume of a QD leads to the relaxation of the phonon selection rule $\Delta q = 0$. As a result, additional transitions will lead to a broadening of the Raman peaks and a concomitant red shift of its mean position. The first-order Raman spectrum of the arrays of QDs is given as

$$I(\omega) \cong \int \frac{d^3 q |C(0,\ q)|^2}{[\omega - \omega(q)]^2 + (\Gamma_0/2)^2} \tag{12.5}$$

where $I(\omega)$ is the Raman intensity, $\Omega(q)$ is the phonon dispersion, Γ_0 is the natural line width, q is the phonon wavevector, and $C(0,\ q)$ are the Fourier coefficients that depend on the shape of the QDs [13]. Assuming one-phonon scattering, spherical QDs and the Gaussian-shaped phonon weighting functions (confined envelopes), the Fourier coefficients for Eq. (12.5) are

$$|C(0,\ q)|^2 \cong e^{-q^2 L^2/4} \tag{12.6}$$

Here, we took L as the characteristic size of the dot and neglected scale factors. We used Eqs. (12.5) and (12.6) for analysis of Raman spectra of Ge quantum dots grown on strained Si.

Although the red shift of optical phonons is frequently observed, it is usually offset by the blue shift due to the strain. This strain is always present in self-assembled QDs grown by MBE using two dissimilar materials with a lattice mismatch. One needs to consider the shape of the Raman peaks in order to separate spatial confinement and strain effects in the Raman spectra of QDs.

In order to describe the phonon assisted optical transitions in semiconductor QDs, Fomin et al. [14] developed a theory that includes the exciton interaction with both adiabatic and Jahn–Teller phonons. The effects of nonadiabaticity of the exciton–phonon system were shown to lead to a significant enhancement of phonon assisted transition probabilities and to multiphonon optical spectra that are considerably different from the Franck–Condon progression. The calculated relative intensity of the phonon satellites and its temperature dependence compare well with the experimental data on the photoluminescence of CdSe quantum dots [14]. In this study, the authors also deduced new selection rules for Raman scattering in QDs of semiconductor materials with a degenerate valence band [15].

Many electronic properties of semiconductor QDs were studied through electron tunneling. The simplest model to describe this process in QDs is the constant-interaction model. This model assumes that the Coulomb interaction between the electrons is independent of their total number N and is determined by the dot capacitance C. In this model, the energy required to put an additional electron, is given by $E_{\text{add}} = e^2/C + \Delta E$, where e is the electron charge, and ΔE is the energy difference between two confined states [see Eq. (12.2)]. Thus, it is important to compare the scaling of Coulomb charging energy e^2/C with the quantum confined energy of a semiconductor QD given by Eq. (12.2). Results of our numerical simulation of the energy scaling in InAs and CdS quantum dots are shown in Fig. 12.1. One can see from this figure that the Coulomb charging energy and the quantum confined energy are comparable for InAs and CdS quantum dots in the range of $10-100°$A. In order to be able to see the single-electron effects in the tunneling current these energies have to be larger than the thermal energy $k_B T$ [16].

12.1.2. Prospects for Applications

Strong spatial confinement effects, which are characteristic for QDs, create an exciting opportunity for reengineering of optical, electronic, and thermal properties of many technologically important semiconductors through the modification of electronic states and phonon dispersion. Recently, a room temperature operation of a GaAs based QD laser was demonstrated at the wavelength of 1.31 μm [17]. The lasing at this frequency was achieved with the threshold current density of $270\ \text{A/cm}^2$ using high-reflectivity facet coating. The results presented in [17] show the potential of QD lasers for use in medium distance fiber interconnects. In another example [18], self-organized $\text{In}_{0.4}\text{Ga}_{0.6}/\text{As}/\text{GaAs}$ quantum-dot single-mode ridge waveguide lasers with intracavity absorber were also grown by MBE. Bistability in the light-current characteristics of a 3-μm single-mode edge-emitting laser was obtained by controlling the intracavity absorber voltage. Self-pulsation was observed with a center frequency of 1.6 GHz and line width < 10 MHz.

The group at UCLA [1–3] observed strong intersubband absorption peaks in boron-doped Ge quantum dots grown on Si. This study suggests the possibility of utilization of QD arrays for infrared (IR) detector applications. The major drawback

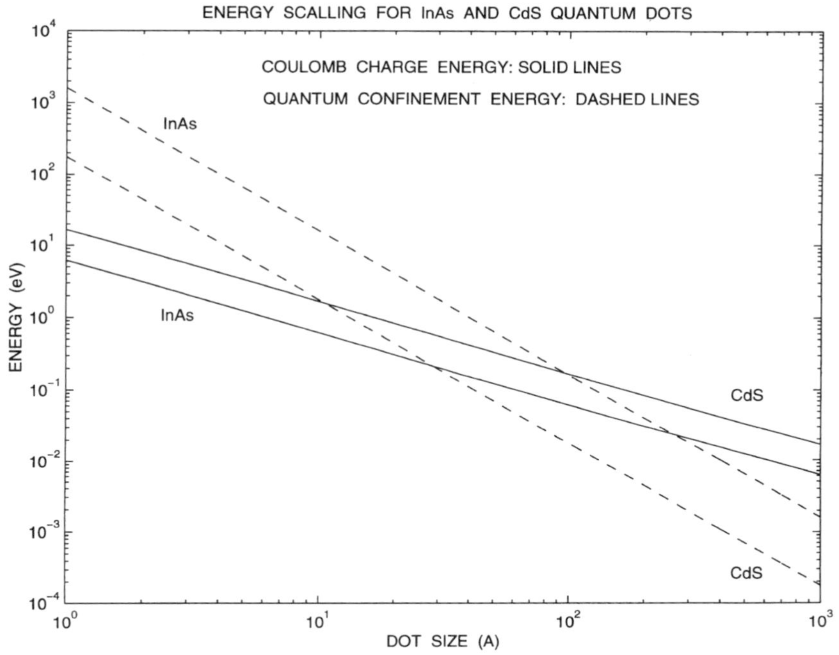

Figure 12.1. Coulomb charge energy and quantum confinement energy scaling in semiconductor QDs.

of the intersubband transitions in multiple quantum well structures is the selection rule of the Γ band, which forbids the detection of normal incident light [19]. Therefore, a grating coupler is needed for the normal incident light in focal plane detector application. Unlike quantum wells, QDs do not have this drawback. They can be effectively used for detection of the normal incident radiation. Another key advantage of the epitaxial grown Ge dots is the possibility of monolithic integration with the mature Si technology. This and other issues of the utilization of semiconductor QDs for mid-IR detection will be discussed in more detail in Section 12.3.

The use of Ge multiple QD arrays on Si substrate for thermoelectric applications was recently suggested by our group [20]. Because of the rather small conduction band offset between Si and Ge, such structures do not strongly deteriorate electron transport while significantly slowing down phonons contributing to thermal conductivity due to strong phonon scattering on the QDs and spatial confinement of acoustic phonon modes. The latter leads to an increase in the thermoelectric figure of merit $ZT = S^2\sigma T/k$ of the multiple QD arrays (T is the temperature, S is the Seebeck coefficient, σ is the electrical conductivity, and k is the thermal conductivity). Utilization of the QD arrays for thermoelectric materials shows great promise for achieving $ZT > 1$ and changing the market of refrigerators and power generators.

A demonstration of two different types of the QD field effect transistors (QD–FET) was recently reported in [21]. These transistors can be used for novel photodetectors, photon-storage devices, and single-electron memory devices controlled by the gate voltage and light illumination. Among other possible applications of QDs, we will discuss in greater detail second-harmonic generation (SHG) and wavelength conversion, nonmagnetic memory devices, and some other electronic devices. A thorough feasibility study of the QD-based implementation of a quantum computing logic gate will also be given in Section 12.3.

12.2. REVIEW OF THE FABRICATION TECHNIQUES

In this section, three of the QD fabrication techniques are given more detailed consideration. They are self-assembly using MBE, nanostamping, and electrochemical self-assembly.

12.2.1. Self-Assembly

A semiconductor QD self-assembled in the MBE systems by the spontaneous islanding in the Stranski–Krastanow growth mode have been around for 6 years or so. The QD formation in this heteroepitaxial growth mode is characterized by the formation of gradually relaxing 3D islands over a fully strained initial two-dimensional (2D) wetting layer through a growth mode transition. Self-assembled QDs can now be grown with good reproducibility and quality control [1–3,22]. The intersubband energy spacing was shown to be easily tunable by adjusting the temperature of the substrate during the growth of QDs and by depositing a cap layer with postgrowth annealing. A number of different material systems, which were used to form self-assembled dots, include groups III A (13)–V A(15) semiconductors (mostly InAs/GaAs and InAs/InP), some groups II B(12)–VI A (16) such as CdSe on GaAs, as well as Ge dots on Si.

More recently, Sharma et al. [22] successfully grew InAs quantum dots on a Si(100) substrate using MBE. Morphological examination by atomic force microscopy (AFM) has shown that the dots had narrow size distribution and high density (see Figs. 12.2 and 12.3). A large lattice mismatch of $\sim 11.5\%$ of the InAs/Si system was suggested as an origin of a large density of dislocation free coherently strained dots with a feature size of 10 nm. Such small dislocation free dots may find applications in silicon based optoelectronics [22]. Self-organized compositional InGaAs quantum dot lasers, grown on Si substrates and operating at 1.013-μm wavelength with a threshold current density of $3.85\,\mathrm{KA/cm^2}$ have already been demonstrated [23].

In most cases, the self-assembled dots grown by MBE are randomly distributed [1–4,22]. Although there were reports of self-ordering of QDs in prepatterned substrates. In [24], the authors reported Ge dots grown on Si(001) partially covered with patterned oxide. On narrow plateaus with well-defined side wall facets, the Ge dots locate preferentially at the edges of the raised Si(001) regions, and the preference is strongest on the narrowest patterns aligned along a $\langle 100 \rangle$ direction. The dots appear to be uniformly spaced along the pattern edges.

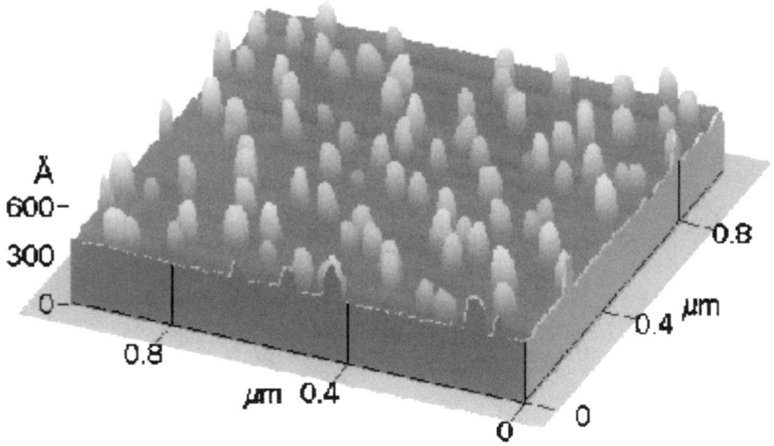

Figure 12.2. The AFM image of InAs quantum dots grown on Si(100). The sample with ~4.6 monolayers of coverage was grown at 450°C. (Courtesy of P. C. Sharma, UCLA).

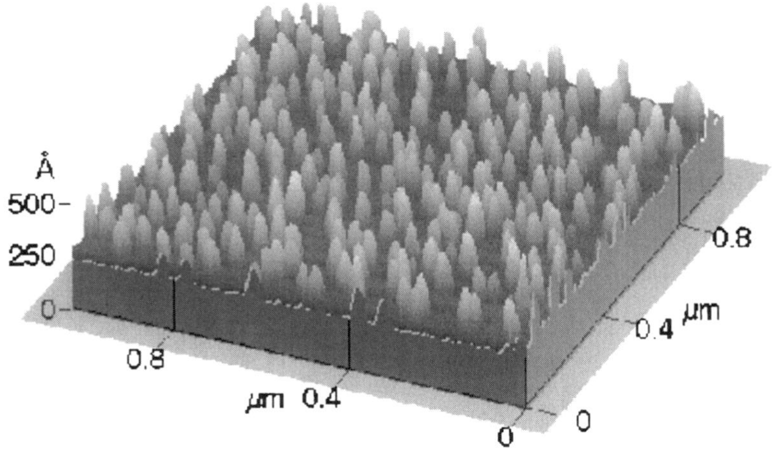

Figure 12.3. The AFM image of InAs quantum dots grown on Si(100). The sample with ~4.1 monolayers of coverage was grown at 450°C. The optimized growth conditions allowed to achieve high densities and uniform dot dimensions. (Courtesy of P. C. Sharma, UCLA).

Fabrication of regimented 3D arrays of QDs was recently reported by Springholz et al. [25]. The self-organization of pyramidal PbSe islands that spontaneously form during strained-layer epitaxial growth of $PbSe/Pb_{1-x}Eu_xTe$ ($x = 0.05-0.1$) superlattices results in the formation of 3D quantum dot crystals [25]. In these crystals, the dots are arranged in a trigonal lattice with a face-centered cubic (fcc) like vertical stacking sequence. The lattice constant of the dot crystal can be tuned continuously by changing the superlattice period. It was also shown that the elastic anisotropy in these artificial dot crystals acts in a manner similar to that of the directed chemical

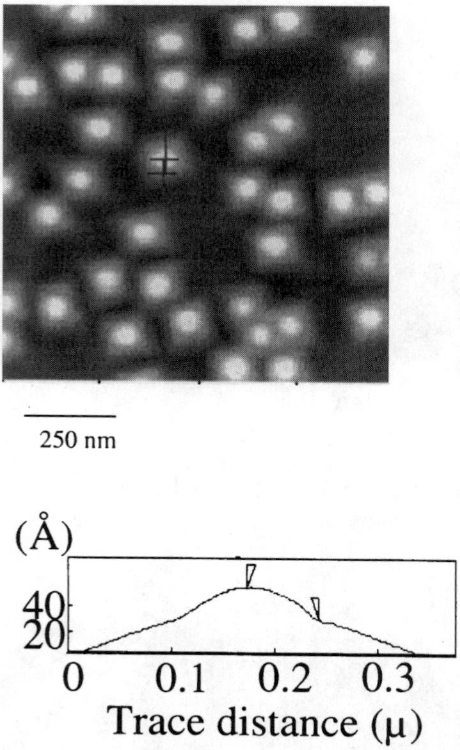

Figure 12.4. The AFM image of Ge quantum dots qrown on Si. (Courtesy of J. L. Liu, UCLA).

bonds of crystalline solids. The narrow size distribution and control of the dot arrangement may be advantageous for optoelectronic device applications.

A majority of optoelectronic applications may not require regimented arrays of QDs but rather randomly distributed dots with an approximately uniform spatial density and a narrow size distribution. On the basis of the Si/Ge system, such types of structures are usually grown using a solid source MBE system. During this process, Si(100) wafers (with resistivity of $\sim 14-22\,\Omega$cm) undergo a standard Shiraki's cleaning procedure, and then are loaded into a MBE system. The protective oxide layer is removed by subsequently heating the substrate to 930°C for 15 min. The substrate temperature is maintained at 650°C during the epitaxial growth. Dot arrays are formed with the deposition of a few monolayers of Ge after an initial growth of a buffer layer. The nominal growth rates for these processes are 0.1 and 0.02 nm/s for Si and Ge, respectively [1–3]. Figure 12.4 shows the top layer of QDs grown by this method. One should also mention here that optoelectronic applications usually require many layers (stacks) of QDs in order to increase the signal-to-noise ratio (see schematic picture on Fig. 12.5). We call these structures a quantum dots superlattice (QDS).

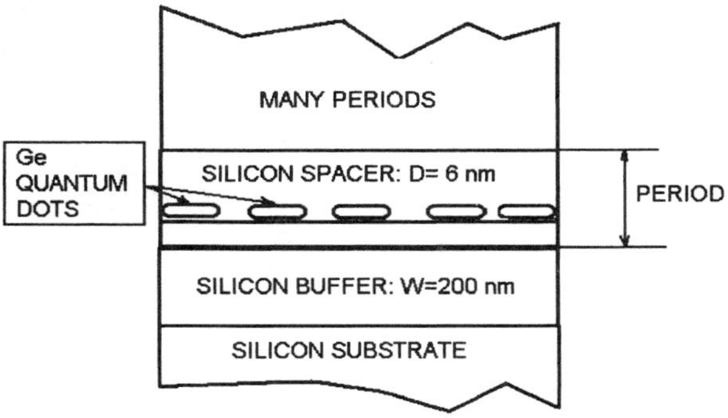

Figure 12.5. Schematic illustration of the QDS grown by MBE. These structures can be used for a variety of applications from thermoelectric cooling devices to IR detectors.

12.2.2. Nanostamping

As we outlined above, many techniques such as e-beam lithography [26], X-ray lithography [27], scanning probe microscopy based lithography [28–30], and others have been used to fabricate structures with sizes $< 100\,\mathrm{nm}$. Although it is possible to generate these feature sizes, the manufacturability of these techniques in a high-throughput microelectronics fabrication environment remains a major challenge.

Microcontact printing is a technique that can generate patterns of self-assembled monolayers (SAMs) of alkanethiolates on surfaces of gold [31], silver [32], copper [33], or of alkysiloxanes on hydroxyl-terminated surfaces [34,35], and others. The patterned SAM can serve as a thin resist film with which the patterns can be transferred to a substrate utilizing an appropriate etching method. Much of the work in this area has been performed on gold. An aqueous solution of ferricynide is used as a wet etchant to transfer the SAM patterns to the thin gold film, which can then serve as a secondary mask to etch underlying semiconductor layers [33,36].

Improved processes are needed before microcontact printing can be considered as a serious candidate for a semiconductor manufacturing process. The issue of particular importance for further development of nanostamping is scaling of the feature size down to a few nanometers. Progress in this direction was recently reported by Kamins et al. [37]. Their approach was to combine MBE self-assembly on a prepatterned substrate with the nanostamping techniques ("nanoimprinting" in their terminology), which uses a mold and etching. When these two techniques are combined, the small patterned features can interact with the self-assembly process, causing the islands to form at the patterned features. The resulting regular array of very small islands may be useful for future devices.

To give a better idea of the process of nanostamping, we describe here the formation of patterned SAM films on SiO_2, crystalline Si, and amorphous Si surfaces. The patterns were transferred successfully to the amorphous Si and crystalline Si surfaces using only the SAM as the resist film and KOH as the etchant. Amorphous Si films

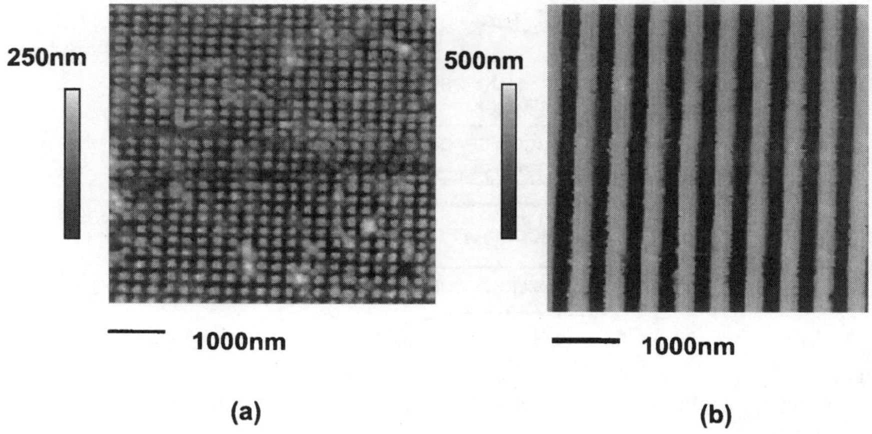

Figure 12.6. The AFM images of a master formed by e-beam lithography, showing the Si array of nanometer scale dots (a) and lines (b). The pitch of the dots is 250 nm and this is ~150 nm. The pitch of the lines is 500 nm and the line width is ~300 nm. The height of the dots and lines is ~350 nm.

resulted in the smallest pillars (~80 nm in diameter). The dimensions of the SAM films and the etched patterns were verified by AFM.

The typical procedure of microcontact printing and subsequent pattern transfer were reported in detail by Wang et al. [38]. Two kinds of master wafers were used to make the polydimethylsiloxane (PDMS) stamps. The first mold is a film made of photoresist patterned by photolithography. The size of the structures here is on the order of microns. The second mold is composed of Si dots and lines formed by e-beam lithography. The dimension of the structures is on the nanometer scale. The AFM images of the Si nanostructures are shown in Fig. 12.6(a) and (b). The pitch of the dots in Fig. 12.6(a) is 250 nm, and the diameter of the dots is ~150 nm. The pitch and width of the lines are 500 and 300 nm, respectively, from the figure. However, due to the AFM tip shape, the actual size of the dots and the line width could be smaller (e.g., 50, 100 nm), respectively. The stamps, which can be used repeatedly, were then cast from these masters.

The substrates used for microcontact imprinting were amorphous Si, crystalline Si, and SiO_2. The formation of the amorphous silicon film (20 nm) was achieved by room temperature e-beam evaporation of silicon onto a SiO_2 substrate. The surface roughness of all the substrates prior to processing was better than a few nanometers. Before contacting with the stamp, the surface was treated in a solution of 30% H_2O_2 and concentrated H_2SO_4 to yield a thin SiO_2 film with a Si–OH group concentration of $\sim 5 \times 10^{14}\,cm^{-2}$ [39]; thus, the surface is highly hydrophilic. A dilute octadecyltrichlorosilane (OTS) solution (one drop in 10 mL hexane) was prepared in a N_2 atmosphere using hexane as the solvent since OTS reacts violently with water. A cotton swab was used to apply the OTS onto the surface of the stamp. The stamp was then brought into contact with the substrate. The Si–Cl bonds in the OTS molecules

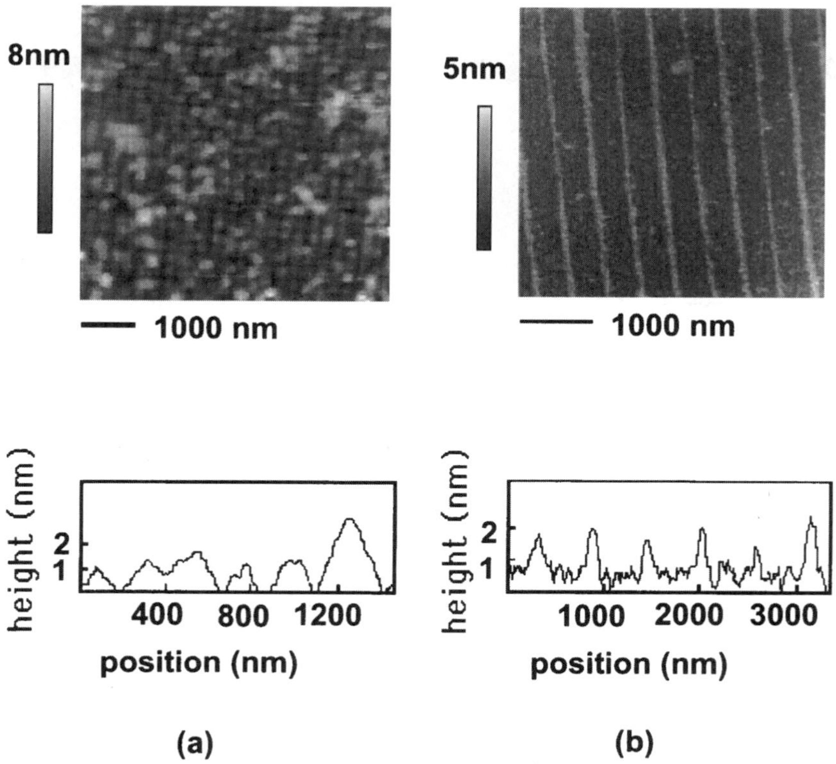

Figure 12.7. The AFM images of nanometer scale SAM patterns formed on the amorphous Si surface by microcontact printing. Their thickness is ~ 2 nm. (a) and (b) are the AFM images of the SAMs formed by the master wafers shown in Fig. 12.2(a) and (b), respectively.

are broken by the OH groups present on the surface of the substrate to form a network of Si–O–Si bonds, resulting in a self-assembled monolayer of OTS on the contact area. A longer contacting time and a stronger applied force result in a more resilient OTS layer, which can better withstand the subsequent processing. In this experiment, the substrate and stamp are kept in contact for more than a few minutes. Figures 12.7(a) and (b) has typical AFM images of the nanometer scale SAM patterns formed on the amorphous Si surface. The height and width of the SAM lines, for the particular example shown in Figure 12.7(b) are ~ 2 and 100 nm, respectively.

Pattern transfer was successful for OTS on both amorphous and crystalline Si by the use of a dilute KOH etching solution. In this process, the wafer is dipped in dilute HF for ~ 8 s to remove the native oxide prior to KOH etching. Figures 12.8(a) and (b) show AFM images of the transferred line patterns on crystalline Si and amorphous Si, depicting a rougher surface for the case of OTS on the crystalline Si surface. The surface roughening is caused by the anisotropic etching of crystalline Si by KOH [see Fig. 12.8(a)]. This problem is circumvented by the use of an amorphous silicon film. As can be seen in Fig. 12.8(b), after the pattern transfer, the

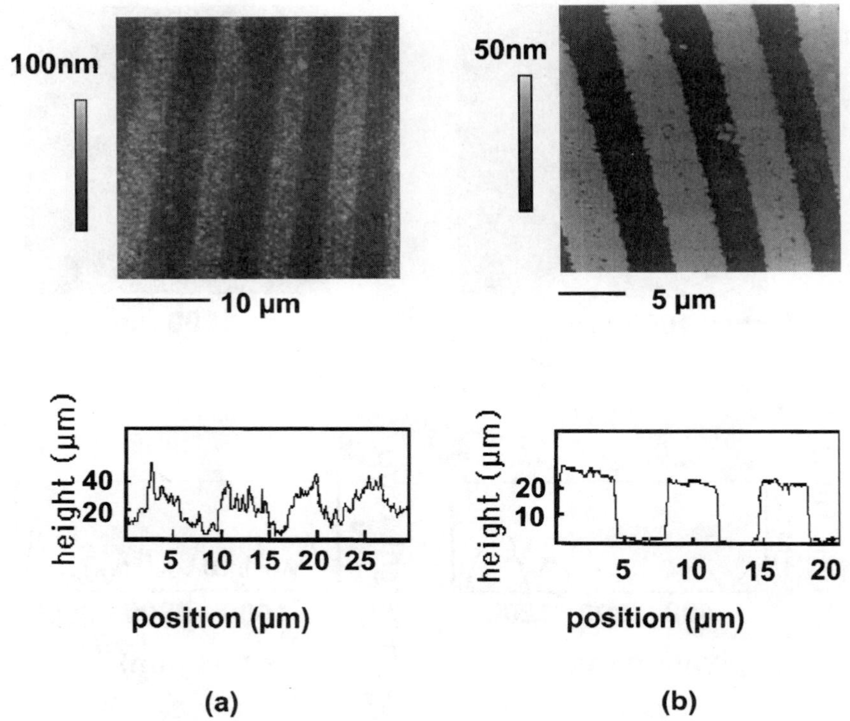

Figure 12.8. The AFM images of line patterns formed on (a) crystalline Si and (b) amorphous Si after dilute KOH etching with the OTS monolayer as masks. The etch depth is ~ 20 nm and the line width is on the order of microns.

amorphous Si surface of the patterns is smoother relative to that in the crystalline Si case. The use of amorphous Si as a transfer layer constitutes a substantial improvement over that of the metal films more commonly used in microcontact printing.

As described in [38], the patterns formed by wet etching of SAMs are the inverse of the original patterns of the master wafer. Since the master wafer, shown in Fig. 12.4(a), has a pillar pattern, the pattern transferred onto the substrates will be holes, with a diameter equal to the base of the pillars. Figures 12.9(a) and (b) shows the AFM images of the holes formed on the amorphous Si surface after etching by dilute KOH. Their pitch is 250 nm, the same as that on the master wafer. When the hole pattern is overetched during pattern transfer to the extent that holes touch each other, a pillar pattern results. Figures 12.9(c) and (d) shows the amorphous pillars with a diameter of 80 nm. With overetching, the size of the pillars can be controlled by the etching time. This amorphous Si pillar pattern can be used as a secondary mask for subsequent patterning if desired. In the case of pattern transfer on a SiO_2 film, the OTS film is not able to withstand etching by HF. Thus, alternative methods for pattern transfer must be used [35].

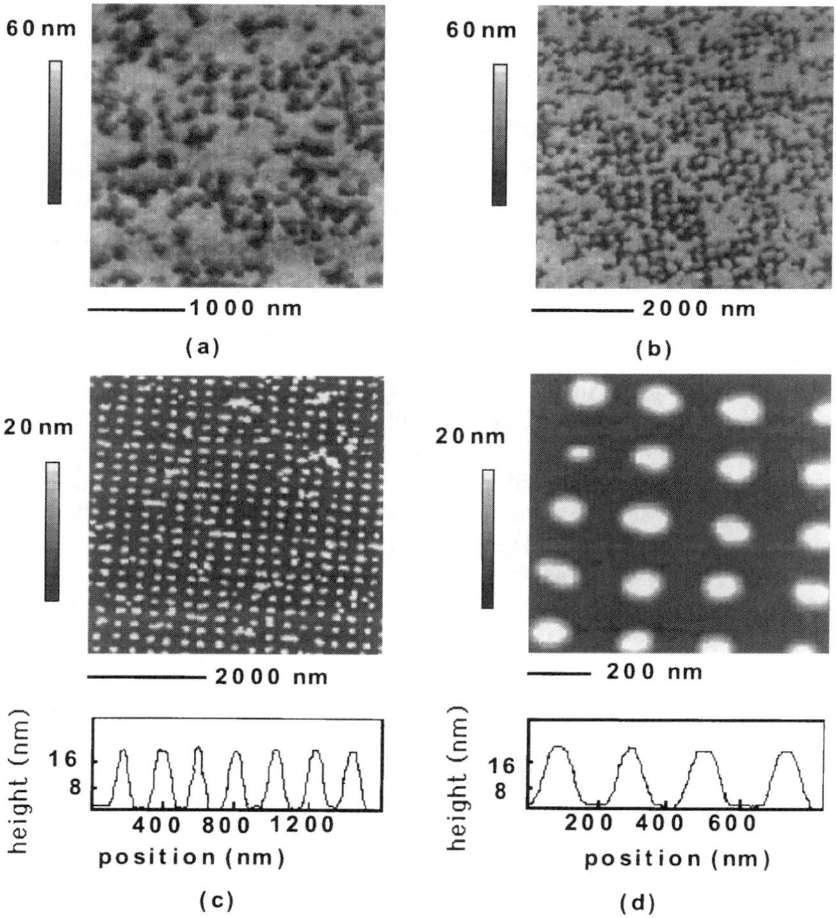

Figure 12.9. The AFM images of the transferred patterns on the amorphous Si surface after etching by dilute KOH: (a) and (b) show the images of the hole patterns. Their pitch is 250 nm, the same as that of the master wafer. The images of the amorphous pillars, with the diameter of 80 nm, are illustrated in (c) and (d). These pillars are obtained after over etching with KOH.

To sum up this section, microcontact printing can be used to form nanometer scale patterns of SAMs on amorphous Si, crystalline Si, and SiO_2 using OTS as the ink and an elastomer as the stamp [38]. The patterns can be subsequently transferred into crystalline Si substrates or amorphous Si films using the SAM of OTS as the resist film. Atomic force microscopy can be used to characterize the quality of the SAM and the resulting patterns. By using a Si pillar structure as the mold, similar pillars < 80 nm in size can be obtained by over etching of the patterned OTS film on amorphous Si using KOH. The size of the resulting amorphous Si pillars can be controlled by the etching time. Further improvement of this technique may allow for fabrication of QDs with strict control over their position. This may be important for electronic applications of QDs.

12.2.3. Electrochemical Self-Assembly

Among other most successful QD (quantum wire) fabrication techniques alternative to MBE is electrochemical self-assembly, which was reported by Bandyopadhyay et al. [8,40] and some others [41]. This technique allows us to produce quasiperiodic 2D arrays of semiconductor QDs embedded in an insulating porous alumina template. Typically, the materials of choice for QDs in this technique are the groups II B(12)–VI A(16) compounds such as CdS or CdSe, although other semiconductors may be deposited inside the pores.

The average physical diameter of the dots produced by this technique is 15 nm, but the effective optical diameter is usually less due to the side depletion. The preparation procedure starts by first anodizing a thin foil of aluminum for a few seconds in 15% H_2SO_4 at a dc current density of 40 mA/cm^2 to produce a nanoporous alumina film on the surface of the foil. These pores form an ordered hexagonal array with an average diameter of 15 nm with a 7% standard deviation. An AFM image of a typical pore assembly is shown in Fig. 12.10. The pores are then selectively filled up with CdS using ac electrodeposition. This is accomplished by immersing

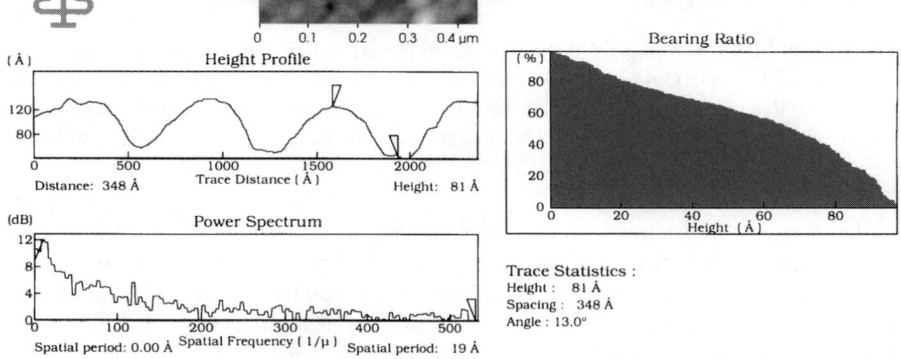

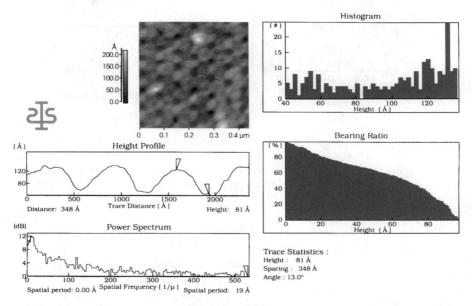

Figure 12.10. The AFM picture of the self-assembled alumina tempate used for synthesis of the arrays of semiconductor QDs. Black spots are the holes that are being filled with different semiconductor materials (CdS, CdSe, etc.). Note the nanoscale feature size of the structures, high degree of periodicity, and density ($>10^{12}$ cm^{-2}) of the template pores.

the alumina film in a boiling aqueous solution of $CdSO_4$. An ac potential of 18 V rms is applied for different durations (5–30 s) using the sample and a graphite rod as electrodes. The Cd^{2+} ion in the solution reacts with the S^{2-} ion left behind in the walls of the pores from the previous anodization step to form CdS. The amount of CdS produced (pore filling) increases superlinearly with the duration of electrodeposition.

The pores have been directly imaged with transmission electron microscopy (TEM) and field-emission SEM [8]. The presence of material in the pores has been verified with cross-section TEM and the stoichiometry was checked with Auger spectroscopy. Optical signatures of CdS were also verified in the past by Raman, photoluminescence, reflection, and absorption [8,40]. These measurements revealed strong quantum confinement effects. In fact, the measured blue shift in the optical spectra indicates that the effective optical diameter of the dots is ~3.5 nm (even though the physical diameter is 15 nm). The size reduction is presumably caused by the side depletion around the periphery of the dots. These are some of the smallest QDs that have shown optical activity.

The variable angle spectroscopic ellipsometry (VASE) was also performed on these samples [42,43] in order to determine (1) the thickness of the alumina layer in which the QDs are embedded, (2) the shape of the CdS dots, (3) the volume fraction of CdS in the alumina layer, and (4) surface roughness and lateral thickness nonuniformity. To obtain information about sample optical constants and structure, ellipsometry data were fitted with a multilayer model based on effective medium approximation. Model parameters were varied (using the Levenberg–Marquardt algorithm) to minimize the mean-square error between two sets of data: measured and model generated. The samples were viewed as arrays of CdS dots in an alumina matrix that could be treated within the Maxwell–Garnett formalism. This allowed us

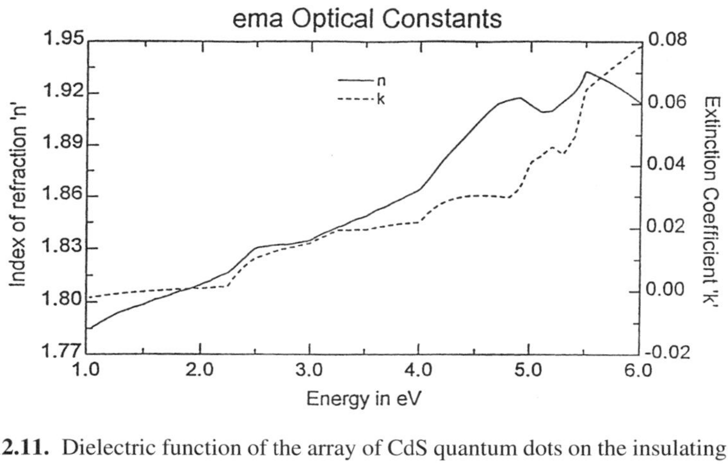

Figure 12.11. Dielectric function of the array of CdS quantum dots on the insulating alumina substrate. The function was determined using VASE data and Maxwell–Garnett model approximation. The shoulder at 2.5 eV corresponds to the fundamental band gap of CdS. (After Balandin [38]).

to determine the volume fraction of CdS in the alumina layer, and the depolarization factor directly related to the dot shape. It was found that the shape of the CdS dots can be best approximated by the rotational ellipsoids with a depolarization factor ~ 0.7. The thickness nonuniformity is $\sim 7\%$; and the volume fraction of CdS for 10-s deposition time was found to be also $\sim 7\%$. Figure 12.11 shows the refractive index of the array of QDs as a function of wavelength extracted from the VASE data. The possibility of *in situ* control of the QD quality and their optical properties (see Fig. 12.11) using VASE can be a crucial factor for future commercial manufacturing and optoelectronic device applications.

12.3. APPLICATION OF QUANTUM DOTS

12.3.1. Quantum Dots for Infrared Detection

Intersubband transitions in semiconductor nanostructures has been a subject of interest from both fundamental physics and applied research point of views. This subject was particularly important for development of the IR photodetectors and lasers. Previously, quantum well structures for detector applications had been studied extensively because of the available mature epitaxial techniques such MBE and chemical vapor deposition (CVD) [44–47]. Compared with 2D quantum well structures, the intersubband absorption in the quasi-0D quantum dot structures have advantages in optical applications because of their sharp delta-like density of states, the reduced intersubband relaxation times and, thus, lower detector noises in their nanostructures [48,49]. With the recent success in fabricating high-quality QDs by the Stranski–Krastanow growth process [50–52], it is timely to examine the intersubband optical transitions of these structures.

Most work to date is based on the electron intersubband absorption in groups III A(13)–V A(15) based QD structures. For example, IR absorption has been reported for doped InGaAs quantum dots for wavelengths $> 20\,\mu m$ [53], and for doped InAs dots in the range of 10–20 μm [54], respectively. Mid-IR photoconductivity at $\sim 3\,\mu m$ has also been studied for delta-doped InAs/AlGaAs quantum dots for subbands to continuum transitions [55].

To date, limited work has been done in SiGe based Ge quantum dot structures. In fact, the large valence band offset of the 0D Ge/Si system, as well as its small light hole effective mass, favor the use of SiGe low-dimensional structures for device applications. The intersubband absorption can be utilized for mid-IR detection. Another key advantage of the epitaxial Ge dot structures, and IR detectors based on this system is the possibility of monolithic integration with the mature Si technology.

In this section, we describe the application of QDs for IR detection in detail following the results of Liu et al. [1,3] and Wu et al. [2]. In both of these works, two samples (A and B) with different Ge layer thickness were prepared in order to study the intersubband optical transitions and energy separation. Both structures consist of 20 periods of boron-doped Ge dot layers and undoped Si barriers. Similar samples without the last Si barrier (with Ge quantum dots grown on the surface) have been

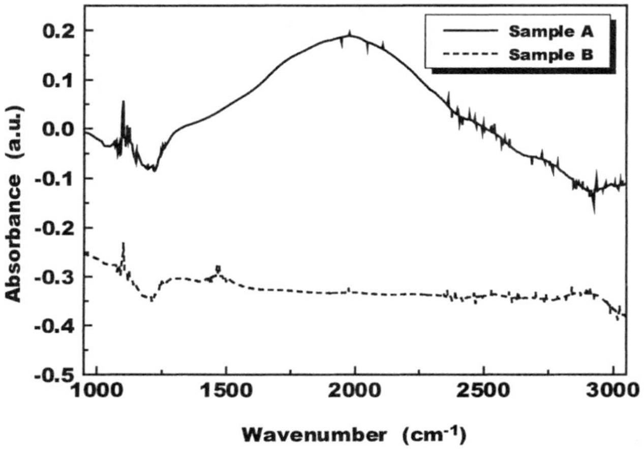

Figure 12.12. The FTIR absorption spectra for two samples with Ge quantum dots taken at room temperature.

examined with AFM indicating estimated base dimensions of 30 and 100 nm, and heights of 10 and 30, for samples A and B, respectively.

These samples were boron-doped, and contained multiple layers of Ge dots (QD superlattices). The IR absorption spectra of the samples were taken at room temperature using a FTIR spectrometer. Standard waveguide structures (10 mm long and 5 mm thick) were prepared with polished backside and polished 45° facets for the IR absorption measurement. The latter allowed for enhanced absorption. A beam condenser was used to focus the IR beam onto the waveguide. An IR polarizer was placed in the path of the IR beam to probe the polarization dependence of the absorption process. A Si substrate waveguide with the same dimensions was used as reference.

The measured absorption spectra of the samples as a function of wavenumber are shown in Fig. 12.12. For sample A, absorption peaks are seen near 2000 cm^{-1} (5 μm) and 1350 cm^{-1} (7.4 μm), where the intersubband absorption processes are supposed to occur in the QDs and Ge wetting layers, respectively. The fwhm of the absorption peak at 5 μm for sample A is ~ 100 meV, and is considerably larger than the intersubband peak width observed in the InGaAs/GaAs quantum dot superlattice (~ 13 meV) [56]. The size uniformity of QDs is a possible explanation for this broadening. The nonparabolicity of the hole bands can also play a significant role in absorption peak broadening. This type of broadening was observed in the quantum well case [45,47]. The absorption peak near 7.4 μm in sample B is believed to be due to the absorption in the wetting layer. The vanishing intersubband absorption in this QD superlattice is due to its larger dot size. As a result, the energy difference between two-hole bound states is so small that it cannot be resolved in the absorption spectra in the investigated energy range. Beyond this energy range, the free carrier absorption tends to obscure the measurement of intersubband transitions. The same absorption peak position in two Ge wetting layers for samples A and B is due to the

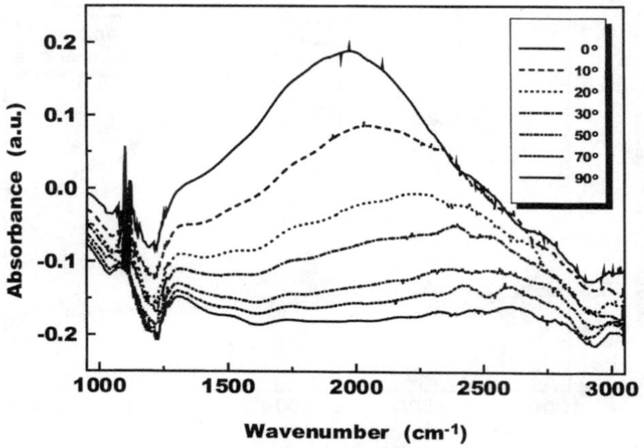

Figure 12.13. Polarization dependent absorption spectra of a QD sample taken at room temperature. The decrease of the absorption with the increasing polarizer angle is due to the reduction of the component of the photon polarization along the axis of the dots.

fact that the thicknesses of wetting layers (usually $3-6$) of samples A and B are about the same under similar growth conditions (e.g., growth temperature) [57]. A sharp peak near $1100\,\text{cm}^{-1}$, which is seen in Fig. 12.13, is mainly due to the strong IR absorption by SiO_2 and H_2O bands.

In order to further confirm the nature of intersubband transitions in the Ge dots, the polarization dependence of the 5-μm peak of sample A is performed. The data is presented in Fig. 12.13. Here, the 0° direction corresponds to 50% polarization along the growth direction of the structure, while the 90° direction is defined as the polarization being parallel to the plane of layers. The absorption of the Ge dots exhibits the maximum at the 0 polarization angle, which corresponds to the direction of the largest confinement in the dots. The absorption strength decreases as the polarization angle increases. This polarization dependence, which has been observed in the *n*-doped InAs/GaAs quantum dot system [58], is similar to the polarization dependence in the intersubband of quantum wells, and is in agreement with the well-known intersubband absorption selection rules.

The details of the study of the intersubband absorption in multiple self-organized Ge dots were reported in [1–3]. This study suggests that boron doped Ge QDS can be used for mid-IR detectors.

12.3.2. Quantum Dots for Second Harmonic Generation

Optical nonlinearities causing frequency conversion are useful in a number of applications such as mixing, switching, limiting, and coupling. Most ordinary solids, however, are not efficient frequency converters because they exhibit extremely small higher order components of dielectric susceptibilties, and also because phase matching, which is needed to optimize frequency conversion process, is difficult to attain in solids that are not birefringent. Furthermore, if the solid has inversion symmetry, then it can exhibit no even-order susceptibility [59] unless the symmetry is broken artificially either by the intentional growth of an asymmetric structure [60], or by an external electric [61] or magnetic [62] field. On rare occasions, the inversion symmetry can be broken spontaneously by built-in fields.

Typically, the second-order optical nonlinearity in a solid has two sources; a bulk contribution and a surface or interface contribution [63]. The inversion symmetry is automatically broken at the interface because of the discontinuity of the crystalline structure and the large gradient of the normal-to-surface component of the electric field of incident radiation. This alone can result in a finite (but small) second-order dielectric susceptibility $\chi^{(2)}$, which causes weak SHG. Other possible mechanisms of SHG in semiconductors include (1) electric quadrupole nonlinearity [64], (2) deformations and stresses in the structure leading to internal fields and nonlinear polarization, (3) lowering of the symmetry of the crystal under the action of the intense electric field of incident radiation, and (4) a high degree of disorder of the crystalline structure (quasiamorphism).

In this section, we follow the results of Balandin et al. [43] and describe the observation of SHG in ordered regimented arrays of 15-nm diameter CdS quantum dots that were prepared by the electrochemical self-assembly process. The details of the preparation of the samples were given by Bandyopadhyay et al. [8,40]. It is most likely that all of the above mentioned nonlinear mechanisms play some role in SHG observed in these samples. However, regardless of its origin, the enhancement of SHG in QDs as compared to the bulk material is very promising. Applications of QD arrays may provide a low cost and convenient way to produce nonlinear optical components. What is also important is that the enhancement was observed for the incident light with an energy smaller than the band-gap energy.

The experimental setup used for the optical SHG measurements consists of a mode-locked Nd:YAG laser producing 10-ns-long pulses with an average power density of $1\ \mathrm{W/cm^2}$ and an absolute peak intensity of $\sim 1\ \mathrm{MW/cm^2}$ at a wavelength of $1.064\ \mu\mathrm{m}$ (photon energy $= 1.32\ \mathrm{eV}$). This corresponds to subband-gap radiation. The bulk band gap of CdS is 2.4 eV and the optically measured band gap, enhanced by quantum confinement, is $\sim 3\ \mathrm{eV}$ for samples prepared by 10-s electrodeposition [8,40,65–67]. The samples were irradiated by a laser at a 20° angle and the spot size was ~ 1 cm, which is approximately equal to the sample size. The second harmonic signal was observed in the reflection mode by collecting all radiation reflected back (in π rad).

A quartz (α-SiO_2) etalon was used as a reference. The nonlinear coefficient of the etalon is $d_{11} = 4.4 \times 10^{-13}\ \mathrm{m/V}$, and the refractive indexes in the spectral region of interest are $\sim n_0 = 1.5350$ and $n_e = 1.5438$ for ordinary and extraordinary rays, respectively. The measured absolute intensity of the SHG of the quartz etalon was $\sim 10^{-15}\ \mathrm{W/cm^2}$. In Fig. 12.14, we show the normalized intensity of the SHG in arrays of CdS quantum dots as a function of the time of deposition. As mentioned before, the time of deposition determines the height of CdS ellipsoids in the alumina pores. The maximum value of the second harmonic intensity, $I(2\omega)$, for our sample with the highest volume fraction of CdS is $\sim 2\%$ of the $I(2\omega)$ of the quartz etalon.

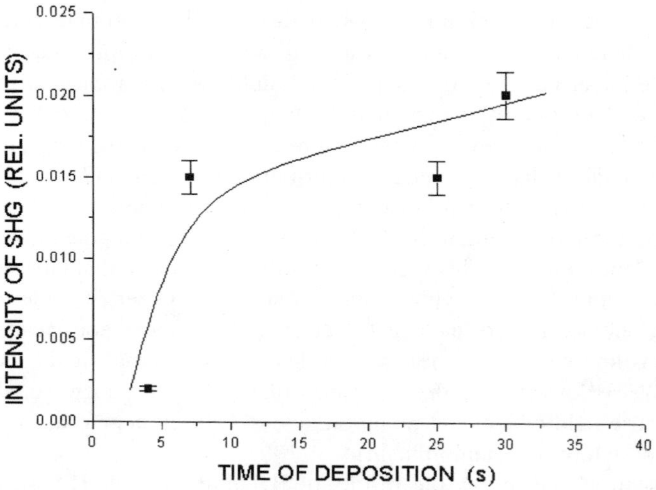

Figure 12.14. The intensity of the SHG in the arrays of CdS quantum dots relative to the intensity of SHG in the quartz etalon. Time of deposition determines the quantity of CdS in the alumina pores (size of the QD).

In order to characterize the process of SHG quantitatively, we use the following formula for the conversion efficiency [68]

$$\frac{I(2\omega)}{I(\omega)} = \frac{2\omega^2 |d_{\text{eff}}|^2 l^2 I(\omega)}{n^3 c^3 \varepsilon_0} \operatorname{sinc}^2(\Delta \boldsymbol{k} \cdot \boldsymbol{l}/2) \tag{12.7}$$

where $\operatorname{sinc}(x) = \sin(x)/x$, $\boldsymbol{l}$ is the interaction length, d_{eff} is the effective material nonlinear coefficient, n is the refractive index of the material, c is the speed of light, $\Delta \boldsymbol{k} = \boldsymbol{k}(2\omega) - 2\boldsymbol{k}(\omega)$ is the wavevector mismatch, and, $I(\omega)$ is the intensity of incident light at the fundamental frequency. Phase matching determines the phase synchronism factor $\operatorname{sinc}^2(\Delta \boldsymbol{k}\boldsymbol{l}/2)$ which is unity at $\Delta \boldsymbol{k}\boldsymbol{l} = 0$. Equation (12.7) is valid in the limit of small conversion and plane wave focusing. It is reasonable to assume that the interaction length, $\boldsymbol{l}$, in our case is approximately equal to the size of a QD. Since the size of the dot is very small (vertical dimension of the dots is of the order of 50 Å) and the difference between refractive indexes at the fundamental and doubled frequencies is insignificant, we can limit our consideration to the case when $\Delta \boldsymbol{k}\boldsymbol{l} \sim 0$. Under such conditions, one can estimate the value of the effective nonlinear coefficient d_{eff} from Eq. (12.7). We find that $d_{\text{eff}} = 8.8 \times 10^{-11}$ m/V, which is about five times larger than that of bulk CdS. Since d_{eff} is proportional to the second-order susceptibility $\chi^{(2)}$, we conclude that our QDs have a value of $\chi^{(2)}$ that is at least five times larger than that of the bulk medium.

As we pointed out, the second-order nonlinearity in our samples is observed at below band-gap frequency. This indicates that the nonlinearity is either related to virtual processes or intraband processes. As far as the latter is concerned, the second harmonic generation may be associated with the intersubband transition dipoles in the conduction band [64]. For samples under consideration, the calculated energy spacing between the electronic subbands is between 1 and 1.5 eV (assuming infinite potential barriers at the interfaces and typical amount of side wall depletion). For the incident photon energy of 1.32 eV, the incident radiation may excite real transitions between two subbands. This mechanism is obviously directly related to quantum confinement and would not be relevant for bulk samples. Other possible origins of the SHG enhancement include electric quadrupole nonlinearity [64], deformations, and stresses in the structure leading to internal fields and nonlinear polarization, lowering of the symmetry of the crystal under the action of the intense electric field of the incident radiation, and high degree of disorder of the crystalline structure (such as amorphism).

Concluding this section, we would like to emphasis that self-assembled semiconductor QD arrays can exhibit strong second-order nonlinearities in their optical response. These nonlinearities are much larger in QDs than in corresponding bulk material. The magnitude of the second-order susceptibility is at least five times larger than that of bulk CdS. It is reasonable to assume that QD arrays fabricated by other techniques (MBE, nanostamping) also will have an enhanced optical nonlinearity. This property of QDs can be used in a number of laser and light source applications as well as for mixing, switching, limiting, and coupling.

12.3.3. Novel Computing Paradigms

It is generally believed that conventional technologies for integrated circuits will approach their limit because of the interconnect bottleneck problem, and because the feature size of the device will eventually reach the atomic scale. These factors will lead to corresponding difficulties with heat management, low gain, and low fanout. At the same time, the manufacturing cost of the conventional patterning process continues to increase. As a result, there have been a number of proposals that examine novel computing paradigms for nanoelectronic logic gates [69,70]. Most of these proposals envision utilization of semiconductor QDs. Enormous difficulties associated with fabrication of a large number of almost identical QDs required for the implementation of Boolean logic have so far slowed down the progress in this area.

Recently, the interest to novel computing architectures based on semiconductor QDs has revived again. The latter was related to the rapid development of the concept of quantum computing [71–75]. Progress in quantum computation schemes has led to many proposals of potentially realizable quantum computers based on various physical systems. Among the most known are quantum computers that utilize laser trapped ions [71], nuclear magnetic resonance (NMR) systems [72], all-optical logic gates [73], Josephson junctions [74], and semiconductor QDs [75]. Successful experimental demonstrations of one and two qubit computers were reported for laser trapped ion systems [76] and NMR systems [77].

The progress in theoretical and experimental development of quantum logic gates on the basis of semiconductor nanostructures has been far short of these systems. This is primarily due to the difficulties in fabrication of high-quality QD arrays with a number of almost identical dots, and the decoherence problem [78], which is inherently more severe for solid-state systems [79]. On the other hand, if these problems are overcome, the quantum computer based on semiconductor QDs offers an attractive alternative to other physical implementations due to its compactness, robustness, and the larger number of qubits that can be realized [80]. Some of these advantages were discussed previously in the context of nanoelectronic architecture for Boolean logic gates [81,82]. Another important advantage of the QD implementation is that further development can make use of well-developed conventional semiconductor technology [83]. It is also generally accepted now that only solid-state (most likely—QD) implementation of a quantum logic gate can be scaled up to the qubit sizes required for useful operation of a quantum computer.

In Section 12.3.3.1 we describe in details a possible implementation of the quantum XOR gate (also called controlled-NOT gate) based on coupled *asymmetrical quantum dots*. We present a feasible, study of this implementation, which includes calculation of the coupling constants and decoherence time [79].

12.3.3.1. Logic Gates Based on Quantum Dots. A classical controlled-NOT gate is a two-bit operator, in which the first bit is the control and the second bit is the target. Its operation or the truth table can be written as follows:

$$\mathrm{CN} := |00\rangle\langle 00| + |01\rangle\langle 01| + |10\rangle\langle 11| + |11\rangle\langle 10| \tag{12.8}$$

Here, if the control bit is in the state $|0\rangle$, the target bit does not change its value after the action of the gate, but if the control bit is in the state $|1\rangle$, the target changes its value after the action of the gate (the last two terms of the equation). Since this gate is logically and physically reversible, it is in principle capable of quantum computing [84]. In order to do this, the gate must act on superpositions of bits (qubits) and coherently transform as in the following example: $\mathrm{a}|00\rangle + \mathrm{b}|11\rangle \rightarrow \mathrm{a}|00\rangle + \mathrm{b}|10\rangle$. Here, each member of the quantum superposition transforms according the classical rule for the gate. Such a gate is conventionally called the quantum XOR gate.

We propose an implementation of the XOR gate with the computational basis states, $\{|0\rangle, |1\rangle\}$ formed by the ground and first excited states of two semiconductor single-electron QDs. Application of a $\pi/2$ pulse or some other fractions of a π pulse [75] will place the gate into a coherent superposition of two states $|0\rangle$ and $|1\rangle$, thus allowing for initial preparation of a qubit. Owing to the asymmetricity of the confining potentials of the two dots and their corresponding uneven charge distributions, the dipole moments p_{ground} and p_{excited} are induced in the ground and excited states $|0\rangle$ and $|1\rangle$, respectively (since the dipoles are parallel, we do not use vector notation). These dipoles are defined as

$$p_i = \int_{-W/2-\Delta}^{W/2+\Delta} \Psi_i z \Psi_i^* \, dz \tag{12.9}$$

where z is the axis along the growth direction (see the inset to Fig. 12.15), W is the size of the dot along the z axis, Φ is the electron wave function, Δ is the maximum

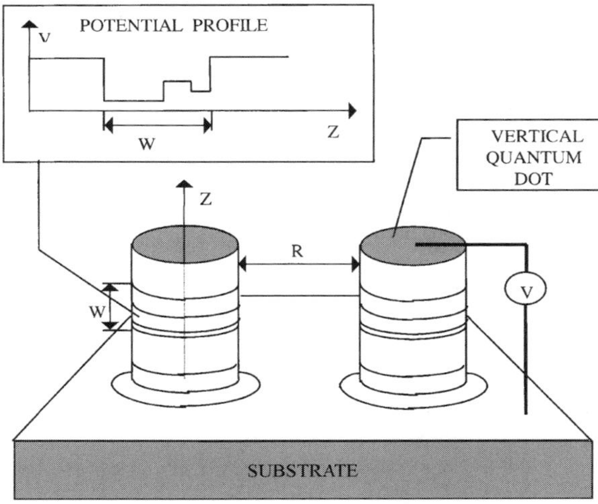

Figure 12.15. Structure of the single-electron asymmetric QDs. Inset shows an example of the potential profile of a QD along the growth direction.

wave function barrier penetration depth, and index i denotes either "ground" or "excited" states of the same dot.

The conditional dynamics of the two-dot structure is achieved via dipole–dipole interaction between two neighboring dots as in [75]. The first QD with the resonant energy $h\nu_c$ (for the transition energy between states $|1\rangle$ and $|0\rangle$) acts as the control qubit while the second dot with the resonant energy $h\nu_t$ acts as the target qubit. The gate is driven optically by application of a proper train of π/N pulses (N is a positive number), which allow us to prepare the qubits, perform Hadamard transforms, and other operations needed for quantum computing [85].

The asymmetry of the potential well has an effect similar to that of applying an electric field [75] while allowing for more degrees of freedom. Control of the potential profile provides better tuning for dipole moments, so that dipoles of the computational basis states can be made exactly equal in the absolute values while opposite in their signs,

$$p_{\text{ground}} = -p_{\text{excited}} \tag{12.10}$$

The latter is required for having desirable Ising-type coupling interaction between the dipoles of the two dots. Moreover, we will argue that it is extremely difficult if not impossible to meet the requirement of Eq. (12.10) in symmetric QDs for realistic values of material parameters and biasing electric field. In principle, conditional quantum dynamics with the resonant frequency of one dot depending on the neighboring dot's state can be built from any such dots brought close together in the Coulomb interaction range [80,86]. But it turns out that in order to have a better resolution of different quantum states and to avoid any undesirable evolution of these states, it is better to construct a system that can be described by an Ising-type interaction [84].

The proposed structure can be implemented using the state-of-the-art technology of QDs, which can be fabricated in a vertical configuration [87,88]. Specifically, we will consider an $A1_xGa_{1-x}As/GaAs$ vertical QD structure containing a tunable number of electrons. Experimental demonstrations of such structures containing from 0 to 50 electrons were reported in [89,90]. By controlling the potential profile of the dots along the growth direction, one can adjust the system in such a way that the condition of Eq. (12.10) holds true [91]. An example of a prototype structure with the desirable potential profile is shown in Fig. 12.15. The gate bias that is required for localization of one electron in the dot structure is on the order of $|V_G| = 0.1$–0.7 V. The electric field induced by this bias will tilt the potential profile shown in Fig. 12.1, but will not introduce any significant change in the values of dipoles as it will be shown in Section 12.4.

12.3.3.2. Operation of the Gate. To a good approximation, the Hamiltonian of noninteracting QDs commutes with the Hamiltonian of the dipole–dipole interaction V_{int}, which can be written as

$$V_{\text{int}}|n,k\rangle = (-1)^{n+k+1}\hbar(J/2)|n,k\rangle \tag{12.11}$$

where $\hbar$ is Plank's constant; $n = 0$ or 1 denotes the state of the control qubit (the first dot); $k = 0$ or 1 denotes the state of the target qubit (the second dot); and the constant of the dipole–dipole interaction J is given by

$$\hbar J = \frac{p_c p_t}{2\pi\varepsilon_0\varepsilon R_r^3} \tag{12.12}$$

where R is the distance between the dots, ε_0 and ε_r are the vacuum and relative dielectric permittivities, respectively. Throughout this chapter, the subscripts c and t refer to the control and target qubits (or dots), respectively. One can see from Eqs. (12.11) and (12.12), that the system of interacting dipoles can be described by the Ising-type interaction where the dipoles associated with the confined QD states act as *effective spins*. This particular form of the interaction Hamiltonian is desirable because all classical bit states are the eigenstates of this interaction, so that there is no undesirable time evolution of the quantum states. Meanwhile, application of an appropriate π pulse (or its fraction) to the target qubit can induce a time evolution that corresponds to the action of the XOR gate.

By using the expression for the Hamiltonian of the dipole–dipole interaction, Eq. (12.11), we can construct an energy diagram for two interacting dots in the absence of the field, which is shown in Fig. 12.16, where we have assumed that $\omega_c > \omega_t$. When the control dot is in state $|1\rangle$, the resonant frequency for the target dot becomes

$$\omega_t' = \frac{E_{11} - E_{10}}{\hbar} = \frac{1}{2}\,(\omega_c + \omega_t - J) - \frac{1}{2}\,(\omega_c - \omega_t + J) = \omega_t - J \tag{12.13}$$

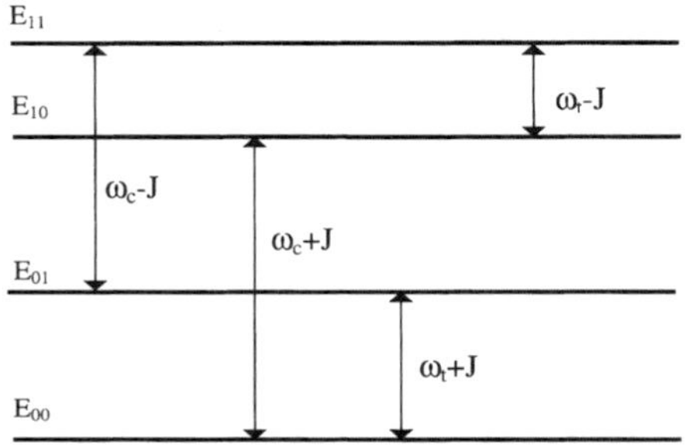

Figure 12.16. Energy levels of two dipole–dipole interacting QDs. It is assumed that $|\omega_c| > |\omega_t|$.

As a result, a π pulse of frequency $\omega_t - J$ causes the transition $|1\rangle \rightarrow |0\rangle$ provided that the control dot is in state $|1\rangle$. In order to be able to determine the state of the system experimentally, the coupling constant has to be relatively large so that

$$\Delta\omega = \omega_{t,c} - J \leq \Gamma \tag{12.14}$$

where Γ is the line broadening due to interaction with phonons, impurities, and so on. The probability of spontaneous transitions must also be small. The latter will be discussed in the Section 12.3.3.3.

12.3.3.3. Coupling between Two Quantum Dots. In order to determine the electron wave functions, confined energies, and dipole moments for each QD, we use the envelope function approximation. The 3D confinement potential can be written as $U(x, y, z) = U_x(x)U_y(y)V(z)$, where $V(z)$ is a finite quantum well profile along the growth direction, and $U_{x,y}$ are assumed to be infinite potential barriers in the normal plane. Neglecting excitonic effects, we may write the envelope function of a conduction band electron confined by the electrostatic potential as

$$\Phi = \phi(x)\phi(y)\chi(z) \tag{12.15}$$

where $\phi(x)$, $\phi(y)$ are regular particle-in-a-box states, and $\chi(z)$ is the envelope along the growth direction to be determined numerically. Since the confinement potential is modified along the z direction, the relevant properties of the dot will be mostly defined by $\chi(z)$. This component of the wave function is found using the transfer matrix method [92], which allows us to take into account the potential profile variation and wave function barrier penetration. The dipoles due to the uneven charge distributions [see Eq. (12.9)] in the two c and t dots can now be rewritten as

$$p_{c,t} = \int_{-W_{c,t}/2-\Delta}^{W_{c,t}/2+\Delta} \chi_{c,t}(z) z \chi_{c,t}^*(z)\, dz \tag{12.16}$$

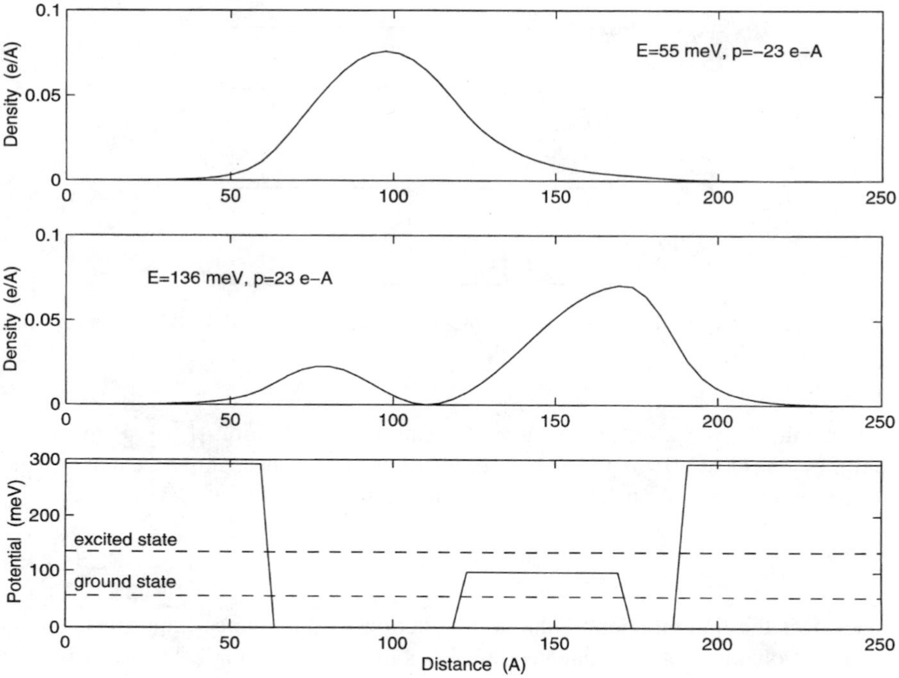

Figure 12.17. Charge density distribution in the asymmetric QD. Results are shown for the ground (upper panel) and first excited states (middle panel). Dipole moments due by the asymmetricity of the distributions are equal in absolute values but opposite in directions. A potential profile, which creates such charge density distributions, is shown below (lower panel).

Material parameters used in simulations correspond to GaAs. The electron effective mass is assumed to be $0.067m_0$, where m_0 is the free electron mass. The relative permittivity of GaAs is taken to be 10.9.

The charge density distributions for the first two confined states together with corresponding confined energies, dipole moments, and the profile of the confinement potential along the growth direction are shown in Fig. 12.17. The height of the potential step approximately corresponds to the heterojunction band offset between GaAs and $Al_{0.2}Ga_{0.8}As$. The dipoles due to the uneven charge distributions along the growth direction are equal in absolute values but opposite in directions ($p_c = -23$ e-A and $p_t = 23$ e-A). The chosen potential profile gives a much better control of the equality $p_{\text{ground}} = -p_{\text{excited}}$ than the simple step well profile. It is interesting to note here that the dot size can be scaled up to a certain extent while keeping the ground-state and excited-state dipoles equal in their absolute value. This is important since (1) the resonance frequencies of the dots should be different, and (2) this allows for the tuning of the dipole–dipole coupling constant J. One should note here that it is almost impossible to meet the condition of Eq. (12.10) for the simple symmetric dot

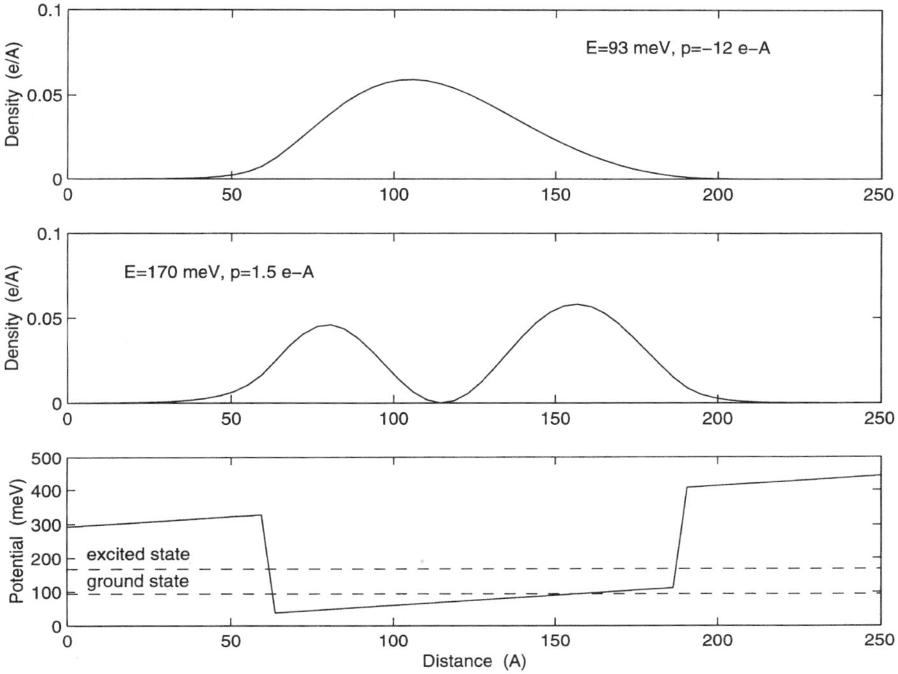

Figure 12.18. Charge density distribution in the symmetric QD biased with electric field. Results are shown for the ground (upper panel) and first excited states (middle panel). Dipole moments induced by the electric field are very different in absolute values. Potential profile is shown below (lower panel).

(well) structure biased with electric field. Figure 12.18 shows such a structure biased with $E = 50$kV/cm electric field. As one can see, the dipoles due to the uneven charge distribution are very different in their values ($p_{\text{ground}} = -12$ e-A and $p_{\text{excited}} = 1.5$ e-A) so that the Ising-like interaction model is not applicable.

By using Eq. (12.16), we calculated the dipole–dipole interaction coupling constant for two asymmetric QDs with the potential profile shown on Fig. 12.17 (the lowest panel). The coupling constant J as a function of the effective size W of the dot is presented in Fig. 12.19. The solid curves correspond to the case when the size of both dots is scaled up, while the dashed curves correspond to the case when the control dot size is fixed at 100 A and only the target dot size is changed. The results are presented for several different dot separation distances. The curves denoted by letters A, B, and C are plotted for dots grown on top of a substrate and, thus separated by air ($\varepsilon_r = 1$). The dash–dotted curve D corresponds to two dots separated by a material with $\varepsilon_r = 10.9$ when both dot sizes vary.

Rather large values of the coupling (as compared to the broadening) allow for experimental resolution of different quantum states. It also reduces requirements for the π/N pulse selectivity. The length of the pulse τ_π has to be in the range specified

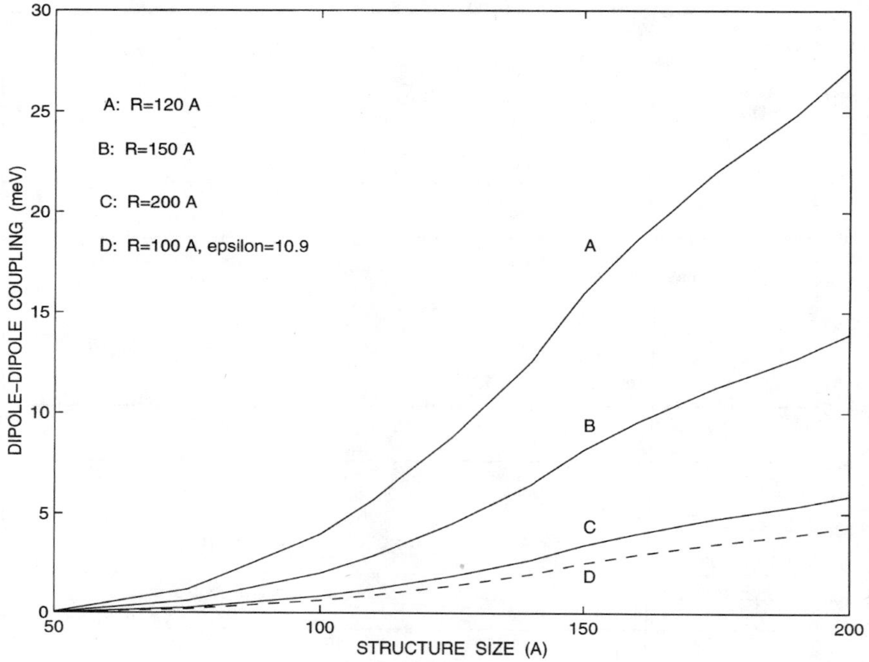

Figure 12.19. Dipole–dipole coupling constant as a function of the dot size along the growth direction. The solid curves correspond to the case when the sizes of both dots are being scaled up. The dashed curves correspond to the case when the size of the control dot is fixed at 100 Å and only that of the target dot varies.

by the condition below

$$1/J < \tau_\pi < \tau_{\rm coh} \tag{12.17}$$

where $\tau_{\rm coh}$ is the quantum coherence time, which depends on the system structure and temperature. For the time allowed for computation ($\tau_{\rm coh}$), we want to use as many driving pulses as possible for increasing the number of computations. Since it is difficult to increase the upper limit of the inequality (12.17), we may try to decrease the lower limit. For the asymmetric dot implementation, the inverse of the coupling constant is on the order of 2.5×10^{-14} s. For comparison, the number obtained in [75] for the gate based on square potential QDs biased with an electric field is on the order of 10^{-12} s. This means that in our XOR gate driven with a femtosecond laser, it is possible to have a sufficient number of qubit manipulation steps. The only technological problem left here is the search for a femtosecond laser in the appropriate frequency range (intersubband transitions).

12.3.3.4. Quantum Coherence. It is well known that a quantum computing algorithm can be implemented only if the quantum coherence of the system is preserved during the time of computation $\tau_{coh} : \tau_{com} > \tau_{coh}$. Here, we estimate the limit of τ_{coh}, considering the dominant decoherence processes relevant to the proposed implementation. Under the assumption of having high-quality structures at low temperatures, we may expect that the decoherence of our system occurs by spontaneous transitions between intersubband states, which serve as the computational basis. These transitions include both radiative and nonradiative processes, which have been well studied for coupled asymmetric quantum well structures [93,94]. In general, for quantum wells, the nonradiative lifetime of the transition ($|1\rangle \rightarrow |0\rangle$) is on the order of 0.1–10 ps, while the radiative lifetime is on the order of 10–100 ns [93]. To our knowledge, the study of relaxation processes in QDs remains incomplete. It is still a subject for debate, particularly regarding the effect of the "phonon bottelneck" on the carrier relaxation rates [95,96]. It is generally believed, however, that the level quantization in QDs slows down carrier relaxation toward the ground state and thus may increase the coherence time of our basic states.

Here, we evaluate the spontaneous radiative lifetime t^r_{spon} and electron–phonon scattering time for a given geometry of our XOR gate structure. The t^r_{spon} can be found as [97]

$$\frac{1}{\tau^r_{\text{spon}}} = \frac{n_r e^2 E^2_{i,j}}{2\pi\varepsilon_0 c^3 m_0 \hbar^2} f_{i,j} \tag{12.18}$$

where n_r is the refractive index, c is the speed of light, $E_{i,j}$ is the intersubband separation energy between states i and j, and $f_{i,j}$ is the oscillator strength given by

$$f_{i,j} = \frac{2m_0 E_{i,j}}{\hbar^2} |\langle j|z|i\rangle|^2 \tag{12.19}$$

We use Eqs. (12.18)–(12.19) to calculate for t^r_{spon} transitions between the ground and the first excited states ($i = 0,\ j = 1$), and to determine how it changes for different potential profiles. The results of the numerical simulations are presented in Table 12.1.

TABLE 12.1. Spontaneous Radiative Lifetime versus Asymmetric Dot Size

W (nm)	$\tau^r_{\text{spon}}(\text{s}) \times 10^9$
5.0	7
10.0	50
12.5	88
15.0	120
17.5	160
20.0	180

We can estimate the lifetime of the basic states limited by the nonradiative transitions using the formalism of [98]. It follows that phonon emission rate by an electron in the state i is given as

$$\frac{1}{\tau_{\mathrm{ph}}} = \frac{2\pi}{\hbar} \sum_{f,q} \alpha^2(q)| < \Psi_f | e^{-iqr} | \Psi_i > |^2 \delta(E_f - E_i + E_q)[N_0(T,\ E_q) + 1] \quad (12.20)$$

where the sum extends over all possible final-electron quantum numbers f and phonon wavevectors $q, N_0(T, E_q)$ is the Bose–Einstein distribution function, E_q is the energy of a phonon with wavevector q, T is the temperature, and α is the coupling constant. An analogous expression can be written for phonon absorption. One has to consider the coupling of electrons to longitudinal acoustic (LA) phonons via deformation potential and the Frohlich interaction between the electron and longitudinal optical (LO) phonons. These two mechanisms are expected to be dominant for our structures, although much weaker than in the bulk. Interaction with longitudinal optical phonons in the QD structure occurs when the level separation is close to the LO phonon energy $\hbar\omega_{\mathrm{LO}}$. In our calculations, we assume that $\hbar\omega_{\mathrm{LO}} \sim 36$ meV. The expressions for the coupling constants are given in [98]. We perform the calculations with the material parameters that correspond to the GaAs/AlGaAs system. The numerical simulation for $T = 4$ K shows that $\tau_{\mathrm{ph}} \sim 10^{-8}$–$10^{-7}$ for a 20-nm wide QD structure, which is comparable to the radiative relaxation time. For higher temperatures, the nonradiative relaxation is stronger and $\tau_{\mathrm{ph}} < \tau^r_{\mathrm{spon}}$. Having these numbers in mind, let us assume that we use 100-fs laser pulses to drive our gates. If we take the average coupling constant from Fig. 12.19 and assume that $\tau_{\mathrm{coh}} \sim 10^{-9} < \min\{\tau_{\mathrm{ph}}, \tau^r_{\mathrm{spon}}\}$, we can see that inequality (12.17) holds: $10^{-14} < 10^{-13} < 10^{-9}$ for our gate. Thus, the strong coupling allows us to use fast laser pulses for optical driving of the quantum XOR gate and perform many computations during the time τ_{coh} provided that the temperature is sufficiently low.

In conclusion, we described an implementation of the quantum XOR gate based on coupled asymmetric QDs. The structure can be realized using the state-of-the-art nanotechnology. The qubit preparation and manipulation are achieved by application of a proper train of π pulses and their fractions. The results of our numerical simulations show that the coupling constant of the dipole–dipole interaction in asymmetric dots can be tuned over a wide range by the choice of the potential profile, separation distances, and material parameters of the dots. This provides conditions for having the Ising-type interaction between the dots. We also examine quantum coherence requirements for the operation of the quantum gate. Finally, we argue that our gate can be driven by a femtosecond laser if one like this is designed for an appropriate frequency range.

12.4. CONCLUSIONS

In this chapter, we described the basic physical properties of QDs and discussed their current and possible future applications. Continuous progress in fabrication techniques for semiconductor QDs resulted in significant achievements in both understanding the physical processes in quasi-0D structures and in their practical applications. Many of these recent developments were addressed in this chapter. Concerning fabrication techniques, we concentrated our attention on MBE self-assemble growth, nanostamping, and electrochemical deposition of semiconductor materials into porous alumina films. Utilization of QDs for IR photodetection, and second-harmonic generation has been discussed. It was shown that arrays of Ge quantum dots on Si offer improved detection capability as compared to quantum wells due to their sensitivity to normal incident light. Enhanced second-order nonlinearity in QDs with respect to the bulk material may lead to the emergence of novel effective nonlinear elements. This and some other aspects of potential applications were also discussed. The second part of this chapter presents a feasibility study of the use of QDs for future nanoelectronic and quantum computing architectures.

ACKNOWLEDGEMENTS

The authors thank Jianlin Liu, P.C. Sharma, Wen-Gang Wu, Yin Sheng Tang, Gaolong Jin, members of the Device Research Laboratory, UCLA, for many discussions on the fabrication of quantum dots and their applications. Special thanks go to P.C. Sharma for reading the manuscript and suggesting some valuable changes. One of the authors (AB) is indebted to S. Bandyopadhyay (UNL) and V. Fomin (University of Antverpen) for their help and encouragement in a process of this work. This work was in part supported by NSF, SRC, and ARO.

REFERENCES

1. J. L. Liu, W. G. Wu, A. Balandin, G. L. Jin, and K. L. Wang, *Appl. Phys. Lett.* ***74***, 185 (1999).
2. W. G. Wu, J. L. Liu, Y. S. Tang, G. L. Jin, and K. L. Wang, in D. C. Houghton and E. A. Fitzgerald, Eds. *Proceedings of SPIE*, Vol. 3630, 5, SPIE, 1999, p. 98.
3. J. L. Liu, Y. S. Tang, K. L. Wang, T. Radetic, and R. Gronsky, *Appl. Phys. Lett* ***74***, 1863 (1999).
4. S. Fafard, Z. R. Wasilewski, C. Ni Allen, D. Picard, P. G. Piva, J. P. McCaffrey, *Superlattices Microstruct.* ***25***, 87 (1999).
5. R. Heitz, N. N. Ledentsov, D. Bimberg, A. Yu. Egorov, *Appl. Phys. Lett.* ***74***, 1701, (1999); A. E. Zhukov, A. R. Kovsh, A. Yu. Egorov, N. A. Maleev, *Fiz. Tekh. Poluprov.* ***33***, 180 (1999) [*Semiconduct.* ***33***, 153 (1999).].
6. L. Kouwenhoven and C. Marcus, *Phys. World* ***35*** (June, 1998).
7. D. Wang, S. G. Tomas, K. L. Wang, Y. Xia, and G. M. Whitesides, *Appl. Phys. Lett.* ***70***, 1593
8. S. Bandyopadhyay, A. E. Miller, and M. Chandrasekhar, in *Proceedings of SPIE*, Vol. 2397, SPIE, 1995, pp. 11–30; R. E. Ricker, A. E. Miller, D.-F. Yue, G. Banerjee, and S. Bandyopadhyay, *J. Electron. Materials* ***25***, 1585 (1996).

9. C. Weisbuch and B. Vinter, *Quantum Semiconductor Structures: Fundamentals and Applications*, Academic, San Diego, CA, 1991; G. Bastard, *Wave Mechanics Applied to Semiconductor Heterostructures*, Halsted, New York, 1988.
10. H. Gotoh, H. Ando, and H. Kanbe, *Appl. Phys. Lett.* ***68***, 2132 (1996).
11. T. Takagahara, *Phys. Rev. B* ***39***, 10206 (1993).
12. L. V. Keldysh, *JETP Lett.* ***29***, 658 (1965).
13. H. Richter, Z. P. Wang, and L. Ley, *Solid State Commun.* ***39***, 625 (1981); I. H. Campbell and P. M. Fauchet, *Solid State Commun.* ***58***, 739 (1986).
14. V. M. Fomin, V. N. Gladilin, J. T. Devreese, and E. P. Pokatilov, *Phys. Rev. B* ***57***, 2415 (1998).
15. V. M. Fomin, E. P. Pokatilov, J. T. Devreese, S. N. Klimin, V. N. Gladilin, and S. N. Balaban, *Solid-State Electron.* 1309 (1998).
16. A. N. Korotkov, in J. Jortner and M. A. Ratner, Eds., *Molecular Electronics*, Blackwell, Oxford, UK.
17. D. L. Huffaker, G. Park, Z. Zou, O. B. Shchekin, and D. G. Deppe, *Appl. Phys. Lett.* ***73***, 2564 (1998).
18. O. Qasaimeh, W.-D. Zhou, J. Phillips, S. Krishna, P. Battacharya, and M. Dutta, *Appl. Phys. Lett.* ***74***, 1654 (1999).
19. K. L. Wang and R. P. G. Karunasiri, in M. O. Manasreh, Eds., *Semiconductor Quantum Wells and Superlattices for Long-Wavelength Infrared Detectors*, Artech House, Norwood, MA, 1993, p. 139.
20. K. L. Wang, A. Balandin, and J. Liu, unpublished results.
21. G. Yusa and H. Sakaki, *Superlattices Microstruct.* ***25***, 247 (1999).
22. P. C. Sharma, K. W. Alt, D. Y. Yeh, D. Wang, and K. L. Wang, *J. Electron. Materials* ***28***, 432 (1999).
23. K. K. Linder, J. Phillips, O. Qasaimeh, X. F. Liu, S. Krishna, P. Bhattacharya, J. C. Jiang, *Appl. Phys. Lett.* ***74***, 1355 (1999).
24. T. I. Kamins and R. S. Williams, *Appl. Phys. Lett.* ***71***, 1201 (1997).
25. G. Springholz, V. Holy, M. Pinczolits, and G. Bauer, *Science* ***282***, 734 (1998).
26. E. A. Dobisz, C. R. K. Marrian, L. M. Shirey, and M. Ancona, *J. Vac. Sci. Technol.* ***B10***, 3067 (1992).
27. A. Moel, M. L. Schattenburg, J. M. Caster, and H. I. Smith, *J. Vac. Sci. Technol.* ***B8***, 1648 (1990).
28. L. Tsau, D. Wang, and K. L. Wang, *Appl. Phys. Lett.* ***64***, 2133 (1994).
29. D. Wang, L. Tsau, and K. L. Wang, *Appl. Phys. Lett.* ***65***, 1914 (1994).
30. D. Wang, L. Tsau, and K. L. Wang, *Appl. Phys. Lett.* ***67***, 1914 (1995).
31. A. Kumar and G. M. Whitesides, *Appl. Phys. Lett.* ***63***, 4 (1993).
32. Y. Xia, E. Kim, and G. M. Whitesides, *J. Electrochem. Soc.* ***143***, 1070 (1996).
33. Y. Xia, M. Mrksich, E. Kim, and G. M. Whitesides, *Chem. Mater.* ***8***, 601 (1996).
34. Y. Xia, M. Mrksich, E. Kim, and G. M. Whitesides, *J. Am. Chem. Soc.* ***117***, 9576 (1995).
35. P. M. St. John and H. G. Craighead, *Appl. Phys. Lett.* ***68***, 7 (1996).
36. E. Kim, A. Kumar, and G. M. Whitesides, *J. Electrochem. Soc.* ***142***, 628 (1995).

37. T. I. Kamins, D. A. A. Ohlberg, R. S. Williams, W. Zhang, W., and S. Y. Chou, *Appl. Phys. Lett.* ***74***, 1773 (1999).

38. D. Wang, S. G. Thomas, K. L. Wang, Y. Xia, and G. M. Whitesides, *Appl. Phys. Lett.* ***70***, 1593 (1997).

39. J. M. Madeley and C. R. Richmond, *Z. Anorg. Allg. Chem.* ***389***, 82 (1972).

40. S. Bandyopadhyay, A. E. Miller, H. C. Chang, G. Banerjee, D-F. Yue, R. E. Ricker, S. Jones, J. A. Eastman, and M. Chandrasekhar, *Nanotechnology* ***7***, 360 (1996).

41. D. Al-Mawlawi, C. Z. Liu, M. Moskovits, *J. Mater. Res.* ***9***, 1014 (1994); J. Heremans, C. M. Thrush, Z. Zhang, X. Sun, M. S. Dresselhaus, J. Y. Ying, and D. T. Morelli, *Phys. Rev. B* ***58***, R10091 (1998).

42. A. Balandin, Ph. D. Theses, University of Notre Dame, 1996.

43. A. Balandin, S. Bandyopadhyay, P. G. Snyder, S. Stefanovich, G. Banerjee, and A. E. Miller, *Phys. Low-Dim. Structur.* ***11/12***, 155 (1997).

44. L. West and S. Eglash, *Appl. Phys. Lett.* ***46***, 1156 (1985).

45. R. P. G. Karunasiri, J. S. Park, and K. L. Wang, *Appl. Phys. Lett.* ***59***, 2588 (1991).

46. P. Kruck, M. Helm, T. Fromherz, G. Bauer, J. F. Nutzel, and G. Abstreiter, *Appl. Phys. Lett.* ***69***, 3372 (1996).

47. S. K. Chum, D. S. Pan, and K. L. Wang, *Phys. Rev. B* ***47***, 15638 (1993).

48. H. Benisty, C. M. Sottomayor-Torres, and C. Weisbuch, *Phys. Rev. B* ***44***, 10945 (1991).

49. U. Bockelmenn, and G. Bastard, *Phys. Rev. B* ***42***, 8947 (1990).

50. K. H. Schmidt, G. Medeiros-Ribeiro, M. Oestreich, P. M. Petroff, and G. H. Dohler, *Phys. Rev. B* ***54***, 11346 (1996).

51. J. Tersoff, C. Teichert, and M. G. Lagally, *Phys. Rev. Lett.* ***76***, 1675 (1996).

52. K. Mukai, N. Ohtsuka, and M. Sugawara, *Appl. Phys. Lett.* ***70***, 2416 (1997).

53. H. Drexler, D. Leonard, W. Hansen, J. P. Kotthaus, G. Medeires-Ribeiro, and P. M. Petroff, *Phys. Rev. Lett* ***73***, 2252 (1994).

54. J. Phillips, K. Kamath, X. Zhou, N. Chervela, and P. Bhattacharya, *Appl. Phys. Lett.* ***71***, 2252 (1994).

55. K. W. Berryman, S. A. Lyon, and M. Segev, *Appl. Phys. Lett.* ***70***, 1861 (1997).

56. D. Pan, Y. P. Zeng, M. Y. Kong, J. Wu, Y. Q. Zhu, C. H. Zheng, J. M. Li, and C. Y. Wang, *Electron. Lett.* ***32***, 1726 (1996).

57. H. Sunamura, S. Fukatsu, N. Usami, and Y. Shiraki, *J. Crystal Growth* ***157***, 265 (1995).

58. S. Sauvage, P. Boucaud, F. H. Julien, J.-M. Gerard and V. Thierry-Mieg, *Appl. Phys. Lett.* ***71***, 2785 (1997).

59. See, for example, P. N. Butcher, and D. Cotter, *The Elements of Nonlinear Optics*, Cambridge University Press, Cambridge, 1990; A. Yariv, *Quantum Electronics*, Wiley, New York, 1989; Y. R. Shen, *The Principles of Nonlinear Optics*, Wiley, New York, 1984.

60. A. Sa'ar, I. Grave, N. Kuze, and A. Yariv, in *Nonlinear Optics: Materials, Phenomena and Devices*, IEEE, New York, 1990 p. 113.

61. M. M. Fejer, S. J. B. Yoo, R. L. Byer, A. Harwit, and J. S. Harris, *Phys. Rev. Lett.* ***62***, 1041 (1989).

62. A. Svizhenko, A. Balandin, and S. Bandyopadhyay, *J. Appl. Phys.* ***81***, 7927 (1997); A. Balandin, A. Svizhenko, and S. Bandyopadhyay, in Zh. Alferov and R. Suris, Eds.,

Nanostructures: Physics and Applications, Russian Academy of Sciences, St. Petersburg, Russia, 1997, p. 294.

63. O. A. Aktsipetrov, A. A. Fedyanin, A. V. Melnikov, J. I. Dadap, X. F. Hu, M. H. Anderson, M. C. Downer, and J. K. Lowell, *Thin Solid Films* ***294***, 231 (1997).
64. V. N. Denisov, B. N. Mavrin, V. B. Podobedov, Kh. Sterin, and B. G. Varshal, *Opt. Spectrosc. (USSR)* ***49***, 221 (1980).
65. S. Bandyopadhyay and A. E. Miller, "Self assembled quantum dots and wires," in S. Nalwa, Ed., *Handbook of Nanostructured Materials and Nanotechnology*, Academic, Japan, 1999.
66. D.-F. Yue, G. Banerjee, A. E. Miller, and S. Bandyopadhyay, *Superlattices and Microstruct.* ***18***, 702 (1995).
67. S. Bandyopadhyay, L. Menon, A. Balandin, D. Zaretsky, A. Varfolomeev, and S. Tereshin, "Electronic bistability in an electrochemically self-assembled array of semiconductor quantum dots," in D. J. Lockwood et al., Eds., *Quantum Confinement V*, The Electrochemical Society, Pennington, NJ, 1999.
68. R. L. Bayer, in P. G. Harper and B. S. Wherrett, Eds., *Non-linear Optics*, Academic, New York, 1977, p. 61.
69. S. Bandyopadhyay and V. P. Roychowdhury, *Superlattices and Microstruct.* ***22***, 411 (1997); S. Bandyopadhyay and V. P. Roychowdhury, *Jpn. J. Appl. Phys., Pt. I* ***35***, 3350 (1996); S. Bandyopadhyay and M. Cahay, "Semiconductor quantum devices," in *Advances in Electronics and Electron Physics*, Vol. 89, Academic, New York, 1994, pp. 93–253.
70. A. N. Korotkov, *Appl. Phys. Lett.* ***67***, 2412 (1995); S. S. Molotkov and S. N. Nazin, *JETP Lett.* ***62***, 272 (1995).
71. J. I. Ciracones P. Zoller, *Phys. Rev. Lett.* ***74***, 4091 (1995).
72. N. A. Gershenfeld and I. Chuang, *Science* ***275***, 350 (1997).
73. G. J. Milburn, *Phys. Rev. Lett.* ***62***, 2124 (1989).
74. A. Shnirman, G. Schoen, and Z. Hermon, *Phys. Rev. Lett.* ***79***, 2371 (1997).
75. A. Barenco, D. Deutsch, A. Eker, and R. Jozsa, *Phys. Rev. Lett.* ***74***, 4083 (1995).
76. C. Monroe, D. M. Meekhof, B. E. King, W. M. Itano, and D. J. Wineland, *Phys. Rev. Lett.* ***75***, 4714 (1995).
77. J. A. Jones and M. Mosca, "Implementation of a quantum algorithm to solve Deutsch's problem on a nuclear magnetic resonance quantum computer," LANL preprint #/980127, 1998.
78. W. G. Unruh, *Phys. Rev. A* ***51***, 992 (1995).
79. A. Balandin and K. L. Wang, *Superlattices. Microstruct.* ***25***, 509 (1999).
80. S. Bandyopadhyay, A. Balandin, F. Roychowdhury, and F. Vatan, *Superlattices. Microstruct.* ***23***, 445 (1998).
81. V. Roychowdhury, D. B. Janes, and S. Bandyopadhyay, *Proceed. IEEE* ***85***, 574 (1997).
82. S. Bandyopadhyay, B. Das, and A. E. Miller, *Nanotechnology* ***5***, 113 (1994).
83. B. E. Kane, *Nature (London)* ***393***, 133 (1998).
84. D. P. DiVincenzo, *J. Appl. Phys.* ***81***, 4602 (1997).
85. S. Lloyd, *Science* ***261***, 1569 (1993).

86. A. Balandin and K. L. Wang, "Implementation of quantum controlled-NOT gates using asymmetric semiconductor quantum dots," in C. P. Williams, Ed., *Lecture Notes in Computer Science*, Vol. 1509, Springer-Verlag, Heidelberg, Germany, 1999.
87. D. G. Austing, T. Honda, and S. Tarucha, *Semicond. Sci. Technol.* ***11***, 388 (1996).
88. S. Tarucha, D. G. Austing, and T. Honda, *Superlattices. Microstruct.* ***18***, 121 (1995).
89. S. Tarucha, D. G. Austing, T. Honda, R. J. van der Hage, and L. P. Kouwenhoven, *Phys. Rev. Lett.* ***77***, 3613 (1996).
90. R. Ashoori, *J. Vacuum Sci. Tech. B* ***15***, 2844 (1997).
91. Y. J. Mii, R. P. G. Karunasiri, and K. L. Wang, *Appl. Phys. Lett.* ***53***, 2050 (1988); K. L. Wang and A. Balandin, unpublished, results.
92. P. F. Yuh and K. L. Wang, *J. Appl. Phys.* ***65***, 4377 (1989); P. F. Yuh and K. L. Wang, *Phys. Rev. B* ***38***, 8377 (1988).
93. F. H. Julien, A. Sa'ar, J. Wang, and J.-P. Leburton, *Electron. Lett.* ***31***, 838 (1995).
94. S. I. Borenstain, and J. Katz, *Appl. Phys. Lett.* ***55***, 654 (1989).
95. K. Mukai, N. Ohtsuka, H. Shoji, and M. Sugawara, *Appl. Phys. Lett.* ***68***, 3013 (1996).
96. S. Grosse, J. H. Sandmann, G. von Plessen, J. Feldmann, H. Lipsanen, M. Sopanen, J. Tulkki, and J. Ahopelto, *Phys. Rev. B* ***55***, 4473 (1997).
97. S. Datta, "Quantum phenomena," in *Modular Series on Solid State Devices*, Vol. II, Addison-Wesley, Readings, MA, 1989.
98. U. Bockelmann and G. Bastard, *Phys. Rev. B* ***42***, 8947 (1990).

INDEX